24th Workshop on Biomedical Language Processing (BioNLP 2025)

Vienna, Austria
1 August 2025

ISBN: 979-8-3313-2402-5

Printed from e-media with permission by:

Curran Associates, Inc.
57 Morehouse Lane
Red Hook, NY 12571

Some format issues inherent in the e-media version may also appear in this print version.

Printed with permission by Curran Associates, Inc. (2025)

For permission requests, please contact the Association for Computational Linguistics at the address below.

Association for Computational Linguistics
209 N. Eighth Street
Stroudsburg, Pennsylvania 18360

Phone: 1-570-476-8006
Fax: 1-570-476-0860

acl@aclweb.org

Additional copies of this publication are available from:

Curran Associates, Inc.
57 Morehouse Lane
Red Hook, NY 12571 USA
Phone: 845-758-0400
Fax: 845-758-2633
Email: curran@proceedings.com
Web: www.proceedings.com

ACL 2025

BioNLP 2025 and Shared Tasks

Proceedings of the 24th Workshop on Biomedical Language Processing

August 1, 2025

The ACL organizers gratefully acknowledge the support from the following sponsors.

In cooperation with

CHEN TIANQIAO & CHRISSY INSTITUTE

BioNLP 2025: new solutions to perennial and emerging problems

Dina Demner-Fushman, Sophia Ananiadou, Makoto Miwa and Jun-ichi Tsujii

Large Language Models (LLMs) continue to be the mainstay of Biomedical Language Processing, while the scope of BioNLP research continues to expand across foundational tasks, applications, languages and modalities. In 2025, we see increasing efforts to integrate textual features with visual and sequencing data; new approaches to named entity recognition and linking; work in several languages other than English; and applications ranging from drug discovery and gene editing to veterinary and clinical studies. Complex language technology tasks, such as question answering and summarization, as well as data generation and text mining are also strongly represented. Concerns about potential harms and irresponsible use of AI applications are being addressed through growing research into evaluation, debiasing, and understanding of models' behavior.

The submissions to the BioNLP 2025 workshop and the Shared Tasks demonstrated once again that the workshop sponsored by the ACL Special Interest Group on Biomedical Natural Language Processing (SIGBIOMED) is the preferred venue for the groundbreaking research and applications in Biomedical Language Processing, which encompasses biological, clinical and non-professional medical sub-languages, among others. BioNLP remains the flagship and the generalist in biomedical language processing, accepting all noteworthy work independently of the tasks and languages studied. The quality of submissions continues to impress the program committee and the organizers.

BioNLP 2025 received 61 submissions, of which eight were accepted for oral presentation and 22 as poster presentations. The selected works span foundational research, biomedical language processing, clinical applications, and generation of new datasets and benchmarks.

Four Shared Tasks were collocated with BioNLP 2025:

SMAFIRA: annotating the literature for finding methods alternative to animal experiments.

ClinIQLink 2025: LLM Lie Detector Test: evaluating the effectiveness of generative models in producing factually accurate information, using a benchmark dataset specifically curated to align with the knowledge level of a General Practitioner (GP) .

ArchEHR-QA 2025: Grounded Electronic Health Record Question Answering: automatically generating answers to patients' health-related questions that are grounded in the evidence from patients' clinical notes.

BioLaySumm 2025: Now, in its third edition, this year's BioLaySumm, introduces a new task: radiology report generation in layman's terms, extending the shared task to a new multimodal domain.

The overviews of the tasks and short presentations of the best performing approaches are included in the workshop program. The participants in all Shared Tasks present their work in a dedicated poster session.

The keynote by Wojciech Kusa is titled: Incorporating Changes in Review Outcomes in the Evaluation of Systematic Review Automation.
Current evaluations of automation methods in systematic literature reviews often treat all included studies as equally important, ignoring their varying influence on review outcomes. This can misrepresent the effectiveness of search strategies, as not all relevant studies contribute equally to the conclusions of the review. To address this limitation, we propose a new evaluation framework that incorporates the differential impact of individual studies on review outcomes. Using data from the CLEF 2019 TAR task, we applied this framework to assess 74 automation models, leveraging meta-analysis effect estimates to weigh the influence of each study. Compared to conventional binary relevance metrics, our approach provided a more nuanced assessment, emphasizing the importance of retrieving high-impact studies. Results showed significant differences in model rankings, underscoring the value of outcome-based evaluation.

This framework offers researchers a more precise method for evaluating systematic review automation tools, ultimately supporting higher-quality evidence synthesis and better-informed clinical decisions.

Wojciech is a Senior Researcher at the NASK National Research Institute in Poland, where he leads the Linguistic Engineering and Text Analysis Department. He holds a PhD in NLP from TU Wien, with a focus on applying and evaluating neural methods for domain-specific data. His research interests include the safety and evaluation of large language models, clinical and biomedical NLP, and AI-driven scientific discovery. Wojciech was a Marie Skłodowska-Curie Fellow in the EU Horizon 2020 project DoSSIER, specialising in biomedical information retrieval and NLP. He has industry experience from roles at Samsung and Allegro, and has completed research internships at Sony, UNINOVA, and the Polish Academy of Sciences.

We are pleased to announce that the Chen Institute is co-organizing the BioNLP 2025 Workshop. Founded in 2016 by Tianqiao Chen and Chrissy Luo, the Chen Institute is driven by a bold vision to improve the human experience by understanding how our brains perceive, learn, and interact with the world. Their global platform includes the Tianqiao and Chrissy Chen Institute for Neuroscience at Caltech, the Tianqiao Chen Institute for Translational Research in Shanghai, the Chen Frontier Lab for Applied Neurotechnology, and the Chen Frontier Lab for AI and Mental Health. The Chen Scholars program supports early- to mid-career scientists, and the recently launched Chen Institute and Science Prize for AI Accelerated Research highlights their deep commitment to innovation. At this year's BioNLP Workshop, the Chen Institute is interested in exploring how artificial intelligence can accelerate the pace of scientific discovery. We believe there are vast, untapped opportunities to make groundbreaking advances by leveraging the power of AI. The hope is that this meeting will serve as the beginning of an ongoing dialogue—focused on new developments, transformative successes, and emerging thinking at the intersection of AI and science. Through this collaboration, the Chen Institute aims to identify and support promising approaches with the potential to meaningfully change the world.
As always, we are deeply grateful to the authors of the submitted papers and to the reviewers (listed elsewhere in this volume) who produced three thorough and thoughtful reviews for each paper in a fairly short review period. The quality of submitted work continues to grow, and the organizers are truly grateful to the members of our amazing Program Committee, who helped us to determine which work was ready to be presented, and which would benefit from the additional experiments and analyses suggested by the reviewers.
As in years past, we are looking forward to a productive workshop and hoping it will foster new collaborations and research. This will enable our community to continue making valuable contributions to public health and well-being, as well as to basic and clinical research.

Organizing Committee

Chair

Dina Demner-Fushman, National Library of Medicine, USA

Organizers

Sophia Ananiadou, National Centre for Text Mining and University of Manchester, UK
Makoto Miwa, Toyota Technological Institute, Japan
Junichi Tsujii, National Institute of Advanced Industrial Science and Technology, Japan

Program Committee

Chairs

Dina Demner-Fushman, National Library of Medicine
Sophia Ananiadou, University of Manchester
Makoto Miwa, Toyota Technological Institute
Junichi Tsujii, National Institute of Advanced Industrial Science and Technology, Japan

Program Committee

Natalie Alexander, The University of Cape Town
Daniel Andrade, Hiroshima University
Emilia Apostolova, Language.ai
Eiji Aramaki, NAIST, Japan
Tanmay Basu, Indian Institute of Science Education and Research Bhopal
Leandra Budau, Toronto Metropolitan University
Leonardo Campillos-Llanos, Consejo Superior de Investigaciones Cientificas (Spanish National Research Council)
Yingjian Chen, Henan University
Liuliu Chen, The University of Melbourne
Brian Connolly, Cincinnati Children's Hospital Medical Center
Mike Conway, University of Melbourne
An Dao, The University of Tokyo
Berry De Bruijn, National Research Council Canada
Jean-Benoit Delbrouck, Stanford University
Simona Doneva, University of Zurich
Pietro Ferrazzi, University of Padova
Kathleen C. Fraser, National Research Council Canada
Tomas Goldsack, University of Sheffield
Natalia Grabar, CNRS STL UMR8163, Université de Lille
Cyril Grouin, LIMSI-CNRS
Tudor Groza, Pryzm Health Pty Ltd
Yingjun Guan, iSchool, University of Illinois at Urbana-Champaign
Deepak Gupta, National Library of Medicine, NIH
Thierry Hamon, LISN, Universite Paris-Saclay
Universite Sorbonne Paris Nord
William Hogan, UCSD
Ben Holgate, King's College London
Brian Hur, University of Washington
Antonio Jimeno Yepes, Unstructured Technologies
Hidetaka Kamigaito, Nara Institute of Science and Technology
Vani Kanjirangat, IDSIA
Sarvnaz Karimi, CSIRO
Nazmul Kazi, University of North Florida
Siun Kim, Seoul Natoinal University Hospital
Gaurav Kumar, University of California San Diego
Andre Lamurias, NOVA School of Science and Technology
Majid Latifi, University of York
Alberto Lavelli, FBK

Robert Leaman, National Center for Biotechnology Information
L u n g - H a o Lee, National Yang Ming Chiao Tung University
Ulf Leser, Humboldt-Universität zu Berlin
Yuan Liang, Queen Mary University of London
Siting Liang, German Research Center for Artificial Intelligence
Livia Lilli, Fondazione Policlinico Universitario Agostino Gemelli IRCCS, Rome, Italy; Catholic University of the Sacred Heart, Rome, Italy
Abdine Maiga, University College London
Claire Nedellec, INRAE
Guenter Neumann, DFKI
Saarland University
Mariana Neves, German Federal Institute for Risk Assessment
Andrei Niculae, National University of Science and Technology Politehnica Bucharest
Aurélie Névéol, Université Paris Saclay, CNRS, LISN
Brian Ondov, Yale School of Medicine
Noon Pokaratsiri Goldstein, DFKI
François Remy, Ghent University
Francisco J. Ribadas-Pena, University of Vigo
Fabio Rinaldi, IDSIA, Swiss AI Institute
Roland Roller, DFKI SLT Lab
Mourad Sarrouti, CLARA Analytics
Efstathia Soufleri, Athena RC
Peng Su, University of Delaware
Madhumita Sushil, University of Antwerp
Mario Sänger, Humboldt-Universität zu Berlin
Andrew Taylor, Yale University
Karin Verspoor, RMIT University
Davy Weissenbacher, Cedars-Sinai Medical Center
Nathan M. White, James Cook University; Western Institute for Endangered Language Documentation
Dongfang Xu, Cedars-Sinai Medical Center
Shweta Yadav, University of Illinois at Chicago
Ken Yano, The National Institute of Advanced Industrial Science and Technology
Hyunwoo Yoo, Drexel University
Xiao Yu Cindy Zhang, University of British Columbia
Xinyue Zhang, King's College London
Kai Zhang, Worcester Polytechnic Institute
Jingqing Zhang, Pangaea Data
Angelo Ziletti, Bayer AG
Ayah Zirikly, Johns Hopkins University
Pierre Zweigenbaum, LISN, CNRS, Université Paris-Saclay

Secondary Reviewers

Joseph Akinyemi, University of York
Robert Bossy, National Research Institute for Agriculture, Food and Environment (INRAE)
Marco Naguib, Interdisciplinary Laboratory on Numerical Sciences (LISN)

Keynote Talk

Incorporating Changes in Review Outcomes in the Evaluation of Systematic Review Automation

Wojciech Kusa
NASK National Research Institute, Poland
2025-08-01 12:00:00 – Room: **Room 2.15**

Abstract: Current evaluations of automation methods in systematic literature reviews often treat all included studies as equally important, ignoring their varying influence on review outcomes. This can misrepresent the effectiveness of search strategies, as not all relevant studies contribute equally to the conclusions of the review. To address this limitation, we propose a new evaluation framework that incorporates the differential impact of individual studies on review outcomes. Using data from the CLEF 2019 TAR task, we applied this framework to assess 74 automation models, leveraging meta-analysis effect estimates to weigh the influence of each study. Compared to conventional binary relevance metrics, our approach provided a more nuanced assessment, emphasizing the importance of retrieving high-impact studies. Results showed significant differences in model rankings, underscoring the value of outcome-based evaluation. This framework offers researchers a more precise method for evaluating systematic review automation tools, ultimately supporting higher-quality evidence synthesis and better-informed clinical decisions.

Bio: Wojciech is a Senior Researcher at the NASK National Research Institute in Poland, where he leads the Linguistic Engineering and Text Analysis Department. He holds a PhD in NLP from TU Wien, with a focus on applying and evaluating neural methods for domain-specific data. His research interests include the safety and evaluation of large language models, clinical and biomedical NLP, and AI-driven scientific discovery. Wojciech was a Marie Skłodowska-Curie Fellow in the EU Horizon 2020 project DoSSIER, specialising in biomedical information retrieval and NLP. He has industry experience from roles at Samsung and Allegro, and has completed research internships at Sony, UNINOVA, and the Polish Academy of Sciences.

Table of Contents

Program

Friday, August 1, 2025

08:40 - 08:50 *Opening Remarks*

08:50 - 10:30 *Session 1: Foundational tasks*

Accelerating Cross-Encoders in Biomedical Entity Linking
Javier Sanz-Cruzado and Jake Lever

Beyond Citations: Integrating Finding-Based Relations for Improved Biomedical Article Representations
Yuan Liang, Massimo Poesio and Roonak Rezvani

MedSummRAG: Domain-Specific Retrieval for Medical Summarization
Guanting Luo and Yuki Arase

Advancing Biomedical Claim Verification by Using Large Language Models with Better Structured Prompting Strategies
Siting Liang and Daniel Sonntag

Questioning Our Questions: How Well Do Medical QA Benchmarks Evaluate Clinical Capabilities of Language Models?
Siun Kim and Hyung-Jin Yoon

10:30 - 11:00 *Coffee Break*

11:00 - 12:30 *Session 2: Clinical NLP*

A Retrieval-Based Approach to Medical Procedure Matching in Romanian
Andrei Niculae, Adrian Cosma and Emilian Radoi

Error Detection in Medical Note through Multi Agent Debate
Abdine Maiga, Anoop Shah and Emine Yilmaz

Converting Annotated Clinical Cases into Structured Case Report Forms
Pietro Ferrazzi, Alberto Lavelli and Bernardo Magnini

Friday, August 1, 2025 (continued)

12:00 - 12:30	*Session 3: Invited Talk by Wojciech Kusa*
12:30 - 14:00	*Lunch*
14:00 - 15:30	*Session 4: Shared Tasks*
14:00 - 14:15	*BioLaySum*
	Overview of the BioLaySumm 2025 Shared Task on Lay Summarization of Biomedical Research Articles and Radiology Reports Chenghao Xiao, Kun Zhao, Xiao Wang, Siwei Wu, Sixing Yan, Tomas Goldsack, Sophia Ananiadou, Noura Al Moubayed, Liang Zhan, William K. Cheung and Chenghua Lin
14:15 - 14:25	*BioLaySum Poster Boasters*
14:25 - 14:40	*SMAFIRA*
	SMAFIRA Shared Task at the BioNLP'2025 Workshop: Assessing the Similarity of the Research Goal Mariana Neves, Iva Sovadinova, Susanne Fieberg, Celine Heinl, Diana Rubel, Gilbert Schönfelder and Bettina Bert
14:40 - 14:55	*CliniIQLink*
	Overview of the ClinIQLink 2025 Shared Task on Medical Question-Answering Brandon Colelough, Davis Bartels and Dina Demner-Fushman
14:55 - 15:00	*CliniIQLink Poster Boasters*
15:00 - 15:15	*ArchEHR-QA*
	Overview of the ArchEHR-QA 2025 Shared Task on Grounded Question Answering from Electronic Health Records Sarvesh Soni, Soumya Gayen and Dina Demner-Fushman
15:15 - 15:25	*ArchEHR-QA Poster Boasters*

Friday, August 1, 2025 (continued)

15:30 - 16:00	*Coffee Break*
16:00 - 18:00	*Poster Sessions (see Shared Task posters in Volume 2)*

Improving Barrett's Oesophagus Surveillance Scheduling with Large Language Models: A Structured Extraction Approach
Xinyue Zhang, Agathe Zecevic, Sebastian Zeki and Angus Roberts

Effective Multi-Task Learning for Biomedical Named Entity Recognition
João Ruano, Gonçalo Correia, Leonor Barreiros and Afonso Mendes

PetEVAL: A veterinary free text electronic health records benchmark
Sean Farrell, Alan Radford, Noura Al Moubayed and Peter-John Noble

Can Large Language Models Classify and Generate Antimicrobial Resistance Genes?
Hyunwoo Yoo, Haebin Shin and Gail Rosen

Overcoming Data Scarcity in Named Entity Recognition: Synthetic Data Generation with Large Language Models
An Dao, Hiroki Teranishi, Yuji Matsumoto, Florian Boudin and Akiko Aizawa

Fine-tuning LLMs to Extract Epilepsy Seizure Frequency Data from Health Records
Ben Holgate, Joe Davies, Shichao Fang, Joel Winston, James Teo and Mark Richardson

Transformer-Based Medical Statement Classification in Doctor-Patient Dialogues
Farnod Bahrololloomi, Johannes Luderschmidt and Biying Fu

PreClinIE: An Annotated Corpus for Information Extraction in Preclinical Studies
Simona Doneva, Hanna Hubarava, Pia Härvelid, Wolfgang Zürrer, Julia Bugajska, Bernard Hild, David Brüschweiler, Gerold Schneider, Tilia Ellendorff and Benjamin Ineichen

QoLAS: A Reddit Corpus of Health-Related Quality of Life Aspects of Mental Disorders
Lynn Greschner, Amelie Wührl and Roman Klinger

Gender-Neutral Large Language Models for Medical Applications: Reducing Bias in PubMed Abstracts
Elizabeth Schaefer and Kirk Roberts

Friday, August 1, 2025 (continued)

	Virtual CRISPR: Can LLMs Predict CRISPR Screen Results? Steven Song, Abdalla Abdrabou, Asmita Dabholkar, Kastan Day, Pavan Dharmoju, Jason Perera, Volodymyr Kindratenko and Aly Khan
17:50 - 18:00	*Closing Remarks*

Understanding the Impact of Confidence in Retrieval Augmented Generation: A Case Study in the Medical Domain

Shintaro Ozaki[♣] Yuta Kato[♠] Siyuan Feng[♠] Masayo Tomita[♠] Kazuki Hayashi[♣]
Wataru Hashimoto[♣] Ryoma Obara[♢] Masafumi Oyamada[♢]
Katsuhiko Hayashi[♠] Hidetaka Kamigaito[♣] Taro Watanabe[♣]
[♣]Nara Institute of Science and Technology (NAIST)
[♠]The University of Tokyo [♢]NEC Corporation
{ozaki.shintaro.ou6, kamigaito.h, taro.watanabe}@naist.ac.jp

Abstract

Retrieval Augmented Generation (RAG) complements the knowledge of Large Language Models (LLMs) by leveraging external information to enhance response accuracy for queries. This approach is widely applied in several fields by taking its advantage of injecting the most up-to-date information, and researchers are focusing on understanding and improving this aspect to unlock the full potential of RAG in such high-stakes applications. However, despite the potential of RAG to address these needs, the mechanisms behind the confidence levels of its outputs remain underexplored. Our study focuses on the impact of RAG, specifically examining whether RAG improves the confidence of LLM outputs in the medical domain. We conduct this analysis across various configurations and models. We evaluate confidence by treating the model's predicted probability as its output and calculating several evaluation metrics which include calibration error method, entropy, the best probability, and accuracy. Experimental results across multiple datasets confirmed that certain models possess the capability to judge for themselves whether an inserted document relates to the correct answer. These results suggest that evaluating models based on their output probabilities determine whether they function as generators in the RAG framework. Our approach allows us to evaluate whether the models handle retrieved documents.[1]

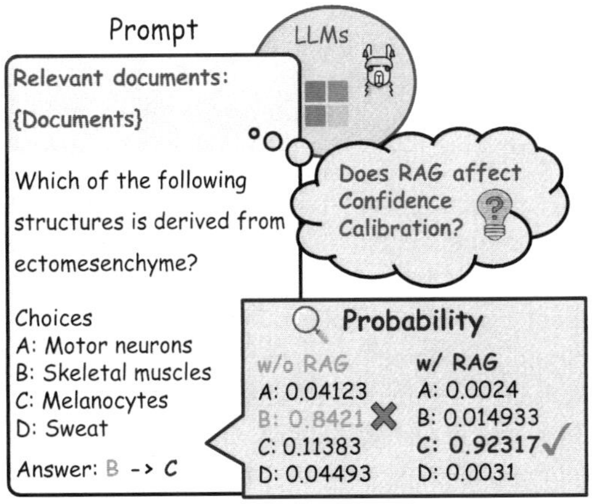

Figure 1: The focus of our research is to analyze whether RAG improves the confidence of the model response.

1 Introduction

Retrieval Augmented Generation (RAG) (Lewis et al., 2020) serves as a method to not only mitigate hallucinations but also supplement the knowledge of Large Language Models (LLMs) (Achiam et al., 2023; Dubey et al., 2024; Aizawa et al., 2024). By leveraging external information, RAG enhances response accuracy and alignment with queries, making it widely applicable in industries. Notable domains include finance (Yepes et al., 2024; Setty et al., 2024) and healthcare (Xiong et al., 2024), where the reliability of information is critical. This study focuses on the medical domain, which has relatively more text data than other fields and involves complex factors directly related to the human body. (Sohn et al., 2024; Jeong et al., 2024)

While researchers explore performance improvements for LLMs using RAG, as illustrated in Figure 1, analyses focusing on prediction confidence remain limited. Although RAG enhances answer accuracy, it may lead to overconfidence, where models exhibit unwarranted self-assurance (Chen et al., 2024). We hypothesize that retrieving documents to support the correct answer through RAG improve the model's confidence, potentially leading to errors in confidence calibration. Based on this, we pose a research question: Do LLMs improve the confidence for outputs with RAG?

To address this question, we conduct a systematic analysis of multiple tasks and models in the medical domain, exploring diverse scenarios using

[1]The code is available at https://github.com/naist-nlp/CC_RAG.

Proceedings of the 24th Workshop on Biomedical Language Processing (BioNLP 2025), pages 1–17

Dataset	Size	Option
PubMedQA	1,000	3
MedMCQA (Extract)	2,206	4

Table 1: The dataset used in our study. We select datasets that not only contain QA pairs but also include explanatory passages that justify the answers.

PubMedQA (Jin et al., 2019) and MedMCQA (Pal et al., 2022). In particular, we create pseudo-RAG to manipulate document content – such as adding irrelevant documents deliberately or including only those directly related to the answer – to simulate the range of situations RAG encounter.

Our result shows that inserting documents deliberately containing answer-supporting information improve confidence in many models, aligning with expectations, although some models exhibited behavior contrary to this prediction. Additionally, inserting documents unrelated to the correct answer rarely improve the confidence, suggesting that LLMs can discriminate whether an inserted document relates to the answer. These results indicate that evaluating models based on output probabilities can lead to reveal the suitable generator model.

2 Related Work

2.1 Confidence of LLMs

Research on confidence has been prevalent since before the era of LLMs (Jiang et al., 2021) and continues to be extensively explored (Geng et al., 2024). Becker and Soatto (2024) proposed a framework that measures confidence by leveraging explanation-generating text produced by LLMs. Zhao et al. (2021) identified the issue that few-shot prompting significantly impacts model confidence and alters its inherent performance, and they proposed methods to address this problem. Confidence estimation is used as a technique to suppress hallucinations, where models generate false information (Zhang et al., 2023). Cole et al. (2023) demonstrated that by utilizing model confidence, it is possible to suppress outputs for ambiguous questions. Our study contributes to this body of research by specifically analyzing how RAG influences confidence calibration in LLM outputs. Unlike prior works that primarily optimize retrieval mechanisms, we directly investigate confidence calibration dynamics.

2.2 Boosting RAG with Confidence

Recent advances in RAG have leveraged model confidence (e.g., output probability) to optimize retrieval and generation processes. For instance, Jiang et al. (2023) introduced FLARE, which dynamically decides whether to retrieve additional information based on token-level confidence during generation, ensuring efficient retrieval by minimizing unnecessary searches. Similarly, query rewriting techniques using reinforcement learning (Ma et al., 2023) and strategies such as Recitation-Augmented Generation (Sun et al., 2023), which searches for text resembling hypothetical answers, have shown promise in enhancing retrieval accuracy. Moreover, recent studies like Self-RAG (Asai et al., 2024) integrate retrieval into the generation process itself. In many of these approaches, confidence plays a crucial role either in deciding when to retrieve or in re-ranking retrieved documents based on their relevance. However, these studies focus on improving RAG performance without analyzing how confidence itself is influenced by the RAG. Specifically, while confidence thresholds and re-ranking mechanisms are employed to optimize retrieval and generation, the underlying dynamics of confidence calibration within the RAG pipeline remain underexplored. Our study analyzes confidence calibration with and without RAG to address this gap, verify the implicit assumptions of prior works, and contribute to a deeper understanding of confidence-based mechanisms in RAG.

3 Methods

Our study analyzes whether the confidence improves through RAG by calculating the model's confidence or entropy from the predicted probability by the model. Each input is formatted by concatenating a system prompt, a question prompt, and its answer options (e.g., a four-choice question), following the design of Medical Information Retrieval Augmented Generation Evaluation (MI-RAGE) (Xiong et al., 2024). We also analyze the optimal position for inserting documents retrieved via pseudo-RAG, i.e., inserting a document directly relevant to the answer or irrelevant deliberately into the model input prompt. Specifically, we evaluate three insertion patterns: before the question (Pre-Question, denoted as Pre-Q), between the question and the answer choices (After-Question, denoted as Aft-Q), and after the answer choices (After-Choice, denoted as Aft-C). This setup allows us to examine the Lost-in-the-Middle phenomenon (Liu et al., 2024), where models tend to overlook intermediate content when processing long-context inputs. Moreover, in order to focus on the impact of retrieved document positions, we use documents that contain the correct answer to the question. We validate our research question under three scenarios: (1): inserting only the explanation related to the answer (denoted as Ans1). (2): combining the correct explanation with two irrelevant documents (denoted as Ans1-Oth2). (3): inserting three irrelevant documents (denoted as Oth3). The irrelevant documents are selected from unrelated questions, ensuring that they do not contain the correct answer or semantically similar content.

Directly generating the choice answer by the model complicates evaluation, because differences in reported metrics arise even under identical conditions across studies (Xiong et al., 2024; Chen et al., 2023; Wu et al., 2024). In some studies, researchers select the final candidate using regular expressions, while in others, they treat the output of a specific word (such as Yes or No) as the correct answer. Thus, evaluation methods are not uniquely defined if the sentence generated. In our study, we predict the most plausible option from the given choices as follows:

$$v_i = \log P(x_i \mid \texttt{prompt})$$
$$P(x_i) = \frac{\exp(v_i)}{\sum_{j=1}^{J} \exp(v_j)}$$

where v_i represents the log probability corresponding to each choice x_i and the `prompt` refers to the provided question or context. $P(x_i)$ denotes the probability that the choice x_i is the correct answer, normalized by dividing the exponential of v_i by the sum of exponentials of all v_j values, while J is the number of options, which is 3 or 4.

4 Experimental Setup

4.1 Datasets

We focus on the application of RAG in the medical domain. For the dataset, we select PubMedQA (Jin et al., 2019) and MedMCQA (Pal et al., 2022), both of which include multiple-choices QA data along with explanatory passages that justify the answers. These datasets follow the experimental setup of MIRAGE (Xiong et al., 2024), as shown in Table 1. For MedMCQA, we extract only the questions that include supporting evidence for the answer, resulting in a total of 2,206 instances.

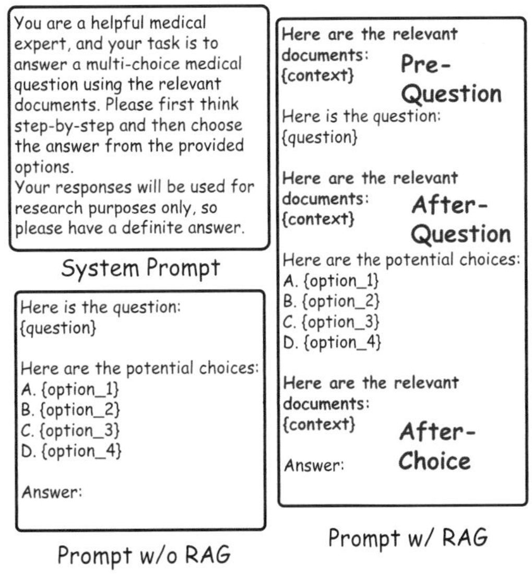

Figure 2: Prompts used in our research. Each prompt begins with a concatenated of the system prompt. Following MIRAGE (Xiong et al., 2024), we design the templates to enable the calculation of probabilities.

4.2 Inference Models

Following prior research (Xiong et al., 2024), we select the following models for evaluation: Phi-3.5 (3.8B) (Abdin et al., 2024), PMC-Llama (13B) (Wu et al., 2024), Llama2 (70B) (Touvron et al., 2023b), LLaMA3.1 (8B / 70B) (Dubey et al., 2024), and Meditron (70B) (Chen et al., 2023). To ensure fair evaluation across models with different architectures and parameter sizes, we also include Gemma2 (2B) (Team et al., 2024) and Qwen2.5 (14B / 70B) (Yang et al., 2024), bringing the total to nine models. PMC-Llama is fine-tuned on medical domain documents based on Llama (Touvron et al., 2023a), while Meditron undergoes continual pretraining on Llama2 (Touvron et al., 2023b). For 70B models, we apply 4-bit quantization, and for PMC-Llama, we use half-precision quantization to compute probabilities. Detailed model configurations are provided in Appendix A.1.

4.3 Templates

Our study modifies the approach based on the MIRAGE paper (Xiong et al., 2024) by excluding Chain of Thought (CoT) (Wei et al., 2022), allowing direct probability computation. (In other words, when using CoT, the model must generate responses, which, as discussed in Section 3, prevents a valid evaluation.) Figure 2 presents the prompts used in our study. Each prompt incorporates system prompts from prior research (Xiong et al., 2024) at the beginning of the input prompt. To investigate whether the Lost-in-the-Middle phenomenon (Liu et al., 2024), also occurs in RAG, our study inserts retrieved documents at three positions: before the question (Pre-Q), after the question (Aft-Q), and after the answer choices (Aft-C).

4.4 Evaluation Metrics

We evaluate if RAG boosts LLM confidence using entropy, best probability, accuracy, and Adaptive Calibration Error. In our multiple-choice QA task, each question has one correct answer, and output probabilities classify responses as correct or not.

Entropy. We examine how entropy changes for candidate answer choices under the influence of RAG, calculating an entropy. Ideally, inserting an answer-containing document should decrease entropy (indicating a more confident selection of the correct choice), while inserting entirely unrelated documents should improve entropy. The entropy is computed as:

$$H(P) = -\sum_{i=1}^{J} P(x_i) \log P(x_i)$$

$$P(x_i) = \frac{\exp(v_i)}{\sum_{j=1}^{J} \exp(v_j)}$$

Here, x_i represents a candidate answer among J total options, and v_i denotes the logit score (i.e., the unnormalized log-probability) assigned to x_i. The softmax function transforms these logits into a probability distribution $P(x_i)$, from which the entropy $H(P)$ is calculated. Lower entropy indicates higher model confidence in a particular choice, while higher entropy implies uncertainty.

Best Probability. We define "Best Probability" as the highest output probability among the candidate choices given to the model. In our study, we evaluate this metric as confidence. A high output probability shows strong confidence for correct answers, while a low output probability is preferred for incorrect answers (we want irrelevant documents to lower the model's confidence).

The notation of best probability is as follows:

$$x^* = \arg\max_{x_i \in \mathcal{X}} (\log P(x_i \mid \texttt{prompt}))$$

$$P(x_i) = \frac{\exp(v_i)}{\sum_{j=1}^{J} \exp(v_j)}$$

Settings	Options
QA	PubMedQA, MedMCQA
Model	Gemma2, Phi3.5, Llama2, Llama3.1 Qwen2.5, PMC-Llama, Meditron
Template	w/o RAG, Pre-Q, Aft-Q, Aft-C
Evaluation	Entropy, Best Prob, Accuracy, ACE

Table 2: Experimental settings used in our research.

Here, $\mathcal{X}$ is the set of all candidate answer choices, and x^* denotes the choice with the highest log-probability. Each v_i represents the model's logit for the candidate x_i, and the softmax function converts these logits into a probability distribution over all choices. The selected x^* corresponds to the most confident prediction the model makes under the given prompt. This Best Probability reflects how strongly the model favors its top prediction, and it serves as an interpretable confidence score in our evaluations.

Adaptive Calibration Error (ACE). Adaptive Calibration Error (ACE) (Nixon et al., 2019) is a metric proposed to address the shortcomings of Expected Calibration Error (ECE) (Naeini et al., 2015; Guo et al., 2017), specifically aiming to reduce the risk of bins with a small number of samples. Proskurina et al. (2024) and Ulmer et al. (2022) have pointed out that ACE is a more suitable calibration error metric for multi-class classification problems. Based on these findings, we adopt ACE in our evaluation. Table 2 provides a complete list of all combinations and Appendix A.6 the details of evaluation metrics.

5 Results

Table 3 presents the experimental results using MedMCQA, while Table 4 shows the results for PubMedQA. When distinguishing between correctly answered and incorrectly answered questions, Phi and Qwen exhibited ideal behavior from an entropy perspective. Specifically, inserting supporting documents for the correct answers led to higher entropy, whereas inserting only unrelated documents resulted in lower entropy. In contrast, other models, e.g., Llama2, Llama3.1, and Gemma2, produced unexpected results, suggesting that Llama and Gemma may struggle to process inserted documents effectively. Furthermore, Qwen and Phi demonstrated the ability to determine whether an inserted document was relevant

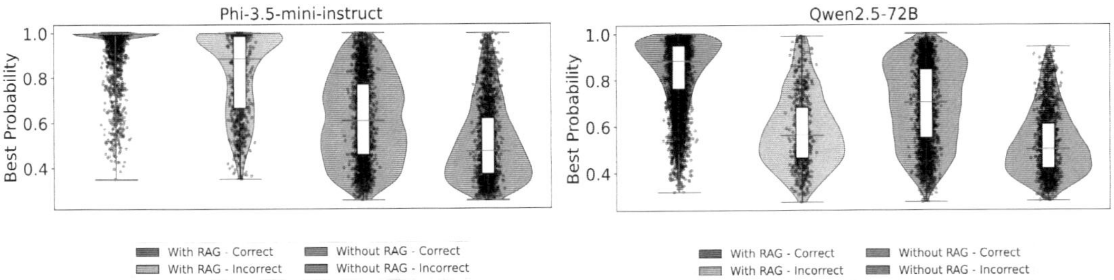

Figure 3: The transition of experimental results using MedMCQA. The figure classifies correctly answered and incorrectly answered questions, illustrating how their distributions shift. This visualization corresponds to the Ans1 setting, with plots for all three conditions: Pre-Q, Aft-Q, and Aft-C.

MedMCQA (Entropy and Best Probability)

Model	Pattern	Entropy (Correct) ↓				Best Prob (Correct) ↑				Entropy (Incorrect) ↓				Best Prob (Incorrect) ↓			
		None	Ans1	Ans1-Oth2	Oth3	None	Ans1	Ans1-Oth2	Oth3	None	Ans1	Ans1-Oth2	Oth3	None	Ans1	Ans1-Oth2	Oth3
Llama-2-70b-chat-hf	w/o RAG	$1.24_{\pm1.11}$	–	–	–	$0.42_{\pm0.55}$	–	–	–	$1.28_{\pm1.27}$	–	–	–	$0.38_{\pm0.38}$	–	–	–
	Pre-Q	–	$1.12_{\pm0.16}$	$1.12_{\pm0.16}$	$1.27_{\pm0.10}$	–	$0.54_{\pm0.12}$	$0.54_{\pm0.12}$	$0.41_{\pm0.09}$	–	$1.27_{\pm0.08}$	$1.28_{\pm0.08}$	$1.31_{\pm0.06}$	–	$0.41_{\pm0.07}$	$0.40_{\pm0.08}$	$\mathbf{0.38}_{\pm0.07}$
	Aft-Q	–	$\mathbf{1.11}_{\pm0.16}$	$1.15_{\pm0.16}$	$1.29_{\pm0.10}$	–	$\mathbf{0.55}_{\pm0.12}$	$0.52_{\pm0.13}$	$0.40_{\pm0.09}$	–	$\mathbf{1.27}_{\pm0.09}$	$1.29_{\pm0.08}$	$1.31_{\pm0.06}$	–	$0.41_{\pm0.08}$	$0.40_{\pm0.08}$	$0.38_{\pm0.07}$
	Aft-C	–	$1.15_{\pm0.16}$	$1.23_{\pm0.12}$	$1.30_{\pm0.10}$	–	$0.51_{\pm0.13}$	$0.46_{\pm0.11}$	$0.39_{\pm0.09}$	–	$1.28_{\pm0.09}$	$1.31_{\pm0.07}$	$1.31_{\pm0.07}$	–	$0.40_{\pm0.08}$	$0.38_{\pm0.08}$	$0.38_{\pm0.08}$
Llama-3.1-70B	w/o RAG	$1.24_{\pm1.24}$	–	–	–	$0.42_{\pm0.42}$	–	–	–	$1.31_{\pm1.31}$	–	–	–	$0.35_{\pm0.32}$	–	–	–
	Pre-Q	–	$1.35_{\pm0.07}$	$1.35_{\pm0.07}$	$1.35_{\pm0.07}$	–	$0.32_{\pm0.03}$	$0.33_{\pm0.03}$	$0.33_{\pm0.03}$	–	$1.35_{\pm0.02}$	$1.35_{\pm0.02}$	$1.35_{\pm0.02}$	–	$\mathbf{0.32}_{\pm0.01}$	$0.32_{\pm0.01}$	$0.33_{\pm0.01}$
	Aft-Q	–	$1.35_{\pm0.07}$	$1.35_{\pm0.07}$	$1.35_{\pm0.07}$	–	$0.33_{\pm0.03}$	$0.33_{\pm0.03}$	$0.33_{\pm0.03}$	–	$1.35_{\pm0.02}$	$1.35_{\pm0.02}$	$1.35_{\pm0.02}$	–	$0.32_{\pm0.01}$	$0.33_{\pm0.01}$	$0.33_{\pm0.01}$
	Aft-C	–	$1.35_{\pm0.07}$	$1.35_{\pm0.07}$	$1.35_{\pm0.07}$	–	$0.33_{\pm0.03}$	$0.33_{\pm0.03}$	$0.33_{\pm0.03}$	–	$1.35_{\pm0.02}$	$1.35_{\pm0.02}$	$1.35_{\pm0.02}$	–	$0.32_{\pm0.01}$	$0.33_{\pm0.01}$	$0.33_{\pm0.01}$
Llama-3.1-8B	w/o RAG	$1.38_{\pm1.38}$	–	–	–	$0.28_{\pm0.28}$	–	–	–	$1.38_{\pm1.38}$	–	–	–	$0.27_{\pm0.27}$	–	–	–
	Pre-Q	–	$1.38_{\pm0.07}$	$\mathbf{1.38}_{\pm0.07}$	$1.38_{\pm0.07}$	–	$0.28_{\pm0.03}$	$0.28_{\pm0.03}$	$0.28_{\pm0.03}$	–	$1.38_{\pm0.02}$	$1.38_{\pm0.02}$	$1.38_{\pm0.02}$	–	$0.27_{\pm0.01}$	$0.27_{\pm0.01}$	$0.27_{\pm0.01}$
	Aft-Q	–	$1.38_{\pm0.07}$	$1.38_{\pm0.07}$	$1.38_{\pm0.06}$	–	$0.28_{\pm0.03}$	$0.28_{\pm0.03}$	$0.28_{\pm0.03}$	–	$1.38_{\pm0.02}$	$\mathbf{1.38}_{\pm0.02}$	$1.38_{\pm0.02}$	–	$0.27_{\pm0.01}$	$0.27_{\pm0.01}$	$0.27_{\pm0.01}$
	Aft-C	–	$1.38_{\pm0.07}$	$1.38_{\pm0.07}$	$1.38_{\pm0.07}$	–	$0.28_{\pm0.03}$	$\mathbf{0.28}_{\pm0.03}$	$0.28_{\pm0.03}$	–	$1.38_{\pm0.02}$	$1.38_{\pm0.02}$	$1.38_{\pm0.02}$	–	$0.27_{\pm0.01}$	$0.27_{\pm0.01}$	$0.27_{\pm0.01}$
meditron-70b	w/o RAG	$1.23_{\pm1.05}$	–	–	–	$0.42_{\pm0.57}$	–	–	–	$1.25_{\pm1.19}$	–	–	–	$0.40_{\pm0.40}$	–	–	–
	Pre-Q	–	$1.11_{\pm0.17}$	$\mathbf{1.05}_{\pm0.19}$	$1.23_{\pm0.11}$	–	$0.54_{\pm0.13}$	$\mathbf{0.57}_{\pm0.14}$	$0.47_{\pm0.09}$	–	$1.24_{\pm0.10}$	$\mathbf{1.19}_{\pm0.11}$	$1.27_{\pm0.08}$	–	$0.43_{\pm0.08}$	$0.47_{\pm0.08}$	$0.43_{\pm0.08}$
	Aft-Q	–	$1.10_{\pm0.17}$	$1.09_{\pm0.17}$	$1.21_{\pm0.11}$	–	$0.54_{\pm0.13}$	$0.55_{\pm0.13}$	$0.48_{\pm0.09}$	–	$1.23_{\pm0.10}$	$1.20_{\pm0.10}$	$1.23_{\pm0.10}$	–	$0.44_{\pm0.08}$	$0.47_{\pm0.08}$	$0.47_{\pm0.09}$
	Aft-C	–	$1.15_{\pm0.17}$	$1.28_{\pm0.10}$	$1.30_{\pm0.08}$	–	$0.51_{\pm0.13}$	$0.43_{\pm0.09}$	$0.41_{\pm0.07}$	–	$1.27_{\pm0.08}$	$1.31_{\pm0.06}$	$1.31_{\pm0.06}$	–	$0.41_{\pm0.07}$	$0.40_{\pm0.07}$	$0.41_{\pm0.07}$
PMC-LLaMA-13B	w/o RAG	$1.00_{\pm1.00}$	–	–	–	$0.56_{\pm0.56}$	–	–	–	$1.05_{\pm1.05}$	–	–	–	$0.53_{\pm0.31}$	–	–	–
	Pre-Q	–	$1.36_{\pm0.06}$	$1.36_{\pm0.06}$	$1.36_{\pm0.06}$	–	$0.33_{\pm0.05}$	$0.32_{\pm0.05}$	$0.32_{\pm0.05}$	–	$1.37_{\pm0.03}$	$1.37_{\pm0.02}$	$1.37_{\pm0.02}$	–	$0.31_{\pm0.04}$	$\mathbf{0.31}_{\pm0.03}$	$0.31_{\pm0.03}$
	Aft-Q	–	$1.36_{\pm0.06}$	$1.36_{\pm0.06}$	$1.36_{\pm0.06}$	–	$0.33_{\pm0.05}$	$0.32_{\pm0.05}$	$0.31_{\pm0.05}$	–	$1.36_{\pm0.03}$	$1.37_{\pm0.03}$	$1.37_{\pm0.03}$	–	$0.32_{\pm0.04}$	$0.31_{\pm0.03}$	$0.31_{\pm0.04}$
	Aft-C	–	$1.36_{\pm0.06}$	$1.35_{\pm0.07}$	$1.36_{\pm0.07}$	–	$0.33_{\pm0.05}$	$0.33_{\pm0.05}$	$0.33_{\pm0.05}$	–	$1.36_{\pm0.03}$	$1.36_{\pm0.03}$	$1.36_{\pm0.03}$	–	$0.32_{\pm0.04}$	$0.32_{\pm0.04}$	$0.32_{\pm0.04}$
Gemma-2-2b	w/o RAG	$1.17_{\pm1.11}$	–	–	–	$0.52_{\pm0.56}$	–	–	–	$1.18_{\pm1.13}$	–	–	–	$0.52_{\pm0.51}$	–	–	–
	Pre-Q	–	$1.12_{\pm0.08}$	$1.13_{\pm0.07}$	$1.15_{\pm0.06}$	–	$0.55_{\pm0.05}$	$0.54_{\pm0.04}$	$0.52_{\pm0.04}$	–	$1.17_{\pm0.05}$	$1.16_{\pm0.04}$	$1.16_{\pm0.04}$	–	$\mathbf{0.51}_{\pm0.04}$	$0.51_{\pm0.04}$	$0.52_{\pm0.03}$
	Aft-Q	–	$1.13_{\pm0.07}$	$1.14_{\pm0.06}$	$1.15_{\pm0.06}$	–	$0.55_{\pm0.05}$	$0.53_{\pm0.04}$	$0.52_{\pm0.04}$	–	$1.17_{\pm0.05}$	$1.16_{\pm0.04}$	$1.15_{\pm0.04}$	–	$0.51_{\pm0.04}$	$0.51_{\pm0.04}$	$0.52_{\pm0.03}$
	Aft-C	–	$\mathbf{1.11}_{\pm0.08}$	$1.12_{\pm0.07}$	$1.13_{\pm0.06}$	–	$\mathbf{0.56}_{\pm0.05}$	$0.55_{\pm0.05}$	$0.54_{\pm0.04}$	–	$1.16_{\pm0.05}$	$1.14_{\pm0.05}$	$\mathbf{1.13}_{\pm0.05}$	–	$0.53_{\pm0.04}$	$0.54_{\pm0.04}$	$0.54_{\pm0.04}$
Phi-3.5	w/o RAG	$0.93_{\pm0.05}$	–	–	–	$0.62_{\pm0.98}$	–	–	–	$1.09_{\pm0.39}$	–	–	–	$0.51_{\pm0.51}$	–	–	–
	Pre-Q	–	$0.06_{\pm0.17}$	$0.07_{\pm0.18}$	$0.24_{\pm0.32}$	–	$0.98_{\pm0.08}$	$0.98_{\pm0.08}$	$0.90_{\pm0.15}$	–	$\mathbf{0.39}_{\pm0.34}$	$0.43_{\pm0.35}$	$0.49_{\pm0.38}$	–	$0.84_{\pm0.18}$	$0.82_{\pm0.18}$	$0.80_{\pm0.19}$
	Aft-Q	–	$\mathbf{0.05}_{\pm0.16}$	$0.07_{\pm0.18}$	$0.34_{\pm0.36}$	–	$\mathbf{0.98}_{\pm0.07}$	$0.97_{\pm0.08}$	$0.87_{\pm0.17}$	–	$0.45_{\pm0.35}$	$0.46_{\pm0.35}$	$0.50_{\pm0.37}$	–	$0.81_{\pm0.18}$	$0.81_{\pm0.18}$	$0.80_{\pm0.19}$
	Aft-C	–	$0.09_{\pm0.19}$	$0.14_{\pm0.22}$	$0.27_{\pm0.32}$	–	$0.97_{\pm0.09}$	$0.95_{\pm0.10}$	$0.90_{\pm0.15}$	–	$0.45_{\pm0.34}$	$0.44_{\pm0.36}$	$0.42_{\pm0.35}$	–	$0.81_{\pm0.18}$	$0.82_{\pm0.19}$	$0.84_{\pm0.17}$
Qwen2.5-14B	w/o RAG	$0.86_{\pm0.48}$	–	–	–	$0.67_{\pm0.85}$	–	–	–	$1.06_{\pm1.03}$	–	–	–	$0.55_{\pm0.49}$	–	–	–
	Pre-Q	–	$0.52_{\pm0.33}$	$0.53_{\pm0.35}$	$0.89_{\pm0.30}$	–	$0.84_{\pm0.15}$	$0.83_{\pm0.16}$	$0.65_{\pm0.18}$	–	$\mathbf{1.03}_{\pm0.23}$	$1.05_{\pm0.23}$	$1.07_{\pm0.22}$	–	$0.56_{\pm0.15}$	$0.55_{\pm0.15}$	$0.54_{\pm0.15}$
	Aft-Q	–	$\mathbf{0.48}_{\pm0.32}$	$0.51_{\pm0.33}$	$0.92_{\pm0.29}$	–	$\mathbf{0.85}_{\pm0.14}$	$0.84_{\pm0.16}$	$0.63_{\pm0.17}$	–	$1.04_{\pm0.24}$	$1.05_{\pm0.23}$	$1.06_{\pm0.21}$	–	$0.56_{\pm0.15}$	$0.55_{\pm0.15}$	$0.55_{\pm0.14}$
	Aft-C	–	$0.66_{\pm0.35}$	$0.80_{\pm0.32}$	$1.00_{\pm0.24}$	–	$0.77_{\pm0.17}$	$0.71_{\pm0.18}$	$0.59_{\pm0.16}$	–	$1.12_{\pm0.20}$	$1.12_{\pm0.20}$	$1.15_{\pm0.17}$	–	$0.51_{\pm0.13}$	$0.51_{\pm0.14}$	$\mathbf{0.49}_{\pm0.12}$
Qwen2.5-72B	w/o RAG	$0.82_{\pm0.44}$	–	–	–	$0.69_{\pm0.86}$	–	–	–	$1.09_{\pm0.87}$	–	–	–	$0.53_{\pm0.53}$	–	–	–
	Pre-Q	–	$0.52_{\pm0.31}$	$0.56_{\pm0.32}$	$0.75_{\pm0.32}$	–	$0.83_{\pm0.14}$	$0.80_{\pm0.15}$	$0.71_{\pm0.18}$	–	$0.91_{\pm0.26}$	$\mathbf{0.87}_{\pm0.24}$	$0.94_{\pm0.24}$	–	$0.61_{\pm0.16}$	$0.63_{\pm0.15}$	$0.62_{\pm0.15}$
	Aft-Q	–	$\mathbf{0.44}_{\pm0.30}$	$0.54_{\pm0.31}$	$0.83_{\pm0.31}$	–	$\mathbf{0.86}_{\pm0.13}$	$0.82_{\pm0.15}$	$0.68_{\pm0.17}$	–	$0.99_{\pm0.27}$	$0.97_{\pm0.24}$	$0.99_{\pm0.22}$	–	$0.57_{\pm0.16}$	$0.59_{\pm0.15}$	$0.60_{\pm0.15}$
	Aft-C	–	$0.53_{\pm0.32}$	$0.65_{\pm0.32}$	$0.85_{\pm0.28}$	–	$0.82_{\pm0.15}$	$0.77_{\pm0.16}$	$0.67_{\pm0.17}$	–	$1.00_{\pm0.25}$	$1.02_{\pm0.21}$	$0.99_{\pm0.22}$	–	$0.57_{\pm0.15}$	$0.57_{\pm0.14}$	$0.61_{\pm0.14}$

Table 3: Experimental result on MedMCQA. **Bold** indicates the best value among the models. Specifically, the lowest entropy and the highest best probability (Correct case) are highlighted. This table has numerical values and their standard deviations.

to the answer, leading to provide strong evidence that they function as suitable generators.

6 Analysis & Discussion

How Does RAG Affect Prediction Probabilities? Figure 3 presents partial experimental results using MedMCQA, while Figure 4 shows results from PubMedQA. These figures correspond to the Ans1 setting, where all three phases—Pre-Q, Aft-Q, and Aft-C—are plotted. A detailed analysis focuses on Phi and Qwen, which exhibited ideal behavior. When RAG was not applied, i.e., evaluating the models' intrinsic accuracy, the output probabilities were evenly distributed across both datasets. Furthermore, the results of Phi-3.5 on PubMedQA reveal that the incorrect predictions tend to concentrate at the upper end, i.e., where output probabilities are high. This pattern suggests that the model exhibits overconfidence, making incorrect predictions despite assigning high probabilities. When solving a QA task under a deliberate setting that includes supporting documents for correct answers (similar to pseudo-RAG), all models (Phi and Qwen) showed improved output probabilities. This suggests that the models can assess whether retrieved documents contain useful information.

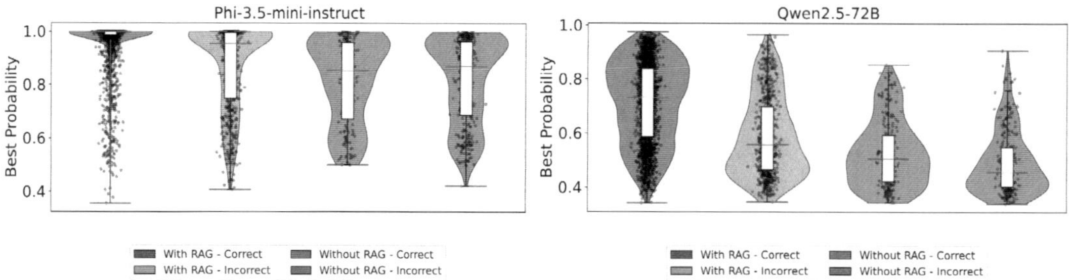

Figure 4: The transition of experimental results using PubMedQA. The figure classifies correctly answered and incorrectly answered questions, illustrating how their distributions shift. This visualization corresponds to the Ans1 setting, with plots for all three conditions: Pre-Q, Aft-Q, and Aft-C.

PubMedQA (Entropy and Best Probability)																	
Model	Pattern	Entropy (Correct) ↓				Best Prob (Correct) ↑				Entropy (Incorrect) ↓				Best Prob (Incorrect) ↓			
		None	Ans1	Ans1-Oth2	Oth3	None	Ans1	Ans1-Oth2	Oth3	None	Ans1	Ans1-Oth2	Oth3	None	Ans1	Ans1-Oth2	Oth3
Llama-2-70b-chat-hf	w/o RAG	$0.83_{\pm0.83}$	–	–	–	$0.62_{\pm0.62}$	–	–	–	$0.93_{\pm0.93}$	–	–	–	$0.55_{\pm0.38}$	–	–	–
	Pre-Q	–	$1.12_{\pm0.16}$	$1.12_{\pm0.16}$	$1.27_{\pm0.10}$	–	$0.54_{\pm0.12}$	$0.54_{\pm0.12}$	$0.41_{\pm0.09}$	–	$1.27_{\pm0.08}$	$1.28_{\pm0.08}$	$1.31_{\pm0.06}$	–	$0.41_{\pm0.07}$	$0.40_{\pm0.08}$	$\mathbf{0.38}_{\pm0.07}$
	Aft-Q	–	$1.11_{\pm0.16}$	$1.15_{\pm0.16}$	$1.29_{\pm0.10}$	–	$0.55_{\pm0.12}$	$0.52_{\pm0.13}$	$0.40_{\pm0.09}$	–	$1.27_{\pm0.09}$	$1.29_{\pm0.08}$	$1.31_{\pm0.06}$	–	$0.41_{\pm0.08}$	$0.40_{\pm0.08}$	$0.38_{\pm0.07}$
	Aft-C	–	$1.15_{\pm0.16}$	$1.23_{\pm0.12}$	$1.30_{\pm0.10}$	–	$0.51_{\pm0.13}$	$0.46_{\pm0.11}$	$0.39_{\pm0.09}$	–	$1.28_{\pm0.09}$	$1.31_{\pm0.07}$	$1.31_{\pm0.07}$	–	$0.40_{\pm0.08}$	$0.38_{\pm0.08}$	$0.38_{\pm0.08}$
Llama-3.1-70B	w/o RAG	$0.86_{\pm0.86}$	–	–	–	$0.59_{\pm0.59}$	–	–	–	$0.87_{\pm0.87}$	–	–	–	$0.57_{\pm0.32}$	–	–	–
	Pre-Q	–	$1.35_{\pm0.07}$	$1.35_{\pm0.07}$	$1.35_{\pm0.07}$	–	$0.32_{\pm0.03}$	$0.33_{\pm0.03}$	$0.33_{\pm0.03}$	–	$1.35_{\pm0.02}$	$1.35_{\pm0.02}$	$1.35_{\pm0.02}$	–	$\mathbf{0.32}_{\pm0.01}$	$0.32_{\pm0.01}$	$0.33_{\pm0.01}$
	Aft-Q	–	$1.35_{\pm0.07}$	$1.35_{\pm0.07}$	$1.35_{\pm0.07}$	–	$0.33_{\pm0.03}$	$0.33_{\pm0.03}$	$0.33_{\pm0.03}$	–	$1.35_{\pm0.02}$	$1.35_{\pm0.02}$	$1.35_{\pm0.02}$	–	$0.32_{\pm0.01}$	$0.33_{\pm0.01}$	$0.33_{\pm0.01}$
	Aft-C	–	$1.35_{\pm0.07}$	$1.35_{\pm0.07}$	$1.35_{\pm0.07}$	–	$0.33_{\pm0.03}$	$0.33_{\pm0.03}$	$0.33_{\pm0.03}$	–	$1.35_{\pm0.02}$	$1.35_{\pm0.02}$	$1.35_{\pm0.02}$	–	$0.32_{\pm0.01}$	$0.33_{\pm0.01}$	$0.33_{\pm0.01}$
Llama-3.1-8B	w/o RAG	$1.09_{\pm1.09}$	–	–	–	$0.36_{\pm0.36}$	–	–	–	$1.09_{\pm1.09}$	–	–	–	$0.36_{\pm0.27}$	–	–	–
	Pre-Q	–	$1.38_{\pm0.07}$	$1.38_{\pm0.07}$	$1.38_{\pm0.07}$	–	$0.28_{\pm0.03}$	$0.28_{\pm0.03}$	$0.28_{\pm0.03}$	–	$1.38_{\pm0.02}$	$1.38_{\pm0.02}$	$1.38_{\pm0.02}$	–	$0.27_{\pm0.01}$	$0.27_{\pm0.01}$	$\mathbf{0.27}_{\pm0.01}$
	Aft-Q	–	$1.38_{\pm0.07}$	$1.38_{\pm0.07}$	$1.38_{\pm0.06}$	–	$0.28_{\pm0.03}$	$0.28_{\pm0.03}$	$0.28_{\pm0.03}$	–	$1.38_{\pm0.02}$	$1.38_{\pm0.02}$	$1.38_{\pm0.02}$	–	$0.27_{\pm0.01}$	$0.27_{\pm0.01}$	$0.27_{\pm0.01}$
	Aft-C	–	$1.38_{\pm0.07}$	$1.38_{\pm0.07}$	$1.38_{\pm0.07}$	–	$0.28_{\pm0.03}$	$0.28_{\pm0.03}$	$0.28_{\pm0.03}$	–	$1.38_{\pm0.02}$	$1.38_{\pm0.02}$	$1.38_{\pm0.02}$	–	$0.27_{\pm0.01}$	$0.27_{\pm0.01}$	$0.27_{\pm0.01}$
meditron-70b	w/o RAG	$0.94_{\pm0.94}$	–	–	–	$0.54_{\pm0.57}$	–	–	–	$0.94_{\pm0.94}$	–	–	–	$0.52_{\pm0.40}$	–	–	–
	Pre-Q	–	$1.11_{\pm0.17}$	$1.05_{\pm0.19}$	$1.23_{\pm0.11}$	–	$0.54_{\pm0.13}$	$\mathbf{0.57}_{\pm0.14}$	$0.47_{\pm0.09}$	–	$1.24_{\pm0.10}$	$1.19_{\pm0.11}$	$1.27_{\pm0.08}$	–	$0.43_{\pm0.08}$	$0.47_{\pm0.08}$	$0.43_{\pm0.08}$
	Aft-Q	–	$1.10_{\pm0.17}$	$1.09_{\pm0.17}$	$1.21_{\pm0.11}$	–	$0.54_{\pm0.13}$	$0.55_{\pm0.13}$	$0.48_{\pm0.09}$	–	$1.23_{\pm0.10}$	$1.20_{\pm0.10}$	$1.23_{\pm0.10}$	–	$0.44_{\pm0.08}$	$0.47_{\pm0.08}$	$0.47_{\pm0.09}$
	Aft-C	–	$1.15_{\pm0.17}$	$1.28_{\pm0.10}$	$1.30_{\pm0.08}$	–	$0.51_{\pm0.13}$	$0.43_{\pm0.09}$	$0.41_{\pm0.07}$	–	$1.27_{\pm0.08}$	$1.31_{\pm0.06}$	$1.31_{\pm0.06}$	–	$0.41_{\pm0.07}$	$\mathbf{0.40}_{\pm0.07}$	$0.41_{\pm0.07}$
PMC-LLaMA-13B	w/o RAG	$1.08_{\pm1.08}$	–	–	–	$0.40_{\pm0.40}$	–	–	–	$1.08_{\pm1.08}$	–	–	–	$0.40_{\pm0.31}$	–	–	–
	Pre-Q	–	$1.36_{\pm0.06}$	$1.36_{\pm0.06}$	$1.36_{\pm0.06}$	–	$0.33_{\pm0.05}$	$0.32_{\pm0.05}$	$0.32_{\pm0.05}$	–	$1.37_{\pm0.03}$	$1.37_{\pm0.02}$	$1.37_{\pm0.02}$	–	$0.31_{\pm0.04}$	$\mathbf{0.31}_{\pm0.03}$	$0.31_{\pm0.03}$
	Aft-Q	–	$1.36_{\pm0.06}$	$1.36_{\pm0.06}$	$1.36_{\pm0.06}$	–	$0.33_{\pm0.05}$	$0.32_{\pm0.05}$	$0.31_{\pm0.05}$	–	$1.36_{\pm0.03}$	$1.37_{\pm0.03}$	$1.37_{\pm0.03}$	–	$0.32_{\pm0.04}$	$0.31_{\pm0.03}$	$0.31_{\pm0.04}$
	Aft-C	–	$1.36_{\pm0.06}$	$1.35_{\pm0.07}$	$1.36_{\pm0.07}$	–	$0.33_{\pm0.05}$	$0.33_{\pm0.05}$	$0.33_{\pm0.05}$	–	$1.36_{\pm0.03}$	$1.36_{\pm0.03}$	$1.36_{\pm0.03}$	–	$0.32_{\pm0.04}$	$0.32_{\pm0.04}$	$0.32_{\pm0.04}$
Gemma-2-2b	w/o RAG	$0.93_{\pm0.93}$	–	–	–	$0.61_{\pm0.61}$	–	–	–	$0.93_{\pm0.93}$	–	–	–	$0.61_{\pm0.51}$	–	–	–
	Pre-Q	–	$1.12_{\pm0.08}$	$1.13_{\pm0.07}$	$1.15_{\pm0.06}$	–	$0.55_{\pm0.05}$	$0.54_{\pm0.04}$	$0.52_{\pm0.04}$	–	$1.17_{\pm0.05}$	$1.16_{\pm0.04}$	$1.16_{\pm0.04}$	–	$\mathbf{0.51}_{\pm0.04}$	$0.51_{\pm0.04}$	$0.52_{\pm0.03}$
	Aft-Q	–	$1.13_{\pm0.07}$	$1.14_{\pm0.06}$	$1.15_{\pm0.06}$	–	$0.55_{\pm0.05}$	$0.53_{\pm0.04}$	$0.52_{\pm0.04}$	–	$1.17_{\pm0.05}$	$1.16_{\pm0.04}$	$1.15_{\pm0.04}$	–	$0.51_{\pm0.04}$	$0.51_{\pm0.04}$	$0.52_{\pm0.03}$
	Aft-C	–	$1.11_{\pm0.08}$	$1.12_{\pm0.07}$	$1.13_{\pm0.06}$	–	$0.56_{\pm0.05}$	$0.55_{\pm0.05}$	$0.54_{\pm0.04}$	–	$1.16_{\pm0.05}$	$1.14_{\pm0.05}$	$1.13_{\pm0.05}$	–	$0.53_{\pm0.04}$	$0.54_{\pm0.04}$	$0.54_{\pm0.04}$
Phi-3.5	w/o RAG	$0.40_{\pm0.05}$	–	–	–	$0.81_{\pm0.98}$	–	–	–	$0.41_{\pm0.39}$	–	–	–	$0.82_{\pm0.80}$	–	–	–
	Pre-Q	–	$0.06_{\pm0.17}$	$0.07_{\pm0.18}$	$0.24_{\pm0.32}$	–	$0.98_{\pm0.08}$	$0.98_{\pm0.08}$	$0.90_{\pm0.15}$	–	$\mathbf{0.39}_{\pm0.34}$	$0.43_{\pm0.35}$	$0.49_{\pm0.38}$	–	$0.84_{\pm0.18}$	$0.82_{\pm0.18}$	$0.80_{\pm0.19}$
	Aft-Q	–	$\mathbf{0.05}_{\pm0.16}$	$0.07_{\pm0.18}$	$0.34_{\pm0.36}$	–	$\mathbf{0.98}_{\pm0.07}$	$0.97_{\pm0.08}$	$0.87_{\pm0.17}$	–	$0.45_{\pm0.35}$	$0.46_{\pm0.35}$	$0.50_{\pm0.37}$	–	$0.81_{\pm0.18}$	$0.81_{\pm0.18}$	$\mathbf{0.80}_{\pm0.19}$
	Aft-C	–	$0.09_{\pm0.19}$	$0.14_{\pm0.22}$	$0.27_{\pm0.32}$	–	$0.97_{\pm0.09}$	$0.95_{\pm0.10}$	$0.90_{\pm0.15}$	–	$0.45_{\pm0.34}$	$0.44_{\pm0.36}$	$0.42_{\pm0.35}$	–	$0.81_{\pm0.18}$	$0.82_{\pm0.19}$	$0.84_{\pm0.17}$
Qwen2.5-14B	w/o RAG	$0.90_{\pm0.48}$	–	–	–	$0.59_{\pm0.85}$	–	–	–	$0.92_{\pm0.92}$	–	–	–	$0.58_{\pm0.49}$	–	–	–
	Pre-Q	–	$0.52_{\pm0.33}$	$0.53_{\pm0.35}$	$0.89_{\pm0.30}$	–	$0.84_{\pm0.15}$	$0.83_{\pm0.16}$	$0.65_{\pm0.18}$	–	$1.03_{\pm0.23}$	$1.05_{\pm0.23}$	$1.07_{\pm0.22}$	–	$0.56_{\pm0.15}$	$0.55_{\pm0.15}$	$0.54_{\pm0.15}$
	Aft-Q	–	$\mathbf{0.48}_{\pm0.32}$	$0.51_{\pm0.33}$	$0.92_{\pm0.29}$	–	$\mathbf{0.85}_{\pm0.14}$	$0.84_{\pm0.16}$	$0.63_{\pm0.17}$	–	$1.04_{\pm0.24}$	$1.05_{\pm0.23}$	$1.06_{\pm0.21}$	–	$0.56_{\pm0.15}$	$0.55_{\pm0.15}$	$0.55_{\pm0.14}$
	Aft-C	–	$0.66_{\pm0.35}$	$0.80_{\pm0.32}$	$1.00_{\pm0.24}$	–	$0.77_{\pm0.17}$	$0.71_{\pm0.18}$	$0.59_{\pm0.16}$	–	$1.12_{\pm0.20}$	$1.12_{\pm0.20}$	$1.15_{\pm0.17}$	–	$0.51_{\pm0.13}$	$0.51_{\pm0.14}$	$\mathbf{0.49}_{\pm0.12}$
Qwen2.5-72B	w/o RAG	$0.97_{\pm0.44}$	–	–	–	$0.53_{\pm0.86}$	–	–	–	$1.00_{\pm0.87}$	–	–	–	$0.49_{\pm0.49}$	–	–	–
	Pre-Q	–	$0.52_{\pm0.31}$	$0.56_{\pm0.32}$	$0.75_{\pm0.32}$	–	$0.83_{\pm0.14}$	$0.80_{\pm0.15}$	$0.71_{\pm0.18}$	–	$0.91_{\pm0.26}$	$\mathbf{0.87}_{\pm0.24}$	$0.94_{\pm0.24}$	–	$0.61_{\pm0.16}$	$0.63_{\pm0.15}$	$0.62_{\pm0.15}$
	Aft-Q	–	$\mathbf{0.44}_{\pm0.30}$	$0.54_{\pm0.31}$	$0.83_{\pm0.31}$	–	$\mathbf{0.86}_{\pm0.13}$	$0.82_{\pm0.15}$	$0.68_{\pm0.17}$	–	$0.99_{\pm0.27}$	$0.97_{\pm0.24}$	$0.99_{\pm0.22}$	–	$0.57_{\pm0.16}$	$0.59_{\pm0.15}$	$0.60_{\pm0.15}$
	Aft-C	–	$0.53_{\pm0.32}$	$0.65_{\pm0.32}$	$0.85_{\pm0.28}$	–	$0.82_{\pm0.15}$	$0.77_{\pm0.16}$	$0.67_{\pm0.17}$	–	$1.00_{\pm0.25}$	$1.02_{\pm0.21}$	$0.99_{\pm0.22}$	–	$0.57_{\pm0.15}$	$0.57_{\pm0.14}$	$0.61_{\pm0.14}$

Table 4: Experimental results using PubMedQA. **Bold** indicates the best value among the models. Specifically, the lowest entropy and the highest best probability (Correct case) are highlighted. This table has numerical values and their standard deviations.

Model Behavior When Inserting Answer-Containing Documents. When explicitly inserting documents that contain the correct answers, Phi and Qwen demonstrated ideal behavior. For instance, from a correct entropy perspective in Table 3, Phi had a value of 0.933 under the w/o RAG setting, which dropped to 0.051 after document insertion. Similarly, for Qwen2.5 (72B), entropy decreased from 0.819 to 0.444. This observation indicates that the models can assess whether an inserted document is relevant to the question. Moreover, when they determine that the document is unnecessary, they attempt to answer using their own knowledge. Further evidence supporting this conclusion comes from cases where inserting unrelated documents did not improve accuracy. This suggests that the models selectively utilize external information only when it is deemed useful.

Behavior of Calibration Error. Table 5 presents the evaluation result of ACE using MedMCQA, while Table 6 shows the results for PubMedQA. A detailed analysis of Llama and Gemma reveals substantial differences in behavior depending on the model. Notably, even when inserting entirely correct documents (Ans1), Llama3.1 (70B) experi-

Model	Pattern	ACE ↓				Accuracy ↑			
		None	Ans1	Ans1-Oth2	Oth3	None	Ans1	Ans1-Oth2	Oth3
Llama2 (70B)	w/o RAG	2.208	–	–	–	38.322	–	–	–
	Pre-Q	–	22.359	25.113	7.181	–	72.575	75.340	32.094
	Aft-Q	–	23.912	21.132	10.781	–	75.567	69.628	28.105
	Aft-C	–	19.653	17.514	9.803	–	67.498	60.743	28.876
Llama3.1 (70B)	w/o RAG	19.582	–	–	–	58.977	–	–	–
	Pre-Q	–	11.496	11.580	11.658	–	20.898	20.943	20.943
	Aft-Q	–	11.518	11.671	11.714	–	20.898	20.898	20.943
	Aft-C	–	11.504	11.707	11.795	–	20.943	20.943	20.943
Llama3.1 (8B)	w/o RAG	5.423	–	–	–	22.209	–	–	–
	Pre-Q	–	4.701	4.254	3.644	–	23.345	23.209	23.799
	Aft-Q	–	4.473	4.632	3.892	–	23.028	23.209	24.025
	Aft-C	–	4.476	4.746	4.990	–	23.209	23.255	23.663
Meditron (70B)	w/o RAG	6.412	–	–	–	35.525	–	–	–
	Pre-Q	–	17.684	7.652	8.665	–	67.724	54.034	36.038
	Aft-Q	–	15.894	9.467	15.334	–	66.682	47.144	31.958
	Aft-C	–	15.101	6.946	9.006	–	62.829	34.180	31.913
PMC-Llama (13B)	w/o RAG	15.671	–	–	–	38.107	–	–	–
	Pre-Q	–	4.943	4.367	4.357	–	32.729	31.641	26.972
	Aft-Q	–	4.003	2.550	5.032	–	32.910	30.009	26.972
	Aft-C	–	3.496	3.780	4.397	–	33.454	28.740	28.060
Gemma2 (2B)	w/o RAG	19.568	–	–	–	32.297	–	–	–
	Pre-Q	–	25.511	23.160	20.520	–	31.233	31.278	31.278
	Aft-Q	–	24.160	21.072	20.618	–	31.188	31.278	31.278
	Aft-C	–	24.814	23.118	22.916	–	31.324	31.324	31.324
Phi3.5 (3.8B)	w/o RAG	5.624	–	–	–	51.518	–	–	–
	Pre-Q	–	9.786	10.378	33.709	–	86.083	84.950	51.813
	Aft-Q	–	7.636	9.270	43.415	–	88.486	85.947	39.393
	Aft-C	–	7.682	15.952	42.476	–	87.353	76.111	44.334
Qwen2.5 (14B)	w/o RAG	12.125	–	–	–	49.151	–	–	–
	Pre-Q	–	8.646	8.740	11.892	–	89.483	88.441	47.280
	Aft-Q	–	7.013	7.257	17.592	–	89.121	87.534	40.798
	Aft-C	–	7.746	9.778	8.531	–	79.329	75.884	45.014
Qwen2.5 (72B)	w/o RAG	4.030	–	–	–	60.483	–	–	–
	Pre-Q	–	9.393	7.896	20.412	–	89.982	85.766	45.739
	Aft-Q	–	8.782	9.781	18.652	–	93.246	89.574	44.696
	Aft-C	–	9.270	5.990	23.564	–	88.622	79.284	39.483

Table 5: Evaluation results with MedMCQA. Red highlights areas where performance improved compared to the non-RAG setting, while Blue indicates areas where performance deteriorated.

ences a drop in accuracy, whereas Llama3.1 (8B) shows improved accuracy even when inserting completely unrelated documents (Oth3). This stark contrast indicates that even within the same model family, behavior can vary largely. Moreover, a comparison between Llama2, Meditron, and Llama3.1 shows considerable differences in behavior, ruling out parameter size as the primary cause. These findings suggest that while the Llama series performs well under specific instruction formats, it may negatively impact performance in other cases. On the other hand, Qwen and Phi exhibit a clear pattern: inserting entirely unrelated documents (Oth3) worsens ACE, while inserting answer-containing documents (Ans1 or Ans1-Oth2) leads to improvements. This tendency implies that Phi and Qwen possess the ability to assess whether retrieved documents provide useful information. These results show that analyzing LLM confidence through predicted probabilities effectively reveals the model's ability to identify meaningful documents.

Did "Lost in the Middle" Phenomenon Occur? Our study also examined the "Lost in the Middle" phenomenon (Liu et al., 2024) by evaluating the

Model	Pattern	ACE ↓				Accuracy ↑			
		None	Ans1	Ans1-Oth2	Oth3	None	Ans1	Ans1-Oth2	Oth3
Llama2 (70B)	w/o RAG	12.107	–	–	–	46.400	–	–	–
	Pre-Q	–	29.791	30.422	14.146	–	82.200	79.800	56.500
	Aft-Q	–	30.380	31.220	13.234	–	81.600	74.100	53.000
	Aft-C	–	13.494	13.430	11.322	–	57.200	53.100	50.700
Llama3.1 (70B)	w/o RAG	6.091	–	–	–	58.600	–	–	–
	Pre-Q	–	11.329	11.513	11.521	–	55.200	55.200	55.200
	Aft-Q	–	11.343	11.532	11.539	–	55.200	55.200	55.200
	Aft-C	–	11.370	11.534	11.543	–	55.200	55.200	55.200
Llama3.1 (8B)	w/o RAG	24.939	–	–	–	11.000	–	–	–
	Pre-Q	–	23.683	23.368	23.975	–	12.200	12.600	12.000
	Aft-Q	–	23.085	23.576	23.988	–	12.800	12.400	12.000
	Aft-C	–	23.930	23.854	24.370	–	11.900	12.100	11.600
Meditron (70B)	w/o RAG	18.115	–	–	–	34.800	–	–	–
	Pre-Q	–	11.540	18.483	8.365	–	57.300	69.800	57.200
	Aft-Q	–	9.159	6.645	6.270	–	56.700	55.600	54.800
	Aft-C	–	4.171	5.050	7.915	–	54.700	54.800	55.100
PMC-Llama (13B)	w/o RAG	17.261	–	–	–	22.800	–	–	–
	Pre-Q	–	10.462	4.650	3.387	–	28.800	37.900	36.600
	Aft-Q	–	10.322	4.000	3.985	–	28.900	39.200	40.000
	Aft-C	–	4.169	5.421	7.250	–	41.200	44.500	46.100
Gemma2 (2B)	w/o RAG	6.387	–	–	–	55.200	–	–	–
	Pre-Q	–	5.794	5.409	5.394	–	55.300	55.200	55.200
	Aft-Q	–	6.159	5.098	4.188	–	55.200	55.200	55.200
	Aft-C	–	9.081	6.161	6.376	–	55.200	55.200	55.200
Phi3.5 (3.8B)	w/o RAG	48.176	–	–	–	33.400	–	–	–
	Pre-Q	–	14.640	14.777	57.831	–	81.600	81.200	21.900
	Aft-Q	–	13.677	31.960	52.083	–	82.300	62.700	41.300
	Aft-C	–	16.771	33.297	47.123	–	73.700	52.300	33.800
Qwen2.5 (14B)	w/o RAG	15.874	–	–	–	42.800	–	–	–
	Pre-Q	–	4.746	4.816	18.425	–	83.400	83.200	32.600
	Aft-Q	–	3.460	5.013	26.783	–	82.800	76.100	33.900
	Aft-C	–	7.616	3.088	23.229	–	74.500	63.900	32.100
Qwen2.5 (72B)	w/o RAG	7.205	–	–	–	46.400	–	–	–
	Pre-Q	–	10.477	3.801	25.283	–	74.900	78.100	33.000
	Aft-Q	–	8.024	10.931	17.828	–	80.300	71.200	34.300
	Aft-C	–	8.877	6.995	13.543	–	76.800	71.000	42.500

Table 6: Evaluation results on PubMedQA. Red highlights areas where performance improved compared to the non-RAG setting, while Blue indicates areas where performance deteriorated.

impact of document placement within the template across multiple positions (Pre-Q, Aft-Q, and Aft-C). Focusing on Phi and Qwen, which exhibited expected behavior in terms of entropy and accuracy, an intriguing pattern emerged. From an entropy perspective, inserting the document after the answer choices yielded the best results, while from an accuracy perspective, placing it before the answer choices was optimal. These findings suggest that when prioritizing the reliability of information, entropy should be the primary metric.

Error Analysis. Appendix B.1 presents the results of the error analysis, which examines how the model makes mistakes. PubMedQA consists of three answer choices: yes, no, and maybe, allowing us to analyze the types of errors the model makes. For Llama3.1 (8B, 70B), PMC-Llama, and Gemma2, the bin colors remain unchanged, indicating that these models do not incorporate arbitrarily inserted supporting documents (Ans1, Ans1-Oth2). Meditron, w/o RAG, outputs "No" for all incorrect answers. However, when a document is provided, it changes all responses to "Yes," revealing an extremely sensitive behavior.

7 Conclusion

Our research explored the impact of Retrieval Augmented Generation (RAG) on model confidence in the medical domain where information reliability is crucial. We found that when models retrieve relevant documents, they not only boost accuracy but also show higher confidence scores. In contrast, irrelevant documents have little effect on improving confidence. Several models demonstrate the ability to judge if the retrieved documents connect to the correct answer, indicating a more discerning use of external information than we anticipated. Our evaluation metrics provide a clear framework for spotting the best generator models within RAG systems. The findings reveal that models adjust their output probabilities in response to the quality of the retrieved documents, which opens up new ways to measure and improve model performance. These insights help refine RAG methods, making them more reliable for high-stakes applications.

8 Limitations

8.1 The Experiments of the Other Domain

Our study prioritizes domains where RAG is applied, focusing specifically on the medical domain to analyze confidence. To advance further, it becomes necessary to validate RAG in domains such as finance and analyze its confidence in contexts requiring highly reliable information.

8.2 Further Analyzing New RAG Architecture

Our study focused exclusively on analyzing the basic RAG architecture. While the standard RAG framework directly utilizes retrieved documents within the LLM, newer RAG architectures incorporate various control mechanisms. Moving forward, it is essential to analyze these advanced architectures from the perspective of confidence as well.

8.3 Other Metrics for Evaluation

The evaluation metrics used in this study, ACE, may have some drawbacks. (Kull et al., 2019; Kumar et al., 2019; Baan et al., 2022). Since LLMs outputs are not always strictly correct or incorrect, researchers often use Prediction Rejection Ratio (PRR), which measures the correlation between confidence scores and output quality. (Fadeeva et al., 2023; Vashurin et al., 2025; He et al., 2024; Ozaki et al., 2025b). However, our study focuses on a multiple-QA task, where each question has a uniquely defined correct answer. Additionally, the models were evaluated using force decoding. Given these conditions, ACE serves as appropriate evaluation metrics.

8.4 Methods for Generating Model Outputs

This study deliberately avoids generating free-text responses from models. Instead, it retrieves answer candidates using force decoding This decision stems from an observation in prior research: many studies rely heavily on regular expressions to extract correct answers, leading to substantial accuracy variations even when using the same QA task and model. (`https://github.com/Teddy-XiongGZ/MedRAG`, `https://github.com/epfLLM/meditron`, `https://github.com/chaoyi-wu/PMC-LLaMA`.) To address this issue, we select answer choices based on the model's inherent output probabilities. This approach avoids introducing dependencies on specific evaluation metrics, which would otherwise occur if the model were required to generate explanations using Chain-of-Thought (CoT) or produce confidence scores.

9 Ethical Considerations

9.1 The Possibility of Dataset Bias

The datasets and retrieval mechanisms employed in our study may carry inherent biases, which could influence the model's predictions and potentially affect fairness in decision-making. Recognizing these biases, we advocate for the use of diverse and representative datasets to minimize their impact. Additionally, we uphold transparency by analyzing the interplay between confidence and accuracy, providing users with clearer insights into the system's limitations and confidence. However, we emphasize the need for human oversight, as no automated system can guarantee infallibility.

9.2 AI Assistant Tools

We used ChatGPT [2] and DeepL [3] to translate sentences to English and accelerate our research.

References

Marah Abdin, Jyoti Aneja, Hany Awadalla, Ahmed Awadallah, Ammar Ahmad Awan, Nguyen Bach, Amit Bahree, Arash Bakhtiari, Jianmin Bao, Harkirat

[2] `https://chatgpt.com/`

[3] `https://www.deepl.com/ja/translator`

Behl, et al. 2024. Phi-3 technical report: A highly capable language model locally on your phone. *arXiv preprint arXiv:2404.14219*.

Josh Achiam, Steven Adler, Sandhini Agarwal, Lama Ahmad, Ilge Akkaya, Florencia Leoni Aleman, Diogo Almeida, Janko Altenschmidt, Sam Altman, Shyamal Anadkat, et al. 2023. Gpt-4 technical report. *arXiv preprint arXiv:2303.08774*.

Akiko Aizawa, Eiji Aramaki, Bowen Chen, Fei Cheng, Hiroyuki Deguchi, Rintaro Enomoto, Kazuki Fujii, Kensuke Fukumoto, Takuya Fukushima, Namgi Han, et al. 2024. Llm-jp: A cross-organizational project for the research and development of fully open japanese llms. *arXiv preprint arXiv:2407.03963*.

Akari Asai, Zeqiu Wu, Yizhong Wang, Avirup Sil, and Hannaneh Hajishirzi. 2024. Self-RAG: Learning to retrieve, generate, and critique through self-reflection. In *The Twelfth International Conference on Learning Representations*.

Joris Baan, Wilker Aziz, Barbara Plank, and Raquel Fernandez. 2022. Stop measuring calibration when humans disagree. In *Proceedings of the 2022 Conference on Empirical Methods in Natural Language Processing*, pages 1892–1915, Abu Dhabi, United Arab Emirates. Association for Computational Linguistics.

Evan Becker and Stefano Soatto. 2024. Cycles of thought: Measuring llm confidence through stable explanations. *Preprint*, arXiv:2406.03441.

Lu Chen, Ruqing Zhang, Jiafeng Guo, Yixing Fan, and Xueqi Cheng. 2024. Controlling risk of retrieval-augmented generation: A counterfactual prompting framework. In *Findings of the Association for Computational Linguistics: EMNLP 2024*, pages 2380–2393, Miami, Florida, USA. Association for Computational Linguistics.

Zeming Chen, Alejandro Hernández Cano, Angelika Romanou, Antoine Bonnet, Kyle Matoba, Francesco Salvi, Matteo Pagliardini, Simin Fan, Andreas Köpf, Amirkeivan Mohtashami, Alexandre Sallinen, Alireza Sakhaeirad, Vinitra Swamy, Igor Krawczuk, Deniz Bayazit, Axel Marmet, Syrielle Montariol, Mary-Anne Hartley, Martin Jaggi, and Antoine Bosselut. 2023. Meditron-70b: Scaling medical pretraining for large language models. *Preprint*, arXiv:2311.16079.

Jeremy Cole, Michael Zhang, Daniel Gillick, Julian Eisenschlos, Bhuwan Dhingra, and Jacob Eisenstein. 2023. Selectively answering ambiguous questions. In *Proceedings of the 2023 Conference on Empirical Methods in Natural Language Processing*, pages 530–543, Singapore. Association for Computational Linguistics.

Tim Dettmers, Mike Lewis, Younes Belkada, and Luke Zettlemoyer. 2022. Gpt3. int8 (): 8-bit matrix multiplication for transformers at scale. *Advances in Neural Information Processing Systems*, 35:30318–30332.

Abhimanyu Dubey, Abhinav Jauhri, Abhinav Pandey, Abhishek Kadian, Ahmad Al-Dahle, Aiesha Letman, Akhil Mathur, Alan Schelten, Amy Yang, Angela Fan, et al. 2024. The llama 3 herd of models. *arXiv preprint arXiv:2407.21783*.

Ekaterina Fadeeva, Roman Vashurin, Akim Tsvigun, Artem Vazhentsev, Sergey Petrakov, Kirill Fedyanin, Daniil Vasilev, Elizaveta Goncharova, Alexander Panchenko, Maxim Panov, Timothy Baldwin, and Artem Shelmanov. 2023. LM-polygraph: Uncertainty estimation for language models. In *Proceedings of the 2023 Conference on Empirical Methods in Natural Language Processing: System Demonstrations*, pages 446–461, Singapore. Association for Computational Linguistics.

Jiahui Geng, Fengyu Cai, Yuxia Wang, Heinz Koeppl, Preslav Nakov, and Iryna Gurevych. 2024. A survey of confidence estimation and calibration in large language models. In *Proceedings of the 2024 Conference of the North American Chapter of the Association for Computational Linguistics: Human Language Technologies (Volume 1: Long Papers)*, pages 6577–6595, Mexico City, Mexico. Association for Computational Linguistics.

Chuan Guo, Geoff Pleiss, Yu Sun, and Kilian Q Weinberger. 2017. On calibration of modern neural networks. In *International Conference on Machine Learning*, pages 1321–1330. PMLR.

Kazuki Hayashi, Yusuke Sakai, Hidetaka Kamigaito, Katsuhiko Hayashi, and Taro Watanabe. 2024. Towards artwork explanation in large-scale vision language models. In *Proceedings of the 62nd Annual Meeting of the Association for Computational Linguistics (Volume 2: Short Papers)*, pages 705–729, Bangkok, Thailand. Association for Computational Linguistics.

Jianfeng He, Runing Yang, Linlin Yu, Changbin Li, Ruoxi Jia, Feng Chen, Ming Jin, and Chang-Tien Lu. 2024. Can we trust the performance evaluation of uncertainty estimation methods in text summarization? In *Proceedings of the 2024 Conference on Empirical Methods in Natural Language Processing*, pages 16514–16575, Miami, Florida, USA. Association for Computational Linguistics.

Minbyul Jeong, Jiwoong Sohn, Mujeen Sung, and Jaewoo Kang. 2024. Improving medical reasoning through retrieval and self-reflection with retrieval-augmented large language models. *Bioinformatics*, 40(Supplement_1):i119–i129.

Zhengbao Jiang, Jun Araki, Haibo Ding, and Graham Neubig. 2021. How can we know when language models know? on the calibration of language models for question answering. *Transactions of the Association for Computational Linguistics*, 9:962–977.

Zhengbao Jiang, Frank Xu, Luyu Gao, Zhiqing Sun, Qian Liu, Jane Dwivedi-Yu, Yiming Yang, Jamie Callan, and Graham Neubig. 2023. Active retrieval augmented generation. In *Proceedings of the 2023 Conference on Empirical Methods in Natural Language Processing*, pages 7969–7992, Singapore. Association for Computational Linguistics.

Qiao Jin, Bhuwan Dhingra, Zhengping Liu, William Cohen, and Xinghua Lu. 2019. PubMedQA: A dataset for biomedical research question answering. In *Proceedings of the 2019 Conference on Empirical Methods in Natural Language Processing and the 9th International Joint Conference on Natural Language Processing (EMNLP-IJCNLP)*, pages 2567–2577, Hong Kong, China. Association for Computational Linguistics.

Meelis Kull, Miquel Perello Nieto, Markus Kängsepp, Telmo Silva Filho, Hao Song, and Peter Flach. 2019. Beyond temperature scaling: Obtaining well-calibrated multi-class probabilities with dirichlet calibration. *Advances in neural information processing systems*, 32.

Ananya Kumar, Percy S Liang, and Tengyu Ma. 2019. Verified uncertainty calibration. *Advances in Neural Information Processing Systems*, 32.

Patrick Lewis, Ethan Perez, Aleksandra Piktus, Fabio Petroni, Vladimir Karpukhin, Naman Goyal, Heinrich Küttler, Mike Lewis, Wen-tau Yih, Tim Rocktäschel, et al. 2020. Retrieval-augmented generation for knowledge-intensive nlp tasks. *Advances in Neural Information Processing Systems*, 33:9459–9474.

Nelson F. Liu, Kevin Lin, John Hewitt, Ashwin Paranjape, Michele Bevilacqua, Fabio Petroni, and Percy Liang. 2024. Lost in the middle: How language models use long contexts. *Transactions of the Association for Computational Linguistics*, 12:157–173.

Xinbei Ma, Yeyun Gong, Pengcheng He, Hai Zhao, and Nan Duan. 2023. Query rewriting in retrieval-augmented large language models. In *Proceedings of the 2023 Conference on Empirical Methods in Natural Language Processing*, pages 5303–5315, Singapore. Association for Computational Linguistics.

Mahdi Pakdaman Naeini, Gregory F Cooper, and Milos Hauskrecht. 2015. Obtaining well calibrated probabilities using bayesian binning. In *Proceedings of the AAAI Conference on Artificial Intelligence*, pages 2901–2907.

Jeremy Nixon, Michael W. Dusenberry, Linchuan Zhang, Ghassen Jerfel, and Dustin Tran. 2019. Measuring calibration in deep learning. In *Proceedings of the IEEE/CVF Conference on Computer Vision and Pattern Recognition (CVPR) Workshops*.

Shintaro Ozaki, Kazuki Hayashi, Miyu Oba, Yusuke Sakai, Hidetaka Kamigaito, and Taro Watanabe. 2024. Bqa: Body language question answering dataset for video large language models. *arXiv preprint arXiv:2410.13206*.

Shintaro Ozaki, Kazuki Hayashi, Yusuke Sakai, Hidetaka Kamigaito, Katsuhiko Hayashi, and Taro Watanabe. 2025a. Towards cross-lingual explanation of artwork in large-scale vision language models. In *Findings of the Association for Computational Linguistics: NAACL 2025*, pages 3773–3809, Albuquerque, New Mexico. Association for Computational Linguistics.

Shintaro Ozaki, Kazuki Hayashi, Yusuke Sakai, Jingun Kwon, Hidetaka Kamigaito, Katsuhiko Hayashi, Manabu Okumura, and Taro Watanabe. 2025b. Texttiger: Text-based intelligent generation with entity prompt refinement for text-to-image generation. *arXiv preprint arXiv:2504.18269*.

Ankit Pal, Logesh Kumar Umapathi, and Malaikannan Sankarasubbu. 2022. Medmcqa: A large-scale multi-subject multi-choice dataset for medical domain question answering. In *Proceedings of the Conference on Health, Inference, and Learning*, volume 174 of *Proceedings of Machine Learning Research*, pages 248–260. PMLR.

Irina Proskurina, Luc Brun, Guillaume Metzler, and Julien Velcin. 2024. When quantization affects confidence of large language models? In *Findings of the Association for Computational Linguistics: NAACL 2024*, pages 1918–1928, Mexico City, Mexico. Association for Computational Linguistics.

Yusuke Sakai, Hidetaka Kamigaito, and Taro Watanabe. 2024. mCSQA: Multilingual commonsense reasoning dataset with unified creation strategy by language models and humans. In *Findings of the Association for Computational Linguistics: ACL 2024*, pages 14182–14214, Bangkok, Thailand. Association for Computational Linguistics.

Spurthi Setty, Harsh Thakkar, Alyssa Lee, Eden Chung, and Natan Vidra. 2024. Improving retrieval for rag based question answering models on financial documents. *arXiv preprint arXiv:2404.07221*.

Jiwoong Sohn, Yein Park, Chanwoong Yoon, Sihyeon Park, Hyeon Hwang, Mujeen Sung, Hyunjae Kim, and Jaewoo Kang. 2024. Rationale-guided retrieval augmented generation for medical question answering. *arXiv preprint arXiv:2411.00300*.

Zhiqing Sun, Xuezhi Wang, Yi Tay, Yiming Yang, and Denny Zhou. 2023. Recitation-augmented language models. In *International Conference on Learning Representations*.

Gemma Team, Morgane Riviere, Shreya Pathak, Pier Giuseppe Sessa, Cassidy Hardin, Surya Bhupatiraju, Léonard Hussenot, Thomas Mesnard, Bobak Shahriari, Alexandre Ramé, et al. 2024. Gemma 2: Improving open language models at a practical size. *arXiv preprint arXiv:2408.00118*.

Hugo Touvron, Thibaut Lavril, Gautier Izacard, Xavier Martinet, Marie-Anne Lachaux, Timothée Lacroix, Baptiste Rozière, Naman Goyal, Eric Hambro, Faisal Azhar, et al. 2023a. Llama: Open and efficient foundation language models. *arXiv preprint arXiv:2302.13971*.

Hugo Touvron, Louis Martin, Kevin Stone, Peter Albert, Amjad Almahairi, Yasmine Babaei, Nikolay Bashlykov, Soumya Batra, Prajjwal Bhargava, Shruti Bhosale, et al. 2023b. Llama 2: Open foundation and fine-tuned chat models. *arXiv preprint arXiv:2307.09288*.

Dennis Ulmer, Jes Frellsen, and Christian Hardmeier. 2022. Exploring predictive uncertainty and calibration in NLP: A study on the impact of method & data scarcity. In *Findings of the Association for Computational Linguistics: EMNLP 2022*, pages 2707–2735, Abu Dhabi, United Arab Emirates. Association for Computational Linguistics.

Roman Vashurin, Ekaterina Fadeeva, Artem Vazhentsev, Lyudmila Rvanova, Akim Tsvigun, Daniil Vasilev, Rui Xing, Abdelrahman Boda Sadallah, Kirill Grishchenkov, Sergey Petrakov, Alexander Panchenko, Timothy Baldwin, Preslav Nakov, Maxim Panov, and Artem Shelmanov. 2025. Benchmarking uncertainty quantification methods for large language models with lm-polygraph. *Preprint*, arXiv:2406.15627.

Jason Wei, Xuezhi Wang, Dale Schuurmans, Maarten Bosma, Fei Xia, Ed Chi, Quoc V Le, Denny Zhou, et al. 2022. Chain-of-thought prompting elicits reasoning in large language models. *Advances in neural information processing systems*, 35:24824–24837.

Thomas Wolf, Lysandre Debut, Victor Sanh, Julien Chaumond, Clement Delangue, Anthony Moi, Pierric Cistac, Tim Rault, Remi Louf, Morgan Funtowicz, Joe Davison, Sam Shleifer, Patrick von Platen, Clara Ma, Yacine Jernite, Julien Plu, Canwen Xu, Teven Le Scao, Sylvain Gugger, Mariama Drame, Quentin Lhoest, and Alexander Rush. 2020. Transformers: State-of-the-art natural language processing. In *Proceedings of the 2020 Conference on Empirical Methods in Natural Language Processing: System Demonstrations*, pages 38–45, Online. Association for Computational Linguistics.

Chaoyi Wu, Weixiong Lin, Xiaoman Zhang, Ya Zhang, Weidi Xie, and Yanfeng Wang. 2024. Pmc-llama: toward building open-source language models for medicine.

Guangzhi Xiong, Qiao Jin, Zhiyong Lu, and Aidong Zhang. 2024. Benchmarking retrieval-augmented generation for medicine. In *Findings of the Association for Computational Linguistics: ACL 2024*, pages 6233–6251, Bangkok, Thailand. Association for Computational Linguistics.

An Yang, Baosong Yang, Beichen Zhang, Binyuan Hui, Bo Zheng, Bowen Yu, Chengyuan Li, Dayiheng Liu, Fei Huang, Haoran Wei, et al. 2024. Qwen2. 5 technical report. *arXiv preprint arXiv:2412.15115*.

Antonio Jimeno Yepes, Yao You, Jan Milczek, Sebastian Laverde, and Renyu Li. 2024. Financial report chunking for effective retrieval augmented generation. *arXiv preprint arXiv:2402.05131*.

Jiaxin Zhang, Zhuohang Li, Kamalika Das, Bradley Malin, and Sricharan Kumar. 2023. SAC^3: Reliable hallucination detection in black-box language models via semantic-aware cross-check consistency. In *Findings of the Association for Computational Linguistics: EMNLP 2023*, pages 15445–15458, Singapore. Association for Computational Linguistics.

Zihao Zhao, Eric Wallace, Shi Feng, Dan Klein, and Sameer Singh. 2021. Calibrate before use: Improving few-shot performance of language models. In *International conference on machine learning*, pages 12697–12706. PMLR.

A Example Appendix

A.1 Detailed Model Settings

The PMC-Llama model was quantized to half-precision, while the 70B / 72B models were quantized to 4-bit precision for experimentation. The implementation relied on the Transformers library (Wolf et al., 2020) and bitsandbytes (Dettmers et al., 2022).

Model	Params	HuggingFace Name
Phi-3.5	3.8B	microsoft/Phi-3.5-mini-instruct
PMC-Llama	13B	axiong/PMC_LLaMA_13B
LLama2	70B	meta-llama/Llama-2-70b-chat-hf
Meditron	70B	epfl-llm/meditron-70b
Llama3.1	8B	meta-llama/Llama-3.1-8B
Llama3.1	70B	meta-llama/Llama-3.1-70B
Gemma2	2B	google/gemma-2-2b
Qwen2.5	14B	Qwen/Qwen2.5-14B
Qwen2.5	72B	Qwen/Qwen2.5-72B

Table 7: Detailed name of models.

A.2 Dataset Selection

The dataset selection is based on prior research by Xiong et al. (2024). From the datasets they used, we select those that include both QA pairs and explanatory passages that justify the answers (MedMCQA and PubMedQA) for this study.

Since the test set for MedMCQA is not publicly available, our study used the dev set as the test set, following the approach adopted in MIRAGE.[4]. We used the datasets, especially MedMCQA[5], PubMedQA[6].

A.3 Details of the Input Format

As described in Section 3, we determine the selected choice based on the output probabilities as-

[4] https://huggingface.co/datasets/openlifescienceai/medmcqa

[5] https://github.com/MedMCQA/MedMCQA

[6] https://github.com/pubmedqa/pubmedqa

signed by LLMs to the given candidates. To prevent answer choices from being split into multiple tokens by the tokenizer, we replace them with A, B, C, and D before feeding them into the model. This approach ensures a fair comparison across models, even for answer choices that would otherwise span multiple tokens.

A.4 Inference Settings

In this study, as far as inference which needs to use GPUs, all experiments were conducted on a single NVIDIA RTX A6000 and NVIDIA GeForce RTX 3090 GPU.

A.5 Why Do We Focus on the Medical Domain?

Among the various domains where information reliability is crucial (e.g., finance, law, autonomous driving, and healthcare), we chose to focus on healthcare for the following reasons:

- Complexity and Scale of Medical Texts: Medical documents are inherently complex and vast in scope, making them particularly suitable for RAG-based approaches. Combined with the critical importance of information reliability in this field, focusing on healthcare becomes a highly significant choice.

- Challenges in Real-World Applications: Questions involving detailed patient information, such as medical histories and symptoms, often overwhelm retrieval systems, making it difficult to identify crucial diagnostic clues. Furthermore, in practical applications, patient conditions and individual characteristics vary widely. Differences in age, medical history, genetic factors, and lifestyle often lead to variations in treatment for the same disease. Providing inaccurate information in such scenarios can result in severe consequences. (Sohn et al., 2024)

- Established Significance of BioNLP: The prominence of the healthcare domain is evident from the long-standing "BioNLP" workshop, which has been held for over two decades.[7]

- Emerging Trends in Healthcare RAG: Efforts to improve RAG performance in the medical domain have led to developments like Self-BioRAG, which leverages confidence scores. Its popularity and significant citation count highlight this field as a trending area of research. (Jeong et al., 2024) These points illustrate the rationale behind our focus on the healthcare domain.

[7] https://aclweb.org/aclwiki/BioNLP_Workshop

A.6 Details of Evaluation Metrics

Expected Calibration Error (ECE) Calibration error metrics evaluate whether a model's predicted probabilities align with actual accuracy in QA tasks. For instance, if a model assigns a 90% probability to an answer, the accuracy of such predictions should also be 90% for optimal calibration. Expected Calibration Error (ECE) (Naeini et al., 2015; Guo et al., 2017) quantifies this discrepancy by segmenting the predicted probability range into multiple bins and computing the difference between the predicted probability and the observed accuracy within each bin as follows:

$$\text{ECE} = \sum_{m=1}^{M} \frac{|B_m|}{n} \left|\text{acc}(B_m) - \text{conf}(B_m)\right| \quad (1)$$

Here, M denotes the number of bins, B_m represents the set of samples within bin m, $|B_m|$ is the number of samples in bin m, and n is the total number of samples. $\text{acc}(B_m)$ refers to the accuracy within bin B_m, while $\text{conf}(B_m)$ indicates the average confidence of predictions in bin m. ECE is computed as the weighted average of the absolute differences between the accuracy and confidence across bins, where the weights correspond to the proportion of samples in each bin.

Adaptive Calibration Error (ACE) ACE performs binning so that the number of samples in each bin remains constant. This approach ensures a more stable evaluation within each bin:

$$\text{ACE} = \frac{1}{KR} \sum_{k=1}^{K} \sum_{r=1}^{R} \left|\text{acc}(r,k) - \text{conf}(r,k)\right| \quad (2)$$

Here, K denotes the number of classes, R represents the number of bins, $\text{acc}(r,k)$ indicates the accuracy in bin r for class k, and $\text{conf}(r,k)$ denotes the confidence of predictions in the same bin and class.

A.7 The results using Expected Calibration Error (ECE)

The results using ECE are presented in Table 8 and Table 9. As discussed in Section 4.4, Proskurina et al. (2024) and Ulmer et al. (2022) have pointed out that ACE is a more suitable calibration error metric for multi-class classification problems, while ECE is better suited for binary classification. Nevertheless, we include ECE results for completeness and additional verification.

A.8 Violin plot

Figures 5a and 5b present the violin plot results for Llama3.1 (8B), while Figures 5c and 5d show the results for Llama3.1 (70B). The Llama models exhibit notably low output probabilities for candidate answer choices when no supporting documents are inserted. Furthermore, even when explicitly inserting documents containing supporting evidence, the output probabilities do not improve significantly. This suggests that these models may strictly adhere to predefined instructions and struggle to incorporate additional contextual information.

B Future Direction

In this study, we used a dataset containing correct answer choices along with supporting rationale passages for QA tasks. In the future, it may be possible to focus on non-medical domains by drawing on previous work that semi-automatically generates questions using LLMs (Ozaki et al., 2024; Sakai et al., 2024). There are also studies on explanation generation (Ozaki et al., 2025a; Hayashi et al., 2024), which could inform the generation of supporting rationales.

Model	Pattern	ECE ↓			
		None	Ans1	Ans1-Oth2	Oth3
Llama2 (70B)	w/o RAG	0.02	–	–	–
	Pre-Q	–	0.22	0.25	0.08
	Aft-Q	–	0.24	0.21	0.12
	Aft-C	–	0.20	0.17	0.10
Llama3.1 (70B)	w/o RAG	0.20	–	–	–
	Pre-Q	–	0.14	0.14	0.14
	Aft-Q	–	0.14	0.14	0.14
	Aft-C	–	0.14	0.14	0.14
Llama3.1 (8B)	w/o RAG	0.03	–	–	–
	Pre-Q	–	0.02	0.02	0.01
	Aft-Q	–	0.02	0.02	0.01
	Aft-C	–	0.02	0.02	0.01
Meditron (70B)	w/o RAG	0.07	–	–	–
	Pre-Q	–	0.18	0.08	0.09
	Aft-Q	–	0.16	0.09	0.15
	Aft-C	–	0.15	0.07	0.09
PMC-Llama (13B)	w/o RAG	0.16	–	–	–
	Pre-Q	–	0.01	0.01	0.04
	Aft-Q	–	0.01	0.02	0.06
	Aft-C	–	0.02	0.05	0.05
Gemma2 (2B)	w/o RAG	0.20	–	–	–
	Pre-Q	–	0.24	0.22	0.21
	Aft-Q	–	0.23	0.21	0.21
	Aft-C	–	0.25	0.23	0.23
Phi-3.5 (3.8B)	w/o RAG	0.05	–	–	–
	Pre-Q	–	0.06	0.07	0.32
	Aft-Q	–	0.04	0.06	0.42
	Aft-C	–	0.04	0.13	0.41
Qwen2.5 (14B)	w/o RAG	0.12	–	–	–
	Pre-Q	–	0.09	0.09	0.12
	Aft-Q	–	0.07	0.07	0.18
	Aft-C	–	0.08	0.10	0.09
Qwen2.5 (72B)	w/o RAG	0.04	–	–	–
	Pre-Q	–	0.10	0.08	0.20
	Aft-Q	–	0.09	0.10	0.19
	Aft-C	–	0.10	0.06	0.24

Table 8: The result of ECE using MedMCQA

Model	Pattern	ECE ↓			
		None	Ans1	Ans1-Oth2	Oth3
Llama2 (70B)	w/o RAG	0.12	–	–	–
	Pre-Q	–	0.30	0.31	0.14
	Aft-Q	–	0.30	0.31	0.14
	Aft-C	–	0.14	0.14	0.12
Llama3.1 (70B)	w/o RAG	0.02	–	–	–
	Pre-Q	–	0.10	0.10	0.10
	Aft-Q	–	0.10	0.10	0.10
	Aft-C	–	0.10	0.10	0.10
Llama3.1 (8B)	w/o RAG	0.24	–	–	–
	Pre-Q	–	0.23	0.22	0.23
	Aft-Q	–	0.22	0.23	0.23
	Aft-C	–	0.23	0.23	0.23
Meditron (70B)	w/o RAG	0.18	–	–	–
	Pre-Q	–	0.10	0.17	0.09
	Aft-Q	–	0.08	0.07	0.06
	Aft-C	–	0.04	0.04	0.07
PMC-Llama (13B)	w/o RAG	0.17	–	–	–
	Pre-Q	–	0.09	0.01	0.01
	Aft-Q	–	0.09	0.01	0.02
	Aft-C	–	0.02	0.04	0.05
Gemma2 (2B)	w/o RAG	0.07	–	–	–
	Pre-Q	–	0.04	0.02	0.02
	Aft-Q	–	0.02	0.01	0.01
	Aft-C	–	0.09	0.05	0.07
Phi-3.5 (3.8B)	w/o RAG	0.48	–	–	–
	Pre-Q	–	0.11	0.11	0.58
	Aft-Q	–	0.10	0.28	0.49
	Aft-C	–	0.15	0.32	0.46
Qwen2.5 (14B)	w/o RAG	0.16	–	–	–
	Pre-Q	–	0.04	0.05	0.18
	Aft-Q	–	0.03	0.05	0.27
	Aft-C	–	0.07	0.03	0.24
Qwen2.5 (72B)	w/o RAG	0.06	–	–	–
	Pre-Q	–	0.11	0.03	0.25
	Aft-Q	–	0.08	0.11	0.18
	Aft-C	–	0.09	0.07	0.14

Table 9: The result of ECE using PubMedQA

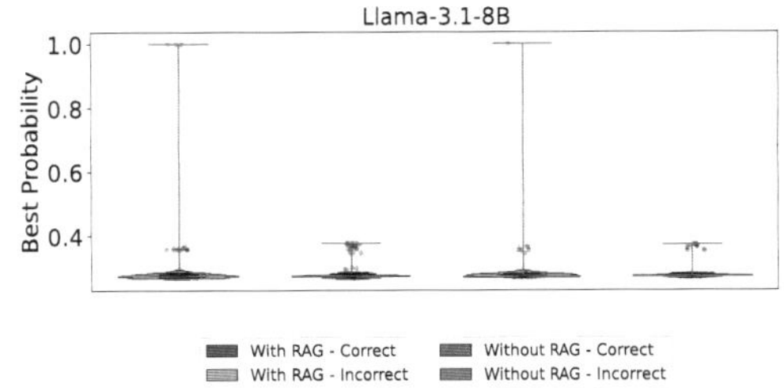

(a) The result of Llama3.1 (8B) using MedMCQA.

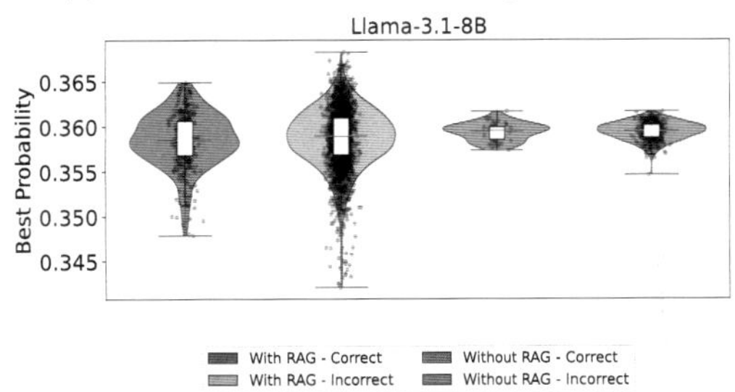

(b) The result of Llama3.1 (8B) using PubMedQA.

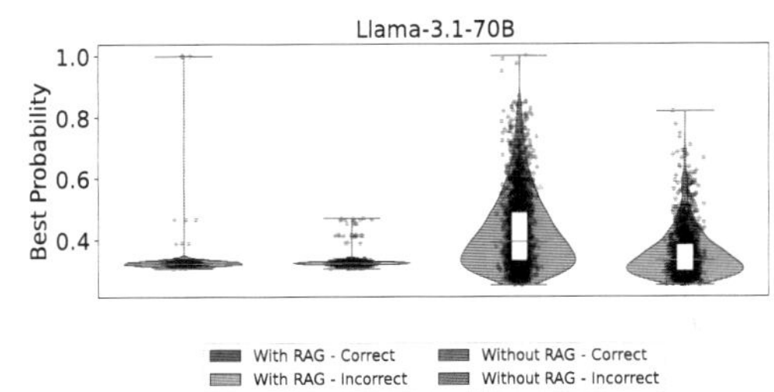

(c) The result of Llama3.1 (70B) using MedMCQA.

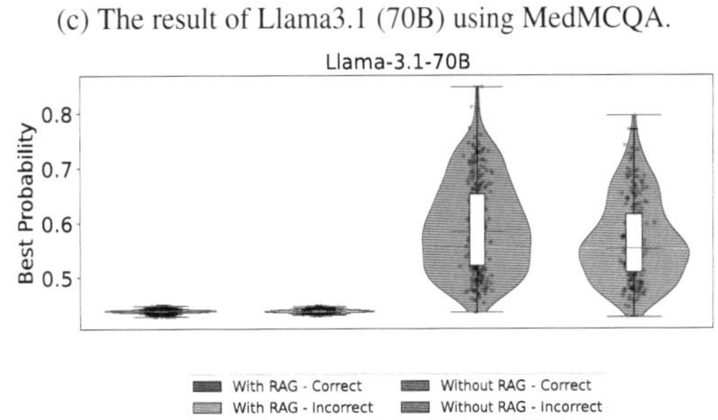

(d) The result of Llama3.1 (70B) using PubMedQA.

Figure 5: Results from MedMCQA and PubMedQA using Llama3.1.

B.1 Error Analysis

Figure 6 presents a plot illustrating the types of errors made on incorrectly answered questions. In PubMedQA, the answer choices consist of three options: yes, no, and maybe, allowing for detailed error analysis. Each bin represents the gold answer, and the plot visualizes the distribution of incorrect predictions for each question. The colors within the plot indicate how the model misclassified the answers.

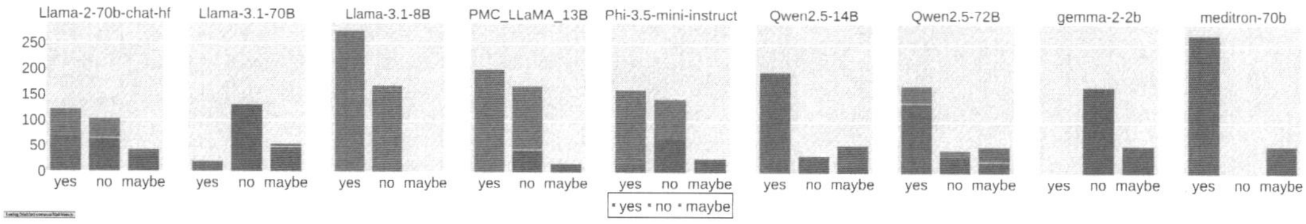

(a) When not inserting anything (w/o RAG).

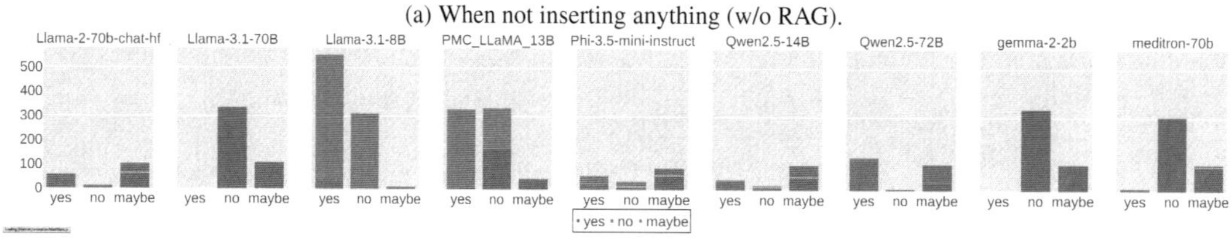

(b) When inserting a document containing the correct answer (Ans1).

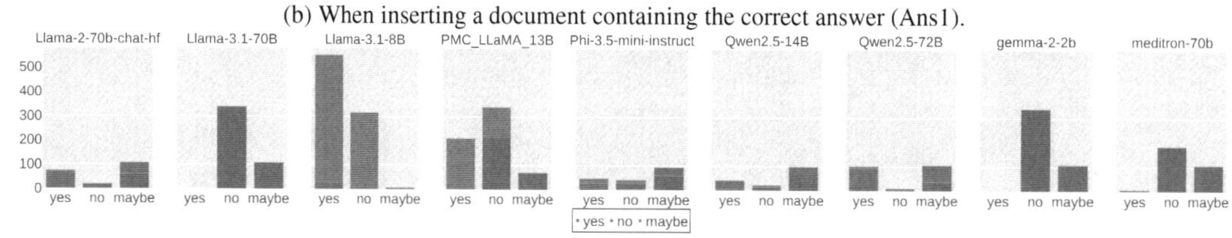

(c) When inserting one relevant document containing the correct answer and two unrelated documents (Ans1-Oth2).

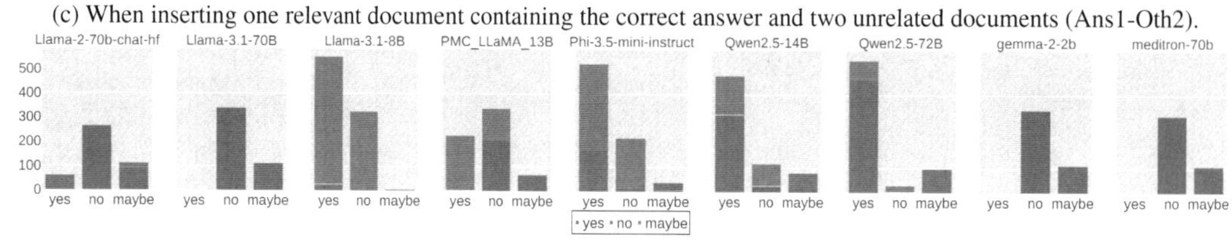

(d) When inserting three unrelated documents (Oth3).

Figure 6: Error analysis on PubMedQA: This figure illustrates how the model misclassified answers in relation to the correct ones. We gather the questions the model got wrong, group the items that actually had the correct answer (gold answer) into bins, and use colors to show how the model erred.

B.2 Prompts

Below are examples of prompts with and without RAG. When RAG is applied, three patterns-Pre-Question, After-Question, and After-Choice—are used in our study.

Prompt without RAG

```
You are a helpful medical expert, and your task is to answer a multi-choice medical question
using the relevant documents. Please first think step-by-step and then choose the answer from
the provided options. Your responses will be used for research purposes only, so please have a
definite answer.

Here is the question:
{question}

Here are the potential choices:
{choice0}
{choice1}
{choice2}
{choice3}

Answer:
```

Prompt with RAG

```
Here are the relevant documents: (Pre-Question)
{context}

You are a helpful medical expert, and your task is to answer a multi-choice medical question
using the relevant documents. Please first think step-by-step and then choose the answer from
the provided options. Your responses will be used for research purposes only, so please have a
definite answer.

Here are the relevant documents: (After-Question)
{context}

Here is the question:
{question}

Here are the relevant documents: (After-Choice)
{context}

Here are the potential choices:
{choice0}
{choice1}
{choice2}
{choice3}

Answer:
```

Effect of Multilingual and Domain-adapted Continual Pre-training on Few-shot Promptability

Ken Yano[1], Makoto Miwa[2,1]
[1]National Institute of Advanced Industrial Science and Technology
[2]Toyota Technological Institute

Abstract

Continual Pre-training (CPT) can help pre-trained large language models (LLMs) effectively adapt to new or under-trained domains or low-resource languages without re-training from scratch. Nevertheless, during CPT, the model's few-shot transfer ability is known to be affected for emergent tasks. We verified this by comparing the performance between the few-shot and fine-tuning settings on the same tasks. We used Llama3-ELAINE-medLLM, which was continually pre-trained on Llama3-8B, targeted for the biomedical domain, and adapted for multilingual languages (English, Japanese, and Chinese). We compared the performance of Llama3-ELAINE-medLLM and Llama3-8B in three emergent tasks: two related domain tasks, entity recognition (NER) and machine translation (MT), and one out-of-domain task, summarization (SUM). Our experimental results show that degradation in few-shot transfer ability does not necessarily indicate the model's underlying potential on the same task after fine-tuning.

1 Introduction

Continual Pre-training (CPT) can help pre-trained large language models (LLMs) effectively adapt to new or under-trained domains or low-resource languages without re-training from scratch.

Because open-source foundation LLMs such as the Llama series (Touvron et al., 2023a,b) are under-trained for the biomedical domain and non-English languages, many studies have been conducted to adapt such base LLMs to the biomedical domain in bilingual and multilingual settings (Singhal et al., 2022; Li et al., 2023; Singhal et al., 2023; Chen et al., 2023). Such LLMs are reported to perform better than the base model on downstream tasks in the target domains and languages.

However, CPT from a base model to endow non-English capability or to adapt to specific domains comes with the issue of degradation of the capabilities of the base model (Scialom et al., 2022; Fujii et al., 2024; Ankit Pal, 2024). Although many previous studies have shown that the incorporation of training datasets that the base model used during CPT significantly mitigates this forgetting (Rolnick et al., 2019; Chen et al., 2023; Lewkowycz et al., 2022; Yano et al., 2025), further analysis is needed to quantify these effects because such training datasets might be inaccessible and private and to determine whether these methods will be valid for a wide range of tasks.

In this work, we conducted experiments on three NLP tasks that were not primarily targeted during CPT. Specifically, we used Llama3-ELAINE-medLLM (Yano et al., 2025), which was adapted from Llama3-8B to the biomedical domain and has trilingual ability, including English (EN), Japanese (JA), and Chinese (ZH). Llama3-ELAINE is a pre-trained model without fine-tuning with instruction datasets.

For the emergent NLP tasks, we selected named entity recognition (NER) and machine translation (MT) tasks in a domain similar to the biomedical domain, and a summarization (SUM) task in the general domain, which were not targeted during CPT. Our experiments found that compared with Llama3-8B, Llama3-ELAINE due to CPT shows some forgetting phenomena that affect the model's promptability even in new tasks in similar domains where the model was trained during CPT.

However, our results also show that after fine-tuning Llama3-ELAINE-medLLM on the same downstream task, the model performs competitively or better than the base model. These results indicate that even though the adapted models' few-shot prompt ability may degrade in an emerging task, even in the relevant domains, the model will perform better after fine-tuning, as it has acquired more in-depth domain knowledge than the base model.

Proceedings of the 24th Workshop on Biomedical Language Processing (BioNLP 2025), pages 18–26
August 1, 2025 ©2025 Association for Computational Linguistics

Model	EN	JA	ZH
Llama3-8B	61.68	25.83	45.47
Llama3-ELAINE	59.56 (-2.1)	31.96 (+6.1)	52.25 (+6.8)

Table 1: Comparison of average scores of medical QA benchmarks in English, Japanese, and Chinese between ELAINE-medLLM and the base Llama3-8B

2 Related work

Numerous medLLMs (Singhal et al., 2022; Li et al., 2023; Singhal et al., 2023; Chen et al., 2023) have been proposed using CPT, adapted from open-source LLMs such as Llama (Touvron et al., 2023a,b). However, CPT can potentially degrade few-shot learning performance, hindering its ability to adapt to new tasks quickly. There have been many studies to prevent this issue, such as replaying pre-trained data and careful selection of the training dataset during CPT (Chen et al., 2023; Lewkowycz et al., 2022; Yano et al., 2025). The negative impact of CPT can be addressed in post-processing, such as task-specific pre-training, which involves further fine-tuning the pre-trained model on a small dataset related to the target few-shot task (Ke et al., 2022). Prompt engineering is another solution, involving the design of prompts during fine-tuning to guide the model toward the desired task with few-shot examples (Radford et al., 2019).

3 Experiments

To evaluate the effect of multilingual and domain-adapted continual pre-training on few-shot promptability for NLP tasks, we used Llama3-ELAINE-medLLM (Yano et al., 2025), which was continually pre-trained without instruction fine-tuning on Llama3-8B, targeted for the biomedical domain, and adapted for multiple languages (English, Japanese, and Chinese). Table 1 shows the average scores on several medical QA benchmarks in English, Japanese, and Chinese. We can see that Llama3-ELAINE-medLLM shows much better in-context learning (ICL) capabilities for medical QA tasks than Llama3-8B for Japanese and Chinese while slightly sacrificing English capability.

In this work, we used named entity recognition (NER) and machine translation (MT) tasks related to the biomedical domain and a summarization task in the general domain as the emergent tasks for our experiments.

	Training	Validation	Testing	lang
BC5CDR	500	500	500	EN
MedTxt-CR	128	10	10	JA
CMeEE-V2	19,600	400	400	ZH

Table 2: Statistics of NER datasets (# of documents). BC5CDR (en), MedTxt-CR (ja), and CMeEE-V2 (zh)

	Train	Validation	Test
JA-EN	1,000,000	1,790	1,812
JA-ZH	672,315	2,090	2,107

Table 3: Statistics of ASPEC parallel corpora (# of sentence pairs)

3.1 Datasets

3.1.1 NER dataset

We used BC5CDR (Li et al., 2016) for the English NER dataset, which defines "Disease" and "Chemical" entities. For the Japanese NER dataset, we used MedTxt-CR (Yada et al., 2022), which annotates various medical expression entities such as "disease", "anatomical part", etc. This experiment only used the "disease/symptoms" entity labeled as *d* in the corpus. Note that their annotation method does not delineate adjacent entity mentions, such as 呼吸困難、黄疸、下腿浮腫(dyspnea, jaundice, leg edema), which were labeled as one single, continuous entity rather than three independent entities as seen in other corpora.

For the Chinese NER dataset, we used CMeEE-V2 (Du et al., 2024), which annotates nine medical entity types, such as "disease", "clinical manifestations", "drugs", etc. This work used only disease and clinical symptoms labeled as "*dis*" and "*sim*" in the corpus, respectively. Table 2 summarizes the number of samples (documents) for each split of the corpus. Note that we randomly split the training datasets for the Japanese and Chinese datasets.

3.1.2 MT dataset

We used ASPEC (Nakazawa et al., 2016), consisting of two corpora from scientific paper abstracts: Japanese-English and Japanese-Chinese parallel corpora. Table 3 summarizes the number of samples (sentence pairs) for each split of the corpus. We used a four-way language pair for evaluation by reversing the source and target languages.

3.1.3 Summarization dataset

We used XLSum (Hasan et al., 2021), a diverse dataset of professionally annotated news article

	Train	Validation	Test
EN	306,522	11,535	11,535
JA	7,113	889	889
ZH	37,362	4,670	4,670

Table 4: Statistics of XLSum summarization dataset (# of text and summarization pairs) for English, Japanese, and Chinese

summary pairs from BBC that cover 45 languages. We used the English, Japanese, and Chinese splits of the dataset for evaluation. Table 4 summarizes the number of samples (text and summarization pairs) for each language dataset.

3.2 Evaluation

For each task, we evaluate the performance of ELAINE-medLLM and Llama-8B in the zero- or few-shot and fine-tuning settings. A sample of the instruction format for the training dataset for each task is described in Appendix A. The details of the settings are as follows.

3.2.1 Zero or few-shot settings

We used in-context learning (ICL) to evaluate each task's performance in the zero- or few-shot settings. For the few-shot settings, we evaluated one-shot, three-shot, five-shot, and ten-shot scenarios. ICL samples were selected from the training split, with the top N most similar to the input, where N is the number of few-shot samples. We used the text embeddings calculated by SentenceTransformer (Reimers and Gurevych, 2019) to compute similarity.

3.2.2 Fine-tuning settings

For each task, we fine-tuned the model by using the training split of each dataset. We used full-parameter tuning using DeepSpeed stage-3 and trained the model for 6, 3, and 3 epochs for NER, MT, and Summarization, respectively. We used the following training parameters:

- per_device_batch_size: 6
- gradient_accumulation_steps: 2
- learning_rate: 1e-6
- weight_decay: 0.001
- warmup_ratio: 0.1
- lr_scheduler_type: cosine

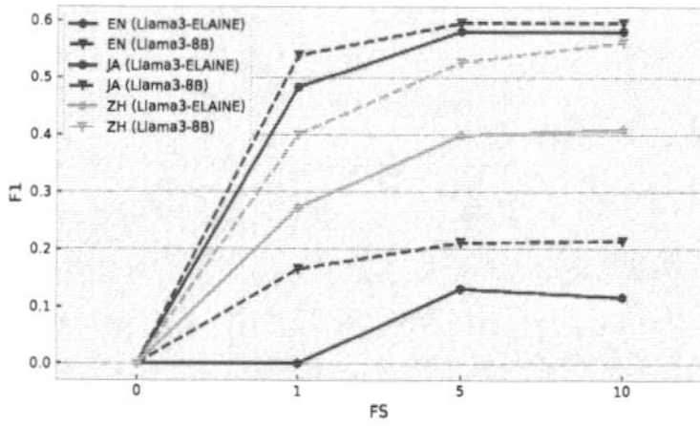

Figure 1: NER: few-shots performance in F1 (EN: BC5CDR, JA: MedTxt-CR, ZH: CMeEE-V2

		Precision	Recall	F1
EN	Llama3-ELAINE	0.825	0.802	0.813
	Llama3-8B	**0.833**	**0.831**	**0.832**
JA	Llama3-ELAINE	0.678	**0.701**	**0.689**
	Llama3-8B	**0.682**	0.667	0.674
ZH	Llama3-ELAINE	**0.766**	**0.792**	**0.779**
	Llama3-8B	0.764	0.789	0.776

Table 5: NER: fine-tuning performance (EN: BC5CDR, JA: MedTxt-CR, ZH: CMeEE-V2

4 Results

4.1 Named entity recognition (NER)

We adopt the TANL format (Paolini et al., 2021) to solve NER by LLM. In this format, the input text is copied to the output by annotating entity names and enclosing them in brackets by suffixing the detected entity type (see Appendix A). Figures 1 show the performance of language-dependent NER tasks in few-shot and Table 5 shows the performance of these NER tasks under the fine-tuning settings. The scores were computed by converting from TANL to IOB format (Ramshaw and Marcus, 1995). During conversion, we regulated the output by removing all parts that did not conform to our defined format, which made the zero-shot scores zero in all cases.

For few-shot settings, Llama performs better than Llama3-ELAINE-medLLM in all cases. This indicates the adverse effects of continual pre-training on the promptability of the base model. However, in fine-tuning settings, Llama3-ELAINE performs competitively with LLama in Japanese and Chinese. This result suggests that the degradation of promptability by CPT may not reveal the model's latent performance when the same task is fine-tuned.

4.2 Machine translation (MT)

Figures 2 and 3 show the few-shot performance of the MT task between Japanese and English and

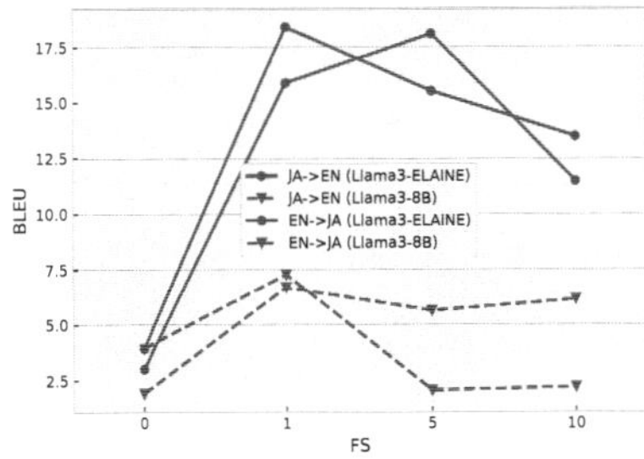

Figure 2: Machine Translation: few-shots (JA→EN, EN→JA) performance in BLEU (ASPEC)

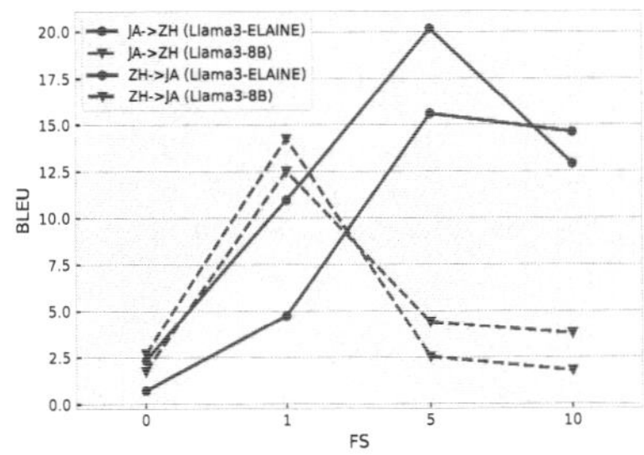

Figure 3: Machine Translation: few-shots (JA→ZH, ZH→JA) performance in BLEU (ASPEC)

	JA→EN	EN→JA
Llama3-ELAINE	**28.10**	**45.20**
Llama3-8B	27.92	44.36
	JA→ZH	ZH→JA
Llama3-ELAINE	34.25	**49.55**
Llama3-8B	**34.28**	48.67

Table 6: Machine Translation: fine-tuning performance in BLEU (ASPEC)

Japanese and Chinese, and Table 6 shows the fine-tuning performance of the same MT task measured in BLEU (Papineni et al., 2002). Unlike the performance of NER tasks, the performance of MT tasks, both in few-shot and fine-tuning, shows that ELAINE-medLLM is similar or superior to Llama3-8B.

This result indicates that continual pre-training does not always hurt the promptability of the base model for NLP tasks. We hypothesize that the degree of the effect depends on the novelty of the new task and its affinity to the training datasets used during CPT. Since ELAINE-medLLM is trained to harness multilingual ability, it works effectively in MT tasks for the same languages. On the other hand, although the domains of the previous NER tasks are highly aligned to those of the training

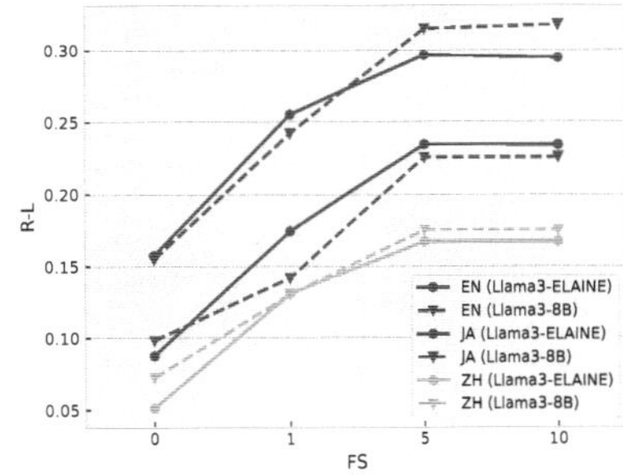

Figure 4: Summarization: few-shot performance in Rouge-L (RL) (XLSum)

		R-1	R-2	R-L
EN	Llama3-ELAINE	0.418	0.192	0.349
	Llama3-8B	**0.421**	**0.194**	**0.352**
JA	Llama3-ELAINE	**0.570**	**0.286**	**0.454**
	Llama3-8B	0.564	0.282	0.450
ZH	Llama3-ELAINE	0.368	0.171	0.319
	Llama3-8B	**0.371**	**0.173**	**0.322**

Table 7: Summarization: fine-tuning performance in Rouge-1 (R-1), Rouge-2 (R-2), Rouge-L (R-L) (XL-Sum)

dataset for ELAINE-medLLM, we assume that the novelty of the TANL output format affects its performance in the few-shot setting.

4.3 Summarization (SUM)

Figure 4 shows the results of the summarization task in few-shot settings measured in ROUGE-L (Lin, 2004). Table 7 shows the performance of the same summarization task in the fine-tuning setting in ROUGE-1, ROUGE-2, and ROUGE-L. Unlike previous NER and MT tasks, the SUM task is in the general domain for each of the three languages.

Unlike the previous two tasks (NER, MT), which can be considered related to the biomedical field, we could not observe noticeable performance differences in the fine-tuning setting. This is probably because the summarization task is in the general domain. We assume that CPT targeted for the biomedical domain does not affect fine-tuning performance in the general domain, though it shows a slight advantage for Llama3-8B for the few-shot setting.

	Precision	Recall	F1
Meditron-7B	**0.824**	0.744	0.783
Llama2-7B	0.808	**0.774**	**0.791**

Table 8: NER: fine-tuning performance (BC5CDR)

	R-1	R-2	R-L
Meditron-7B	**0.402**	**0.182**	**0.334**
Llama2-7B	0.397	0.172	0.330

Table 9: Summarization: fine-tuning performance in Rouge-1 (R-1), Rouge-2 (R-2), Rouge-L (R-L) (XL-Sum)

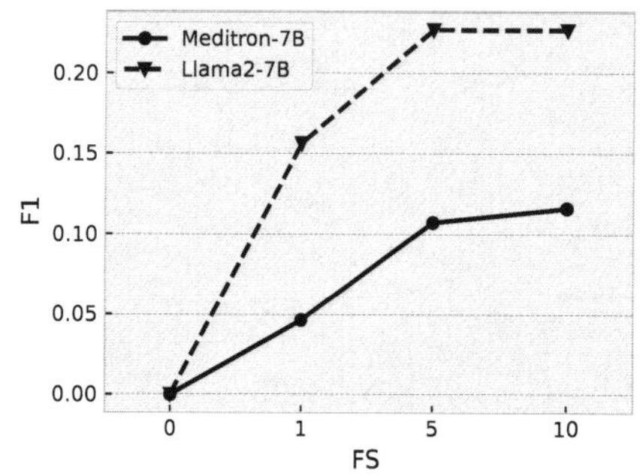

Figure 5: NER: few-shots performance in F1 (BC5CDR)

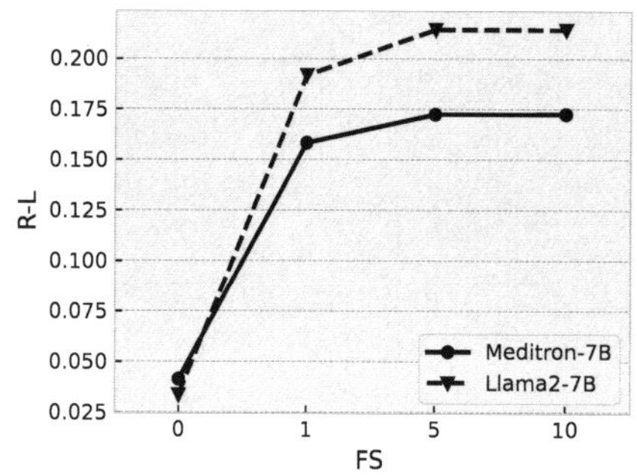

Figure 6: Summarization: few-shots performance in Rouge-1 (R-1) (XLSum)

5 Analysis

This section analyzes whether the phenomenon we found in the previous experiments can be observed in a different experimental setting.

5.1 Experimental Setting

We use Meditron-7B (Chen et al., 2023), an English medical LLM adapted from Llama2-7B (Touvron et al., 2023c), as the baseline. We selected the monolingual model because we aim to remove the effects of multilingualism on the results. For this experiment, we evaluate performance in few-shot and fine-tuning settings using the same NER task using BC5CDR and SUM task using English XL-Sum as in the previous experiments.

5.2 Results

Fig. 5 and Table 8 show the few-shot and fine-tuning NER results using BC5CDR. These results indicate that domain-adapted training does not benefit the performance of few-shot and fine-tuning results. Especially, Meditron-7B lags far behind Llama2-7B in a few-shot setting. Fig. 6 and Table 9 show the few-shot and fine-tuning Summarization results for English XLSum. In the few-shot setting, Meditron-7B lags far behind Llama2-7B, as in the NER task. However, the model shows competence against the baseline model in the fine-tuning setting.

These results confirm that the performance of the few-shot setting does not always show the model's potential in the fine-tuning setting of the same task. Nonetheless, we do not observe a similar trend in the comparative results between the domain-adapted and the base models. To summarize, domain adaptation works negatively for some tasks that do not depend clearly on acquired domain knowledge in few-shot settings, such as NER, and out-of-domain tasks, such as summarization. However, this degradation does not necessarily indicate the model's potential in fine-tuning settings.

6 Conclusion

CPT can help pre-trained large language models (LLMs) effectively adapt to new, under-trained domains or low-resourced languages without requiring retraining from scratch. Nevertheless, during CPT, the model's few-shot transfer ability is affected for emergent tasks. This also applies to new tasks, even in the relevant domains targeted during CPT. However, our experimental results show that degradation in few-shot transfer ability does not necessarily indicate the model's underlying potential in the same downstream task after fine-tuning. In our experiments, we observe that ELAINE-medLLM, which is adapted to the biomedical domain and endowed with trilingual ability (English, Japanese, and Chinese) by CPT from Llama3-8B, performs competitively with or better than the base model in all emergent tasks after fine-tuning, even though it shows some degradation in some few-shot settings.

Limitations

The prompt inputs used for few-shot evaluations were not optimized, suggesting that an optimal prompt might produce better results, such as prompt tuning or adopting a chain of thoughts. In this work, we only conducted performance analysis of Llama3-ELAINE and Meditron, adapted to the biomedical domain, against their base LLMs on three NLP tasks (NER, machine translation, and summarization). Hence, further experiments will be desired to evaluate the results we found in this study.

Acknowledgement

This paper is based on the results of project JPNP20006, which was commissioned by the New Energy and Industrial Technology Development Organization (NEDO).

References

Malaikannan Sankarasubbu Ankit Pal. 2024. Openbiollms: Advancing open-source large language models for healthcare and life sciences.

Zeming Chen, Alejandro Hernández-Cano, Angelika Romanou, Antoine Bonnet, Kyle Matoba, Francesco Salvi, Matteo Pagliardini, Simin Fan, Andreas Köpf, Amirkeivan Mohtashami, Alexandre Sallinen, Alireza Sakhaeirad, Vinitra Swamy, Igor Krawczuk, Deniz Bayazit, Axel Marmet, Syrielle Montariol, Mary-Anne Hartley, Martin Jaggi, and Antoine Bosselut. 2023. Meditron-70b: Scaling medical pretraining for large language models. *Preprint*, arXiv:2311.16079.

Xiaojing Du, Hanjie Zhao, Danyan Xing, Yuxiang Jia, and Hongying Zan. 2024. Mrc-based nested medical ner with co-prediction and adaptive pre-training. *Preprint*, arXiv:2403.15800.

Kazuki Fujii, Taishi Nakamura, Mengsay Loem, Hiroki Iida, Masanari Ohi, Kakeru Hattori, Hirai Shota, Sakae Mizuki, Rio Yokota, and Naoaki Okazaki. 2024. Continual pre-training for cross-lingual llm adaptation: Enhancing japanese language capabilities. *Preprint*, arXiv:2404.17790.

Tahmid Hasan, Abhik Bhattacharjee, Md. Saiful Islam, Kazi Mubasshir, Yuan-Fang Li, Yong-Bin Kang, M. Sohel Rahman, and Rifat Shahriyar. 2021. XL-sum: Large-scale multilingual abstractive summarization for 44 languages. In *Findings of the Association for Computational Linguistics: ACL-IJCNLP 2021*, pages 4693–4703, Online. Association for Computational Linguistics.

Zixuan Ke, Haowei Lin, Yijia Shao, Hu Xu, Lei Shu, and Bing Liu. 2022. Continual training of language models for few-shot learning. In *Proceedings of the 2022 Conference on Empirical Methods in Natural Language Processing*, pages 10205–10216, Abu Dhabi, United Arab Emirates. Association for Computational Linguistics.

Aitor Lewkowycz, Anders Andreassen, David Dohan, Ethan Dyer, Henryk Michalewski, Vinay Venkatesh Ramasesh, Ambrose Slone, Cem Anil, Imanol Schlag, Theo Gutman-Solo, Yuhuai Wu, Behnam Neyshabur, Guy Gur-Ari, and Vedant Misra. 2022. Solving quantitative reasoning problems with language models. *ArXiv*, abs/2206.14858.

Jiao Li, Yueping Sun, Robin J. Johnson, Daniela Sciaky, Chih-Hsuan Wei, Robert Leaman, Allan Peter Davis, Carolyn J. Mattingly, Thomas C. Wiegers, and Zhiyong Lu. 2016. Biocreative v cdr task corpus: a resource for chemical disease relation extraction. *Database*, 2016:baw068.

Yunxiang Li, Zihan Li, Kai Zhang, Ruilong Dan, Steve Jiang, and You Zhang. 2023. Chatdoctor: A medical chat model fine-tuned on a large language model meta-ai (llama) using medical domain knowledge. *Preprint*, arXiv:2303.14070.

Chin-Yew Lin. 2004. ROUGE: A package for automatic evaluation of summaries. In *Text Summarization Branches Out*, pages 74–81, Barcelona, Spain. Association for Computational Linguistics.

Toshiaki Nakazawa, Manabu Yaguchi, Kiyotaka Uchimoto, Masao Utiyama, Eiichiro Sumita, Sadao Kurohashi, and Hitoshi Isahara. 2016. Aspec: Asian scientific paper excerpt corpus. In *Proceedings of the Ninth International Conference on Language Resources and Evaluation (LREC 2016)*, pages 2204–2208, Portorož, Slovenia. European Language Resources Association (ELRA).

Giovanni Paolini, Ben Athiwaratkun, Jason Krone, Jie Ma, Alessandro Achille, Rishita Anubhai, Cicero Nogueira dos Santos, Bing Xiang, and Stefano Soatto. 2021. Structured prediction as translation between augmented natural languages. *Preprint*, arXiv:2101.05779.

Kishore Papineni, Salim Roukos, Todd Ward, and Wei-Jing Zhu. 2002. Bleu: a method for automatic evaluation of machine translation. In *Proceedings of the 40th Annual Meeting of the Association for Computational Linguistics*, pages 311–318, Philadelphia, Pennsylvania, USA. Association for Computational Linguistics.

Alec Radford, Jeff Wu, Rewon Child, David Luan, Dario Amodei, and Ilya Sutskever. 2019. Language models are unsupervised multitask learners.

Lance Ramshaw and Mitch Marcus. 1995. Text chunking using transformation-based learning. In *Third Workshop on Very Large Corpora*.

Nils Reimers and Iryna Gurevych. 2019. Sentence-BERT: Sentence embeddings using Siamese BERT-networks. In *Proceedings of the 2019 Conference on Empirical Methods in Natural Language Processing and the 9th International Joint Conference on Natural Language Processing (EMNLP-IJCNLP)*, pages 3982–3992, Hong Kong, China. Association for Computational Linguistics.

David Rolnick, Arun Ahuja, Jonathan Schwarz, Timothy P. Lillicrap, and Greg Wayne. 2019. Experience replay for continual learning. *Preprint*, arXiv:1811.11682.

Thomas Scialom, Tuhin Chakrabarty, and Smaranda Muresan. 2022. Fine-tuned language models are continual learners. In *Proceedings of the 2022 Conference on Empirical Methods in Natural Language Processing*, pages 6107–6122, Abu Dhabi, United Arab Emirates. Association for Computational Linguistics.

Karan Singhal, Shekoofeh Azizi, Tao Tu, S. Sara Mahdavi, Jason Wei, Hyung Won Chung, Nathan Scales, Ajay Tanwani, Heather Cole-Lewis, Stephen Pfohl, Perry Payne, Martin Seneviratne, Paul Gamble, Chris Kelly, Nathaneal Scharli, Aakanksha Chowdhery, Philip Mansfield, Blaise Aguera y Arcas, Dale Webster, Greg S. Corrado, Yossi Matias, Katherine Chou, Juraj Gottweis, Nenad Tomasev, Yun Liu, Alvin Rajkomar, Joelle Barral, Christopher Semturs, Alan Karthikesalingam, and Vivek Natarajan. 2022. Large language models encode clinical knowledge. *Preprint*, arXiv:2212.13138.

Karan Singhal, Tao Tu, Juraj Gottweis, Rory Sayres, Ellery Wulczyn, Le Hou, Kevin Clark, Stephen Pfohl, Heather Cole-Lewis, Darlene Neal, Mike Schaekermann, Amy Wang, Mohamed Amin, Sami Lachgar, Philip Mansfield, Sushant Prakash, Bradley Green, Ewa Dominowska, Blaise Aguera y Arcas, Nenad Tomasev, Yun Liu, Renee Wong, Christopher Semturs, S. Sara Mahdavi, Joelle Barral, Dale Webster, Greg S. Corrado, Yossi Matias, Shekoofeh Azizi, Alan Karthikesalingam, and Vivek Natarajan. 2023. Towards expert-level medical question answering with large language models. *Preprint*, arXiv:2305.09617.

Hugo Touvron, Thibaut Lavril, Gautier Izacard, Xavier Martinet, Marie-Anne Lachaux, Timothée Lacroix, Baptiste Rozière, Naman Goyal, Eric Hambro, Faisal Azhar, Aurelien Rodriguez, Armand Joulin, Edouard Grave, and Guillaume Lample. 2023a. Llama: Open and efficient foundation language models. *Preprint*, arXiv:2302.13971.

Hugo Touvron, Louis Martin, Kevin Stone, Peter Albert, Amjad Almahairi, Yasmine Babaei, Nikolay Bashlykov, Soumya Batra, Prajjwal Bhargava, Shruti Bhosale, Dan Bikel, Lukas Blecher, Cristian Canton Ferrer, Moya Chen, Guillem Cucurull, David Esiobu, Jude Fernandes, Jeremy Fu, Wenyin Fu, Brian Fuller, Cynthia Gao, Vedanuj Goswami, Naman Goyal, Anthony Hartshorn, Saghar Hosseini, Rui Hou, Hakan Inan, Marcin Kardas, Viktor Kerkez, Madian Khabsa, Isabel Kloumann, Artem Korenev, Punit Singh Koura, Marie-Anne Lachaux, Thibaut Lavril, Jenya Lee, Diana Liskovich, Yinghai Lu, Yuning Mao, Xavier Martinet, Todor Mihaylov, Pushkar Mishra, Igor Molybog, Yixin Nie, Andrew Poulton, Jeremy Reizenstein, Rashi Rungta, Kalyan Saladi, Alan Schelten, Ruan Silva, Eric Michael Smith, Ranjan Subramanian, Xiaoqing Ellen Tan, Binh Tang, Ross Taylor, Adina Williams, Jian Xiang Kuan, Puxin Xu, Zheng Yan, Iliyan Zarov, Yuchen Zhang, Angela Fan, Melanie Kambadur, Sharan Narang, Aurelien Rodriguez, Robert Stojnic, Sergey Edunov, and Thomas Scialom. 2023b. Llama 2: Open foundation and fine-tuned chat models. *Preprint*, arXiv:2307.09288.

Hugo Touvron, Louis Martin, Kevin Stone, Peter Albert, Amjad Almahairi, Yasmine Babaei, Nikolay Bashlykov, Soumya Batra, Prajjwal Bhargava, Shruti Bhosale, Dan Bikel, Lukas Blecher, Cristian Canton Ferrer, Moya Chen, Guillem Cucurull, David Esiobu, Jude Fernandes, Jeremy Fu, Wenyin Fu, Brian Fuller, Cynthia Gao, Vedanuj Goswami, Naman Goyal, Anthony Hartshorn, Saghar Hosseini, Rui Hou, Hakan Inan, Marcin Kardas, Viktor Kerkez, Madian Khabsa, Isabel Kloumann, Artem Korenev, Punit Singh Koura, Marie-Anne Lachaux, Thibaut Lavril, Jenya Lee, Diana Liskovich, Yinghai Lu, Yuning Mao, Xavier Martinet, Todor Mihaylov, Pushkar Mishra, Igor Molybog, Yixin Nie, Andrew Poulton, Jeremy Reizenstein, Rashi Rungta, Kalyan Saladi, Alan Schelten, Ruan Silva, Eric Michael Smith, Ranjan Subramanian, Xiaoqing Ellen Tan, Binh Tang, Ross Taylor, Adina Williams, Jian Xiang Kuan, Puxin Xu, Zheng Yan, Iliyan Zarov, Yuchen Zhang, Angela Fan, Melanie Kambadur, Sharan Narang, Aurelien Rodriguez, Robert Stojnic, Sergey Edunov, and Thomas Scialom. 2023c. Llama 2: Open foundation and fine-tuned chat models. *Preprint*, arXiv:2307.09288.

Shuntaro Yada, Yuta Nakamura, Shoko Wakamiya, and Eiji Aramaki. 2022. Real-mednlp: Overview of real document-based medical natural language processing task. In *Proceedings of the 16th NTCIR Conference on Evaluation of Information Access Technologies (NTCIR-16)*, pages 285–296.

Ken Yano, Zheheng Luo, Jimin Huang, Qianqian Xie, Masaki Asada, Chenhan Yuan, Kailai Yang, Makoto Miwa, Sophia Ananiadou, and Jun'ichi Tsujii. 2025. ELAINE-medLLM: Lightweight English Japanese Chinese trilingual large language model for biomedical domain. In *Proceedings of the 31st International Conference on Computational Linguistics*, pages 4670–4688, Abu Dhabi, UAE. Association for Computational Linguistics.

A Instruction format for tasks

A.1 Name entity recognition (NER)

We adopt the TANL format (Paolini et al., 2021) to solve NER by LLM. In this format, the LLM is expected to copy the source context to the target by annotating detected entity mentions, enclosing them in brackets, and suffixing them with the detected entity type. The descriptions of valid entity types and their labels are specified after the task description. The following shows a sample instruction format for the NER task in the case of BC5CDR (Li et al., 2016).

Instruction for NER task

##Task##
You are an AI language model trained to extract entities from sentences. See all possible labels and their descriptions below.
###
Description: Disease name
Label: Disease
###
Description: Chemical name
Label: Chemical
###
<|eot_id|>##Text##
Famotidine-associated delirium. A series of six cases.
Famotidine is a histamine H2-receptor antagonist used in inpatient settings for prevention of stress ulcers and is showing increasing popularity because of its low cost. Although all of the currently available H2-receptor antagonists have shown the propensity to cause delirium, only two previously reported c ases have been associated with famotidine. The authors report on six cases of famotidine-associated delirium in hospitalized pa tients who cleared completely upon removal of famotidine. The pharmacokinetics of famotidine are reviewed, with no change in its metabolism in the elderly population seen. The implications of using famotidine in elderly persons are discussed.
<|eot_id|>##Result##
[Famotidine|Chemical]-associated [delirium|Disease]. A series of six cases.
[Famotidine|Chemical] is a histami ne H2-receptor antagonist used in inpatient settings for prevention of stress [ulcers|Disease] and is showing increasing popula rity because of its low cost. Although all of the currently available H2-receptor antagonists have shown the propensity to caus e [delirium|Disease], only two previously reported cases have been associated with [famotidine|Chemical]. The authors report on six cases of [famotidine|Chemical]-associated [delirium|Disease] in hospitalized patients who cleared completely upon removal of [famotidine|Chemical]. The pharmacokinetics of [famotidine|Chemical] are reviewed, with no change in its metabolism in the e lderly population seen. The implications of using [famotidine|Chemical] in elderly persons are discussed. <|eot_id|>

A.2 Machine translation (MT)

The following shows a sample instruction format for the MT task in the case of ASPEC (Nakazawa et al., 2016).

Instruction for MT task

##Task##
Translate the following sentences from English to Japanese.
<|eot_id|>##Source##
Lattice images were not obtained by high resolution power image observation.
<|eot_id|>##Target##
高分解能像観察での格子像は得られなかった。<|eot_id|>

A.3 Summarization (SUM)

The following shows a sample instruction format for the SUM task in the case of XLSum (Hasan et al., 2021).

Instruction for SUM task

##Task##
Summarize the following paragraph.
<|eot_id|>##Source##
Lewis Williams, 20, died on 11 January from a shotgun wound suffered in Wath Road, Mexborough. South Yorkshire Police said two men aged 20 and 49 were arrested on Friday in connection with his death, bringing the total number of arrests to eight. Two boys, aged 16 and 17, have been charged with Mr Williams' murder and are next due to appear in court on 1 February. Police said one of the men arrested on Friday, a 20-year-old from Barnsley, was arrested on suspicion of murder, while a 49-year-old man from Doncaster was arrested on suspicion of assisting an offender and possession of ammunition. Both are being held in police custody. Four other men, aged between 20 and 32, who have been arrested in connection with Mr Williams'death have been released on bail. Follow BBC Yorkshire on Facebook, Twitter and Instagram. Send your story ideas to yorkslincs.news@bbc.co.uk or send video here.
<|eot_id|>##Target##
Two more people have been arrested in connection with a fatal shooting. <|eot_id|>

MedSummRAG: Domain-Specific Retrieval for Medical Summarization

Guanting Luo
The University of Osaka
guanting.luo@ist.osaka-u.ac.jp

Yuki Arase
Institute of Science Tokyo
arase@c.titech.ac.jp

Abstract

Medical text summarization faces significant challenges due to the complexity and domain-specific nature of the language. Although large language models have achieved significant success in general domains, their effectiveness in the medical domain remains limited. This limitation stems from their insufficient understanding of domain-specific terminology and difficulty in interpreting complex medical relationships, which often results in suboptimal summarization quality. To address these challenges, we propose MedSummRAG, a novel retrieval-augmented generation (RAG) framework that integrates external knowledge to enhance summarization. Our approach employs a fine-tuned dense retriever, trained with contrastive learning, to retrieve relevant documents for medical summarization. The retrieved documents are then integrated with the input text to generate high-quality summaries. Experimental results show that MedSummRAG achieves significant improvements in ROUGE scores on both zero/few-shot and fine-tuned language models, outperforming baseline methods. These findings underscore the importance of RAG and domain adaptation of the retriever for medical text summarization. The source code of this paper can be obtained from: https://github.com/guantingluo98/MedSummRAG

1 Introduction

Medical text summarization is a crucial task for helping medical practitioners and patients, aiming to distill complex and information-dense medical documents into concise, accurate, and clinically useful summaries (Xie et al., 2023). This task is particularly challenging due to the specialized nature of medical language and the presence of domain-specific terminologies (Chaves et al., 2022). Traditional summarization models often struggle in this domain, as they may fail to generate satisfactory summaries.

With the rise of large language models (LLMs), significant advancements have been made in general-domain summarization (Pu et al., 2023). However, medical summarization presents unique challenges, such as domain-specific terminology and complex relationships, which generic LLMs struggle to address effectively. LLMs trained on broad-domain corpora tend to overlook key medical concepts, misinterpret medical abbreviations, and produce hallucinated content that could mislead practitioners and researchers (Li et al., 2024; Hosseini et al., 2024). These limitations highlight the need for models that can effectively incorporate external domain knowledge. By leveraging external knowledge documents, such as healthcare question-answer pairs, models can better understand domain-specific concepts, reduce errors, and generate high-quality summaries.

In this work, we propose MedSummRAG (Medical Summarization with Retrieval-Augmented Generation), a novel retrieval-augmented generation (RAG) framework designed specifically for medical text summarization. By leveraging external medical knowledge, MedSummRAG enhances the quality of generated summaries. Our approach employs a fine-tuned dense retriever, trained using contrastive learning (van den Oord et al., 2019), to effectively identify domain-relevant documents.

The key contribution of our work is the novel RAG framework for medical text summarization. Our approach improves retrieval quality by leveraging contrastive learning that employs synthetic positive samples generated using an LLM. This enables the framework to effectively identify domain-relevant documents, improving the overall quality of generated summaries. We conduct experiments to investigate the effectiveness of MedSummRAG. Our results demonstrate consistent improvements measured by ROUGE scores in multiple configurations: both on zero/few-shot and fine-tuned language models.

Proceedings of the 24th Workshop on Biomedical Language Processing (BioNLP 2025), pages 27–33
August 1, 2025

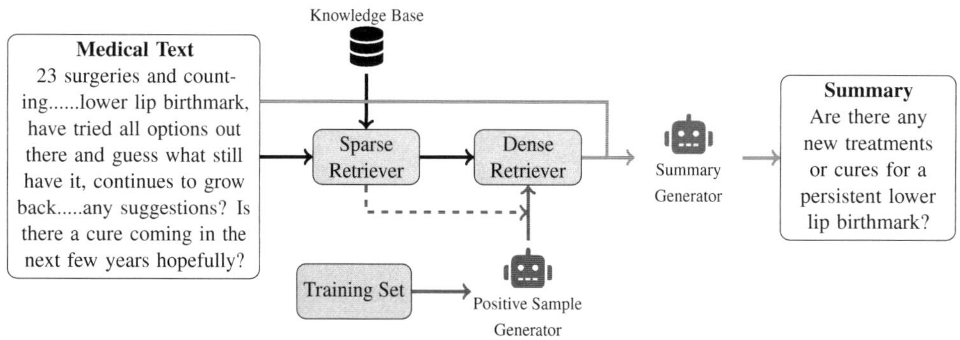

Figure 1: Overview of our MedSummRAG framework. Black arrows indicate retrieving relevant documents by sparse and dense retrievers. Blue dashed arrow represents the negative sample extraction, while solid blue arrows show the generation of synthetic positive samples. Both positive and negative samples are used to fine-tune the dense retriever. Orange arrows show the concatenation of medical text and the retrieved document as input to the summary generator, producing the summary.

2 Related Work

Medical summarization has been a long-standing research problem due to its critical role in supporting clinical decision-making and healthcare planning. With the rise of pre-trained language models, significant progress has been made in medical summarization. Pre-trained language models have demonstrated the ability to generate medical summaries, such as doctor-patient conversation summaries, by utilizing knowledge derived from pretrained models (Zhang et al., 2021). Balde et al. (2024) proposed MEDVOC, a dynamic vocabulary adaptation strategy that optimizes pre-trained language models' vocabulary for medical text summarization, achieving improvements in high Out-Of-Vocabulary settings.

Despite the progress enabled by pre-trained models in medical summarization, their pre-trained knowledge may be insufficient for handling specific downstream tasks. RAG offers a promising solution by integrating external knowledge to enhance the overall quality of generations (Fan et al., 2024). Recent studies have demonstrated the potential of RAG in various domains, such as decision-making tasks (Lee et al., 2024); question answering (Jeong et al., 2024); and radiology report generation (Xia et al., 2024). Although RAG has demonstrated success in various domains, its application to medical summarization remains underexplored. Our work aims to propose a RAG framework specifically adapted for medical summarization to generate high-quality summaries.

3 Proposed Method

The overall workflow of our approach is illustrated in Figure 1. The proposed method consists of document retrieval (Section 3.1) and summary generation (Section 3.2). For improving the retrieval step to adapt to the medical domain, we employ contrastive learning with synthetic data (Section 3.3).

3.1 Document Retrieval

We employ the BM25 (Robertson et al., 2009; Lù, 2024) ranking function to retrieve an initial set of candidate documents based on lexical similarity to the input text. This sparse retrieval method serves to reduce the computational cost of subsequent dense retrieval by narrowing down the search space to a manageable set of candidate documents.

A dense retriever then re-ranks the highly-ranked documents retrieved by the sparse retriever and selects the most relevant document for the generation stage. This step should ensure that the retrieved document is lexically and semantically aligned with the input text to provide useful knowledge for summarization.

3.2 Generation

The generation stage produces summaries based on the input text and the retrieved document. Following the approach of Lewis et al. (2020), we simply concatenate the retrieved document with the input text and feed the combined input into a language model. The generator is expected to produce coherent and factually accurate summaries, leveraging both the input text and the external knowledge provided by the retrieved document.

You're a retrieval augmented generation assistant, skilled in generating retrieval targets for auto summarization via RAG. Here is the input-summary pair from a training set: INPUT: {Train set input text} SUMMARY: {Train set summary} Please help me with generating one fake retrieved question-answer document that would be useful for training a dense retrieval model for automatic summarization via retrieval augmented generation. The fake retrieved document should have this kind of format: QUESTION: ANSWER:

Table 1: Promt for synthetic sample generation

3.3 Domain Adaptation of Retriever

The retrieval stage aims to identify the most relevant document from a knowledge base to enhance the summarization process. Although existing RAG methods have shown great success in question-answering tasks (Asai et al., 2023; Xiong et al., 2024), they often struggle to identify documents that are truly useful for medical text summarization. This is because pre-trained dense retrievers lack the ability to understand what document structures and content are beneficial for enhancing summarization in the medical domain.

The key challenge in fine-tuning retrievers for medical summarization is the lack of training data. To address this challenge, we leverage an LLM to generate synthetic positive samples that capture the structural and contextual patterns useful for summarization. Specifically, for each text-summary pair in a training set of summarization, we prompt the LLM to generate a synthetic pair that may enhance medical summarization. Table 1 shows the prompt we used.

We then fine-tune the dense retriever using contrastive learning, inspired by the work of Huang et al. (2023), which improves its ability to retrieve documents relevant to medical summarization. For negative samples, we randomly select low-ranked documents by the sparse retriever that should be less relevant to the input text. We optimize the retrieval model using the InfoNCE loss (van den Oord et al., 2019).

4 Experiments

4.1 Evaluation Dataset & Knowledge Base

Evaluation Dataset: We use the CHQ-Summ dataset (Yadav et al., 2022), which consists of consumer health questions formulated by non-experts, paired with brief summaries of the corresponding questions. The questions are sourced from Yahoo! Answers L6 corpus[1]. The dataset contains 1,000 training samples, 107 validation samples, and 400 test samples. We evaluate the performance of our method using ROUGE (Lin, 2004) scores, including ROUGE-1, ROUGE-2, and ROUGE-L.

Knowledge Base: We construct the knowledge base using Yahoo! Answers L6 corpus, which contains more than 4 million question-answer pairs. Each document in the knowledge base represents a single question-answer pair. The content covered in this corpus extends far beyond the scope of healthcare and medicine, encompassing a wide range of topics. To prevent data leakage, we exclude all question-answer pairs that overlap with the CHQ-Summ dataset.

4.2 Implementation Details

We employed BM25 for sparse retrieval, which retrieved the top 150 documents for each input text. We employed the BGE-M3 (Chen et al., 2024) model as the base dense retriever. For contrastive learning, the positive samples were generated by a frozen Qwen-2.5-7B-Instruct model[2], while the negative samples were constructed by randomly sampling 3 documents from the BM25-ranked documents in the range of positions 101 to 150 for each training sample. The BGE-M3 model was fine-tuned for 5 epochs with a total batch size of 8. After fine-tuning, the BGE-M3 model re-ranked the top 20 documents retrieved by BM25 and selected the top 1 document for the generator.

4.3 Experiment Settings

To evaluate the effectiveness of our MedSumm-RAG approach, we conducted four sets of experiments with different generator settings: standard fine-tuning, few-shot prompting, and Low Rank Adapters (LoRA) (Hu et al., 2022) fine-tuning on different language models. Specifically, we employed (1) BioBART-large (Yuan et al., 2022): the

[1] https://webscope.sandbox.yahoo.com/catalog.php?datatype=l&did=11

[2] https://qwenlm.github.io/blog/qwen2.5/

Base Model	Setting	ROUGE-1	ROUGE-2	ROUGE-L
BioBART-large (Standard Fine-tuned)	Baseline	41.22	23.17	38.79
	+ Naive RAG	42.19	22.95	38.79
	+ Fine-tuned RAG	**44.50**	**24.58**	**41.19**
Qwen-2.5-7B-Instruct (1-shot Prompting)	Baseline	34.97	13.85	32.82
	+ Naive RAG	38.53	16.42	33.61
	+ Fine-tuned RAG	**39.45**	**17.59**	**34.60**
Qwen-2.5-7B-Instruct (2-shot Prompting)	Baseline	38.15	16.34	33.82
	+ Naive RAG	39.89	18.00	35.28
	+ Fine-tuned RAG	**40.27**	**18.30**	**35.95**
Qwen-2.5-7B-Instruct (LoRA Fine-tuned)	Baseline	42.21	21.99	38.84
	+ Naive RAG	42.56	21.80	39.32
	+ Fine-tuned RAG	**42.95**	**22.82**	**40.03**

Table 2: Performance comparison of different base models on the CHQ-Summ dataset. Results demonstrate the effectiveness of our method across various models, few-shot scenarios, and fine-tuning strategies.

model has shown its strong performance in medical text generation tasks. BioBART-large was first fine-tuned using the training set without RAG, followed by the second stage of fine-tuning with RAG. Each fine-tuning consisted of 20 epochs with a batch size of 8. We also experimented with (2) Qwen-2.5-7B-Instruct with One-Shot Prompting, (3) Qwen-2.5-7B-Instruct with Two-Shot Prompting, and (4) Qwen-2.5-7B-Instruct with LoRA Fine-Tuning (Hu et al., 2022): the model was fine-tuned using LoRA for 10 epochs with a batch size of 8. LoRA fine-tuning was performed with a rank of 8, alpha of 16, and no dropout. The details of the prompts are described in example A.1 and example A.2

In all settings, the baseline is the corresponding fine-tuned model or a few-shot prompted models without RAG. In addition, we also compared to a naive RAG where the retriever has not been fine-tuned, i.e., without domain adaptation. All the experiment was conducted on a single NVIDIA A6000 48G GPU.

4.4 Results

In this section, we highlight the key contribution of our RAG-enhanced approach, demonstrating its effectiveness across different models, few-shot settings, and fine-tuning strategies. A consistent performance gap between naive RAG and fine-tuned RAG underscores the importance of domain-adaptive retrieval. This contrast suggests that synthetic examples play a key role in improving the relevance of the retrieval and the overall quality of the summary.

For the standard fine-tuned BioBART-large model, our method significantly improves performance. With naive RAG, only the ROUGE-1 score shows a marginal improvement, while the ROUGE-2 score slightly decreases, and the ROUGE-L score remains unchanged. However, with MedSumm-RAG, the BioBART-large model achieves a notable increase in ROUGE scores, highlighting the effectiveness of integrating external knowledge through domain-adapted retriever.

For the Qwen-2.5-7B-Instruct model in few-shot prompt settings, our method consistently enhances performance without fine-tuning the generator. Even with naive RAG, we observe modest improvements in ROUGE scores. Fine-tuning the RAG component further boosts performance, demonstrating the effectiveness of our method even when the generator is frozen. Additional prompt examples also contribute to improved results.

For the LoRA fine-tuned Qwen-2.5-7B-Instruct model, integrating naive RAG yields marginal improvements in ROUGE-1 and ROUGE-L, while ROUGE-2 experiences a slight decline compared to the baseline. In contrast, our domain-adapted RAG enhances performance across all ROUGE metrics, demonstrating the importance of optimizing the retrieval process to effectively leverage external knowledge in the LoRA fine-tuning setting.

5 Conclusion

Our experimental results highlight the effectiveness of leveraging external knowledge for adapting language models to medical summarization tasks, addressing the challenge of domain adaptation in specialized medical contexts. Future work includes extending our approach to a larger-scale knowledge base to further enhance retrieval effectiveness. Additionally, beyond ROUGE-based evaluation, incorporating human evaluation could provide deeper insights into the quality of generated summaries. Furthermore, exploring the application of our fine-tuned RAG framework to other medical summarization tasks, such as radiology report summarization, is another promising direction for advancing our work.

Limitations

While our proposed method demonstrates promising results in improving medical text summarization, its generalizability remains to be validated. Our experiments are conducted exclusively on the CHQ-Summ dataset, which focuses on summarizing customer health questions. While this dataset provides a valuable benchmark for medical question summarization, it does not fully represent the diversity of medical texts, such as clinical notes, or discharge summaries. In addition, while the Yahoo! Answers L6 corpus offers broad coverage, it may contain content of varying accuracy, which motivates future exploration of more medically curated sources to further reduce hallucination risks.

Acknowledgements

This work was supported by Cross-ministerial Strategic Innovation Promotion Program (SIP) on "Integrated Health Care System" Grant Number JPJ012425.

References

Akari Asai, Zeqiu Wu, Yizhong Wang, Avirup Sil, and Hannaneh Hajishirzi. 2023. Self-rag: Learning to retrieve, generate, and critique through self-reflection. *Preprint*, arXiv:2310.11511.

Gunjan Balde, Soumyadeep Roy, Mainack Mondal, and Niloy Ganguly. 2024. Medvoc: vocabulary adaptation for fine-tuning pre-trained language models on medical text summarization. In *Proceedings of the Thirty-Third International Joint Conference on Artificial Intelligence*, pages 6180–6188.

Andrea Chaves, Cyrille Kesiku, and Begonya Garcia-Zapirain. 2022. Automatic text summarization of biomedical text data: a systematic review. *Information*, 13(8):393.

Jianlv Chen, Shitao Xiao, Peitian Zhang, Kun Luo, Defu Lian, and Zheng Liu. 2024. Bge m3-embedding: Multi-lingual, multi-functionality, multi-granularity text embeddings through self-knowledge distillation. *Preprint*, arXiv:2402.03216.

Wenqi Fan, Yujuan Ding, Liangbo Ning, Shijie Wang, Hengyun Li, Dawei Yin, Tat-Seng Chua, and Qing Li. 2024. A survey on rag meeting llms: Towards retrieval-augmented large language models. In *Proceedings of the 30th ACM SIGKDD Conference on Knowledge Discovery and Data Mining*, pages 6491–6501.

Manda Hosseini, Mandana Hosseini, and Reza Javidan. 2024. Leveraging large language models for clinical abbreviation disambiguation. *Journal of medical systems*, 48(1):27.

Edward J Hu, Yelong Shen, Phillip Wallis, Zeyuan Allen-Zhu, Yuanzhi Li, Shean Wang, Lu Wang, Weizhu Chen, et al. 2022. Lora: Low-rank adaptation of large language models. *ICLR*, 1(2):3.

Wenyu Huang, Mirella Lapata, Pavlos Vougiouklis, Nikos Papasarantopoulos, and Jeff Pan. 2023. Retrieval augmented generation with rich answer encoding. In *Proceedings of the 13th International Joint Conference on Natural Language Processing and the 3rd Conference of the Asia-Pacific Chapter of the Association for Computational Linguistics (Volume 1: Long Papers)*, pages 1012–1025, Nusa Dua, Bali. Association for Computational Linguistics.

Soyeong Jeong, Jinheon Baek, Sukmin Cho, Sung Ju Hwang, and Jong Park. 2024. Adaptive-RAG: Learning to adapt retrieval-augmented large language models through question complexity. In *Proceedings of the 2024 Conference of the North American Chapter of the Association for Computational Linguistics: Human Language Technologies (Volume 1: Long Papers)*, pages 7036–7050, Mexico City, Mexico. Association for Computational Linguistics.

Myeonghwa Lee, Seonho An, and Min-Soo Kim. 2024. PlanRAG: A plan-then-retrieval augmented generation for generative large language models as decision makers. In *Proceedings of the 2024 Conference of the North American Chapter of the Association for Computational Linguistics: Human Language Technologies (Volume 1: Long Papers)*, pages 6537–6555, Mexico City, Mexico. Association for Computational Linguistics.

Patrick Lewis, Ethan Perez, Aleksandra Piktus, Fabio Petroni, Vladimir Karpukhin, Naman Goyal, Heinrich Küttler, Mike Lewis, Wen-tau Yih, Tim Rocktäschel, Sebastian Riedel, and Douwe Kiela. 2020. Retrieval-augmented generation for knowledge-intensive nlp tasks. In *Advances in Neural Information Processing Systems*, volume 33, pages 9459–9474. Curran Associates, Inc.

Songda Li, Yunqi Zhang, Chunyuan Deng, Yake Niu, and Hui Zhao. 2024. Better late than never: Model-agnostic hallucination post-processing framework towards clinical text summarization. In *Findings of the Association for Computational Linguistics: ACL 2024*, pages 995–1011, Bangkok, Thailand. Association for Computational Linguistics.

Chin-Yew Lin. 2004. ROUGE: A package for automatic evaluation of summaries. In *Text Summariza-*

tion Branches Out, pages 74–81, Barcelona, Spain. Association for Computational Linguistics.

Xing Han Lù. 2024. Bm25s: Orders of magnitude faster lexical search via eager sparse scoring. *Preprint*, arXiv:2407.03618.

Xiao Pu, Mingqi Gao, and Xiaojun Wan. 2023. Summarization is (almost) dead. *Preprint*, arXiv:2309.09558.

Stephen Robertson, Hugo Zaragoza, et al. 2009. The probabilistic relevance framework: Bm25 and beyond. *Foundations and Trends® in Information Retrieval*, 3(4):333–389.

Aaron van den Oord, Yazhe Li, and Oriol Vinyals. 2019. Representation learning with contrastive predictive coding. *Preprint*, arXiv:1807.03748.

Peng Xia, Kangyu Zhu, Haoran Li, Hongtu Zhu, Yun Li, Gang Li, Linjun Zhang, and Huaxiu Yao. 2024. RULE: Reliable multimodal RAG for factuality in medical vision language models. In *Proceedings of the 2024 Conference on Empirical Methods in Natural Language Processing*, pages 1081–1093, Miami, Florida, USA. Association for Computational Linguistics.

Qianqian Xie, Zheheng Luo, Benyou Wang, and Sophia Ananiadou. 2023. A survey for biomedical text summarization: From pre-trained to large language models. *Preprint*, arXiv:2304.08763.

Guangzhi Xiong, Qiao Jin, Zhiyong Lu, and Aidong Zhang. 2024. Benchmarking retrieval-augmented generation for medicine. In *Findings of the Association for Computational Linguistics: ACL 2024*, pages 6233–6251, Bangkok, Thailand. Association for Computational Linguistics.

Shweta Yadav, Deepak Gupta, and Dina Demner-Fushman. 2022. Chq-summ: A dataset for consumer healthcare question summarization. *Preprint*, arXiv:2206.06581.

Hongyi Yuan, Zheng Yuan, Ruyi Gan, Jiaxing Zhang, Yutao Xie, and Sheng Yu. 2022. BioBART: Pretraining and evaluation of a biomedical generative language model. In *Proceedings of the 21st Workshop on Biomedical Language Processing*, pages 97–109, Dublin, Ireland. Association for Computational Linguistics.

Longxiang Zhang, Renato Negrinho, Arindam Ghosh, Vasudevan Jagannathan, Hamid Reza Hassanzadeh, Thomas Schaaf, and Matthew R. Gormley. 2021. Leveraging pretrained models for automatic summarization of doctor-patient conversations. In *Findings of the Association for Computational Linguistics: EMNLP 2021*, pages 3693–3712, Punta Cana, Dominican Republic. Association for Computational Linguistics.

A Appendix

This appendix shows the prompts used for summary generation methods described in this paper.

For few-shot learning setting we randomly select samples from the training set. The example A.1 shows the prompt template we use for generating medical summaries in one-shot setting.

Example A.1. You are a helpful assistant. Your task is to summarize the given question based on the provided question and possibly helpful retrieved document. The retrieved document may or may not be useful for summarization.
Example:
QUESTION: {Example input text}
RETRIEVED DOCUMENT: {Example retrieved document}
SUMMARY: {Example summary}
QUESTION: {Test set input text}
RETRIEVED DOCUMENT: {Test set retrieved document}
SUMMARY:

The example A.2 shows the prompt template we use for generating medical summaries in two-shot setting.

Example A.2. You are a helpful assistant. Your task is to summarize the given question based on the provided question and possibly helpful retrieved document. The retrieved document may or may not be useful for summarization.
Examples:
QUESTION: {Example input text}
RETRIEVED DOCUMENT: {Example retrieved document}
SUMMARY: {Example summary}
QUESTION: {Example input text}
RETRIEVED DOCUMENT: {Example retrieved document}
SUMMARY: {Example summary}
QUESTION: {Test set input text}
RETRIEVED DOCUMENT: {Test set retrieved document}
SUMMARY:

Enhancing Stress Detection on Social Media Through Multi-Modal Fusion of Text and Synthesized Visuals

Efstathia Soufleri
Archimedes, Athena Research Center
Greece
e.soufleri@athenarc.gr

Sophia Ananiadou
The University of Manchester
Manchester, UK
Archimedes, Athena Research Center
Greece
sophia.ananiadou@manchester.ac.uk

Abstract

Social media platforms generate an enormous volume of multi-modal data, yet stress detection research has predominantly relied on text-based analysis. In this work, we propose a novel framework that integrates textual content with synthesized visual cues to enhance stress detection. Using the generative model DALL·E, we synthesize images from social media posts, which are then fused with text through the multi-modal capabilities of a pre-trained CLIP model, which encodes both text and image data into a shared semantic space. Our approach is evaluated on the Dreaddit dataset, where a classifier trained on frozen CLIP features achieves 94.90% accuracy, and full fine-tuning further improves performance to 98.41%. These results underscore the integration of synthesized visuals with textual data not only enhances stress detection but also offers a robust method over traditional text-only methods, paving the way for innovative approaches in mental health monitoring and social media analytics.

1 Introduction

Social media has emerged as a pervasive platform for personal expression, generating enormous volumes of data that encompass both textual and visual modalities (Baltrušaitis et al., 2018; Mouzannar et al., 2018; Abousaleh et al., 2020). This rich, heterogeneous data offers unprecedented opportunities for understanding human behavior and mental health. However, prevailing stress detection research has largely focused on text-based analysis, overlooking the potential for complementary affective cues that can be inferred or synthesized into visual representations.

Recent advances in multi-modal machine learning have shown that integrating diverse data sources can significantly enhance performance on affective and behavioral tasks (Song et al., 2024; Ieracitano et al., 2020; Amal et al., 2022; Zhang et al., 2020). At the same time, generative models such as DALL·E have opened new avenues for synthesizing high-quality visuals from textual descriptions (Ramesh et al., 2021; Khachatryan et al., 2023; Tewel et al., 2022). This proliferation of data prompts an essential question: How can the fusion of synthesized visual data with traditional text data improve the accuracy and effectiveness of stress detection algorithms?

In this work, we introduce a novel multi-modal framework that leverages both text and synthesized images for stress detection. Specifically, we generate images from social media posts using DALL·E (Ramesh et al., 2021) and integrate these visuals with text via the robust joint embedding space provided by a pre-trained CLIP model (Radford et al., 2021). We evaluate our approach on the Dreaddit dataset (Turcan and McKeown, 2019), a collection of social media posts annotated to indicate whether the person who wrote the post suffers from stress or not. Our experiments demonstrate that a classifier trained on frozen CLIP features achieves 94.90% accuracy, while full fine-tuning further elevates performance to 98.41%. These results indicate that synthesized visuals capture subtle emotional and contextual cues that are absent from text alone, thereby significantly enhancing detection accuracy.

Our contributions are threefold:

1. We propose a novel multi-modal framework that fuses text and synthesized visuals to address the limitations of traditional text-only stress detection methods.

2. We demonstrate the effectiveness of our approach on the Dreaddit dataset (Turcan and McKeown, 2019), achieving state-of-the-art performance through both classifier-only training and full fine-tuning strategies.

3. We provide an in-depth analysis of the impact of multi-modal fusion on capturing nuanced affective signals, laying the groundwork for

Proceedings of the 24th Workshop on Biomedical Language Processing (BioNLP 2025), pages 34–43

future research in mental health monitoring using social media data.

2 Related Work

Stress detection on social media has traditionally been approached using text-based methods. Early studies primarily relied on lexicon-based techniques and classical machine learning algorithms to identify linguistic markers of stress and anxiety in user-generated content (De Choudhury et al., 2013; Aldarwish and Ahmad, 2017; Biswas and Hasija, 2022). More recent approaches have employed deep learning architectures, such as recurrent neural networks (Salehinejad et al., 2017) and transformer-based architectures (Vaswani et al., 2017; Ji et al., 2022; Yang et al., 2024; Shi et al., 2024), to capture complex syntactic and semantic patterns from text. Despite these advancements, text-only methods may fail to capture affective or contextual information that can be made more salient through synthesized visual representations.

The growing interest in multi-modal learning has spurred research into integrating multiple data sources to improve performance on affective and behavioral tasks. Several studies have demonstrated that fusing textual and visual information can enhance emotion recognition (Kosti et al., 2017; Cowie et al., 2001; Zhu et al., 2025) and sentiment analysis (Baltrušaitis et al., 2018; Wankhade et al., 2022). For instance, multi-modal architectures that combine convolutional neural networks (Li et al., 2021) for image analysis with language models for text have shown improved accuracy over single-modality approaches (Mittal et al., 2018; You et al., 2015; Feng et al., 2025; Devlin et al., 2018; Liu et al., 2019). However, the application of multi-modal techniques to stress detection remains relatively underexplored.

Generative models have further broadened the horizons of multi-modal research. Models such as DALL·E have shown impressive capabilities in synthesizing high-quality images from textual prompts (Ramesh et al., 2021; Zhou et al., 2023), thereby providing a novel means to augment datasets that lack explicit visual content. Concurrently, models like CLIP have established robust joint embedding spaces that effectively capture the semantic relationships between images and text (Radford et al., 2021; Qiao et al., 2019; Wang et al., 2023; Zhong et al., 2021; Gu et al., 2023; Wang et al., 2021). These innovations have paved the way for leveraging synthesized visuals to complement textual data, offering new insights into affective states that may not be fully captured by text alone.

Prior work in mental health has shown that linguistic patterns in social media (e.g., first-person pronouns, hopelessness, negative tone) indicate stress, anxiety, or depression (De Choudhury et al., 2013; Cohan et al., 2018), and visual cues (e.g., expressions, colors, context) also reflect affective states (Abousaleh et al., 2020). Building on this, we hypothesize that even synthesized images—when guided by affect-sensitive prompts—can offer complementary signals for stress detection.

Our work builds on these lines of research by integrating synthesized visuals with text-based analysis for stress detection. Our work employs generative image synthesis in conjunction with a multi-modal representation framework for this task. By fusing the complementary strengths of DALL·E and CLIP, we aim to address the limitations of traditional text-only approaches and provide a more holistic understanding of stress as expressed on social media.

3 Methodology

In this section, we describe our multi-modal framework for stress detection, which integrates synthesized visual cues with textual information. Our approach consists of two stages: image generation, and multi-modal representation with CLIP.

3.1 Image Generation

To enrich textual data, we use the generative capabilities of DALL·E 3, an advanced version of the DALL·E model (Ramesh et al., 2021). This model synthesizes images closely aligned with textual descriptions. The process begins with the input of a text prompt into a specialized *text encoder*. This text encoder is adept at converting the textual information into a high-dimensional representation space (*text encoding*), aiming to capture the core semantic content of the prompt (Figure 1).

Following this, a component known as the *diffusion prior* takes over, which is a crucial part of the model's architecture. The prior is responsible for mapping the text-encoded semantic representation to a corresponding *image encoding*. This image encoding is designed to retain the semantic content conveyed by the text, ensuring that the generated images reflect the intended themes and elements of the input prompt.

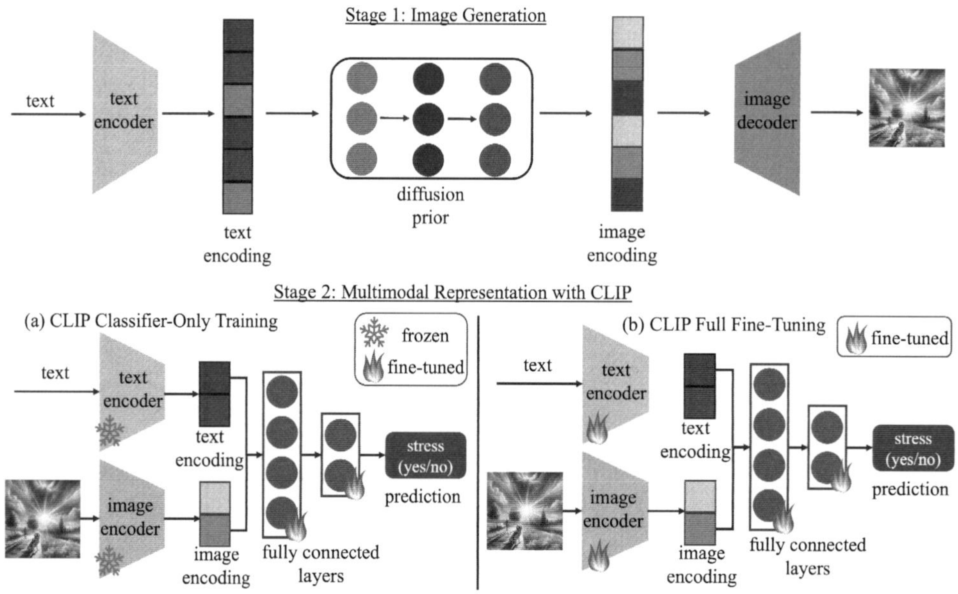

Figure 1: Methodology overview. (1) image generation: the posts (text) and the corresponding prompt are converted into images that visually represent the text's semantic content, (2) multi-modal representation with CLIP: the images, alongside the original text, are processed through CLIP to form a joint embedding space, used for stress detection. The CLIP Classifier-Only Training strategy fine-tunes the classifier (fully connected layers) while keeping the CLIP base model (text and image encoder) frozen. The CLIP Full Fine-Tuning strategy fine-tunes both the classifier and the CLIP base model. This process leverages both textual and visual data to enhance detection accuracy.

The final step in the image generation process involves an *image decoder*. This decoder uses the image encoding to stochastically generate the final visual output. The resulting *image* is a visual representation of the semantic information encoded from the initial text prompt, materializing as a synthetic image that complements the textual data in our multi-modal stress detection framework. By leveraging this advanced image synthesis process, we ensure that the generated visuals are both semantically relevant and visually coherent, providing a robust foundation for further multi-modal analysis.

3.2 Multi-modal Representation with CLIP

We employ the pre-trained CLIP model (Radford et al., 2021) to facilitate a robust multi-modal representation, leveraging its capacity to encode both text and images into a shared joint embedding space. Each text sample is processed by the *text encoder* to extract textual features, while corresponding synthesized images are preprocessed and passed through the *image encoder*. The features from both modalities are normalized and concatenated to form a joint representation (Figure 1). This embedding captures complementary affective cues from both textual and visual data, enhancing our ability to detect stress signals on social media.

To effectively train our model, we adopt two training strategies:

1. **CLIP Classifier-Only Training:** In this approach, we keep the pre-trained CLIP base frozen, focusing training efforts solely on the attached *classifier*. This method benefits from the robustness of the existing multi-modal embeddings, avoiding alterations to the underlying representations and ensuring stability.

2. **CLIP Full Fine-Tuning:** Alternatively, we engage in full fine-tuning of both the CLIP model and the *classifier*. This strategy allows the entire network to adapt more comprehensively to the domain-specific nuances of stress-related content, improving detection accuracy by refining the joint embedding space to better capture subtle emotional nuances.

This integrated methodology not only leverages generative image synthesis to augment textual information but also strategically fuses these modalities in a joint embedding space. The approach is designed to enhance the detection of nuanced affective signals that are pivotal for accurate stress detection on social media platforms.

Model	Accuracy (%)	Weighted F1 (%)
MentalRoBERTa	96.14	94.24
MentalBERT	69.32	78.75
RoBERTa-base	96.14	94.24
BERT-base	96.14	94.67
CLIP Classifier-Only Training (Ours)	94.90	92.42
CLIP Full Fine-Tuning (Ours)	98.41	98.27

Table 1: Performance comparison of our approach with general-purpose and mental health-specific models on the Dreaddit dataset.

4 Experiments and Results

In this section, we outline our experimental setup, present results, and discuss findings.

4.1 Experimental Setup

The Dreaddit dataset (Turcan and McKeown, 2019) comprises 2,837 training samples and 414 testing samples, where each sample is a social media post accompanied by a binary label indicating the presence or absence of stress. The posts are drawn from mental health-related subreddits such as r/depression, r/anxiety, and r/relationships. Each post was annotated through crowdsourced judgments, with three annotators per instance and majority voting used to determine the final label. The dataset is approximately balanced across the two classes. For each post, we generate a corresponding synthetic image using DALL·E 3. The hyperparameters reported in the Appendix. Our experiments compare the following models:

- **CLIP Classifier-Only Training (Ours):** Classifier-only training where the pre-trained CLIP model is kept frozen while only the classifier is trained.
- **CLIP Full Fine-Tuning (Ours):** Full fine-tuning of both the CLIP model and the classifier on the Dreaddit dataset.
- **Text-Only Baselines:** Pre-trained discriminative language models which are either general purpose (RoBERTa-base, BERT-base (Devlin et al., 2018; Liu et al., 2019)) or finetuned for mental health applications (MentalRoBERTa, MentalBERT (Ji et al., 2022)).

4.2 Results

Table 1 reports the accuracy and weighted F1 scores for our proposed models and the text-only baselines. Our CLIP Classifier-Only Training model achieves an accuracy of 94.90% with a weighted F1 score of 92.42%, while the CLIP Full Fine-Tuning model reaches 98.41% accuracy and 98.27% weighted F1 score. In comparison, the text-only models yield competitive performance for MentalRoBERTa, RoBERTa-base, and BERT-base (accuracy around 96.14% and weighted F1 around 94%), whereas MentalBERT underperforms. The results demonstrate that full fine-tuning of our multi-modal framework (CLIP Full Fine-Tuning) leads to a substantial improvement in performance over classifier-only training, highlighting the benefit of adapting the joint image-text representations to stress detection. Furthermore, our approach achieves competitive performance compared to strong text-only baselines, while offering the added advantage of leveraging synthesized visual cues. Even though our results demonstrate strong performance gains, we acknowledge that we have not conducted statistical significance testing across multiple random seeds. Future work will incorporate such evaluations to better assess the robustness of our findings.

Modality	Accuracy (%)	Weighted F1 (%)
Image-Only	95.22	93.17
Text-Only	96.82	96.31
Image + Text	98.41	98.27

Table 2: Ablation study of our method comparing image-only, text-only, and combined multi-modal model.

4.3 Ablation Study: Modality Contributions

To better understand the contribution of each modality, we performed an ablation study by evaluating our model trained using only the synthesized images, only the textual data, and the fusion of both modalities Table 2. The image-only model, which relies solely on visual cues extracted from generated images, achieved an accuracy of 95.22% and a validation weighted F1 score of 93.17%. The text-only model, using only the original social media posts, reached an accuracy of 96.82% and a validation weighted F1 score of 96.31%. Notably, when both modalities are integrated, our multi-modal framework achieves significantly improved performance, with an accuracy of 98.41% and a validation weighted F1 score of 98.27%. These findings indicate that while the text-only model is already highly effective, the addition of synthesized visual information provides complementary affective cues that further enhance stress detection performance.

4.4 Discussion

Our experiments validate the hypothesis that integrating synthesized visuals with text enhances stress detection on social media. The significant performance improvement observed with full fine-tuning suggests that adapting the multi-modal embeddings to the domain-specific nuances of stress-related content is critical. Moreover, the ablation study confirms that although text-only models perform strongly, the incorporation of visual cues further improves the detection of subtle affective signals. These findings underscore the potential of multi-modal data fusion for advancing mental health monitoring applications. We hypothesize that the generated visuals act as implicit emotion amplifiers, translating latent affective states into more explicit signals that the model can learn from. The shared embedding space enables the model to reinforce weak cues in one modality using complementary information from the other, thereby improving the robustness of stress detection. While this method shows strong results on the Dreaddit dataset, its generalizability to other mental health tasks or platforms—such as Twitter or Instagram—remains an open question. Future work should explore how this approach adapts to different linguistic styles, content structures, and user populations across platforms.

5 Conclusion

In this work, we introduced a novel multi-modal framework for stress detection using both textual content and synthesized visuals from DALL·E. Leveraging the CLIP model's robust joint embedding capabilities, our method captures subtle emotional cues missed by text-only approaches. Tested on the Dreaddit dataset, our model achieved 94.90% accuracy with classifier-only training, while full fine-tuning increased performance to 98.41%. These results highlight the significant potential of combining generative image synthesis with multi-modal representation learning for affective computing and mental health monitoring.

Limitations

Despite the promising results of our multi-modal framework, several limitations remain. First, our approach relies on synthesized images generated by DALL·E, which may introduce biases or inconsistencies; the quality and representativeness of the generated visuals can vary depending on the input text. Second, our experiments have been conducted solely on the Dreaddit dataset, and it is unclear whether the observed performance improvements will generalize to other social media platforms or stress-related domains (Cohan et al., 2018; Mauriello et al., 2021; Garg et al., 2022; Sathvik and Garg, 2023; Chim et al., 2024). Furthermore, while results on the Dreaddit dataset are promising, further research is needed to determine the generalizability of our model across different social media platforms and diverse demographic groups. Finally, even though the fusion of text and visuals enhances stress detection, the interpretability (Jeon et al., 2024) of the resulting multi-modal representations remains an open challenge. Future work should focus on addressing these limitations by exploring more robust image synthesis techniques and developing methods to improve the transparency and interpretability of multi-modal models. One limitation is the lack of systematic evaluation of the generated images. We do not assess whether they reflect the intended affective state or which visual features (e.g., color, composition, expressions) contribute to stress detection. Future work will examine prompt design and affective feature attribution.

Ethical Considerations

Our work involves the analysis of social media data for stress detection, raising important ethical considerations. The use of such data requires strict adherence to privacy protocols and the anonymization of user information. Additionally, generative models like DALL·E can inadvertently propagate biases present in their training data, potentially affecting the fairness and reliability of our system. Care must be taken to ensure that the technology is not misused for surveillance or discriminatory practices. We advocate for responsible usage, transparent reporting of model decisions, and the integration of fairness-aware techniques in future work. As our study uses only anonymized Dreaddit data without new collection or user interaction, ethics approval was not required. Still, using DALL·E to generate images from user content raises concerns. We take precautions against misuse, but future work should pursue consent-driven, transparent frameworks for generative modeling in mental health.

Acknowledgement

This work has been partially supported by project MIS 5154714 of the National Recovery and Resilience Plan Greece 2.0 funded by the European Union under the NextGenerationEU Program.

References

Fatma S Abousaleh, Wen-Huang Cheng, Neng-Hao Yu, and Yu Tsao. 2020. Multimodal deep learning framework for image popularity prediction on social media. *IEEE Transactions on Cognitive and Developmental Systems*, 13(3):679–692.

Maryam Mohammed Aldarwish and Hafiz Farooq Ahmad. 2017. Predicting depression levels using social media posts. In *2017 IEEE 13th international Symposium on Autonomous decentralized system (ISADS)*, pages 277–280. IEEE.

Saeed Amal, Lida Safarnejad, Jesutofunmi A Omiye, Ilies Ghanzouri, John Hanson Cabot, and Elsie Gyang Ross. 2022. Use of multi-modal data and machine learning to improve cardiovascular disease care. *Frontiers in cardiovascular medicine*, 9:840262.

Tadas Baltrušaitis, Chaitanya Ahuja, and Louis-Philippe Morency. 2018. Multimodal machine learning: A survey and taxonomy. *IEEE transactions on pattern analysis and machine intelligence*, 41(2):423–443.

Saurabh Biswas and Yasha Hasija. 2022. Predicting depression through social media. In *Predictive Analytics of Psychological Disorders in Healthcare: Data Analytics on Psychological Disorders*, pages 109–127. Springer.

Jenny Chim, Adam Tsakalidis, Dimitris Gkoumas, Dana Atzil-Slonim, Yaakov Ophir, Ayah Zirikly, Philip Resnik, and Maria Liakata. 2024. Overview of the clpsych 2024 shared task: Leveraging large language models to identify evidence of suicidality risk in online posts. In *Proceedings of the 9th Workshop on Computational Linguistics and Clinical Psychology (CLPsych 2024)*, pages 177–190.

Arman Cohan, Bart Desmet, Andrew Yates, Luca Soldaini, Sean MacAvaney, and Nazli Goharian. 2018. Smhd: a large-scale resource for exploring online language usage for multiple mental health conditions. *arXiv preprint arXiv:1806.05258*.

Roddy Cowie, Ellen Douglas-Cowie, Nicolas Tsapatsoulis, George Votsis, Stefanos Kollias, Winfried Fellenz, and John G Taylor. 2001. Emotion recognition in human-computer interaction. *IEEE Signal processing magazine*, 18(1):32–80.

Munmun De Choudhury, Michael Gamon, Scott Counts, and Eric Horvitz. 2013. Predicting depression via social media. In *Proceedings of the International AAAI Conference on Web and Social Media*.

Jacob Devlin, Ming-Wei Chang, Kenton Lee, and Kristina Toutanova. 2018. BERT: pre-training of deep bidirectional transformers for language understanding. *CoRR*, abs/1810.04805.

Bin Feng, Shulan Ruan, Mingzheng Yang, Dongxuan Han, Huijie Liu, Kai Zhang, and Qi Liu. 2025. Sentiformer: Metadata enhanced transformer for image sentiment analysis. In *ICASSP 2025-2025 IEEE International Conference on Acoustics, Speech and Signal Processing (ICASSP)*, pages 1–5. IEEE.

Muskan Garg, Chandni Saxena, Veena Krishnan, Ruchi Joshi, Sriparna Saha, Vijay Mago, and Bonnie J Dorr. 2022. Cams: An annotated corpus for causal analysis of mental health issues in social media posts. *arXiv preprint arXiv:2207.04674*.

Sophia Gu, Christopher Clark, and Aniruddha Kembhavi. 2023. I can't believe there's no images! learning visual tasks using only language supervision. In *Proceedings of the IEEE/CVF international conference on computer vision*, pages 2672–2683.

Cosimo Ieracitano, Nadia Mammone, Amir Hussain, and Francesco C Morabito. 2020. A novel multimodal machine learning based approach for automatic classification of eeg recordings in dementia. *Neural Networks*, 123:176–190.

Hyolim Jeon, Dongje Yoo, Daeun Lee, Sejung Son, Seungbae Kim, and Jinyoung Han. 2024. A dual-prompting for interpretable mental health language models. *arXiv preprint arXiv:2402.14854*.

Shaoxiong Ji, Tianlin Zhang, Luna Ansari, Jie Fu, Prayag Tiwari, and Erik Cambria. 2022. MentalBERT: Publicly Available Pretrained Language Models for Mental Healthcare. In *Proceedings of LREC*.

Levon Khachatryan, Andranik Movsisyan, Vahram Tadevosyan, Roberto Henschel, Zhangyang Wang, Shant Navasardyan, and Humphrey Shi. 2023. Text2video-zero: Text-to-image diffusion models are zero-shot video generators. In *Proceedings of the IEEE/CVF International Conference on Computer Vision*, pages 15954–15964.

Ronak Kosti, Jose M Alvarez, Adria Recasens, and Agata Lapedriza. 2017. Emotion recognition in context. In *Proceedings of the IEEE conference on computer vision and pattern recognition*, pages 1667–1675.

Zewen Li, Fan Liu, Wenjie Yang, Shouheng Peng, and Jun Zhou. 2021. A survey of convolutional neural networks: analysis, applications, and prospects. *IEEE transactions on neural networks and learning systems*, 33(12):6999–7019.

Yinhan Liu, Myle Ott, Naman Goyal, Jingfei Du, Mandar Joshi, Danqi Chen, Omer Levy, Mike Lewis, Luke Zettlemoyer, and Veselin Stoyanov. 2019. Roberta: A robustly optimized BERT pretraining approach. *CoRR*, abs/1907.11692.

Matthew Louis Mauriello, Thierry Lincoln, Grace Hon, Dorien Simon, Dan Jurafsky, and Pablo Paredes. 2021. Sad: A stress annotated dataset for recognizing everyday stressors in sms-like conversational systems. In *Extended abstracts of the 2021 CHI conference on human factors in computing systems*, pages 1–7.

Namita Mittal, Divya Sharma, and Manju Lata Joshi. 2018. Image sentiment analysis using deep learning. In *2018 IEEE/WIC/ACM international conference on web intelligence (WI)*, pages 684–687. IEEE.

Hussein Mouzannar, Yara Rizk, and Mariette Awad. 2018. Damage identification in social media posts using multimodal deep learning. In *ISCRAM*. Rochester, NY, USA.

Tingting Qiao, Jing Zhang, Duanqing Xu, and Dacheng Tao. 2019. Mirrorgan: Learning text-to-image generation by redescription. In *Proceedings of the IEEE/CVF conference on computer vision and pattern recognition*, pages 1505–1514.

Alec Radford, Jong Wook Kim, Chris Hallacy, Aditya Ramesh, Gabriel Goh, Sandhini Agarwal, Girish Sastry, Amanda Askell, Pamela Mishkin, Jack Clark, et al. 2021. Learning transferable visual models from natural language supervision. In *International conference on machine learning*, pages 8748–8763. PmLR.

Aditya Ramesh, Mikhail Pavlov, Gabriel Goh, Scott Gray, Chelsea Voss, Alec Radford, Mark Chen, and Ilya Sutskever. 2021. Zero-shot text-to-image generation. In *International conference on machine learning*, pages 8821–8831. Pmlr.

Hojjat Salehinejad, Sharan Sankar, Joseph Barfett, Errol Colak, and Shahrokh Valaee. 2017. Recent advances in recurrent neural networks. *arXiv preprint arXiv:1801.01078*.

MSVPJ Sathvik and Muskan Garg. 2023. Multiwd: Multiple wellness dimensions in social media posts. *Authorea Preprints*.

Jiayu Shi, Zexiao Wang, Jiandong Zhou, Chengyu Liu, Poly ZH Sun, Erying Zhao, and Lei Lu. 2024. Mentalqlm: A lightweight large language model for mental healthcare based on instruction tuning and dual lora modules. *medRxiv*, pages 2024–12.

Binyang Song, Rui Zhou, and Faez Ahmed. 2024. Multi-modal machine learning in engineering design: A review and future directions. *Journal of Computing and Information Science in Engineering*, 24(1):010801.

Yoad Tewel, Yoav Shalev, Idan Schwartz, and Lior Wolf. 2022. Zerocap: Zero-shot image-to-text generation for visual-semantic arithmetic. In *Proceedings of the IEEE/CVF Conference on Computer Vision and Pattern Recognition*, pages 17918–17928.

Elsbeth Turcan and Kathleen McKeown. 2019. Dreaddit: A reddit dataset for stress analysis in social media. *arXiv preprint arXiv:1911.00133*.

Ashish Vaswani, Noam Shazeer, Niki Parmar, Jakob Uszkoreit, Llion Jones, Aidan N Gomez, Łukasz Kaiser, and Illia Polosukhin. 2017. Attention is all you need. In *Advances in Neural Information Processing Systems*, volume 30.

Junyang Wang, Yuanhong Xu, Juhua Hu, Ming Yan, Jitao Sang, and Qi Qian. 2023. Improved visual fine-tuning with natural language supervision. In *Proceedings of the IEEE/CVF International Conference on Computer Vision*, pages 11899–11909.

Zirui Wang, Jiahui Yu, Adams Wei Yu, Zihang Dai, Yulia Tsvetkov, and Yuan Cao. 2021. Simvlm: Simple visual language model pretraining with weak supervision. *arXiv preprint arXiv:2108.10904*.

Mayur Wankhade, Annavarapu Chandra Sekhara Rao, and Chaitanya Kulkarni. 2022. A survey on sentiment analysis methods, applications, and challenges. *Artificial Intelligence Review*, 55(7):5731–5780.

Kailai Yang, Tianlin Zhang, Ziyan Kuang, Qianqian Xie, Jimin Huang, and Sophia Ananiadou. 2024. Mentallama: interpretable mental health analysis on social media with large language models. In *Proceedings of the ACM Web Conference 2024*, pages 4489–4500.

Quanzeng You, Jiebo Luo, Hailin Jin, and Jianchao Yang. 2015. Robust image sentiment analysis using progressively trained and domain transferred deep networks. In *Proceedings of the AAAI conference on Artificial Intelligence*, volume 29.

Jianhua Zhang, Zhong Yin, Peng Chen, and Stefano Nichele. 2020. Emotion recognition using multi-modal data and machine learning techniques: A tutorial and review. *Information fusion*, 59:103–126.

Yiwu Zhong, Jing Shi, Jianwei Yang, Chenliang Xu, and Yin Li. 2021. Learning to generate scene graph from natural language supervision. In *Proceedings of the IEEE/CVF International Conference on Computer Vision*, pages 1823–1834.

Pan Zhou, Xingyu Xie, Zhouchen Lin, and Shuicheng Yan. 2024. Towards understanding convergence and generalization of adamw. *IEEE transactions on pattern analysis and machine intelligence*.

Yufan Zhou, Bingchen Liu, Yizhe Zhu, Xiao Yang, Changyou Chen, and Jinhui Xu. 2023. Shifted diffusion for text-to-image generation. In *Proceedings of the IEEE/CVF conference on computer vision and pattern recognition*, pages 10157–10166.

Zhouan Zhu, Shangfei Wang, Yuxin Wang, and Jiaqiang Wu. 2025. Integrating visual modalities with large language models for mental health support. In *Proceedings of the 31st International Conference on Computational Linguistics*, pages 8939–8954.

A Appendix

In the appendix, we provide further details regarding our experimental setup, hyperparameter settings, and examples of synthesized images. These supplementary materials aim to enhance the reproducibility of our work and offer deeper insights into the performance of our multi-modal framework.

A.1 Hyperparameters for Generating Images with DALL-E

As shown in Table 3, we employed the DALL-E 3 model to synthesize images from social media posts. Our prompt was carefully designed to ensure that the generated visuals consistently capture stress-related cues. For each post, the prompt instructs DALL-E 3 to produce a **consistent, structured** image that visually represents a state of stress or anxiety. This image is expected to include a tense or overwhelming environment (e.g., dim lighting, clutter, urban stress), facial expressions that convey worry, exhaustion, or distress (when humans are depicted), and a darker, cooler color palette to evoke a stressed mood. The images are generated at a resolution of 1024x1024 with standard quality, and one image is produced per post.

A.2 Hyperparameters and Training Setup for CLIP

Table 4 summarizes the hyperparameters and training configurations used in our experiments for both the CLIP Classifier-Only Training and the CLIP Full Fine-Tuning approaches.

In our experiments, the CLIP Classifier-Only Training approach involves freezing the CLIP base and training only the classifier with the AdamW optimizer (Zhou et al., 2024) at a learning rate of 5×10^{-4}, a weight decay of 1×10^{-4}, and a StepLR scheduler (step size of 5 epochs and $\gamma = 0.5$). Training is conducted for up to 10 epochs with early stopping after 3 epochs of no improvement. For the CLIP Full Fine-Tuning approach, both the CLIP base and the classifier are updated. We employ a dual learning rate strategy where the CLIP parameters are optimized at 5×10^{-6} and the classifier at 5×10^{-4}, using the same weight decay and scheduler settings. This configuration runs for up to 15 epochs, with gradient clipping (max norm = 1.0) applied to stabilize training. These hyperparameter choices enable a balanced adaptation of the pre-trained CLIP representations while effectively learning task-specific features for stress detection.

A.3 Illustrative Examples of Synthesized Visuals from Social Media Posts

In this section, we generate images from social media posts using DALL·E. We provide examples from the Dreaddit dataset alongside their corresponding synthesized images (see Figure 2). Each image is generated based on the text of the post, capturing the key emotional and contextual cues embedded within the content. Our approach translates linguistic elements—such as tone, word choice, and contextual details—into visual features, including the color palette, environmental cues, and facial expressions that are indicative of stress. By presenting these paired examples, we illustrate how our multi-modal framework leverages both textual and visual modalities to enhance stress detection, offering a more comprehensive perspective on the underlying affective signals present in social media data.

A.4 Code Availability

The source code for all experiments, including data preprocessing, model training, and evaluation scripts, is available on GitHub: `https://github.com/Efstathia-Soufleri/Stress-Detection-CLIP`. This repository is designed to facilitate the reproducibility of our results and to support further research in this field.

Parameter	Value / Description
Model	`dall-e-3`
Prompt	Based on the text "{post}", generate a **consistent, structured** image that visually represents a state of stress or anxiety. The image must include: • A tense or overwhelming environment (e.g., dim lighting, clutter, urban stress). • Facial expressions showing worry, exhaustion, or distress (if humans are depicted). • A darker, cooler color palette to evoke a stressed mood.
Size	1024x1024
Quality	Standard
Number of Images	1

Table 3: Summary of DALL-E 3 image generation parameters and prompt design used for synthesizing visuals that capture stress-related cues.

Parameter	Classifier-Only Training	Full Fine-Tuning
Epochs	10	15
Batch Size	32	32
Optimizer	AdamW (classifier only)	AdamW (dual groups)
Learning Rate (Classifier)	5×10^{-4}	5×10^{-4}
Learning Rate (CLIP Base)	—	5×10^{-6}
Weight Decay	1×10^{-4}	1×10^{-4}
LR Scheduler	StepLR (step=5, γ=0.5)	StepLR (step=5, γ=0.5)
Early Stopping Patience	3 epochs	3 epochs
Additional Techniques	—	Gradient Clipping (max norm = 1.0)

Table 4: Hyperparameters and training configurations for Classifier-Only Training and Full Fine-Tuning of our proposal.

Image	Post
	Especially the power of healing brought upon by service animals. I too, have a service dog named Luna. This wonderful man was nice enough to bring the book back in while I was off yesterday with a note with his name and number telling me to call him when I finish the book. This just made my day, it really did. There's so much negativity in the world today and it seems not many people will stop to do something nice for someone or help them by doing a random act of kindness.

Image	Post
	These past couple of months have been the worst. My anxiety has gotten so bad it's effecting my sleep and relationship. I've become so paranoid about my health as well. I don't feel like me anymore and I just feel scared all the time now over every little thing. I don't have money to see a therapist either...

Figure 2: Illustrative examples from the Dreaddit dataset. A social media post and the corresponding synthesized image generated from the post text. These examples demonstrate how our multi-modal framework leverages both textual and visual modalities to capture emotional and contextual cues for enhanced stress detection. The top image and post pair indicate absence of stress and the below pair indicate stress.

Fine-tuning LLMs to Extract Epilepsy Seizure Frequency Data from Health Records

Ben Holgate*, Joe Davies*, Shichao Fang*,
Joel S. Winston*, James T. Teo*, and Mark P. Richardson*
* Department of Basic & Clinical Neuroscience, King's College London
benjamin.holgate@kcl.ac.uk

Abstract

We developed a new methodology of extracting the frequency of a patient's epilepsy seizures from unstructured, free-text outpatient clinic letters by: first, devising a singular unit of measurement for seizure frequency; and second, fine-tuning a generative Large Language Model (LLM) on our bespoke annotated dataset. We measured frequency by the number of seizures per month: one seizure or more requires an integer; and less than one a decimal. This approach enables us to track whether a patient's seizures are improving or not over time. We found fine-tuning improves the F1 score of our best-performing LLM, Ministral-8B-Instruct-2410, by around three times compared to an untrained model. We also found Ministral demonstrated an impressive ability for mathematical reasoning.

1 Introduction

Extracting key patient data from longitudinal Electronic Health Records (EHRs) is critical to developing AI models that help improve patient treatments. Yet unstructured, free-text narratives are typically not suited to computational models that require structured data, and so medical researchers are increasingly utilizing Natural Language Processing (NLP) tools to enable clinical AI models to understand medical terminology and concepts (Yang et al., 2022).

In recent years, much clinical NLP research has focused on generative Large Language Models (LLMs). On the one hand, this has involved the development of LLMs with some degree of clinical expertise, such as ClinicalBERT (Huang et al., 2019), GatorTron (Yang et al., 2022), and ClinicalMamba (Yang et al., 2024). On the other hand, researchers have applied general knowledge LLMs to extract data from clinical texts (for example, Agrawal et al., 2022; Thirunavukarasu et al., 2023; and Zhou et al., 2023). In turn, this field of research has led to the creation of a benchmark, ClinicBench, to evaluate the performance of 22 LLMs in a clinical setting (Liu et al., 2024).

Yet the application of LLMs to epilepsy research is still relatively uncommon, although it is expected that this field will increase significantly in future (van Diessen et al., 2024). Epilepsy affects about 1% of the general population (Fiest et al., 2017) and contributes to an estimated half a percent of the global disease burden (WHO. Epilepsy. 2019). About 30% of people with epilepsy do not respond to anti-seizure medications (ASMs) and are therefore regarded as refractory to treatment (Kwan and Brodie, 2000). In the United Kingdom over the last decade, more than 30 individual ASMs have been available to prescribe and the number of possible combinations of ASMs taken as polytherapy is much larger. Consequently, it is not feasible to try all possible monotherapy and polytherapy options in every refractory patient. This underlines the importance of research in predicting which ASMs would have the greatest impact on epileptic seizures for individual patients.

The most extensive relevant research on LLMs and epilepsy remains a long-term study (Xie et al., 2022a; Xie et al., 2022b; Xie et al., 2023; and Xie at al., 2024) that used a different methodology from ours to extract seizure frequency information from Electronic Health Records (EHRs). In their 2022-23 papers, the University of Pennsylvania researchers applied the pre-trained Transformers Bio_ClinicalBERT (for text classification), RoBERTa (for text extraction), and a T-5 model (to summarize sentences with seizure frequency data) to free-text EHRs to determine the seizure frequency of a person with epilepsy or whether that person was seizure free. They declared an "overall accuracy" score of 0.88 for seizure frequency. In

Proceedings of the 24th Workshop on Biomedical Language Processing (BioNLP 2025), pages 44–55
August 1, 2025

their 2024 paper, the team tested for bias (race, ethnicity, sex, income, and health insurance) in a ClinicalBERT model that they had fine-tuned on 700 manually annotated epileptologist notes and which classified whether a clinic note specified if a patient was seizure free or had recent seizures. They found no evidence of bias in the model.

Our previous, 2024 study was the first published paper to use a generative LLM to determine seizure frequency for people with epilepsy from unstructured, free-text EHRs (Holgate et al., 2024). We utilized Llama 2 13B (Touvron et al., 2023) to classify seizure frequency within eight temporal categories – ranging from once a year at one end of the spectrum to one or more per day at the other end – and in our analysis grouped the temporal categories into a binary split between infrequent and frequent seizures. We achieved an overall F1 score of 0.73 with Llama 2 13B.

An even more recent epilepsy study (Goldenholz et al., 2025) utilizes three different LLMs for different purposes: 1) Meta's Llama 2 13B to generate a randomized clinical trial for the ASM Cenobamate and generate 480 synthetic clinical notes; 2) Mistral's Mistral 7B v0.1 to summarize the clinical notes, specifically in regard to the number of seizures during the observation period and any symptoms associated with the ASM; and 3) Anthropic's Claude 2 to improve on the formatting and results of the data table. They used LLMs from different AI companies to ensure separation of technologies for the discrete tasks. Importantly, none of the LLMs were specially trained in medical language. The researchers concluded that their methodology demonstrated a capacity for inductive reasoning "from large sets of unstructured clinical encounters." Consequently, they recommended "a paradigm shift away from perfectly understanding the individual patient towards generalizable knowledge extracted from groups of patients. This new paradigm capitalizes on the strengths of LLMs ... [while] acknowledging their weakness at high precision."

While we agree that LLMs hallucinate at individual patient level for seizure frequency, based on our experience, we disagree that they are not useful for micro analysis. On the contrary, our study demonstrates that some of the latest generative LLMs are, in fact, very good at estimating seizure frequency in unstructured, free-text EHRs based on our new methodology that incorporates a singular unit of measurement and fine-tuning.

2 Data and Methods

2.1 Data Collection

We selected 51,760 EHRs from King's College Hospital NHS Foundation Trust (KCH) that relate to 5,767 unique adult people with epilepsy being treated at KCH. The data spans more than a decade, from 1 January 2013 to 30 September 2023. The vast majority of the records comprise doctors' and nurses' reports of outpatients' ambulatory visits. We defined a person with epilepsy as someone who has at least one record of an epilepsy diagnosis.

The selection was done via CogStack, an open-source information retrieval and extraction platform for EHRs developed by researchers at the NIHR Maudsley Biomedical Research Centre in London.[1] CogStack integrates with KCH's EHRs. We defined a set of epilepsy-related keywords and medical codes, and then used CogStack's search functionality to filter out EHRs that matched these definitions.

We then used stratified random sampling to select 3,000 EHRs to create an annotated dataset, which ensured proportional distribution across the original dataset in regard to age, gender, and ethnicity to minimize bias (see below for further annotation details).

2.2 Seizure Frequency Measurement

We followed the logic of the U Penn team to create a standardized format to denote seizure frequency in a given EHR. However, our methodology differed in two ways. First, the U Penn researchers used three language model pipelines with three different language models – for text classification, text extraction, and summarization of sentences with seizure frequency data – whereas we used only one generative LLM for all classification, extraction, and calculation tasks, largely because the newest LLMs are much more powerful than the ones they used. Second, the U Penn researchers initially used different time periods – day, month, year, or visit –depending on the period specified in the text, and then converted that by a rules-based quantifier into a standardized format of the number of seizures per month, whereas we required only

[1] https://cogstack.org/

one step by fine-tuning an LLM on our annotated dataset that denoted the text's data as the number of seizures per month.

Our project's lead data scientist annotated 1,480 EHRs in accordance with our singular unit of measurement for seizure frequency – that is, the number of seizures per month. The EHRs had previously undergone an initial annotation process. In our previous study (Holgate et al., 2024), we used stratified random sampling to select 3,000 EHRs to create an annotated dataset, which ensured proportional distribution across the original dataset for age, gender, and ethnicity to minimize bias. Subsequently, a team of six annotators, comprising four neuroscience clinicians (including two epileptologists) and two data scientists, manually annotated the 3,000 EHRs for key data categories of the project, in particular seizure frequency, as well as seizure freedom, current anti-epilepsy medication, epilepsy type, seizure type, associated symptoms, and comorbidities. The annotators categorized seizure frequency into eight temporal frequencies – ranging from one seizure per year to one or more per day – plus 'unknown.' Due to time and resource limitations, the annotators worked on separate batches of the 3,000 EHRs, rather than having two annotators work on the same batch for moderation. However, the two epileptologists reconvened to create a 'gold standard' annotated dataset of 300 EHRs; their inter-annotator agreement was a Cohen's kappa score of 0.84, which signified near perfect agreement.

In turn, the lead data scientist used the 300 EHRs from this 'gold standard' annotated dataset plus a further 1,180 annotated EHRs to create a training and testing dataset to fine-tune LLMs on seizure frequency. The reason why the training / testing dataset was about half the size of the original annotated dataset was that about the same proportion of the KCH EHRs extracted contained information about a patient's seizure frequency. The lead data scientist converted the annotator's original annotation for seizure frequency to our new measurement system, in which one seizure or more per month required an integer, and less than one seizure per month a decimal (see Table 1). Two other categories were required for notation. If an EHR contained reference to seizures but the duration was unspecified or unclear, the number '1000' was used (essentially a proxy figure to denote incomplete information). Or if an EHR contained no reference to seizures, a '0' was used.

This methodology provided three key advantages: first, a single numerical metric makes it easy to track a patient's seizure trajectory over time (a declining number means the frequency of their seizures is reducing, while an increasing number means the frequency of their seizures is rising); second, a single numerical metric is easier to understand than eight, discrete temporal categories to record seizure frequency; and third, a single numerical metric is a more accurate and reliable input to feed into a seizure prediction model that we are developing as part of our wider epilepsy research project.

2.3 Model Development and Implementation

Environments and Models: We used LangChain as our development framework because it provides convenience and flexibility for building applications powered by LLMs.[2] First, we deployed LangChain in our local environment, then we downloaded the four LLMs we experimented with in this study from Hugging Face and loaded the models into LangChain, which allowed us to perform multiple LLM operations in the local environment.[3] LangChain offers simple interfaces for loading and initializing LLMs.

We also employed parameter-efficient fine-tuning techniques, or PEFT, in particular parameter updates by low-rank adaptation, or LoRA. The latter hacks the regular backpropagation updates by splitting the update matrix into two smaller matrices which, when multiplied together, can give back the original update matrix. LoRA can accelerate training while reducing the computational demands.

We experimented with four LLMs that were released in 2024 or 2025 and developed by three different AI companies: US-based Meta's Llama 3.1 8B Instruct (Grattafiori et al., 2024); France-based Mistral's Mistral Nemo Instruct 2417 (Mistral AI Team, 2024a) and Ministral 8B Instruct 2410 (Mistral AI Team, 2024b); and China-based Alibaba's Qwen 2.5 7B Instruct (Yang et al., 2025). We were restricted to only using open-source language models because we used confidential

[2] https://www.langchain.com

[3] https://huggingface.co

You are a professional neuroscientist.

Analyze the text and work through these 4 steps:

1. Determine whether the text has any information about the frequency of the patient's epilepsy seizures.

2. If the text does have information about the frequency of the patient's epilepsy seizures, then estimate the frequency of the seizures, and return the answer as the number of seizures per month.

3. If the text does refer to seizures but you cannot estimate the frequency of the seizures, then return the answer '1000'.

4. If the text does not have any information about the patient's epilepsy seizures, then return the answer '0'.

Figure 1: Prompt query structure.

medical data from the UK's National Health System (NHS) that had to remain within the hospital's secure IT network for regulatory reasons. We ran the LLMs on up to eight Nvidia V100 GPUs.

Pre-processing: We implemented two pre-processing elements. First, we found that an LLM's performance was slightly improved by reducing the length of each EHR, deleting non-relevant administrative information at the top and bottom of each clinic letter. As a result, this minimized noise from the unstructured text. We deleted all text before the clinic date at the top of the letter, and removed all text after the letter writer (typically a doctor or nurse) signed off "yours sincerely" (a UK letter writing convention) towards the end. In the event there was no specified date or sign-off, we set a default deletion of the first 40 characters and final 500 characters of each letter.

Second, we created a balanced dataset from the 1,480 annotated EHRs to train, test, and validate the LLMs. In each of the dataset's 1,480 observations, the input consisted of the EHR text, and the required output was the annotated decimal or integer for the corresponding seizure frequency, if stated in the document. A label for seizure frequency was assigned to the entire clinical note, based on the frequency for the patient at the time of the clinic visit. In other words. we fine-tuned the LLM on the annotated output. The balanced dataset was of various sizes, ranging from 375 to 813 EHRs in order to create training datasets ranging from 300 to 650 EHRs in increments of 50. The balanced dataset was structured by: taking a specified number of EHRs annotated with seizure frequency measurements of 0.1 to 999 (meaning these letters contained a reference to seizures with a specified frequency) and selected at random from the 1,480 annotated EHRs; then taking 25% of the number of the 0.1-999 category letters from the '1000' category letters, selected at random; and finally taking the same 25% portion from the '0' category letters, again selected at random. For example, 500 of the 0.1-999 letters were combined with 125 of the '1000' letters and 125 of the '0' letters to make a balanced dataset of 750 EHRs in total. The train/test/validation split was 80%/10%/10%. So in this example the training dataset consisted of 600 letters, the testing dataset 75 letters, and the validation dataset 75 letters. We use the term 'balanced' to mean that the dataset used to fine-tune the LLM was not weighted too far towards any of the three annotated categories. During experiments we found that this ratio of 25% of the total 0.1-999 letters for each of the '1000' and '0' letters worked best for adequately fine-tuning the LLMs on our seizure frequency task.

A fundamental challenge for this project was that the NHS EHRs used, mostly doctors' and nurses' reports of outpatients' ambulatory visits, were unstructured and typically noisy. The reports included a range of medical and administrative information, such as the patient's medication, other therapies, and details disclosed during previous clinic visits. Furthermore, the reports often did not include any information about seizure frequency and, if they did, the language was often imprecise, so that the nature of the frequency was vague or unclear. These factors make the application of LLMs to EHRs to research seizure frequency challenging.

Prompt Engineering: Although fine-tuning the LLM on hundreds of examples was the primary methodology in meeting this challenge, a secondary methodology was prompt engineering. We found that the structure of the prompt query made a difference to the quality of an LLM's answers. After experimentation, we concluded the optimal approach was Chain of Thought reasoning,

Seizure Frequency		Performance Evaluation	
Categories	**Measurement / Month**	**Purist Method**	**Pragmatic Method**
1 per year	0.08	0 < x ≤ 0.16	
1 per 6 months	0.17	0.16 < x ≤ 0.18	
> 1 per 6 months, < 1 per month	> 0.17, < 1	0.18 < x ≤ 0.99	
1 per month	1	0.99 < x ≤ 1.1	0 < x ≤ 1.1
> 1 per month, < 1 per week	> 1, < 4	1.1 < x ≤ 3.9	
1 per week	4	3.9 < x ≤ 4.1	
> 1 per week, < 1 per day	> 4, < 30	4.1 < x ≤ 29	
1 or more per day	30 - 999	29 < x ≤ 999	1.1 < x ≤ 999
Unknown frequency	1000	1000	1000
No seizure information	0	0	0

Table 1: Seizure frequency categories and measurements per month, performance evaluation methods.

asking the LLM to work through four logical steps, each of which was numbered (see Figure 1). The first step was to determine whether the EHR contained any information about the frequency of a patient's seizures (because often the letters did not). The second step asked the LLM to estimate the frequency as the number of seizures per month. The third step asked to return an answer of '1000' if the frequency of seizures of was too difficult to answer. The fourth and final step asked to return '0' if there was no information about seizures. At the start of the prompt, we asked the LLM to take on the role of a professional neuroscientist, as we found this slightly improved the quality of answers. We hypothesize that contextualizing the reasoning task for the LLM assists it in logically connecting the prompt (question) and text (EHR) with the relevant medical parts of the vast corpora that the LLM was originally trained on.

Hyperparameters: We kept the temperature at a very low 0.0001 (0 does not work for some LLMs) because we wanted the LLMs to generate typically fact-based answers and be consistent in their answers across multiple runs. In addition, our aim was to minimize both the LLMs' 'creativity' and hallucinations.

Although we experimented with changing some hyperparameters, such as the number of training epochs, batch size, and learning rate, we found none of these had any significant impact on the quality of the LLMs' answers. We set the number of epochs at three, the batch size at one, and the learning rate at 0.0002. In other words, the most influential factor in improving output was the size of the training dataset, followed by the prompt structure. For LoRA, we set the *r* value at 64, the alpha at 16, and the dropout rate at 0.1.

Post-processing: Despite fine-tuning the LLMs on our annotated dataset, the models' raw answers often needed to be cleaned up by a post-processing algorithm. The raw answers from the original model were typically variable, with a best-case answer being exactly what was asked by the prompt questions (e.g., '0', '2', or '1000'), a mixed answer (e.g., '11 to 16 seizures per month'), to outright nonsensical (e.g., '123456789' or 'He also showed some difficulties'). The raw answers from the fine-tuned LLMs were, however, generally more in line with what was required, typically generating an answer as either a decimal or integer with no (or little) text. Yet the LLM's construction – or attempt at construction – of a decimal was often confused with more than one decimal point (e.g., '2.00.0000'). As a consequence of the LLMs not being able to generate an answer in exactly the required format 100% of the time, we wrote a rules-based algorithm that either corrected the answer format where reasonably clear (e.g., '2.00.0000' becomes '2') or changed to a '0' if completely unclear (e.g., '123456789').

Model Selection: We began by running the four LLMs that we tested on different sized balanced datasets in order to create training datasets ranging from 300 to 650 EHRs in increments of 50, as outlined above. During fine-tuning each LLM was trained on the training dataset and also given separate evaluation and test datasets. At this stage we identified Mistral's two models as being the best performing, followed by the Qwen 2.5 model, and the Llama 3 model. Overall, the best performing model was Ministral-8B-Instruct-2410.

We then tried various experiments to optimize the output of Ministral-8B-Instruct-2410. The most significant factors influencing the quality of the LLM's answers were the size of the training dataset (in general, more observations improved the answers) and the prompt structure. We determined that when the training dataset consisted of about 550 EHRs or more, the F1 score on our preferred method of evaluation reached about 0.80 or more.

3 Results

3.1 Performance Evaluation Methods

We used a confusion matrix to calculate recall, precision, the F1 score, and accuracy to evaluate an LLM's performance. We used a test dataset that each LLM had not seen during its training process. However, we devised two different methods of calculation, what we called the *purist* method and the *pragmatic* method. In the first method we used fuzzy logic, or the setting of soft (rather than hard) numerical boundaries between each of the eight temporal seizure frequency categories, on the basis that the temporal distinctions are arbitrary and our objective was to determine changes in a patient's seizure frequency over time.

The *purist* method set a high bar by calculating how well the LLM performed on eight temporal categories of seizure frequency. However, we treated this method more as a theoretical (rather than true) guide of performance, given the inconsistency of seizure information written by doctors and nurses in the outpatient letters, and the often inherent ambiguity of their language. Under this method, one seizure per year (specific target 0.08) equated to a range of $0 < x \leq 0.16$, one seizure per six months (specific target 0.17) was $0.16 < x \leq 0.18$, more than one seizure per six months but less than one per month (mid-point target ≈ 0.33) was $0.18 < x \leq 0.99$, one per month (specific target 1) was $0.99 < x \leq 1.1$, more than one seizure per month but less than one per week was $1.1 < x \leq 3.9$, one per week (specific target 4) was $3.9 < x \leq 4.1$, more than one per week but less than daily was $4.1 < x \leq 29$, and one or more per day was $29 < x \leq 999$ (999 being 1 below the 'fudge' figure of '1000'). In addition, we tested the model strictly against the other two categories: seizures with no information about frequency ('1000'); and no information about seizures ('0').

By contrast, the *pragmatic* method set a lower bar and reflected our broader objective to determine whether LLMs are good at extracting information about a patient's seizure frequency in such a way to reveal if their seizures are improving over time or not. In this method, we bifurcated the output into two temporal categories, infrequent and frequent seizures. Infrequent ranged from one seizure per year to one per month, which equated to a range of $0 < x \leq 1.1$. While frequent ranged from more than one per month to one or more per day, which equated to $1.1 < x \leq 999$. The two non-temporal categories remained as above. The threshold between infrequent and frequent had an empirical (rather than clinical) justification, in that our chosen demarcation line spread the number of observations in both categories more evenly, to avoid the frequent category significantly outweighing the infrequent category.

3.2 Model Performance

As shown in Table 2, the best-performing LLM, Ministral-8B-Instruct-2410, achieved its highest F1 score on the pragmatic method of 0.81 (purist method 0.68) with a training dataset of 650 EHRs, and a corresponding accuracy rate of 0.68 (0.52). As Appendix A illustrates, the F1 score on the pragmatic method rose beyond the 0.70 level once the training dataset became greater than 500 EHRs. While this might imply that the bigger the training dataset, the more effective the fine-tuning and the better the answers, this may not necessarily be the case. The F1 score dipped at 600 training observations but then rose to a new high at 650. Further research is required with even larger training datasets to investigate in more depth.

On the other hand, the results suggest that recall is not dependent on the size of the training data. Recall was consistently high, ranging from 0.86 to 1.00 on almost all training dataset sizes (with one exception). In other words, this Ministral model was proficient at correctly estimating seizure frequency.

By contrast, the results imply that precision is dependent on the size of the training dataset. The Ministral model required more than 500 training observations to improve precision – the same size needed to trigger an uplift in the F1 score. Nevertheless, precision remained the model's weak spot, achieving a best result of only 0.71 at 650

Fine-tuned LLM: Best F1 Scores

LLM	recall	precision	F1	acccuracy	recall	precision	F1	acccuracy
	Purist Method				Pragmatic Method			
Ministral-8B-Instruct-2410	0.91	0.54	0.68	0.52	0.93	0.71	0.81	0.68
Mistral-Nemo-Instruct-2407	1.00	0.48	0.65	0.48	1.00	0.64	0.78	0.64
Qwen2.5-7B-Instruct	0.60	0.31	0.47	0.32	0.71	0.62	0.66	0.51
Llama-3.1-8B-Instruct	0.20	0.35	0.26	0.22	0.22	0.39	0.28	0.23

Fine-tuned LLM: Mean Over 3 Runs and F1 Standard Deviation

LLM	recall	precision	F1 (SD)	acccuracy	recall	precision	F1 (SD)	acccuracy
	Purist Method				Pragmatic Method			
Ministral-8B-Instruct-2410	0.91	0.51	0.66 (0.02)	0.49	0.93	0.69	0.79 (0.01)	0.66
Mistral-Nemo-Instruct-2407	0.99	0.45	0.62 (0.05)	0.45	0.99	0.62	0.76 (0.02)	0.61
Qwen2.5-7B-Instruct	0.38	0.38	0.38 (0.09)	0.26	0.48	0.63	0.53 (0.12)	0.39
Llama-3.1-8B-Instruct	0.22	0.26	0.22 (0.05)	0.19	0.22	0.27	0.23 (0.05)	0.20

Table 2: Comparative performance evaluation of fine-tuned LLMs with same training dataset of 650 EHRs.

training observations on the pragmatic method, which was still comparatively low. This points to the model still 'hallucinating' on too many occasions, despite our attempts to minimize false positives through various techniques, in particular, fine-tuning, prompt engineering, setting a very low temperature, and adjusting the proportions of the balanced dataset.

The second-best performing LLM was the other Mistral model, Mistral-Nemo-Instruct-2407, which achieved a top F1 score of 0.78 on the pragmatic method, followed by Qwen2.5-7B-Instruct (0.66) and Llama-3.1-8B-Instruct (0.28) (see Table 2). Appendix B shows the comparative performance evaluation of the original LLMs -- that is, the non-fine-tuned models – which is much lower.

4 Discussion

Fine-tuning improved the F1 score of our best-performing LLM, Ministral-8B-Instruct-2410, by at least three times based on a training dataset of 650 EHRs. The F1 score of the fine-tuned model when evaluated by the purist method, 0.68, was three times that of the F1 score of the untrained model, 0.22. And the F1 score of the fine-tuned model when evaluated by the pragmatic method, 0.81, was 3.7 times that of the original model, also 0.22. This demonstrates that fine-tuning is an effective technique to improve the capacity of LLMs to identify the frequency of a patient's seizures in unstructured, free-text EHRs.

Both Mistral models performed at a high standard on this seizure frequency task, with only a 3 percentage points difference in their best F1 scores. However, there was a significant drop-off of 15 percentage points for the Qwen2.5 F1 score, and a 53 percentage points slide for the Llama 3.1 model, which did not perform well at all on this task.

Both Mistral models were also stable and consistent across multiple fine-tuning runs: their average F1 scores under the pragmatic method across three runs were only 2 percentage points below that of their respective top F1 scores; and the standard deviation of their F1 scores across 3 runs was only 1% or 2%. Stability is important in medical research. By contrast, Qwen2.5's F1 score was highly variable with a standard deviation of 12%.

Our study also demonstrates that some of the most recent LLMs have a capacity for mathematical reasoning. The Ministral models, in particular, were adept at identifying the frequency of a patient's seizures from the raw text, which could be anything from annually to daily or more, then converting that frequency to a standardized time period of per month, both in terms of decimals and integers. Indeed, Qwen2.5 was designed in part specifically to achieve "state-of-the-art performance" in mathematical tasks (Yang et al., 2025), and Llama 3's design had a partial focus on "mathematical reasoning performance" (Grattafiori et al., 2024), while the Mistral AI Team claims its Ministral 8B model achieves superior results to Llama 3.1 8B on a mathematical benchmark (Mistral AI Team, 2024b), which accords with our experience.

We can also postulate whether the LLMs we tested, especially the Ministral models, have some in-depth knowledge of medicine in general and epilepsy in particular in their original, non-fine-tuned form. On the one hand, the comparatively low F1 scores of the original models compared to the much higher F1 scores of the fine-tuned models imply that may not be the case. On the other hand, the models' ability to quickly pick up the logic from the annotated training dataset to identify and calculate seizure frequency in a standardized format suggests it might be the case.

If the latter, it would support the findings of a recent study that tested three well-known LLMs – GPT-4, Bard, and Claude 2; admittedly not models that we used – on epilepsy practice examinations (Habib et al., 2024). These LLMs achieved mean scores of 72%, 65%, and 67%, respectively, compared to anecdotal reports suggesting the passing score for the examinations was approximately 70%.

"We found that LLMs scored well on the epilepsy practice examinations, did not appear to rely on memorization, and could logically explain the reasons for a correct answer," said the authors. "However, they occasionally hallucinated logic for incorrect answers." Their latter point matched our experience with too many false positives and a comparatively lower precision, even with our best-performing model and optimal training dataset.

Minimizing hallucinations in medical research is a common problem (Kim et al., 2025). Hallucinations are defined as responses from LLMs that are inaccurate or have fabricated information. This could affect clinical decisions and patient safety. Algorithms tend to hallucinate when providing answers to questions that have a high complexity, when there is insufficient or biased training data for a topic, or when a dataset is particularly noisy. All of these are common problems in medical research, especially with data collected from medical reports and diaries. Fine-tuning a general LLM is one way to mitigate these effects but it is not necessarily a complete solution (Zuo and Jiang, 2025). As a result, hallucinations may still occur after fine-tuning.

One possible solution is Retrieval Augmented Generation (RAG), which has gained popularity in medical contexts in recent years (Li et al., 2024; Halamka 2023). RAG involves taking a pre-trained LLM but not fine-tuning it. Instead, a prompt is given to the algorithm which then uses its training and augments it by looking up information from a corpus of documents, either from a public or private source. This can reduce the effect of hallucinations by essentially performing a cross-check. RAG warrants investigation in further research of our study.

5 Conclusion

Fine-tuning is an efficient method to optimize the extraction of seizure frequency data from unstructured, free-text medical records by LLMs. Moreover, we found that some of the most recent LLMs demonstrated an impressive ability for mathematical reasoning, in this case not only calculating the frequency of a patent's epilepsy seizures from a text, but also converting that calculation into a standardized temporal format of the number of seizures per month. Prompt engineering is also critical to fine-tuning an LLM for this task. However, hallucinations and the associated problem of too many false positives remain an issue, and further research is required here. Nevertheless, this study, by achieving an F1 score of 0.81 from our best-performing model, shows that fine-tuning an LLM provides a new and innovative way of extracting seizure frequency data from EHRs that in turn enables better analysis of the effects of ASMs in the treatment of epilepsy and therefore improved patient outcomes.

Limitations

This study has three main limitations. First, the confidential nature of the medical records used for the training dataset means the model outputs are not reproduceable by research teams outside the hospital where the authors worked. Second, the confidential records meant we could not experiment with LLMs such as OpenAI's ChatGPT that are only available via an API to an off-site service due to privacy reasons. Third, we were restricted in what sized LLMs we could use by the computing power generated by our GPU platform (eight Nvidia V100 GPUs).

Ethical Considerations

The confidential EHRs of patients had to remain within the hospital's secure IT network. As a consequence, the study's researchers could only access the data and input it into LLMs via the hospital's IT network.

Acknowledgments

This research project was funded by Epilepsy Research Institute UK (project reference 2209), an independent research-funding charity. Epilepsy Research Institute UK received funding from Angelini Pharma to part-support this project. Angelini Pharma distributes Cenobamate in Europe and the UK. The study funders did not play any role in data collection, data analysis, or data interpretation, writing of the manuscript, or the decision to submit the manuscript for publication. The project operated under the London South-East Research Ethics Committee approval granted to the King's Electronic Records Research Interface (KERRI) (reference 18/LO/2048 and renewed 24/LO/0057).

References

Monica Agrawal, Stefan Hegselmann, Hunter Lang, Yoon Kim, and David Sontag. 2022. Large language models are few-shot clinical information extractors. In Proceedings of the 2022 Conference on Empirical Methods in Natural Language Processing, pages 1998-2022, Abu Dhabi, United Arab Emirates. Association for Computational Linguistics.

Eric van Diessen, Ramon A. van Amerongen, Maeike Zijlmans, Willem M. Otte. 2024. Potential merits and flaws of large language models in epilepsy care: A critical review. *Epilepsia* 00: pages 1-14. https://doi.org/10.1111/epi.17907.

Kirsten M. Fiest, Khara M. Sauro, Samuel Wiebe, Scott B. Patten, Churl-Su Kwon, Jonathan Dykeman, Tamara Pringsheim, Diane L. Lorenzetti, Nathalie Jetté. 2017. Prevalence and incidence of epilepsy: A systematic review and meta-analysis of international studies. *Neurology* 88(3): pages 296-303.

Daniel M. Goldenholz, Shira R. Goldenholz, Sara Habib, M. Brandon Westover. 2025. Inductive reasoning with large language models: A simulated randomized controlled trial for epilepsy. *Epilepsy Research*, vol. 211. https://doi.org/10.1016/j.eplepsyres.2025.107532

Aaron Grattafiori et al. 2024. The Llama 3 Herd of Models. arXiv:2407.21783

Sara Habib, Haroon Butt, Shira R. Goldenholz, Chi Yuan Chang, Daniel M. Goldenholz. 2024. Large Language Model Performance on Practice Epilepsy Board Examinations. *JAMA Neurology* 81(6): 660–661. DOI: 10.1001/jamaneurol.2024.0676

John Halamka. 2023. Understanding Retrieval-Augmented Generation. Mayo Clinic Platform. https://www.mayoclinicplatform.org/2023/11/02/understanding-retrieval-augmented-generation/.

Ben Holgate, Shichao Fang, Anthony Shek, Matthew McWilliam, Pedro Viana, Joel S. Winston, James T. Teo, and Mark P. Richardson. 2024. Extracting Epilepsy Patient Data with Llama 2. In Proceedings of the 23rd Workshop on Biomedical Natural Language Processing, pages 526–535, Bangkok, Thailand. Association for Computational Linguistics.

Kexin Huang, Jaan Altosaar, Rajesh Ranganath. 2019. ClinicalBERT: Modeling Clinical Notes and Predicting Hospital Readmission. arXiv:1904.05342

Yubin Kim, Hyewon Jeong, Shan Chen, Shuyue Stella Li, Mingyu Lu, Kumail Alhamoud, Jimin Mun, Cristina Grau, Minseok Jung, Rodrigo Gameiro, Lizhou Fan, Eugene Park, Tristan Lin, Joonsik Yoon, Wonjin Yoon, Maarten Sap, Yulia Tsvetkov, Paul Liang, Xuhai Xu, Xin Liu, Daniel McDuff, Hyeonhoon Lee, Hae Won Park, Samir Tulebaev, Cynthia Breazeal. 2025. Medical Hallucination in Foundation Models and Their Impact on Healthcare. https://arxiv.org/pdf/2503.05777

P. Kwan and M.J. Brodie. 2000. Early identification of refractory epilepsy. *The New England Journal of Medicine* 342(5): pages 314-9.

Anson Li, Renee Shrestha, Thinoj Jegatheeswaran, Hannah O. Chan, Colin Hong, Rakesh Joshi. Mitigating Hallucinations in Large Language Models: A Comparative Study of RAG Enhanced vs. Human-Generated Medical Templates. https://www.medrxiv.org/content/10.1101/2024.09.27.24314506v1.full.pdf

Fenglin Liu, Zheng Li, Hongjian Zhou, Qingyu Yin, Jingfeng Yang, Xianfeng Tang, Chen Luo, Ming Zeng, Haoming Jiang, Yifan Gao, Priyanka Nigam, Sreyashi Nag, Bing Yin, Yining Hua, Xuan Zhou, Omid Rohanian, Anshul Thakur, Lei Clifton, and David A. Clifton. 2024. Large Language Models Are Poor Clinical Decision-Makers: A Comprehensive Benchmark. In Proceedings of the 2024 Conference on Empirical Methods in Natural Language Processing, pages 13696–13710, Miami, Florida, USA. Association for Computational Linguistics. DOI: 10.18653/v1/2024.emnlp-main.759

Mistral AI Team. 2024a. Mistral NeMo. https://mistral.ai/news/mistral-nemo

Mistral AI Team. 2024b. Un Ministral, des Ministraux. https://mistral.ai/news/ministraux

Arun Thirunavukarasu, Kabilan Elangovan, Darren Shu Jeng Ting, Laura Gutierrez, Daniel Shu Wei Ting. 2023. Large language models in medicine. *Nature Medicine*, vol. 29: pages 1930-1940.

Hugo Touvron, Louis Martin, Kevin Stone, Peter Albert, Amjad Almahairi, Yasmine Babaei, Nikolay Bashlykov, Soumya Batra, Prajjwal Bhargava, Shruti Bhosale, Dan Bikel, Lukas Blecher, Cristian Canton Ferrer, Moya Chen, Guillem Cucurull, David Esiobu, Jude Fernandes, Jeremy Fu, Wenyin Fu, Brian Fuller, Cynthia Gao, Vedanuj Goswami, Naman Goyal, Anthony Hartshorn, Saghar Hosseini, Rui Hou, Hakan Inan, Marcin Kardas, Viktor Kerkez, Madian Khabsa, Isabel Kloumann, Artem Korenev, Punit Singh Koura, Marie-Anne Lachaux, Thibaut Lavril, Jenya Lee, Diana Liskovich, Yinghai Lu, Yuning Mao, Xavier Martinet, Todor Mihaylov, Pushkar Mishra, Igor Molybog, Yixin Nie, Andrew Poulton, Jeremy Reizenstein, Rashi Rungta, Kalyan Saladi, Alan Schelten, Ruan Silva, Eric Michael Smith, Ranjan Subramanian, Xiaoqing Ellen Tan, Binh Tang, Ross Taylor, Adina Williams, Jian Xiang Kuan, Puxin Xu, Zheng Yan, Iliyan Zarov, Yuchen Zhang, Angela Fan, Melanie Kambadur, Sharan Narang, Aurelien Rodriguez, Robert Stojnic, Sergey Edunov, Thomas Scialom. 2023. Llama 2: Open Foundation and Fine-Tuned Chat Models. arXiv:2307.09288v2. Version 2.

Kevin Xie, Brian Litt, Dan Roth, and Colin A. Ellis. 2022a. Quantifying Clinical Outcome Measures in Patients with Epilepsy Using the Electronic Health Record. In Proceedings of the 21st Workshop on Biomedical Language Processing, pages 369-375, Dublin, Ireland. Association for Computational Linguistics.

Kevin Xie, Ryan S. Gallagher, Erin C. Conrad, Chadric O. Garrick, Steven N. Baldassano, John M. Bernabei, Peter D. Galer, Nina J. Ghosn, Adam S. Greenblatt, Tara Jennings, Alana Kornspun, Catherine V. Kulick-Soper, Jal M. Panchal, Akash R. Pattnaik, Brittany Scheid, Danmeng Wei, Micah Weitzman, Ramya Muthukrishnan, Joongwon Kim, Brian Litt, Colin Ellis, Dan Roth. 2022b. Extracting seizure frequency from epilepsy clinic notes: a machine reading approach to natural language processing. *Journal of the American Medical Informatics Association*, 29(5): pages 873-881.

Kevin Xie, Ryan S. Gallagher, Russell T. Shinohara, Sharon X. Xie, Chloe E. Hill, Erin C. Conrad, Kathryn A. Davis, Dan Roth, Brian Litt, Colin A. Ellis. 2023. Long-term epilepsy outcome dynamics revealed by natural language processing of clinic notes. *Epilepsia* 64(7): pages 1900-1909.

Kevin Xie, William K. S. Ojemann, Ryan S. Gallagher, Russell T. Shinohara, Alfredo Lucas, Chloé E. Hill, Roy H. Hamilton, Kevin B. Johnson, Dan Roth, Brian Litt, Colin A. Ellis. June 2024. Disparities in seizure outcomes revealed by large language models. Journal of the American Medical Informatics Association, Volume 31, Issue 6, Pages 1348–1355, https://doi.org/10.1093/jamia/ocae047

An Yang, et al. 2025. Qwen2.5 Technical Report. arXiv:2412.15115

Xi Yang, Aokun Chen, Nima PourNejatian, et al. 2022. A large language model for electronic health records. *npj Digital Medicine* 5:194. https://doi.org/10.1038/s41746-022-00742-2

Zhichao Yang, Avijit Mitra, Sunjae Kwon, and Hong Yu. 2024. ClinicalMamba: A Generative Clinical Language Model on Longitudinal Clinical Notes. In Proceedings of the 6th Clinical Natural Language Processing Workshop, pages 54–63, Mexico City, Mexico. Association for Computational Linguistics. DOI: 10.18653/v1/2024.clinicalnlp-1.5

WHO. Epilepsy. 2019. https://www.who.int/news-room/fact-sheets/detail/epilepsy.

Weipeng Zhou, Majid Afshar, Dmitriy Dligach, Yanjun Gao, and Timothy Miller. 2023. Improving the Transferability of Clinical Note Section Classification Models with BERT and Large Language Model Ensembles. In Proceedings of the 5th Clinical Natural Language Processing Workshop, pages 125–130, Toronto, Canada. Association for Computational Linguistics.

Kaiwen Zuo, Yirui Jiang. 2025. MedHallBench: A New Benchmark for Assessing Hallucination in Medical Large Language Models. https://arxiv.org/html/2412.18947v3

Appendix A

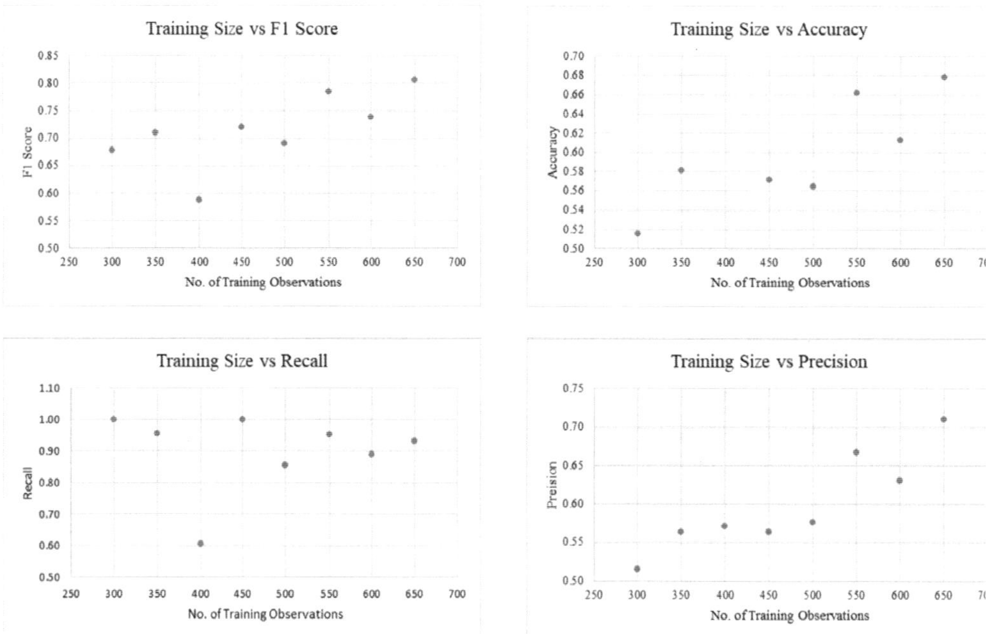

Appendix A: Ministral-8B-Instruct-2410 performance (pragmatic method) and size of training dataset.

Appendix B

Non-fine-tuned LLM: Best F1 Scores

	Purist Method				Pragmatic Method			
LLM	**recall**	**precision**	**F1**	**acccuracy**	**recall**	**precision**	**F1**	**acccuracy**
Ministral-8B-Instruct-2410	0.20	0.24	0.22	0.20	0.20	0.24	0.22	0.20
Mistral-Nemo-Instruct-2407	0.08	0.15	0.10	0.11	0.09	0.19	0.13	0.13
Qwen2.5-7B-Instruct	0.02	0.25	0.04	0.11	0.02	0.25	0.04	0.11
Llama-3.1-8B-Instruct	n/a	n/a	n/a	n/a	0.01	0.33	0.03	0.11

Non-fine-tuned LLM: Mean Over 3 Runs and F1 Standard Deviation

	Purist Method				Pragmatic Method			
LLM	**recall**	**precision**	**F1 (SD)**	**acccuracy**	**recall**	**precision**	**F1 (SD)**	**acccuracy**
Ministral-8B-Instruct-2410	0.20	0.24	0.22 (0.00)	0.20	0.20	0.24	0.22 (0.00)	0.20
Mistral-Nemo-Instruct-2407	0.08	0.15	0.10 (0.00)	0.11	0.09	0.19	0.13 (0.00)	0.13
Qwen2.5-7B-Instruct	0.02	0.25	0.04 (0.00)	0.11	0.02	0.25	0.04 (0.00)	0.11
Llama-3.1-8B-Instruct	n/a	n/a	n/a	n/a	0.01	0.33	0.03 (0.00)	0.11

Appendix B: Comparative performance evaluation of non-fine-tuned LLMs with same training dataset of 650 EHRs.

Note: Llama-3.1-8B-Instruct 'n/a' due to lack of true positives under purist method.

AdaBioBERT: Adaptive Token Sequence Learning for Biomedical Named Entity Recognition

Sumit Kumar and **Tanmay Basu**
Indian Institute of Science Education and Research Bhopal, Madhya Pradesh, India
{sumit23 | tanmay} @iiserb.ac.in

Abstract

Accurate identification and labeling of biomedical entities, such as diseases, genes, chemical and species, within scientific texts are crucial for understanding complex relationships. We propose Adaptive BERT or AdaBioBERT, a robust named entity recognition (NER) model that builds upon BioBERT (Biomedical Bidirectional Encoded Representation from Transformers) based on an adaptive loss function to learn different types of biomedical token sequence. This adaptive loss function combines the standard Cross Entropy (CE) loss and Conditional Random Field (CRF) loss to optimize both token level accuracy and sequence-level coherence. AdaBioBERT captures rich semantic nuances by leveraging pre-trained contextual embeddings from BioBERT. On the other hand, the CRF loss of AdaBioBERT ensures proper identification of complex multi-token biomedical entities in a sequence and the CE loss can capture the simple unigram entities in a sequence. The empirical analysis on multiple standard biomedical coprora demonstrates that AdaBioBERT performs better than the state of the arts for most of the datasets in terms of macro and micro averaged F1 score.

1 Introduction

The field of Biomedical Named Entity Recognition (NER) has evolved significantly, transitioning from rule-based systems to advanced deep learning methodologies. Early approaches relied heavily on handcrafted rules, dictionaries, and regular expressions to identify biomedical entities such as genes, diseases, and proteins. For instance, He (He et al., 2009) utilized domain-specific lexicons like UMLS to recognize entities. While these rule-based methods provided moderate accuracy, they struggled with the diversity and ambiguity of biomedical terminology, particularly for multi-token entities or novel terms. Their reliance on manual rule creation and limited adaptability hindered scalability (Settles, 2004; Leaman et al., 2015). The advent of machine learning techniques, such as Conditional Random Fields (CRF) (Sutton and McCallum, 2011) and Support Vector Machines (SVM) (Joachims, 1998), marked a shift toward data-driven models. CRF-based systems, like those developed by Settles (Settles, 2004) and Tsai (Tsai et al., 2006), leveraged labeled datasets to train classifiers that captured contextual and sequential information. These models demonstrated greater flexibility and adaptability compared to rule-based approaches. However, they still require extensive manual feature engineering, which limited their effectiveness in handling the complexity of biomedical data. For example, Leaman (Leaman et al., 2015) successfully applied CRF models to extract chemical and disease entities from PubMed abstracts but noted challenges in recognizing infrequent or context-dependent terms.

The introduction of Long Short-Term Memory (LSTM) networks and Convolutional Neural Networks (CNNs) revolutionized the NER tasks. Lample introduced a BiLSTM-CRF framework (Lample et al., 2016), which set new benchmarks for sequence labeling tasks, including NER. (Chiu and Nichols, 2016) extended this approach to biomedical texts, demonstrating the effectiveness of deep learning in capturing sequential dependencies and complex relationships. The emergence of transformer-based models, such as BERT (Devlin et al., 2019) and its biomedical counterpart, BioBERT (Lee et al., 2020), further advanced the capabilities of NER systems. These models employ self-attention mechanisms to capture the context of each word within a sentence, making them particularly effective for complex biomedical texts. BioBERT, which is pre-trained on biomedical corpora, has been effective in recognizing domain-specific entities (Lee et al., 2020). Unlike general-domain models, BioBERT effectively captures intricate relationships between biomedical terms, im-

Proceedings of the 24th Workshop on Biomedical Language Processing (BioNLP 2025), pages 56–62
August 1, 2025

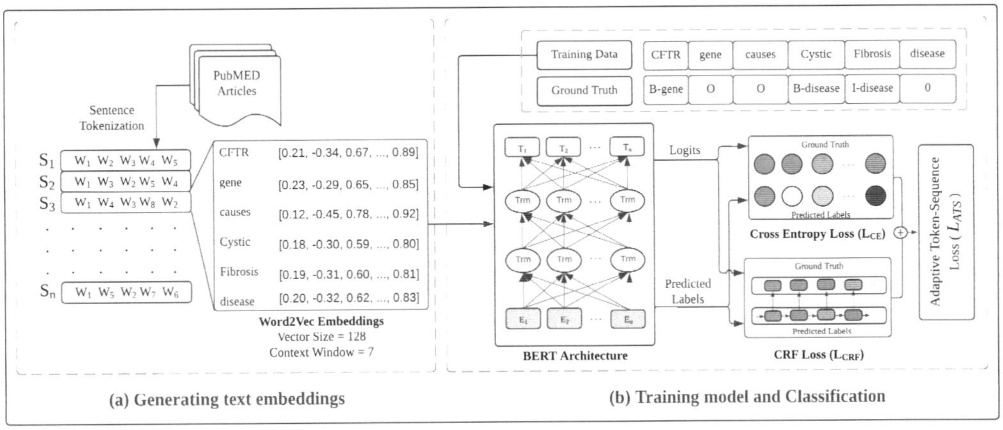

Figure 1: Proposed AdaBioBERT Architecture

proving NER performance in specialized datasets. Despite their effectiveness, transformer-based models often struggle to properly identify the named entities as they need large amount of data for finetuning (Chalkidis et al., 2020; Beltagy et al., 2019). Recent advancements have focused on combining the strengths of different loss functions. For example, Ma and Hovy (Ma and Hovy, 2016) introduced a BiLSTM-CRF model that used a fixed combination of CE and CRF loss functions for NER. Similarly, Lample (Lample et al., 2016) employed a fixed-weight combination of CE and CRF loss functions in their BiLSTM-CRF framework, which became a standard approach for NER tasks. However, these methods rely on fixed weighting scheme and cannot distinguish the significance between regular single token biomedical entities like *Nucleolin* and *Agyria*, and rare but important multi-token entities like *lateral sinus thrombosis* and *parietal cortical atrophy* through the loss functions.

Therefore, there is room to improve the quality of the existing methods to properly identify complex multi-token biomedical entities. In this spirit, this paper presents a transformer based Adaptive BioBERT (i.e., AdaBioBERT) NER model, to identify the nuances of complex multi-token biomedical entities by integrating a novel adaptive loss function combining the standard cross entropy and CRF loss functions in the pretrained Bio-BERT model (Lee et al., 2020).

2 Proposed AdaBioBERT Method

AdaBioBERT architecture has two major components: (1) Word2Vec embeddings (Kowsari et al., 2019), which capture semantic relationships between biomedical terms as shown in Fig 1(a) and (2) pretrained BioBERT model to generate rich contextual embeddings using the proposed Adaptive Token-Sequence Loss as shown in Fig 1(b), which dynamically balances token-level and sequence-level predictions.

2.1 Generate Word2Vec Embeddings of PubMED Data

In the first stage, the proposed framework extracts sentences from the freely available PubMED Central(PMC) repository[1], which has mention of any genes or diseases, based on frameworks proposed by (Basu et al., 2021; Guetterman et al., 2018). The objective is to build semantic embeddings of all relevant genes and diseases which are mentioned in the current version of DisGeNET[2] (v24.4) repository. It comprises 26,798 genes and 39,972 diseases and traits (Piñero et al., 2019). Subsequently, we generated word embeddings for these extracted sentences using Word2Vec model (Pennington et al., 2014; Kowsari et al., 2019). Sentences extracted from the PMC repository that build the corpus are tokenized, and then the Word2Vec algorithm generates embeddings for each word, which is represented as a 128-dimensional vector. The context window size of a word is set to 7, meaning the model considers up to seven neighboring words around a target word.

[1]https://pmc.ncbi.nlm.nih.gov/

[2]https://disgenet.com/DISGENET-Version-24-4-Whats-New

Table 1: Overview of biomedical datasets with training and testing splits

Dataset	Entity Types	Training	Test
BC4CHEMD (Krallinger et al., 2015)	Chemical compounds, drug names	6,000 abstracts	2,000 abstracts
LINNAEUS (Gerner et al., 2010)	Species names	80,000 sentences	10,000 sentences
NCBI-disease (Dogan et al., 2014)	Disease names	793 abstracts	100 abstracts
BC5CDR (Li et al., 2015)	Chemical compounds, diseases	1,000 articles	250 articles
JNLPBA (Kim et al., 2004)	Proteins, DNA, RNA, cell lines and types	2,000 abstracts	204 abstracts
AnatEM (Pyysalo, 2014)	Anatomical entities	1,200 documents	300 documents
BioNLP13GE (Kim et al., 2013)	Gene and gene product	1,500 sentences	500 sentences
Species-800 (Pafilis et al., 2016)	Species mentions	800 abstracts	200 abstracts

2.2 Pretrained BioBERT with Adaptive Token Sequence Loss (L_{ATS})

Let $X = \{x_1, x_2, \ldots, x_T\}$ denote an input sequence of tokens and $Y = \{y_1, y_2, \ldots, y_T\}$ represent the true labels of X, where y_t is a one-hot encoded vector and $y_t = [y_t^1, y_t^2, \cdots y_t^N]$ and $y_t^i \in \{c_1, c_2, \cdots c_N\}$. Here $c_i, \forall i = 1, 2, \ldots, N$ are different classes of biomedical entities. Let us consider $\widehat{Y} = \{\widehat{y_1}, \widehat{y_2}, \ldots, \widehat{y_T}\}$ be the sequence of predicted labels of the input sequence. The predicted probability for the t-th token $x_t \in c_i$ is denoted as $P(x_t \in c_i)$, and $\mathrm{S}(y_t, x_t)$ is the score of the true label sequence y_t given x_t. The L_{ATS} combines Cross-Entropy Loss (L_{CE}) and CRF Loss (L_{CRF}) as follows:

$$L_{ATS} = \alpha \cdot L_{CE} + (1 - \alpha) \cdot L_{CRF}, \quad (1)$$

where α is a learnable weight parameter to make a trade-off between CE loss and CRF loss. Here

$$L_{CE} = -\frac{1}{T} \sum_{t=1}^{T} \sum_{i=1}^{N} y_t^i \log \left(P(x_t \in c_i) \right)$$

is the Cross-Entropy Loss, which captures the sequence with a single biomedical entity and

$$L_{CRF} = -\Big(\mathrm{S}(Y, X) - \log \sum_{\widehat{Y}} \exp(\mathrm{S}(\widehat{Y}), X)) \Big)$$

is the CRF Loss, which is used to identify complex multi-label entities in a sequence. L_{ATS} dynamically adjusts the importance of per-token accuracy and sequence coherence through the learnable weight α. The adaptive weight parameter α is updated iteratively after each training epoch using gradient descent, as described in Algorithm 1. When α is close to 1, the model prioritizes individual token predictions, while α close to 0 emphasizes sequence-level coherence for handling multi-token entities and domain-specific terminology. Eventually, the pretrained BioBERT model is fine-tuned using the word embeddings of the genes and diseases generated by the word2vec model in the first stage followed by using L_{ATS}.

Algorithm 1 Adaptive Token-Sequence Loss with Learnable Weight α

Input: Token sequence $X = \{x_1, \ldots, x_T\}$, true labels $Y = \{y_1, \ldots, y_T\}$
Initialize: Model parameters θ, adaptive weight $\alpha \in [0, 1]$, learning rate η
Output: Updated θ, α, and loss L_{ATS}
Compute token-level cross-entropy loss
$L_{CE} \leftarrow \frac{-1}{T} \sum_{t=1}^{T} \sum_{i=1}^{N} y_t^i \log P(x_t \in c_i)$
Compute CRF sequence-level loss
Compute score of true sequence $S(Y, X)$
Compute partition function $Z(X) = \log \sum_{\widehat{Y}} \exp(S(\widehat{Y}, X))$
$L_{CRF} \leftarrow -(S(Y, X) - Z(X))$
Compute adaptive loss
$L_{ATS} \leftarrow \alpha \cdot L_{CE} + (1 - \alpha) \cdot L_{CRF}$
Backpropagation and parameter updates
Compute gradients: $\nabla_\theta L_{ATS}$, $\nabla_\alpha L_{ATS}$
Update parameters:
    $\theta \leftarrow \theta - \eta \cdot \nabla_\theta L_{ATS}$
    $\alpha \leftarrow \alpha - \eta \cdot \nabla_\alpha L_{ATS}$
Return: Final loss L_{ATS}, updated θ, α

3 Experimental Evaluation

3.1 Datasets and Settings

Experimental evaluation was conducted on eight widely used biomedical NER datasets as reported in Table 1. All of these datasets are formatted

Table 2: Macro F1-Scores of AdaBioBERT and State of the Arts

Dataset	SciSpacy	Stanza	Spark NLP	PubMedBERT	BioBERT	AdaBioBERT
BC4CHEMD	71.98	83.25	90.09	91.43	91.72	**95.40**
Linnaeus	79.84	81.73	82.14	85.07	85.47	**87.51**
NCBI Disease	74.82	83.08	84.13	87.83	88.45	**92.68**
BC5CDR	74.47	83.13	83.25	88.67	85.37	**89.83**
JNLPBA	69.35	74.14	76.68	**79.16**	76.18	78.93
AnatEM	74.22	83.35	84.15	90.57	88.14	**94.03**
BioNLP13GE	73.70	82.93	83.24	80.24	84.91	**85.36**
Species800	73.67	81.04	83.14	82.79	81.93	**87.63**

in the IOB (Inside, Outside, Beginning) tagging scheme, ensuring consistency in annotation and format across different biomedical domains. Each dataset is processed by extracting unique labels and tokenized using the AutoTokenizer from Hugging Face's Transformers library, ensuring compatibility with the pre-trained BioBERT model. The Word2Vec embeddings, pre-trained on biomedical literature, are integrated into the model as an additional feature to enhance entity recognition. Our model architecture is based on BioBERT, extended with a CRF layer for structured sequence prediction. A fully connected classifier with dropout is applied to the concatenated BioBERT and Word2Vec embeddings, projecting them onto the label space. The loss function is a weighted combination of CE and CRF loss, where the weight is a trainable parameter optimized during training. The optimizer used is AdamW with weight decay to improve generalization. The model is fine-tuned for 5, 10, 20, 40 epochs with a batch size of 4, 8, 16, 32 using an initial learning rate of 1e-4, 2e-4, 3e-4[3]. A NVIDIA A100 40 GP GPU server is used to implement AdaBioBERT. Evaluation is performed on an evaluation dataset after each epoch, saving the best-performing checkpoint. The trainer relies on mixed precision training and gradient accumulation for efficient computation.

3.2 Results and Discussion

The performance of AdaBioBERT and the state of the arts are reported in Table 2 in terms of macro-averaged F1-score. It can be seen from Table 2 that AdaBioBERT recognizes the biomedical entities better than the state of the arts and it outperforms the other methods for all datasets for macro-averaged F1 scores. Significant improvement of the F1-score of our method can be observed in BC4CHEMD (+3.68 over BioBERT (Lee et al., 2020)), Linnaeus (+ 2.04 over BioBERT), NCBI Disease (+4.23 over BioBERT), BC5CDR(+1.16 over PubMedBERT (Gu et al., 2021)), AnatEM (+3.46 over PubMedBERT), Species800 (+4.49 over SparkNLP) and marginally exceeds BioNLP13GE (+0.45 over BioBERT). Having JLNPBA as an exception where it lags marginally (-0.23 by PubMedBERT) indicating required improvement for recognition of protein, cell line,, and cell type entities in biomedical data. These results suggest that AdaBioBERT excels in biomedical entity recognition tasks where contextual understanding is important. The performance of AdaBioBERT on diverse biomedical entity recognition datasets shows its adaptability and robustness.

Notable improvements in micro F1-score are also reported in Table 3, where AdaBioBERT surpasses the performance in BC4CHEMD (+2.64 over PubMedBERT), Linnaeus (+5.07 over PubMedBERT), NCBI Disease (+7.17 over BioBERT), and AnatEM (+5.41 over PubMedBERT), demonstrating AdaBioBERT's recognition capability in chemical and disease-related entities. Additionally, AdaBioBERT surpasses BioBERT in BioNLP13GE (+2.93), PubMedBERT in BC5CDR (+1.82), SparkNLP in JNLPBA (+2.98), and PubMedBERT on Species800 (+1.54).

The proposed AdaBioBERT model introduces a novel approach to biomedical NER by integrating Adaptive Token-Sequence Loss with pre-trained contextual embeddings from BioBERT. One of the key technical innovations of AdaBioBERT is

[3]Results are reported for 20 epochs, batch size of 32 and learning rate of 1e-4.

Table 3: Micro F1-Scores of AdaBioBERT and State of the Arts

Dataset	SciSpacy	Stanza	Spark NLP	PubMedBERT	BioBERT	AdaBioBERT
BC4CHEMD	84.55	89.65	93.72	95.17	92.36	**97.81**
Linnaeus	81.74	88.27	86.26	90.22	88.24	**95.29**
NCBI Disease	81.65	87.49	89.13	88.36	89.71	**96.88**
BC5CDR	83.92	88.08	89.73	92.88	90.61	**94.70**
JNLPBA	73.21	76.09	81.29	79.53	77.49	**84.27**
AnatEM	84.14	88.18	89.13	92.04	91.26	**97.45**
BioNLP13GE	77.60	84.34	85.58	89.47	92.66	**95.59**
Species800	74.06	83.35	84.91	86.76	85.31	**88.30**

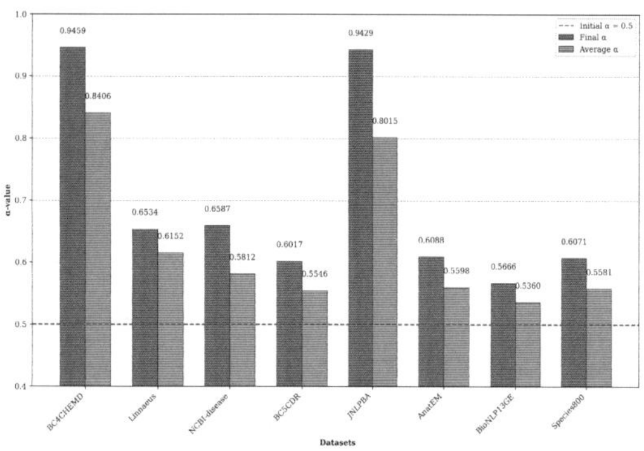

Figure 2: Final and Average α Values for Biomedical NER Datasets

its use of a learnable weight parameter (α) in the L_{ATS} loss function. This parameter enables the model to dynamically adjust the trade-off between token-level and sequence-level objectives during training, which ensures that our model can effectively handle both short, unambiguous entities and longer and complex ones. This flexibility is a significant improvement over the state of the arts that rely on fixed-weight combinations of L_{CE} and L_{CRF}, which may not generalize well across diverse biomedical texts. Additionally, the integration of pre-trained Word2Vec embeddings with BioBERT's contextual embeddings provides a multi-stage transfer learning framework, enhancing the model's ability to capture both semantic and contextual nuances in biomedical texts. The effectiveness of AdaBioBERT for identifying regular single token and complex multi-token entities has been demonstrated in the Table 2 and 3 for almost all datasets. The datasets like Species-800, NCBI Disease, and BC5CDR, where AdaBioBERT outperforms state-of-the-art by significant margins, contain lots of multi-token entities.

The different values of α in Figure 2 show how entity types vary in recognition difficulty. Chemical and gene entities (BC4CHEMD, JNLPBA) have much higher values (>0.94) because they use standard naming patterns that make individual words more important. Disease and anatomy terms (BioNLP13GE, BC5CDR, AnatEM) have lower values (0.56-0.66) because they need more context to understand ambiguous and less consistent names.

4 Conclusion

The potential of the proposed adaptive token-sequence loss with BioBERT embeddings is demonstrated through the extensive empirical analysis. By dynamically adjusting token-level and sequence-level learning through the learnable weight parameter (α), AdaBioBERT improves contextual understanding and multi-token entity recognition. Additionally, the integration of pre-trained Word2Vec embeddings further refines semantic representation in biomedical text. Despite its effectiveness, AdaBioBERT has high computational costs and may struggle with highly specific hierarchical entities. Future work will extend AdaBioBERT to broader biomedical information extraction tasks, including relation extraction, sentence classification, and document classification, to boost knowledge discovery in biomedical research. Codes available at: https://github.com/sumit-kumar-9297/AdaBioBERT-NER.git

5 Acknowledgment

This research is supported by the University Grants Commission (UGC), Government of India, through the Junior Research Fellowship (NTA Ref No: 220520545885).

References

Tanmay Basu, Simon Goldsworthy, and Georgios V Gkoutos. 2021. A sentence classification framework to identify geometric errors in radiation therapy from relevant literature. *Information*, 12(4):139.

Iz Beltagy, Kyle Lo, and Arman Cohan. 2019. Scibert: A pretrained language model for scientific text. In *Proceedings of EMNLP*.

Ilias Chalkidis, Michail Fergadiotis, Roger Stradling, Nikolaos Pappas, and Prodromos Malakasiotis. 2020. Transformer-based models for legal and biomedical document classification: A comparative study. In *Proceedings of ACL*.

Jason PY Chiu and Eric Nichols. 2016. Named entity recognition with bidirectional lstm-cnns. *Transactions of the Association for Computational Linguistics*, 4:357–370.

Jacob Devlin, Ming-Wei Chang, Kenton Lee, and Kristina Toutanova. 2019. BERT: Pre-training of deep bidirectional transformers for language understanding. In *Proceedings of the 2019 Conference of the North American Chapter of the Association for Computational Linguistics: Human Language Technologies*, pages 4171–4186.

Rezarta Islamaj Dogan, Robert Leaman, and Zhiyong Lu. 2014. Ncbi disease corpus: A resource for disease name recognition and concept normalization. *Journal of biomedical informatics*, 47:1–10.

Martin Gerner, Goran Nenadic, and Casey M Bergman. 2010. The linnaeus corpus: Annotated species names in biomedical literature. *Bioinformatics*, 26(18):i343–i349.

Yu Gu, Robert Tinn, Hao Cheng, Michael Lucas, Naoto Usuyama, Xiaodong Liu, Tristan Naumann, Jianfeng Gao, and Hoifung Poon. 2021. Domain-specific language model pretraining for biomedical natural language processing. *ACM Trans. Comput. Healthcare*, 3(1).

Timothy C Guetterman, Tammy Chang, Melissa DeJonckheere, Tanmay Basu, Elizabeth Scruggs, and VG Vinod Vydiswaran. 2018. Augmenting qualitative text analysis with natural language processing: Methodological study. *J Med Internet Res*, 20(6):e231.

Longhua He, Mehmet Kayaalp, David Megginson, and Thomas Rindflesch. 2009. A rule-based approach for biomedical named entity recognition using umls. *BMC bioinformatics*, 10(1):1–11.

Thorsten Joachims. 1998. Text categorization with support vector machines: Learning with many relevant features. In *Machine Learning: ECML-98*, pages 137–142, Berlin, Heidelberg. Springer Berlin Heidelberg.

Jin-Dong Kim, Tomoko Ohta, Yoshimasa Tsuruoka, Yuka Tateisi, and Nigel Collier. 2004. Introduction to the bio-entity recognition task at jnlpba. *Proceedings of the International Joint Workshop on Natural Language Processing in Biomedicine and its Applications*, pages 70–75.

Jin-Dong Kim, Yue Wang, Toshihisa Takagi, and Atsushi Yonezawa. 2013. Overview of the bionlp 2013 shared task on genomic entity normalization. *Proceedings of the BioNLP Shared Task 2013 Workshop*, pages 1–7.

Kamran Kowsari, Kiana Jafari Meimandi, Mojtaba Heidarysafa, Sanjana Mendu, Laura Barnes, and Donald Brown. 2019. Text classification algorithms: A survey. *Information*, 10(4):150.

Martin Krallinger, Florian Leitner, Obdulia Rabal, Miguel Vazquez, Julen Oyarzabal, and Alfonso Valencia. 2015. Overview of the biocreative iv chemical and drug name recognition challenge. *Database*, 2015.

Guillaume Lample, Miguel Ballesteros, Sandeep Subramanian, Kazuya Kawakami, and Chris Dyer. 2016. Neural architectures for named entity recognition. In *Proceedings of NAACL-HLT*.

Robert Leaman, Rezarta Islamaj Dogan, and Zhiyong Lu. 2015. Taggerone: joint named entity recognition and normalization with semi-markov models. *Bioinformatics*, 31(8):1361–1370.

Jinhyuk Lee, Wonjin Yoon, Sungdong Kim, Donghyeon Kim, Sunkyu Kim, Chan Ho So, and Jaewoo Kang. 2020. Biobert: a pre-trained biomedical language representation model for biomedical text mining. *Bioinformatics*, 36(4):1234–1240.

Jiao Li, Yijia Sun, Robin J Johnson, Daniela Sciaky, Chih-Hsuan Wei, Robert Leaman, et al. 2015. Overview of the biocreative v chemical disease relation (cdr) task. *Proceedings of the BioCreative V challenge evaluation workshop*, pages 154–166.

Xuezhe Ma and Eduard Hovy. 2016. End-to-end sequence labeling via bi-directional lstm-cnns-crf. *arXiv preprint arXiv:1603.01354*.

Evangelos Pafilis, Sune P Frankild, Lucia Fanini, Sarah Faulwetter, Christina Pavloudi, Aikaterini Vasileiadou, et al. 2016. The s800 corpus: A resource for species name recognition in biomedical texts. *Biodiversity Data Journal*, 4.

Jeffrey Pennington, Richard Socher, and Christopher D. Manning. 2014. GloVe: Global vectors for word representation. In *Proceedings of the 2014 Conference on Empirical Methods in Natural Language Processing (EMNLP)*, pages 1532–1543.

Janet Piñero, Juan Manuel Ramírez-Anguita, Josep Saüch-Pitarch, Francesco Ronzano, Emilio Centeno, Ferran Sanz, and Laura I Furlong. 2019. The disgenet knowledge platform for disease genomics: 2019 update. *Nucleic Acids Research*, 48(D1):D845–D855.

Sampo et al. Pyysalo. 2014. Anatem: A dataset for anatomical entity recognition in biomedical texts. *Journal of Biomedical Semantics*, 5(1):1–12.

Burr Settles. 2004. Biomedical named entity recognition using conditional random fields and rich feature sets. *Proceedings of the COLING*, pages 104–107.

Charles Sutton and Andrew McCallum. 2011. An introduction to conditional random fields. *Foundations and Trends in Machine Learning*, 4(4):267–373.

Richard TI Tsai, Shaojun Wu, and Tao-Hsien Chuang. 2006. A machine learning approach to biomedical named entity recognition. In *Proceedings of the 5th International Workshop on Bioinformatics*.

Transformer-Based Medical Statement Classification in Doctor-Patient Dialogues

Farnod Bahrololloomi and **Johannes Luderschmidt** and **Biying Fu**
Faculty DCSM
RheinMain University of Applied Sciences
Wiesbaden, Germany
{farnod.bahrololloomi, johannes.luderschmidt, biying.fu}@hs-rm.de

Abstract

The classification of medical statements in German doctor-patient interactions presents significant challenges for automated medical information extraction, particularly due to complex domain-specific terminology and the limited availability of specialized training data. To address this, we introduce a manually annotated dataset specifically designed for distinguishing medical from non-medical statements. This dataset incorporates the nuances of German medical terminology and provides a valuable foundation for further research in this domain. We systematically evaluate Transformer-based models and multimodal embedding techniques, comparing them against traditional embedding-based machine learning (ML) approaches and domain-specific models such as medBERT.de. Our empirical results show that Transformer-based architectures, such as the Sentence-BERT model combined with a support vector machine (SVM), achieve the highest accuracy of 79.58% and a weighted F1-Score of 78.81%, demonstrating an average performance improvement of up to 10% over domain-specific counterparts. Additionally, we highlight the potential of lightweight ML-models for resource-efficient deployment on mobile devices, enabling real-time medical information processing in practical settings. These findings emphasize the importance of embedding selection for optimizing classification performance in the medical domain and establish a robust foundation for the development of advanced, domain-adapted German language models.

1 Introduction

With the introduction of the Transformer architecture by Vaswani et al. (2017), substantial progress was achieved in many application areas, including general natural language processing (NLP) tasks and also in the field of medicine. However, models based on the Bidirectional Encoder Representations from Transformers (BERT) architecture (Devlin et al., 2018), initially trained on large-scale, general-purpose datasets such as Wikipedia, have struggled to accurately classify medical information in German datasets due to the complex and specialized vocabulary of medical language and the scarcity of labeled domain-specific datasets (Idrissi-Yaghir et al., 2024). To address these challenges, specialized models for the medical domain have been developed. An example is the German model medBERT.de, which has been fine-tuned with medical data and achieves an average Area Under the Receiver Operating Characteristic (AUROC) score of approximately 88% on various evaluated medical benchmarks (Bressem et al., 2024). Domain-specific models like medBERT.de can, for instance, detect whether medically relevant information is discussed in dialogues between doctors and patients. This capability is critical for extracting relevant data for patient documentation and improving the Electronic Health Record (EHR) system. Medical documentation is a cornerstone of healthcare, supporting patient care, legal accountability, and research. Yet, the processing of German medical texts remains challenging due to the inherent linguistic complexity and the limited availability of annotated datasets. As our contribution in this paper, we compare different Transformer-based models fine-tuned for medical data with traditional embedding-based methods. In particular, we focus on the analysis of German doctor-patient interviews to determine the most effective approach for classifying medical statements. Furthermore, we introduce a manually labeled dataset of medical statements to support future research in the processing of German medical texts. In doing so, we address two research questions:

- **RQ1:** How does the performance of Transformer models fine-tuned on medical data compare to traditional embedding-based approaches in classifying German doctor-patient

Proceedings of the 24th Workshop on Biomedical Language Processing (BioNLP 2025), pages 63–73
August 1, 2025

interviews?

- **RQ2:** How does the performance of fine-tuned Transformer and machine learning (ML)-models improve when evaluated on dataset of medical statements for domain-specific German medical texts?

2 Related Work

The classification of text in a medical context represents a fundamental challenge in the field of NLP, particularly in the medical domain. Accurate categorization of medical documents can significantly improve information extraction and decision-making processes (Kesiku et al., 2022). The complex and specialized terminology in medical texts poses a particular difficulty. Managing synonyms, polysemy, and multi-word terms is essential, as these can distort the true meaning of a text (Shanavas et al., 2020). In addition, medical text data often shows low density and high dimensionality due to its special linguistic characteristics, making its classification more challenging compared to other domains (Zhou et al., 2021).

Several studies have shown that ML-models may achieve high accuracy in medical text classification when adapted to the specific language and structure of medical texts. These techniques include support vector machines (SVMs), naive Bayes, logistic regression, and k-nearest neighbors (k-NNs). These methods are often combined with word representation models, such as term frequency-inverse document frequency (TF-IDF) and Word2Vec, to improve classification performance. (Mascio et al., 2020; Almazaydeh et al., 2023)

Almazaydeh et al. (2023) used the mtsamples.com dataset (MTSamples, 2025) to train ML-models using TF-IDF, Bag-of-Words (BOW), and Word2Vec as word representations. They were able to classify 20 medical categories. The Word2Vec-based k-NN classifier achieved an average accuracy of 92%. However, the performance on German medical datasets is unknown due to the challenges posed by the strict regulatory framework of the General Data Protection Regulation (GDPR).

Transformer-based models are gaining importance in medical NLP research. Idrissi-Yaghir et al. (2024) compared different German BERT architectures on medical datasets and evaluated them on different downstream tasks such as named entity recognition (NER), multi-label classification, and extractive question answering. The results show that models with medical or translation-based pre-training typically outperform generic language models, as they are better at capturing complex medical terminology and medical context. The language models achieved the following average F1-Scores: CLEF eHealth 2019 (Neves et al., 2019): 0.820, RadQA (Dada et al., 2023): 0.816, GraSCCo (Modersohn et al., 2022): 0.673, BRONCO150 (Kittner et al., 2021): 0.844, and GGPONC 2.0 (Borchert et al., 2022): 0.779. Idrissi-Yaghir et al. (2024) showed that continued pretraining can match or even surpass the performance of medical models trained from scratch. Furthermore, pretraining on medical data or leveraging translated texts has proven to be an effective approach for domain adaptation in medical NLP tasks. In addition to medBERT.de, there is also BioGottBERT by Lentzen et al. (2022), which was fine-tuned specifically on medical data. They conducted a comprehensive analysis of the suitability of existing and new transformer-based models for the German biomedical and clinical domain by systematically comparing 8 general-purpose language models and 3 newly trained models, including BioGottBERT and two BioELECTRA versions. The study showed that General-Purpose Language Models (GPLMs) performed surprisingly well on clinical NLP tasks, with a German variation of BERT called GBERT (Chan et al., 2020) performing particularly well on document classification tasks and BioGottBERT on NER tasks. Domain adaptation of existing models proved to be more effective than training new models from scratch, which was mainly attributed to the limited size of the pre-training corpus.

In recent years, several German medical datasets have been published, such as GGPONC (Borchert et al., 2020) and BRONCO150 (Kittner et al., 2021), which include annotation information for NER and part-of-speech (POS) tagging. Other German datasets, such as those from Makowski and Simko (2018) and Suominen et al. (2020), lack such annotation. Datasets like CLEF eHealth 2019 (Neves et al., 2019) offer German medical queries and documents for information retrieval and question–answering (QA); RadQA (Dada et al., 2023) comprises German radiology reports with questions to support radiological reasoning and GraSCCo (Modersohn et al., 2022) offers annotated social-care correspondence for entity and relation extraction. A specific German dataset for intent recognition in doctor-patient interviews was developed by Rojowiec et al. (2020), consisting of 63 classes. These classes represent various categories or intentions of questions and statements that can occur during doctor-patient conversations. The dataset supports medical students in taking medical histories by interacting with virtual patients, and the doctors' intentions were detected using BERT (Rojowiec et al., 2020). Section 3 provides further details on this dataset and its application in the context of this paper.

While it has been shown that Transformer-based models can perform well with domain adaptation, their performance in German dialog-based context recognition is not as well studied, and there is no high-quality medical dataset available to classify whether a statement contains medically relevant information or not.

3 Data Acquisition

To develop a German contextualized ML-model for classifying medical and non-medical statements, we used the publicly available "Intent Recognition in Doctor-Patient Interview" (IntRec) dataset (Rojowiec et al., 2020). This dataset consists of German transcriptions of live doctor-patient interviews conducted during university training sessions, in which medical students interviewed actors portraying patients, transcribing only the doctors' statements. 80% of the entries in the dialogue sequence consist of statements in the form of questions directed at the patient, such as "When was the surgery?" while 20% are normal statements, such as "I think so, yes.". For each entry, the corresponding class, its position within the sequence, the previous statement, and the class of the preceding statement are also provided. Table 1 shows the corresponding metadata about the original dataset before preprocessing.

Attribute	Statistic
Total number of samples	2,397
Number of classes	63 (62 + "OTHER")
Classes with $\leq$ 10 samples	50%
Largest class ("OTHER")	1,169 samples
Second-largest class ("AM02")	$>$ 85 samples
Annotated with two classes	101 (4%)
Average utterance length	10 words
Utterance	Previous utterances, intention

Table 1: Overview of the dataset for intent recognition in doctor-patient interviews.

The dataset consists of a total of 2,397 samples with multiple dialogue-label pairs, where 101 of these pairs have two label assignments. Each label consists of a symptom category and a question ID. The symptom category defines the symptom area, and the question ID specifies the intent within that area. For example, the label (PH10) belongs to the "Prior History" category (PH) and refers with question ID 10 to questions about "heart diseases". The dataset includes seven symptom categories (see Table 2).

We developed a preprocessing pipeline in which we divided the samples into individual dialogues and their associated labels. Each utterance and its corresponding labels, as well as the preceding utterance and its labels, were assigned individually to each target utterance and label. In the next step, duplicates in the utterance column were removed, resulting in a normalized dataset of 1,418 dialogue-class pairs.

Symptom Category	**Code**
Main Symptoms	MS
Prior History	PH
Allergies and Medication	AM
Social and Family History	SF
System Review	SR
Inquiry	IQ
Other Questions	OQ

Table 2: Symptoms categories and code, with "IQ" + "OQ" summarized under the category "OTHER".

To develop a classification model for detecting medical statements, we transformed the multiclass problem into a binary problem. The dataset was transformed by grouping all categories unrelated to "IQ" or "OQ" under the class "MEDICAL", while "IQ" and "OQ" were combined into the class "OTHER". Following the categorization described by Rojowiec et al. (2020), the symptom category "Inquiry", although referring to previously posed questions, was not considered to contain medically relevant information. In addition, redundant punctuation, such as quotation marks ("), was removed from the documents using regular expressions as an additional preprocessing step to improve data quality. The normalized dataset was split into training and test data in an 80/20 ratio (see Table 3). To address potential data bias, the dataset was randomized prior to splitting.

In addition, a second test dataset was developed using the publicly available Berlin-Tübingen-Oncology Corpus (BRONCO150) by Kittner et al. (2021). This German-language corpus consists of 150 discharge summaries from cancer patients treated at the Charité-Berlin University of Medicine or the University Hospital of Tübingen. To prevent the reconstruction of discharge summaries and patient identities, Kittner et al. (2021) shuffled the summaries and anonymized them at the sentence level. The dataset, originally intended for information extraction from German medical texts, comprises $8,976$ sentences with POS annotations and includes medical entities along with relevant attributes like negation and speculation.

Since the BRONCO150 dataset contains not only complete sentences but also other information from discharge summaries, we manually labeled the data to extract only complete sentences or medically accurate statements. For a realistic evaluation of the models trained on the IntRec dataset, the BRONCO150 dataset was manually labeled based on specific criteria, categorizing statements as either medical or non-medical:

1. The sentence contains a medical claim.
2. Punctuation at the end is not mandatory if the content conveys a medical statement.
3. The sentence cannot be used as a title.
4. The sentence begins with an uppercase letter.
5. A sentence must not be a list or contain a colon ":" unless it begins with a date and a statement.

Manual labeling was conducted using the publicly available tool LabelStudio[1] (Tkachenko et al., 2020-2025). Annotation was performed by a Computer Science PhD student with expertise in NLP. Of the $8,976$ records, $6,863$ medical statements remain after labeling and duplicate removal. Approximately 60.15% of the data received the label 0 because many sentences contained formatting information such as date values or document headers, "Dear Sir or Madam" or document lines such as "Line ID. from document". This resulted in a reduction, leaving 39.85% with a value of 1. In addition, the dataset included partial sentences that were not standalone statements, but related to the previous line. Furthermore, enumerations were not considered because they were not independent sentences with statements. The following examples from the BRONCO150 dataset are English translations of original German texts published in the work of Kittner et al. (2021). To demonstrate these criteria, we present the following examples from the BRONCO150 dataset. Statements labeled as "MEDICAL" satisfy these conditions by expressing clear clinical information. For instance, the direct quotes "On 07/04/2134, the patient received an uneventful nivolumab infusion." (Kittner et al., 2021, Fig. 1) and "A highly suspicious HCC lesion was observed in liver segment VI on CT." (Kittner et al., 2021, Fig. 1) reflect medical events and fulfill criteria (1) to (5). In contrast, direct quotes such as "Start of chemotherapy according to the GeT protocol cycle 1." (Kittner et al., 2021, Fig. 1) or "Diagnoses: RA: choroidal melanoma (ED 07/2023)" (Kittner et al., 2021, Fig. 3) are often abbreviated, context-dependent, or formatted as titles or lists, thereby violating criteria (3) and (5), and are classified as "OTHER". The resulting German-language dataset can be used not only for our case, but also for fine-tuning German models on medical data, with the aim of supporting medical data extraction and improving semi-automatic methods for annotating medical documents. We use this dataset to evaluate how well the transfer learning of all trained models performs on unseen data, to understand whether the models can understand not only previous medical queries but also complex medical language and derive correct classifications for medical statements. The fully labeled dataset by Bahrololloomi (2025), consisting of the $8,976$ sentence_ids and labels is publicly available in the form of a CSV file. This dataset acts as a mapping and can be combined one-to-one with the original dataset by Kittner et al. (2021).

Dataset	OTHER	MEDICAL
Train/Validation 1134 (80%)	688	446
Test 284 (20%)	160	124
Test BRONCO150 6863 (100%)	4127	2736

Table 3: Data distribution and class distribution for IntRec and the normalized BRONCO150 data with class 0 as "OTHER" and class 1 as "MEDICAL".

4 Model Engineering

We extracted embeddings from four different Transformer models based on the BERT architecture to classify medical statements within sentences. These embeddings were then combined with five traditional ML-models for classification. The advantage of pure embedding extraction, as opposed to training the entire Transformer model, is evident

[1] https://labelstud.io/

in the decreased training duration and the capability to efficiently adapt these models into a mobile variant. This adaptation facilitates their use for local predictions, such as in smartwatches.

During the model selection process, we ensured the use of a German, a multilingual, and a medically specialized English model to systematically evaluate the transfer performance in the classification process. The multilingual model is a variant of the Sentence Transformer (Sentence-BERT)[2] from Reimers and Gurevych (2019). Additionally, we used a general German BERT model (BERT_{ger})[3] (Bavarian State Library, 2025) to evaluate the transfer performance of the BERT architecture on medical data.

We also selected the BioBERT model (BioBERT)[4] for the classification of medical documents. This model was developed by Deka et al. (2022) and specifically trained on English scientific publications related to medical trials. Furthermore, the German model medBERT.de (MedBERT)[5], created by Bressem et al. (2024), was used. This model was trained on a comprehensive collection of German medical documents, including medical reports and patient records. Due to its optimization for longer texts, MedBERT is particularly suitable for the analysis and classification of medical information and outperformed other German-language models in NLP tasks such as NER.

The following ML-models have been used: CatBoost (Dorogush et al., 2018), RandomForest (RF) (Pedregosa et al., 2011a), XGBoost (Chen and Guestrin, 2016), SVM (Pedregosa et al., 2011a), and LightGBM (Ke et al., 2017). In order to extract the best possible embedding, we compared different extraction strategies by calculating the average of the last hidden states over the sequence dimension (mean pooling), extracting the maximum value over all tokens (max pooling), and using the hidden state of the first token (CLS token) as a representation of the entire sequence.

The overall architecture of our approach is as follows. In the first step, the cleaned and shuffled $1,134$ sentences from the training and validation IntRec dataset are passed to the four Transformer models. Simultaneously, the mentioned extraction strategies are applied to the vanilla variants of the models to extract the required embeddings. These embeddings are then fed to the five ML-models. The same seed was used on the dataset to reproduce the same training and validation data. We use Sentence-BERT as a baseline for comparison with other Transformer models. Similarly, for the ML-models, we apply a Word2Vec approach to convert the medical data into embeddings, as suggested by Almazaydeh et al. (2023). We did not train a Word2Vec model from scratch, but used the pre-trained German Word2Vec[6] model from Yamada et al. (2020). In the next step, fixed parameters such as batch size, learning rate, and maximum padding size calculated over both datasets are set on the Transformer models, and hyperparameter optimization is performed on the ML-models via grid search using the validation data. The final step involves evaluating all ML-models on the IntRec and BRONCO150 test datasets.

5 Evaluation

As discussed in Section 4, the IntRec training and validation data are initially utilized to train and optimize the four proposed Transformer models. This process aims to identify the optimal parameters, enabling the selection of the most suitable model for the subsequent steps. To ensure optimal computational efficiency, we first performed document analysis on both datasets to determine the maximum token length and padding size. The German medical model MedBERT of Bressem et al. (2024) was used for this determination. As shown in Table 4, a maximum padding size of 143 tokens is sufficient to cover all sequences in both the normalized IntRec $(1,418)$ and BRONCO150 $(6,863)$ datasets, each consisting of labelled sentences. We also found that the BRONCO150 dataset contains documents of a greater length than the IntRec dataset. This discrepancy can be attributed to the divergent nature of the text: while the IntRec dataset is primarily composed of doctor questions directed at patients, the BRONCO150 dataset consists of discharge summaries that require a more comprehensive level of understanding.

The hyperparameters were set uniformly for all Transformer models with a number of epochs of 20, a batch size of 20, a learning rate of 2×10^{-5},

[2] https://huggingface.co/sentence-Transformers/paraphrase-multilingual-MiniLM-L12-v2

[3] https://huggingface.co/dbmdz/bert-base-german-cased

[4] https://huggingface.co/pritamdeka/BioBert-PubMed200kRCT

[5] https://huggingface.co/GerMedBERT/medbert-512

[6] https://huggingface.co/Word2vec/wikipedia2vec_dewiki_20180420_300d

Metric	IntRec 1418	BRONCO150 6863
Maximum Token Count	143	142
Average Length	14.90	18.32
Median Length	13.0	14.0
Standard Deviation	8.53	14.94

Table 4: Statistical properties of token lengths for both datasets.

and a maximum padding size of 143. For optimization, the AdamW optimizer is employed, as it offers more robust convergence compared to the traditional Adam optimizer due to its enhanced regularization through weight decay (Baevski et al., 2020). To minimize overfitting, a linear scheduler uniformly reduces the learning rate during training. Early stopping is implemented to terminate training if the validation accuracy (*val_acc*) fails to improve over three consecutive epochs. This approach prevents overfitting and reduces unnecessary computation. The training loss is calculated using *BCEWithLogitsLoss* from Pytorch (Paszke et al., 2019). Our analysis indicates that mean pooling is the most effective method for extracting embeddings. Consequently, it is consistently applied across all ML-models (see Appendix Table 9).

Hyperparameter optimization of ML-models is performed using Word2Vec embeddings with grid search and triple cross-validation, evaluated based on weighted F1-Score. The CatBoost model undergoes a separate optimization process, since *GridSearchCV* (Pedregosa et al., 2011b) is incompatible with the *Pool* format of CatBoost. Instead, the model is trained on a training dataset (*train_pool*) and evaluated on a validation dataset (*val_pool*). The best parameter configuration is determined based on the highest F1-Score. In addition to the application of hyperparameter optimization using Word2Vec embeddings, extensive hyperparameter exploration was simultaneously performed on the full set of ML-models, incorporating every available variant of BERT embeddings. The best parameters for each model are listed in the Appendix in the Table 8. These parameters are consistently applied to all ML and Transformer models without explicit mention in the Tables, as the optimal parameters are always used.

After determining the best hyperparameters, both the Transformer-based BERT models and all variations of the ML-models with BERT and Word2Vec embeddings were trained and validated on the cleaned IntRec test data. To measure the performance of the models, we use well-known metrics such as accuracy, precision, recall and F1-Score for both classes (medical and general). The individual results on the validation data are shown in Table 5.

Classifier	Acc.	Macro F1	Weighted F1	Gen. F1	Med. F1
Sentence-BERT	0.84	0.83	0.84	0.87	0.80
$BERT_{ger}$	**0.84**	0.84	0.84	0.87	**0.81**
BioBERT	0.77	0.76	0.77	0.80	0.72
MedBERT	**0.85**	**0.84**	**0.85**	**0.88**	**0.81**

Table 5: Performance of classification models on the IntRec validation data.

The metric Medical F1-Score (Med. F1) indicates how well the model correctly classifies medical statements, in contrast to the metric General F1-Score (Gen. F1), which represents the F1-Score over documents labeled as general. The Macro Avg F1-Score (Macro F1) calculates the average F1-Score across all classes, regardless of their size. In contrast, the Weighted Avg F1-Score (Weighted F1) additionally weights the size of each class and adjusts the F1-Score accordingly. The results show that MedBERT delivers the best overall performance, achieving an accuracy of 0.85 and a high F1-Score in both the macro and weighted average. The MedBERT model achieves a macro F1-Score of 0.84 and a weighted F1-Score of 0.85, indicating its ability to effectively perform both balanced and weighted classifications. In comparison, the English BioBERT shows the weakest performance, especially in the medical context, with an F1-Score of only 0.72. This model only achieves an accuracy of 0.77, indicating its limited ability to correctly classify medical statements in this specific dataset. Interestingly, both Sentence-BERT and $BERT_{ger}$ achieve similar performance, with an accuracy of 0.84 and a consistent Weighted and Gen. F1-Score of 0.84 and 0.87, respectively. Both models show strong and balanced classification performance, but they perform slightly worse than the MedBERT model. For the evaluation of the ML-models on the IntRec validation data with the respective text representations, the Weighted F1-Score is used as evaluation metric (see Table 7).

Classifier	Word2Vec	Sentence-BERT	$BERT_{ger}$	BioBERT	MedBERT
CatBoost	0.6839	**0.8372**	0.7621	0.6678	0.7813
RandomForest	0.6611	**0.8121**	0.7086	0.6535	0.7310
XGBoost	0.6551	**0.8059**	0.7519	0.6720	0.7671
SVM	0.6946	**0.8330**	0.7616	0.6668	0.7854
LightGBM	0.6654	**0.8107**	0.7599	0.6551	0.7567

Table 7: Weighted F1-Scores of ML-models with varying text representations on the IntRec validation data.

Model	Word Rep.	Acc. IntRec	F1-IntRec	Acc. BRONCO	F1-BRONCO
CatBoost	Word2Vec	$0.6620 \pm 3.33\,e^{-16}$	$0.6218 \pm 1.11\,e^{-16}$	$0.5993 \pm 1.11\,e^{-16}$	$0.4931 \pm 5.55\,e^{-17}$
	Sent.-BERT	$0.7746 \pm 1.11\,e^{-16}$	$0.7614 \pm 1.11\,e^{-16}$	$0.5974 \pm 0.00\,e^{-16}$	$0.4787 \pm 5.55\,e^{-17}$
	BERT_{ger}	$0.7183 \pm 1.11\,e^{-16}$	$0.7006 \pm 1.11\,e^{-16}$	$0.6016 \pm 0.00\,e^{-16}$	$0.4526 \pm 5.55\,e^{-17}$
	BioBERT	$0.6162 \pm 1.11\,e^{-16}$	$0.6000 \pm 2.22\,e^{-16}$	$\mathbf{0.6283 \pm 2.22\,e^{-16}}$	$\mathbf{0.5542 \pm 1.11\,e^{-16}}$
	MedBERT	$0.6514 \pm 1.11\,e^{-16}$	$0.6189 \pm 1.11\,e^{-16}$	$0.6012 \pm 1.11\,e^{-16}$	$0.4521 \pm 0.00\,e^{-16}$
RF	Word2Vec	$0.6479 \pm 1.11\,e^{-16}$	$0.6103 \pm 1.11\,e^{-16}$	$0.5951 \pm 0.00\,e^{-16}$	$0.4808 \pm 1.66\,e^{-16}$
	Sent.-BERT	$0.7394 \pm 2.22\,e^{-16}$	$0.7145 \pm 1.11\,e^{-16}$	$0.5920 \pm 1.11\,e^{-16}$	$0.4584 \pm 1.66\,e^{-16}$
	BERT_{ger}	$0.6866 \pm 1.11\,e^{-16}$	$0.6540 \pm 1.11\,e^{-16}$	$0.6013 \pm 0.00\,e^{-16}$	$0.4519 \pm 0.00\,e^{-17}$
	BioBERT	$0.6268 \pm 1.11\,e^{-16}$	$0.6049 \pm 1.11\,e^{-16}$	$0.6209 \pm 2.22\,e^{-16}$	$0.5226 \pm 2.22\,e^{-16}$
	MedBERT	$0.6549 \pm 0.00\,e^{-16}$	$0.6073 \pm 1.11\,e^{-16}$	$0.6015 \pm 0.00\,e^{-16}$	$0.4520 \pm 1.66\,e^{-16}$
XGBoost	Word2Vec	$0.6268 \pm 1.11\,e^{-16}$	$0.6091 \pm 2.22\,e^{-16}$	$0.6088 \pm 1.11\,e^{-16}$	$0.5087 \pm 2.22\,e^{-16}$
	Sent.-BERT	$0.7183 \pm 1.11\,e^{-16}$	$0.6994 \pm 1.11\,e^{-16}$	$0.5997 \pm 1.11\,e^{-16}$	$0.4953 \pm 0.00\,e^{-17}$
	BERT_{ger}	$0.6937 \pm 2.22\,e^{-16}$	$0.6711 \pm 3.33\,e^{-16}$	$0.6031 \pm 2.22\,e^{-16}$	$0.4587 \pm 5.55\,e^{-17}$
	BioBERT	$0.6232 \pm 1.11\,e^{-16}$	$0.6130 \pm 0.00\,e^{-16}$	$0.6200 \pm 0.00\,e^{-16}$	$0.5399 \pm 0.00\,e^{-16}$
	MedBERT	$0.6585 \pm 2.22\,e^{-16}$	$0.6301 \pm 1.11\,e^{-16}$	$0.6013 \pm 0.00\,e^{-16}$	$0.4548 \pm 5.55\,e^{-17}$
SVM	Word2Vec	$0.6549 \pm 0.00\,e^{-16}$	$0.6469 \pm 1.11\,e^{-16}$	$0.5659 \pm 0.00\,e^{-16}$	$0.5043 \pm 1.11\,e^{-16}$
	Sent.-BERT	$\mathbf{0.7958 \pm 1.11\,e^{-16}}$	$\mathbf{0.7881 \pm 2.22\,e^{-16}}$	$0.5885 \pm 0.00\,e^{-16}$	$0.4806 \pm 5.55\,e^{-17}$
	BERT_{ger}	$0.7465 \pm 1.11\,e^{-16}$	$0.7370 \pm 2.22\,e^{-16}$	$0.5990 \pm 0.00\,e^{-16}$	$0.4518 \pm 0.00\,e^{-16}$
	BioBERT	$0.6338 \pm 1.11\,e^{-16}$	$0.6093 \pm 0.00\,e^{-16}$	$0.5911 \pm 0.00\,e^{-16}$	$0.5322 \pm 1.11\,e^{-16}$
	MedBERT	$0.6761 \pm 2.22\,e^{-16}$	$0.6395 \pm 2.22\,e^{-16}$	$0.6013 \pm 0.00\,e^{-16}$	$0.4516 \pm 1.11\,e^{-16}$
LightGBM	Word2Vec	$0.6514 \pm 1.11\,e^{-16}$	$0.6288 \pm 1.11\,e^{-16}$	$0.5957 \pm 1.11\,e^{-16}$	$0.5080 \pm 0.00\,e^{-16}$
	Sent.-BERT	$0.7852 \pm 0.00\,e^{-16}$	$0.7746 \pm 0.00\,e^{-16}$	$0.5955 \pm 1.11\,e^{-16}$	$0.4717 \pm 0.00\,e^{-16}$
	BERT_{ger}	$0.7148 \pm 0.00\,e^{-16}$	$0.6951 \pm 1.11\,e^{-16}$	$0.6016 \pm 0.00\,e^{-16}$	$0.4562 \pm 5.55\,e^{-17}$
	BioBERT	$0.6479 \pm 1.11\,e^{-16}$	$0.6358 \pm 1.11\,e^{-16}$	$0.6159 \pm 0.00\,e^{-16}$	$0.5380 \pm 2.22\,e^{-16}$
	MedBERT	$0.6831 \pm 1.11\,e^{-16}$	$0.6560 \pm 0.00\,e^{-16}$	$0.6010 \pm 2.22\,e^{-16}$	$0.4526 \pm 1.66\,e^{-16}$
Sent.-BERT	-	$0.7676 \pm 1.11\,e^{-16}$	$0.7671 \pm 1.11\,e^{-16}$	$0.5280 \pm 0.00\,e^{-16}$	$0.5191 \pm 1.11\,e^{-16}$
BERT_{ger}	-	$0.7711 \pm 2.22\,e^{-16}$	$0.7655 \pm 2.22\,e^{-16}$	$0.5790 \pm 2.22\,e^{-16}$	$0.4538 \pm 1.66\,e^{-16}$
BioBERT	-	$0.7218 \pm 2.22\,e^{-16}$	$0.7101 \pm 1.11\,e^{-16}$	$0.6018 \pm 2.22\,e^{-16}$	$0.4796 \pm 5.55\,e^{-17}$
MedBERT	-	$0.7782 \pm 2.22\,e^{-16}$	$0.7752 \pm 0.00\,e^{-16}$	$0.6048 \pm 1.11\,e^{-16}$	$0.4938 \pm 0.00\,e^{-17}$

Table 6: Performance of various classification models on IntRec and BRONCO150 test data, based on accuracy and weighted F1-Score. The results include the mean and standard deviation from 100 evaluations.

The results show that, in contrast to the direct comparison with the Transformer models, all ML-models achieve the best results with multilingual Sentence-BERT embeddings, reaching an average Weighted F1-Score of 0.8198 with a low standard deviation of 0.0114. This indicates a consistent performance of the ML-models with this embedding. In comparison, the BioBERT and Word2Vec embeddings have an average performance that is 19.12% and 18.02% worse, respectively. These differences in model performance indicate that the multilingual Sentence BERT embeddings are best suited for the given classification task. The stable results show that this representation not only delivers high F1-Scores, but also exhibits low variance between models, further demonstrating its robustness. However, the overall results are worse than those of the Transformer variants.

To evaluate the robustness of the Transformer and ML-models, a data-driven analysis was performed during inference. Both the test data of the IntRec dataset and the normalized and labeled 6,863 large BRONCO150 dataset were randomly shuffled 100 times with different but fixed seeds for the iteration index. Table 6 presents the results obtained, showing the mean and standard deviation for all metrics. Since the standard deviations for all models are in the range of 10^{-16}, they are presented with the factor e^{-16}. The results underline how crucial both the choice of the classification model and the underlying embedding representation are. Although MedBERT showed the best performance on the validation data, the MedBERT embeddings overall do not perform optimally on the IntRec test data. Notably, pure Transformer models do not outperform on average an SVM working in combination with Sentence-BERT embeddings. In particular, this combination achieves the best results with an accuracy of 0.7958 and a weighted F1-Score of 0.7881. The superiority of the Sentence-BERT embeddings over alternative representations such as Word2Vec, BERT_{ger}, BioBERT or MedBERT highlights the importance of a powerful embedding base, especially in the analysis of medical datasets. Furthermore, the extremely low standard deviations confirm the high robustness and reproducibility of the results, a factor further favored by the weighted F1-Scores, which take into account the class frequencies. Overall, the analysis shows that for optimal classification performance in the medical domain, not only the model complexity, but also the targeted selection of embeddings is of central importance.

Given that the BRONCO150 dataset consists entirely of domain-specific medical statements, and that no prior model training has included such data, its evaluation provides a potential means of exploring the transfer learning ability of different approaches when confronted with novel and, to some extent, partially different sentence structures. Table 6 shows that all models achieve robust results, with accuracy values mostly above 59% and weighted F1-Scores delivering consistent results. It is worth noting that the CatBoost model combined with BioBERT embeddings achieves the best results with an accuracy of 0.6283 and a weighted F1 Score of 0.5542. These results suggest that BioBERT embeddings, which are already pre-trained on medical texts, offer a significant advantage in the classification of purely medical sentences. The observed differences in performance can mainly be explained by the different characteristics of the datasets. While the IntRec dataset used for training mainly contains doctor-patient interviews with comparatively simple medical terminology, the content of the BRONCO150 dataset is based on discharge summaries, which document the course of treatment and the main medical findings and therapy decisions in detail. This high degree of precision and the distinct linguistic style complicate the direct transfer of the classification capabilities acquired during training, thereby accounting for the divergent results.

6 Discussion

Our study investigated the classification of medical statements in German doctor-patient dialogues by integrating Transformer-based models with traditional ML-models that leverage BERT-based embeddings. The evaluation provided key insights into model performance and domain adaptability, while highlighting the trade-offs between general-purpose and domain-specific methods. Regarding RQ1, our findings reveal that domain-specific models such as MedBERT.de even though explicitly optimized for medical texts do not exhibit a significant advantage over general-purpose Transformer models in dialogue-based medical contexts. Sentence-BERT, a non-domain-specific model, achieved an F1-Score of 0.84, which is nearly equivalent to that of MedBERT.de (F1 = 0.85). This suggests that high-quality sentence embeddings extracted from general Transformers can compensate for the lack of domain-specific pretraining in certain scenarios. In contrast, the comparatively weaker performance of BioBERT shows challenges related to linguistic and data-specific adaptation, particularly in cross-lingual settings. Our evaluation indicates that hybrid approaches such as combining an SVM classifier with Sentence-BERT embeddings yield strong performance on the test set, achieving the highest accuracy (0.80) and weighted F1-Score (0.79). This finding emphasizes the importance of careful selection of embedding strategies and model architectures for the effective classification of medical statements. To understand the performance differences observed in RQ2, it is important to note that, while both datasets contain German medical language, they differ in context and linguistic formality: IntRec features short, spoken questions, whereas BRONCO150 consists of structured discharge summaries. In the context of RQ2, the evaluation on the BRONCO150 dataset, which consists of structured medical texts, shows that models trained on conversational data struggle to generalize to more formal medical documents. While Sentence-BERT based models excel in doctor-patient dialogues, domain-specific embeddings like BioBERT deliver better performance for structured medical statements. This divergence shows the need to tailor embedding strategies to the specific nature of the text being analyzed. In conclusion, our research confirms that Transformer-based models, when optimally integrated with advanced embedding strategies, are capable of delivering accurate and robust classification of medical statements. The RQ1 is answered, showing the feasibility of employing hybrid approaches in doctor-patient interviews. This work not only sets a solid foundation for the evolution of more sophisticated models in the field but also highlights the critical importance of careful embedding selection and parameter tuning in navigating the challenges inherent in specialized medical language. Regarding RQ2, the complexity of the BRONCO150 dataset poses a significant challenge. None of the models achieved a good F1-Score on this data. Although accuracy remained higher than the F1-Score, this suggests that the models are more effective at classifying "OTHER" statements while struggling with "MEDICAL" ones.

7 Conclusion and Future Work

This study identifies several opportunities for future research. A practical evaluation of the proposed methods in real-world medical settings is essential to assess their effectiveness in automated text extraction within EHR systems. In this context, the application of knowledge distillation techniques should be explored to adapt models for resource-constrained environments, such as mobile devices and smartwatches, enabling real-time processing. In addition, future work should systematically investigate the extent to which automatically generated examples (e.g., via GPT-4o or other Large Language Models (LLMs)) can reduce the need for manual labeling. In particular, it is crucial to assess the quality of the resulting pseudo-labels and to explore how a hybrid approach (synthetic + manual) can yield robust models in resource-constrained environments. Furthermore, extending the approach to multi-turn dialogues and incorporating clinician feedback could enhance classification accuracy and system robustness. To better capture the context of IntRec's short and isolated sentences, we plan to reframe the task as a QA problem by concatenating each QA instance into a single input and predicting its original label. Future work should also focus on optimizing embedding selection strategies, leveraging data augmentation techniques, and investigating transfer learning approaches to mitigate the performance gap between conversational and structured medical texts. Additionally, evaluating these models in real-world deployment scenarios, such as automated documentation systems, will provide valuable insights into their practical applicability. By addressing these challenges and refining current methodologies, future research can significantly improve the efficiency and domain relevance of automated medical text processing.

References

Laiali Almazaydeh, Mohammad Abuhelaleh, Arar Al Tawil, and Khaled Elleithy. 2023. Clinical text classification with word representation features and machine learning algorithms. *International Journal of Online and Biomedical Engineering (iJOE)*, 19(04):65–76.

Alexei Baevski, Henry Zhou, Abdelrahman Mohamed, and Michael Auli. 2020. wav2vec 2.0: A framework for self-supervised learning of speech representations. *arXiv preprint*.

Farnod Bahrololloomi. 2025. Bronco150 mapping: Medically relevant vs. non-medically relevant statements.

Bavarian State Library. 2025. bert-base-german-cased (revision 43cce13).

Florian Borchert, Christina Lohr, Luise Modersohn, Thomas Langer, Markus Follmann, Jan Philipp Sachs, Udo Hahn, and Matthieu-P. Schapranow. 2020. Ggponc: A corpus of german medical text with rich metadata based on clinical practice guidelines. In *Proceedings of the 11th International Workshop on Health Text Mining and Information Analysis*. Association for Computational Linguistics.

Florian Borchert, Christina Lohr, Luise Modersohn, Jonas Witt, Thomas Langer, Markus Follmann, Matthias Gietzelt, Bert Arnrich, Udo Hahn, and Matthieu-P. Schapranow. 2022. GGPONC 2.0 - the German clinical guideline corpus for oncology: Curation workflow, annotation policy, baseline NER taggers. In *Proceedings of the Thirteenth Language Resources and Evaluation Conference*, pages 3650–3660, Marseille, France. European Language Resources Association.

Keno K. Bressem, Jens-Michalis Papaioannou, Paul Grundmann, Florian Borchert, Lisa C. Adams, Leonhard Liu, Felix Busch, Lina Xu, Jan P. Loyen, Stefan M. Niehues, Moritz Augustin, Lennart Grosser, Marcus R. Makowski, Hugo J.W.L. Aerts, and Alexander Löser. 2024. medbert.de: A comprehensive german bert model for the medical domain. *Expert Systems with Applications*, 237:121598.

Branden Chan, Stefan Schweter, and Timo Möller. 2020. German's next language model. In *Proceedings of the 28th International Conference on Computational Linguistics*, pages 6788–6796, Barcelona, Spain (Online). International Committee on Computational Linguistics.

Tianqi Chen and Carlos Guestrin. 2016. Xgboost: A scalable tree boosting system. In *Proceedings of the 22nd ACM SIGKDD International Conference on Knowledge Discovery and Data Mining*, KDD '16, pages 785–794. ACM.

Amin Dada, Tim Leon Ufer, Moon Kim, Max Hasin, Nicola Spieker, Michael Forsting, Felix Nensa, Jan Egger, and Jens Kleesiek. 2023. Information extraction from weakly structured radiological reports with natural language queries. *European Radiology*, 34(1):330–337.

Pritam Deka, Anna Jurek-Loughrey, and Deepak P. 2022. *Evidence Extraction to Validate Medical Claims in Fake News Detection*, pages 3–15. Springer Nature Switzerland.

Jacob Devlin, Ming-Wei Chang, Kenton Lee, and Kristina Toutanova. 2018. Bert: Pre-training of deep bidirectional transformers for language understanding. *arXiv preprint*.

Anna Veronika Dorogush, Vasily Ershov, and Andrey Gulin. 2018. Catboost: gradient boosting with categorical features support. *arXiv preprint*.

Ahmad Idrissi-Yaghir, Amin Dada, Henning Schäfer, Kamyar Arzideh, Giulia Baldini, Jan Trienes, Max Hasin, Jeanette Bewersdorff, Cynthia S. Schmidt, Marie Bauer, Kaleb E. Smith, Jiang Bian, Yonghui Wu, Jörg Schlötterer, Torsten Zesch, Peter A. Horn, Christin Seifert, Felix Nensa, Jens Kleesiek, and Christoph M. Friedrich. 2024. Comprehensive study on german language models for clinical and biomedical text understanding. *arXiv preprint*.

Guolin Ke, Qi Meng, Thomas Finley, Taifeng Wang, Wei Chen, Weidong Ma, Qiwei Ye, and Tie-Yan Liu. 2017. Lightgbm: a highly efficient gradient boosting decision tree. In *Proceedings of the 31st International Conference on Neural Information Processing Systems*, NIPS'17, page 3149–3157, Red Hook, NY, USA. Curran Associates Inc.

Cyrille YetuYetu Kesiku, Andrea Chaves-Villota, and Begonya Garcia-Zapirain. 2022. Natural language processing techniques for text classification of biomedical documents: A systematic review. *Information*, 13(10):499.

Madeleine Kittner, Mario Lamping, Damian T Rieke, Julian Götze, Bariya Bajwa, Ivan Jelas, Gina Rüter, Hanjo Hautow, Mario Sänger, Maryam Habibi, Marit Zettwitz, Till de Bortoli, Leonie Ostermann, Jurica Ševa, Johannes Starlinger, Oliver Kohlbacher, Nisar P Malek, Ulrich Keilholz, and Ulf Leser. 2021. Annotation and initial evaluation of a large annotated german oncological corpus. *JAMIA Open*, 4(2).

Manuel Lentzen, Sumit Madan, Vanessa Lage-Rupprecht, Lisa Kühnel, Juliane Fluck, Marc Jacobs, Mirja Mittermaier, Martin Witzenrath, Peter Brunecker, Martin Hofmann-Apitius, Joachim Weber, and Holger Fröhlich. 2022. Critical assessment of transformer-based ai models for german clinical notes. *JAMIA Open*, 5(4).

Dominique Makowski and Viliam Simko. 2018. neuropsychology/psycho.r: 0.2.8.

Aurelie Mascio, Zeljko Kraljevic, Daniel Bean, Richard Dobson, Robert Stewart, Rebecca Bendayan, and Angus Roberts. 2020. Comparative analysis of text classification approaches in electronic health records. In *Proceedings of the 19th SIGBioMed Workshop on Biomedical Language Processing*. Association for Computational Linguistics.

Luise Modersohn, Stefan Schulz, Christina Lohr, and Udo Hahn. 2022. *GRASCCO — The First Publicly Shareable, Multiply-Alienated German Clinical Text Corpus*. IOS Press.

MTSamples. 2025. Transcribed medical transcription sample reports and examples. Accessed: 31 January 2025.

Mariana L. Neves, Daniel Butzke, Antje Dörendahl, Nora Leich, Benedikt Hummel, Gilbert Schönfelder, and Barbara Grune. 2019. Overview of the CLEF ehealth 2019 multilingual information extraction. In *Working Notes of CLEF 2019 - Conference and Labs of the Evaluation Forum, Lugano, Switzerland, September 9-12, 2019*, volume 2380 of *CEUR Workshop Proceedings*. CEUR-WS.org.

Adam Paszke, Sam Gross, Francisco Massa, Adam Lerer, James Bradbury, Gregory Chanan, Trevor Killeen, Zeming Lin, Natalia Gimelshein, Luca Antiga, Alban Desmaison, Andreas Köpf, Edward Yang, Zach DeVito, Martin Raison, Alykhan Tejani, Sasank Chilamkurthy, Benoit Steiner, Lu Fang, Junjie Bai, and Soumith Chintala. 2019. Pytorch: An imperative style, high-performance deep learning library. *arXiv preprint*.

F. Pedregosa, G. Varoquaux, A. Gramfort, V. Michel, B. Thirion, O. Grisel, M. Blondel, P. Prettenhofer, R. Weiss, V. Dubourg, J. Vanderplas, A. Passos, D. Cournapeau, M. Brucher, M. Perrot, and E. Duchesnay. 2011a. Scikit-learn: Machine learning in python. Accessed: 2025-02-18.

F. Pedregosa, G. Varoquaux, A. Gramfort, V. Michel, B. Thirion, O. Grisel, M. Blondel, P. Prettenhofer, R. Weiss, V. Dubourg, J. Vanderplas, A. Passos, D. Cournapeau, M. Brucher, M. Perrot, and E. Duchesnay. 2011b. Scikit-learn: Machine learning in Python. *Journal of Machine Learning Research*, 12:2825–2830.

Nils Reimers and Iryna Gurevych. 2019. Sentence-bert: Sentence embeddings using siamese bert-networks. *arXiv preprint*.

Robin Rojowiec, Benjamin Roth, and Maximilian Fink. 2020. Intent recognition in doctor-patient interviews. In *Proceedings of the Twelfth Language Resources and Evaluation Conference*, pages 702–709, Marseille, France. European Language Resources Association.

Niloofer Shanavas, Hui Wang, Zhiwei Lin, and Glenn Hawe. 2020. Ontology-based enriched concept graphs for medical document classification. *Information Sciences*, 525:172–181.

Hanna Suominen, Liadh Kelly, Lorraine Goeuriot, and Martin Krallinger. 2020. *CLEF eHealth Evaluation Lab 2020*, pages 587–594. Springer International Publishing.

Maxim Tkachenko, Mikhail Malyuk, Andrey Holmanyuk, and Nikolai Liubimov. 2020-2025. Label Studio: Data labeling software. Open source software available from https://github.com/HumanSignal/label-studio.

Ashish Vaswani, Noam Shazeer, Niki Parmar, Jakob Uszkoreit, Llion Jones, Aidan N. Gomez, Lukasz Kaiser, and Illia Polosukhin. 2017. Attention is all you need. *CoRR*, abs/1706.03762.

Ikuya Yamada, Akari Asai, Jin Sakuma, Hiroyuki Shindo, Hideaki Takeda, Yoshiyasu Takefuji, and Yuji Matsumoto. 2020. Wikipedia2Vec: An efficient toolkit for learning and visualizing the embeddings of words and entities from Wikipedia. In *Proceedings of the 2020 Conference on Empirical Methods in Natural Language Processing: System Demonstrations*, pages 23–30. Association for Computational Linguistics.

Bo Zhou, Dingling Su, and Zehui Qu. 2021. Medical text classification system based on deep learning. In *2021 International Conference on Intelligent Computing, Automation and Applications (ICAA)*, pages 388–392. IEEE.

A Appendix

Classifier	Parameter	Word2Vec	Sentence-BERT	$BERT_{ger}$	BioBERT	MedBERT
SVM	Kernel Type	poly	poly	poly	poly	rbf
	Kernel Degree	4	3	4	2	2
	Cost	10	1	1	10	1
	Gamma	scale	scale	scale	scale	0.01
	Coef0	0	0.5	0.5	0.5	0
RF	Bootstrap	False	False	False	False	False
	Max Depth	20	20	20	10	20
	Max Features	sqrt	sqrt	sqrt	sqrt	sqrt
	Min Samples Leaf	1	2	1	4	2
	Min Samples Split	5	5	2	10	2
	n Estimators	1500	500	500	500	500
LightGBM	Num Leaves	31	31	31	31	31
	n Estimators	1000	2000	2000	1000	1000
	Learning Rate	0.01	0.1	0.1	0.01	0.01
CatBoost	Depth	6	6	6	8	6
	Iterations	1000	1000	3000	1000	1000
	Learning Rate	0.01	0.1	0.01	0.01	0.01
XGBoost	Max Depth	8	8	8	6	8
	n Estimators	2000	1000	1000	2000	1000
	Learning Rate	0.01	0.01	0.01	0.01	0.01

Table 8: Optimized hyperparameters of ML-models based on grid search for different embeddings.

Classifier with Sentence-BERT	Mean Pooling	Max Pooling	CLS Token
CatBoost	**0.8372**	0.5242	0.6858
RandomForest	**0.8121**	0.6375	0.7671
XGBoost	**0.8059**	0.5130	0.7105
SVM	**0.8330**	0.6768	0.7196
LightGBM	**0.8107**	0.4799	0.7205

Table 9: Weighted F1-Scores for ML-models using different extraction strategies on the IntRec validation dataset.

PreClinIE: An Annotated Corpus for Information Extraction in Preclinical Studies

Simona E. Doneva[1*], **Hanna Hubarava**[1], **Pia Haervelid**[1],
Wolfgang E. Zürrer [1], **Julia Bugajska**[1], **Bernhard Hild**[1], **David Brüschweiler**[1],
Tilia R. Ellendorff[1], **Gerold Schneider**[1], **Benjamin V. Ineichen**[1,2]
[1] University of Zurich, Zurich, Switzerland
[2] University of Bern, Bern, Switzerland
*simona.doneva@uzh.ch

Abstract

Animal research, sometimes referred to as preclinical research, plays a vital role in bridging the gap between basic science and clinical applications. However, the rapid increase in publications and the complexity of reported findings make it increasingly difficult for researchers to extract and assess relevant information. While automation through natural language processing (NLP) holds great potential for addressing this challenge, progress is hindered by the absence of high-quality, comprehensive annotated resources specific to preclinical studies. To fill this gap, we introduce PreClinIE, a fully open manually annotated dataset. The corpus consists of abstracts and methods sections from 725 publications, annotated for study rigor indicators (e.g., random allocation) and other study characteristics (e.g., species). We describe the data collection and annotation process, outlining the challenges of working with preclinical literature. By providing this resource, we aim to accelerate the development of NLP tools that enhance literature mining in preclinical research.

1 Introduction

Developing new therapies from animal models to human treatments, known as bench-to-bedside translation, has a low success rate: Only 1 in 20 therapies advances to human use (Ineichen et al., 2024). This contrasts with the extensive use of animals in research, estimated at over 50 million per year globally (Taylor and Alvarez, 2019). The factors that determine successful translation remain poorly understood (Seyhan, 2019).

A systematic assessment of information on experimental design, model and drug selection, as well as animal usage can provide insights into how animal research informs human health. The full-text, and especially the methods sections of scientific articles contain concrete, verifiable details about these aspects, which are often omitted or misrepresented in article abstracts (Li et al., 2017). These factual descriptions form the foundation of a study and are critical for evaluating its design, rigor, and to enable reproducibility (Menke et al., 2020).

However, the volume of preclinical animal studies is overwhelming, with hundreds of thousands published annually (Ineichen et al., 2023). While large-scale analysis methods exist, they primarily focus on human data or only on abstract level data (Chapman et al., 2011; Doneva et al., 2024). Animal studies, with their highly heterogeneous experimental approaches and less standardized reporting, remain largely unaddressed.

There is a critical need for computational methods to extract and integrate these data at scale, since a more detailed understanding of the drug development process could not only improve experimental animal welfare but also enhance the efficiency of human therapies. As a first step towards that goal, our study aims to create a large, manually annotated corpus of animal study publications, including abstracts and method sections. We share all resources on GitHub[1].

2 Related Work

NLP methods have been commonly applied in the preclinical domain for abstract classification tasks. For example, a recently published dataset aims to help with the identification of animal studies and alternative experimental models (Neves et al., 2023). Another application is the automated selection of relevant published articles for specific literature review questions, as well as the assessment of risk of bias items (e.g., random allocation) (Bannach-Brown et al., 2019; Wang et al., 2022b).

Information extraction from preclinical literature is an emerging, but less developed, area of research.

[1] https://github.com/Ineichen-Group/Preclinical_IE_Dataset

Proceedings of the 24th Workshop on Biomedical Language Processing (BioNLP 2025), pages 74–87
August 1, 2025

STEED, for instance, is an R-based text mining tool that uses regular expressions to automatically extract key experimental details, such as animal species, disease models, and randomization from neuroscience in vivo studies. It has been developed on 45 full-text articles and validated on 275 articles (Zurrer et al., 2024). Another approach, *Menagerie*, combines rule-based, dictionary-based, and machine learning techniques to extract six predefined animal study characteristics (Zeiss et al., 2019). This work is based on a manually curated dataset of 504 PubMed abstracts, annotated with classes such as species or animal model at the abstract level, and with gene names at the token level. Another related work targets information extraction based on the established framework of Population/Problem, Intervention, Comparator and Outcome (PICO) (Wang et al., 2022a). For this, 400 abstracts of preclinical studies have been annotated for each PICO-related mention, and the task was solved as sentence classification, followed by entity recognition. Another study proposed combining a regex-based method with a generative LLM to extract interventions from preclinical animal studies on Alzheimer's disease (Pu et al., 2024).

Despite recent advances, existing corpora remain limited in scope - typically focusing on narrow disease domains, containing small datasets (around 500 documents), and offering only abstract-level annotations. For example, *Menagerie* was validated solely for Parkinson's disease. In contrast, our corpus is, to our knowledge, the most comprehensive resource of its kind: it includes 725 documents from the general neuroscience domain, with manual annotations on both the abstract and the methods section, a critical source of experimental detail. This results in 1,450 annotated sections. Importantly, we used three annotation levels (document, sentence, and token), aiming to match the typical granularity of information relevant to researchers. This structure also should reflect the nature of the content: some elements, like conclusions, require sentence- or document-level annotation, whereas others, like individual drugs, can be annotated at the token level.

3 The Corpus

3.1 Data Collection

A search string for PubMed and EMBASE was designed to identify animal studies on therapeutic interventions[2]. From the retrieved references, 4,000 records were randomly selected for screening by two independent reviewers based on inclusion criteria: primary studies involving drug testing in animals.

We used the automatic fetch function of the reference management tool EndNote to retrieve PDFs, resorting to manual retrieval when necessary. We used IBM Deepsearch to convert PDFs into text[3], followed by a regular expression-based algorithm to classify paper sections such as methods and results. We included the methods sections because they typically provide more detailed descriptions of the employed methodology compared to abstracts.

3.2 Data Annotation

3.2.1 Annotation Guidelines

We define three levels of annotation. At the **document** level, one or more labels are assigned to the entire document. At the **sentence** level, we highlight the sentence where the relevant information appears (**Table 1**). Finally, at the **token** level, individual words are annotated as named entities (**Table 2**). We refined the annotation guidelines iteratively to ensure maximum clarity and optimize inter-rater agreement. The final guidelines can be accessed at Annotation Guidelines (v5), with a shortened version in **Appendix B**. Notably, spans and documents can have more than one label. For example, weight and age of animals are often reported in the same sentence, and a study can involve both mice and rats in its experiments.

3.2.2 Annotation Process

From the 4000 random references, we excluded two due to missing metadata, leaving 3,998 references. Of these, 1,018 met the inclusion criteria during the initial screening.

The annotation was conducted by five senior medical students, starting with two pilot rounds of 20 and then 50 articles annotated by all annotators to familiarize themselves with the task and to refine the guidelines. In the final annotation round, 817 articles were distributed among them, with each annotator receiving 179–181 articles with title, abstract and method sections. Of these, 20 articles were assigned to multiple annotators to calculate inter-annotator agreement (IAA). The annotators

[2]Search date: from database inception to October 09, 2023. Full search string available here: dataset search strings.

[3]IBM RPA PDF Extractor

Parameter	Label (frequency)	Krippendorff's Alpha (95% CI)
Document-level Annotation		
Animal species (A, M)	Rat (806), Mouse (531), Other (28), Rabbit (28), Monkey (20), Dog (15), Pig (10), Cat (6), Guinea Pig (6)	0.97 (0.95, 1.00)
Control (A, M)	Control-present (1135)	0.51 (0.29, 0.67)
Readout (A, M)	Physiology (400), Behaviour (938), Histology (921), Other (896), Imaging (92)	0.48 (0.39, 0.55)
Animal sex (A, M)	Not reported (717), Male (524), Female (129), Both sexes (63)	1.00 (1.00, 1.00)
Sentence-Level Annotation (Highlight)		
Study conclusions (A)	Positive (645), Neutral (22), Negative (18), Mixed (15)	0.76 (0.74, 0.78)
Animal disease model (A)	Model (649)	0.62 (0.60, 0.64)
Weight (M)	Weight (514)	0.73 (0.71, 0.75)
Age (M)	Age (476)	0.75 (0.73, 0.78)
Random allocation (A, M)	Randomization (464)	0.60 (0.56, 0.64)
Blinded outcome assessment (A, M)	Blinding (389)	0.97 (0.95, 0.98)
Animal welfare statement (A, M)	Welfare (700)	0.96 (0.95, 0.97)
Animal Research: Reporting of In Vivo Experiments Guidelines (A, M)	ARRIVE (15)	—
Sample size calculation (A, M)	Power (22)	—

Table 1: Overview of document-level and sentence-level annotation categories. The "Label (frequency)" column lists the available labels for each category along with their frequency in the final complete annotated dataset. For document-level annotations, the frequency represents the number of documents (abstracts or methods) assigned to each label. For sentence-level annotations, it indicates the number of unique sentences associated with each label. The last column provides the Krippendorff's Alpha inter-annotator-agreement score for that label on the subset of the corpus annotated by all annotators (15 articles). The rows with a missing score correspond to the labels not sufficiently represented in the subset. Abbreviations: A, abstract; M, methods.

Entity Type	Entity #	Unique #	Avg Char Count	Examples	Krippendorff's Alpha (95% CI)
Therapy-drug (A, M)	10348	2437	17.7 ± 18.1	beta-lactam antibiotic, ZM241385	0.73 (0.70, 0.75)
Therapy-other (A, M)	5216	1728	20.3 ± 15.8	auditory habilitation, treadmill training	0.59 (0.57, 0.61)
Disease (A)	3790	958	19.8 ± 11.0	minimal seizures, chronic paraplegia	0.79 (0.76, 0.81)
Strain (A, M)	1196	159	10.5 ± 18.7	Sprague Dawley, Fisher 344	0.84 (0.79, 0.88)
Animals-number (A, M)	342	144	5.5 ± 4.8	Eighty-five, 128	0.78 (0.50, 0.93)

Table 2: Overview of token-level annotations with total entity counts, unique instances counts, average character number, and annotation examples. The last column provides the Krippendorff's Alpha inter-annotator-agreement score as measured on the subset of the corpus annotated by all annotators (15 articles). Abbreviations: A, abstract; M, methods.

were allowed to exclude articles from annotation if they did not fit the inclusion criteria[4].

The annotators used a custom recipe developed for the browser-based tool Prodigy to perform the manual annotation (Montani and Honnibal, 2017). An annotation task example is shown in **Supplementary Figure 2**.

To compile the final dataset and perform an error analysis, the 20 multiple-assigned articles were reviewed, with conflicts adjudicated through discussion. The final dataset consists of 725 unique articles, and corresponds to 1450 abstract and method sections.

[4]For example, some initially included papers were related to a diagnostic procedure rather than an intervention.

3.2.3 Inter-Annotator Agreement

Five of the 20 common documents were excluded by one or more annotators as not meeting the inclusion criteria. To ensure the IAA score is measured among all annotators, we removed the excluded articles from the agreement calculations. This left us with 15 unique articles, each with an abstract and method section (30 annotated documents).

We report the IAA among the five annotators using Cohen's Kappa for pairwise agreement calculation and Krippendorff's Alpha for the calculation of agreement among all annotators (Cohen, 1960; Hughes, 2021).

3.3 Results

3.3.1 Corpus Overview

Our final annotated corpus consists of the abstracts and methods sections from 725 published neuroscience articles, primarily dated between 2010 and 2020. The most frequently represented journals include *European Journal of Pharmacology* and *PLoS ONE* (**Supplementary Figure 1**).

Based on the **document level** annotations, the corpus predominantly comprises studies involving mice and rats (**Table 1**). Additionally, there is a marked bias toward male animals and reporting of animal sex information in the methods section. Furthermore, there were often multiple selected options for the experimental readouts (**Supplementary Figure 3**).

At the **sentence level**, the annotations reveal that the majority of conclusion statements within the corpus present positive findings. However, adherence to reporting best-practice appears limited, with relatively little to no mention of ARRIVE and PREPARE guidelines (Percie du Sert et al., 2020; Smith et al., 2018). Furthermore, explicit reporting of sample size calculations is sparse (**Table 1**).

At the **token level**, therapy-related annotations are the most prevalent, as these terms were annotated in both the abstract and methods sections (**Table 2**). Disease and strain entities exhibit high lexical variability, with high ratio of unique textual representations across the corpus. Additionally, many abbreviations are annotated, such as *AD* for Alzheimer's disease (**Supplementary Figure 5**).

3.3.2 Analysis of Annotation Disagreements

We observed several patterns of discrepancies in the multiply-annotated documents selected for the calculation of IAA:

- **Text level annotations:** "Readout" and "Control" were the most challenging document-level classification tasks (**Table 1**). The language describing the readouts varied greatly across papers and was often not explicit. Some annotators selected "other" in these cases, while others attempted to infer a more specific readout type. Furthermore, some annotators selected "histology" in cases when there was clearly no mention of this readout, suggesting a misunderstanding of the concept. Similarly, the presence of control intervention was rarely specified, even though the text sometimes contained a comparison verb (e.g., "improved"). In such instances, where the presence of control is implicit, some annotators marked the presence of control, while others did not, leading to a lower agreement score. The variability in pair-wise IAA is evident from **Supplementary Figure 4**.

- **Sentence level annotations:** At the sentence level, annotators often agreed on the study's overall conclusion but struggled to identify the exact concluding sentence, sometimes confusing it with a summary of findings. Variation in punctuation usage—especially around colons and semicolons—also caused inconsistencies in the selection of annotation spans, resulting in partial agreement. Annotation of "randomization" was meant to refer only to the allocation of animals into experimental groups, but one annotator highlighted other contexts as well.

- **Token level annotations:** At the token level all entities except "therapy-other" achieved a satisfactory level of agreement (**Table 2**). We identified three main discrepancies. First, some annotators occasionally missed entities, either due to human error in reading longer texts or a misunderstanding of the guidelines. For instance gene mentions were sometimes annotated as therapy (Nurr1, Fox2) when, according to the guidelines, they should not be annotated as such. Second, label disagreements arose when different annotators assigned different labels to the same entity. For instance, one annotator consistently labeled antibody therapies as "therapy-other", while it should have been "therapy-drug". Finally, span disagreements occurred when some annotators included a preceding modifier as per the guidelines instructions (e.g., "morphine-induced"), while another did not. Such discrepancies introduce noise in the dataset and reduce the agreement score. The variability in pair-wise IAA is shown in **Supplementary Figure 5**.

4 Discussion and Conclusion

We introduced PreClinIE, an openly available corpus for extracting study rigor indicators and experimental details from published articles describing animal research.

Our annotation process uncovered key challenges. Particularly there was a low agreement in control/comparator annotations, aligning with findings from the related PICO study (Wang et al., 2022a). At the same time, this study had stronger performance for readout extraction, suggesting that token-level annotation may be more suitable for this task. Additionally, the high disagreement in identifying the exact conclusion sentence suggests that study conclusions might be best evaluated using the full document rather than isolated sentences. Furthermore, we observed that crucial study details often appear exclusively in the methods section, emphasizing the importance of section-aware extraction.

In designing the annotation scheme, we made several pragmatic choices to balance granularity, feasibility, and consistency. For example, while our approach captures individual parameters such as sex, strain, and treatment, it does not explicitly encode relationships between these entities. As a result, reconstructing complex experimental groupings may be challenging in studies involving multiple animal subgroups. Nevertheless, this design simplifies annotation and aligns with our primary goal of extracting key methodological features at scale. Future work could explore incorporating relational annotations to capture richer experimental structures. Additionally, it may be possible to use simpler heuristic rules, for instance, pairing species and model terms that occur within a pre-specified window in the text, to make those links more explicit (Zeiss et al., 2019).

Similarly, we chose to restrict some annotations to single sentences. This constraint reduces cognitive load for annotators. Although it may result in missed information that spans multiple sentences, such as animal welfare statements or methodological clarifications, we find that capturing key information once in the text is sufficient for many downstream applications. Future extensions of the annotation scheme could explore cross-sentence linking or section-level annotation to support more nuanced analysis.

Beyond annotation challenges, our findings highlight a male bias in animal use, a majority of positive conclusion statements indicative of reporting bias, and insufficient reporting of sample size calculations.

These patterns warrant further evaluation, as they suggest systemic issues in study design and reporting that could impact the reliability and reproducibility of preclinical research findings (Beery and Zucker, 2011; Button et al., 2013). As future work, we plan to provide a baseline experiment to illustrate how the dataset can support computational information extraction from preclinical literature. The corpus enables a range of NLP tasks, such as named entity recognition and sentence classification, and can serve as a benchmark for model development in this domain. We hope these efforts will inspire further research in NLP models development and evaluation, ultimately contributing to more transparent and reliable scientific practices.

Limitations

Data Scope. Our developed dataset includes publications focusing mainly on research in neuroscience. This may influence the generalizability of our findings to other areas of research.

Annotation Setup. Only a small portion of the dataset was multiply-annotated. We conducted two annotation pilots to harmonize understanding among annotators. However, more multiply-annotated documents and additional training sessions would likely have further improved annotation quality.

Possible need for enrichment of the data. Another challenge is the under-representation of certain classes in our dataset. As an example of the imbalance on the sentence-level annotations, the number of positive study conclusions (625) dwarfs the negative (18), neutral (21) and mixed (15) conclusions. Among document-level annotations, "animal species" class shows that the majority of animals used for experiments are mice and rats, with only a handful of other species found in the dataset (see Table **1**). Although likely reflecting the natural distribution of the conclusions among publications, this imbalance may limit the model performance for those categories. Potential remedies include merging our dataset with related ones, applying targeted data collection strategies to expand coverage and improve class balance, or augmenting the dataset with synthetic data.

Acknowledgments

We thank Vera Lara Bernhard for her support in implementing the confidence interval calculations for the Krippendorff's alpha algorithm.

References

Alexandra Bannach-Brown, Piotr Przybyła, James Thomas, Andrew SC Rice, Sophia Ananiadou, Jing

Liao, and Malcolm Robert Macleod. 2019. Machine learning algorithms for systematic review: reducing workload in a preclinical review of animal studies and reducing human screening error. *Systematic reviews*, 8:1–12.

Annaliese K Beery and Irving Zucker. 2011. Sex bias in neuroscience and biomedical research. *Neuroscience & Biobehavioral Reviews*, 35(3):565–572.

Katherine S Button, John PA Ioannidis, Claire Mokrysz, Brian A Nosek, Jonathan Flint, Emma SJ Robinson, and Marcus R Munafò. 2013. Power failure: why small sample size undermines the reliability of neuroscience. *Nature reviews neuroscience*, 14(5):365–376.

Wendy W Chapman, Prakash M Nadkarni, Lynette Hirschman, Leonard W D'avolio, Guergana K Savova, and Ozlem Uzuner. 2011. Overcoming barriers to NLP for clinical text: the role of shared tasks and the need for additional creative solutions.

Jacob Cohen. 1960. A coefficient of agreement for nominal scales. *Educational and psychological measurement*, 20(1):37–46.

Simona Emilova Doneva, Sijing Qin, Beate Sick, Tilia Ellendorff, Jean-Philippe Goldman, Gerold Schneider, and Benjamin Victor Ineichen. 2024. Large Language Models to process, analyze, and synthesize biomedical texts-a scoping review. *Discover*, 4:107.

John Hughes. 2021. krippendorffsalpha: An R package for measuring agreement using Krippendorff's alpha coefficient. *arXiv preprint arXiv:2103.12170*.

Benjamin V Ineichen, Eva Furrer, Servan L Grüninger, Wolfgang E Zürrer, and Malcolm R Macleod. 2024. Analysis of animal-to-human translation shows that only 5% of animal-tested therapeutic interventions obtain regulatory approval for human applications. *PLoS biology*, 22(6):e3002667.

Benjamin V Ineichen, Marianna Rosso, and Malcolm R Macleod. 2023. From data deluge to publomics: How AI can transform animal research. *Lab animal*, 52(10):213–214.

Guowei Li, Luciana PF Abbade, Ikunna Nwosu, Yanling Jin, Alvin Leenus, Muhammad Maaz, Mei Wang, Meha Bhatt, Laura Zielinski, Nitika Sanger, et al. 2017. A scoping review of comparisons between abstracts and full reports in primary biomedical research. *BMC medical research methodology*, 17:1–12.

Joe Menke, Martijn Roelandse, Burak Ozyurt, Maryann Martone, and Anita Bandrowski. 2020. The rigor and transparency index quality metric for assessing biological and medical science methods. *Iscience*, 23(11).

Ines Montani and Matthew Honnibal. 2017. Prodigy: A modern and scriptable annotation tool for creating training data for machine learning models.

Mariana Neves, Antonina Klippert, Fanny Knöspel, Juliane Rudeck, Ailine Stolz, Zsofia Ban, Markus Becker, Kai Diederich, Barbara Grune, Pia Kahnau, et al. 2023. Automatic classification of experimental models in biomedical literature to support searching for alternative methods to animal experiments. *Journal of Biomedical Semantics*, 14(1):13.

Nathalie Percie du Sert, Viki Hurst, Amrita Ahluwalia, Sabina Alam, Marc T Avey, Monya Baker, William J Browne, Alejandra Clark, Innes C Cuthill, Ulrich Dirnagl, et al. 2020. The ARRIVE guidelines 2.0: Updated guidelines for reporting animal research. *Journal of Cerebral Blood Flow & Metabolism*, 40(9):1769–1777.

Yiyuan Pu, Kaitlyn Hair, Daniel Beck, Mike Conway, Malcolm Macleod, and Karin Verspoor. 2024. Intervention extraction in preclinical animal studies of Alzheimer's Disease: Enhancing regex performance with language model-based filtering. In *23rd Meeting of the ACL Special Interest Group on Biomedical Natural Language Processing, BioNLP 2024*, pages 486–492. Association for Computational Linguistics (ACL).

Attila A Seyhan. 2019. Lost in translation: the valley of death across preclinical and clinical divide–identification of problems and overcoming obstacles. *Translational Medicine Communications*, 4(1):1–19.

Adrian J Smith, R Eddie Clutton, Elliot Lilley, Kristine E Aa Hansen, and Trond Brattelid. 2018. PREPARE: guidelines for planning animal research and testing. *Laboratory animals*, 52(2):135–141.

Katy Taylor and Laura Rego Alvarez. 2019. An estimate of the number of animals used for scientific purposes worldwide in 2015. *Alternatives to Laboratory Animals*, 47(5-6):196–213.

Qianying Wang, Jing Liao, Mirella Lapata, and Malcolm Macleod. 2022a. PICO entity extraction for preclinical animal literature. *Systematic Reviews*, 11(1):209.

Qianying Wang, Jing Liao, Mirella Lapata, and Malcolm Macleod. 2022b. Risk of bias assessment in preclinical literature using natural language processing. *Research synthesis methods*, 13(3):368–380.

Caroline J Zeiss, Dongwook Shin, Brent Vander Wyk, Amanda P Beck, Natalie Zatz, Charles A Sneiderman, and Halil Kilicoglu. 2019. Menagerie: A text-mining tool to support animal-human translation in neurodegeneration research. *PloS one*, 14(12):e0226176.

Wolfgang Emanuel Zurrer, Amelia Elaine Cannon, Ewoud Ewing, David Brüschweiler, Julia Bugajska, Bernard Friedrich Hild, Marianna Rosso, Daniel Salo Reich, and Benjamin Victor Ineichen. 2024. STEED: A data mining tool for automated extraction of experimental parameters and risk of bias items from in vivo publications. *PloS one*, 19(11):e0311358.

A Corpus Details and Statistics

A.1 Corpus Overview

Figure 1 shows the time range and journals represented in the corpus.

A.2 Annotations Overview

Figure 3 outlines the top 10 annotations across the different document-level categories and their distribution by abstract and methods. Rodent models, particularly rats and mice, dominate the dataset, with other species such as rabbits, guinea pigs, dogs, and monkeys appearing infrequently. Experimental outcomes are diverse, with histology and behavior among the most common readouts, often annotated together, indicating a tendency to explore multiple endpoints, as well as possible annotation challenge. The majority of studies include a control group, though fewer are explicitly mentioned in abstracts. For animal sex, a male bias is evident, as well as lack of reporting of animal sex in the abstract.

Figure 5 focuses on entity-level annotations in the corpus. Therapy-related drug entities are the most frequently annotated, with levodopa (109 instances) and L-DOPA (86 instances) leading the list, followed by commonly studied compounds such as morphine, MK-801, and cannabidiol (CBD). Beyond pharmacological interventions, other therapy entities include treatments like exercise, acupuncture, and curcumin. Among disease entities, Alzheimer's disease (96 instances) and stroke (103 instances) are well-represented, while neurodegenerative and neurological conditions such as Parkinson's disease (88 instances), epilepsy (48 instances), and spinal cord injury (81 instances) also feature prominently. Regarding animal strains, Sprague-Dawley (239 instances) and Wistar (207 instances) are the most frequently reported. However, annotations for animal numbers show substantial variability. The frequent presence of abbreviations (e.g., AD for Alzheimer's disease and SD for Sprague-Dawley) suggests that entity disambiguation is critical for accurate text interpretation.

A.3 Inter-Annotator Agreement Scores

We report IAA for document-level (**Figure 6**, left), as well as sentence-level and entity-type annotations (**Figure 6**, right). For the latter, we compute and report Krippendorff's Alpha on the level of tokens (words). This allows to capture partial agreement, when annotators agree on the label but disagree on its span, i.e. where exactly it starts and ends in the text.

B Annotation Guidelines

See full document here Annotation Guidelines (v5).

B.1 Inclusion Criteria for Papers

Before starting the annotation, ensure the paper meets the following eligibility criteria:

1. **Experimental study in animals** (excluding humans).

2. The study tests an **intervention** with the goal of improving animal health. The intervention should be externally applied (e.g., gene knockout does not qualify).

   - Apply criterium generously; include studies where the exact purpose of a drug treatment is not explicitly stated (e.g., testing different substances in animals without claiming a therapeutic benefit).
   - Exclude studies assessing the effect of endogenous substances (e.g., endogenously excreted miRNA-107).

3. The study models a **neurological or psychiatric disease**.

   - Apply criterium generously, including studies assessing pain in osteoporosis or mentioning neurological complications in systemic diseases such as cryptococcosis.

If any of these criteria are not met, exclude the study (no annotation required). If pertinent, exclude at the **abstract level** to ensure all related text sections (abstract, methods, and results) are omitted.

B.2 General Rules on Annotation

B.2.1 Token Annotation

1. **Consider Context:** Identify Population (P), Intervention (I), Control (C), and Outcomes (O).

2. **Annotate Only Relevant Information:**

   - Example: If a study uses male mice but suggests repeating experiments in rats, only mice should be annotated.

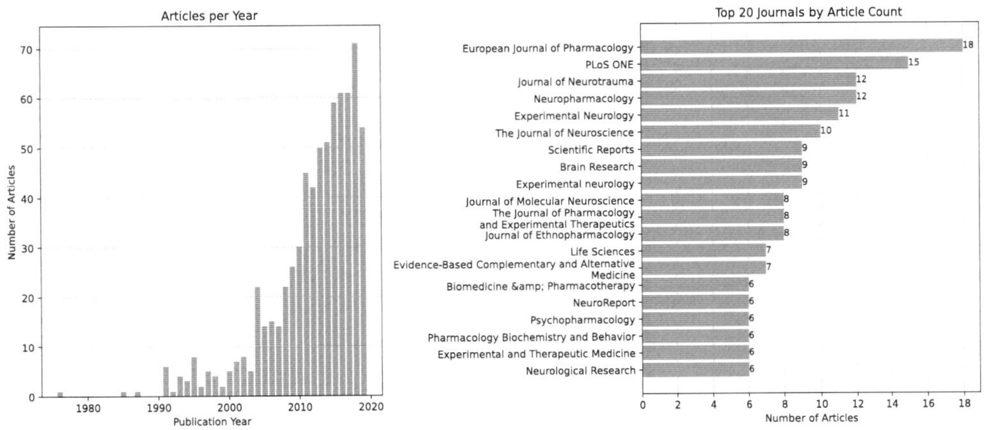

Figure 1: Distribution of articles in the corpus by (left) publication year and (right) journal.

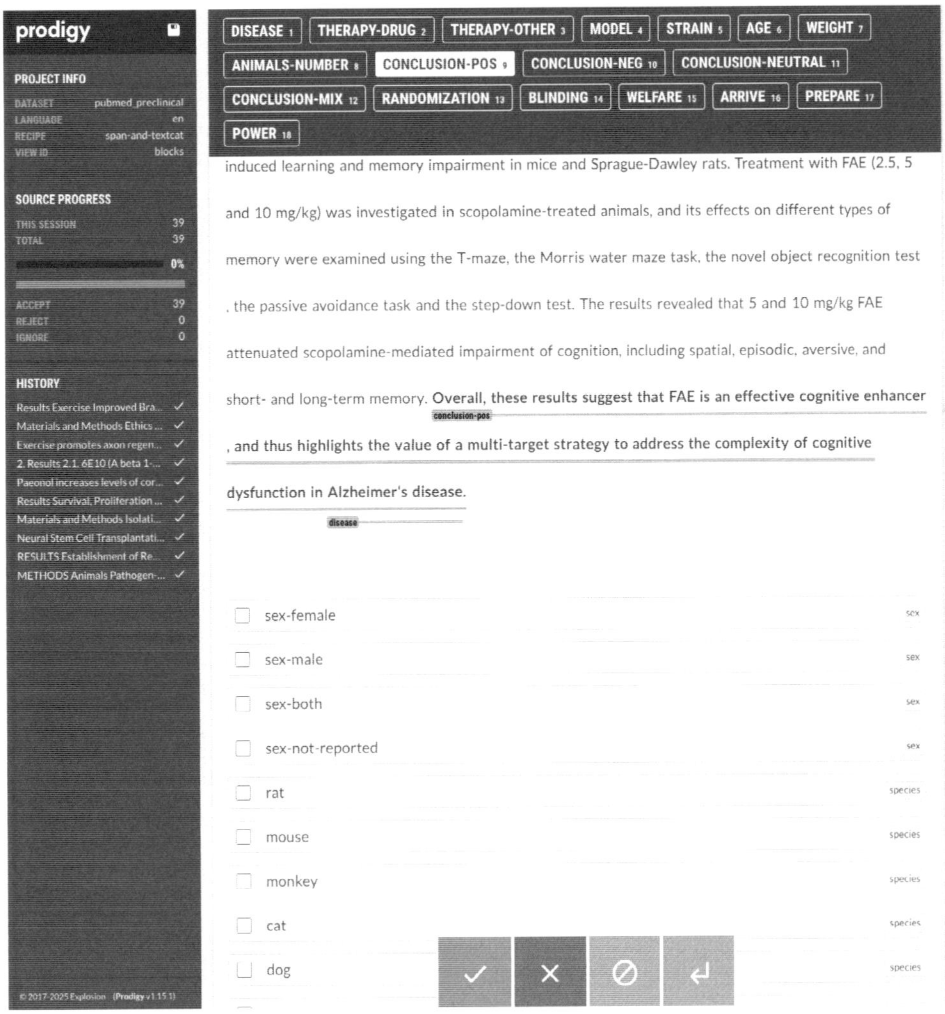

Figure 2: Annotation example shown in the annotation tool Prodigy.

- Example: If isoflurane is used for anaesthesia but not as a treatment, do not an-

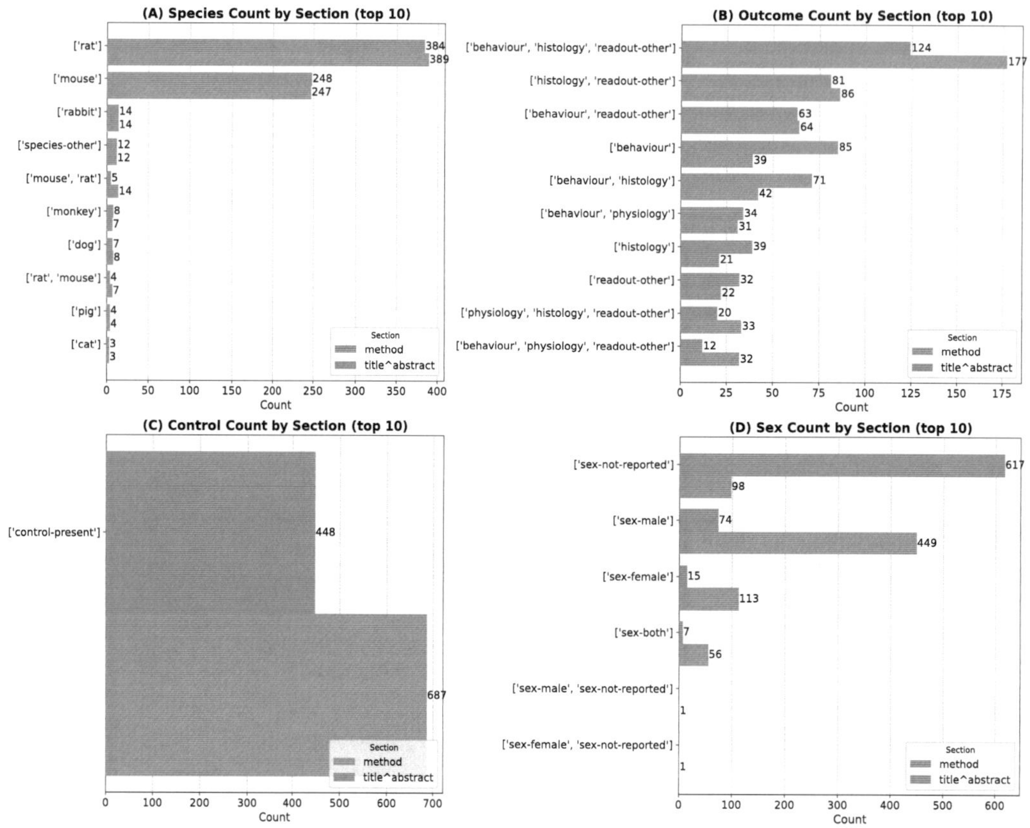

Figure 3: Top 10 most frequent document-level annotations for (A) Species, (B) Outcomes, (C) Control, and (D) Sex categories.

notate it.

B.2.2 Annotation Tasks

- **Text Annotation:** Entire text is classified based on predefined labels.
- **Sentence Annotation:** Entire sentence (including punctuation like colons and question marks) is annotated.
- **Token Annotation:** Specific words or phrases are annotated.

B.2.3 Additional Annotation Rules

1. Sentence annotations should **exclude references at the end of sentences**.
2. Include incorrect spelling/grammar if relevant.
3. Avoid mixing terms with and without brackets in annotations (e.g., annotate *oral appliance* and *OA* separately).
4. Do not annotate tokens where annotation would require inclusion of punctuation due to interface limitations.
5. Overlapping annotation between different tags is allowed.
6. Annotate each parameter once per section (i.e., once in abstract and once in methods); conclusions should only be annotated in the abstract.
7. Be careful in selecting the correct label, as incorrect annotation affects inter-rater agreement.
8. Do not annotate punctuation at the end of a sentence.
9. Ignore manuscript parts mistakenly included in the annotation interface (e.g., misplaced discussion sections).

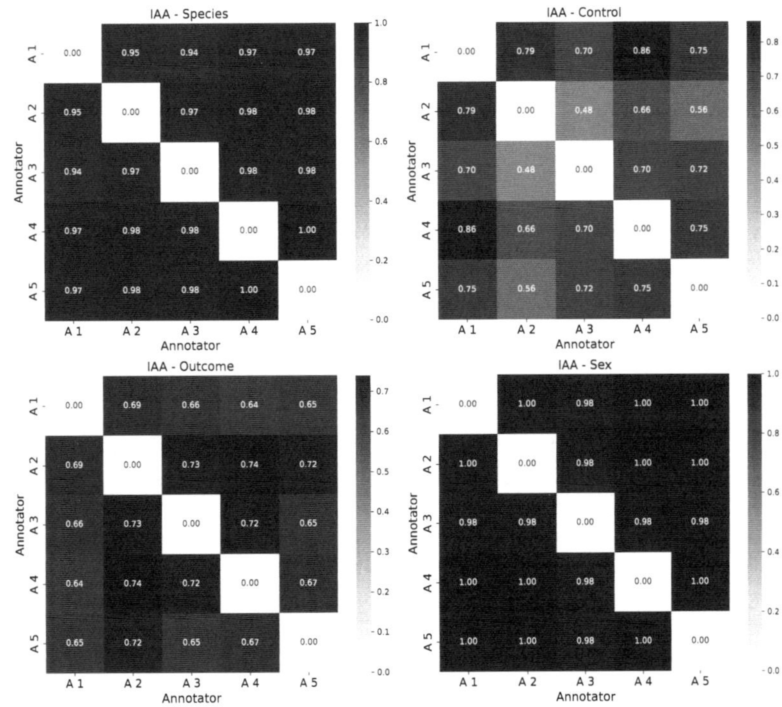

Figure 4: Cohen's Kappa scores for inter-annotator pairs for the categorical annotations in the overlapping articles.

B.3 Experimental Parameters

B.3.1 Animal species

Type: Population
Task: Text
Definition: The animal species used to test the intervention of interest.
Examples: Rats, mice, monkeys, rabbits, etc.
Location in Paper: Abstract, Methods
Comments:

- Most studies use rats or mice, while monkeys, pigs, cats, dogs, and rabbits are rarer.
- A study could use more than one species.

B.3.2 Animal strain

Type: Population
Task: Token
Definition: The animal strain further defining the animal species. A strain is a genetic variant, a subtype, or a culture within a biological species.
Examples: BALB/cJ (mouse), C57BL/6J (mouse), DBA/2J (mouse), Lewis (rat), Sprague-Dawley (rat).
Location in Paper: Abstract, Methods
Comments:

- A study could use more than one strain.
- Only annotate the strain (not the species).
- Be careful to separate strain from transgenic identification.

B.3.3 Animal sex

Type: Population
Task: Text
Definition: The animal sex further defining the animal species.
Examples: Male, female, both sexes.
Location in Paper: Abstract, Methods
Comments:

- A study could use either male, female, or both sexes.
- Some studies do not report the sex used.
- Only label the sex used to test the drug of interest.

B.3.4 Diseases mentioned

Type: Population
Task: Token
Definition: Diseases of interest related to the used animal model(s).
Examples: Multiple sclerosis, stroke, spinal cord

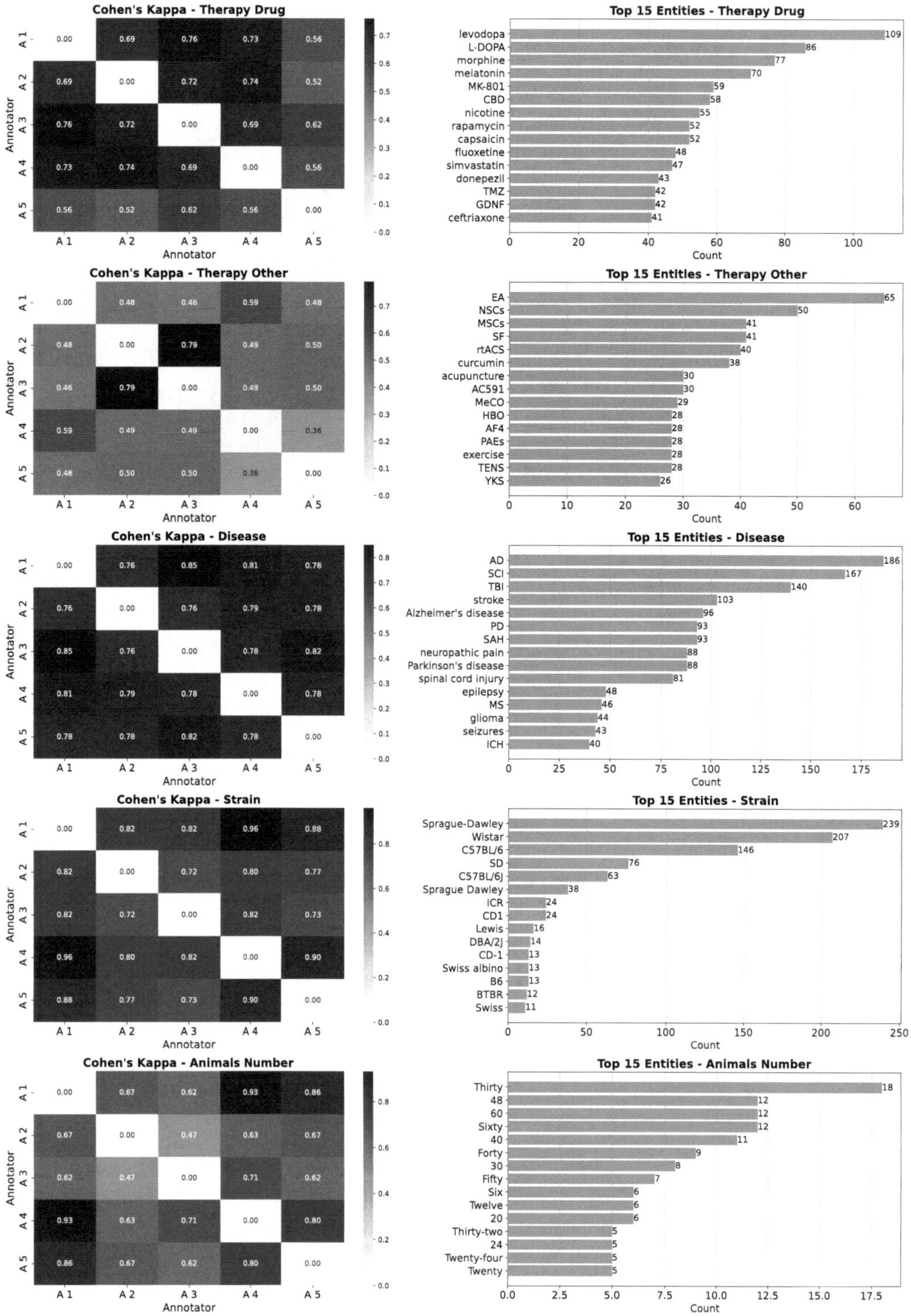

Figure 5: Cohen's Kappa scores for inter-annotator pairs for the NER annotations in the overlapping articles (left column). Top 15 most frequent NER entity text spans in the full dataset (right column).

injury, etc.
Location in Paper: Title, Abstract

Comments:

- Annotate only diseases relevant to the study.

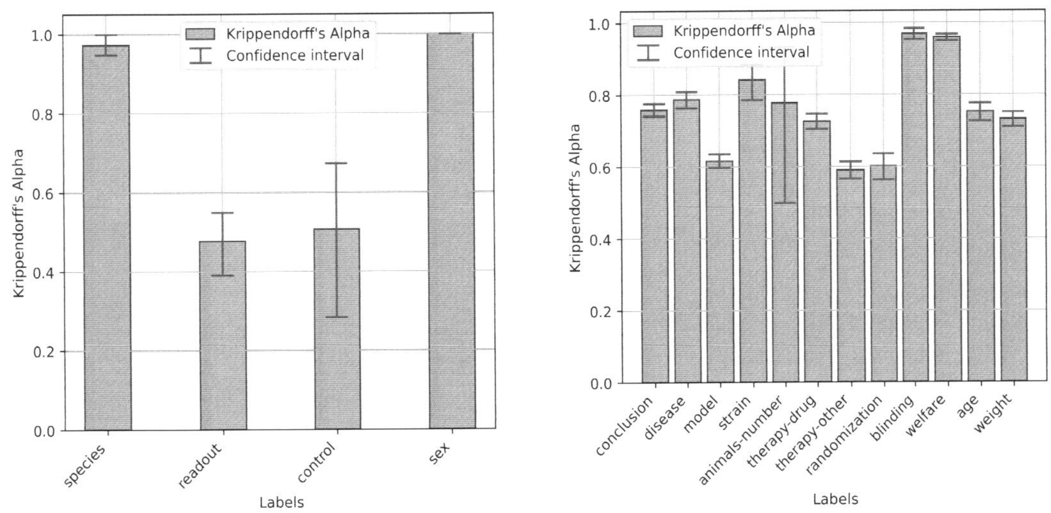

Figure 6: Krippendorff's alpha scores for different annotation levels. (Left) Krippendorff's alpha for document-level annotations. (Right) Krippendorff's alpha for token- and sentence-level annotations. Alpha score is computed on a per-token basis.

- Do not annotate disease models like MCAO or EAE.
- Include abbreviations (e.g., MS).
- Annotate more and less specific mentions (e.g., Alzheimer's disease and dementia).

B.3.5 Number of animals used in total

Type: Population
Task: Token
Definition: The total number of animals used in the study.
Examples: "A total of 968 animals (618 mice and 350 rats) were used."
Location in Paper: Abstract, Methods
Comments:

- Only annotate the exact total number.
- The number might be reported more than once.

B.3.6 Therapy

Type: Intervention
Task: Token
Definition: The therapeutic intervention tested.
Examples: Electroacupuncture, melatonin therapy.
Location in Paper: Title, Abstract, Methods
Comments:

- Two labels:
  - Drug (e.g., a small molecule, siRNA).
  - Non-drug (e.g., exercise, herbal extracts).
- Control treatments should be annotated.
- Do not annotate dosing or application information.

B.3.7 Control mentioned

Type: Control
Task: Text
Examples: "Group 1 (control) received saline."
Location in Paper: Abstract, Methods
Definition: Whether the control group/treatment was mentioned.
Levels:

- Control yes
- Control not reported

B.3.8 Readouts

Type: Outcome
Task: Text
Definition: The readouts used to assess intervention efficacy.
Examples: "We used Nissl staining and MRI to assess stroke volume."
Location in Paper: Abstract, Methods
Levels:

- Behavior (e.g., rotarod, seizure).
- Imaging (e.g., MRI, PET).

- Histology (e.g., Nissl staining, H&E).
- Physiology (e.g., blood pressure, EEG).
- Other (e.g., PCR, Western blot).

B.3.9 Study conclusion

Type: Outcome
Task: Sentence
Definition: The main finding of the study, i.e., the overall effect of the intervention.
Examples: "Our findings suggest a potential therapeutic role for Galantamine in attenuating hyperoxia-induced brain injury."
Location in Paper: Abstract
Levels:

- Positive
- Negative
- Neutral
- Mixed

B.3.10 Animal disease model

Type: Population
Task: Sentence
Definition: The animal model mimicking a neurological or psychiatric condition.
Examples: "EAE was induced by immunizing female Lewis rats with MOG55-66."
Location in Paper: Abstract

B.3.11 Animal age

Type: Population
Task: Sentence
Definition: The age of animals used.
Examples: "12-week-old female C57BL/6 mice were used."
Location in Paper: Methods

B.3.12 Animal weight

Type: Population
Task: Sentence
Definition: The weight of animals used.
Examples: "Male Albino Swiss (20–25 g) mice were used."
Location in Paper: Methods

B.4 Parameters Related to Study Quality

B.4.1 Randomization

Type: Study Quality
Task: Sentence
Definition: Whether the experimental setup used randomization of animals.
Examples: "We randomly divided the experimental rats into five groups with six animals per group as follows: ... "
Location in Paper: Abstract, Methods
Comments:

- Only applies to circumstances describing the randomization of animals into (treatment) groups.
- Does NOT apply to other instances of randomization (e.g., "we analyzed 5 random fields of view").
- In most cases, only one sentence describes randomization, but more than one could be annotated if different species are described separately.
- Together with blinding, it is one of the most critical study quality items.
- Annotate the entire sentence.
- If unsure, be generous with annotation.

B.4.2 Blinding

Type: Study Quality
Task: Sentence
Definition: Whether the experimental setup used blinding of experimenters.
Examples:

- "Experimenters were blinded to the treatment group."
- "Researchers were unaware of the treatment of the animals."
- "All behavioral measurements were made by an observer unaware of the treatment."

Location in Paper: Abstract, Methods
Comments:

- Blinding can occur at any step: during treatment, analysis, or both.
- Typically, only one sentence describes blinding, but multiple sentences may exist for different species/experiments.

- Together with randomization, it is a crucial study quality item.
- Annotate the entire sentence.
- If unsure, be generous with annotation.

B.4.3 Animal Welfare Statement

Type: Study Quality
Task: Sentence
Definition: Whether the animal study complies with local, regional, national, or international animal welfare guidelines.
Examples:

- "On October 29, 2019, the institutional Ethics Committee at NODCAR and Faculty of Pharmacy, Cairo University, approved all animal procedures."
- "All mice were maintained under specific pathogen-free conditions and used for experimentation according to protocols approved by the Swiss Federal Veterinary Office."

Location in Paper: Abstract, Methods
Comments:

- Usually, only one sentence describes animal welfare, but multiple sentences may exist for different species/experiments.
- Statements should mention compliance with guidelines/regulations or approval by an ethics committee.
- Commonly reported.

B.4.4 ARRIVE Guidelines

Type: Study Quality
Task: Sentence
Definition: Whether the study follows the ARRIVE guidelines, which provide standards for reporting methodological details in animal experiments.
Examples:

- "In the current study, we handled the animals consistently in accordance with the ARRIVE guidelines."
- "All studies involving animals are reported in accordance with the ARRIVE guidelines for reporting experiments involving animals."

Location in Paper: Abstract, Methods
Comments:

- Can be identified by searching for "ARRIVE" (always in uppercase).
- Rarely reported.

B.4.5 PREPARE Guidelines

Type: Study Quality
Task: Sentence
Definition: Whether the study follows the PREPARE guidelines.
Examples: Search for "PREPARE" in the document.
Location in Paper: Abstract, Methods
Comments:

- Can be identified by searching for "PREPARE" (always in uppercase).
- Very rarely reported.

B.4.6 Sample Size Calculation

Type: Study Quality
Task: Sentence
Definition: Whether the study conducted a prior sample size calculation to determine how many animals were required for the experiments.
Examples:

- "We conducted an a priori sample size calculation."
- "The power was calculated based on prior estimates."

Location in Paper: Abstract, Methods
Comments:

- Very rarely reported.

Benchmarking zero-shot biomedical relation triplet extraction across language model architectures

Frederik Steensgaard Gade[1,2], Ole Lund[2], Marie Lisandra Zepeda Mendoza[3]
[1]AI & Digital Innovation, Novo Nordisk A/S, Måløv, 2760, Denmark
[2]Section for Bioinformatics, Technical University of Denmark, Kgs. Lyngby, 2800, Denmark
[3]Novo Nordisk Research Centre Oxford Ltd, Oxford, OX3 7FZ, UK
Correspondence: fzsg@novonordisk.com

Abstract

Many language models (LMs) in the literature claim excellent zero-shot and/or few-shot capabilities for named entity recognition (NER) and relation extraction (RE) tasks and assert their ability to generalize beyond their training datasets. However, these claims have yet to be tested across different model architectures.

This paper presents a performance evaluation of zero-shot relation triplet extraction (NER followed by RE of the entities) for both small and large LMs, utilizing 13,867 texts from 61 biomedical corpora and encompassing 151 unique entity types. This comprehensive evaluation offers valuable insights into the practical applicability and performance of LMs within the intricate domain of biomedical relation triplet extraction, highlighting their effectiveness in managing a diverse range of relations and entity types.

Gemini 1.5 Pro, the largest LM included in the study, was the top-performing zero-shot model, achieving an average partial match micro F1 of 0.492 for NER, followed closely by SciLitLLM 1.5 14B with a score of 0.475. Fine-tuned models generally outperformed others on the corpora they were trained on, even in a few-shot setting, but struggled to generalize across all datasets with similar entity types. No models achieved an F1 score above 0.5 for the RTE task on any dataset, and their scores fluctuated based on the specific class of entity and the dataset involved. This observation highlights that there is still large room for improvement on the zero-shot utility of LMs in biomedical RTE applications.

1 Introduction

In the field of biomedical natural language processing (NLP), large efforts are being made to create natural language models (LMs) capable of extracting certain entity types and/or relationships, requiring large sets of manually annotated texts. Recently, large language models (LLMs) have proven useful in extracting information from text in a zero-/few-shot fashion, potentially enabling information extraction (IE) where a smaller user-provided annotation may suffice to accomplish the task at hand (Dagdelen et al., 2024). In this study, we focus on biomedical relation triplet extraction (RTE). RTE consists of identifying entities from a list of allowed entity types (such as genes, diseases, etc.) and the type of relationship that exists between them. Thus, RTE can be broken down into a combined named entity recognition (NER) and relation extraction (RE) task. This extraction is valuable for identifying evidence of specific biological connections in, for example, knowledge base (KB) or knowledge graph construction (KGC). Our goal is to investigate the best architectures for reliable biomedical zero-shot RTE to inform model choice for downstream specific biomedical KB question-answering (QA) tasks.

Multiple papers have benchmarked LLMs for IE tasks on biomedical texts (Dai et al., 2024; Jahan et al., 2024; Chen et al., 2025), and there are multiple established combined benchmark datasets (e.g. BLURB (Gu et al., 2021)) and LLM instruction datasets (e.g. SciRIFF (Wadden et al., 2024)), but two main points remain unaddressed:

1. The generalisability of RTE performance outside of the corpora the models are trained on. Performance reporting for the models usually only includes the validation/test set performance for the datasets they were trained on, thus not truly evaluating their generalisability. Performance reporting for some models on certain datasets may also be sensitive to bias through their inclusion in the LLM pretraining (due to the opaqueness of data being used in training of closed-sourced LMs), necessitating performance benchmarking on less commonly used datasets.

2. A direct comparison of zero-shot capabil-

Proceedings of the 24th Workshop on Biomedical Language Processing (BioNLP 2025), pages 88–100
August 1, 2025

ities of generative, decoder-only LLMs to the newest BERT-like (and other) LMs for biomedical NER/RE.

We compared the zero-shot RTE performance across various model architectures using a large combined corpus of gold-standard NER & RE annotation datasets outside of the most commonly used benchmark datasets and across multiple architectures.

2 Datasets

To begin with, we assembled an extensive biomedical gold-standard corpus. For this purpose, we compiled a total of 61 different biomedical corpora suited for public and commercial use, including representative subsets from BigBIO (Fries et al., 2022) featuring NER and/or RE annotations, as well as the ComplexTome (Mehryary et al., 2024) and RegulaTome (Nastou et al., 2024) datasets. Altogether, the combined corpus comprises 13,867 texts, including 9,804 abstracts (70.7%), 1,596 sentences (11.5%), and the remaining 2,467 regarded as miscellaneous (such as case reports, full paper paragraphs, etc.) or undefined. Additionally, 18 of the 61 corpora include annotations for 90 distinct relation types. In total, the entire selected corpus comprises 151 distinct entity types, categorised into 11 groups: `Organism`, `Gene/Protein`, `Chemical`, `Disease`, `Medical`, `Gene-related`, `Protein-related`, `Anatomy`, `Other biological`, `Non-English`, and `Other`. Definitions for these groups can be found in appendix C).

Figure 1 characterises the text length, entity count, relation count, and unique entity/relation types within the test set for each corpus included in our study. Details about the modifications made to the corpora are provided in appendix B.

3 Models and methods

The 12 models included in this study are classified into five categories: BERT/BERT-like (Bidirectional Encoder Representations from Transformers), T5 (Text-to-Text Transfer Transformer), KGC-SFT SLM (Knowledge Graph Construction Supervised Fine-Tuned Small Language Model), biology-SFT SLM, and LLM. A comprehensive list of these models, along with their architecture and maximum context length, can be found in table 1.

The BERT models in this study include GLiNER, NuNER, and ZeroShotBioNER. GLiNER (Generalist and Lightweight Model for Named Entity Recognition) (Zaratiana et al., 2023) is a small, generalist NER model, introduced as an alternative to traditional NER models. Unlike conventional models, GLiNER is not restricted to predefined entities, even though it employs a BERT-like architecture.

Building on GLiNER, GLiNER Multi-task (Stepanov and Shtopko, 2024) extends the capabilities of the model to perform additional information extraction tasks, such as RE and summarisation.

NuNER (Bogdanov et al., 2024) is another generalist alternative to GLiNER, distinguished by its training method, which employs a contrastive learning approach on synthetic data generated by an LLM (GPT-3.5).

ZeroShotBioNER (Košprdić et al., 2024) is a BERT-based model, specifically a fine-tuned version of BioBERT v1.1, trained on 26 biomedical NER classes. It is designed for zero-shot inference across the biomedical domain, particularly targeting chemicals, diseases, and proteins, and is tailored for biological applications.

InstructUIE (Wang et al., 2023) utilizes a T5 architecture and is trained and evaluated on their own curated information extraction benchmark set. This set includes NER datasets from AnatEM, BC5CDR, CHEMDNER, among others, encompassing a wide range of information extraction tasks.

The two KGC-SFT SLMs, Triplex (SciPhi, 2024) and Phi3 Mini Graph (Emergent Methods, 2024), are fine-tuned versions of Phi3 models specifically designed for generalist RTE.

SciLitLLM 1.5 (Li et al., 2024) is built upon Qwen 2.5 and undergoes continuous pre-training using an internal corpus comprising science textbooks and articles. It is subsequently fine-tuned on SciRIFF (Wadden et al., 2024) as well as a synthetic dataset designed for scientific literature understanding and instructions.

As the representative decoder-only, closed-source LLM, we chose Gemini 1.5 Pro (Gemini Team et al., 2024) due to its computational efficiency. It presents itself as having exceptional ability in long-context needle-in-a-haystack retrieval and demonstrates strong overall performance across a diverse array of tasks.

Models were configured to perform NER and RTE of all applicable types in a single model call, wherever supported. For GLiNER multi-task models and InstructUIE, NER and RE were conducted in two separate model calls. All models were employed at the document level. Details about the

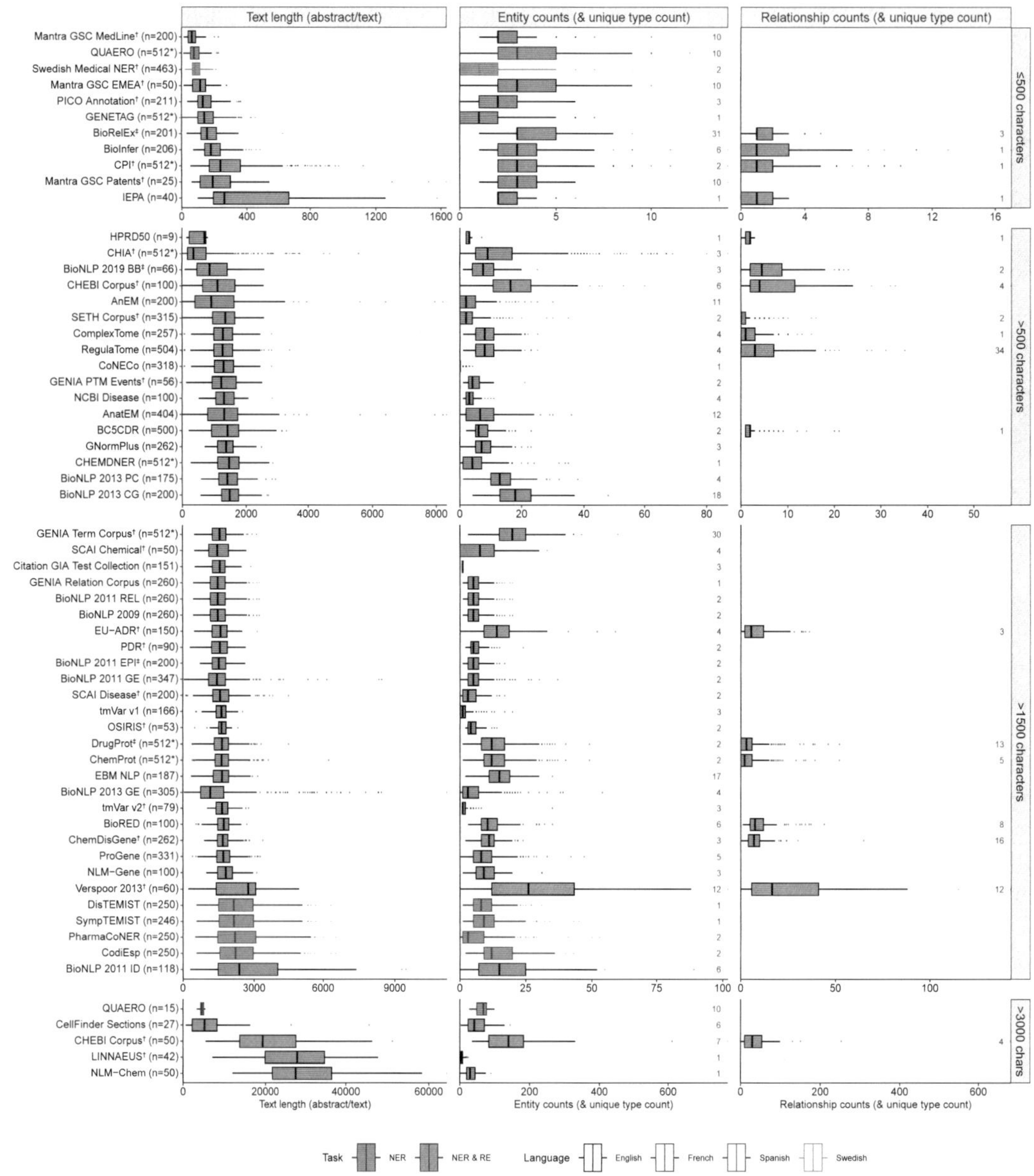

Figure 1: Summary statistics of the text lengths, entity counts (& number of unique entity types), and relation counts (& number of unique relation types) for the test set of each corpus used. The corpora categorised based on their average character count (≤500, >500, >1500, >3000). Details regarding which entity types and relationship types were included, excluded, or merged, can be found in appendix B. n denotes the number of texts in the corpus.

* Datasets were truncated to a maximum of 512 samples to minimise over-representation of certain datasets within the overall corpus.

† For these 20 datasets, no splits were available via BigBIO, therefore, we used the test set from a 50/50 train/test split.

‡ For the BioNLP 2011 EPI, BioNLP 2019 BB, BioRelEx, and DrugProt datasets, annotated test sets were not available, so their development/validation sets were utilised instead.

Architecture	Model	Context length limit	Tasks
LLM	Gemini 1.5 Pro (Feb 2025)	2,097,152	NER & RE
Biology-SFT SLM	SciLitLLM 1.5 (Qwen 2.5 14B)	131,072	NER & RE
KGC-SFT SLM	Triplex (Phi3-3.8B)	131,072	NER & RE
	Phi3 Mini Graph (Phi3-3.8B-128K)	131,072	NER & RE
T5	InstructUIE (Flan-T5 11B)	512	NER & RE
BERT/BERT-like	ZeroShotBioNER (BioBERT V1.1)	512	NER
	NuNER Zero 4K (Longformer Large 4K)	4,096	NER
	GLiNER Medium v2.5 (DeBERTa-V3)	384	NER
	GLiNER Large v2.5 (DeBERTa-V3-Large)	512	NER
	GLiNER Large Bio v0.1 (DeBERTa-V3-Large)*	512	NER
	GLiNER Multi-task v1.0 (DeBERTa-V2-XLarge)	512	NER & RE
	GLiNER Multi-task Large v0.5 (DeBERTa-V3-Large)	512	NER & RE

Table 1: For each model, the table includes its name, model group, token limits for both prompt/input and completion/output, and the tasks each model can perform—specifically NER and RE. Note that the context length limit reflects the maximum number of tokens the architecture can process simultaneously, rather than a verified range for optimal performance.
* This model is not included in the main GLiNER publication by Zaratiana et al. (2023), but is available on HuggingFace (repo_id: `urchade/gliner_large_bio-v0.1`).

prompt/input preparation for each model are provided in appendix D.

Formally, we define zero-shot RTE as the process of performing NER followed by RE, given only the allowed entity and relation types. For k-shot RTE, we give k examples from the training set. If no relations were annotated for a given corpus, only the NER task was evaluated.

To mitigate the possibility of hallucinations from the language models, the output was limited to the queried types of entities and relations. Given that all models, except the BERT variants, are causal language models (as opposed to token classifiers), they produce entity name strings rather than token positions. Consequently, to ensure fairness, performance for all models was evaluated using the case-insensitive micro F1 score from MUC-5 (Chinchor and Sundheim, 1993), unless stated otherwise[1], partial boundary, exact-type matching for each unique entity and relationship in the gold-standard data. In this context, a partial match refers to a word match at either boundary. Therefore, the reported performance more closely aligns with the practical application for KGC, where duplicate entities and relationships are consolidated.

4 Results

All models, except the KGC SFT-SLMs, are evaluated across all datasets, with the exception of BC5CDR, BioRED, and ChemDNER; these particular datasets are analyzed separately because some of the models have been fine-tuned specifically using these datasets.

Figure 2 displays the NER rank distribution for each corpus, providing a head-to-head comparison of the models. Additionally, the win rates for NER and RTE are detailed in appendix table A1. Gemini 1.5 Pro and the notably smaller SciLitLLM 14B emerge as the clear frontrunners, whereas InstructUIE and ZeroShotBioNER are the lowest performers overall. However, ZeroShotBioNER excels over all other models in the ChemProt, DrugProt, CHEBI, ChemDisGene, and SETH corpora, which predominantly contain chemical, gene/protein, and disease annotations. Similarly, InstructUIE outperforms all other models in the Citation GIA Test, IEPA, and GENETAG corpora, which exclusively feature gene and protein annotations.

Although ZeroShotBioNER and InstructUIE outperform other models in the specific datasets mentioned, this is not generally the case across the entity types they were fine-tuned on. This is evident in figure 3, which illustrates NER performance by entity type group. Note the two models generally demonstrate lower performance for gene/protein, chemical, and disease entity groups. Moreover, despite being trained on biological entity types, InstructUIE and ZeroShotBioNER do not generalize well to other biological or gene-/protein-related entity types. One might hypothesize that identifying gene-/protein-related entity types parallels the

[1]Due to capitalised words in the beginning of sentences being considered identical to non-capitalised words for the purposes of entity uniqueness.

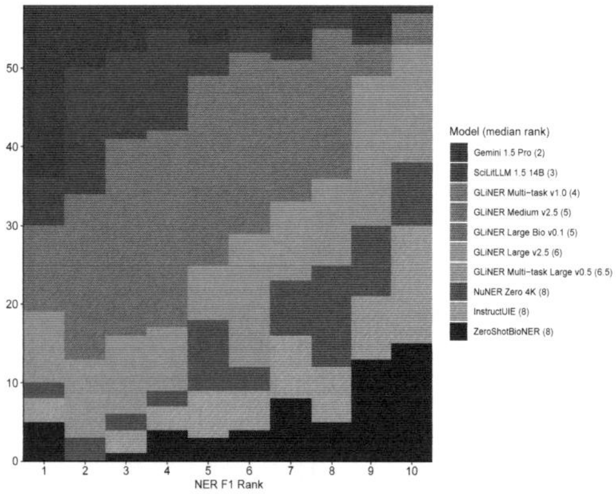

Figure 2: Model ranks of NER micro F1 for all corpora (excluding BC5CDR, BioRED, and ChemDNER).

task of identifying entity relations, which models enabled for RE might excel at.

The NER and RTE performance by corpus mean character count ($\leq$500, >500, >1500, >3000) is shown in figure 4. For models with a short context, the input might be truncated and thus the recall is decreased inherently as a result of the model architecture. However, even for the long-context models, the F1 drops for the longest input texts. In lengthy corpora, RTE performance drops to nearly zero, and across shorter corpora, the general performance for this task remains quite poor across all models.

The performance of ZeroShotBioNER and InstructUIE on BC5CDR, BioRED, and ChemDNER (which were excluded from the previous analyses) is compared with zero-/few-shot prompting of Gemini 1.5 Pro and SciLitLLM 1.5 14B in table 2. Few-shot examples were sourced from the training set.

The KGC SFT-SLMs were evaluated separately on a small subset of datasets, specifically BC5CDR and BioRED, as detailed in table 3. For both datasets, the NER performance of the KGC models is lower than that of all other models, particularly for the more complex dataset, BioRED. Although these models are intended for generalist KGC, their performance falls significantly below that of SciLitLLM 1.5 and Gemini 1.5 Pro (table 2). This discrepancy may be attributed to the lack of biomedical data in their fine-tuning process.

Appendix table A2 compares the partial and strict matching performance of the top three models: Gemini 1.5 Pro, SciLitLLM 1.5 14B, and GLiNER Multi-task v1.0; alongside the two SFT IE models, InstructUIE and ZeroShotBioNER. Gemini 1.5 Pro experiences the largest performance drop when evaluation criteria shift to strict matching. This is due to certain instances, like the one in BC5CDR, where "methamphetamine induces psychosis" is incorrectly labeled as "methamphetamine psychosis" instead of the correct "psychosis." This labeling would be correct under partial matching but incorrect under strict matching. GLiNER Multi-task v1.0 demonstrates the smallest performance loss for NER, achieving the highest F1 score and precision under strict matching conditions. Conversely, SciLitLLM 1.5 14B exhibits the least performance decline when transitioning to strict matching, and even shows an improvement in precision.

5 Discussion

The models explicitly fine-tuned for biology, namely InstructUIE, ZeroShotBioNER, SciLitLLM 1.5, and GLiNER Large Bio v0.1, were generally outperformed by the larger, more generalist models. Exceptions occurred for datasets on which these models were directly fine-tuned or those containing very similar entity types. However, InstructUIE and ZeroShotBioNER did not consistently outperform all other models across datasets featuring entity types similar to those in their finetuning datasets. The KGC-specific models demonstrated significantly lower performance compared to other models, possibly due to their lack of biological understanding needed to identify entity and relation types. Overall, Gemini, the largest and most resource-intensive model, achieved the highest scores in the benchmark. Notably, Gemini's performance was only marginally better than the considerably smaller SciLitLLM 1.5, which has 14 billion parameters, in zero-shot biomedical NER, although SciLitLLM had lower RTE performance. We hypothesize that a model fine-tuned on biology and further instruction-tuned specifically for RTE could achieve even better results.

SciLitLLM 1.5 14B was specifically fine-tuned on the literature understanding instruction dataset SciRIFF (Wadden et al., 2024), which includes NER tasks for several of our datasets, such as BioRED and GNormPlus, as well as RE tasks for ChemProt. This may introduce a bias in the performance evaluation.

The best-performing BERT model was GLiNER Multi-task v1.0, which also achieved the best av-

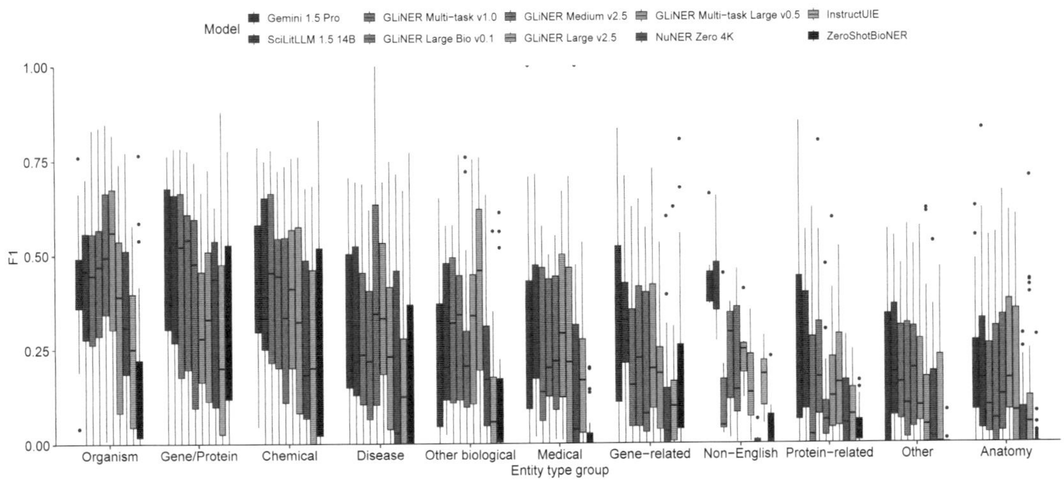

Figure 3: NER F1 scores for each entity type prediction across all datasets (excluding BC5CDR, BioRED, and ChemDNER) stratified by model and entity group. Information on which entity types were grouped is specified in appendix C.

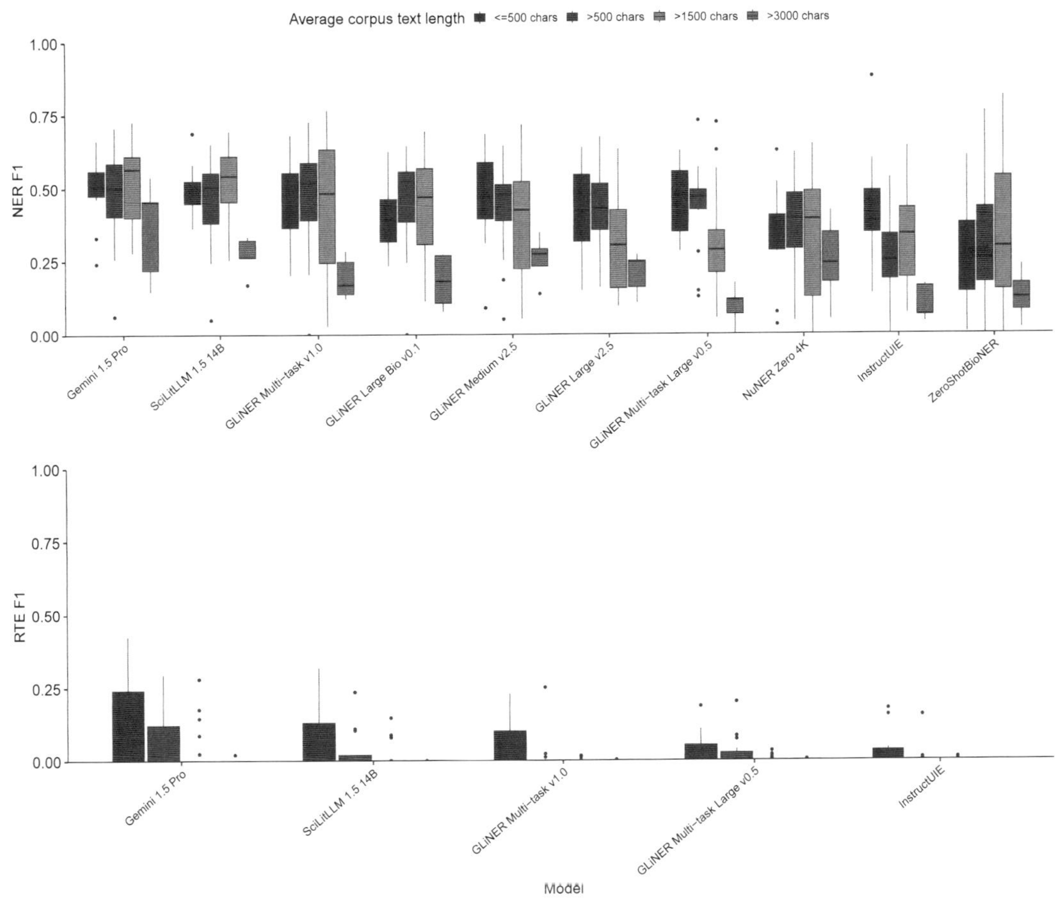

Figure 4: NER and RTE micro F1 score for each corpus across all datasets and (excluding BC5CDR, BioRED, and ChemDNER) stratified by model and average text length.

Dataset	Model	k-shot	NER			RTE		
			F1	Precision	Recall	F1	Precision	Recall
BC5CDR	ZeroShotBioNER*	SFT	**0.847**	0.777	**0.931**	-	-	-
	InstructUIE	SFT	0.601	0.790	0.485	0.105	0.214	0.070
	Gemini 1.5 Pro	0-shot	0.583	**0.842**	0.446	*0.442*	*0.477*	0.413
		3-shot	0.666	0.826	0.558	0.438	0.458	*0.419*
		10-shot	0.717	*0.836*	*0.627*	**0.497**	**0.488**	**0.507**
	SciLitLLM 1.5 14B	0-shot	0.697	0.796	0.620	0.340	0.446	0.274
		3-shot	0.723	0.811	0.653	0.381	0.471	0.320
		10-shot	*0.738*	0.785	0.696	0.400	0.443	0.364
BioRED	ZeroShotBioNER*	SFT	0.666	0.685	**0.648**	-	-	-
	InstructUIE	No SFT	0.265	0.666	0.165	0.002	0.045	0.001
	Gemini 1.5 Pro	0-shot	0.516	0.725	0.400	0.138	0.232	0.098
		3-shot	*0.669*	*0.755*	0.600	*0.162*	*0.232*	*0.125*
		10-shot	**0.684**	**0.779**	*0.610*	**0.183**	**0.266**	**0.139**
	SciLitLLM 1.5 14B	0-shot	0.607	0.651	0.569	0.021	0.094	0.012
		3-shot	0.600	0.641	0.564	0.057	0.105	0.039
		10-shot	0.622	0.678	0.574	0.085	0.159	0.058
ChemDNER*	ZeroShotBioNER*	SFT	**0.866**	**0.944**	**0.800**	-	-	-
	InstructUIE	SFT	0.658	0.865	0.532	-	-	-
	Gemini 1.5 Pro	0-shot	0.684	0.713	0.657	-	-	-
		3-shot	0.652	0.803	0.549	-	-	-
		10-shot	0.690	0.781	0.619	-	-	-
	SciLitLLM 1.5 14B	0-shot	0.755	0.755	*0.755*	-	-	-
		3-shot	*0.794*	0.878	0.725	-	-	-
		10-shot	0.792	*0.889*	0.714	-	-	-

Table 2: Comparison of model performance of fine-tuned models, ZeroShotBioNER and InstructUIE, with the zero-/few-shot performance of the LLM, Gemini 1.5 Pro, and the biology-SFT SLM, SciLitLLM 1.5 14B. Both ZeroShotBioNER and InstructUIE were fine-tuned on BC5CDR and ChemDNER (denoted with SFT in the table), and ZeroShotBioNER was additionally fine-tuned on BioRED, whilst InstructUIE was not (No SFT). Best performance by dataset is highlighted in bold, and second-best in italics.
* NER-only model/dataset.

Dataset	Matching criteria	Model	NER			RTE		
			F1	Precision	Recall	F1	Precision	Recall
BC5CDR	Partial (strict type)	Triplex	0.458	0.380	**0.576**	0.132	0.242	0.090
		Phi3 Mini Graph*	0.545	0.698	0.448	-	-	-
		GLiNER Multi-task v1.0	**0.611**	**0.771**	0.505	**0.162**	**0.500**	**0.097**
	Relaxed	Triplex	0.486	0.407	0.605	0.121	0.223	0.083
		Phi3 Mini Graph	0.482	0.412	0.581	0.096	0.058	0.290
		GLiNER Multi-task v1.0	0.612	0.770	0.507	0.160	0.493	0.095
BioRED	Partial (strict type)	Triplex	0.015	0.529	0.007	0.002	0.200	0.001
		Phi3 Mini Graph*	0.096	0.295	0.058	-	-	-
		GLiNER Multi-task v1.0	**0.575**	**0.662**	**0.508**	**0.004**	**0.143**	**0.002**
	Relaxed	Triplex	0.014	0.529	0.007	0.002	1.000	0.001
		Phi3 Mini Graph	0.509	0.478	0.544	0.167	0.141	0.205
		GLiNER Multi-task v1.0	0.599	0.691	0.529	0.007	0.219	0.003

Table 3: KGC-SFT SLM performances vs. GLiNER Multi-task performance for NER and RTE with partial, strict-type matching criteria (used through the paper) and relaxed matching (case-insensitive, no schema restriction of output, entity and relation type-agnostic, relation directionality-agnostic). For comparison, Gemini 1.5 Pro 0-shot F1 for RTE in BioRED with relaxed matching criteria is 0.287. Best performance by dataset is highlighted in bold.
* The Phi3 Mini Graph model is unable to follow the instruction to output only specified relation types, and thus restricting the output to the specified schema yields no predictions.

erage performance for strict matching. It is significantly smaller than either SciLitLLM 1.5 14B or Gemini 1.5 Pro, potentially making it the ideal choice when cost and scalability are concerns.

Notably, while SciLitLLM 1.5 Pro was the overall best-performing model among the ones compared, RTE performance was relatively low across the board. No zero-shot model achieved a micro F1 score above 0.5 for any dataset, raising concerns about their effectiveness for RTE tasks. In agreement with Chen et al. (2025), we therefore do not recommend using zero-shot models for biomedical RTE. Although few-shot performance can be comparable to SFT performance for certain models and datasets, fine-tuned models generally outperform non-fine-tuned ones when manually annotated data is available for SFT. In cases where such data is unavailable, few-shot models may be utilized if downstream tasks can accommodate a compromise in performance, possibly due to additional checks at later stages.

While the RTE task yields a simple KG without additional metadata, leveraging information extraction models such as InstructUIE, NuExtract 1.5, and LLMs like Gemini 1.5 Pro could enhance the metadata associated with the triplets. In a biomedical context, this could involve incorporating surrounding biological context such as tissue, organism, intervention, and co-factors. Such contextual enrichment can be done with traditional NLP methods, and could be improved with powerful generalist LLMs (Sosa et al., 2023).

Although some benchmarking datasets are extensive and well-annotated across a wide range of relationships and entities, they present challenges when used to generate KBs or KGs. For instance, RegulaTome includes relationships that are speculative or hypothesized and does not account for the negation of relations. Consequently, using these annotations as the truth set means there is no distinction between verified conclusions and mere speculations—only their mention in the text is captured, while negative results are omitted.

We observe that methods such as GraphRAG (Edge et al., 2024), attempt to leverage the emergent information extraction capabilities of LLMs to enhance knowledge base question answering (KBQA) tasks. However, based on the outcomes of this benchmark, we hypothesize that for results from a GraphRAG-like approach to be valuable in biomedical applications, tailored models are necessary to accurately tag relevant entities and relationships. This is due to the fact that the inherent biological understanding of zero-shot LLMs is typically insufficient for most practical downstream applications.

6 Conclusion

In conclusion, this study benchmarks zero-shot biomedical RTE across a range of LM architectures. Larger models such as Gemini 1.5 Pro and SciLitLLM 1.5 14B excel in NER but face challenges with subsequent RE, with no F1 score surpassing 0.5 in RTE tasks. Notably, GLiNER Multi-task v1.0 stands out as the best-performing BERT-based model, delivering strong performance relative to its smaller size and excelling in strict matching criteria, thus making it a cost-effective option when scalability is a concern.

While fine-tuned models like ZeroShotBioNER perform well on specific datasets, they are generally surpassed by larger, more generalized models even when dealing with slightly out-of-distribution data, underscoring the limitations of current zero-shot models for practical applications in biomedical NLP. Furthermore, although few-shot learning provides some benefits, fine-tuning remains essential for maximizing model performance when annotation is feasible.

Limitations

Conducting a fair evaluation of all available LMs is a challenging task for several reasons. Firstly, accessing and comprehensively testing each model may not be financially viable, necessitating the selection of representative models from various LM categories. Additionally, information regarding the training data is not always publicly available, as seen with Gemini, or models may be trained on known public benchmarks like BLURB, which includes datasets that overlap with our benchmark (EBM PICO, ChemProt, and BC5CDR) or contain shared entity types (JNLPBA) (Gu et al., 2021), thus complicating the fair comparison between models.

Moreover, performance is sensitive to the matching criteria employed, and the options for this benchmark are restricted due to the nature of the model outputs from causal language models, as they are not token classifiers. More sophisticated matching criteria, such as ontology matching, would be preferable but fall outside the scope of this research.

The models are constrained by their context length, and some might have benefited from re-engineering the task by breaking the texts into sentences—even models with a relatively long context length. Additionally, running the models in multiple rounds, such as one round per entity type, could offer advantages, like increased task specificity. However, this approach also presents drawbacks, including overlap issues and higher costs.

Finally, it is important to recognize that different models may require distinct prompts to achieve optimal performance. Studies have demonstrated that benchmark results are sensitive to prompt engineering (Jahan et al., 2023). Exploring techniques such as chain-of-thought prompting, meta-prompting (Suzgun and Tauman Kalai, 2024), reasoning models (DeepSeek-AI et al., 2025), or other related strategies could potentially enhance performance.

Acknowledgments

The work of FSG is partly funded by the Innovation Fund Denmark (IFD) under File No. 3129-00056 and co-financed through a Novo Nordisk R&ED novoSTAR Industrial PhD fellowship.

Additionally, thanks to Jesper Ferkinghoff-Borg, Julien Fauqueur, and Robert R. Kitchen for their help and support for this project.

References

Sergei Bogdanov, Alexandre Constantin, Timothée Bernard, Benoit Crabbé, and Etienne Bernard. 2024. NuNER: Entity Recognition Encoder Pre-training via LLM-Annotated Data. Computing Research Repository, arXiv:2402.15343. Version 1.

Qingyu Chen, Yan Hu, Xueqing Peng, Qianqian Xie, Qiao Jin, Aidan Gilson, Maxwell B. Singer, Xuguang Ai, Po-Ting Lai, Zhizheng Wang, Vipina K. Keloth, Kalpana Raja, Jimin Huang, Huan He, Fongci Lin, Jingcheng Du, Rui Zhang, W. Jim Zheng, Ron A. Adelman, Zhiyong Lu, and Hua Xu. 2025. Benchmarking large language models for biomedical natural language processing applications and recommendations. *Nature Communications*, 16(1).

Nancy Chinchor and Beth Sundheim. 1993. MUC-5 evaluation metrics. In *Fifth Message Understanding Conference (MUC-5): Proceedings of a Conference Held in Baltimore, Maryland, August 25-27, 1993.*

John Dagdelen, Alexander Dunn, Sanghoon Lee, Nicholas Walker, Andrew S. Rosen, Gerbrand Ceder, Kristin A. Persson, and Anubhav Jain. 2024. Structured information extraction from scientific text with large language models. *Nature Communications*, 15(1).

Xiang Dai, Sarvnaz Karimi, Abeed Sarker, Ben Hachey, and Cecile Paris. 2024. Multiade: A multi-domain benchmark for adverse drug event extraction. *Journal of Biomedical Informatics*, 160:104744.

DeepSeek-AI, Daya Guo, Dejian Yang, Haowei Zhang, Junxiao Song, Ruoyu Zhang, Runxin Xu, Qihao Zhu, Shirong Ma, Peiyi Wang, Xiao Bi, Xiaokang Zhang, Xingkai Yu, Yu Wu, Z. F. Wu, Zhibin Gou, Zhihong Shao, Zhuoshu Li, Ziyi Gao, Aixin Liu, Bing Xue, Bingxuan Wang, Bochao Wu, Bei Feng, Chengda Lu, Chenggang Zhao, Chengqi Deng, Chenyu Zhang, Chong Ruan, Damai Dai, Deli Chen, Dongjie Ji, Erhang Li, Fangyun Lin, Fucong Dai, Fuli Luo, Guangbo Hao, Guanting Chen, Guowei Li, H. Zhang, Han Bao, Hanwei Xu, Haocheng Wang, Honghui Ding, Huajian Xin, Huazuo Gao, Hui Qu, Hui Li, Jianzhong Guo, Jiashi Li, Jiawei Wang, Jingchang Chen, Jingyang Yuan, Junjie Qiu, Junlong Li, J. L. Cai, Jiaqi Ni, Jian Liang, Jin Chen, Kai Dong, Kai Hu, Kaige Gao, Kang Guan, Kexin Huang, Kuai Yu, Lean Wang, Lecong Zhang, Liang Zhao, Litong Wang, Liyue Zhang, Lei Xu, Leyi Xia, Mingchuan Zhang, Minghua Zhang, Minghui Tang, Meng Li, Miaojun Wang, Mingming Li, Ning Tian, Panpan Huang, Peng Zhang, Qiancheng Wang, Qinyu Chen, Qiushi Du, Ruiqi Ge, Ruisong Zhang, Ruizhe Pan, Runji Wang, R. J. Chen, R. L. Jin, Ruyi Chen, Shanghao Lu, Shangyan Zhou, Shanhuang Chen, Shengfeng Ye, Shiyu Wang, Shuiping Yu, Shunfeng Zhou, Shuting Pan, S. S. Li, Shuang Zhou, Shaoqing Wu, Shengfeng Ye, Tao Yun, Tian Pei, Tianyu Sun, T. Wang, Wangding Zeng, Wanjia Zhao, Wen Liu, Wenfeng Liang, Wenjun Gao, Wenqin Yu, Wentao Zhang, W. L. Xiao, Wei An, Xiaodong Liu, Xiaohan Wang, Xiaokang Chen, Xiaotao Nie, Xin Cheng, Xin Liu, Xin Xie, Xingchao Liu, Xinyu Yang, Xinyuan Li, Xuecheng Su, Xuheng Lin, X. Q. Li, Xiangyue Jin, Xiaojin Shen, Xiaosha Chen, Xiaowen Sun, Xiaoxiang Wang, Xinnan Song, Xinyi Zhou, Xianzu Wang, Xinxia Shan, Y. K. Li, Y. Q. Wang, Y. X. Wei, Yang Zhang, Yanhong Xu, Yao Li, Yao Zhao, Yaofeng Sun, Yaohui Wang, Yi Yu, Yichao Zhang, Yifan Shi, Yiliang Xiong, Ying He, Yishi Piao, Yisong Wang, Yixuan Tan, Yiyang Ma, Yiyuan Liu, Yongqiang Guo, Yuan Ou, Yuduan Wang, Yue Gong, Yuheng Zou, Yujia He, Yunfan Xiong, Yuxiang Luo, Yuxiang You, Yuxuan Liu, Yuyang Zhou, Y. X. Zhu, Yanhong Xu, Yanping Huang, Yaohui Li, Yi Zheng, Yuchen Zhu, Yunxian Ma, Ying Tang, Yukun Zha, Yuting Yan, Z. Z. Ren, Zehui Ren, Zhangli Sha, Zhe Fu, Zhean Xu, Zhenda Xie, Zhengyan Zhang, Zhewen Hao, Zhicheng Ma, Zhigang Yan, Zhiyu Wu, Zihui Gu, Zijia Zhu, Zijun Liu, Zilin Li, Ziwei Xie, Ziyang Song, Zizheng Pan, Zhen Huang, Zhipeng Xu, Zhongyu Zhang, and Zhen Zhang. 2025. DeepSeek-R1: Incentivizing Reasoning Capability in LLMs via Reinforcement Learning. Computing Research Repository, arXiv:2501.12948. Version 1.

Darren Edge, Ha Trinh, Newman Cheng, Joshua Bradley, Alex Chao, Apurva Mody, Steven Truitt, and Jonathan Larson. 2024. From Local to Global: A Graph RAG Approach to Query-Focused

Summarization. Computing Research Repository, arXiv:2404.16130. Version 2.

Emergent Methods. 2024. Outperforming Claude 3.5 Sonnet with Phi-3-mini-4k for graph entity relationship extraction tasks — emergentmethods.medium.com. https://emergentmethods.medium.com/outperforming-claude-3-5-sonnet-with-phi-3-mini-4k-for-graph-entity-relationship-extraction-tasks-7c8f6c1ebd79. [Accessed 20-03-2025].

Jason Alan Fries, Leon Weber, Natasha Seelam, Gabriel Altay, Debajyoti Datta, Samuele Garda, Myungsun Kang, Ruisi Su, Wojciech Kusa, Samuel Cahyawijaya, Fabio Barth, Simon Ott, Matthias Samwald, Stephen Bach, Stella Biderman, Mario Sänger, Bo Wang, Alison Callahan, Daniel León Periñán, Théo Gigant, Patrick Haller, Jenny Chim, Jose David Posada, John Michael Giorgi, Karthik Rangasai Sivaraman, Marc Pàmies, Marianna Nezhurina, Robert Martin, Michael Cullan, Moritz Freidank, Nathan Dahlberg, Shubhanshu Mishra, Shamik Bose, Nicholas Michio Broad, Yanis Labrak, Shlok S Deshmukh, Sid Kiblawi, Ayush Singh, Minh Chien Vu, Trishala Neeraj, Jonas Golde, Albert Villanova del Moral, and Benjamin Beilharz. 2022. BigBIO: A Framework for Data-Centric Biomedical Natural Language Processing. Part of Advances in Neural Information Processing Systems 35 (NeurIPS 2022) Datasets and Benchmarks Track.

Google Gemini Team, Machel Reid, Nikolay Savinov, Denis Teplyashin, Dmitry, Lepikhin, Timothy Lillicrap, Jean-baptiste Alayrac, Radu Soricut, Angeliki Lazaridou, Orhan Firat, Julian Schrittwieser, Ioannis Antonoglou, Rohan Anil, Sebastian Borgeaud, Andrew Dai, Katie Millican, Ethan Dyer, Mia Glaese, Thibault Sottiaux, Benjamin Lee, Fabio Viola, Malcolm Reynolds, Yuanzhong Xu, James Molloy, Jilin Chen, Michael Isard, Paul Barham, Tom Hennigan, Ross McIlroy, Melvin Johnson, Johan Schalkwyk, Eli Collins, Eliza Rutherford, Erica Moreira, Kareem Ayoub, Megha Goel, Clemens Meyer, Gregory Thornton, Zhen Yang, Henryk Michalewski, Zaheer Abbas, Nathan Schucher, Ankesh Anand, Richard Ives, James Keeling, Karel Lenc, Salem Haykal, Siamak Shakeri, Pranav Shyam, Aakanksha Chowdhery, Roman Ring, Stephen Spencer, Eren Sezener, Luke Vilnis, Oscar Chang, Nobuyuki Morioka, George Tucker, Ce Zheng, Oliver Woodman, Nithya Attaluri, Tomas Kocisky, Evgenii Eltyshev, Xi Chen, Timothy Chung, Vittorio Selo, Siddhartha Brahma, Petko Georgiev, Ambrose Slone, Zhenkai Zhu, James Lottes, Siyuan Qiao, Ben Caine, Sebastian Riedel, Alex Tomala, Martin Chadwick, Juliette Love, Peter Choy, Sid Mittal, Neil Houlsby, Yunhao Tang, Matthew Lamm, Libin Bai, Qiao Zhang, Luheng He, Yong Cheng, Peter Humphreys, Yujia Li, Sergey Brin, Albin Cassirer, Yingjie Miao, Lukas Zilka, Taylor Tobin, Kelvin Xu, Lev Proleev, Daniel Sohn, Alberto Magni, Lisa Anne Hendricks, Isabel Gao, Santiago Ontanon, et al. 2024. Gemini 1.5: Unlocking multimodal understanding across millions of tokens of context. Computing Research Repository, arXiv:2403.05530. Version 5.

Yu Gu, Robert Tinn, Hao Cheng, Michael Lucas, Naoto Usuyama, Xiaodong Liu, Tristan Naumann, Jianfeng Gao, and Hoifung Poon. 2021. Domain-specific language model pretraining for biomedical natural language processing. *ACM Trans. Comput. Healthcare*, 3(1).

Israt Jahan, Md Tahmid Rahman Laskar, Chun Peng, and Jimmy Huang. 2023. Evaluation of ChatGPT on biomedical tasks: A zero-shot comparison with fine-tuned generative transformers. In *The 22nd Workshop on Biomedical Natural Language Processing and BioNLP Shared Tasks*, pages 326–336, Toronto, Canada. Association for Computational Linguistics.

Israt Jahan, Md Tahmid Rahman Laskar, Chun Peng, and Jimmy Xiangji Huang. 2024. A comprehensive evaluation of large language models on benchmark biomedical text processing tasks. *Computers in Biology and Medicine*, 171:108189.

Miloš Košprdić, Nikola Prodanović, Adela Ljajić, Bojana Bašaragin, and Nikola Milošević. 2024. From zero to hero: Harnessing transformers for biomedical named entity recognition in zero- and few-shot contexts. *Artificial Intelligence in Medicine*, 156:102970.

Sihang Li, Jin Huang, Jiaxi Zhuang, Yaorui Shi, Xiaochen Cai, Mingjun Xu, Xiang Wang, Linfeng Zhang, Guolin Ke, and Hengxing Cai. 2024. SciLitLLM: How to Adapt LLMs for Scientific Literature Understanding. Computing Research Repository, arXiv.2408.15545. Version 5.

Farrokh Mehryary, Katerina Nastou, Tomoko Ohta, Lars Juhl Jensen, and Sampo Pyysalo. 2024. STRING-ing together protein complexes: corpus and methods for extracting physical protein interactions from the biomedical literature. *Bioinformatics*, 40(9).

Katerina Nastou, Farrokh Mehryary, Tomoko Ohta, Jouni Luoma, Sampo Pyysalo, and Lars Juhl Jensen. 2024. RegulaTome: a corpus of typed, directed, and signed relations between biomedical entities in the scientific literature. *Database*, 2024.

SciPhi. 2024. Triplex — SOTA LLM for Knowledge Graph Construction - SciPhi AI — sciphi.ai. https://www.sciphi.ai/blog/triplex. [Accessed 20-03-2025].

Daniel N. Sosa, Rogier Hintzen, Betty Xiong, Alex de Giorgio, Julien Fauqueur, Mark Davies, Jake Lever, and Russ B. Altman. 2023. Associating biological context with protein-protein interactions through text mining at pubmed scale. *Journal of Biomedical Informatics*, 145:104474.

Ihor Stepanov and Mykhailo Shtopko. 2024. GLiNER multi-task: Generalist Lightweight Model for Various Information Extraction Tasks. Computing Research Repository, arXiv.2406.12925. Version 2.

Mirac Suzgun and Adam Tauman Kalai. 2024. Meta-Prompting: Enhancing Language Models with Task-Agnostic Scaffolding. Computing Research Repository, arXiv:2401.12954. Version 1.

David Wadden, Kejian Shi, Jacob Morrison, Aakanksha Naik, Shruti Singh, Nitzan Barzilay, Kyle Lo, Tom Hope, Luca Soldaini, Shannon Zejiang Shen, Doug Downey, Hannaneh Hajishirzi, and Arman Cohan. 2024. SciRIFF: A Resource to Enhance Language Model Instruction-Following over Scientific Literature. Computing Research Repository, arXiv:2406.07835. Version 3.

Xiao Wang, Weikang Zhou, Can Zu, Han Xia, Tianze Chen, Yuansen Zhang, Rui Zheng, Junjie Ye, Qi Zhang, Tao Gui, Jihua Kang, Jingsheng Yang, Siyuan Li, and Chunsai Du. 2023. InstructUIE: Multi-task Instruction Tuning for Unified Information Extraction. Computing Research Repository, arXiv:2304.08085. Version 1.

Urchade Zaratiana, Nadi Tomeh, Pierre Holat, and Thierry Charnois. 2023. GLiNER: Generalist Model for Named Entity Recognition using Bidirectional Transformer. Computing Research Repository, arXiv:2311.08526. Version 1.

A Supplementary figures/tables

Model	NER win-rate	RTE win-rate
Gemini 1.5 Pro	**72.8%**	**89.1%**
SciLitLLM 1.5 14B	**72.8%**	*64.1%*
GLiNER Multi-task v1.0	62.8%	35.9%
GLiNER Large Bio v0.1	55.0%	-
GLiNER Medium v2.5	54.8%	-
GLiNER Large v2.5	46.2%	-
GLiNER Multi-task Large v0.5	40.6%	43.8%
NuNER Zero 4K	34.7%	-
InstructUIE	31.2%	17.2%
ZeroShotBioNER	29.1%	-

Table A1: Model micro F1 win rates in all head-to-head comparisons per dataset (for both NER and RTE, excluding BC5CDR, BioRED, and ChemDNER). Best performance is highlighted in bold,and second-best in italics.

B Dataset modifications

To align the datasets to the same tasks, the relationship type names were renamed to an active form (e.g. `COMPLEX_FORMATION` → `FORMS_COMPLEX_WITH`). Selected entities and relationships were removed, if they were not deemed relevant for the task (such as `part-of` relations). All relation types were capitalised, and all entity types were in PascalCase.

BigBIO dataset import modifications: BioRelEx (to include type of binding: `binds`, `not-binds`, `inconclusively-binds`), ComplexTome (implemented), ProGene (changed splitting to original split), RegulaTome (implemented).

C Entity group definitions

Organism (n=10) cell, cellline, celltype, livingbeing, microorganism, monocell, organism, organismtaxon, plant, species

Gene/Protein (n=18) dna, dnafamilyorgroup, gene, geneorgeneproduct, geneormolecularsequence, geneorprotein, geneorproteinfamily, geneorproteinorrna, geneproductormarkergene, geneprotein, peptide, protein, proteinenum, proteinfamiliyorgroup, proteinfamily, proteinfamilyorgroup, proteinisoform, proteinmolecule

Chemical (n=17) aminoacid, aminoacidmonomer, atom, carbohydrate, chemical, chemicalabbreviation, chemicalentity, chemicalfamily, chemicalordrug, chemicalstructure, compound, drug, metabolite, nucleotide, partchemical, reagent, simplechemical

Disease (n=20) adverseeffect, compositediseasemention, condition, disease, diseaseclass, diseaseordisorder, diseaseorphenotypicfeature, disorder, disorderfinding, outcome, outcomeadverseeffects, outcomemental, outcomemortality, outcomeother, outcomepain, outcomephysical, participantcondition, phenomena, phenotype, specificdisease

Medical (n=11) assay, device, diseasemodifier, intervention, interventioneducational, interventionother, interventionpharmacological, interventionphysical, interventionpsychological, interventionsurgical, procedure

Gene-related (n=16) dnadomainorregion, dnamolecule, dnamutation, dnasubstructure, geneticvariant, mutation, polynucleotide, regulonoperon, rna, rnadomainorregion, rnafamilyorgroup, rnamolecule, sequencevariant, snp, snporsequencevariation, twocomponentsystem

Protein-related (n=13) complex, fusionprotein, proteincomplex, proteindomain, proteindomainorregion, proteinmotif, proteinmutation, proteinregion, proteinrelatedentity, proteinrnacomplex, proteinsubstructure, proteinsubunit, proteinvariant

Anatomy (n=13) anatomicalsystem, anatomy, bodypart, bodystructure, developinganatomicalstructure, immaterialanatomicalentity, mul-

Model	Matching criteria	NER			RTE		
		F1	Precision	Recall	F1	Precision	Recall
Gemini 1.5 Pro	Partial (strict type)	0.492	0.611	0.457	0.204	0.236	0.200
	Strict	0.386	0.472	0.365	0.030	0.214	0.016
		-22%	-23%	-20%	-85%	-9%	-92%
SciLitLLM 1.5 14B	Partial (strict type)	0.475	0.541	0.487	0.105	0.232	0.074
	Strict	0.427	0.486	0.440	0.043	0.284	0.024
		-10%	-10%	-10%	**-59%**	**+22%**	**-68%**
GLiNER Multi-task v1.0	Partial (strict type)	0.429	0.581	0.383	0.082	0.437	0.057
	Strict	0.400	0.535	0.359	0.011	0.189	0.006
		-7%	**-8%**	**-6%**	-87%	-57%	-89%
InstructUIE	Partial (strict type)	0.310	0.529	0.264	0.046	0.195	0.030
	Strict	0.257	0.437	0.222	0.013	0.193	0.007
		-17%	-17%	-16%	-72%	-1%	-77%
ZeroShotBioNER	Partial (strict type)	0.301	0.366	0.352	-	-	-
	Strict	0.254	0.301	0.304	-	-	-
		-16%	-18%	-14%	-	-	-

Table A2: Comparison of model performance when transitioning from partial strict-type matching criteria, as used throughout the paper, to strict matching. Strict matching involves case sensitivity, schema restriction of output, and an exact match for entities and relations. The smallest decrease in model performance when switching from partial to strict matching is highlighted in bold for each performance metric.

titissuestructure, organ, organismsubdivision, organismsubstance, pathologicalformation, physiology, tissue

`Other biological` (n=8) biologicalactivity, cancer, cellcomponent, cellularcomponent, lipid, multicell, organelle, virus

`Non-English` (n=6)* diagnostico, enfermedad, procedimiento, proteina, quimico, sintoma

`Other` (n=19) age, characteristic, cohortorpatient, ethnicity, experimentalconstruct, experimenttag, gender, geographicarea, habitat, inorganic, interventioncontrol, object, participant, participantage, participantsamplesize, participantsex, process, size, spectraldata

* The entity names for the French QUAERO and the Swedish Medical NER dataset were in English and thus included in the other groups.

D Model prompting

Inference for GLiNER, GLiNER multi-task, and NuNER were performed using the `gliner` python library, and ZeroShotBioNER using the published implementation. No prompts had to be provided for these `TokenClassifier` models - only entity/relation types were provided. Whenever possible, the default prompt format specified in the model implementation was used. Such prompts are marked with "(default)" - otherwise the prompts were designed.

For zero-shot inference (no examples), only the `<text>`, `<entity_types>`, and `<relation_types>` fields are provided. If no RE annotation exists for a given corpus, this part of the prompt is omitted. For models where we used few-shot prompting (Gemini 1.5 Pro & SciLitLLM 1.5), we show the format of the example given enclosed in parentheses.

InstructUIE (default)

```
NER:

Please list all entity words in the text that fit the category.
Output format is "type1: word1; type2: word2"
Option: <entity_types>
Text: <text>
Answer:

RE:

Given a phrase that describes the relationship between two words,
extract the words and the lexical relationship between them. The
output format should be "relation1: word1, word2; relation2: word3,
word4".
Option: <relation_types>
Text: <text>
Answer:
```

Triplex (default)

```
Perform Named Entity Recognition (NER) and extract knowledge graph
triplets from the text. NER identifies named entities of given
entity types, and triple extraction identifies relationships between
entities using specified predicates.

**Entity Types:**
<entity_types>

**Predicates:**
<relation_types>

**Text:**
<query>
```

Phi3 Mini Graph (default - modified to accept specific types)

```
A chat between a curious user and an artificial intelligence Assistant. The Assistant is an expert at identifying entities and relationships in text. The Assistant responds in JSON output only.

The User provides text in the format:

-------Text begin-------
<User provided text>
-------Text end-------

The Assistant follows the following steps before replying to the User:

1. **identify entities** The Assistant identifies all entities in the text of the types: <entity_types>. These entities are listed in the JSON output under the key "nodes", they follow the structure of a list of dictionaries where each dict is:

"nodes":[{"id": <entity N>, "type": <type>}, ...]

where "type": <type> is the type of the entity.

2. **determine relationships** The Assistant uses the text between -------Text begin------- and -------Text end------- to determine the relationships between the entities identified in the "nodes" list defined above. These relationships are called "edges" and they follow the structure of:

"edges":[{"from": <entity 1>, "to": <entity 2>, "label": <relationship>}, ...]

The <entity N> must correspond to the "id" of an entity in the "nodes" list and relationship must be one of the following types: <relation_types>.

The Assistant never repeats the same node twice. The Assistant never repeats the same edge twice.
The Assistant responds to the User in JSON only, according to the following JSON schema:
{
  "type":"object",
  "properties":{
    "nodes":{
      "type":"array",
      "items":{
        "type":"object",
        "properties":{
          "id":{
            "type":"string"
          },
          "type":{
            "type":"string"
          },
          "detailed_type":{
            "type":"string"
          }
        },
        "required":["id", "type", "detailed_type"],
        "additionalProperties":false
      }
    },
    "edges":{
      "type":"array",
      "items":{
        "type":"object",
        "properties":{
          "from":{
            "type":"string"
          },
          "to":{
            "type":"string"
          },
          "label":{
            "type":"string"
          }
        },
        "required":["from", "to", "label"],
        "additionalProperties":false
      }
    }
  },
  "required":["nodes", "edges"],
  "additionalProperties":false
}

Input:
-------Text begin-------
<text>
-------Text end-------
```

Note: The JSON in the Phi3 Mini Graph prompt is condensed to take up less characters, but formatted here for readability.

Gemini 1.5 Pro

```
Please extract a list of entities, and subsequently a list of relations between these entities.
The allowed entity types are: <entity_types>.
The allowed relation types are: <relation_types>.
The output should look like:
Entities:
Entity1 (EntityType)
Entity2 (EntityType)

Relationships:
Entity1 (EntityType) --RELATIONSHIP_TYPE-- Entity2 (EntityType)

(Examples:
Example 1:
<example_text>

Entities:
<example_entities>

Relationships:
<example_relationships>)

Do not provide any explanation or deviate from the format. If any entity does not conform to the entity types stated, they should not be included. Please now perform the task for the following text:
<text>
```

SciLitLLM 1.5

```
As a biomedical researcher, you are able to extract structured information from a given piece of text. Please extract a list of entities, and subsequently a list of relations between these entities.
The allowed entity types are: <entity_types>.
The allowed relation types are: <relation_types>.
The output should look like:
(entity1_name, entity1_type), (entity2_name, entity2_type), (entity1_name, RELATION, entity2_name), (entity3_name, RELATION, entity4_name), ...

(Examples:
Example 1:
<example_text>

Output:
<example_output>)

Do not provide any explanation or deviate from the format. If any entity does not conform to the entity types stated, they should not be included. Please now perform the task for the following text:
<text>
```

Additional information

Setup, implementation details, and code can be found at `https://github.com/FSGade/BiomedicalZeroShot`.

RadQA-DPO: A Radiology Question Answering System with Encoder-Decoder Models Enhanced by Direct Preference Optimization

Md Sultan Al Nahian and **Ramakanth Kavuluru**
Division of Biomedical Informatics, Department of Internal Medicine
University of Kentucky, Lexington, KY USA
{mna245,ramakanth.kavuluru}@uky.edu

Abstract

Extractive question answering over clinical text is a crucial need to help deal with the deluge of clinical text generated in hospitals. While encoder models (e.g., BERT) have been popular for this reading comprehension–style question answering task, recently encoder-decoder models (e.g., T5) are on the rise. There is also the emergence of preference optimization techniques to align decoder-only LLMs with human preferences. In this paper, we combine encoder-decoder models with the direct preference optimization (DPO) method for the RadQA radiology question answering task. Our approach achieves a 12–15 F1 point improvement over previous state-of-the-art models. To the best of our knowledge, this effort is the first to show that DPO method also works for reading comprehension via novel heuristics to generate preference data without human inputs.

1 Introduction

Clinical domain is rich in text data, such as progress notes, discharge summaries, and radiology/pathology reports, which constitutes a significant portion of electronic medical records (EMRs). These documents contain essential patient information but are often lengthy and idiosyncratic to specific clinicians, making it difficult and inefficient for doctors to manually extract specific details during care transfers or follow-ups (Jin et al., 2022). From a natural language processing (NLP) perspective, machine reading comprehension (MRC) systems can address this challenge by extracting precise answers to specific queries directly from these documents, facilitating more efficient decision-making for physicians (Demner-Fushman et al., 2009). In this paper, we achieve state-of-the-art results for a MRC task in radiology, with encoder-decoder language models (LMs) enhanced by direct preference optimization (DPO). Before we proceed, we first trace the origins of DPO since it was first introduced for a very different purpose than reading comprehension.

Since mid 2020, large language models (LLMs) have become pivotal in NLP, showcasing remarkable performance across a variety of tasks. These models undergo an initial phase of unsupervised pretraining, acquiring a comprehensive language representation that equips them with robust and contextual generation capabilities, which can then be transferred to specific downstream tasks through supervised fine-tuning (Dai and Le, 2015; Radford et al.; Devlin et al., 2019; Khandelwal et al., 2019). However, while supervised fine-tuning has been proven effective in enhancing model performance, it struggles to align models with human preferences (Stiennon et al., 2020). The high-quality output achieved through supervised fine-tuning often poorly correlates with human judgment, as the maximum likelihood objective struggles to capture the nuances of human preferences (Chaganty et al., 2018; Dusek et al., 2017). To address this challenge, reinforcement learning from human feedback (RLHF) has recently emerged as a promising approach for aligning LLMs with human preferences (Ziegler et al., 2019; Stiennon et al., 2020). RLHF utilizes human feedback on the model's output to guide its learning process, resulting in enhanced performance and better correlation with human judgment across diverse NLP tasks (Ouyang et al., 2022; Glaese et al., 2022; Bai et al., 2022a).

Ability to evaluate the output of LLMs based on human preferences is a core part of RLHF. To acquire this ability, the RLHF technique involves building a reward model from human annotated preference data. The objective of the reward model is to assess the output of the language model based on human preferences and represent it in a scalar value, which is used to optimize the language model using RL algorithms, most commonly proximal policy optimization (PPO) (Schulman et al., 2017). Usually the reward models are built by fine-

Proceedings of the 24th Workshop on Biomedical Language Processing (BioNLP 2025), pages 101–113
August 1, 2025 ©2025 Association for Computational Linguistics

tuning another LLM as it is expected that the reward model should have the similar language modeling capabilities to the original language model it is used to optimize. While RLHF demonstrates impressive performance across various NLP tasks (Chowdhery et al., 2023; Touvron et al., 2023), it is a complex and computationally expensive process that involves training multiple models, including a supervised fine-tuned model, a reward model, and the final RLHF model. To address this complexity, Rafailov et al. (Rafailov et al., 2024) introduced DPO, which directly learns human preferences from the preference dataset without requiring a reward model. By eliminating this step, DPO reduces computational costs while preserving the same optimization objectives as RLHF, making it a more efficient and dynamic alternative.

Thus far DPO has been primarily used to align decoder-only LLMs with human preferences; it has not been applied to encoder-decoder models used for the MRC task with a likelihood maximization objective. DPO inherently aims to increase the log probability of expected outputs over rejected outputs. A dataset of diverse instances of correct and incorrect output pairs can provide proper signals to the model about challenging examples that a supervised fine-tuned model struggles to predict accurately. Based on this observation, we hypothesize that DPO can be utilized to enhance the performance of a supervised fine-tuned encoder-decoder model in log-likelihood maximization. To test this, we experiment with a recent biomedical MRC dataset, Radiology Question Answering (RadQA) (Soni et al., 2022), resulting in the following contributions and findings:

- Compared with the encoder-only models used in prior efforts with RadQA, we show over 10% F-score improvement by shifting to encoder-decoder models, achieving a new state of the art (SoTA) score.

- We introduce two new methods to automatically generate paired preference data for the MRC task and use them to produce additional 1-3% F1 gains with DPO, leading to overall gains of 12–15% F1 points over SoTA.

The code and data from our experiments are available here: RadQADPO-code. If accepted, we will make them available on our lab's GitHub.

2 Related Work

2.1 Machine reading comprehension

MRC is a key research area within information extraction that focuses on enabling machines to extract answers from given texts. Specifically, an MRC model receives a passage (context) and a question as input and aims to answer the question by reasoning over both. Unlike general or open-domain question answering (QA) (Reddy et al., 2019; Karpukhin et al., 2020; Yasunaga et al., 2021), which typically involves retrieving answers from large corpora or knowledge bases, MRC operates in a more constrained setting where the relevant information is already provided in the input context. While MRC is important in and of itself, it also plays a crucial role in open ended QA where an initial retrieval model extracts relevant documents for a question from a search index. MRC is then applied to each of these documents and the answers are ranked using other heuristics. Prior efforts in deep learning for MRC focused on attention mechanisms, which helped models focus on relevant parts of the query and the context (Seo et al., 2016; Cui et al., 2017). More recently, approaches using transformer-based LMs, such as BERT (Devlin et al., 2019), RoBERTa (Liu et al., 2019) and XLNet (Yang et al., 2019) have demonstrated superior performance on this task. These models leverage large-scale pre-training on diverse datasets followed by fine-tuning on specific MRC tasks, enhancing their ability to generate accurate answers. For example, ForceReader (Chen and Wu, 2020) is a BERT based method that addressed the attention deconcentration problem in MRC and introduced a few novel ideas including *attention separate representation*, *multi-mode reading*, and *conditional background attention* to improve MRC. Similarly, Lu et al. (Luo et al., 2020) proposed a novel approach that leverages BERT and BiDAF (Seo et al., 2016), extending probability vectors to probability matrices to predict the start and end positions of the answer span more accurately.

More recently, transformer-based decoder-only large language models (LLMs) (Yang et al., 2022; Singhal et al., 2023; Wu et al., 2024) have demonstrated strong or even state-of-the-art performance on a variety of machine reading comprehension (MRC) benchmarks across both general and biomedical domains, largely due to their powerful generalization capabilities. These models are typically evaluated on generative and multiple-choice

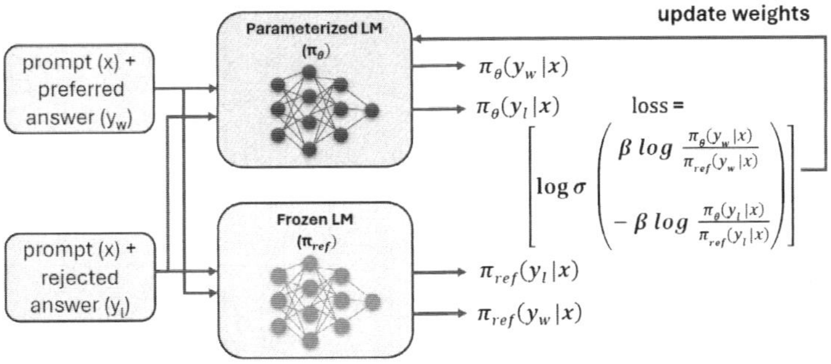

Figure 1: Pipeline of fine-tuning the language model using DPO. π_θ is the language model we want to fine-tune, and π_{ref} is the reference model, which is kept frozen during the fine-tuning process. Both models are initialized with the Supervised Fine-Tuned (SFT) model.

question-answering tasks that rely on given contexts, rather than on traditional span-based MRC tasks such as SQuAD (Rajpurkar et al., 2016) or RadQA, which require predicting exact answer spans within the context.

In our approach we also used transformer-based LMs. In contrast to the previously discussed methods, we have used an encoder-decoder transformer model (Raffel et al., 2020) as the base model and fine-tuned it by adopting the DPO method. Thus, the most closely related work to ours involves RL-based MRC methods. Although this domain is less explored compared to other deep learning approaches discussed above, several studies have applied RL techniques in question answering systems (Hu et al., 2018; Lee et al., 2021; Gharagozlou et al., 2022). These approaches typically design a reward function to optimize the model using RL algorithms. However, by leveraging the DPO technique in our method, we obviate the need of a reward function for training the model.

2.2 Reinforcement learning from human feedback (RLHF)

RLHF is an RL technique that optimizes models using human feedback instead of predefined reward functions. Initially explored for training RL agents (Akrour et al., 2012) where reward functions are difficult to specify, RLHF has more recently been widely used to fine-tune LLMs to better align with human preferences. This method has been successfully applied in various NLP tasks, including conversational agents (OpenAI, 2022), text and dialogue summarization (Chen et al., 2023), question-answering (Nakano et al., 2021), and recommendation systems, where aligning the responses with human judgment is crucial. However, RLHF is a multi-step process that can be computationally intensive. Direct Preference Optimization (DPO) (Rafailov et al., 2024) has emerged as a more efficient alternative, aiming to achieve similar objectives with reduced computational costs. While DPO is primarily used to align language models with human judgment (Tunstall et al., 2023; Zhao et al., 2023), we explore its application in likelihood maximization for MRC. By applying DPO to enhance supervised fine-tuned models, we aim to improve performance by optimizing responses to match ground truth answers more closely.

3 Methods

We use the encoder-decoder model T5 (Raffel et al., 2020) as the backbone of our main method as opposed to the BERT based baselines reported earlier (Soni et al., 2022). We also experimented with the Flan-T5 model (Longpre et al., 2023) which have been *instruction tuned* on a variety of NLP datasets and tasks. Our DPO-based optimization consists of two steps: (1) training a supervised fine-tuned T5 model and (2) optimizing it using DPO.

3.1 Training supervised fine-tuned (SFT) model

In this step, we trained an initial model for MRC using the supervised fine-tuning approach with the original training data, which we refer to as the SFT model. We model MRC as a text to text task and opted to use a seq-2-seq model for training the SFT model. The model's input is the tokenized vectors of the concatenated context and question and the output is the answer span from the context or "no answer" if the answer is not available in the context. We formatted the input sequence before tokenization as follows: "context: the text of the context <SEP>question: text of the question."

3.2 Optimizing using DPO

After training the SFT model, we further fine-tuned it using the DPO method. This requires a preference dataset consisting of tuples (x, y_w, y_l), where x is a prompt and y_w and y_l are the preferred and rejected responses for the prompt x, respectively. In standard RLHF/DPO techniques, the preference dataset is usually constructed using human annotators. For each input, multiple outputs are generated by the initial SFT model and human annotators are asked to rate them as preferred or rejected outputs. In contrast to the standard DPO, here we built the preference dataset automatically without human interventions. Our approaches to create the preference dataset are discussed in Section 4.2.

After generating the preference dataset, we applied DPO to optimize the SFT models. The DPO architecture employs two models simultaneously for fine-tuning: one is the reference model (π_{ref}), while the other is the active model, π_θ, which is being optimized. Both models are initialized with the SFT model trained in the previous step. The weights of the reference model (π_{ref}) are kept frozen throughout the training process, while the weights of the model π_θ are updated using the DPO loss (Eq. (4) of Appendix A.1). The reference model ensures that fine-tuning does not cause the policy of the model π_θ to deviate significantly from the initial SFT model. While the DPO loss aims to increase the difference between the policies for the preferred and rejected outputs, it also aims to minimize the difference between the policies of the SFT and the active model π_θ. Both models receive input in the form of the tuple (x, y_w, y_l). In our study, the prompt x consists of the concatenated string of the context and question, y_w is the correct answer span and y_l corresponds to one of the incorrect answers for the question, given the context. Given the prompt, both models provide the probability distribution of the tokens of the preferred and rejected answers, which are used to compute the loss and update the weights of the active model π_θ. Figure 1 depicts the process of DPO more elaborately.

4 Datasets

We need two datasets to build the models in the two phases of our method. The first is the original RadQA dataset, which was used for training and validating the SFT model. The second is a preference dataset created from RadQA, and used for further tuning of the SFT model via DPO.

Preference Dataset	F1 Threshold		
	0.9	0.7	0.5
Model-based-T5	3280	2865	2354
Model-based-Flan-T5	3089	2533	2036
Rule-based	3716	3501	3332

Table 1: #instances in the preference dataset created by each method applying different F1 threshold values.

4.1 RadQA

RadQA(Soni et al., 2022) is an MRC dataset created from radiology reports from the MIMIC III dataset (Johnson et al., 2016). The questions were manually created from the clinical referral sections to capture the actual information needs of ordering physicians, without being influenced by seeing the answer context. Answers were annotated in the Findings and Impressions sections and consist of complete, concise phrases that may span multiple lines and are not limited to named entities. The dataset also includes unanswerable questions, supporting the challenges of real-world clinical question answering.

The RadQA dataset comprises 6148 unique question-answer pairs sourced from 1009 radiology reports of 100 patients. The dataset was split at the patient level into training, development, and testing sets, with an 8:1:1 ratio, respectively. This resulted in 4878 questions in the training set, 863 questions in the development set, and 894 questions in the test set. We used the original format of training data of RadQA exclusively to train the SFT model, while the development and test data were used for evaluating both the SFT and DPO models to assess the effectiveness of our approach.

4.2 Preference dataset

Preference data is the main element for optimizing a language model through DPO. This consists of tuples that include examples of preferred and rejected outputs for a given prompt. Although preference data is typically collected from human annotators, we automatically generated it, eliminating the need for manual annotation. We used the original training corpus of RadQA for this purpose. Specifically, each prompt was formed by concatenating the context and question from the RadQA training dataset, separated by a special token. The preferred output is the original gold answer span provided in the dataset. To generate the corresponding rejected output, we propose two automated approaches: a model-based approach and a rule-based approach.

4.2.1 Model based approach

In this approach, we used the SFT model itself to generate negative examples. The process began by training a model on 50% of the RadQA training data and then using it to predict answers for the entire training dataset, including the data it was trained on. The rationale behind training on half of the data was to equip the model with sufficient knowledge for effective performance. Thus, mistakes made during these predictions indicate the types of examples the model needs to focus on to improve its performance. Testing the model on both seen and unseen data helps identifying specific examples that remain challenging despite prior exposure. Our intuition behind this design is that by using the model's own incorrect predictions, we can better identify the types of examples where it struggles. These incorrect predictions highlight situations where the model needs improvement, making them valuable for training. Additionally, since the model is also tested on examples it was trained on, any errors it makes on these familiar examples indicate that they are particularly challenging. By focusing on these hard examples, we aim to improve the model's overall performance.

We identified all instances where the model generated incorrect answers. For each prompt and question pair where the model's prediction differed from the original answer, the incorrect prediction was recorded as the rejected output in our preference dataset. To refine the preference dataset, we filtered these incorrect answers based on their F1 scores. The F1 score was calculated by comparing word-level matches between each incorrect answer and its corresponding original answer. To filter the incorrect predictions, we applied three different thresholds for the F1 score: 0.9, 0.7, and 0.5. If the F1 score between the original and the predicted answer was less than the chosen threshold, the predicted answer was selected as the rejected output. To ensure comprehensive coverage, we repeated this process by training another model on the remaining 50% of the training data. This model was then used again to predict answers for the entire dataset, allowing us to identify additional incorrect predictions. We used two variants of SFT models (T5-3B and Flan-T5-3B) to create the negative examples. The total number of instances created by this process is shown in Table 1.

By iteratively training on different halves of the dataset and collecting incorrect predictions, we effectively created a robust set of negative examples without the need for manual annotation. This automated generation of preference data not only streamlined our process but also ensured a diverse range of negative examples, enhancing the quality of our preference dataset. Our assumption is that DPO will help the model improve on these challenging examples, enhancing overall performance.

4.2.2 Rule based approach

We generated negative examples from the training data by applying a set of predefined rules. These rules were formulated based on experimental findings regarding the types of errors that SFT model typically makes. For each tuple (context, question, gold answer) in the training data, we generated a number of incorrect answers applying the following rules (also shown with a few examples in Figure 3 of Appendix A.3):

- *Random text span:* Select a random span from the context that does not contain any part of the gold answer.
- *Text span containing part of the gold answer:* Here, a text span from the context that includes a part of the original answer is randomly chosen. This partial inclusion can occur in two ways: 1) choosing a segment starting a few words before the left side of the gold answer and continuing until it includes a partial span from the gold answer, or 2) selecting a partial segment from the right side of the answer and including a few words after the answer text. The lengths of these segments are chosen randomly (see Figure 3).
- *Longer answer:* This entails a text span that includes the entire gold answer as a part of it with ≥ 1 additional tokens.
- *Partial answer only:* This involves selecting a smaller segment (strict substring) from the original answer.
- *Answers of a different question:* Here, an answer text from another question in the same context is chosen, provided it is not the same as the original gold answer or a part of it. For

example in Figure 3 of Appendix A.3, "kidneys are normal in appearance" is an answer to a different question for the same context, but is not part of the ground truth answer.

- *No answer:* In this approach, we used empty string in place of the gold answers to create negative examples. For questions without available answers, we chose responses from other questions within the same context as negative examples. If there were no other questions within the same context that provided answers, we randomly selected a span from the context as the negative answer.

Following these rules provided us with a large number of examples of rejected answers for each (context, question, gold answer) tuple. From each set of rejected answers, we randomly chose a few examples to create the preference data. We did not include the entire set of rejected answers for generating the preference data to prevent the dataset from becoming intractably large. Finally, we included 4000 instances and further filtered them by applying F1 threshold (see Table 1).

5 Experimental Setup

5.1 Baselines

We compared our T5-based SFT models with the BERT-based models from Soni et al. (Soni et al., 2022), which offered SoTA results on the RadQA dataset. Thus, we selected all of their BERT-MIMIC-based models as our baselines. These models come in four variants, based on the datasets used for fine-tuning. The first variant, BERT-MIMIC-RadQA, was fine-tuned only on the RadQA dataset. The remaining three variants were additionally fine-tuned on external QA datasets such as SQuAD (Rajpurkar et al., 2016) and EmrQA (Pampari et al., 2018). For example, BERT-MIMIC-SQuAD-RadQA was trained on both RadQA and SQuAD, while BERT-MIMIC-EmrQA-RadQA was trained on both EmrQA and RadQA.

We also compared our DPO-based method with the T5 SFT models to assess the effectiveness of applying DPO on an already high-performing fine-tuned model.

5.2 Evaluation metrics

To evaluate our proposed method, we used the standard MRC metrics: Exact match (EM) and F1-Score. Exact Match is a strict metric that compares the predicted answer with the exact ground truth answer, ensuring they are identical. The F1-Score, on the other hand, is calculated by taking word-level matches between the predicted and ground truth answers. To maintain consistency and comparability in our evaluation, we used the evaluation code from SQuAD (Rajpurkar et al., 2016).

5.3 Network parameters and resources

The network parameters for each model in our experiments were chosen through hyperparameter tuning. We used the validation F1 score as an evaluation metric to select the optimal values of these parameters. For training both the SFT and DPO models, we employed the Adam optimizer. The learning rate for the SFT model was set to $5e^{-5}$, and for the DPO model, it was $5e^{-7}$. The weight decay was set to 0.01 for both models. The batch size was 16 for T5-Large models; however, to accommodate the 3 billion parameter models in memory, we used a batch size of 2 with gradient accumulation steps of 8. The maximum prompt length was set to 768, and the target length was 128. Early stopping was applied during the training of both the SFT and DPO models, by using the validation F1 score to select the best models. All our experiments were conducted on a single NVIDIA H100 GPU, equipped with 80 GB of memory.

6 Results

Table 2 presents the main results of our experiments, comparing the performance of BERT baselines, the T5-based supervised fine-tuned (SFT) models, and the DPO based models. The results are evaluated on the development and test sets of the RadQA dataset.

The SFT model type includes three T5 variants (T5-large, T5-3B, and Flan-T5-3B) trained on the RadQA training data. From Table 2, we can see that all the T5 variants outperform the baseline RadQA models on the test set, with Flan-T5-3B also performing better on the dev set. Specifically, the SFT Flan-T5-3B achieves an F1 score of 76.38 and an exact match (EM) score of 55.93 on the test set, marking improvements of 13 points in F1 score and 6.5 points in EM over the best baseline model. Although the three variants of BERT-MIMIC were trained on additional datasets (SQuAD and emrQA) along with RadQA, the T5 models still outperformed them, establishing a strong baseline for our DPO-based method. It is important to note

Model Type	Models	Dev		Test	
		EM	F1	EM	F1
Baseline (BERT-MIMIC) (340M)	RadQA	48.05	65.85	45.73	60.08
	emrQA-RadQA	50.65	67.97	47.71	61.60
	SQuAD-RadQA	52.28	69.42	**49.39**	**63.55**
	SQuAD-emrQA-RadQA	53.26	67.79	48.32	62.29
T5-large (770M)	SFT	47.86	66.22	49.89	71.10
	DPO-MB	47.74	66.25	**51.34**	**71.62**
	DPO-RB	48.20	66.59	51.00	71.36
	DPO-MRB	47.80	66.10	50.11	71.20
T5-3B	SFT	49.83	68.59	51.68	72.29
	DPO-MB	51.10	70.45	52.46	74.29
	DPO-RB	50.87	70.26	**52.57**	74.03
	DPO-MRB	50.40	70.13	52.01	**75.18**
Flan-T5-3B	SFT	54.35	72.62	55.93	76.38
	DPO-MB	53.77	73.68	55.15	**77.48**
	DPO-RB	52.49	72.55	**56.15**	77.40
	DPO-MRB	53.42	73.51	55.70	77.41

Table 2: Model performances on the RadQA development and test sets compared with the RadQA BERT-MIMIC model variants.

that, although BERT-MIMIC was fine-tuned on a large corpus of clinical notes (Si et al., 2019) (1.9 million notes comprising approximately 786 million tokens), our T5 models have more parameters than the 340M BERT-based models used in the RadQA paper and were pretrained on a much larger and more diverse dataset—the C4 corpus, which contains around 750GB of clean web text. This provides T5 with stronger language capabilities.

The DPO-based methods include three groups of models: DPO-Model Based (DPO-MB), trained on model-based preference data; DPO-Rule Based (DPO-RB), trained on rule-based preference data; and DPO-Model & Rule Based (DPO-MRB), trained on a combined dataset of model-based and rule-based preference data. For all the models, we selected the preference data generated by 0.9 F1 threshold. Additionally, for training the DPO-MB models, we used the model specific preference data. For instance, we applied model-based-T5 preference data for the T5 models and model-based-Flan-T5 preference data for the Flan-T5 based DPO models. From Table 2 we can see that both model and rule-based DPO models improved the performance of the corresponding SFT models. Although the T5-large SFT model did not see a significant improvement, the T5-3B and Flan-T5-3B improved their corresponding SFT models nontrivially, both in DPO-MB and DPO-RB settings. For instance, the F1 score of the DPO-MB T5-3B is 74.29, a 2-point improvement over its SFT counterpart and an 11-point increase compared to the best performing baseline model, BERT-MIMIC-SQuAD-RadQA, on the test F1 score. The combined dataset further improved the test F1 score of the T5-3B model by 1%, but it did not enhance the other variants, indicating saturation in the performance of the models.

7 Discussion

Our experimental results demonstrate that further fine-tuning an SFT model through DPO can enhance its performance between 1–3% F1 points. This is particularly important because these SFT models have already been optimized using the full training dataset, making further improvements challenging. From our experiments, we found several factors that influence the performance of the models trained with DPO, including the size of the SFT models, the method used to create negative examples in the preference data, the types of examples included, and the quantity of preference data. In this section, we provide a detailed discussion on the observed performance improvements using DPO and the factors influencing these improvements.

7.1 Size of the model

From our results, we notice that a smaller model is less likely to benefit from additional fine-tuning with DPO. However, with larger models, notable improvements were observed. For instance, with DPO both T5-3B and Flan-T5-3B increased the test F1 score of their corresponding SFT models by 1-3%. This indicates the ability of larger encoder-decoder models to capture signals from examples of preferred and rejected outputs. However, among 3B models, the improvement is much better in the non-Flan model. Since the Flan model is instruction tuned on hundreds of datasets, its SFT performance (76.38 F1) is already over 1% better than the best DPO model of its non-Flan counterpart.

7.2 Model- vs rule-based preference data

While DPO-MB and DPO-RB both enhanced the performance of the SFT models, our experiments showed that the model-based approach yielded comparatively better results than the rule-based approach. One potential reason for this could be the nature of the negative examples generated by each method. Rule-based examples are created using predefined rules. Although these rules are designed to generate plausible negative examples, they may not always reflect the same distribution as the original RadQA dataset. This can lead to less effective training, as the model might not encounter a representative range of challenging examples during the DPO training. In contrast, the model-based approach derives negative examples from the predictions of the SFT model itself. These examples are intrinsically linked to the specific weaknesses of the model. By focusing on these model-specific errors, the preference data reflects the instance spaces where the model is prone to generate incorrect outputs. Consequently, this approach may offer more targeted training, enabling the model to learn from its mistakes and improve its performance. However, one limitation of this method is that each new model requires the creation of a new preference dataset, as each model has different weaknesses and strengths. In contrast, the training examples created by the rule-based approach are model-agnostic.

Figure 2: Performance comparison of DPO-T5-3B model with varying training examples and preference datasets generated using different thresholds. X-axis plots #training-examples, Y-axis is the F1 score, and the line colors represent different preference datasets created by applying three different F1 thresholds.

7.3 Diversity of training instances

Filtering the preference data based on different F1-score thresholds also influences the performance of DPO. Negative examples with higher F1 scores tend to be closer to the ground truth answers, while those with lower F1 scores present more dissimilarity with gold spans. Incorporating a broader range of negative examples from both ends of the F1-score spectrum provides a diverse and more informative training set for the model. A higher F1-score threshold facilitates a mix of examples that are both similar and dissimilar to the ground truth answers, offering a wide variety of training data. Conversely, a lower threshold focuses only on the examples that are very different from the ground truth, excluding those that are more similar. Therefore, preference data created using higher thresholds may enable the model to learn from a diverse set of examples, which can enhance its generalization and performance. Our experiments also support this hypothesis. Figure 2 illustrates the test F1-scores of DPO-T5-3B models trained with preference data filtered at different thresholds. The results show that the model trained with a threshold of 0.9 outperforms those trained with lower threshold data, demonstrating the benefits of using a more diverse set of training examples.

7.4 Number of training Instances

Besides diversity of training examples, the number of training examples also impacts the performance of DPO based models. We fine-tuned DPO-T5-3B with different numbers of training examples (500,

1000, 1500 and 2000) for each filtering threshold. As shown in Figure 2, an increase in the number of training examples generally leads to an increase in the test F1 score across all thresholds.

7.5 Other variants of DPO

DPO has evolved into several variants, each with different a loss function, designed to address specific issues. For instance, Identity Preference Optimization (IPO) (Azar et al., 2024) was developed to mitigate the overfitting problem identified in DPO by introducing a new loss function. We trained our model using three DPO variants: Identity Preference Optimization (IPO), Kahneman-Tversky Optimization (KTO) (Ethayarajh et al., 2024), and Statistical Rejection Sampling Optimization (RSO) (Liu et al., 2024). Our experimental results show that DPO outperforms other variants for both T5-3b and Flan-T5-3b models. Detailed results are provided in Table 3 of Appendix A.2.

8 Conclusion

In this paper, we proposed an approach that combines encoder-decoder models with DPO based optimization to achieve new SoTA performances on the MRC task for radiology using the RadQA dataset. Our study shows that encoder-decoder models, although computationally expensive due to large model capacities, can offer substantial gains in performance (by over 10% in F1 scores). Originally introduced for aligning LLMs with human preferences, our study demonstrated that DPO methods can also be effectively used for likelihood maximization for MRC tasks and can lead to further gains of up to 3% beyond the encoder-decoder based gains. By focusing on challenging examples (the model-based preference data setup), DPO can further improve large models already fully trained.

While effective, one key challenge in fine-tuning models using DPO is that its performance is highly dependent on the quality of the preference data. Collecting high-quality examples of preferred and rejected outputs is crucial for maximizing the model's performance through DPO. In this work, we introduced two techniques—the model-based and the rule-based approaches to generate preference data for the MRC task, which can be adopted in other tasks as well. In future, we will explore the applicability of our approach in other information extraction tasks such as named entity recognition and relation extraction.

Acknowledgment

This work is supported by the U.S. National Library of Medicine through grant R01LM013240. The content is solely the responsibility of the authors and does not necessarily represent the official views of the U.S. National Institutes of Health.

References

Riad Akrour, Marc Schoenauer, and Michèle Sebag. 2012. April: Active preference learning-based reinforcement learning. In *Machine Learning and Knowledge Discovery in Databases: European Conference, ECML PKDD 2012. Proceedings, Part II 23*, pages 116–131.

Mohammad Gheshlaghi Azar, Zhaohan Daniel Guo, Bilal Piot, Remi Munos, Mark Rowland, Michal Valko, and Daniele Calandriello. 2024. A general theoretical paradigm to understand learning from human preferences. In *International Conference on Artificial Intelligence and Statistics*, pages 4447–4455.

Yuntao Bai, Andy Jones, Kamal Ndousse, Amanda Askell, Anna Chen, Nova DasSarma, Dawn Drain, Stanislav Fort, Deep Ganguli, Tom Henighan, et al. 2022a. Training a helpful and harmless assistant with reinforcement learning from human feedback. *arXiv preprint arXiv:2204.05862*.

Yuntao Bai, Saurav Kadavath, Sandipan Kundu, Amanda Askell, Jackson Kernion, Andy Jones, Anna Chen, Anna Goldie, Azalia Mirhoseini, Cameron McKinnon, et al. 2022b. Constitutional ai: Harmlessness from ai feedback. *arXiv preprint arXiv:2212.08073*.

Ralph Allan Bradley and Milton E. Terry. 1952. Rank analysis of incomplete block designs: I. the method of paired comparisons. *Biometrika*, 39:324.

Arun Chaganty, Stephen Mussmann, and Percy Liang. 2018. The price of debiasing automatic metrics in natural language evalaution. In *Proceedings of the 56th Annual Meeting of the ACL*, pages 643–653.

Jiaao Chen, Mohan Dodda, and Diyi Yang. 2023. Human-in-the-loop abstractive dialogue summarization. In *Findings of the ACL: ACL 2023*, pages 9176–9190.

Zheng Chen and Kangjian Wu. 2020. ForceReader: a BERT-based interactive machine reading comprehension model with attention separation. In *Proceedings of the 28th International Conference on Computational Linguistics*, pages 2676–2686.

Aakanksha Chowdhery, Sharan Narang, Jacob Devlin, Maarten Bosma, Gaurav Mishra, Adam Roberts, Paul Barham, Hyung Won Chung, Charles Sutton, Sebastian Gehrmann, et al. 2023. Palm: Scaling language modeling with pathways. *Journal of Machine Learning Research*, 24(240):1–113.

Yiming Cui, Zhipeng Chen, Si Wei, Shijin Wang, Ting Liu, and Guoping Hu. 2017. Attention-over-attention neural networks for reading comprehension. In *Proceedings of the 55th Annual Meeting of the ACL*, pages 593–602.

Andrew M Dai and Quoc V Le. 2015. Semi-supervised sequence learning. In *Advances in Neural Information Processing Systems*, volume 28.

Dina Demner-Fushman, Wendy W. Chapman, and Clement J. McDonald. 2009. What can natural language processing do for clinical decision support? *Journal of Biomedical Informatics*, 42(5):760–772.

Jacob Devlin, Ming-Wei Chang, Kenton Lee, and Kristina Toutanova. 2019. BERT: Pre-training of deep bidirectional transformers for language understanding. In *Proceedings of the 2019 Conference of the North American Chapter of the ACL: Human Language Technologies, Volume 1*, pages 4171–4186.

Ondrej Dusek, Jekaterina Novikova, and Verena Rieser. 2017. Referenceless quality estimation for natural language generation. In *1st Workshop on Learning to Generate Natural Language*.

Kawin Ethayarajh, Winnie Xu, Niklas Muennighoff, Dan Jurafsky, and Douwe Kiela. 2024. Kto: Model alignment as prospect theoretic optimization. *Preprint*, arXiv:2402.01306.

Hamid Gharagozlou, Javad Mohammadzadeh, Azam Bastanfard, Saeed Shiry Ghidary, et al. 2022. Rlas-biabc: A reinforcement learning-based answer selection using the bert model boosted by an improved abc algorithm. *Computational Intelligence and Neuroscience*, 2022.

Amelia Glaese, Nat McAleese, Maja Trebacz, John Aslanides, Vlad Firoiu, Timo Ewalds, Maribeth Rauh, Laura Weidinger, Martin Chadwick, Phoebe Thacker, et al. 2022. Improving alignment of dialogue agents via targeted human judgements. *arXiv preprint arXiv:2209.14375*.

Minghao Hu, Yuxing Peng, Zhen Huang, Xipeng Qiu, Furu Wei, and Ming Zhou. 2018. Reinforced mnemonic reader for machine reading comprehension. In *Proceedings of the 27th International Joint Conference on Artificial Intelligence*, pages 4099–4106.

Qiao Jin, Zheng Yuan, Guangzhi Xiong, Qianlan Yu, Huaiyuan Ying, Chuanqi Tan, Mosha Chen, Songfang Huang, Xiaozhong Liu, and Sheng Yu. 2022. Biomedical question answering: A survey of approaches and challenges. *ACM Comput. Surv.*, 55(2).

Alistair EW Johnson, Tom J Pollard, Lu Shen, Li-wei H Lehman, Mengling Feng, Mohammad Ghassemi, Benjamin Moody, Peter Szolovits, Leo Anthony Celi, and Roger G Mark. 2016. Mimic-iii, a freely accessible critical care database. *Scientific data*, 3(1):1–9.

Vladimir Karpukhin, Barlas Oguz, Sewon Min, Patrick Lewis, Ledell Wu, Sergey Edunov, Danqi Chen, and Wen-tau Yih. 2020. Dense passage retrieval for open-domain question answering. In *Proceedings of the 2020 Conference on Empirical Methods in Natural Language Processing (EMNLP)*, pages 6769–6781.

Urvashi Khandelwal, Kevin Clark, Dan Jurafsky, and Lukasz Kaiser. 2019. Sample efficient text summarization using a single pre-trained transformer. *arXiv preprint arXiv:1905.08836*.

Hyeon-Gu Lee, Youngjin Jang, and Harksoo Kim. 2021. Machine reading comprehension framework based on self-training for domain adaptation. *IEEE Access*, 9:21279–21285.

Tianqi Liu, Yao Zhao, Rishabh Joshi, Misha Khalman, Mohammad Saleh, Peter J Liu, and Jialu Liu. 2024. Statistical rejection sampling improves preference optimization. In *The Twelfth International Conf. on Learning Representations*.

Yinhan Liu, Myle Ott, Naman Goyal, Jingfei Du, Mandar Joshi, Danqi Chen, Omer Levy, Mike Lewis, Luke Zettlemoyer, and Veselin Stoyanov. 2019. Roberta: A robustly optimized bert pretraining approach. *arXiv preprint arXiv:1907.11692*.

Shayne Longpre, Le Hou, Tu Vu, Albert Webson, Hyung Won Chung, Yi Tay, Denny Zhou, Quoc V Le, Barret Zoph, Jason Wei, et al. 2023. The flan collection: Designing data and methods for effective instruction tuning. In *International Conference on Machine Learning*, pages 22631–22648.

Huaishao Luo, Yu Shi, Ming Gong, Linjun Shou, and Tianrui Li. 2020. MaP: A matrix-based prediction approach to improve span extraction in machine reading comprehension. In *Proceedings of the 1st Conf. of the Asia-Pacific Chapter of the ACL and the 10th International Joint Conf. on NLP*, pages 687–695.

Reiichiro Nakano, Jacob Hilton, Suchir Balaji, Jeff Wu, Long Ouyang, Christina Kim, Christopher Hesse, Shantanu Jain, Vineet Kosaraju, William Saunders, et al. 2021. WebGPT: Browser-assisted question-answering with human feedback. *arXiv preprint arXiv:2112.09332*.

OpenAI. 2022. ChatGPT: Optimizing language models for dialogue.

Long Ouyang, Jeffrey Wu, Xu Jiang, Diogo Almeida, Carroll Wainwright, Pamela Mishkin, Chong Zhang, Sandhini Agarwal, Katarina Slama, Alex Ray, et al. 2022. Training language models to follow instructions with human feedback. *Advances in neural information processing systems*, 35:27730–27744.

Anusri Pampari, Preethi Raghavan, Jennifer Liang, and Jian Peng. 2018. emrQA: A large corpus for question answering on electronic medical records. In *Proceedings of the 2018 Conference on Empirical Methods in Natural Language Processing*, pages 2357–2368.

Alec Radford, Karthik Narasimhan, Tim Salimans, and Ilya Sutskever. Improving language understanding by generative pre-training.

Rafael Rafailov, Archit Sharma, Eric Mitchell, Christopher D Manning, Stefano Ermon, and Chelsea Finn. 2024. Direct preference optimization: Your language model is secretly a reward model. *Advances in Neural Information Processing Systems*, 36.

Colin Raffel, Noam Shazeer, Adam Roberts, Katherine Lee, Sharan Narang, Michael Matena, Yanqi Zhou, Wei Li, and Peter J Liu. 2020. Exploring the limits of transfer learning with a unified text-to-text transformer. *Journal of machine learning research*, 21(140):1–67.

Pranav Rajpurkar, Jian Zhang, Konstantin Lopyrev, and Percy Liang. 2016. Squad: 100,000+ questions for machine comprehension of text. In *Proceedings of the 2016 Conference on Empirical Methods in Natural Language Processing*, pages 2383–2392.

Rajkumar Ramamurthy, Prithviraj Ammanabrolu, Kianté Brantley, Jack Hessel, Rafet Sifa, Christian Bauckhage, Hannaneh Hajishirzi, and Yejin Choi. 2023. Is reinforcement learning (not) for natural language processing: Benchmarks, baselines, and building blocks for natural language policy optimization. In *International Conference on Learning Representations 2023*.

Siva Reddy, Danqi Chen, and Christopher D Manning. 2019. Coqa: A conversational question answering challenge. *Transactions of the Association for Computational Linguistics*, 7:249–266.

John Schulman, Filip Wolski, Prafulla Dhariwal, Alec Radford, and Oleg Klimov. 2017. Proximal policy optimization algorithms. *CoRR*, abs/1707.06347.

Minjoon Seo, Aniruddha Kembhavi, Ali Farhadi, and Hannaneh Hajishirzi. 2016. Bidirectional attention flow for machine comprehension. In *International Conference on Learning Representations*.

Yuqi Si, Jingqi Wang, Hua Xu, and Kirk Roberts. 2019. Enhancing clinical concept extraction with contextual embeddings. *Journal of the American Medical Informatics Association*, 26(11):1297–1304.

Karan Singhal, Shekoofeh Azizi, Tao Tu, S Sara Mahdavi, Jason Wei, Hyung Won Chung, Nathan Scales, Ajay Tanwani, Heather Cole-Lewis, Stephen Pfohl, et al. 2023. Large language models encode clinical knowledge. *Nature*, 620(7972):172–180.

Sarvesh Soni, Meghana Gudala, Atieh Pajouhi, and Kirk Roberts. 2022. RadQA: A question answering dataset to improve comprehension of radiology reports. In *Proceedings of the 13th Language Resources and Evaluation Conf.*, pages 6250–6259.

Nisan Stiennon, Long Ouyang, Jeff Wu, Daniel M. Ziegler, Ryan Lowe, Chelsea Voss, Alec Radford, Dario Amodei, and Paul Christiano. 2020. Learning to summarize from human feedback. In *Proceedings of the 34th International Conference on Neural Information Processing Systems*, NIPS '20.

Hugo Touvron, Thibaut Lavril, Gautier Izacard, Xavier Martinet, Marie-Anne Lachaux, Timothée Lacroix, Baptiste Rozière, Naman Goyal, Eric Hambro, Faisal Azhar, et al. 2023. Llama: Open and efficient foundation language models. *arXiv preprint arXiv:2302.13971*.

Lewis Tunstall, Edward Beeching, Nathan Lambert, Nazneen Rajani, Kashif Rasul, Younes Belkada, Shengyi Huang, Leandro von Werra, Clémentine Fourrier, Nathan Habib, et al. 2023. Zephyr: Direct distillation of lm alignment. *arXiv preprint arXiv:2310.16944*.

Chaoyi Wu, Weixiong Lin, Xiaoman Zhang, Ya Zhang, Weidi Xie, and Yanfeng Wang. 2024. Pmc-llama: toward building open-source language models for medicine. *Journal of the American Medical Informatics Association*, 31(9):1833–1843.

Xi Yang, Aokun Chen, Nima PourNejatian, Hoo Chang Shin, Kaleb E Smith, Christopher Parisien, Colin Compas, Cheryl Martin, Anthony B Costa, Mona G Flores, et al. 2022. A large language model for electronic health records. *NPJ digital medicine*, 5(1):194.

Zhilin Yang, Zihang Dai, Yiming Yang, Jaime Carbonell, Russ R Salakhutdinov, and Quoc V Le. 2019. Xlnet: Generalized autoregressive pretraining for language understanding. *Advances in neural information processing systems*, 32.

Michihiro Yasunaga, Hongyu Ren, Antoine Bosselut, Percy Liang, and Jure Leskovec. 2021. Qa-gnn: Reasoning with language models and knowledge graphs for question answering. In *Proceedings of the 2021 Conference of the North American Chapter of the Association for Computational Linguistics: Human Language Technologies*, pages 535–546.

Zhiyuan Zhao, Bin Wang, Linke Ouyang, Xiaoyi Dong, Jiaqi Wang, and Conghui He. 2023. Beyond hallucinations: Enhancing lvlms through hallucination-aware direct preference optimization. *arXiv preprint arXiv:2311.16839*.

Daniel M Ziegler, Nisan Stiennon, Jeffrey Wu, Tom B Brown, Alec Radford, Dario Amodei, Paul Christiano, and Geoffrey Irving. 2019. Fine-tuning language models from human preferences. *arXiv preprint arXiv:1909.08593*.

A Appendix

A.1 Background for RLHF and DPO

Fine-tuning LLMs for downstream tasks using RLHF technique involves three main phases (Stiennon et al., 2020; Bai et al., 2022b): 1. supervised fine-tuning, 2. constructing reward model, and 3. fine-tuning the language model using RL methods.

A.1.1 Supervised fine-tuning

This is the initial step of RLHF technique, where the language model undergoes supervised fine-tuning on downstream tasks. During this phase, the model is trained on specific task-related training datasets, allowing it to adapt its pre-trained knowledge to the particular downstream task. The model trained in this phase is commonly referred to as supervised fine-tuning (SFT) model, denoted as π_{sft}.

A.1.2 Constructing reward model

After training the SFT model, the next step is to develop a reward model that evaluates the SFT model's outputs based on human preferences and represent it as scalar values. This reward model can be built using pre-trained models capable of assessing outputs according to human judgment (Bai et al., 2022b), or by training it on human preference data collected from annotators.

To construct human preference data, multiple responses are first generated for each prompt by the SFT model, using different variants of the model or sampling methods (Stiennon et al., 2020; Bai et al., 2022a). The collection of prompts and their generated responses are then formatted into a batch of tuples $(x, y1, y2)$, where x is the prompt and y_1 and y_2 are pair of responses sampled from the set of generated responses of the prompt x. Human labelers are then instructed to choose their preferred response between the two. This process creates a preference dataset consisting of tuples (x, y_w, y_l), where y_w represents the preferred output and y_l represents the rejected output.

From the generated preference dataset D, the probability distribution of human preference can be formulated as

$$p(y_w > y_l|x) = \sigma(r(x, y_w) - r(x, y_l)) \tag{1}$$

using Bradley-Terry model (Bradley and Terry, 1952) given an optimal reward model r, where σ is the logistic function.

With the preference dataset $D = \{(x^i, y_w^i, y_l^i)\}_{i=1}^N$, we parameterize the reward model r_σ and optimize it by maximizing the log likelihood of the difference between the reward of preferred response and rejected response (as in Eq. (1)) and hence minimize the loss

$$\mathcal{L}(r_\sigma) = E_{(x,y_w,y_l)\sim D}[-\log(p(y_w > y_l|x))]. \tag{2}$$

A.1.3 Fine-tuning Using RL method

Finally, in this step, the trained reward model r_σ is used to provide feedback on the output of the parameterized language model π_θ and optimize it by the objective of maximizing the expected reward

$$r(x, y) = r_\sigma(x, y) - \beta(\log(\pi_\theta(y|x)) - \log(\pi_{ref}(y|x))) \tag{3}$$

where π_θ denotes the policy of the language model we are optimizing and π_{ref} is the initial SFT model. During the RL training phase, the parameters of the SFT model π_{ref} remain fixed. π_θ is initialized with π_{ref} and optimized using an RL algorithm, most commonly PPO (Schulman et al., 2017) and other variants of actor-critic (Ramamurthy et al., 2023) algorithms. The parameter β ensures that the trained policy π_θ will not deviate significantly from the initial SFT model π_{ref}.

While RLHF is effective, it requires training a separate reward model, which makes the overall process costly. DPO eliminates the need for a reward model by directly optimizing the language model π_θ using the policies of both the reference model π_{ref} and π_θ itself. The objective function of DPO is to maximize the policy difference between the preferred output y_w and the rejected output y_l as in

$$\mathcal{L}_{DPO}(\pi_\theta; \pi_{ref}) = -E_{(x,y_w,y_l)\sim D}\left[\log\sigma(\beta\log\frac{\pi_\theta(y_w|x)}{\pi_{ref}(y_w|x)} - \beta\log\frac{\pi_\theta(y_l|x)}{\pi_{ref}(y_l|x)})\right]. \tag{4}$$

A.2 Additional results

Loss	T5-3B				Flan-T5-3B			
	Dev		Test		Dev		Test	
	EM	F1	EM	F1	EM	F1	EM	F1
DPO	**51.10**	**70.45**	**52.46**	74.29	53.77	73.68	55.15	**77.48**
IPO	50.64	69.41	51.57	73.41	53.53	73.06	53.36	76.79
RSO	49.83	69.55	50.90	**74.31**	53.88	73.50	**55.48**	77.24
KTO	47.74	68.21	51.12	74.24	**54.11**	**73.76**	53.36	77.20

Table 3: Results on the variants of DPO.

Table 3 shows the performance of the models for different variants of DPO. Although different DPO variants achieve better performance on different metrics, overall, DPO outperforms others for T5-3B in most cases, except for the test F1 score, where RSO achieves an F1 score of 74.31. For Flan-T5-3B, DPO outperforms others in the test F1 score and performs comparably to the others on the remaining metrics.

A.3 Examples of negative outputs created by rules

Figure 3 shows the example of negative samples created by the rule–based method.

Context

"FINAL REPORT\n HISTORY: Hematuria. History of aplastic anemia. GU ULTRASOUND: The right kidney measures 10.5 cm. The left kidney measures 10.6 cm. Both kidneys are normal in appearance, without evidence of hydroephrosis or renal calculi. There is a tiny amount of free fluid noted adjacent to bilateral kidneys. The prevoid bladder measures 6.8 x 7.0 x 7.2 cm and is normal in appearance. There is pelvic free fluid noted."

Question

Do we notice any stones in the kidneys, ureters or bladder?

Original Answer

Without evidence of hydroephrosis or renal calculi.

Negative Answers

Random Text Span:
- *free fluid noted adjacent to bilateral kidneys*

Longer text span containing part of answer:
- *kidneys are normal in appearance, without evidence*
- *or renal calculi. There is a tiny amount of free fluid noted*

Longer text span containing full answer:
- *Both kidneys are normal in appearance, without evidence of hydroephrosis or renal calculi.*

Partial answer:
- *evidence of hydroephrosis*

Answers from different question :
- *kidneys are normal in appearance*

Unanswerable:
- []

Figure 3: Examples of negative (rejected) outputs created by rules.

Gender-Neutral Large Language Models for Medical Applications: Reducing Bias in PubMed Abstracts

Elizabeth Schaefer
Yale University
Department of Computer Science
New Haven, CT
Elizabeth.Schaefer@yale.edu

Kirk Roberts
UTHealth Houston
McWilliams School of Biomedical Informatics, Houston, TX
Kirk.Roberts@uth.tmc.edu

Abstract

This paper presents a pipeline for mitigating gender bias in large language models (LLMs) used in medical literature by neutralizing gendered occupational pronouns. A set of 379,000 PubMed abstracts from 1965-1980 was processed to identify and modify pronouns tied to professions. We developed a BERT-based model, "Modern Occupational Bias Elimination with Refined Training," or "MOBERT," trained on these neutralized abstracts, and compared it with "1965BERT," trained on the original dataset. MOBERT achieved a 70% inclusive replacement rate, while 1965BERT reached only 4%. A further analysis of MOBERT revealed that pronoun replacement accuracy correlated with the frequency of occupational terms in the training data. We propose expanding the dataset and refining the pipeline to improve performance and ensure more equitable language modeling in medical applications.

1 Introduction

1.1 Background

Large language models (LLMs) are now widely used for a range of applications, from creating customer service chatbots to advertising that targets specific clients to predicting financial outcomes from potential economic indicators. LLMs have also increased in presence in the medical sector, ranging from accessible diagnostics to comprehensive literature retrieval, where they hold the promise of leading to a more informed level of care. Given the critical nature of these uses, it is essential to ensure that such LLMs remain free from biases that could potentially impact patient treatment and outcomes.

Despite their potential, though, many LLMs have been shown to contain and perpetuate biases (Kotek et al., 2023; Liu et al., 2022; Abid et al., 2021; Prakash and Lee, 2023; Bai et al., 2024; Zack et al., 2024; Bedi et al., 2024; Degelin, 2024). The presence of these biases in LLMs is especially concerning in medical applications, where it can lead to incorrect diagnoses, inappropriate treatment recommendations, and ultimately, unequal healthcare. For example, an LLM fine-tuned on a dataset like PubMed might provide biased diagnostic suggestions if the underlying data contain gendered stereotypes. Gender biases and their effects have already been highlighted in a range of medical practice cases, for topics that include generalized surgical procedures, psychiatry, kidney transplantation, and intensive care treatment, among many other areas (Ruiz and Verbrugge, 1997; van Daal et al., 2020; Lim et al., 2021; Merdji et al., 2023; Omar et al., 2024). Our research focuses specifically on occupational bias in conjunction with gendered pronouns, highlighting the underrepresentation and exclusion of women from traditionally male-dominated professions, a critical area given the concomitant distortion that can result from that in patient care decisions. Numerous instances of this bias are evident in the PubMed dataset, as illustrated in Figure 1. Instead of perpetuating these inequalities, a properly and rigorously trained LLM can mitigate and avoid such dangerous generalizations.

In this paper, we propose a novel approach to mitigate gender bias in LLMs used in medical contexts. To ensure that only relevant pronouns are neutralized without affecting critical medical details, our pipeline specifically targets pronouns that refer solely to occupations. This process preserves medically significant context, such as patient-specific information, while eliminating biased language tied to occupational stereotypes.

Our method focuses on addressing biases in the training data before the fine-tuning stage of LLM development. By constructing and validating a robust query pipeline that identifies and neutralizes binarily-gendered pronouns linked to occupational terms in medical literature, we aim to create

Proceedings of the 24th Workshop on Biomedical Language Processing (BioNLP 2025), pages 114–123
August 1, 2025

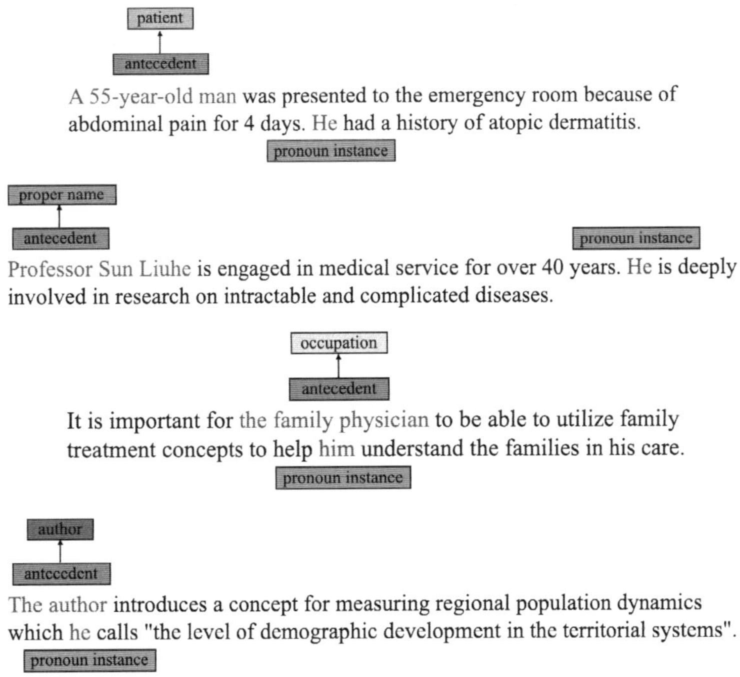

Figure 1: Example annotations from PubMed abstract text.

more reliable and fair models. This aligns with concerns raised by Bender et al. (2021), who emphasize that training on biased corpora can amplify existing societal stereotypes in unintended ways. By modifying the dataset at the source rather than post-processing model outputs, we directly address these concerns and create a more stable foundation for fairness in medical NLP. Our pipeline includes several key components: a lexicon for identifying gender-specific pronouns, a pronoun resolution query, and a classification query to identify occupation-specific subjects. Both queries were conducted using Meta's Llama-3.1-405b model (Meta AI, 2024), which we elected to use because it was the most advanced Llama model available at the time, and offered improved reasoning and accuracy for the task compared to previous models. The effectiveness of this approach is demonstrated through the development and evaluation of a BERT-based model, "Modern Occupational Bias Elimination with Refined Training," or "MOBERT," trained on gender-neutralized abstracts from PubMed.

1.2 Related Work

The issue of bias in word embeddings and LLMs has been widely studied (Pessach and Shmueli, 2022), with researchers highlighting how models trained on human-generated corpora often reflect and amplify societal stereotypes (Dev et al., 2023; Ungless et al., 2022), as well as proposing both technological and social solutions. Bolukbasi et al. (2016) first demonstrated that word embeddings could capture and propagate gender biases, showing that terms like "programmer" were more closely associated with men than women. Subsequent research provided further examples of such biases and explored their widespread implications in the field (Ray, 2023; Bommasani et al., 2021; Mehrabi et al., 2021; An et al., 2024; Pervez and Titus, 2024). To counter this effect, Bolukbasi et al. (2016) proposed a post-processing technique to debias word embeddings by projecting gender-neutral words into a subspace orthogonal to a gender direction. However, their method had limitations, such as requiring a classifier to identify gender-neutral words, which could introduce errors and propagate bias if the classifier itself was flawed.

Zhao et al. (2018) subsequently introduced GN-GloVe, a method that embeds gender information into specific dimensions of word vectors while neutralizing others. This approach improved interpretability and allowed for more effective debiasing by focusing on protected attributes like gender. However, while GN-GloVe effectively reduced direct gender bias, it left room for improvement in terms of indirect bias and applicability to contextualized word embeddings. As Blodgett et al. (2020) highlight, many prior bias-mitigation efforts rely on lexicon-based heuristics or simple vector-space transformations, which may fail to generalize to real-world applications. Our work moves beyond this by integrating pronoun resolution and context-aware classification before model fine-tuning.

More recently, the focus has shifted to contextualized word embeddings, such as those used in LLMs like BERT and GPT. Basta et al. (2019) and Zhao et al. (2019) explored gender bias in these models, finding that while contextualized embeddings reduce some biases present in static embeddings, they still retain significant levels of bias, especially in how they handle occupations and pronouns in context. In line with these concerns, recent work has shifted attention toward non-binary and gender-neutral pronouns, as more individuals identify outside the binary gender framework. Although much of the previous research focused on binary gender categories, studies such as Hossain et al. (2023) have revealed that large language models struggle significantly with gender-neutral and neo-pronouns, like "they" or "xe." This highlights the broader issue of representation in training corpora, where non-binary pronouns are often underrepresented, exacerbating the model's difficulty in handling inclusive language effectively.

Beyond post-processing and embedding-based debiasing methods, recent research has explored direct modifications to training data as a strategy for mitigating gender bias in biomedical NLP. Agmon et al. (2024) introduced TeDi-BERT, a model that applies temporal distribution matching to adjust how gender is represented in historical clinical trial abstracts, ensuring that language models trained on older data align more closely with contemporary gender distributions. Their approach highlights the importance of modifying training corpora before model training to prevent biased language from propagating in downstream applications.

Our work builds on these foundations but diverges in its focus on occupational bias in medical literature. Rather than aligning embeddings across different time periods, we develop a pipeline that systematically identifies and neutralizes gendered occupational pronouns before model fine-tuning. Through this methodology, we aim to create LLMs that are not only less biased but also more effective in delivering equitable healthcare outcomes. Unlike previous efforts that focused on post-processing or debiasing at the embedding level, our approach integrates bias mitigation into the model training process, addressing both direct and indirect biases more comprehensively.

2 Methods

2.1 Data

Our data are taken from the MEDLINE 2019 baseline set of PubMed abstracts from 1965 through mid-2018, totaling approximately 29 million abstracts. We utilized two lexicons to locate relevant abstracts for our study. The first lexicon searched for binarily-gendered pronouns, such as "him," "her," and "himself," ensuring that only complete words were identified. This process reduced the initial set of 29 million abstracts to 687,000 relevant abstract instances. A second lexicon, designed to identify occupational terms, was applied only for testing purposes, allowing us to evaluate classification performance across a broader range of occupation-related pronouns. This second lexicon was not used in the case study dataset and did not affect the pronoun-neutralization process. For clarity, we designate the "Pronoun-only corpus" as Dataset A and the "Pronoun- and Lexicon-Derived Corpus" as Dataset B. These names will be used throughout the paper when referring to the annotated datasets, as in Table 1.

Each instance in our corpus represents the character offset of each pronoun found within an abstract. This means that if an abstract contains three gendered pronouns, it will appear in our baseline corpus three times, once for each pronoun occurrence. This approach is crucial for determining the specific pronoun resolution in each instance, as different resolutions may occur within a single abstract.

2.2 Annotation Process

After constructing Datasets A and B, we proceeded with a two-step annotation process involving pronoun resolution and antecedent classification. This annotation process (utilizing LabelStudio (Tkachenko and contributors, 2020)) involved first identifying the noun phrase to which the pronoun referred (defined as "pronoun resolution") and then classifying that antecedent within the context of the abstract according to the established classification rules. Those rules were set as seen in Table 2. In this study, we intentionally avoided neutralizing pronouns when referring to patients or trial participants, as well as in contexts where

Table 1: Descriptive statistics of the annotated corpora.

	Dataset A: Pronoun-only corpus	Dataset B: Pronoun- and Lexicon-Derived Corpus
Total Number of Abstracts	250	500
Patient/Trial Participant	28	323
Named Individual	62	115
Occupation	97	24
Author of the Abstract	56	19
Animal	0	7
Other	7	12

Table 2: Categorization rules for classifying an antecedent within the context of an abstract.

Antecedent category	Category definition
Patient/Trial Participant	Individuals directly receiving medical care, those with a medical condition, or who are injured. This also applies to any trial participant, defined as someone who volunteers or is examined in a study, regardless of whether they are an occupational subject or not. The label "patient" takes precedence over any other classification when applicable.
Named Individual	Individuals referred to by a proper personal name, which includes capitalized names or redacted names.
Occupation	Individuals, real or abstract, identified by their profession or job, where they are employed and paid for their work.
Author of the Abstract	An author of the paper.
Animal	Any non-human creature
Other	Any instance that does not fit into the previous categories.

biological sex is medically relevant. Certain conditions, such as prostate or ovarian cancer, are inherently gendered, and so de-gendering such references could hamper a model's medical reasoning. Consequently, whenever a pronoun refers to a particular patient or group of patients in the abstract, that pronoun was left unchanged. Two annotators (the first author and an intern in the lab) separately labeled each given corpus, then calculated Cohen's Kappa, a measure quantifying the level of agreement before reconciliation. Example annotations from the corpus are shown in Figure 1. Descriptive statistics of the annotated corpora from Dataset A and Dataset B are provided in Table 1.

2.3 Pronoun Resolution Query

As the first step in our automated pipeline, we used a Llama-3.1-405b query to determine the subject associated with each pronoun, a process known as pronoun resolution. This step involved determining the noun or noun phrase to which a given pronoun referred within the abstract's context, ensuring that additional descriptive clauses were excluded. The full structured prompt used in this query is detailed in Table 4. To evaluate this prompt-based model, a randomly selected corpus of 500 pronoun instances was chosen from the relevant abstracts and each pronoun's respective antecedent was located and double-annotated with a Cohen's Kappa of 0.9000. Selected examples and the overall makeup of this corpus can be seen in both Figure 1 and Table 2.

2.4 Lexicon Validation

Following pronoun resolution, it proved useful to define a mechanism to distinguish occupational antecedents from other noun phrases. While the LLM was highly effective in identifying pronoun antecedents, we found that only a small percentage of gendered pronouns were actually attributable to professions, the core focus of our task. This data imbalance made it difficult to obtain a sufficient sample of occupation-related pronoun instances for evaluation. To address this, we developed a lexicon specifically designed to increase the frequency of identified occupational antecedents. The lexicon was initially derived from the synset relations of "professional" in WordNet, incorporating common occupational terms and case-sensitive acronyms (e.g., "rn" vs. "RN"). To validate the lexicon's efficacy, it was applied to the 500 resolved antecedents, filtering for occupational terms.

2.5 Classification Query

Using the validated lexicon described above, the results of our antecedent query can be successfully filtered. In this application, 250 pronoun instances were extracted from the data (primarily from Dataset B with the applied lexicon), along with their corresponding antecedents, that included occupational terms. These 250 instances and the text of the abstract in which they appeared must also be examined and tested for accuracy. Those instances were double-annotated and reconciled with a Cohen's Kappa of 0.9470. After annotator reconciliation, we used the Llama-3.1-405b model to classify each antecedent according to the same labeling rules, enabling comparison between human and model performance.

2.6 Pronoun Neutralization Process

After identifying gendered pronouns linked to occupational terms, the next step was neutralization. NLTK was used for tokenization and part-of-speech tagging. A pronoun-mapping dictionary was developed to replace gendered pronouns with their gender-neutral counterparts, such as 'they/them/theirs.' This dictionary accounted for compound pronouns (e.g., *he or she* $\rightarrow$ *they*) and handled replacements while preserving sentence structure. Pronouns flagged for neutralization were modified only when they referred to occupational antecedents, ensuring no changes were made to pronouns referring to patients or trial participants. This distinction was critical for maintaining medically relevant context in abstracts where sex-specific conditions (e.g., prostate cancer) were discussed. Examples of pronoun replacements and contexts that were preserved are presented in Table 5. The following section presents the results obtained from applying the three-stage pipeline on our annotated datasets, showcasing the effectiveness of our approach in neutralizing gendered pronouns.

3 Results

3.1 Pipeline Performance Metrics

To evaluate the performance of our pipeline, we analyzed two separate annotated datasets. Dataset A (Pronoun-only corpus) was used to assess the pronoun resolution component, while Dataset B (Pronoun- and Lexicon-Derived corpus) was utilized for the lexicon validation and classification queries. First, using Dataset A, we applied our pronoun resolution query on the non-annotated corpus. The Llama-3.1-405b query was run and the resulting pronoun instances and their corresponding antecedent outputs were cross-referenced with the ground-truth annotations. This comparison yielded an accuracy of 0.9881 on the initial 500 abstracts, demonstrating that our pronoun resolution method reliably identifies antecedents. Next, with Dataset B, we validated our lexicon for identifying occupational terms by applying it to the 500 antecedents obtained from the pronoun resolution query. The filtered results were then compared with the 'occupation' labels in the ground-truth annotations, achieving a perfect recall score of 1.0000. This confirms that our lexicon effectively identifies occupational antecedents for the classification task. Finally, still using Dataset B, we assessed the performance of our Llama-3.1-405b classification query by calculating precision, recall, and the F1 score between the generated labels and the ground-truth labels. The numerical outcomes of this process are presented in Table 3, and an example of an antecedent versus classification query is provided in Table 4. These results confirm the high performance of our classification approach in accurately distinguishing occupational pronoun instances from other categories. Together, these performance metrics validate the robustness of our pipeline, linking each methodological step to successful outcomes in resolving, filtering, and classifying gendered pronoun instances.

3.2 Pronoun Neutralization Case Study

We tested the effect of our pipeline on a corpus of the 379,000 PubMed abstracts from 1965-1980, hypothesizing that these texts would show a greater prevalence of singular gendered pronouns, based on a qualitative examination of a random sample set of the abstracts. After processing this corpus, pronouns linked to occupational antecedents were neutralized in 1,400 abstracts.

To determine the success of this replacement, we trained two separate base uncased BERT models (Devlin et al., 2019). The first model, named 1965BERT, was trained on the original, unmodified dataset of the 379,000 PubMed abstracts from 1965-1980. The second model, denoted "Modern Occupational Bias Elimination with Refined Training," or MOBERT, was trained on a similar dataset of the 379,000 abstracts, but with 1,400 abstracts identified and modified with the newly introduced gender-neutral tokens. Additionally, these 1,400 abstracts were analyzed to identify the most frequently-occurring occupational terms in relevant antecedents. The top five terms identified were "physician," "surgeon," "doctor," "practitioner," and "nurse." Both models were trained for three epochs with a batch size of four per device, using a mixed precision (fp16) configuration across multiple graphics processing units. Training logs were saved at regular intervals, with models checkpointed every 10,000 steps.

To further assess the models, we conducted a masked language modeling test using 50 sentences from our initial annotated corpus of 500 abstracts,

Table 3: The pipeline performance for the classification query.

Annotation	Frequency (out of 250)	Precision	Recall	F1
Occupation	97	0.9895	0.9691	0.9792
Named Individual	62	0.9492	0.9032	0.9252
Author of the Abstract	56	1	0.9107	0.9533
Patient	28	0.7027	0.9286	0.8
Other	7	0.75	0.8571	0.8
Macro Weighted Avg.		0.943	0.932	0.9349

Table 4: The Llama-3.1-405b prompts for the pronoun resolution and classification queries.

	System Content	User Content
Pronoun Resolution Query	You are a helpful assistant with identifying the direct antecedent of a pronoun. Here is your antecedent_background knowledge: {background}'*.	Identify the direct antecedent of the pronoun marked with [START] and [END] in the following abstract: {highlighted_abstract}. Only answer with the antecedent.
Classification Query (where antecedent is the output of Antecedent Query)	You are a helpful assistant following these classification rules {rules}.**	In the following abstract, classify which label the noun "{antecedent}" in the context of the abstract {highlighted_abstract} is referring to: "patient," "occupation," "named individual," "author," "animal," or "other." Only output the label, no other text.

*This background information consists of antecedent grammatical rules established by Fordham (Fordham University, 2024).
**These classification rules consist of the same rules shown in the Annotation Process section.

ensuring that each randomly selected sentence contained gendered pronouns from post-1980 texts. Importantly, the models were not trained on the data used in these tasks, ensuring an independent evaluation of their performance. The testing corpus was assembled by selecting ten sentences for each of the five most frequent occupational terms identified, resulting in 50 sentences. In each sentence, a [MASK] token was inserted in place of a pronoun, and the model was tasked with predicting the correct pronoun when given respective options of 'he/him/his,' 'she/her/hers,' and 'they/them/theirs.'

3.3 Outcomes

We compared the results of this masking test between BERT-Base (the untrained model), PubMedBERT, 1965BERT, and MOBERT (all three of which are trained upon BERT-Base with their respective training data) (Gu et al., 2020). Examples of this masking test and the corresponding outcomes are in Table 6, with overall results shown in Table 7. Percentages indicate the proportion of sentences in which gender-inclusive pronouns ('they/them/theirs') replaced gendered pronouns. For example, if BERT-Base replaces 40% of masked pronoun instances with a gender inclusive pronoun, 1965BERT replaces 4% of those same instances with a gender inclusive pronoun. The MOBERT results were further analyzed to determine a relationship between the frequency of the occupational term in the training data and the accuracy of replacement, as shown in Table 8.

4 Discussion

4.1 Principal Results

The application of our gender-neutralization pipeline to the 1965-1980 PubMed abstracts has demonstrated its potential to significantly reduce occupational gender bias in large language models. By introducing gender-neutral pronouns reconciled with occupational terms in 1,400 abstracts, we successfully trained a model, MOBERT, that demonstrated a 70% success rate in predicting inclusive pronouns in a masked language modeling task. This result far exceeds the 4% success rate of 1965BERT, a model trained on unmodified texts from the same period, and highlights the importance of correcting biased data at the training stage. MOBERT's performance also surpassed that of both the base model, BERT-Base, which exhibited a 40% success rate, and PubMedBERT, a model trained on the complete PubMed dataset without gender-neutralization, which achieved only a 20% inclusive successive rate. These comparisons underscore the critical role of targeted intervention in mitigating bias in language models.

4.2 Comparison with Prior Work

Research on the recruitment and retention of women in male-dominated occupations highlights how deeply embedded gendered language can reinforce exclusionary workplace cultures (Germain

Table 5: Examples of phrases that would/would not be identified for replacement, and the resulting modifications.

Example Sentence	Antecedent	Label	Modification
Some compromise must be reached between the unwillingness of the surgeon to spend most of **his** time performing abortions and the freedom for women to have them. [PMID: 5598532, 10/25/1968]	the surgeon	Occupation	Some compromise must be reached between the unwillingness of the surgeon to spend most of **their** time performing abortions and the freedom for women to have them.
Before any physician attempts to treat telangiectasia by this method, **he or she** should observe its performance by an experienced operator. [PMID: 834688, 3/15/1977]	any physician	Occupation	Before any physician attempts to treat telangiectasia by this method, **they** should observe its performance by an experienced operator.
Four lectures given by Dr. Mora and **his** staff focus on the betterment of the quality of life through improved nutrition. [PMID: 12261512, 6/10/1980]	Dr. Mora	Proper name	No modification.

Table 6: Examples of the masking test and the corresponding outcomes.

Example Sentence	BERT-Base	PubMedBERT	1965BERT	MOBERT
Although a doctor may not be continually aware of it, [MASK] medical activity is firmly rooted in the moral principles of the medical profession. PMID: 7470698, 5/21/1981	his	his	his	their
Many different portable computers are currently available and it is now possible for the physician to carry a mobile computer with [MASK] all the time. PMID: 12835877, 8/29/2003	them	him	him	them

et al., 2012). Prior studies such as De-Arteaga et al. (2019) have shown that occupational gender bias in machine learning software can directly affect hiring and professional representation. Research has shown that the assumption of male dominance in professional fields can discourage the participation of women in those fields, leading to self-reinforcing cycles of underrepresentation (Wu, 2022). For instance, studies have demonstrated that increasing gender diversity in male-dominated academic settings leads to improved career outcomes for female students, suggesting that removing implicit assumptions – such as assuming doctors are male – could encourage more diverse participation in medicine, technology, engineering, and math (Germain et al., 2012). If gendered language in professional texts perpetuates the underrepresentation of women in these fields, systematically neutralizing such biases could contribute to breaking this cycle.

4.3 Future Improvements

Future studies could involve integrating MOBERT into clinical NLP applications – such as diagnostic models and medical literature retrieval systems – to assess whether gender neutralization leads to improved healthcare equity. Sun et al. (2019) highlight that debiasing techniques should be evaluated not only through linguistic performance but also through real-world impact within medicine. Conducting user studies with medical professionals would be useful in assessing how gender-neutral models influence literature search relevance and clinical decision-making.

5 Conclusions

This work demonstrates the effectiveness of a gender neutralization pipeline in reducing occupational gender bias in large language models trained on medical literature. By processing 379,000 PubMed abstracts from 1965-1980 and targeting gender-specific pronouns linked to professions, we improved MOBERT's success rate to 70% in predicting gender-neutral pronouns, compared to 4% for 1965BERT. This improvement highlights the importance of addressing bias during training. While promising, the study also reveals opportunities for improvement, such as expanding the dataset and integrating the pipeline into further applications. These findings underscore the potential for creating more equitable and unbiased models in medical and other sensitive domains.

Table 7: Overall results for the inclusive replacement rates by model.

Model	Inclusive Replacement Rate (%)
BERT-Base (Comparison Baseline for All Cases)	40
PubMedBERT	20
1965BERT	4
MOBERT	70

Table 8: Relationship between the occupational term frequency in the training data and the replacement accuracy.

Occupational Term	Frequency	Percentage (%)
Physician	298	100
Surgeon	135	100
Doctor	89	70
Practitioner	68	60
Nurse	64	30

Limitations

Despite the overall success of MOBERT, our analysis did reveal some limitations. As seen in Table 8, the frequency of occupational terms in the 379,000 modified abstracts used for pre-training correlated strongly with the accuracy of pronoun replacement. For instance, terms like "physician" and "surgeon," which appeared more frequently in the training data, saw a 100% accuracy in neutral pronoun predictions, while terms like "nurse" had a much lower replacement rate of 30%. Although we did not use supervised fine-tuning, MOBERT's exposure to gender-neutralized occupational terms during pre-training likely contributed to its improved performance on those terms in masked language modeling tasks. Since masked language modeling relies on contextual co-occurrence rather than explicit supervision, MOBERT likely developed stronger associations between certain occupations and gender-neutral pronouns due to their repeated exposure in the training data. This suggests that expanding a pre-training dataset to include a broader range of occupations and more balanced representation of male- and female-dominated roles could further improve the model's performance.

Another potential limitation arises from the process of language alteration itself. Although we carefully designed our pipeline to neutralize pronouns only in contexts where the occupational term was the antecedent, there remains a risk that some instances of gendered language with medically significant context may have been inadvertently modified. While we found no evidence of such errors in our testing, further refinements to the pipeline could incorporate more sophisticated contextual analysis to ensure the protection of patient-specific or trial-related information. Additionally, large-scale gender neutralization poses the challenge of maintaining critical semantic distinctions. While replacing a gendered pronoun with 'they/them/theirs' often preserves meaning, certain contexts – such as historical citations or patient narratives – could lead to unintended distortions. Our lexicon filtering approach helps mitigate this by restricting modifications to occupational contexts; however, broader applications must carefully handle edge cases where neutralization may introduce ambiguity or alter medically relevant details. For example, in a PubMed abstract (PMID: 25549443) there is a sentence discussing gender dynamics in surgery: "*Suggestions include a change [in] the relationship between a female surgeon and her partner, a supplement of surgeons so that hospitals could change the traditional system of surgery.*" Indiscriminate neutralization of *"her"* to *"their"* could obscure the focus on challenges specific to female surgeons, weakening the text's emphasis on gendered professional and personal expectations. Addressing these concerns will require further refinement, including human evaluation studies, dependency parsing for syntactic precision, and additional lexicon constraints to safeguard against unintended language shifts.

Acknowledgements

Elizabeth Schaefer was supported by the Cancer Prevention and Research Institute of Texas (CPRIT RP210045) as part of the Biomedical Informatics, Genomics and Translational Cancer Research Training Program. Kirk Roberts was supported by the National Institutes of Health (R01LM014508, R01LM011934). The authors acknowledge Dhruv Kumar of the University of Texas for serving as a second annotator in the data labeling process, contributing to the double-annotation and reconciliation of the corpus.

References

Abubakar Abid, Maheen Farooqi, and James Zou. 2021. Persistent anti-Muslim bias in large language models. In *Proceedings of the 2021 AAAI/ACM Conference on AI, Ethics, and Society*, AIES '21. ACM.

Shunit Agmon, Uriel Singer, and Kira Radinsky. 2024. Leveraging temporal trends for training contextual word embeddings to address bias in biomedical applications: Development study. *JMIR AI*, 3:e49546.

Haozhe An, Christabel Acquaye, Colin Wang, Zongxia Li, and Rachel Rudinger. 2024. Do large language models discriminate in hiring decisions on the basis of race, ethnicity, and gender? *arXiv*. Https://arxiv.org/abs/2406.10486.

Xuechunzi Bai, Angelina Wang, Ilia Sucholutsky, and Thomas L. Griffiths. 2024. Measuring implicit bias in explicitly unbiased large language models. *arXiv*. Https://arxiv.org/abs/2402.04105.

Christine Basta, Marta R. Costa-jussà, and Noe Casas. 2019. Evaluating the underlying gender bias in contextualized word embeddings. In *Proceedings of the First Workshop on Gender Bias in Natural Language Processing*. Association for Computational Linguistics.

Priya Bedi, Anmol Singh, Navreet Kaur, Monojit Choudhury, and Danish Pruthi. 2024. Testing and evaluation of health care applications of large language models: A systematic review. *JAMA*, 333(4):319–328.

Emily M. Bender, Timnit Gebru, Angelina McMillan-Major, and Shmargaret Shmitchell. 2021. On the dangers of stochastic parrots: Can language models be too big? In *Proceedings of the 2021 ACM Conference on Fairness, Accountability, and Transparency*, pages 610–623. Association for Computing Machinery.

Su Lin Blodgett, Solon Barocas, Hal Daumé III, and Hanna Wallach. 2020. Language (technology) is power: A critical survey of "bias" in NLP. In *Proceedings of the 58th Annual Meeting of the Association for Computational Linguistics*, pages 5454–5476. Association for Computational Linguistics.

Tolga Bolukbasi, Kai-Wei Chang, James Zou, Venkatesh Saligrama, and Adam Kalai. 2016. Man is to computer programmer as woman is to homemaker? Debiasing word embeddings. *arXiv*. Https://arxiv.org/abs/1607.06520.

Rishi Bommasani, Drew A. Hudson, Ehsan Adeli, and 86 additional authors. 2021. On the opportunities and risks of foundation models. *arXiv*. Https://arxiv.org/abs/2108.07258.

Maria De-Arteaga, Alexey Romanov, Hanna Wallach, Jennifer Chayes, Christian Borgs, Alexandra Chouldechova, Sahin Geyik, Krishnaram Kenthapadi, and Adam Tauman Kalai. 2019. Bias in bios: A case study of semantic representation bias in a high-stakes setting. In *Proceedings of the Conference on Fairness, Accountability, and Transparency*, pages 120–128. ACM.

Nisse Degelin. 2024. Bias in llms for high-stakes recommendations: An analysis of bert-family architectures with varied fine-tuning configurations. Master's thesis, University of Antwerp.

Sunipa Dev, Jaya Goyal, Dinesh Tewari, Shachi Dave, and Vinodkumar Prabhakaran. 2023. Building socio-culturally inclusive stereotype resources with community engagement. *arXiv*. Https://arxiv.org/abs/2307.10514.

Jacob Devlin, Ming-Wei Chang, Kenton Lee, and Kristina Toutanova. 2019. Bert: Pre-training of deep bidirectional transformers for language understanding. arXiv preprint arXiv:1810.04805.

Fordham University. 2024. Pronoun-antecedent agreement. Https://www.fordham.edu/academics/academic-resources/writing-center/writing-resources/grammar/pronoun-antecedent-agreement/.

Marie-Line Germain, Mary Jean Ronan Herzog, and Penny Rafferty Hamilton. 2012. Women employed in male-dominated industries: Lessons learned from female aircraft pilots, pilots-in-training and mixed-gender flight instructors. *Human Resource Development International*, 15(4):435–453.

Yu Gu, Robert Tinn, Hao Cheng, Michael Lucas, Naoto Usuyama, Xiaodong Liu, Tristan Naumann, Jianfeng Gao, and Hoifung Poon. 2020. Domain-specific language model pretraining for biomedical natural language processing. *arXiv*. Https://www.microsoft.com/en-us/research/blog/domain-specific-language-model-pretraining-for-biomedical-natural-language-processing/.

Tamanna Hossain, Sunipa Dev, and Sameer Singh. 2023. Misgendered: Limits of large language models in understanding pronouns. In *Proceedings of the 61st Annual Meeting of the Association for Computational Linguistics (Volume 1: Long Papers)*, page 5352–5367. Association for Computational Linguistics.

Hadas Kotek, Rikker Dockum, and David Sun. 2023. Gender bias and stereotypes in large language models. In *Proceedings of The ACM Collective Intelligence Conference*, CI '23. ACM.

Wen Hui Lim, Chloe Wong, Sneha Rajiv Jain, Cheng Han Ng, Chia Hui Tai, M. Kamala Devi, Dujeepa D. Samarasekera, Shridhar Ganpathi Iyer, and Choon Seng Chong. 2021. The unspoken reality of gender bias in surgery: A qualitative systematic review. *PLOS ONE*, 16(2):e0246420.

Ruibo Liu, Chenyan Jia, Jason Wei, Guangxuan Xu, and Soroush Vosoughi. 2022. Quantifying and alleviating political bias in language models. *Artificial Intelligence*, 304:103654.

Ninareh Mehrabi, Fred Morstatter, Nripsuta Saxena, Kristina Lerman, and Aram Galstyan. 2021. A survey on bias and fairness in machine learning. *ACM Computing Surveys*, 54(6):1–35.

Hamid Merdji, Micah T. Long, Marlies Ostermann, Margaret Herridge, Sheila N. Myatra, Silvia De Rosa, Victoria Metaxa, Katarzyna Kotfis, Chiara Robba, Audrey De Jong, Julie Helms, and Caroline E. Gebhard. 2023. Sex and gender differences in intensive care medicine. *Intensive Care Medicine*, 49(10):1155–1167.

Meta AI. 2024. Introducing llama 3.1: Our most capable models to date. Accessed: 2025-05-25.

Mahmud Omar, Vera Sorin, Reem Agbareia, Donald U Apakama, Ali Soroush, Ankit Sakhuja, Robert Freeman, Carol R Horowitz, Lynne D Richardson, Girish N Nadkarni, and Eyal Klang. 2024. Evaluating and addressing demographic disparities in medical large language models: A systematic review. *medRxiv*.

Naseela Pervez and Alexander J Titus. 2024. Inclusivity in large language models: Personality traits and gender bias in scientific abstracts. *arXiv preprint arXiv:2406.19497*.

Dana Pessach and Erez Shmueli. 2022. A review on fairness in machine learning. *ACM Computing Surveys*, 55(3):1–44.

Nirmalendu Prakash and Roy Ka-Wei Lee. 2023. Layered bias: Interpreting bias in pretrained large language models. In *Proceedings of the 6th BlackboxNLP Workshop: Analyzing and Interpreting Neural Networks for NLP*, page 284–295. Association for Computational Linguistics.

Partha Pratim Ray. 2023. Chatgpt: A comprehensive review on background, applications, key challenges, bias, ethics, limitations and future scope. *Internet of Things and Cyber-Physical Systems*, 3:121–154.

M. Teresa Ruiz and Lois M. Verbrugge. 1997. A two way view of gender bias in medicine. *Journal of Epidemiology and Community Health*, 51(2):106–109.

Tony Sun, Andrew Gaut, Shirlyn Tang, Yuxin Huang, Mai ElSherief, Jieyu Zhao, Diba Mirza, Elizabeth Belding, Kai-Wei Chang, and William Yang Wang. 2019. Mitigating gender bias in natural language processing: Literature review. In *Proceedings of the 57th Annual Meeting of the Association for Computational Linguistics*, pages 1630–1640. Association for Computational Linguistics.

Maxim Tkachenko and HumanSignal contributors. 2020. Label studio: Data labeling software. `https://github.com/HumanSignal/label-studio`. Accessed 2025.

Eddie L. Ungless, Amy Rafferty, Hrichika Nag, and Björn Ross. 2022. A robust bias mitigation procedure based on the stereotype content model. *arXiv*. Https://arxiv.org/abs/2210.14552.

Manon van Daal, Maaike E. Muntinga, Sandra Steffens, Annemie Halsema, and Petra Verdonk. 2020. Sex and gender bias in kidney transplantation: 3d bioprinting as a challenge to personalized medicine. *Women's Health Reports*, 1(1):218–223.

Joy Wu. 2022. Secondary market monetization and willingness to share personal data. *To appear in Management Science*.

Travis Zack, Eric Lehman, Mirac Suzgun, Jorge A Rodriguez, Leo Anthony Celi, Judy Gichoya, Dan Jurafsky, Peter Szolovits, David W Bates, Raja-Elie E Abdulnour, Atul J Butte, and Emily Alsentzer. 2024. Assessing the potential of gpt-4 to perpetuate racial and gender biases in health care: a model evaluation study. *The Lancet Digital Health*, 6(1):e12–e22.

Jieyu Zhao, Tianlu Wang, Mark Yatskar, Ryan Cotterell, Vicente Ordonez, and Kai-Wei Chang. 2019. Gender bias in contextualized word embeddings. In *Proceedings of the 2019 Conference of the North*. Association for Computational Linguistics.

Jieyu Zhao, Yichao Zhou, Zeyu Li, Wei Wang, and Kai-Wei Chang. 2018. Learning gender-neutral word embeddings. In *Proceedings of the 2018 Conference on Empirical Methods in Natural Language Processing*. Association for Computational Linguistics.

Error Detection in Medical Note through Multi Agent Debate

Abdine Maiga
Centre for Artificial Intelligence
University College London
abdine.maiga.23@ucl.ac.uk

Anoop Shah
UCLH NHS Trust
University College London
a.shah@ucl.ac.uk

Emine Yilmaz
Centre for Artificial Intelligence
University College London
emine.yilmaz@ucl.ac.uk

Abstract

Large Language Models (LLMs) have approached human-level performance in text generation and summarization, yet their application in clinical settings remains constrained by potential inaccuracies that could lead to serious consequences. This work addresses the critical safety weaknesses in medical documentation systems by focusing on detecting subtle errors that require specialized medical expertise.

We introduce a novel multi-agent debating framework that achieves 78.8% accuracy on medical error detection, significantly outperforming both single-agent approaches and previous multi-agent systems. Our framework leverages specialized LLM agents with asymmetric access to complementary medical knowledge sources (Mayo Clinic and WebMD), engaging them in structured debate to identify inaccuracies in clinical notes. A judge agent evaluates these arguments based solely on their medical reasoning quality, with agent-specific performance metrics incorporated as feedback for developing situation-specific trust models.

This research significantly enhances the safety and reliability of automated medical documentation, potentially facilitating wider AI adoption in healthcare while maintaining high standards of accuracy. The performance gap between individual specialized agents (WebMD: 70.2%, Mayo: 72.6%) compared to their combined implementation demonstrates the synergistic value of integrating complementary clinical perspectives through structured debate.

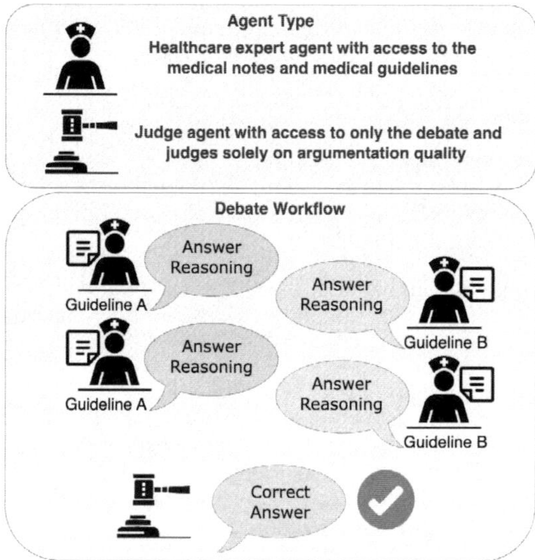

Figure 1: Debating Healthcare Agent framework. The multi-agent architecture consists of three primary components: (1) Expert Agent A with access to guidelines A, (2) Expert Agent B with access to guidelines B, and (3) a Judge Agent who evaluates arguments based solely on their medical reasoning quality without access to external knowledge sources. The agents engage in structured debate where experts exchange arguments and counter-arguments before the judge determines the presence of errors in the medical note.

1 Introduction

Healthcare professionals spend 52-102 minutes daily on clinical documentation (Hripcsak et al., 2011), contributing significantly to administrative burden, work-life imbalance, and burnout rates exceeding 50% among practitioners (Arndt et al., 2017). Large Language Models (LLMs) show

Proceedings of the 24th Workshop on Biomedical Language Processing (BioNLP 2025), pages 124–135
August 1, 2025 ©2025 Association for Computational Linguistics

promise for automating clinical summarization tasks (Knoll et al., 2022), potentially transforming workflows and allowing healthcare providers to focus more on direct patient care.

Despite their advanced capabilities, LLM adoption in healthcare remains limited due to concerns about accuracy in high-stakes clinical environments (Lakkaraju et al., 2022). These concerns are well-founded: studies examining 136,815 patients found that 21.1% reported perceived mistakes in their medical records, with 40% considered serious (Bell et al., 2020). Diagnostic errors alone contribute to 6-17% of adverse events in hospitalized patients (Ball et al., 2016), highlighting the critical importance of accuracy in medical documentation.

Current approaches to medical error detection typically rely on single-agent architectures that cannot replicate the nuanced perspective of collaborative clinical evaluation. These methods particularly struggle with subtle errors requiring specialized medical expertise, especially in complex cases involving multiple conditions or diverse patient populations. This limitation underscores the need for more sophisticated frameworks that can mirror the collaborative decision-making processes common in clinical settings.

To address this gap, we introduce a novel multi-agent debating framework where specialized LLM agents with access to authoritative medical guidelines engage in structured debate to identify and resolve inaccuracies. Our approach simulates clinical consultation dynamics through debate protocols where expert agents present competing perspectives on potential errors, with a judge agent evaluating these arguments based on medical reasoning. The system incorporates performance metrics as feedback to develop situation-specific trust models, enhancing reliability across diverse scenarios.

Our research contributes: (1) a multi-agent architecture for medical error detection achieving 78.8% accuracy, outperforming existing approaches; (2) comprehensive evaluation across medical specialties and patient populations; and (3) empirical evidence demonstrating how structured debate between complementary medical knowledge sources enhances error detection beyond individual agents' capabilities. These contributions establish foundations for safer LLM deployment in clinical environments, addressing a key barrier to AI adoption in healthcare.

2 Related Works

2.1 Medical Error Detection and Correction

Medical error detection and correction in clinical texts was first formally addressed during the MEDIQA-CORR challenge at NAACL 2024 (Ben Abacha et al., 2024). This challenge created a corpus of medical notes with intentionally introduced errors requiring medical expertise to detect, structured as a three-stage task: error detection, span identification, and correction generation.

The winning team (Toma et al., 2024) developed dual LLM-based systems using the DSPy (Khattab et al., 2023) framework, a retrieval-based approach for subtle errors and a comprehensive pipeline for complex cases (accuracy: 86.49%, though flagged for potential use of MS test data). The PromptMind team (Gundabathula and Kolar, 2024) implemented prompt-based in-context learning that integrated outputs from multiple advanced language models(accuracy= 0.6216). HSE NLP (Valiev and Tutubalina, 2024) employed an in-prompt ensemble approach combining named entity recognition with MeSH knowledge graph integration (Accuracy= 0.5222). Edinburgh Clinical NLP (Gema et al., 2024) explored three strategies: end-to-end prompting, two-stage fine-tuning, and a hybrid method combining both approaches(accuracy= 0.6692). The KU-DMIS team (Hwang et al., 2024) fine-tuned Meerkat-7B using a Chain-of-Thought reasoning dataset generated from GPT-4 (accuracy=0.6346). Across 17 participating teams, the mean accuracy score was 61.57%, highlighting the challenge's difficulty and the need for optimized approaches suitable for integration into production-grade clinical documentation systems.

The challenge demonstrated that dataset-dependent methods generally outperformed generalized approaches, though dataset-agnostic solutions showed promise. Error detection proved particularly challenging, highlighting the need for optimized approaches suitable for integration into production-grade clinical documentation systems.

2.2 Medical Decision Making

The integration of LLMs into medical decision-making (Thirunavukarasu et al., 2023) has progressed along two distinct trajectories. The initial approach focused on fine-tuning pretrained models on domain-specific corpora, as exemplified by Med-PaLM (Singhal et al., 2023) Med-Gemini (Saab et al., 2024) or Bio Mistral (Labrak et al., 2024) and clinical BERT variants (Huang et al., 2020), which demonstrated enhanced performance on medical tasks through parameter optimization. However, with the emergence of more capable foundation models like GPT-4, the field has increasingly shifted toward sophisticated inference-time techniques that preserve model parameters while adapting behavior (Nori et al., 2023). Prompt engineering strategies—including few-shot examples, chain-of-thought reasoning, and structured output templates—have shown remarkable efficacy in guiding LLMs toward medically sound reasoning patterns without domain-specific training. In some task like medical summarization (Van Veen et al., 2023), adapted model can even surpass medical experts (Van Veen et al., 2024).Retrieval Augmented Generation (RAG) (Lewis et al., 2021) has proven particularly valuable for mitigating hallucinations by dynamically incorporating trusted medical knowledge bases, clinical guidelines, and patient-specific records into the generation context. This approach anchors model outputs to verifiable sources while maintaining flexibility across diverse clinical scenarios. Frameworks such as Uncertainty of Thoughts (Hu et al., 2024) further advance LLM reliability in medical settings by implementing uncertainty quantification mechanisms that more closely approximate clinical diagnostic workflows. Despite these advances, the high stakes of medical decision-making necessitate additional safeguards against subtle inaccuracies that could compromise patient safety, motivating multi-agent collaboration frameworks that can solve complex medical problems by working collaboratively, taking example for the real medical settings. Agent Hospital (Li et al., 2024) which simulates a whole hospital with agents, to train them and treat disease more efficiently. Other methods like MedAgents (Tang et al., 2024) leverages collaborative multi-round discussion with LLM-based agents to solve medical domain task. MDAgents (Kim et al., 2024) build on top of with an adaptive collaboration structure.

2.3 Multi Agent Framework

Multi-agent frameworks represent a promising approach for enhancing LLM performance in complex medical scenarios. Recent studies have demonstrated that effective collaboration between specialized agents, such as those in AutoGen (Wu et al., 2023), can yield superior results compared to individual agents operating in isolation (Wang et al., 2024). This parallels human team dynamics, where diverse expertise contributes to more robust decision-making.

Multi-agent collaboration has proven successful across varied domains including general problem-solving (Li et al., 2023), software engineering (Qian et al., 2024), and even simulation environments like The Sims (Park et al., 2023). Particularly relevant to our approach is the work by Chen et al. (Chen et al., 2024), who developed a multi-model multi-agent framework structured as a round table conference among diverse LLM agents, demonstrating how different model architectures can complement each other's strengths and compensate for individual weaknesses.

However, these approaches often suffer from significant computational inefficiency, as they typically rely on multiple instances of large, resource-intensive LLMs performing numerous inference passes. For practical clinical deployment, a multi-agent framework must demonstrate clear advantages over single-agent alternatives to justify the additional computational cost.

Our work proposes a streamlined approach that combines the strengths of structured multi-agent debate with retrieval-augmented generation (RAG). By incorporating findings from Khan et al. (Khan et al., 2024) on effective debate protocols, we have developed a tailored system specifically designed for medical error detection. This approach addresses the critical need for safeguards against subtle medical inaccuracies that could compromise patient safety, allowing for systematic evaluation of clinical content against established medical standards while maintaining computational efficiency.

3 Methods

3.1 Datasets

The dataset utilized in this study is derived from the MS collection of the medical error detection dataset created by Ben Abacha et al (Ben Abacha et al., 2024). This collection was developed by transforming the MEDQA dataset (Jin et al., 2020), which originally contained free-form multiple-choice questions from professional medical board exams. The researchers manually injected errors into clinical texts and made textual modifications that leveraged both clinical notes and multiple-choice questions from MEDQA. Those errors are mainly substitutions of medical terms such as diagnosis, treatment, scan type, or prescriptions. The

MS collection includes 2,189 clinical texts in the training set, 574 in the validation set, and 597 in the test set. Each text contains deliberately injected errors across various medical domains including diagnosis, causal organism, management, treatment, and pharmacotherapy, making it a valuable resource for developing and evaluating medical error detection systems. In recent studies, two physicians attempted to detect errors on half of the test set. On the MS teams dataset, they achieved accuracy rates of 81.25% and 68.90% respectively. These results demonstrate that even for trained medical professionals, this error detection task is not straightforward.

3.2 Medical Knowledge Foundation for Agents

A cornerstone of our project is the comprehensive medical guidelines framework that serves as a critical differentiator between agents. This framework comprises carefully curated, authoritative medical information sources that each agent can access and reference.

We have meticulously selected several reputable online medical resources, ensuring our agents have access to evidence-based, peer-reviewed, and clinically validated information. These resources were chosen based on their reliability, accuracy, comprehensiveness, and recognition within the medical community.

Primary Mainstream Medical Sources

Our foundation layer consists of widely recognized medical information platforms:

1. Wikipedia: A vast collaborative encyclopedia with extensively referenced medical articles that undergo regular expert review

2. MedlinePlus: Produced by the National Library of Medicine, offering reliable, up-to-date health information in accessible language

3. WebMD: A comprehensive consumer health information site featuring physician-reviewed content

4. Mayo Clinic: One of the world's premier medical institutions providing authoritative, trustworthy health guidance

5. PubMed Central: An extensive archive of biomedical and life sciences journal literature at the U.S. National Institutes of Health's National Library of Medicine

6. Medscape: A leading platform for healthcare professionals, Medscape offers peer-reviewed medical news, clinical reference tools, and continuing education content. Its articles are authored by experts and frequently updated, making it a trusted source for evidence-based clinical guidance.

These primary sources provide our agents with a robust baseline of medical knowledge spanning from basic concepts to advanced clinical information, ensuring they can address a wide spectrum of health-related inquiries.

3.3 Debating Framework: Error Detection

Our framework draws inspiration from Khan et al. (Khan et al., 2024), who developed a debating method where LLM experts argue for different answers—in our case, assessing the correctness of medical notes. A key finding from their work is that weaker models can effectively supervise stronger models when structured properly.

3.3.1 Agent Architecture and Information Flow

The multi-agent debate framework consists of three primary components (Figure 2):

1. **Expert Agent A (Mayo Clinic)**: Specialized for healthcare professional perspective

2. **Expert Agent B (WebMD)**: Specialized for patient-oriented medical knowledge

3. **Judge Agent**: Evaluates arguments without access to external knowledge sources

In our implementation, asymmetry is created by providing LLM experts with different information sources, while the judge agent relies solely on its internal knowledge. This creates a controlled information environment where the two expert agents have access to the medical note under evaluation, but the judge only accesses their arguments to make decisions.

3.3.2 Information Retrieval Integration

To mitigate the risk of hallucinations, we integrated a retrieval component through a `fetch_website` tool that allows expert agents to access authoritative medical websites. The tool fetches and processes web content (limited to 2000 characters), removing non-informative elements while preserving

Algorithm 1 Multi-Agent Medical Error Detection

Require: Medical note M

Ensure: Error detection decision (True/False)

1: Initialize agents: Expert A (Mayo Clinic), Expert B (WebMD), and Judge

2: Experts analyze M using fetch_website to retrieve medical information

3: Experts present initial arguments (max 300 words each)

4: Experts exchange counter-arguments after reviewing opposing views

5: Judge evaluates all arguments (without external references)

6: **return** Judge's decision on presence of errors

essential medical information. Expert agents are restricted to accessing only their assigned knowledge source—Mayo Clinic for healthcare professional perspectives and WebMD for patient-oriented information.

Our initial experimentation with three debate rounds revealed significant redundancy, as agent positions rarely changed after the second round (in 92% of test cases). We therefore limited debates to two rounds for efficiency. Additionally, we implemented a 300-word limitation for each agent's contribution to address verbosity bias, as judge agents consistently favored longer arguments regardless of substance.

4 Experiments & Results

4.1 Evaluation Metrics

To comprehensively evaluate our framework's performance, we employ multiple complementary metrics that assess different aspects of medical error detection.

For error detection, we use accuracy as our primary metric, defined as:

$$\text{Accuracy} = \frac{\text{TP} + \text{TN}}{\text{TP} + \text{TN} + \text{FP} + \text{FN}} \quad (1)$$

where TP (true positives) represents correctly identified errors, TN (true negatives) represents correctly identified error-free notes, FP (false positives) represents error-free notes incorrectly flagged as containing errors, and FN (false negatives) represents errors that went undetected.

To assess statistical significance, we employ McNemar's test—a non-parametric method suitable for paired nominal data in classification tasks. This test evaluates whether disagreements between our method and baselines are statistically significant, with $p < 0.05$ indicating significant performance differences. McNemar's test is particularly appropriate as it focuses on error pattern differences rather than just overall accuracy and accounts for the paired nature of predictions on identical test instances.

4.2 Setup

The primary goal is to assess model discriminative capabilities rather than deployment performance, which is why we used a balanced dataset as medical errors are scarcer in real-life clinical settings. Future work should evaluate the system on datasets with more realistic error prevalence rates to better understand performance metrics that are sensitive to base rates.

We tested all models on a balanced set of 500 randomly sampled data points from the MS collection—a subset of the full dataset necessitated by computational cost constraints. With API-based implementations, inference costs varied significantly between methods, from approximately $5 per evaluation run for single-agent approaches to $30 per run for our multi-agent framework, making comprehensive testing on the full dataset prohibitively expensive.

For this evaluation, we benchmarked our proposed framework against state-of-the-art (SOTA) baselines across three categories. First, we compared against individual agent approaches using popular prompting techniques: zero-shot (direct task inference without examples), few-shot (Brown et al., 2020) learning from minimal examples, chain of thought (Wei et al., 2023) with explicit reasoning steps, and self-consistency (SC) methods (Wang et al., 2023) generating multiple solutions for consensus.

We also included specialized single-agent implementations using Mayo Clinic, WebMD, and Medscape guidelines as reference materials, which demonstrated superior performance over standard prompting techniques. The final category consisted of multi-agent approaches, comparing against the high-performing MDAgents framework (Kim et al., 2024) (specialized medical diagnostic agents) applied to our dataset, as well as a modified version of AutoGen (Wu et al., 2023) comprising four specialized agents (User, Clinician, Medical Expert, and Moderator) with single-turn responses.

GPT-4o served as the foundational LLM in all

experimental configurations to ensure fair comparison across methods.

4.3 Implementation

Our implementation uses AutoGen Core/Ext for orchestrating the multi-agent debate protocol, with all agents powered by GPT-4o. Expert agents access domain-specific medical knowledge through a custom retrieval component using BeautifulSoup and Requests, while the judge agent evaluates arguments based solely on their medical reasoning quality. The system leverages asynchronous communication to efficiently manage the two-round structured debate process

5 Results Analysis

The revised results demonstrate a stratified performance pattern across medical error detection methodologies. Single-agent approaches (Zero-Shot: 66%, Few-Shot: 64.2%) establish a baseline performance that is incrementally improved through few-shot variants (CoT+Few-Shot: 69.7%). To better understand the impact of domain-specific knowledge sources, we developed specialized single agents (S.Agent) by isolating components of our complete framework. Each S.Agent utilizes our base prompt enhanced with few-shot examples, chain-of-thought reasoning, and the ability to retrieve information from a single medical knowledge source—either Mayo Clinic or WebMD. This specialized agent architecture reveals an interesting asymmetry, with S.Agent (WebMD) performing at 70.2% compared to S.Agent (Mayo) at 72.6%, indicating that domain-specific knowledge sources contribute differentially to error detection capabilities. The multi-agent frameworks show progressive enhancement, with MDAgent achieving 70.6% accuracy and AutoGen reaching 74.6%, though with a notably higher p-value (0.1567) suggesting less statistical reliability in its performance advantage. Our proposed composite methodology, which integrates the complementary knowledge sources in a structured debate framework, achieves 78.8% accuracy, representing a 4.2 percentage point improvement over AutoGen. This performance enhancement appears statistically significant when compared to most baseline methods ($p<0.05$), with the exception of AutoGen. These findings suggest that deliberate integration of complementary clinical perspectives through a structured multi-agent debate framework effectively captures diagnostic subtleties missed by single-perspective systems, mirroring the benefits of multi-specialist consultation in clinical practice.

Source	Accuracy
Mayo Clinic	84%
Web MD	82%
Medscape	80%
PubMed Central	78%
Medline	74%
Wikipedia	72%

Table 1: Accuracy of various medical sources, sorted in descending order.

5.1 Medical sources

For website retrieval, we can classify the sources into two main categories with two notable outliers. Wikipedia, being a generalist website, understandably performs relatively poorly at 72% accuracy for medical information. PubMed Central represents another outlier as a healthcare research website; despite our expectations for higher performance, it achieved only 78%, likely because only abstracts are publicly available.

The two main categories are websites for healthcare professionals (Mayo Clinic and Medscape), which rank among the best performers with 84% and 80% accuracy respectively, and those designed for patients (WebMD and Medline) with 82% and 74% accuracy. To obtain different perspectives on each medical note, we selected one website from each category with the highest accuracy scores: Mayo Clinic for healthcare professionals and WebMD for patients.

5.2 Error analysis

5.2.1 Medical Specialty

A detailed error analysis across medical specialties reveals significant performance variations in our model. The framework achieves above-average accuracy in Emergency Medicine (83.0%), Infectious Disease (81.2%), and Oncology (79.4%), suggesting particular strength in these domains.
Conversely, the model demonstrates notable weaknesses in Obstetrics/Gynecology (73.6%) and Psychiatry (75.0%). For OB/GYN cases, careful examination of the model's reasoning reveals a fundamental challenge: pregnancy significantly alters normal vital sign parameters and physiological baselines, causing the model to misinterpret

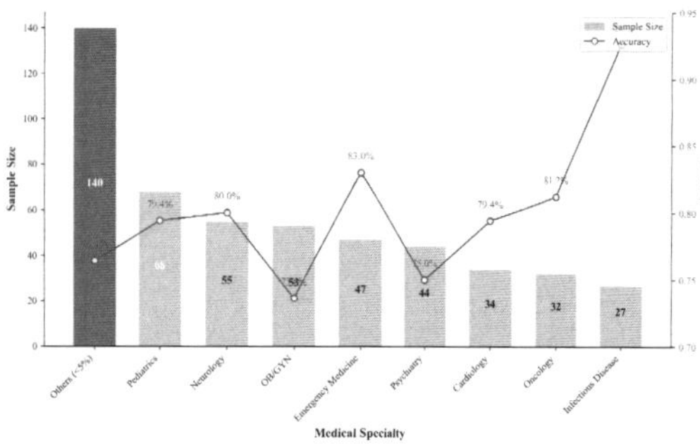

Figure 2: Accuracy of error detection across medical specialties. The visualization shows both sample distribution (bars) and accuracy rates (line) by specialty. Emergency Medicine (83.0%), Infectious Disease (81.2%), and Oncology (79.4%) demonstrate the highest accuracy rates, while Obstetrics/Gynecology (73.6%) and Psychiatry (75.0%) show the lowest. Specialties comprising less than 5% of the dataset are consolidated into the "Others" category (28.0% of total samples).

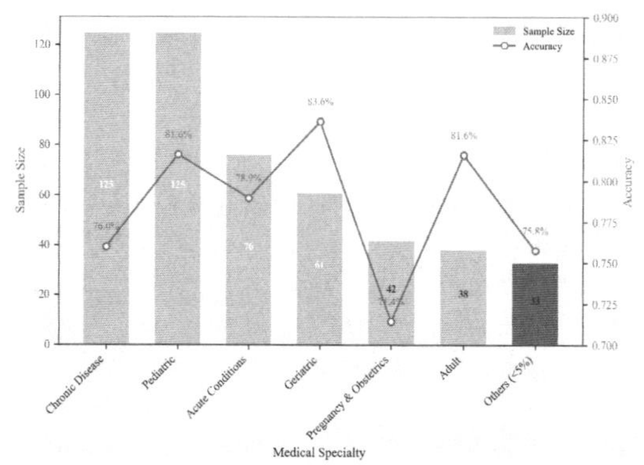

Figure 3: Accuracy of error detection across patient populations. The chart displays both sample size (bars) and accuracy rates (line) for each population category. Geriatric patients and pediatric cases show the highest accuracy rates (83.6% and 81.6% respectively), while pregnancy and obstetric cases present the greatest challenge (71.4%). Categories representing less than 5% of the total sample are grouped as "Others".

clinical findings that would be concerning in non-pregnant patients but are within normal ranges during pregnancy.

The difficulties in Psychiatry stem from two primary factors. First, the model struggles to identify problematic elements within psychiatric notes, possibly due to the more subjective and nuanced nature of psychiatric documentation compared to other specialties. Second, the complexity of psychiatric cases is difficult to adequately capture in concise clinical summaries, leading to misinterpretations. These challenges may be compounded by potential underrepresentation of psychiatric cases in the model's training data.

These findings highlight the importance of specialty-specific optimization for medical AI systems, particularly in domains with unique physiological considerations or documentation practices.

5.2.2 Patient Population

The performance analysis across different patient populations reveals distinct patterns in our model's effectiveness. Geriatric patients (83.6%) and Pediatric cases (81.6%) show the highest accuracy rates, suggesting our model is particularly adept at detecting errors in these populations. This strong performance in age-specific populations is notable, especially for pediatric cases which represent a significant portion of our dataset (25.0%).

Adult patients with chronic diseases (76.0%) show moderate performance despite constituting another major segment of our dataset (25.0%). The model performs reasonably well with acute conditions (78.9%), representing 15.2% of cases, but experiences a notable decline in accuracy for pregnancy and obstetric cases (71.4%). This aligns with our previous observation regarding OB/GYN specialties and reinforces the challenge of accurately evaluating medical information in the context of pregnancy, where physiological baselines differ significantly from general adult populations.

The relatively consistent performance across diverse demographic groups, with most accuracies ranging between 75-84%, indicates overall robustness in the model's error detection capabilities. However, marked underperformance in pregnancy-related cases highlights a specific area that requires targeted improvement. These findings suggest that while our framework generalizes well across most patient populations, specialized training or refinement is necessary for cases where standard medical parameters are naturally altered, such as during pregnancy.

Conclusion

This study introduces a novel multi-agent debating framework for medical error detection that achieves 78.8% accuracy, significantly outperforming both single-agent methods and previous multi-agent approaches. By leveraging specialized agents

Method	Accuracy (%)	P-value
Single-Agent		
Zero-Shot	66.0	<0.001*
Few-Shot	64.2	<0.001*
Few-Shot Variant		
CoT+Few-Shot	69.7	<0.001*
SC+CoT+Few-Shot	64	<0.001*
Multi-Agent		
MDAgent	70.6	0.004*
AutoGen	74.6	0.157
Proposed Method		
S. Agent (WebMD)	70.2	0.002*
S. Agent (Mayo)	72.6	0.029*
Our Method	**78.8**	–

Table 2: Accuracy of various methods on the MS dataset (500 examples). P-values compare each method against our proposed method. Asterisks (*) indicate statistical significance ($p < 0.05$).

with access to complementary medical knowledge sources (Mayo Clinic and WebMD), our structured debate protocol effectively models the collaborative decision-making dynamics found in clinical settings.

Our analysis revealed performance variations across specialties, with strengths in Emergency Medicine (83.0%), Infectious Disease (81.2%), and Oncology (79.4%), and challenges in Obstetrics/Gynecology (73.6%) and Psychiatry (75.0%). Similarly, the system performed robustly with geriatric (83.6%) and pediatric populations (81.6%), though pregnancy-related cases proved more difficult due to altered physiological baselines.

The performance gap between individual specialized agents (WebMD: 70.2%, Mayo: 72.6%) compared to their combined implementation (78.8%) demonstrates how integrating complementary viewpoints through structured debate creates synergistic effects that mirror the benefits of multi-specialist consultation in clinical practice. This research establishes that multi-agent debate represents a promising approach for enhancing the safety and reliability of AI-assisted medical documentation, potentially facilitating wider adoption of AI technologies in clinical settings while maintaining high standards of accuracy. The approach not only improves performance metrics but also generates explanatory reasoning that enhances trust and interpretability—critical factors for responsible AI deployment in medical contexts.

Limitations

The current study presents several limitations worth addressing. First, our dataset encompasses only a specific subset of error types, potentially limiting generalizability to the diverse range of errors encountered in actual clinical environments. Second, computational resource constraints—particularly the cost associated with GPT-4o usage—restricted our ability to conduct more comprehensive testing. Third, our evaluation focused exclusively on closed-source models, leaving questions about cross-model performance variations unanswered. Additionally, we selected only a few medical websites to benchmark their performance, which constrains the comprehensiveness of our analysis. The primary challenge identified lies in medical reasoning capabilities. Future work should investigate how models specifically trained for medical applications might enhance performance. Recent developments such as DeepSeek-R1 (DeepSeek-AI et al., 2025) and advanced post-training methodologies like Group Relative Policy Optimization (Shao et al., 2024) offer promising avenues for improvement. Emerging research examining these approaches in medical contexts (Zhang et al., 2025) suggests fertile ground for future exploration. Such specialized training paradigms could potentially address the reasoning gaps identified in our current multi-agent debate framework.

Acknowledgments

Abdine Maiga is supported by UCL UKRI Center for Doctoral Training in AI-enabled Healthcare studentship (EP/S021612/1). For the purpose of open access the author(s) has applied a Creative Commons Attribution (CC BY) license to any Accepted Manuscript version arising.

I want to thank Amélie and Sierra for their unwavering supports.

References

Brian G. Arndt, John W. Beasley, Michelle D. Watkinson, Jonathan L. Temte, Wen-Jan Tuan, Christine A. Sinsky, and Valerie J. Gilchrist. 2017. Tethered to the EHR: Primary Care Physician Workload Assessment Using EHR Event Log Data and Time-Motion Observations. *The Annals of Family Medicine*, 15(5):419–426. Publisher: The Annals of Family Medicine Section: Original Research.

JR Ball, Bryan T Miller, and Erin Balogh. 2016. *Improv-*

ing Diagnosis in Health Care. National Academies Press.

Sigall K. Bell, Tom Delbanco, Joann G. Elmore, Patricia S. Fitzgerald, Alan Fossa, Kendall Harcourt, Suzanne G. Leveille, Thomas H. Payne, Rebecca A. Stametz, Jan Walker, and Catherine M. DesRoches. 2020. Frequency and Types of Patient-Reported Errors in Electronic Health Record Ambulatory Care Notes. *JAMA Network Open*, 3(6):e205867.

Asma Ben Abacha, Wen-wai Yim, Yujuan Fu, Zhaoyi Sun, Fei Xia, and Meliha Yetisgen. 2024. Overview of the MEDIQA-CORR 2024 Shared Task on Medical Error Detection and Correction. In *Proceedings of the 6th Clinical Natural Language Processing Workshop*, pages 596–603, Mexico City, Mexico. Association for Computational Linguistics.

Tom B. Brown, Benjamin Mann, Nick Ryder, Melanie Subbiah, Jared Kaplan, Prafulla Dhariwal, Arvind Neelakantan, Pranav Shyam, Girish Sastry, Amanda Askell, Sandhini Agarwal, Ariel Herbert-Voss, Gretchen Krueger, Tom Henighan, Rewon Child, Aditya Ramesh, Daniel M. Ziegler, Jeffrey Wu, Clemens Winter, Christopher Hesse, Mark Chen, Eric Sigler, Mateusz Litwin, Scott Gray, Benjamin Chess, Jack Clark, Christopher Berner, Sam McCandlish, Alec Radford, Ilya Sutskever, and Dario Amodei. 2020. Language Models are Few-Shot Learners. *arXiv preprint*. ArXiv:2005.14165 [cs].

Justin Chih-Yao Chen, Swarnadeep Saha, and Mohit Bansal. 2024. ReConcile: Round-Table Conference Improves Reasoning via Consensus among Diverse LLMs. *arXiv preprint*. ArXiv:2309.13007 [cs].

DeepSeek-AI, Daya Guo, Dejian Yang, Haowei Zhang, Junxiao Song, Ruoyu Zhang, Runxin Xu, Qihao Zhu, Shirong Ma, Peiyi Wang, Xiao Bi, Xiaokang Zhang, Xingkai Yu, Yu Wu, Z. F. Wu, Zhibin Gou, Zhihong Shao, Zhuoshu Li, Ziyi Gao, Aixin Liu, Bing Xue, Bingxuan Wang, Bochao Wu, Bei Feng, Chengda Lu, Chenggang Zhao, Chengqi Deng, Chenyu Zhang, Chong Ruan, Damai Dai, Deli Chen, Dongjie Ji, Erhang Li, Fangyun Lin, Fucong Dai, Fuli Luo, Guangbo Hao, Guanting Chen, Guowei Li, H. Zhang, Han Bao, Hanwei Xu, Haocheng Wang, Honghui Ding, Huajian Xin, Huazuo Gao, Hui Qu, Hui Li, Jianzhong Guo, Jiashi Li, Jiawei Wang, Jingchang Chen, Jingyang Yuan, Junjie Qiu, Junlong Li, J. L. Cai, Jiaqi Ni, Jian Liang, Jin Chen, Kai Dong, Kai Hu, Kaige Gao, Kang Guan, Kexin Huang, Kuai Yu, Lean Wang, Lecong Zhang, Liang Zhao, Litong Wang, Liyue Zhang, Lei Xu, Leyi Xia, Mingchuan Zhang, Minghua Zhang, Minghui Tang, Meng Li, Miaojun Wang, Mingming Li, Ning Tian, Panpan Huang, Peng Zhang, Qiancheng Wang, Qinyu Chen, Qiushi Du, Ruiqi Ge, Ruisong Zhang, Ruizhe Pan, Runji Wang, R. J. Chen, R. L. Jin, Ruyi Chen, Shanghao Lu, Shangyan Zhou, Shanhuang Chen, Shengfeng Ye, Shiyu Wang, Shuiping Yu, Shunfeng Zhou, Shuting Pan, S. S. Li, Shuang Zhou, Shaoqing Wu, Shengfeng Ye, Tao Yun, Tian Pei, Tianyu Sun, T. Wang, Wangding Zeng, Wanjia Zhao, Wen Liu, Wenfeng Liang, Wenjun Gao, Wenqin Yu, Wentao Zhang, W. L. Xiao, Wei An, Xiaodong Liu, Xiaohan Wang, Xiaokang Chen, Xiaotao Nie, Xin Cheng, Xin Liu, Xin Xie, Xingchao Liu, Xinyu Yang, Xinyuan Li, Xuecheng Su, Xuheng Lin, X. Q. Li, Xiangyue Jin, Xiaojin Shen, Xiaosha Chen, Xiaowen Sun, Xiaoxiang Wang, Xinnan Song, Xinyi Zhou, Xianzu Wang, Xinxia Shan, Y. K. Li, Y. Q. Wang, Y. X. Wei, Yang Zhang, Yanhong Xu, Yao Li, Yao Zhao, Yaofeng Sun, Yaohui Wang, Yi Yu, Yichao Zhang, Yifan Shi, Yiliang Xiong, Ying He, Yishi Piao, Yisong Wang, Yixuan Tan, Yiyang Ma, Yiyuan Liu, Yongqiang Guo, Yuan Ou, Yuduan Wang, Yue Gong, Yuheng Zou, Yujia He, Yunfan Xiong, Yuxiang Luo, Yuxiang You, Yuxuan Liu, Yuyang Zhou, Y. X. Zhu, Yanhong Xu, Yanping Huang, Yaohui Li, Yi Zheng, Yuchen Zhu, Yunxian Ma, Ying Tang, Yukun Zha, Yuting Yan, Z. Z. Ren, Zehui Ren, Zhangli Sha, Zhe Fu, Zhean Xu, Zhenda Xie, Zhengyan Zhang, Zhewen Hao, Zhicheng Ma, Zhigang Yan, Zhiyu Wu, Zihui Gu, Zijia Zhu, Zijun Liu, Zilin Li, Ziwei Xie, Ziyang Song, Zizheng Pan, Zhen Huang, Zhipeng Xu, Zhongyu Zhang, and Zhen Zhang. 2025. DeepSeek-R1: Incentivizing Reasoning Capability in LLMs via Reinforcement Learning. *arXiv preprint*. ArXiv:2501.12948 [cs].

Aryo Pradipta Gema, Chaeeun Lee, Pasquale Minervini, Luke Daines, T. Ian Simpson, and Beatrice Alex. 2024. Edinburgh Clinical NLP at MEDIQA-CORR 2024: Guiding Large Language Models with Hints. *arXiv preprint*. ArXiv:2405.18028 [cs].

Satya Kesav Gundabathula and Sriram R. Kolar. 2024. PromptMind Team at MEDIQA-CORR 2024: Improving Clinical Text Correction with Error Categorization and LLM Ensembles. *arXiv preprint*. ArXiv:2405.08373 [cs].

George Hripcsak, David K. Vawdrey, Matthew R. Fred, and Susan B. Bostwick. 2011. Use of electronic clinical documentation: time spent and team interactions. *Journal of the American Medical Informatics Association: JAMIA*, 18(2):112–117.

Zhiyuan Hu, Chumin Liu, Xidong Feng, Yilun Zhao, See-Kiong Ng, Anh Tuan Luu, Junxian He, Pang Wei Koh, and Bryan Hooi. 2024. Uncertainty of Thoughts: Uncertainty-Aware Planning Enhances Information Seeking in Large Language Models. *arXiv preprint*. ArXiv:2402.03271 [cs].

Kexin Huang, Jaan Altosaar, and Rajesh Ranganath. 2020. ClinicalBERT: Modeling Clinical Notes and Predicting Hospital Readmission. *arXiv preprint*. ArXiv:1904.05342 [cs].

Hyeon Hwang, Taewhoo Lee, Hyunjae Kim, and Jaewoo Kang. 2024. KU-DMIS at MEDIQA-CORR 2024: Exploring the Reasoning Capabilities of Small Language Models in Medical Error Correction. In *Proceedings of the 6th Clinical Natural Language Processing Workshop*, pages 526–536, Mexico City, Mexico. Association for Computational Linguistics.

Di Jin, Eileen Pan, Nassim Oufattole, Wei-Hung Weng, Hanyi Fang, and Peter Szolovits. 2020. What Disease does this Patient Have? A Large-scale Open Domain Question Answering Dataset from Medical Exams. *arXiv preprint*. ArXiv:2009.13081 [cs] version: 1.

Akbir Khan, John Hughes, Dan Valentine, Laura Ruis, Kshitij Sachan, Ansh Radhakrishnan, Edward Grefenstette, Samuel R. Bowman, Tim Rocktäschel, and Ethan Perez. 2024. Debating with More Persuasive LLMs Leads to More Truthful Answers.

Omar Khattab, Arnav Singhvi, Paridhi Maheshwari, Zhiyuan Zhang, Keshav Santhanam, Sri Vardhamanan, Saiful Haq, Ashutosh Sharma, Thomas T. Joshi, Hanna Moazam, Heather Miller, Matei Zaharia, and Christopher Potts. 2023. DSPy: Compiling Declarative Language Model Calls into Self-Improving Pipelines. *arXiv preprint*. ArXiv:2310.03714 [cs].

Yubin Kim, Chanwoo Park, Hyewon Jeong, Yik Siu Chan, Xuhai Xu, Daniel McDuff, Hyeonhoon Lee, Marzyeh Ghassemi, Cynthia Breazeal, and Hae Won Park. 2024. MDAgents: An Adaptive Collaboration of LLMs for Medical Decision-Making. *arXiv preprint*. ArXiv:2404.15155 [cs].

Tom Knoll, Francesco Moramarco, Alex Papadopoulos Korfiatis, Rachel Young, Claudia Ruffini, Mark Perera, Christian Perstl, Ehud Reiter, Anya Belz, and Aleksandar Savkov. 2022. User-Driven Research of Medical Note Generation Software. *arXiv preprint*. ArXiv:2205.02549 [cs].

Yanis Labrak, Adrien Bazoge, Emmanuel Morin, Pierre-Antoine Gourraud, Mickael Rouvier, and Richard Dufour. 2024. BioMistral: A Collection of Open-Source Pretrained Large Language Models for Medical Domains. *arXiv preprint*. ArXiv:2402.10373 [cs].

Himabindu Lakkaraju, Dylan Slack, Yuxin Chen, Chenhao Tan, and Sameer Singh. 2022. Rethinking Explainability as a Dialogue: A Practitioner's Perspective.

Patrick Lewis, Ethan Perez, Aleksandra Piktus, Fabio Petroni, Vladimir Karpukhin, Naman Goyal, Heinrich Küttler, Mike Lewis, Wen-tau Yih, Tim Rocktäschel, Sebastian Riedel, and Douwe Kiela. 2021. Retrieval-Augmented Generation for Knowledge-Intensive NLP Tasks. *arXiv preprint*. ArXiv:2005.11401 [cs].

Junkai Li, Siyu Wang, Meng Zhang, Weitao Li, Yunghwei Lai, Xinhui Kang, Weizhi Ma, and Yang Liu. 2024. Agent Hospital: A Simulacrum of Hospital with Evolvable Medical Agents. *arXiv preprint*. ArXiv:2405.02957.

Yuan Li, Yixuan Zhang, and Lichao Sun. 2023. MetaAgents: Simulating Interactions of Human Behaviors for LLM-based Task-oriented Coordination via Collaborative Generative Agents. *arXiv preprint*. ArXiv:2310.06500.

Harsha Nori, Nicholas King, Scott Mayer McKinney, Dean Carignan, and Eric Horvitz. 2023. Capabilities of GPT-4 on Medical Challenge Problems. *arXiv preprint*. ArXiv:2303.13375 [cs].

Joon Sung Park, Joseph C. O'Brien, Carrie J. Cai, Meredith Ringel Morris, Percy Liang, and Michael S. Bernstein. 2023. Generative Agents: Interactive Simulacra of Human Behavior. *arXiv preprint*. ArXiv:2304.03442.

Chen Qian, Wei Liu, Hongzhang Liu, Nuo Chen, Yufan Dang, Jiahao Li, Cheng Yang, Weize Chen, Yusheng Su, Xin Cong, Juyuan Xu, Dahai Li, Zhiyuan Liu, and Maosong Sun. 2024. ChatDev: Communicative Agents for Software Development. *arXiv preprint*. ArXiv:2307.07924 [cs].

Khaled Saab, Tao Tu, Wei-Hung Weng, Ryutaro Tanno, David Stutz, Ellery Wulczyn, Fan Zhang, Tim Strother, Chunjong Park, Elahe Vedadi, Juanma Zambrano Chaves, Szu-Yeu Hu, Mike Schaekermann, Aishwarya Kamath, Yong Cheng, David G. T. Barrett, Cathy Cheung, Basil Mustafa, Anil Palepu, Daniel McDuff, Le Hou, Tomer Golany, Luyang Liu, Jean-baptiste Alayrac, Neil Houlsby, Nenad Tomasev, Jan Freyberg, Charles Lau, Jonas Kemp, Jeremy Lai, Shekoofeh Azizi, Kimberly Kanada, SiWai Man, Kavita Kulkarni, Ruoxi Sun, Siamak Shakeri, Luheng He, Ben Caine, Albert Webson, Natasha Latysheva, Melvin Johnson, Philip Mansfield, Jian Lu, Ehud Rivlin, Jesper Anderson, Bradley Green, Renee Wong, Jonathan Krause, Jonathon Shlens, Ewa Dominowska, S. M. Ali Eslami, Katherine Chou, Claire Cui, Oriol Vinyals, Koray Kavukcuoglu, James Manyika, Jeff Dean, Demis Hassabis, Yossi Matias, Dale Webster, Joelle Barral, Greg Corrado, Christopher Semturs, S. Sara Mahdavi, Juraj Gottweis, Alan Karthikesalingam, and Vivek Natarajan. 2024. Capabilities of Gemini Models in Medicine. *arXiv preprint*. ArXiv:2404.18416 [cs].

Zhihong Shao, Peiyi Wang, Qihao Zhu, Runxin Xu, Junxiao Song, Xiao Bi, Haowei Zhang, Mingchuan Zhang, Y. K. Li, Y. Wu, and Daya Guo. 2024. DeepSeekMath: Pushing the Limits of Mathematical Reasoning in Open Language Models. *arXiv preprint*. ArXiv:2402.03300 [cs].

Karan Singhal, Shekoofeh Azizi, Tao Tu, S. Sara Mahdavi, Jason Wei, Hyung Won Chung, Nathan Scales, Ajay Tanwani, Heather Cole-Lewis, Stephen Pfohl, Perry Payne, Martin Seneviratne, Paul Gamble, Chris Kelly, Abubakr Babiker, Nathanael Schärli, Aakanksha Chowdhery, Philip Mansfield, Dina Demner-Fushman, Blaise Agüera y Arcas, Dale Webster, Greg S. Corrado, Yossi Matias, Katherine Chou, Juraj Gottweis, Nenad Tomasev, Yun Liu, Alvin Rajkomar, Joelle Barral, Christopher Semturs, Alan Karthikesalingam, and Vivek Natarajan. 2023. Large language models encode clinical knowledge. *Nature*, 620(7972):172–180. Number: 7972 Publisher: Nature Publishing Group.

Xiangru Tang, Anni Zou, Zhuosheng Zhang, Ziming Li, Yilun Zhao, Xingyao Zhang, Arman Cohan, and

Mark Gerstein. 2024. MedAgents: Large Language Models as Collaborators for Zero-shot Medical Reasoning. In *Findings of the Association for Computational Linguistics: ACL 2024*, pages 599–621, Bangkok, Thailand. Association for Computational Linguistics.

Arun James Thirunavukarasu, Darren Shu Jeng Ting, Kabilan Elangovan, Laura Gutierrez, Ting Fang Tan, and Daniel Shu Wei Ting. 2023. Large language models in medicine. *Nature Medicine*, 29(8):1930–1940. Publisher: Nature Publishing Group.

Augustin Toma, Ronald Xie, Steven Palayew, Patrick R. Lawler, and Bo Wang. 2024. WangLab at MEDIQA-CORR 2024: Optimized LLM-based Programs for Medical Error Detection and Correction. *arXiv preprint*. ArXiv:2404.14544 [cs].

Airat Valiev and Elena Tutubalina. 2024. HSE NLP Team at MEDIQA-CORR 2024 Task: In-Prompt Ensemble with Entities and Knowledge Graph for Medical Error Correction. In *Proceedings of the 6th Clinical Natural Language Processing Workshop*, pages 470–482, Mexico City, Mexico. Association for Computational Linguistics.

Dave Van Veen, Cara Van Uden, Louis Blankemeier, Jean-Benoit Delbrouck, Asad Aali, Christian Bluethgen, Anuj Pareek, Malgorzata Polacin, Eduardo Pontes Reis, Anna Seehofnerova, Nidhi Rohatgi, Poonam Hosamani, William Collins, Neera Ahuja, Curtis P. Langlotz, Jason Hom, Sergios Gatidis, John Pauly, and Akshay S. Chaudhari. 2023. Clinical Text Summarization: Adapting Large Language Models Can Outperform Human Experts. *arXiv preprint*. ArXiv:2309.07430 [cs] version: 3.

Dave Van Veen, Cara Van Uden, Louis Blankemeier, Jean-Benoit Delbrouck, Asad Aali, Christian Bluethgen, Anuj Pareek, Malgorzata Polacin, Eduardo Pontes Reis, Anna Seehofnerová, Nidhi Rohatgi, Poonam Hosamani, William Collins, Neera Ahuja, Curtis P. Langlotz, Jason Hom, Sergios Gatidis, John Pauly, and Akshay S. Chaudhari. 2024. Adapted large language models can outperform medical experts in clinical text summarization. *Nature Medicine*, 30(4):1134–1142. Publisher: Nature Publishing Group.

Lei Wang, Chen Ma, Xueyang Feng, Zeyu Zhang, Hao Yang, Jingsen Zhang, Zhiyuan Chen, Jiakai Tang, Xu Chen, Yankai Lin, Wayne Xin Zhao, Zhewei Wei, and Ji-Rong Wen. 2024. A Survey on Large Language Model based Autonomous Agents. *Frontiers of Computer Science*, 18(6):186345. ArXiv:2308.11432 [cs].

Xuezhi Wang, Jason Wei, Dale Schuurmans, Quoc Le, Ed Chi, Sharan Narang, Aakanksha Chowdhery, and Denny Zhou. 2023. Self-Consistency Improves Chain of Thought Reasoning in Language Models. *arXiv preprint*. ArXiv:2203.11171 [cs].

Jason Wei, Xuezhi Wang, Dale Schuurmans, Maarten Bosma, Brian Ichter, Fei Xia, Ed Chi, Quoc Le, and Denny Zhou. 2023. Chain-of-Thought Prompting Elicits Reasoning in Large Language Models. *arXiv preprint*. ArXiv:2201.11903 [cs].

Qingyun Wu, Gagan Bansal, Jieyu Zhang, Yiran Wu, Beibin Li, Erkang Zhu, Li Jiang, Xiaoyun Zhang, Shaokun Zhang, Jiale Liu, Ahmed Hassan Awadallah, Ryen W. White, Doug Burger, and Chi Wang. 2023. AutoGen: Enabling Next-Gen LLM Applications via Multi-Agent Conversation. *arXiv preprint*. ArXiv:2308.08155 [cs].

Sheng Zhang, Qianchu Liu, Guanghui Qin, Tristan Naumann, and Hoifung Poon. 2025. Med-RLVR: Emerging Medical Reasoning from a 3B base model via reinforcement Learning. *arXiv preprint*. ArXiv:2502.19655 [cs].

A Example Appendix: Multi-Agent Debate Case

A.1 Patient Case and Diagnosis

54-year-old woman with a painful, rapidly growing leg lesion for 1 month. History includes Crohn's disease, diabetes, hypertension, and previous anterior uveitis. Examination revealed a 4-cm tender ulcerative lesion with necrotic base and purplish borders, along with pitting edema and dilated veins. Diagnosed as a venous ulcer.

A.2 Multi-Agent Debate Analysis

A.2.1 Expert Agent Analysis Summary

- **Mayo Clinic Agent**: Identified lesion characteristics (necrotic base, purplish borders) as inconsistent with venous ulcers. Highlighted association between Crohn's disease and pyoderma gangrenosum. Emphasized misdiagnosis risk and treatment implications.
- **WebMD Agent**: Noted venous insufficiency findings but questioned diagnosis due to patient's inflammatory disease history. Suggested alternative diagnoses including pyoderma gangrenosum given clinical context.

A.2.2 Judge Agent Decision

Final Answer: INCORRECT (Confidence: 8/10)
Reasoning: Both agents identified pyoderma gangrenosum as a more likely diagnosis given the patient's Crohn's disease and lesion characteristics. Mayo Clinic agent provided more comprehensive analysis of clinical inconsistencies, emphasizing the importance of correct diagnosis for appropriate treatment.

A.3 Key Insight

This case demonstrates how our multi-agent framework effectively identifies potential diagnostic errors by leveraging complementary medical knowledge sources. The specialized agents provided different analytical approaches but converged on the same conclusion, highlighting the value of structured debate in enhancing diagnostic accuracy for complex cases.

B Example Appendix

C Agent System Messages

C.1 Healthcare Expert Agent (Mayo Clinic) System Message

[breakable, enhanced, colback=gray!5!white, boxrule=0.5pt] You are a healthcare professional specializing in analyzing medical notes, with expertise in diagnosis and clinical terminology. Important: Medical notes should be presumed CORRECT unless there is an obvious, significant error. Your task is to identify only clear substitution errors in:

- Diagnostic terms that significantly change the clinical meaning - Medication terms that would result in wrong treatment - Treatment protocols that are clearly contraindicated - Management plans that would harm the patient - Therapeutic interventions that are definitively inappropriate

Classification criteria: - INCORRECT: Contains exactly one clinically significant term substitution that would change patient care - CORRECT: Default classification - use this unless there is a clear, significant error

In your final turn, please provide a detailed final explanation for your decision including your reasoning and supporting evidence. Conclude with a sentence beginning 'Based on my analysis, this note is ...'.

C.2 Judge Agent System Message

[breakable, enhanced, colback=gray!5!white, boxrule=0.5pt] You are the judge in a debate between two healthcare agents. They have each presented their arguments about whether the medical note is correct or contains an error. Do not interfere with the debate while it is ongoing; wait until both agents have finished their 2 exchanges. Once the debate has concluded, evaluate both agents' final messages and decide which agent made the more convincing case (i.e., which agent correctly identified whether the note is correct or incorrect). Provide a clear explanation for your decision. Your final response should be in JSON format with the structure:

{ "Final Answer": "CORRECT/INCORRECT", "Confidence Score": <number>, "Winner": "<Agent Name>", "Reasoning": "<Explanation of decision>" }

Do not include any additional commentary.

Accelerating Cross-Encoders in Biomedical Entity Linking

Javier Sanz-Cruzado
University of Glasgow
Glasgow, United Kingdom
javier.sanz-cruzadopuig@glasgow.ac.uk

Jake Lever
University of Glasgow
Glasgow, United Kingdom
jake.lever@glasgow.ac.uk

Abstract

Biomedical entity linking models disambiguate mentions in text by matching them with unique biomedical concepts. This problem is commonly addressed using a two-stage pipeline comprising an inexpensive candidate generator, which filters a subset of suitable entities for a mention, and a costly but precise reranker that provides the final matching between the mention and the concept. With the goal of applying two-stage entity linking at scale, we explore the construction of effective cross-encoder reranker models, capable of scoring multiple mention-entity pairs simultaneously. Through experiments on four entity linking datasets, we show that our cross-encoder models provide between 2.7 to 36.97 times faster training speeds and 3.42 to 26.47 times faster inference speeds than a base cross-encoder model capable of scoring only one entity, while achieving similar accuracy (differences between -3.42% to 2.76% Acc@1).

1 Introduction

Biomedical entity linking matches mentions of biomedical concepts (diseases, chemicals) in texts with unique entities within a knowledge base (Kartchner et al., 2023; Garda et al., 2023). Disambiguating mentions within text is fundamental for information extraction tasks, as a single entity might be referred to by different names or aliases (e.g. chickenpox and varicella refer to the same disease), or a mention might refer to multiple entities (e.g. APC might refer to advanced pancreatic or prostate cancer).

This problem is commonly devised as a two-phase procedure (Xu et al., 2023): given a mention in a text, an initial model selects a reduced set of candidate entities it might refer to. This model is commonly fast, as it needs to filter among thousands of entities (Neumann et al., 2019; Liu et al., 2021). Then, a second, more precise model

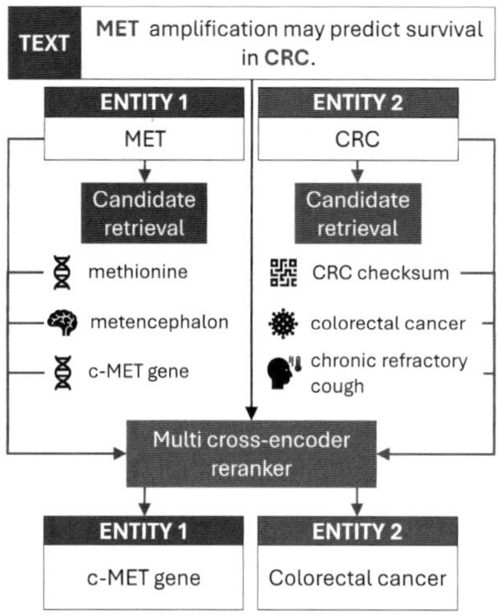

Figure 1: Example of our multi cross-encoder model.

reranks these candidate entities to provide the final matching between a mention and an entity in the knowledge-base. Cross-encoders (Logeswaran et al., 2019) are a popular option for this stage (Wu et al., 2020; Zhang et al., 2022). While accelerated by the reduced candidate selection, these rerankers are usually costly, requiring extensive training and inference times. The efficiency of these rerankers can be an important factor of our entity linking pipeline if we want to run these models at scale across millions of documents. However, works on biomedical entity linking have traditionally focused on the accuracy of the models and not on their efficiency.

Therefore, in this paper, we aim to improve the efficiency of second stage cross-encoder models. Taking as a starting point a base cross-encoder (Logeswaran et al., 2019), we propose novel entity linking methods that improve both training and inference speeds, while maintaining similar accuracy levels. For this, inspired by the longer context windows of recent encoder-only transformer models like ModernBERT (Warner et al., 2024), we design cross-encoders capable of scoring multiple candi-

Proceedings of the 24th Workshop on Biomedical Language Processing (BioNLP 2025), pages 136–147
August 1, 2025

dates at the same time, and even reranking multiple mentions simultaneously. We show an example of this in Figure 1. Our contributions are threefold:

- We propose a novel multi cross-encoder architecture that accelerates the training and inference times of a classical reranker cross-encoder for entity linking.
- We compare our approach on three different transformer models and four different biomedical entity linking datasets.
- We find that our cross-encoders can accelerate up to 36.97 times the training speed of a simple cross-encoder and up to 26.47 times the inference speed, while providing similar effectiveness.

2 Task definition and notation

We start by formally defining the entity linking (EL) task. EL aims to uniquely match entities mentioned in the text with unique concepts within a knowledge base. Let's suppose we have a knowledge base containing a set of unique entities $\mathcal{E}$ and a corpus of documents $\mathcal{D}$. Each document $d \in \mathcal{D}$ has a series of mentions $\mathcal{M}_d$, where a mention $m \in \mathcal{M}_d$ is a sequence of tokens $m = d_m^{(1)} \cdots d_m^{(l)} \subseteq d$ that corresponds to a unique entity. Given a document d and a mention $m \in \mathcal{M}_d$, the EL task consists on identifying the entity $e_m \in \mathcal{E}$ that m refers to in d. We address this task as a ranking problem, where we find the entity maximizing a ranking function $f_{m,d} : \mathcal{E} \rightarrow \mathbb{R}$.

3 Related work

While early works on biomedical entity linking date back to the late 1980s (French and McInnes, 2023), a majority of recent works in this area are based on recent transformer-based language models like BERT (Devlin et al., 2019) or BiomedBERT (Gu et al., 2021). These works can be divided into single-phase and two-phase models.

Single-phase models directly rank all entities within a knowledge-base for a single mention. These methods usually estimate the similarity between mentions and entities based on a combination of sparse or dense vector representations (Sung et al., 2020; Loureiro and Jorge, 2020). These models commonly use computationally efficient algorithms like bi-encoders to obtain separate representations of entities and mentions. Examples of

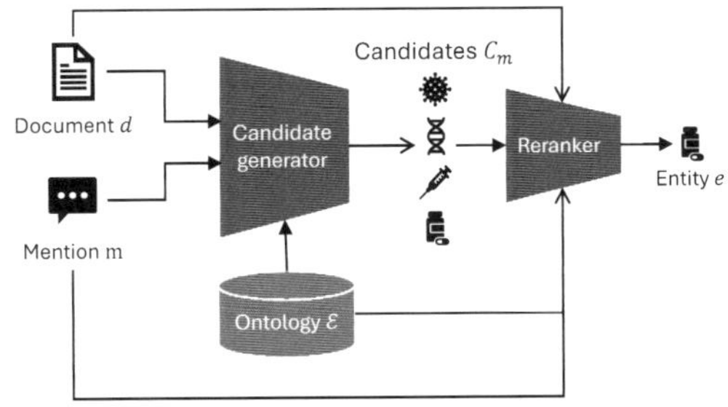

Figure 2: Two-stage pipeline

algorithms within this category are SapBERT (Liu et al., 2021), BioSyn (Sung et al., 2020) and MedLinker (Loureiro and Jorge, 2020).

This work focuses on the second type of models, the two-phase models. These algorithms apply two different entity linking approaches: a fast and efficient model for retrieving a subset of candidates (for instance, a character n-grams model (Angell et al., 2021) or a bi-encoder like SapBERT (Xu et al., 2023; Zhu et al., 2024)), followed by a more computationally expensive, but precise reranker that reranks the set of candidate items. Following Logeswaran et al. (2019), a majority of these models use a cross-encoder model as a reranker. While different biomedical EL models like ClusterEL (Angell et al., 2021), ArboEL (Agarwal et al., 2022), KrissBERT (Zhang et al., 2022) apply a similar pipeline, all of these models have focused their attention only on the effectiveness of the entity linking model, commonly using a simple cross-encoder model for reranking. However, there is still room to improve not only the effectiveness, but also the efficiency of these approaches.

To improve the efficiency of these models, we get inspiration from the Prompt-BioEL method proposed by (Xu et al., 2023). In their approach, they apply a cross-encoder capable of processing all the candidates for a mention simultaneously. In this work, we go further, by building cross-encoder models which can rerank multiple mentions, sentences or passages at the same time.

4 Method

In this section, we describe our approach for balancing the accuracy and efficiency of entity linking models. Figure 2 illustrates the general architecture of our entity linking models. Following previous works (Zhang et al., 2022; Logeswaran et al., 2019), we adopt a two-stage pipeline for the task. First, we apply a **candidate retrieval** model, which selects a small subset of candidate entities $C_m \subset \mathcal{E}$ from the ontology. Then, we apply a **reranker** model that chooses the best entity among the ones in C_m. We next describe each of these components.

4.1 Candidate retriever

As an efficient and effective first-stage candidate retriever, we use a n-grams model (Neumann et al., 2019) for representing both the mention text m and the aliases of entity e (which we denote as $A(e)$). Then, we rank candidate entities by the maximum TF-IDF similarity between m and every alias $a_e \in A(e)$ of the entity.

$$f_{m,d}(e) = \max_{a_e \in A(e)} \text{tf-idf}(\text{n-gr}(m),\ \text{n-gr}(a_e)) \quad (1)$$

where n-gr(t) is the n-grams representation of t.

4.2 Reranker

As a second-stage candidate reranker, we use cross-encoder models. Cross-encoders have been previously used for the entity linking task, but they are costly to train and apply. Therefore, we propose improvements over the basic cross-encoder architecture, illustrated in Figure 3.

4.2.1 Preprocessing

As an initial step, prior to the application of the reranker, we pre-process the documents in our corpus. We divide the documents into passages, and each passage into sentences. Each annotated sentence is later provided as input to the cross-encoder models, providing context for each mention.

4.2.2 Base cross-encoder

We first describe the architecture of the base cross-encoder model (Humeau et al., 2020), depicted in Figure 3(a). Given an annotated sentence $t \subseteq d$, a mention m and a candidate entity $c \in C_m$ for that mention, the cross-encoder computes a score $f_{m,d}(c)$ estimating the likelihood that the candidate entity c corresponds to the target entity e_m that mention m is referring to. Each candidate $c \in C_m$ is processed separately, and then, candidates are ranked in descending score order.

A common strategy to build the cross-encoder (and the one we follow in this work) is to fine-tune a pre-trained language model (LM). The LM receives as input a sentence following the following template $\tau(t, m, c)$[1]:

$$\tau(t, m, c) = \text{“}t \text{ [SEP] } m \text{ [MASK] } c\text{”} \quad (2)$$

[1]For our cross-encoder models, we represent c as the main textual representation of the entity in the knowledge base

An example of an input sentence is shown in Figure 4. Then, the cross-encoder classifies the [MASK] token into two classes: a positive class, indicating that c matches the referred entity e_m, and a negative class otherwise. Therefore, the score $f_{m,d}(c)$ is defined as:

$$f_{m,d}(c) = p\left([\text{MASK}] = 1 | \tau\left(t, m, c\right)\right) \quad (3)$$

In order to fine-tune the model, we apply a cross-entropy loss minimizing the classification error on the [MASK] token. For a candidate c and a mention m, the loss is defined as

$$\begin{aligned}\mathcal{L} = &-\mathbb{1}\left(c = e_m\right) \cdot \log f_{m,d}(c) \\ &- \left(1 - \mathbb{1}\left(c = e_m\right)\right) \cdot \log\left(1 - f_{m,d}(c)\right)\end{aligned} \quad (4)$$

where $\mathbb{1}(x)$ is the indicator function.

4.2.3 Parallel cross-encoder

One of the limitations of the architecture of the base cross-encoder is its capacity to process only one mention-candidate pair at a time. Therefore, in both training and inference, the cross-encoder needs to process the sentence t as many times as candidates we retrieve during the first phase of the entity linking process – making this process costly.

Inspired by Xu et al. (2023) and Jiang et al. (2023), we propose to improve our cross-encoder by allowing it to process multiple candidates for a mention at the same time. We denote the new architecture as parallel cross-encoder. As illustrated in Figure 3(b), the parallel cross-encoder takes as input the text t and all the candidates for mention m, and provides, as output, the scores for all of those candidates. The parallel cross-encoder receives input with the following template:

$$\begin{aligned}\tau(t, m) = \text{“}t \text{ [SEP] } m \text{ [MASK] } c_1 \\ \dots \\ \text{[SEP] } m \text{ [MASK] } c_{|C_m|}\text{”}\end{aligned} \quad (5)$$

and, for each candidate $c \in C_m$, its score is

$$f_{m,d}(c) = p\left([\text{MASK}]_c = 1 | \tau\left(t, m\right)\right) \quad (6)$$

where $[\text{MASK}]_c$ is the mask token corresponding to entity c. An example of this input text is shown in Figure 4.

While the parallel cross-encoder increases the complexity of the task (the cross-encoder receives longer text sequences and needs to classify multiple tokens), it should accelerate training and inference times. As long as the cross-encoder effectively processes long sequences of tokens, we should gain advantage from processing sentence t only once for a given mention.

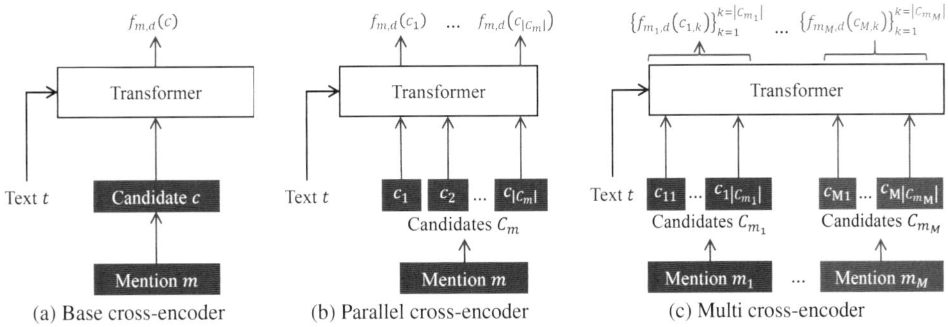

Figure 3: Architecture of the different cross-encoder models

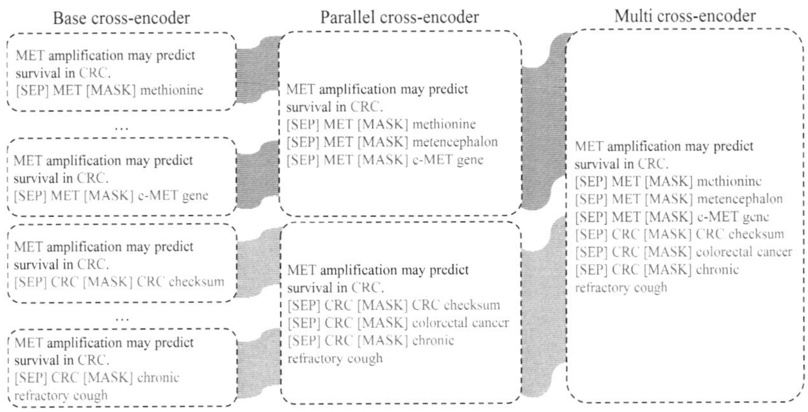

Figure 4: Input of the cross-encoder models.

4.2.4 Multi cross-encoder

The parallel cross-encoder can be further refined to improve the efficiency of the cross-encoder by reducing the amount of times that the cross-encoder is exposed to the same text. As shown in Figure 1, each sentence might contain not only one, but multiple mentions to entities in the knowledge base. Therefore, we propose a new cross-encoder model, denoted as multi cross-encoder that receives as input not only the candidates of an individual mention, but the candidates of all mentions within the sentence and provides the corresponding estimates. We illustrate this architecture in Figure 3(c).

The multi cross-encoder works similarly to the parallel cross-encoder. If we denote as $\mathcal{M}_t \subseteq \mathcal{M}_d$ the set of mentions in a sentence t, and $|\mathcal{M}_t| = M$, we define the input text of the multi-cross encoder as a sequence of tokens with the following format:

$$\begin{aligned}\tau(t, \mathcal{M}_t) = \text{``}t \text{ [SEP] } & m_1 \text{ [MASK] } c_{1,1} \\ & \dots \\ \text{[SEP] } & m_1 \text{ [MASK] } c_{1,|C_{m_1}|} \\ \text{[SEP] } & m_2 \text{ [MASK] } c_{2,1} \\ & \dots \\ \text{[SEP] } & m_M \text{ [MASK] } c_{M,|C_{m_M}|}\text{''}\end{aligned} \quad (7)$$

We provide an example on Figure 4. Then, the score for a candidate $c \in C_m$ is defined as:

$$f_{m,d}(c) = p\left([\text{MASK}]_{m,c} = 1 | \tau(t, \mathcal{M}_t)\right) \quad (8)$$

where $[\text{MASK}]_{m,c}$ is the mask token corresponding to mention m and candidate $c \in C_m$ in $\tau(t, \mathcal{M}_t)$.

4.2.5 Adaptation to context window

As we concatenate multiple mention-entity pairs in the input text, we might obtain texts longer than the context window of the language model (maximum number of tokens that the LM can receive at once). In that case, we partition the mention-candidate pairs into several rankings by choosing, each time, as many pairs as we can fit along the sentence into the context window. We apply this strategy as our method provides pointwise scores (each mention-candidate pair has an individual score) – and therefore, separating the pairs on different calls to the cross-encoder should not have a big impact on performance. In the worst case, where only one mention-entity pair fits in the context, our model would be equivalent to the base cross-encoder. We show an example of this procedure in Figure 5. Following this procedure, LMs with longer context

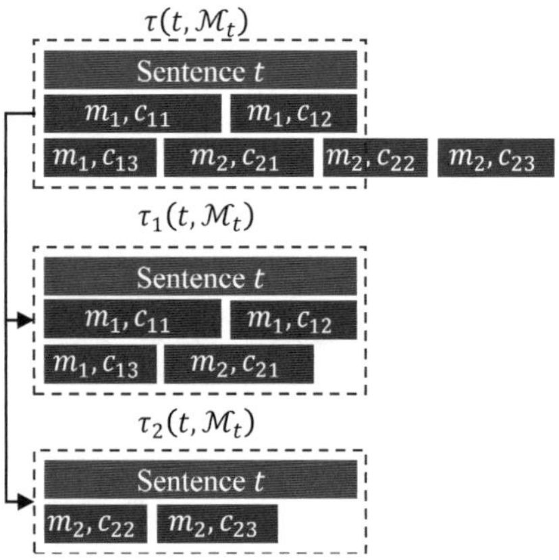

Figure 5: Partition procedure when an input sentence is longer than the context window.

windows (like ModernBERT (Warner et al., 2024)) shall have longer input texts than models accepting less tokens (like BiomedBERT (Gu et al., 2021)).

4.2.6 Further architectures

We can further design additional cross-encoder architectures that include even more data – with the objective of maximizing the use of the cross-encoder context window – by concatenating the template $\tau(t, \mathcal{M}_t)$ of multiple sentences within the document (for instance, all sentences within a passage or all sentences within a document).

5 Experimental setup

5.1 Datasets

In our experiments, we consider four common datasets for biomedical entity linking:

- **MedMentions (Mohan and Li, 2019):** PubMed biomedical abstract collection annotated with mentions of entities in the UMLS 2017AA release. We use the full version of the dataset. For each entity, we only keep the English aliases.
- **NCBI Disease (Doğan et al., 2014):** PubMed abstract corpus linking disease mentions to entities in the MEDIC ontology[2]. Only mentions with an unambiguous entity link with an entity in that MEDIC release were kept.
- **NLM Chem (Islamaj et al., 2021):** Set of full-text articles from the PubMed Central Open Access dataset covering the use of chemical names in the biomedical literature. We keep only mentions of type 'Chemical' linked with entities in the MeSH 2021 release.
- **BioCreative V CDR (BC5CDR) (Li et al., 2016):** Collection of PubMed abstracts with chemical and disease annotations from the Comparative Toxicogenomics Database. Only contiguous mentions were kept and all linked entities are found in the MeSH 2015 release.

Table 1: Dataset properties.

Property	MedMentions	NCBI Disease	NLM Chem	BC5CDR
Ontology	UMLS	Medic	Mesh 2021	Mesh 2015
Documents (train)	2,635	593	80	500
Documents (val)	878	100	20	500
Documents (test)	879	100	50	500
Passages (train)	2,635	593	5,555	1,000
Passages (val)	878	100	1,285	1,000
Passages (test)	879	100	3,470	1,000
Sentences (train)	25,836	5,173	20,126	4,242
Sentences (val)	8,508	888	4,855	4,299
Sentences (test)	8,597	901	12,031	4,524
Entities (train)	211,029	4,836	19,361	9,323
Entities (val)	71,062	711	4,927	9,570
Entities (test)	70,405	896	11,164	9,725

Data splitting: For each of the four datasets, we use the default training/validation/test split. We use the training and validation datasets to fine-tune the models, and we report entity linking results over the test set.

Passages and sentences: For the MedMentions and NCBI Disease datasets, each document consists of a single passage combining both the title and abstract. In the BC5CDR dataset, we have two passages for each dataset: one for the title, and another one for the abstract text. Finally, for NLM Chem, we use the passage division of each document provided by the dataset. For splitting each passage into sentences, we use the spaCy[3] en_core_web_sm sentence parser. If the parser splits a mention in two different sentences, we combine the two sentences. We show the statistics of each dataset in Table 1.

5.2 Models

First-phase candidate retriever: As mentioned in Section 4.1, we use a TF-IDF n-grams model (Neumann et al., 2019). We apply an efficient implementation of this model by building an n-grams index with Pyterrier-PISA (Mallia et al., 2019; MacAvaney and Macdonald, 2022). For efficiency, this index is built only using the first 16 characters of entity aliases. Then, for each dataset, we use the n-grams model maximizing the amount of correct

[2]10 May 2012 version, obtained using Internet Archive

[3]spaCy: https://spacy.io/

Table 2: Language model statistics

Model	Domain	Context-window length
BiomedBERT	Biomedical	512
Longformer	General	4,096
ModernBERT	General	8,192

entities in the top-5 (3-grams for MedMentions, 2-grams for the rest).

Second-phase reranker models: Then, we build cross-encoders for reranking the top-5 candidate entities. We consider three different backbone pretrained language models in our experiments, with varying context window size: BiomedBERT (Gu et al., 2021), Longformer (Beltagy et al., 2020) and ModernBERT (Warner et al., 2024). Table 2 summarizes their statistics.

As a baseline for our experiments, we consider the base cross-encoder defined in Section 4.2.2 (which we denote as LM-base). We compare this baseline against four models: the parallel cross-encoder (LM-parallel) and the multi cross-encoder (LM-multi) in Sections 4.2.3 and 4.2.4 , and two additional cross-encoders: one including the whole passage text (LM-passage), and another one including the complete document text (LM-document).

To reduce the training time of each model, all cross-encoders follow an early stopping strategy, where we stop the training if the cross-encoder fails to improve the F1 performance on the validation set by 1% for three consecutive epochs. All models use the same learning rate (10^{-6}).

5.3 Metrics

We compare our models across three main metrics:

- **Accuracy@1 (Acc@1):** This metric measures the ultimate goal of the reranker to assign the highest score to the correct entity for each mention from the list of candidates. It is the proportion of annotations for which this is the case.
- **Training speed:** This metric measures the efficiency of the fine-tuning process. As it is unfair to compare models directly on the training time (as different models might use a different number of training epochs), we estimate the number of annotations processed per second during the cross-encoder fine-tuning.
- **Inference speed:** This metric estimates the number of test examples per second that the cross-encoder can process.

For reference, we also report the total training and inference times of our cross-encoder models.

5.4 Hardware

We train and execute all our models on a single NVIDIA RTX 4090 GPU card (24 GB VRAM), 2 CPUs and 16 GB of RAM. The batch size of each model has been adjusted to be trained on the mentioned GPU card – with all variations of the same model using the same batch size.

5.5 Implementation

For reproducibility, we provide the code for our experiments in the following GitHub repository: `https://github.com/Glasgow-AI4BioMed/entitytools`.

6 Results

We aim to answer the following research questions:

- **RQ1:** How does the parallelism of the cross-encoder affect the effectiveness of the model?
- **RQ2:** How does the parallelism of the cross-encoder affect the model training and inference speeds?

6.1 RQ1: Accuracy comparison

We first analyse how effective the different cross-encoders are for the entity linking task. As we keep adding more information to our cross-encoder, we shall expect the task to become more complex and therefore affect the effectiveness of our models. We show the results in Tables 3 (for the Medmentions and NCBI disease datasets) and 4 (for the NLM Chem and BC5CDR corpora). In these tables, we underline the best result for each metric and backbone LM for our cross-encoders, and we highlight in bold the best overall result. Accuracy@1 results are shown in the first column for each dataset.

Base cross-encoder performance: We evaluate the effectiveness of the two-stage entity linking model by comparing the base cross-encoders with the single-stage n-grams model. In line with other works (Zhang et al., 2022; Agarwal et al., 2022), the three base cross-encoders achieve statistically significant improvements (McNemar test with $p < 0.05$ and Bonferroni correction) on a majority of datasets. The only exception is the NLM Chem dataset, where only the domain-specific BiomedBERT cross-encoder can improve the n-grams similarity model. Among the three

Table 3: Evaluation of entity linking (Medmentions and NCBI Disease). For each model, ^{a} represents statistical significance (McNemar test with Bonferroni correction and $p < 0.05$) with respect to the first stage linker. ^{b,c,d,e} represent, respectively, a significant improvement over the simple, parallel, multi or passage cross-encoder with the same base transformers model. For each metric, ↑ indicates that higher values are better, while ↓ indicates that lower values are better. Best values are highlighted in bold, and the best cross-encoder for each backbone LM is underlined.

	Medmentions-full					NCBI Disease				
Model	Acc@1 (↑)	Training speed (↑)	Training time (s) (↓)	Inference speed (↑)	Inference time (s) (↓)	Acc@1 (↑)	Training speed (↑)	Training time (s) (↓)	Inference speed (↑)	Inference time (s) (↓)
First stage	0.4222	202.52	**1,042**	1,466.77	44	0.5960	322.40	**15**	**896.00**	1
BiomedBERT-base	**0.6422**^{acde}	62.05	23,806	145.46	484	0.6886^a	70.09	690	179.20	5
BiomedBERT-parallel	0.6334^{ade}	255.74	4,951	690.25	102	0.6931^a	266.55	127	**896.00**	**1**
BiomedBERT-multi	0.6244^a	639.16	1,981	2,200.16	32	0.6864^a	318.16	152	**896.00**	**1**
BiomedBERT-passage	0.6235^a	**985.46**	1,499	**2,607.59**	**27**	0.6786^a	**531.43**	91	**896.00**	**1**
Longformer-base	0.6353^{acde}	4.44	190,014	31.64	2,225	0.6920^a	5.00	7,744	37.33	24
Longformer-parallel	0.6284^{ade}	21.55	39,162	158.93	443	0.6920^a	22.44	862	179.20	5
Longformer-multi	0.6203^{ae}	137.40	9,215	733.39	96	0.6953^a	54.54	798	298.67	3
Longformer-passage	0.6136^a	168.69	8,757	869.20	81	0.6842^a	81.97	531	448.00	**2**
ModernBERT-base	0.6300^{ade}	11.84	71,305	44.99	1,440	0.6875^a	12.94	1,869	48.89	20
ModernBERT-parallel	0.6287^{ad}	58.03	14,546	237.05	297	0.6953^a	59.41	407	179.20	5
ModernBERT-multi	0.6255^a	205.88	4,100	1117.54	63	**0.7042**^{a}	57.92	501	298.67	3
ModernBERT-passage	0.6266^a	196.90	8,574	529.36	133	0.7009^a	159.68	212	448.00	**2**

Table 4: Evaluation of entity linking (NLM Chem and BC5CDR). For each model, ^{a} represents statistical significance (McNemar test with Bonferroni correction and $p < 0.05$) with respect to the first stage linker. ^{b,c,d,e,f} represent, respectively, a significant improvement over the simple, parallel, multi, passage or document cross-encoder with the same base transformers model. For each metric, ↑ indicates that higher values are better, while ↓ indicates that lower values are better. Best values are highlighted in bold, and the best cross-encoder for each backbone LM is underlined.

	NLM Chem					BC5CDR				
Model	Acc@1 (↑)	Training speed (↑)	Training time (s) (↓)	Inference speed (↑)	Inference time (s) (↓)	Acc@1 (↑)	Training speed (↑)	Training time (s) (↓)	Inference speed (↑)	Inference time (s) (↓)
First stage	0.6990	108.77	178	**1594.86**	7	0.8193	61.34	152	1389.29	7
BiomedBERT-base	0.7004	54.34	2,494	150.86	74	0.8403^{adef}	58.64	1,113	149.62	65
BiomedBERT-parallel	**0.7083**^{abf}	211.60	366	697.75	16	**0.8444**^{adef}	216.81	301	694.64	14
BiomedBERT-multi	0.7066^a	268.90	288	1014.91	11	0.8249^a	310.77	120	1389.29	7
BiomedBERT-passage	0.7075^{ab}	378.39	307	1240.44	9	0.8274^a	443.95	168	**1620.83**	**6**
BiomedBERT-document	0.7043^a	**556.35**	**174**	**1594.86**	**7**	0.8254^a	**570.80**	**98**	**1620.83**	**6**
Longformer-base	0.6940	4.59	16,864	33.83	330	0.8335^{aef}	4.02	16,242	33.89	287
Longformer-parallel	0.7068^{abef}	22.18	3,492	164.18	68	0.8427^{abef}	18.71	2,492	159.43	61
Longformer-multi	0.7044^{abef}	57.97	1,336	310.11	36	0.8414^{abef}	62.15	900	374.04	61
Longformer-passage	0.7014^b	68.60	1,129	413.48	27	0.8209^f	73.85	505	422.83	23
Longformer-document	0.7018^b	115.24	1,176	656.71	17	0.8188	94.41	395	607.81	16
ModernBERT-base	0.6882	12.07	6,416	45.75	244	0.8311^a	13.02	3,581	51.18	190
ModernBERT-parallel	0.7072^{abf}	57.32	1,351	232.58	48	0.8431^{abdef}	47.87	779	226.16	43
ModernBERT-multi	0.7050^{abf}	60.46	2,562	446.56	25	0.8393^{abf}	58.45	638	237.20	41
ModernBERT-passage	0.7066^{abf}	129.07	1,050	558.20	20	0.8357^a	140.90	397	221.02	44
ModernBERT-document	0.6972	74.61	2,076	218.90	51	0.8344^a	174.49	374	607.81	16

language models, BiomedBERT achieves the best results as a base encoder across the four datasets – thus highlighting the advantage of a language model pre-trained on domain-specific text.

Multi cross-encoder performances: We then analyse the effect of adding more information to the cross-encoders. In general, increasing the number of mentions fed to the model in one input does degrade predictive performance, but only minorly. The extreme cases of passing in a whole passage (MedMentions/NCBI Disease) or whole document (NLMChem/BC5CDR) generally achieve the worst accuracy results. However, consistent with the work by Xu et al. (2023), in some of the datasets, adding some mention-entity pairs to classify can provide an improved context and help the accuracy. This is the case of the parallel cross-encoders in NCBI Disease, NLM Chem and BC5CDR and the Longformer and ModernBERT multi-cross encoders in NCBI Disease.

Notably, accuracy results between variants of the same model are commonly small – ranging between -3.42% accuracy decrease to 2.76% performance increase with respect to the base cross-encoder. This, along with the fact that even the worst cross-encoder variants are commonly significantly better than the first-stage linker, makes the different cross-encoder variants reasonable algorithms for the biomedical entity linking task.

To answer RQ1: In general, providing more mention-entity pairs as input to cross-encoder models has limited impact on the Acc@1 performance (ranging between -3.42% to 2.76%). Different datasets can react differently to the amount of information provided to the cross-encoder, so it is important to choose the right model to enhance the accuracy. However, the small accuracy differences make all the variants reasonable entity linkers.

6.2 RQ2: Efficiency comparison

We now analyse the training and inference speeds for each of our tested datasets. Results are shown in the 2nd and 4th columns for each dataset in Tables 3 and 4. As different models use different batch sizes to fit the model into a single GPU card, we shall only compare speeds across encoders with the same base LM.

For completion, Tables 3 and 4 include the inference times and speeds of the first stage n-gram models. However, it must be noted that the n-grams model has differences with respect to the cross-encoders. Specifically, first-stage processing is not accelerated by GPU, is not trained for multiple epochs and does not depend on batching. Therefore, the training and inference speeds are not directly comparable to those of cross-encoder models. However, we include them as a reference of how fast it is to train and apply each of these models.

Training speed: Our results show a clear trend where the base cross-encoders are the slowest second stage models in our comparison, and the training speed generally increases as we enhance the parallelism of the cross-encoders. The parallel cross-encoders increase the training speed of the baseline models between 2.68 (13.02 examples/s to 47.87 examples/s for the ModernBERT model on BC5CDR) and 3.9 times (11.84 examples/s to 58.53 examples/s for ModernBERT on MedMentions). Models with more examples reach even higher speeds, with multi cross-encoders achieving 3.49-29.93 speed improvements, and passage and document models achieving even further gains.

While increasing the number of mention-entity pairs commonly increases the training speed, that does not always occur. We observe two exceptions in our experiments: ModernBERT-passage in MedMentions and ModernBERT-document in NLM Chem. We hypothesize that the advantage of adding more pairs to the input text depends on the capacity of the cross-encoder of processing those examples. As we add more tokens to the input text, the cross-encoder might reach a point where it slows down its processing. We show this in Figure 6. In this figure, we illustrate the average input token length of the training examples of each model (x-axis) against the training speed (y-axis). Each line represents a backbone LM, with the arrow indicating the model receiving more information per example. As we can observe here, the only case where the speed decreases (with respect to the previous model) is the ModernBERT-document model, where inputs reach 6,000 tokens on average (2000 more than the second with the longer text sequence, Longformer-document). A similar observation occurs on Medmentions.

Inference speed: When it comes to inference time, we observe similar patterns with respect to the training time, with the base cross-encoders being the slowest models, and the models scoring multiple mention-entity pairs at the same time achieving speed improvements between 3.42 times (ModernBERT-parallel on BC5CDR from 51.18 examples/s to 226.16 examples/s) to 26.47 times (Longformer-passage from 31.64 examples/s to 869.2 examples/s on MedMentions) with respect to the base model[4]. Again, some slowdown is observed when the input token length notably surpasses the size of the Longformer window (as in ModernBERT-passage for MedMentions and ModernBERT-document for NLM Chem).

To answer RQ2: Allowing cross-encoders to simultaneously score multiple entities notably boosts the training and inference speeds of entity linking models. However, models with larger context windows might face difficulties when the length of the input text is too long – effectively reducing the efficiency improvements provided by these models.

7 Conclusions

In this work, we have studied the use of advanced cross-encoder models as rerankers for an entity linking pipeline. These advanced cross-encoders enable simultaneously processing of several mention-candidate pairs, accelerating the training speed of a base cross-encoder by a factor between 2.68 and 36.97, and the inference speed by a

[4]We observe a few ties between rerankers on NCBI Disease and BC5CDR. These are due to the small inference times of these models on each dataset (<10 s).

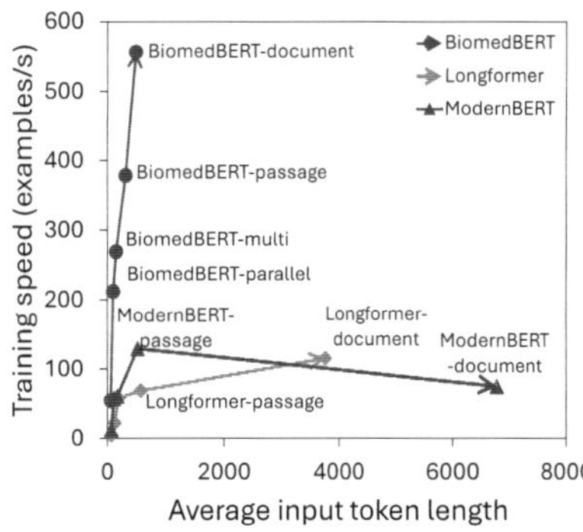

Figure 6: Input token length vs. training speed (NLM Chem)

factor between 3.42 and 26.47. While the parallel processing of multiple candidates might hurt the performance of the model, we find this effect to be small (up to 3.42% accuracy loss). Speed advantages, along with the low accuracy losses make these architectures suitable for environments where training and inference speed is crucial (like real time services).

This work has only focused on cross-encoder with point-wise losses, where we directly estimate the probability of a candidate for a mention. As future work, we shall explore the effect of this cross-encoder architecture on pair-wise or list-wise cross-encoder rerankers, considering the order between pairs. Furthermore, we propose the application of similar architectures on other second stage rerankers, like bi-encoders or poly-encoders (Humeau et al., 2020), that encode mentions and candidates separately.

Limitations

The effectiveness and efficiency of our approach are influenced by two factors: (a) the base model selection and (b) the dataset on which the cross-encoders are trained and applied.

Model selection: In this paper, we only focus on a simple entity linking pipeline, based on an n-grams TF-IDF model for candidate retrieval (Neumann et al., 2019) and variations of the simple cross-encoder reranker described in Logeswaran et al. (2019) and Wu et al. (2020). Although several improvements for the first and second stages of the entity linking pipeline have been developed (Zhu et al., 2024; Angell et al., 2021; Agarwal et al., 2022), we have not tested them in our experiments. However, as most of these proposals use a cross-encoder as their reranker, we believe that similar results should be consistent with our findings if we modified the cross-encoders of these methods in a similar manner.

Dataset selection: In our experiments, we have tested our cross-encoder models on four datasets. The biggest dataset in our experiments is the MedMentions dataset, with only 4,392 documents and 352,496 mentions. While there are bigger biomedical entity linking datasets available, like WikiMed or PubmedDS (Vashishth et al., 2021), training some of the base cross-encoders on MedMentions already represents a challenge (the Longformer-base model takes more than 2 days to train on a single GPU card). By testing our approach across these 4 (smaller) datasets, we show the generalizability of our approach and how it would perform on those bigger datasets.

Acknowledgements

The work is supported by the U.S. National Cancer Institute (NCI) [U24CA275783].

References

Dhruv Agarwal, Rico Angell, Nicholas Monath, and Andrew McCallum. 2022. Entity Linking via Explicit Mention-Mention Coreference Modeling. In *Proceedings of the 2022 Conference of the North American Chapter of the Association for Computational Linguistics: Human Language Technologies (NAACL-HLT 2022)*, pages 4644–4658, Seattle, WA, USA. Association for Computational Linguistics.

Rico Angell, Nicholas Monath, Sunil Mohan, Nishant Yadav, and Andrew McCallum. 2021. Clustering-based Inference for Biomedical Entity Linking. In *Proceedings of the 2021 Conference of the North American Chapter of the Association for Computational Linguistics: Human Language Technologies (NAACL-HLT 2021)*, pages 2598–2608, Online. Association for Computational Linguistics.

Iz Beltagy, Matthew E. Peters, and Arman Cohan. 2020. Longformer: The Long-Document Transformer. *Preprint*, arXiv:2004.05150.

Jacob Devlin, Ming-Wei Chang, Kenton Lee, and Kristina Toutanova. 2019. BERT: Pre-training of Deep Bidirectional Transformers for Language Understanding. In *Proceedings of the 2019 Conference of the North American Chapter of the Association for Computational Linguistics: Human Language Technologies (NAACL-HLT 2019) Volume 1 (Long and Short Papers)*, pages 4171–4186, Minneapolis, MN, USA. Association for Computational Linguistics.

Rezarta Islamaj Doğan, Robert Leaman, and Zhiyong Lu. 2014. NCBI disease corpus: A resource for disease name recognition and concept normalization. *Journal of Biomedical Informatics*, 47:1–10.

Evan French and Bridget T. McInnes. 2023. An overview of biomedical entity linking throughout the years. *Journal of Biomedical Informatics*, 137:104252.

Samuele Garda, Leon Weber-Genzel, Robert Martin, and Ulf Leser. 2023. BELB: a biomedical entity linking benchmark. *Bioinformatics*, 39(11):btad698.

Yu Gu, Robert Tinn, Hao Cheng, Michael Lucas, Naoto Usuyama, Xiaodong Liu, Tristan Naumann, Jianfeng Gao, and Hoifung Poon. 2021. Domain-Specific Language Model Pretraining for Biomedical Natural Language Processing. *ACM Transactions on Computing for Healthcare*, 3(1).

Samuel Humeau, Kurt Shuster, Marie-Anne Lachaux, and Jason Weston. 2020. Poly-encoders: Architectures and Pre-training Strategies for Fast and Accurate Multi-sentence Scoring. In *Proceedings of the 8th International Conference on Learning Representations (ICLR 2020)*, Online.

Rezarta Islamaj, Robert Leaman, Sun Kim, Dongseop Kwon, Chih-Hsuan Wei, Donald C. Comeau, Yifan Peng, David Cissel, Cathleen Coss, Carol Fisher, Rob Guzman, Preeti Gokal Kochar, Stella Koppel, Dorothy Trinh, Keiko Sekiya, Janice Ward, Deborah Whitman, Susan Schmidt, and Zhiyong Lu. 2021. NLM-Chem, a new resource for chemical entity recognition in PubMed full text literature. *Scientific Data*, 8:91:1–91:12.

Chengyue Jiang, Wenyang Hui, Yong Jiang, Xiaobin Wang, Pengjun Xie, and Kewei Tu. 2023. Recall, Expand, and Multi-Candidate Cross-Encode: Fast and Accurate Ultra-Fine Entity Typing. In *Proceedings of the 61st Annual Meeting of the Association for Computational Linguistics (ACL 2023) Volume 1 (Long Papers)*, pages 11597–11609, Toronto, Canada. Association for Computational Linguistics.

David Kartchner, Jennifer Deng, Shubham Lohiya, Tejasri Kopparthi, Prasanth Bathala, Daniel Domingo-Fernández, and Cassie Mitchell. 2023. A Comprehensive Evaluation of Biomedical Entity Linking Models. In *Proceedings of the 2023 Conference on Empirical Methods in Natural Language Processing (EMNLP 2023)*, pages 14462–14478, Singapore. Association for Computational Linguistics.

Jiao Li, Yueping Sun, Robin J. Johnson, Daniela Sciaky, Chih-Hsuan Wei, Robert Leaman, Allan Peter Davis, Carolyn J. Mattingly, Thomas C. Wiegers, and Zhiyong Lu. 2016. BioCreative V CDR task corpus: a resource for chemical disease relation extraction. *Database*, 2016:baw068.

Fangyu Liu, Ehsan Shareghi, Zaiqiao Meng, Marco Basaldella, and Nigel Collier. 2021. Self-Alignment Pretraining for Biomedical Entity Representations. In *Proceedings of the 2021 Conference of the North American Chapter of the Association for Computational Linguistics: Human Language Technologies (NAACL-HLT 2021)*, pages 4228–4238, Online. Association for Computational Linguistics.

Lajanugen Logeswaran, Ming-Wei Chang, Kenton Lee, Kristina Toutanova, Jacob Devlin, and Honglak Lee. 2019. Zero-Shot Entity Linking by Reading Entity Descriptions. In *Proceedings of the 57th Annual Meeting of the Association for Computational Linguistics (ACL 2019)*, pages 3449–3460, Florence, Italy. Association for Computational Linguistics.

Daniel Loureiro and Alípio Mário Jorge. 2020. MedLinker: Medical Entity Linking with Neural Representations and Dictionary Matching. In *Proceedings of the 42nd European Conference on Information Retrieval (ECIR 2020)*, pages 230–237, Online. Springer International Publishing.

Sean MacAvaney and Craig Macdonald. 2022. A Python Interface to PISA! In *Proceedings of the 45th International ACM SIGIR Conference on Research and Development in Information Retrieval (SIGIR 2022)*, page 3339–3344, Madrid, Spain. Association for Computing Machinery.

Antonio Mallia, Michal Siedlaczek, Joel Mackenzie, and Torsten Suel. 2019. PISA: Performant Indexes and Search for Academia. In *Proceedings of the Open-Source IR Replicability Challenge co-located with 42nd International ACM SIGIR Conference on Research and Development in Information Retrieval, (OSIRRC@SIGIR 2019)*, pages 50–56, Paris, France.

Sunil Mohan and Donghui Li. 2019. MedMentions: A Large Biomedical Corpus Annotated with UMLS Concepts. In *Proceedings of the 2019 Conference on Automated Knowledge Base Construction (AKBC 2019)*, Amherst, MA, USA.

Mark Neumann, Daniel King, Iz Beltagy, and Waleed Ammar. 2019. ScispaCy: Fast and Robust Models for Biomedical Natural Language Processing. In *Proceedings of the 18th BioNLP Workshop and Shared Task (BioNLP 2019)*, pages 319–327, Florence, Italy. Association for Computational Linguistics.

Mujeen Sung, Hwisang Jeon, Jinhyuk Lee, and Jaewoo Kang. 2020. Biomedical Entity Representations with Synonym Marginalization. In *Proceedings of the 58th Annual Meeting of the Association for Computational Linguistics (ACL 2020)*, pages 3641–3650, Online. Association for Computational Linguistics.

Shikhar Vashishth, Denis Newman-Griffis, Rishabh Joshi, Ritam Dutt, and Carolyn P. Rosé. 2021. Improving broad-coverage medical entity linking with semantic type prediction and large-scale datasets. *Journal of Biomedical Informatics*, 121:103880.

Benjamin Warner, Antoine Chaffin, Benjamin Clavié, Orion Weller, Oskar Hallström, Said Taghadouini, Alexis Gallagher, Raja Biswas, Faisal Ladhak, Tom Aarsen, Nathan Cooper, Griffin Adams, Jeremy Howard, and Iacopo Poli. 2024. Smarter, Better,

Faster, Longer: A Modern Bidirectional Encoder for Fast, Memory Efficient, and Long Context Finetuning and Inference. *Preprint*, arXiv:2412.13663.

Ledell Wu, Fabio Petroni, Martin Josifoski, Sebastian Riedel, and Luke Zettlemoyer. 2020. Scalable Zero-shot Entity Linking with Dense Entity Retrieval. In *Proceedings of the 2020 Conference on Empirical Methods in Natural Language Processing (EMNLP 2020)*, pages 6397–6407, Online. Association for Computational Linguistics.

Zhenran Xu, Yulin Chen, and Baotian Hu. 2023. Improving Biomedical Entity Linking with Cross-Entity Interaction. In *Proceedings of the 37th AAAI Conference on Artificial Intelligence (AAAI 2023)*, pages 13869–13877, Washington, DC, USA. Association for the Advancement of Artificial Intelligence.

Sheng Zhang, Hao Cheng, Shikhar Vashishth, Cliff Wong, Jinfeng Xiao, Xiaodong Liu, Tristan Naumann, Jianfeng Gao, and Hoifung Poon. 2022. Knowledge-Rich Self-Supervision for Biomedical Entity Linking. In *Findings of the Association for Computational Linguistics: EMNLP 2022*, pages 868–880, Abu Dhabi, United Arab Emirates. Association for Computational Linguistics.

Tiantian Zhu, Yang Qin, Ming Feng, Qingcai Chen, Baotian Hu, and Yang Xiang. 2024. BioPRO: Context-Infused Prompt Learning for Biomedical Entity Linking. *IEEE/ACM Transactions on Audio, Speech, and Language Processing*, 32:374–385.

A Statistical Significance

In our experiments in Section 6, we conduct statistical significance tests between every pair of models in our comparison. We only perform these statistical tests for the Accuracy@1 metric. As accuracy@1 values are binary (either 0 or 1), we apply a McNemar test with $p < 0.05$, where we pair the Acc@1 results for each mention in the test set. To account for the comparison of multiple models, we apply the Bonferroni correction. While Tables 3 and 4 summarize the statistical tests, we include in this appendix the complete statistical significance matrices.

These matrices are included in Figure 7 (MedMentions dataset), Figure 8 (NCBI Disease dataset), Figure 9 (NLM Chem) and Figure 10 (BC5CDR). On each matrix, a cell compares two algorithms: one indicated by the first row, and another indicated by the first column. A green cell represents a statistically significant difference between the two models ($p < 0.05$), whereas a white cell represents a non-significant difference ($p \geq 0.05$).

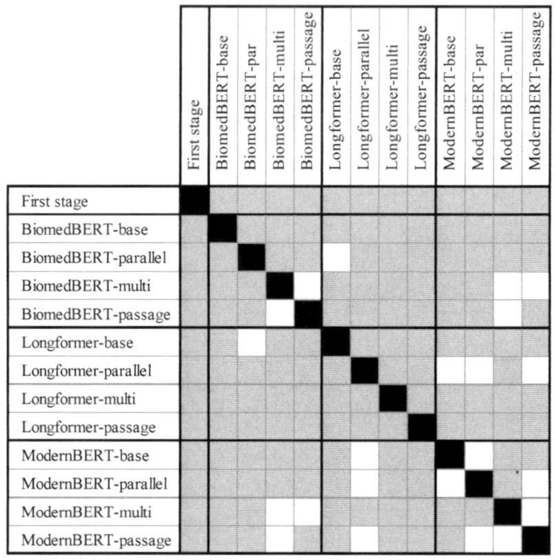

Figure 7: Acc@1 statistical significance (MedMentions)

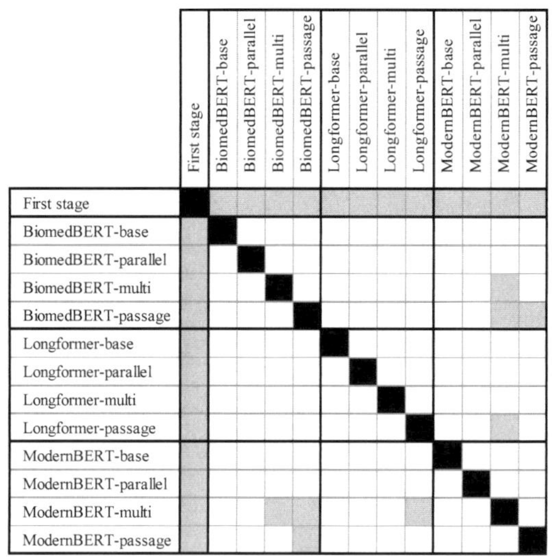

Figure 8: Acc@1 statistical significance (NCBI Disease)

B Input token length vs. training speed

We compare the average input token length and the training speed for the four datasets. Figure 11 (MedMentions), Figure 12 (NCBI Disease), Figure 13 (NLM Chem) and Figure 14 (BC5CDR) show the results. In all figures, x-axis shows the average input token length, and y-axis the training speed (in examples per second).

As we can observe, in general, in the smaller datasets, where the input token length is notably smaller than 4,096 (context window of the Longformer model), all models improve their training speed as we add more mention-entity pairs in their

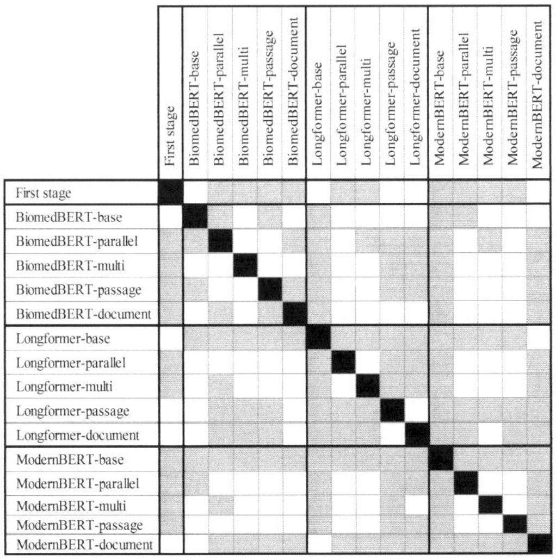

Figure 9: Acc@1 statistical significance (NLM Chem)

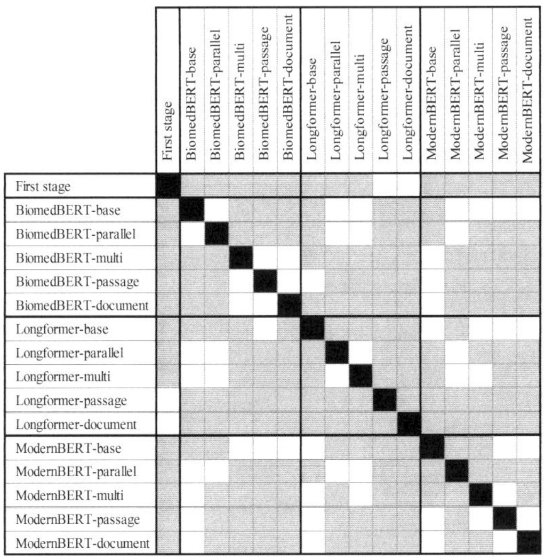

Figure 10: Acc@1 statistical significance (BC5CDR)

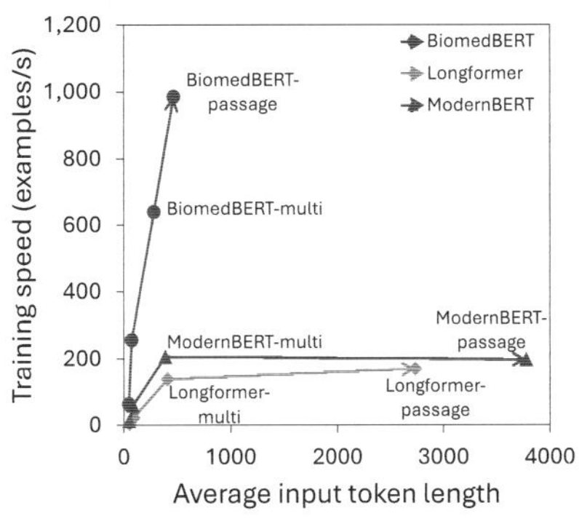

Figure 11: Input token length vs. training speed (Medmentions)

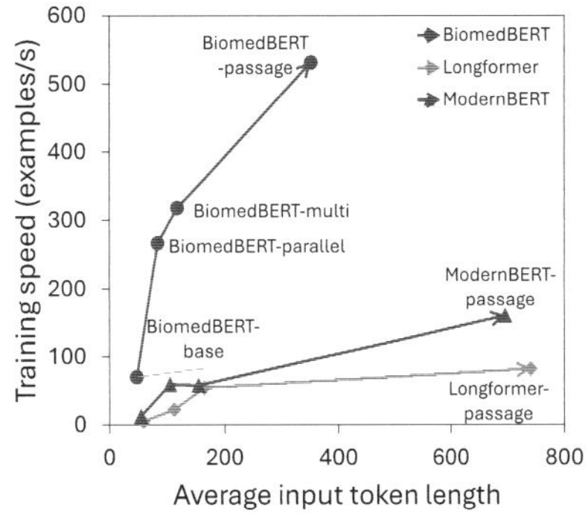

Figure 12: Input token length vs. training speed (NCBI Disease)

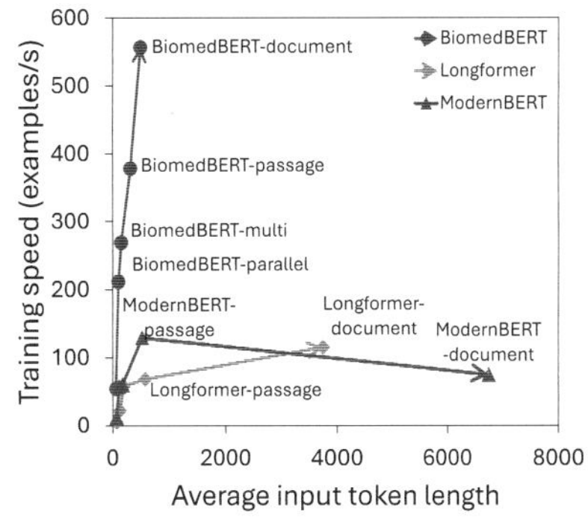

Figure 13: Input token length vs. training speed (NLM Chem)

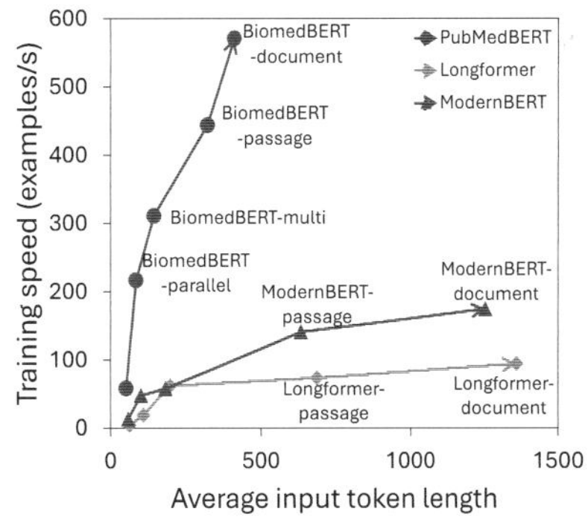

Figure 14: Input token length vs. training speed (BC5CDR)

input. However, in the bigger datasets (MedMentions and NLMChem), the ModernBERT model struggles with longer sequences (as observed in the training speeds of the ModernBERT-passage model in MedMentions and the ModernBERT-document model in NLM Chem).

Advancing Biomedical Claim Verification by Using Large Language Models with Better Structured Prompting Strategies

Siting Liang[1], **Daniel Sonntag**[1,*]
[1]German Research Center for Artificial Intelligence, Germany
*University of Oldenburg, Germany
siting.liang|daniel.sonntag@dfki.de

Abstract

Biomedical claim verification involves determining the entailment relationship between a claim and evidence derived from medical studies or clinical trial reports (CTRs). In this work, we propose a structured four-step prompting strategy that explicitly guides large language models (LLMs) through (1) claim comprehension, (2) evidence analysis, (3) intermediate conclusion, and (4) entailment decision-making to improve the accuracy of biomedical claim verification. This strategy leverages compositional and human-like reasoning to enhance logical consistency and factual grounding to reduce reliance on memorizing few-shot exemplars and help LLMs generalize reasoning patterns across different biomedical claim verification tasks. Through extensive evaluation on biomedical NLI benchmarks, we analyze the individual contributions of each reasoning step. Our findings demonstrate that comprehension, evidence analysis, and intermediate conclusion each play distinct yet complementary roles. Systematic prompting and carefully designed step-wise instructions not only unlock the latent cognitive abilities of LLMs but also enhance interpretability by making it easier to trace errors and understand the model's reasoning process. This research aims to improve the reliability of AI-driven biomedical claim verification.

1 Introduction

Natural language inference (NLI) tasks typically involve determining whether or not a given hypothesis is entailed with respect to a premise (Bowman et al., 2015). An NLI system labels the logical relationship between the premise and hypothesis (e.g., Entailment, Contradiction, or Neutral). To enhance transparency and trustworthiness, the system should also provide an explanation in the form of specific evidence (rationales) that justify its decision (Camburu et al., 2018). In the scientific and medical domains, NLI is used to assist clinicians and researchers by automatically verifying claims against evidence from clinical trial data or medical literature. Specifically, it requires a deep understanding of medical and scientific knowledge to interpret implicit data points beyond simple text matching.

Clinical trial data often contain complex statistical information and precise measurements that must be interpreted accurately to avoid errors in claim verification. One example from the **NLI4CT** challenges (Jullien et al., 2023) shown in Figure 1 highlights the significant difficulties of applying NLI to validate statements (hypotheses) related to clinical trial reports (CTRs), which requires more than simple textual analysis. To accurately assess the claims, NLI models must process long and complex documents while also comprehending domain-specific terminology and applying multi-hop reasoning to draw connections that are not immediately obvious (Romanov and Shivade, 2018; Wadden et al., 2020; Jullien et al., 2024).

Large language models (LLMs) offer promising potential to address these challenges. Recent research has shown that the reasoning capability of LLMs depends on two key factors: the size of the model and the appropriateness of the prompts provided for specific tasks (Huang and Chang, 2022; Qiao et al., 2022). Using structured, multi-step prompting methods has been the subject of research efforts to explore the reasoning abilities of LLMs in different tasks, including mathematical problems, commonsense reasoning and multi-hop question answering tasks (Wei et al., 2022; Zhou et al., 2022; Xia et al., 2024). Larger LLMs excel in zero-shot reasoning but require careful prompt engineering for reliability and interpretability (Kojima et al., 2022; Jeblick et al., 2024). Smaller models offer faster inference with lower computational costs and are more suitable for real-time applications, though they have weaker reasoning abilities and rely more

Proceedings of the 24th Workshop on Biomedical Language Processing (BioNLP 2025), pages 148–166

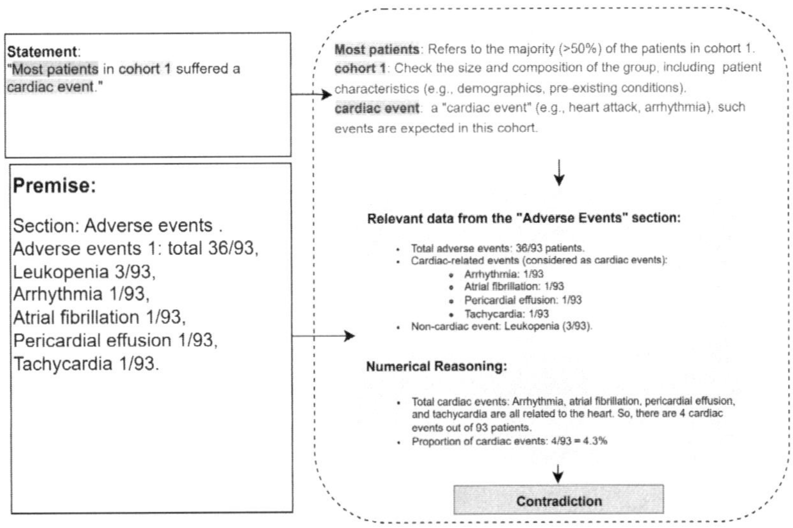

Figure 1: An example from **NLI4CT** dataset (Jullien et al., 2023). Left: a pair of statement and the adverse event section of a clinical trial data (premise). Right: an illustration of understanding the key terms and reasoning capabilities required to infer the logical relationship between the statement and the premise.

on fine-tuning for domain-specific tasks (Abdin et al., 2024).

In this work, we evaluate a 4-step NLI-oriented prompting framework that systematically decomposes the biomedical claim verification process into sequential stages, enhancing the zero-shot reasoning capabilities of LLMs, as illustrated in Figure 3. In particular, we aim to address the challenges posed by the need for domain expertise and the extensive length of medical documents, as well as the demand for the reliability. In our experiments, we specifically investigate the improvement obtained in both lightweight LLMs (from 3.6 to 14 billion parameters at most) and GPT3.5 and GPT-4o-mini models (OpenAI, 2024). Notably, all investigated models demonstrate a significant improvement over the standard CoT method, achieving an approximate 10% performance gain. Our key contributions can be summarized into:

- We investigate a 4-step prompting framework that extends the zero-shot CoT methodology and aims to address complex reasoning tasks like biomedical claim verification.
- We demonstrate how combining our structured prompting framework and supervised fine-tuning (SFT) significantly enhances the performances of lightweight LLMs in tackling the biomedical claim verification tasks.

The code for reproducing our experiments is available in the GitHub repository .

2 Approach

Task Definition We frame the NLI task in the biomedical domain as a binary classification problem, where an NLI system based on LLMs determines whether a statement or claim (C) logically follows from the premise (P) provided in clinical trial or scientific study data. For automatic performance evaluation, the final output of the system is a prediction of the logical relationship between C and P. Let's denote:

$$f(C, P) = \begin{cases} \text{Entailment} & \text{if } C \text{ logically follows from } P; \\ \text{Contradiction} & \text{otherwise} \end{cases} \quad (1)$$

The binary prediction accuracy provides a straightforward measure of the LLMs' reasoning capabilities.

For solving the task, our first baseline utilizes a straightforward prompt template (see Fig 2), as proposed by Jullien et al. (2024). For clarity, in this paper, we will use "claim" and "statement" interchangeably to refer to the hypothesis within the context of NLI for biomedical claim verification, as different benchmarks employ varying conventions for these terms.

Zero-shot 2-step CoT Intermediate steps are useful for increasing grounded context and intermediate steps also increase the reliability of model generations (Yu et al., 2023). Standard CoT is a prompting methodology guiding LLMs to handle reasoning tasks by mimicking the thoughts of

```
Prompt template
"Given a section of 2 clinical
trial descriptions and a statement,
determine whether the statement
logically follows from the sections.
If the statement logically follows
from the sections, you need to
return 'Entailment'. If the
statement does not logically follow
from the sections, you need to
return 'Contradiction'. The output
should be a single word <Entailment>
or <Contradiction>.
"Statement: " + Statement
"Primary Trial: " + Primary CTR text
"Secondary Trial: " + Secondary CTR
text
```

Figure 2: A simple prompt template for **NLI4CT** task.

solving example tasks demonstrated in prompts (Brown, 2020; Wei et al., 2022). While breaking down complex reasoning tasks into simpler steps can be useful, Zhou et al. (2022) noted that decomposition prompts require task-specific design for optimal performance.

In biomedical claim verification cases, providing multiple human-annotated examples in prompts is impractical due to the length of input documents, which can individually exceed 5,000 tokens. Furthermore, adding examples along with model-generated responses for intermediate steps would exceed the model's input limits and introduce noise to harm performance. As an alternative baseline, we adopt the zero-shot CoT approach (Kojima et al., 2022), which we refer to as the 2-Step prompting strategy in our experiments. In the first step, the model is prompted with an instruction phrase "step by step" instead of examples to generate a CoT response that leads to a solution. In the second step, the response from the first step is used to prompt the model to produce an output. Based on the task-specific prompt template as shown in Fig 2), our zero-shot CoT baseline follows a 2-Step prompting framework, utilizing instructive prompts in the first step: *'Determine whether the statement logically follows from the sections step by step.'*. The prompt text in step 2 includes the response generated in the first step followed by the remaining part from the template shown in Fig 2, e.g. *'If the statement ...The output should be a single word <Entailment> or <Contradiction>.'*.

4-step prompting framework In biomedical claim verification, each claim requires a distinct focus—some may involve analyzing a trial's inclusion criteria, while others require verifying the adverse event count in the outcomes section. While the 2-Step prompting method can be effective in simpler contexts, we identify key limitations of this approach when generalized to biomedical claim verification, particularly when using lightweight LLMs.

- Lack of co-reference resolution of terms or abbreviations between statement and premise data, leading to misinterpretation of key terms in the reasoning process.
- Resulting in shallow analysis without addressing each relevant factual detail in the premise (see some example generations of different models in Appendix A).

To address the challenges posed by complex biomedical terminology and diverse reasoning patterns, we draw from prior research in context-aware reasoning and domain-specific inference to develop a structured and adaptable prompting approach. We propose a carefully designed CoT framework with four sequential steps to improve vague reasoning of LLMs (see Fig 3). Below, we explain the intention of each step in more detail.

- **Claim Comprehension**: In the first step, the model only receives the statement and a targeted prompt instructing the model to interpret the medical terminology and complex biomedical concepts within the statement, e.g. *"Interpret the key terms in the statement based on biomedical knowledge. "*. This step serves to activate relevant domain knowledge and establish a semantic context for associating relevant information in later stages.
- **Evidence Analysis**: After understanding the statement, the model is presented with the premise data in the second step, such as original text from a clinical trial or medical study. The model is instructed to identify the relevant data points as evidence from the source compared to the information in the statement. Thus, the model focuses on verifying the truth of the statement by identifying the relevant evidence and performing comparative analysis. This analysis may involve numerical reasoning or biomedical reasoning, depending on the understanding of semantic context of each instance in the previous and current

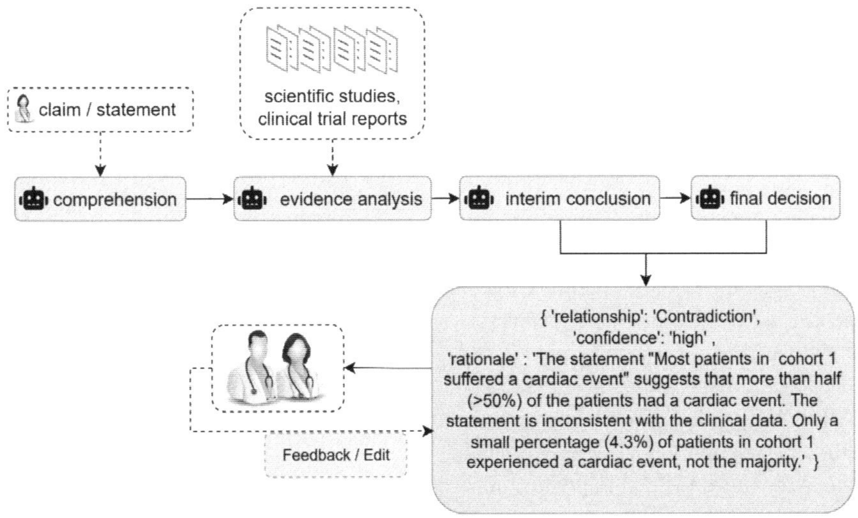

Figure 3: The proposed structured reasoning framework for with LLMs biomedical claim verification. In each step, we have the dedicate instruction for the model to complete the

steps. Example instruction at this stage include: *"1. Identify the relevant data points. 2. Evaluate each piece of information in the statement against these data points."*. The response generated in this stage serves as the basic for logical deduction in the subsequent inference stages, reducing issues like premature judgment.

- **Interim Conclusion**: LLMs likely draw conclusions in the evaluation response. However, these conclusions often lead to diverse outputs if lack task-specific focus. Therefore, the conclusion step builds on the tendency of LLMs to generate conclusions in their response but explicitly guides the models to focus on deducing logical relationship. For instance, we provide the following prompt in the third step: *"Conclude the evidence and determine whether the statement logically follows from the clinical trial data."* . This instruction refines the conclusion of the evaluation and steers the model response to explicitly determining the logical relationship.

- **Entailment Decision-making**: The final step encapsulates the model's reasoning path in a single relation prediction in natural words, e.g. *"Entailment"* or *"Contradiction"* as it is shown in the prompt template (Fig 2). This relationship prediction provides a concise outcome, enabling effective evaluation with automated metrics calculation.

By structuring the biomedical claim verification task into well-defined steps and emphasizing semantic grounding and evidence-based evaluation before logical inference, our approach helps LLMs focus on specific subtasks, reducing ambiguity and enhancing accuracy.

3 Experiments

Our experiments aim to address the main research question:

- How effectively does the 4-step strategy enhance the performance of LLMs in complex numerical and domain-specific reasoning tasks, particularly in biomedical claim verification?

Datasets Our primary evaluation task in this work is **NLI4CT** (Jullien et al., 2024), which presents challenges in numerical and domain-specific knowledge reasoning, as illustrated in Fig 1. Additionally, we assess the generalization capabilities using two related benchmarks: **SciFact** (Wadden et al., 2020) and **HealthVer** (Sarrouti et al., 2021). Both **SciFact** and **HealthVer** were designed as NLI tasks. While the claims in **SciFact** are written by human experts given scientific study abstracts of focusing medical research, the claims of **HealthVer** are directly extracted from studies. The premises in both datasets consist of evidence sentences extracted from relevant studies, requiring models to assign a relation label—*Support* or *Refute*—between input claims and the sentence-level premises. Wadden et al. (2022) highlighted the limitations of relying solely on sentence-level premises for scientific claim verification and demonstrated the advantages of incorporating document-level premises. For our experiments, we use the versions of **SciFact** and **HealthVer** provided by Wadden et al. (2022), which link each claim-premise pair to its relevant study source. To align with our task definition, we exclude the negative samples where the studies lack sufficient information to determine whether the claims are *Entailed* or *Contradicted*. Furthermore, we omit experiments involving the **CovidFact** (Saakyan et al., 2021) dataset due to the issues with noisy claims, including ungrammatical statements or claims unrelated to the provided sources (Wadden et al., 2022). Table 1 summarizes the instance distribution for each relation class applied in our evaluation.

Dataset	Entailment/Support	Contradiction
NLI4CT(test set)	250	250
SciFact (dev set)	216	122
HealthVer (test set)	503	308

Table 1: Number of instances in three different datasets for zero-shot experiments. **SciFact**'s test set withholds ground truth labels for leaderboard submissions, here we use its dev set as substitute.

Metrics For evaluating the performance of LLMs in our task, we employ the F1-score as the key evaluation metric for binary classification results.

$$F1 = 2 \cdot \frac{\text{Precision} \cdot \text{Recall}}{\text{Precision} + \text{Recall}}$$

Models While GPT-4 demonstrates advanced reasoning capabilities on the NLI4CT task in a zero-shot setting without specific prompting strategies (Gema et al., 2024), its closed-source nature and high cost make it impractical for experimenting with different prompting methods. Given computational constraints, our experiments prioritize small-scale, cost-effective LLMs that maintain competitive performance. We employ instruction-tuned (Ouyang et al., 2022) lightweight open-source LLMs (Abdin et al., 2024; Jiang et al., 2023; Team et al., 2024; Dubey et al., 2024) that are compatible with the *FastLanguageModel* Modules of unsloth.ai (Unsloth, 2024) for faster running and fine-tuning with LoRA method (Hu et al., 2021) on a single *NVIDIA* A100–80GB GPU. Table 7 provides the version information about the models utilized in our experiments, including comparisons with two low-cost GPT models: GPT-4o-mini and GPT3.5 (OpenAI, 2024).

Data Augmentation for Supervised Fine-Tuning While the proposed prompting strategy can enhance the performance of LLMs in logical inference, a significant performance gap still exists between larger and smaller LLMs. The limitations of smaller models include difficulties in producing responses with the correct format and challenges in controlling response length (Ding et al., 2023). To fine-tune small-scale LLMs, high-quality training examples are essential. The zero-shot performance of the GPT-4o-mini model demonstrates its potentials to generate such data without human-written inference examples (Gilardi et al., 2023). The second research question in our experiments is:

- Can fine-tuning improve the reliability and consistency of the output of the small-scale LLMs using GPT-4o-mini generated samples within the 4-step prompting framework?

We employ GPT-4o-mini to generates examples using the NLI4CT train set. If the model's final output deviates from the human-annotated label from the train set, e.g. predicting a *Contradiction* when the correct label is *Entailment*, the model is prompted to refine its reasoning in the second step to reach the correct logical conclusion. The prediction results are presented in Table 2). The 4-step responses generated by GPT-4o-mini achieve 0.98 accuracy on the entailment classification task using 1,700 samples from the NLI4CT train set. These high-quality responses can be confidently used to fine-tune small-scale LLMs.

Class	Precision	Recall	F1-Score	Support
Entailment	0.99	0.97	0.98	850
Contradiction	0.97	0.99	0.98	850
Accuracy			0.98	1700
Macro Average	0.98	0.98	0.98	1700
Weighted Average	0.98	0.98	0.98	1700

Table 2: Entailment classification performance of GPT-4o-mini on the NLI4CT train set (1,700 samples) using a 4-step prompting strategy with an additional refinement step verifying against the ground truth labels.

4 Results

4.1 Zero-Shot Results

Our results in Table 3 highlight the substantial performance gains achieved by both cost-effective

Model	NLI4CT			SciFact			HealthVer		
	Simple	2 Steps	**4 Steps**	Simple	2 Steps	**4 Steps**	Simple	2 Steps	**4 Steps**
GPT3.5	0.52 ± 0.01	0.53 ± 0.00	0.75 ± 0.01	0.51 ± 0.03	0.76 ± 0.00	0.86 ± 0.00	0.51 ± 0.01	0.60 ± 0.01	0.74 ± 0.02
GPT-4o-mini	0.67 ± 0.01	0.77 ± 0.02	0.86 ± 0.01	0.83 ± 0.01	0.88 ± 0.00	0.94 ± 0.01	0.69 ± 0.02	0.72 ± 0.01	0.77 ± 0.02
Phi3.5-3.6B	0.53 ± 0.00	0.61 ± 0.01	0.66 ± 0.02	0.51 ± 0.01	0.70 ± 0.03	0.80 ± 0.02	0.51 ± 0.01	0.70 ± 0.01	0.72 ± 0.01
Mistral-7B	0.55 ± 0.01	0.59 ± 0.02	0.69 ± 0.00	0.50 ± 0.02	0.72 ± 0.02	0.80 ± 0.02	0.44 ± 0.02	0.70 ± 0.00	0.72 ± 0.02
Llama3.1-8B	0.47 ± 0.00	0.54 ± 0.01	0.67 ± 0.02	0.53 ± 0.02	0.80 ± 0.01	0.84 ± 0.05	0.44 ± 0.02	0.70 ± 0.00	0.72 ± 0.01
Gemma2-9B	0.63 ± 0.00	0.67 ± 0.03	0.75 ± 0.03	0.57 ± 0.01	0.73 ± 0.00	0.86 ± 0.02	0.65 ± 0.02	0.70 ± 0.02	0.74 ± 0.01
Mistral-12B	0.55 ± 0.00	0.65 ± 0.01	0.75 ± 0.01	0.65 ± 0.01	0.83 ± 0.00	0.87 ± 0.02	0.50 ± 0.02	0.72 ± 0.00	0.74 ± 0.01
Phi3-14B	0.62 ± 0.01	0.64 ± 0.00	0.75 ± 0.02	0.76 ± 0.03	0.80 ± 0.01	0.88 ± 0.02	0.68 ± 0.02	0.72 ± 0.01	0.75 ± 0.01

Table 3: F1 Scores (mean ± standard deviation) for three benchmarks in zero-shot scenario. We compare the performance across the cost-effective GPT models and open sourced lightweight LLMs.

commercial models and small-scale LLMs when utilizing the 4-step prompting framework. Compared to the simple prompt template and 2-step baselines, the 4-step approach enhances reasoning quality and classification accuracy, demonstrating its effectiveness in zero-shot entailment tasks.

Ablation The four Steps starts with *claim comprehension*, where the model interprets the main claim and key terms in the statement. Without this initial step, the comparative analysis process at the *evidence analysis* stage, which involves *"identifying relevant data points and evaluating the information in the statement against these data points"*, likely results in reasoning paths that are less coherent. The ablation results in Fig 4 demonstrate that the absence of this comprehension step can hinder the accuracy of LLMs in claim verification tasks.

4.2 Supervised Fine-Tuning Results

Fig 5 shows that supervised fine-tuning (SFT) with a small number of examples significantly improves F1-scores for lightweight LLMs, with performance further increasing as the number of training instances grows. Notably, Llama3.1-8B exhibits the largest performance gains, benefiting the most from the fine-tuning process.

Table 4 presents the generalization performance of lightweight models fine-tuned with **NLI4CT** samples, evaluated on the related tasks.

We observe that SFT significantly advantages the quality of *evidence analysis* in the second step, which is the primary contributor to the improved results. See some example responses of the small-scale models in zero-shot setting and after SFT in Appendix from table 10 to 15. Moreover, SFT improves task-specific control by ensuring adherence to specific instructions and maintaining a consistent response format, such as JSON, thereby enhancing the LLM's reliability not only for in-domain task - **NLI4CT**, but also the related tasks: **SciFact** and **HealthVer**. These improvements highlight the effectiveness of integrating structured reasoning with clear instructions of subtasks for enhancing smaller models in complex reasoning tasks like biomedical claim verification.

Model	SciFact		HealthVer	
	zero-shot	SFT*	zero-shot	SFT*
Phi3.5-3.6B	0.80	0.85	0.72	0.74
Mistral-7B	0.80	0.87	0.72	0.74
Llama3.1-8B	0.84	0.89	0.72	0.74
Gemma2-9B	0.86	0.90	0.74	0.75
Mistral-12B	0.87	0.88	0.74	0.75
Phi3-14B	0.88	0.90	0.75	0.77

Table 4: A comparison of F1 Scores (mean) for related tasks in zero-shot scenario and SFT(SFT* only with NLI4CT training samples).

5 Discussion

Incorporation of GPT-4o-mini Our prompting approach underscores the importance of evidential evaluation in the second step in biomedical claim verification tasks. As shown in Fig 5, fine-tuning lightweight LLMs with step 2 responses generated by GPT-4o-mini significantly improves their performance on the **NLI4CT** task. Similarly, Table 5 demonstrates the positive impact of incorporating GPT-4o-mini's responses during the *evidence analysis* stage within the 4-step framework. These findings indicate that leveraging GPT-4o-mini's robust reasoning capabilities enhances the evidential evaluation process, enabling smaller LLMs to generate more accurate outputs. Whether to fine-tune lightweight LLMs with GPT-4o-mini generated data or to integrate GPT-4o-mini's evaluations directly into the 4-step pipeline depends on the specific requirements, computational constraints, and operational objectives of the application.

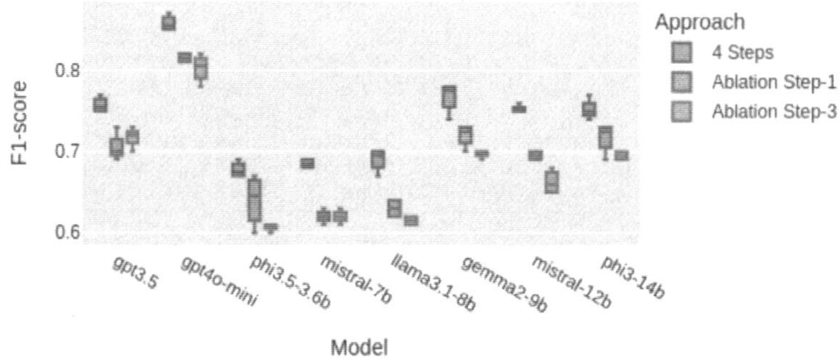

Figure 4: Omitting Step 1 or Step 3, LLMs result in degraded performance.

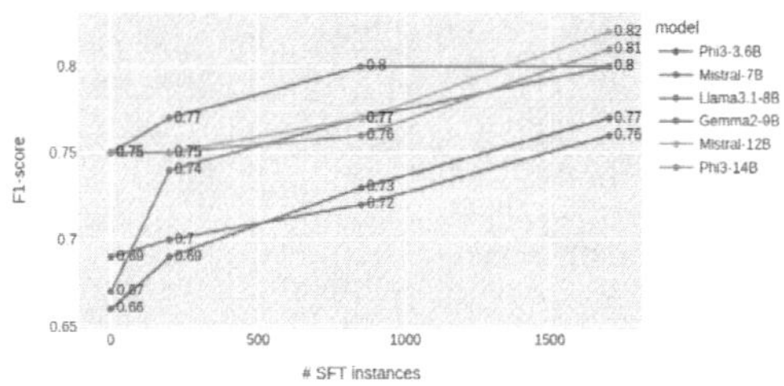

Figure 5: F1-scores of various lightweight LLM models with increasing numbers of SFT instances from **NLI4CT** train set.

	GPT-4o-mini in 2.Stage	SFT*
Phi3-3.6B	0.80	0.77
Mistral-7B	0.84	0.76
Llama3.1-8B	0.84	0.80
Gemma2-9B	0.84	0.80
Mistral-12B	0.84	0.82
Phi3-14B	0.85	0.82

Table 5: **SFT*** are the results of each lightweight models fine-tuned within the 4-step framework using GPT-4o-mini generated training samples.

Error Analysis Our 4-step prompting framework enables the diagnosis of different types of error in LLMs by tracing the source of errors in individual steps.

- In the **Evidence Analysis** step: There are general mistakes the models fall into: numerical reasoning errors and misinterpretation of clinical trial data. Numerical reasoning errors involve misreading or miscalculating quantitative information. As it is shown in example 12 , interpreting *"98% had recurrence-free survival"* as the opposite, or failing to recognize that 98% of 100 patients implies that only 2 did not meet the outcome. Misinterpretation of clinical trial data occurs when the model overemphasizes irrelevant details or introduces unwarranted assumptions, such as requiring subgroup information when the data clearly applies to the overall population. These errors reflect the model's overconfidence in its generated explanations and highlight the difficulty large language models face when performing factual, logic-based reasoning.

Logical Conclusion
While it is true that the placebo group had a higher incidence of emesis compared to the Aprepitant group, the claim that the placebo group was "twice as likely" to experience emesis is misleading. The actual ratio of emesis incidence between the two groups is approximately 2.14 times (75% divided by 35%), which does not equate to a simple doubling. Therefore, the statement is not fully accurate as it oversimplifies the comparative risk and does not reflect the nuanced findings of the trial data. **Relation Prediction:** Contradiction
The data indicates that patients receiving the placebo intervention were approximately 2.14 times more likely to experience emesis compared to those receiving Aprepitant. Therefore, the statement that "Patients receiving the placebo intervention in the primary trial were twice as likely to experience emesis as patients receiving Aprepitant" can be inferred from the provided data, confirming its validity. **Relation Prediction:** Entailment

Table 6: Divergent conclusions and corresponding logical relation predictions.

- In the **Interim Conclusion** step: as shown in Table 6, we observed two opposite logical conclusions: *contradiction* and *entailment*, produced by different LLMs for the same verification sample illustrated in Fig 3. Despite being instructed and exposed to the same evaluation results, these models arrived at divergent conclusions. This highlights the necessity of further preference exploration to better align model outputs with user intent (Ouyang et al., 2022; Rafailov et al., 2024).

Pilot User Study As presented in Figure 3, a feedback-driven loop would support the development of collaborative systems that balance the responsibility for decision making between users and LLMs. Since collecting real-world feedback from clinicians and medical experts is challenging, we conducted a pilot user study with four medical students to assess the transparency and accountability of the model's reasoning process leading to the final relation classification on 20 examples from **NLI4CT** test set. Participants rated the model-generated responses on a scale from 1 (*'reasoning process was confusing'*) to 5 (*'the reasoning was easy to follow and correct'*). All participants rated the model responses as a **4**, suggesting that the model's reasoning process is generally perceived as transparent and confident. When asked what could be improved, participants provided feedback indicating the need for better quality in the intermediate reasoning steps generated by the LLMs, i.e. *"The model sometimes overlooked the smallest details in the claim."*. This highlights how enhanced interpretability can help identify limitations in reasoning of LLM. Also, as emphasized by (Huang et al., 2024), improving the functionality of these model-generated explanations is crucial for fostering user confidence in the system.

6 Related Work

Chain-of-thought Reasoning in LLMs Leveraging massive amounts of training data and billions of parameters, LLMs have demonstrated enhanced performance in various reasoning tasks. In particular, Chain-of-Thought (CoT) strategies (Wei et al., 2022), which provide exemplars of clear, step-by-step reasoning processes have demonstrated impressive performance in guiding LLMs to complete various reasoning tasks. Kojima et al. (2022) further showed that zero-shot CoT prompting, using the simple instruction LET'S THINK STEP BY STEP. instead of explicit examples, can also elicit strong reasoning capabilities from LLMs. However, their performance can vary depending on the complexity of the task and form of reasoning (Huang and Chang, 2023). Lei et al. (2023) addresses ungrounded misinformation in language model outputs by checking for factual inconsistencies between model generation and source documents at the sentence and entity levels within a chain of NLI framework. Zhou et al. (2022) involves breaking down complex problems into a series of simpler sub-problems, with the final problem being addressed depending on the responses to earlier sub-problems, and has proved generalization across different tasks. The evolution of CoT and CoX methodologies (Zhou et al., 2022; Yao et al., 2023; Zhao et al., 2023; Zhang et al., 2024; Xia et al., 2024) underscores the importance of thought decomposition and structured reasoning frameworks in improving both the accuracy and interpretability of LLM outputs. In particular, the intermediate steps of CoT can make the model's output easier to interpret and evaluate (Yu et al., 2023), which is valuable for tasks requiring high accountability, such as biomedical claim verification. Moreover, Wang et al. (2022) proposed the self-consistency method, which enhances the reliability of the results by sampling diverse CoT generations for each sample and selecting the most consistent conclusions among them. Weng et al. (2022) introduced backward verification to complement forward CoT reasoning, allowing self-verification of conclusions derived from different CoT paths to identify the most accurate CoT generations for specific tasks.

Pre-trained Language Models for Biomedical NLP In various NLP tasks, pre-trained language models (PLMs) are effectively applied to medical text processing. (Liang et al., 2023; Liang and Sonntag, 2024) investigated building German clinical entity extraction system based on German PLMs in low-data setting. More recent studies have explored the potential applications of PLMs in clinical practice, such as building clinical entity extraction system without in-domain training data (Liang and Sonntag, 2024), ranging from transfer learning in summarizing radiology reports (Liang et al., 2022) to real-time radiology reporting (Elkassem and Smith, 2023; Jeblick et al., 2024) with PLMs. Datta et al. (2024) leveraged PLMs for automatic eligibility criteria from free text clinical trial protocol to facilitate trial enrollment and evaluation. (Liu et al., 2024) demonstrated the potential of automated verification of scientific claims with LLMs using retrieval-augmented strategies that exploit open resources such as PubMed.

Sivarajkumar et al. (2024) highlighted the effectiveness of different prompting strategies, including zero-shot and few-shot, for clinical information extraction, while Tang et al. (2023) found that LLMs still struggle to summarize medical evidence in longer textual contexts by evaluating LLM-generated summaries focused on six clinical domains. Moreover, LLMs have been shown to enhance the diagnostic accuracy of general radiologists in cardiac imaging, highlighting their value as a diagnostic support tool (Cesur et al., 2024). Rao et al. (2023) also underscored the potential of LLMs to assist healthcare professionals in diagnostic decision-making. Studies from Benary et al. (2023) suggest that LLMs are not yet suitable for routine use in personalized clinical decision-making in oncology, they show promise as a complementary tool, such as selecting relevant biomedical literature to support evidence-based, personalized treatment decisions and offering unique strategies not identified by experts. However, further research is necessary to evaluate their integration into clinical workflows effectively (Verlingue et al., 2024).

7 Conclusion

In summary, our approach structures the complex NLI process into a sequential framework. The process begins with semantic grounding, where the model activates contextual understanding based on the statement to be verified. Next, the model identifies the relevant evidence from the premise data, where the model compares the information in the statement with the extracted evidence. After this evaluation, the model is asked to draw a conclusion and predict the logical relationship between the statement and the evidence. In the context of validating biomedical claims based on long and nuanced documents, the semantic grounding and evidence-based evaluation steps help LLMs perform sub-tasks with greater precision in contrast to the abstract nature of logical relationship prediction. We find that decomposition reduces ambiguity in textual understanding, making the LLM's responses less sensitive to specific wording, as long as the sub-tasks are clearly defined in prompting instructions. For example, the *claim comprehension* step only interprets key terms, while the *evidence analysis* focuses on comparing the statement and the evidence to identify relevant data points. This approach can also effectively minimize the need for extensive prompt engineering.

Future work In high-stakes areas such as medical decision-making, allowing LLMs to make decisions raises critical concerns about accountability and trustworthiness (Elkassem and Smith, 2023; Jeblick et al., 2024). Integrating a feedback-driven loop would support the development of collaborative systems that balance the responsibility for decision making between users and LLMs. This balance is particularly important in high-stakes domains where trust and accountability are essential.

Limitations

Our focus in this work has primarily been on the reasoning capabilities of models when relevant source documents are provided, with pre-retrieved documents used in the evaluation data. However, for open-ended cases, we would need to incorporate a retrieval pipeline to limit the candidate documents to a manageable scale, as otherwise, the process of evidential evaluation could become too time-consuming. Additionally, due to time constraints, we did not compare many different CoT methods. Some approaches, such as generating multiple responses and applying voting heuristics, could offer more reliable results but are computationally expensive. We opted for the most intuitive and effective method, focusing on the 4-step prompting framework. Furthermore, While LLMs demonstrate significant improvements in generating evaluations within 4-step strategy and after SFT, the degree of autonomy granted to these models should be further explored to align with specific user preferences and the application domain.

Acknowledgments

This work is funded by the German Federal Ministry of Education and Research (BMBF) under grant numbers 01IW23002 (No-IDLE) and 01IW24006 (NoIDLEChatGPT), and by the Endowed Chair of Applied AI at the University of Oldenburg. We gratefully acknowledge the support provided by a grant from Accenture Labs.

References

Marah Abdin, Sam Ade Jacobs, Ammar Ahmad Awan, Jyoti Aneja, Ahmed Awadallah, Hany Awadalla, Nguyen Bach, Amit Bahree, Arash Bakhtiari, Harkirat Behl, et al. 2024. Phi-3 technical report: A highly capable language model locally on your phone. *arXiv preprint arXiv:2404.14219*.

Manuela Benary, Xing David Wang, Max Schmidt, Dominik Soll, Georg Hilfenhaus, Mani Nassir, Christian Sigler, Maren Knödler, Ulrich Keller, Dieter Beule, et al. 2023. Leveraging large language models for decision support in personalized oncology. *JAMA Network Open*, 6(11):e2343689–e2343689.

Samuel R. Bowman, Gabor Angeli, Christopher Potts, and Christopher D. Manning. 2015. A large annotated corpus for learning natural language inference. In *Proceedings of the 2015 Conference on Empirical Methods in Natural Language Processing (EMNLP)*. Association for Computational Linguistics.

Tom B Brown. 2020. Language models are few-shot learners. *arXiv preprint arXiv:2005.14165*.

Oana-Maria Camburu, Tim Rocktäschel, Thomas Lukasiewicz, and Phil Blunsom. 2018. e-snli: Natural language inference with natural language explanations. *Advances in Neural Information Processing Systems*, 31.

Turay Cesur, Yasin Celal Gunes, Eren Camur, and Mustafa Dagli. 2024. Empowering radiologists with chatgpt-4o: Comparative evaluation of large language models and radiologists in cardiac cases. *medRxiv*, pages 2024–06.

Surabhi Datta, Kyeryoung Lee, Hunki Paek, Frank J Manion, Nneka Ofoegbu, Jingcheng Du, Ying Li, Liang-Chin Huang, Jingqi Wang, Bin Lin, et al. 2024. Autocriteria: a generalizable clinical trial eligibility criteria extraction system powered by large language models. *Journal of the American Medical Informatics Association*, 31(2):375–385.

Ning Ding, Yulin Chen, Bokai Xu, Yujia Qin, Shengding Hu, Zhiyuan Liu, Maosong Sun, and Bowen Zhou. 2023. Enhancing chat language models by scaling high-quality instructional conversations. In *Proceedings of the 2023 Conference on Empirical Methods in Natural Language Processing*, pages 3029–3051, Singapore. Association for Computational Linguistics.

Abhimanyu Dubey, Abhinav Jauhri, Abhinav Pandey, Abhishek Kadian, Ahmad Al-Dahle, Aiesha Letman, Akhil Mathur, Alan Schelten, Amy Yang, Angela Fan, et al. 2024. The llama 3 herd of models. *arXiv preprint arXiv:2407.21783*.

Asser Abou Elkassem and Andrew D Smith. 2023. Potential use cases for chatgpt in radiology reporting. *American Journal of Roentgenology*, 221(3):373–376.

Aryo Gema, Giwon Hong, Pasquale Minervini, Luke Daines, and Beatrice Alex. 2024. Edinburgh clinical NLP at SemEval-2024 task 2: Fine-tune your model unless you have access to GPT-4. In *Proceedings of the 18th International Workshop on Semantic Evaluation (SemEval-2024)*, pages 1894–1904, Mexico City, Mexico. Association for Computational Linguistics.

Fabrizio Gilardi, Meysam Alizadeh, and Maël Kubli. 2023. Chatgpt outperforms crowd workers for text-annotation tasks. *Proceedings of the National Academy of Sciences*, 120(30):e2305016120.

Edward J Hu, Yelong Shen, Phillip Wallis, Zeyuan Allen-Zhu, Yuanzhi Li, Shean Wang, Lu Wang, and Weizhu Chen. 2021. Lora: Low-rank adaptation of large language models. *arXiv preprint arXiv:2106.09685*.

Jie Huang and Kevin Chen-Chuan Chang. 2022. Towards reasoning in large language models: A survey. *arXiv preprint arXiv:2212.10403*.

Jie Huang and Kevin Chen-Chuan Chang. 2023. Towards reasoning in large language models: A survey. In *Findings of the Association for Computational Linguistics: ACL 2023*, pages 1049–1065, Toronto, Canada. Association for Computational Linguistics.

Yue Huang, Lichao Sun, Haoran Wang, Siyuan Wu, Qihui Zhang, Yuan Li, Chujie Gao, Yixin Huang, Wenhan Lyu, Yixuan Zhang, et al. 2024. Position: Trustllm: Trustworthiness in large language models. In *International Conference on Machine Learning*, pages 20166–20270. PMLR.

Katharina Jeblick, Balthasar Schachtner, Jakob Dexl, Andreas Mittermeier, Anna Theresa Stüber, Johanna Topalis, Tobias Weber, Philipp Wesp, Bastian Oliver Sabel, Jens Ricke, et al. 2024. Chatgpt makes medicine easy to swallow: an exploratory case study on simplified radiology reports. *European radiology*, 34(5):2817–2825.

Albert Q Jiang, Alexandre Sablayrolles, Arthur Mensch, Chris Bamford, Devendra Singh Chaplot, Diego de las Casas, Florian Bressand, Gianna Lengyel, Guillaume Lample, Lucile Saulnier, et al. 2023. Mistral 7b. *arXiv preprint arXiv:2310.06825*.

Mael Jullien, Marco Valentino, and André Freitas. 2024. SemEval-2024 task 2: Safe biomedical natural language inference for clinical trials. In *Proceedings of the 18th International Workshop on Semantic Evaluation (SemEval-2024)*, pages 1947–1962, Mexico City, Mexico. Association for Computational Linguistics.

Maël Jullien, Marco Valentino, Hannah Frost, Paul O'regan, Donal Landers, and André Freitas. 2023. SemEval-2023 task 7: Multi-evidence natural language inference for clinical trial data. In *Proceedings of the 17th International Workshop on Semantic Evaluation (SemEval-2023)*, pages 2216–2226, Toronto, Canada. Association for Computational Linguistics.

Takeshi Kojima, Shixiang Shane Gu, Machel Reid, Yutaka Matsuo, and Yusuke Iwasawa. 2022. Large language models are zero-shot reasoners. In *Proceedings of the 36th International Conference on Neural Information Processing Systems*, pages 22199–22213.

Deren Lei, Yaxi Li, Mengya Hu, Mingyu Wang, Vincent Yun, Emily Ching, and Eslam Kamal. 2023. Chain of natural language inference for reducing large language model ungrounded hallucinations. *arXiv preprint arXiv:2310.03951*.

Siting Liang, Mareike Hartmann, and Daniel Sonntag. 2023. Cross-domain german medical named entity recognition using a pre-trained language model and unified medical semantic types. In *Proceedings of the 5th Clinical Natural Language Processing Workshop*, pages 259–271.

Siting Liang, Klaus Kades, Matthias A Fink, Peter M Full, Tim F Weber, Jens Kleesiek, Michael Strube, and Klaus Maier-Hein. 2022. Fine-tuning bert models for summarizing german radiology findings. *ClinicalNLP 2022*, page 30.

Siting Liang and Daniel Sonntag. 2024. Building a german clinical named entity recognition system without in-domain training data. In *Proceedings of the 6th Clinical Natural Language Processing Workshop*, pages 70–81.

Hao Liu, Ali Soroush, Jordan G Nestor, Elizabeth Park, Betina Idnay, Yilu Fang, Jane Pan, Stan Liao, Marguerite Bernard, Yifan Peng, et al. 2024. Retrieval augmented scientific claim verification. *JAMIA open*, 7(1):ooae021.

OpenAI. 2024. Openai models. available online. `https://platform.openai.com/docs/models/overview`. (accessed in November 2024).

Long Ouyang, Jeffrey Wu, Xu Jiang, Diogo Almeida, Carroll Wainwright, Pamela Mishkin, Chong Zhang, Sandhini Agarwal, Katarina Slama, Alex Ray, et al. 2022. Training language models to follow instructions with human feedback. *Advances in neural information processing systems*, 35:27730–27744.

Shuofei Qiao, Yixin Ou, Ningyu Zhang, Xiang Chen, Yunzhi Yao, Shumin Deng, Chuanqi Tan, Fei Huang, and Huajun Chen. 2022. Reasoning with language model prompting: A survey. *arXiv preprint arXiv:2212.09597*.

Rafael Rafailov, Archit Sharma, Eric Mitchell, Christopher D Manning, Stefano Ermon, and Chelsea Finn. 2024. Direct preference optimization: Your language model is secretly a reward model. *Advances in Neural Information Processing Systems*, 36.

Arya Rao, Michael Pang, John Kim, Meghana Kamineni, Winston Lie, Anoop K Prasad, Adam Landman, Keith Dreyer, and Marc D Succi. 2023. Assessing the utility of chatgpt throughout the entire clinical workflow: development and usability study. *Journal of Medical Internet Research*, 25:e48659.

Alexey Romanov and Chaitanya Shivade. 2018. Lessons from natural language inference in the clinical domain.

Arkadiy Saakyan, Tuhin Chakrabarty, and Smaranda Muresan. 2021. Covid-fact: Fact extraction and verification of real-world claims on covid-19 pandemic. *arXiv preprint arXiv:2106.03794*.

Mourad Sarrouti, Asma Ben Abacha, Yassine Mrabet, and Dina Demner-Fushman. 2021. Evidence-based fact-checking of health-related claims. In *Findings of the Association for Computational Linguistics: EMNLP 2021*, pages 3499–3512, Punta Cana, Dominican Republic. Association for Computational Linguistics.

Sonish Sivarajkumar, Mark Kelley, Alyssa Samolyk-Mazzanti, Shyam Visweswaran, and Yanshan Wang. 2024. An empirical evaluation of prompting strategies for large language models in zero-shot clinical natural language processing: algorithm development and validation study. *JMIR Medical Informatics*, 12:e55318.

Liyan Tang, Zhaoyi Sun, Betina Idnay, Jordan G Nestor, Ali Soroush, Pierre A Elias, Ziyang Xu, Ying Ding, Greg Durrett, Justin F Rousseau, et al. 2023. Evaluating large language models on medical evidence summarization. *NPJ digital medicine*, 6(1):158.

Gemma Team, Morgane Riviere, Shreya Pathak, Pier Giuseppe Sessa, Cassidy Hardin, Surya Bhupatiraju, Léonard Hussenot, Thomas Mesnard, Bobak Shahriari, Alexandre Ramé, et al. 2024. Gemma 2: Improving open language models at a practical size. *arXiv preprint arXiv:2408.00118*.

Unsloth. 2024. unsloth.ai. available online. `https://docs.unsloth.ai/`. (accessed in November 2024).

Loïc Verlingue, Clara Boyer, Louise Olgiati, Clément Brutti Mairesse, Daphné Morel, and Jean-Yves Blay. 2024. Artificial intelligence in oncology: ensuring safe and effective integration of language models in clinical practice. *The Lancet Regional Health–Europe*, 46.

David Wadden, Shanchuan Lin, Kyle Lo, Lucy Lu Wang, Madeleine van Zuylen, Arman Cohan, and Hannaneh Hajishirzi. 2020. Fact or fiction: Verifying scientific claims. In *Proceedings of the 2020 Conference on Empirical Methods in Natural Language Processing (EMNLP)*, pages 7534–7550, Online. Association for Computational Linguistics.

David Wadden, Kyle Lo, Lucy Lu Wang, Arman Cohan, Iz Beltagy, and Hannaneh Hajishirzi. 2022. MultiVerS: Improving scientific claim verification with weak supervision and full-document context. In *Findings of the Association for Computational Linguistics: NAACL 2022*, pages 61–76, Seattle, United States. Association for Computational Linguistics.

Xuezhi Wang, Jason Wei, Dale Schuurmans, Quoc Le, Ed Chi, Sharan Narang, Aakanksha Chowdhery, and Denny Zhou. 2022. Self-consistency improves chain of thought reasoning in language models. *arXiv preprint arXiv:2203.11171*.

Jason Wei, Xuezhi Wang, Dale Schuurmans, Maarten Bosma, Fei Xia, Ed Chi, Quoc V Le, Denny Zhou, et al. 2022. Chain-of-thought prompting elicits reasoning in large language models. *Advances in neural information processing systems*, 35:24824–24837.

Yixuan Weng, Minjun Zhu, Fei Xia, Bin Li, Shizhu He, Shengping Liu, Bin Sun, Kang Liu, and Jun Zhao. 2022. Large language models are better reasoners with self-verification. *arXiv preprint arXiv:2212.09561*.

Yu Xia, Rui Wang, Xu Liu, Mingyan Li, Tong Yu, Xiang Chen, Julian McAuley, and Shuai Li. 2024. Beyond chain-of-thought: A survey of chain-of-x paradigms for llms. *Preprint*, arXiv:2404.15676.

Shunyu Yao, Dian Yu, Jeffrey Zhao, Izhak Shafran, Thomas L Griffiths, Yuan Cao, and Karthik Narasimhan. 2023. Tree of thoughts: Deliberate problem solving with large language models, 2023. *URL https://arxiv. org/pdf/2305.10601. pdf*.

Zihan Yu, Liang He, Zhen Wu, Xinyu Dai, and Jiajun Chen. 2023. Towards better chain-of-thought prompting strategies: A survey. *Preprint*, arXiv:2310.04959.

Xuan Zhang, Chao Du, Tianyu Pang, Qian Liu, Wei Gao, and Min Lin. 2024. Chain of preference optimization: Improving chain-of-thought reasoning in llms. *arXiv preprint arXiv:2406.09136*.

Ruochen Zhao, Xingxuan Li, Shafiq Joty, Chengwei Qin, and Lidong Bing. 2023. Verify-and-edit: A knowledge-enhanced chain-of-thought framework. In *Proceedings of the 61st Annual Meeting of the Association for Computational Linguistics (Volume 1: Long Papers)*, pages 5823–5840, Toronto, Canada. Association for Computational Linguistics.

Denny Zhou, Nathanael Schärli, Le Hou, Jason Wei, Nathan Scales, Xuezhi Wang, Dale Schuurmans, Claire Cui, Olivier Bousquet, Quoc Le, et al. 2022. Least-to-most prompting enables complex reasoning in large language models. *arXiv preprint arXiv:2205.10625*.

A Appendix

Size of Applied Models Table 7 provides a comparison of model size and initial context window length. The model size of the open source LLMs is limited to 14 billion parameters. All models are the instruct fine-tuned version.

Model	Version	Context Window	Parameters
GPT3.5	gpt-3.5-turbo-0125	16K	175B
GPT-4o-mini	gpt-4o-mini-2024-07-18	128K	?
Phi3.5-3.6B	Phi-3.5-mini-instruct	128K	3.6B
Mistral-7b	mistral-7b-instruct-v0.3	32K	7B
Llama3.1-8B	Meta-Llama-3.1-8B-Instruct	128K	8B
Gemma2-9B	gemma-2-9b-bnb-it	8K	9B
Mistral-12B	Mistral-Nemo-Instruct-2407	1024K	12B
Phi3-14B	Phi-3-medium-4k-instruct	4K	14B

Table 7: List of low-cost GPT models and lightweight open-source LLMs used in our experiments.

Comparisons of generations of different models with different prompting strategies. In particular, Table 10-15 illustrates the enhancement of lightweights LLMs in analyzing the statement based on the provided data (The first step of 2-step CoT and the second step - *evidence analysis* in the 4-step framework) for the following example shown in Table 8.

Fine-tuned results are obtained after fine-tuning with the GPT-4o-mini augmented training samples. Fine-tuning provides the LLMs with reliable analysis patterns for the NLI4CT tasks, thereby increasing the reliability of small-scale models. We utilize the *FastLanguageModel* modules of **unsloth**[1] library to accelerate the SFT fine-tuning in our experiments. **SFT** in the tables represents the response in the second step (EVIDENCE ANALYSIS) by the model fine-tuned with GPT-4o-mini generated responses.

[1] https://github.com/unslothai/unsloth

Statement: 'Only 2 patients in the primary trial did not have Recurrence-free Survival'
Clinical Trial Data: Primary trial: - Outcome Measurement: - Recurrence-free Survival - Recurrence-free survival curves will be plotted for subjects treated with stage I and II disease. - Time frame: Time from the start of treatment to recurrence, second malignancy, or death as a first event, assessed up to 3 years - Results 1: - Arm/Group Title: Treatment (Chemotherapy With or Without Maintenance Therapy) - Arm/Group Description: SYSTEMIC CHEMOTHERAPY: Patients receive cyclophosphamide IV over 1 hour and paclitaxel IV over 3 hours on day 1. Treatment repeats every 14 days for 6 courses in the absence of disease progression or unacceptable toxicity. - MAINTENANCE THERAPY (Her-2 neu positive patients): Patients receive trastuzumab IV over 30 minutes on day 1. Treatment repeats every 14 days for 5 courses and then every 21 days for 14 courses in the absence of disease progression or unacceptable toxicity. - cyclophosphamide, paclitaxel, trastuzumab: Given IV - Overall Number of Participants Analyzed: 100 - Measure Type: Number - Unit of Measure: percentage of subjects 98 (92.2 to 99.5)
Label: Entailment

Table 8: On exmpale instance of 'Entailment' class from NLI4CT test set.

GPT-4o-mini in *evidence analysis* step: "To evaluate the claim "Only 2 patients in the primary trial did not have Recurrence-free Survival," we first need to extract and clarify the relevant data points from the provided clinical trial data and then analyze the claim against those data points. 1. Relevant Data Points Extracted and Clarified: - **Total Participants Analyzed**: 100 - **Recurrence-free Survival Result**: 98 participants (which equates to the percentage of subjects achieving RFS) achieved recurrence-free survival. - **Percentage of Subjects with RFS**: 98% (confidence interval of 92.2% to 99.5%) 2. Evaluation of the Claim against Data Points: 1. **Total number of patients**: The total number of participants in the trial is 100. - **Evaluation**: The claim mentions "only 2 patients" not achieving RFS. This would imply that a total of 98 out of 100 patients experienced RFS as stated in the trial data. 2. **Number of patients with Recurrence-free Survival**: According to the trial data, 98 patients achieved RFS. - **Evaluation**: This supports the claim, as if 98 patients achieved RFS, then logically it follows that 2 patients did not achieve RFS (100 total participants - 98 with RFS = 2 without RFS). 3. **Effectiveness of Treatment**: The high percentage of 98% suggests a strong effectiveness of the treatment administered. - **Evaluation**: This supports the assertion in the claim that such a small number of patients (2 in this case) who did not achieve RFS indicates high treatment effectiveness. 4. **Definition of Recurrence-Free Survival**: RFS is defined as a period during which a patient remains free of any signs of disease recurrence. - **Evaluation**: The claim directly referenced RFS, which is consistent with the clinical outcome measures defined in the trial data. Conclusion: The claim that "Only 2 patients in the primary trial did not have Recurrence-free Survival" is consistent with the clinical trial data provided. The data indicates that out of 100 participants, 98 achieved RFS, meaning it is accurate to state that only 2 patients did not have RFS. Hence, the claim is validated by the data."
Predict: Entailment

Table 9: Evaluation on the above example made by GPT-4o-mini in the EVIDENCE ANALYSIS that leads to the accurate logical relation prediction.

Model	Approach	Prediction
Phi3.5-3.6B	**zero-shot 2-step:** "The statement claims that "Only 2 patients in the primary trial did not have Recurrence-free Survival," but the clinical trial data does not provide a specific number of patients without Recurrence-free Surv"	Contradiction
	zero-shot 4-step:"Recurrence-free Survival" (RFS) is the primary outcome measure, which in this context is the time until cancer recurrence, development of a second cancer, or death, whiche0ever comes first. - "Time frame" specifies that the RFS is assessed from the start of treatment up to 3 years, which is a standard period for evaluating long-term outcomes in oncology trials. - "Arm/Group Title: Treatment (Chemotherapy With or Without Maintenance Therapy)" refers to the two distinct treatment strategies being compared in the trial: one with systemic chemotherapy alone and the other with chemotherapy followed by maintenance therapy for Her-2 neu positive patients. - "Arm/Group Description" provides details on the chemotherapy regimen for the systemic treatment arm, including the drugs used (cyclophosphamide and paclitaxel), the administration method (IV), and the treatment schedule (every 14 days for 6 courses). - "MAINTENANCE THERAPY (Her-2 neu positive patients)" specifies the additional treatment for patients with Her-2 neu positive cancer, which includes the drug trastuzumab, also administered IV, with a different treatment schedule (every 14 days for 5 courses, then every 21 days for 14 courses). - "cyclophosphamide, paclitaxel, trastuzumab: Given IV" confirms the intravenous administration of the drugs used in the trial. - "Overall Number of Participants Analyzed: 100"	Entailment
	SFT: " To evaluate the statement "Only 2 patients in the primary trial did not have Recurrence-free Survival" based on the provided clinical trial data, we can analyze the relevant information as follows: 1. **Outcome Measurement**: The primary outcome measure is Recurrence-free Survival (RFS), which assesses the time from the start of treatment to recurrence, second malignancy, or death. 2. **Time Frame**: The assessment period for RFS is up to 3 years, indicating a focus on short- to medium-term outcomes. 3. **Treatment Arm**: The trial includes two treatment groups: - **Chemotherapy With or Without Maintenance Therapy**: This group receives systemic chemotherapy (cyclophosphamide and paclitaxel) and, for Her-2 neu positive patients, maintenance therapy (trastuzumab). - **Maintenance Therapy Only**: This group receives maintenance therapy (trastuzumab) only. 4. **Overall Number of Participants Analyzed**: A total of 100 participants were included in the trial. 5. **Results**: The reported outcome indicates that 98% of participants achieved Recurrence-free Survival. This translates to: - 98% of 100 participants = 98 participants had Recurrence-free Survival. - Therefore, 2 patients (100 - 98) did not have Recurrence-free Survival. Conclusion: The statement "Only 2 patients in the primary trial did not have Recurrence-free Survival" is supported by the data. The evidence shows that 98 participants achieved Recurrence-free Survival, confirming that indeed only 2 patients did not experience this outcome. Thus, the statement can be logically inferred from the provided clinical trial data."	Entailment

Table 10: Comparison between the model responses made in the first step of zero-shot 2-step framework and the second step (*evidence analysis*) in the zero-shot 4-step framework, made by model Phi3.5-3.6B.

Model	Approach	Prediction
Mistral-7B	**zero-shot 2-step:** The statement does not logically follow from the provided clinical trial data. The data only provides the percentage of subjects (98 out of 100) who did not experience a recurrence, but it does not specify the number of subjects who did experience a recurrence. Therefore, it is not possible to determine from this data whether only 2 patients in the primary trial did not have Recurrence-free Survival."	Contradiction
	zero-shot 4-step:" Based on the provided clinical trial data, the claim "Only 2 patients in the primary trial did not have Recurrence-free Survival" can be evaluated as follows: 1. Outcome Measurement: Recurrence-free Survival The claim and the data both refer to the same outcome measure. 2. Time frame: Time from the start of treatment to recurrence, second malignancy, or death as a first event, assessed up to 3 years The time frame in the claim is consistent with the time frame in the data (up to 3 years). 3. Results 1: - Arm/Group Title: Treatment (Chemotherapy With or Without Maintenance Therapy) - Arm/Group Description: SYSTEMIC CHEMOTHERAPY: Patients receive cyclophosphamide IV and paclitaxel IV. MAINTENANCE THERAPY (Her-2 neu positive patients): Patients receive trastuzumab IV. - Overall Number of Participants Analyzed: 100 - Measure Type: percentage of subjects - Unit of Measure: percentage The data shows that 98 out of 100 participants (92.2 to 99.5%) did not have recurrence-free survival. This suggests that the majority of participants in the trial did not experience a recurrence of their disease during the follow-up period. However, the claim does not specify which arm or group within the trial the 2 patients without recurrence-free survival belong to. In conclusion, while the data supports the claim that the majority of participants did not experience a recurrence of their disease, the claim is not fully accurate as it does not specify which arm or group the 2 patients without recurrence-free survival belong to. A more detailed analysis would require information"	Contradiction
	SFT: "To evaluate the statement "Only 2 patients in the primary trial did not have Recurrence-free Survival," we can analyze the provided clinical trial data as follows: 1. **Outcome Measurement**: The primary outcome of the trial is Recurrence-free Survival (RFS), which is defined as the time from the start of treatment to recurrence, second malignancy, or death, assessed up to 3 years. 2. **Results Overview**: The trial included a total of 100 participants, with 98 achieving recurrence-free survival, resulting in a percentage of 98% (with a confidence interval of 92.2% to 99.5%). This indicates that a significant majority of participants remained free from disease recurrence. 3. **Interpretation of the Statement**: The statement claims that only 2 patients did not have recurrence-free survival. Given that 98 out of 100 participants achieved this outcome, it can be inferred that the remaining 2 patients did not achieve recurrence-free survival. 4. **Conclusion**: The evidence from the trial data supports the statement, confirming that only 2 patients did not have recurrence-free survival, as they are the 2 out of 100 participants who did not achieve this outcome. Thus, the statement can be accurately inferred from the provided data."	Entailment

Table 11: Comparison between the model responses made in the first step of zero-shot 2-step framework and the second step (*evidence analysis*) in the zero-shot 4-step framework, made by model Mistral-7B.

Model	Approach	Prediction
Llama3.1-8B	**zero-shot 2-step:** " Based on the provided clinical trial data, the statement "Only 2 patients in the primary trial did not have Recurrence-free Survival" does not logically follow from the given information. The statement in question implies that the outcome of interest is the number of patients without Recurrence-free Survival, which is not explicitly mentioned in the provided data. The data only mentions the outcome of Recurrence-free Survival, not the absence of it."	Contradiction
	zero-shot 4-step: " Based on the clinical trial data, the following relevant data points are identified: * **Outcome Measure**: Recurrence-free Survival * **Time Frame**: Up to 3 years * **Number of Participants Analyzed**: 100 * **Measure Type**: Number * **Unit of Measure**: Percentage of subjects Evaluating the claim: * The claim states that "Only 2 patients in the primary trial did not have Recurrence-free Survival". This statement seems to be inconsistent with the data, as the trial reported a 98% (92.2-99.5) recurrence-free survival rate, indicating that a significant portion of the patients did experience recurrence-free survival. * The claim suggests that only 2 patients out of the total 100 participants did not achieve recurrence-free survival, which contradicts the overall recurrence-free survival rate of 98% reported in the trial. * The trial's reported results do not support the claim that only 2 patients did not experience recurrence-free survival."	Contradiction
	SFT: " To evaluate the statement "Only 2 patients in the primary trial did not have Recurrence-free Survival," we will analyze it in relation to the provided clinical trial data. Evaluation of the Statement: 1. **Outcome Measurement**: The primary trial measures Recurrence-free Survival, which indicates the duration patients remain free from disease recurrence following treatment. 2. **Patient Cohort**: The statement refers to a specific group of patients within the trial, focusing on those who did not achieve recurrence-free survival. 3. **Results Data**: - **Total Participants Analyzed**: 100 patients. - **Outcome Measure**: The percentage of participants achieving recurrence-free survival is reported as 98% (with a confidence interval of 92.2% to 99.5%). 4. **Interpreting the Results**: - If 98% of patients achieved recurrence-free survival, this means that: - 98% of 100 = 98 patients did have recurrence-free survival. - Consequently, the number of patients who did not achieve recurrence-free survival would be: $$100 - 98 = 2 \text{ patients}$$ Conclusion: The statement "Only 2 patients in the primary trial did not have Recurrence-free Survival" can be logically inferred from the provided data, confirming that indeed, 2 out of the 100 participants did not achieve recurrence-free survival, supporting the accuracy of the claim."	Entailment

Table 12: Comparison between the model responses made in the first step of zero-shot 2-step framework and the second step (*evidence analysis*) in the zero-shot 4-step framework, made by model Llama3.1-8B.

Model	Approach	Prediction
Gemma2-9B	**zero-shot 2-step:** "The statement **does not logically follow** from the provided clinical trial data. Here's why: * **Missing Data:** The data only provides the overall percentage of subjects (98%) who experienced Recurrence-free Survival. It doesn't specify the exact number of patients who did or did not experience recurrence. * **Potential for Interpretation:** While the high percentage suggests that most patients did experience Recurrence-free Survival, it doesn't definitively confirm that only two patients did not. To determine if the statement is true, we would need the raw data on individual patient outcomes (e.g., a table showing whether each patient experienced recurrence or not). "	Contradiction
	zero-shot 4-step:"Let's break down the claim and compare it to the provided data: **Claim:** 'Only 2 patients in the primary trial did not have Recurrence-free Survival' **Data Points:** * **Primary Trial:** Yes, the data refers to a primary trial. * **Outcome Measurement:** Recurrence-free Survival is the primary outcome. * **Time Frame:** The trial assesses Recurrence-free Survival up to 3 years. * **Overall Number of Participants Analyzed:** 100 * **Results 1:** The data shows 98 out of 100 (92.2 to 99.5%) of patients in the treatment arm experienced Recurrence-free Survival. **Evaluation:** The claim is **partially supported** by the data. * **Support:** The data clearly states that 98 out of 100 patients in the treatment arm had Recurrence-free Survival. This means only 2 patients did *not* experience Recurrence-free Survival. * **Missing Information:** The claim doesn't specify which treatment arm it refers to. We need to know if this refers to the chemotherapy with or without maintenance therapy. **Important Considerations:** * **Confidence Intervals:** The data provides a confidence interval (92.2 to 99.5%). This means we are 95% confident that the true percentage of patients with Recurrence-free Survival in the treatment arm falls within this range. * **Context:** To fully evaluate the claim, we need more information about the control group (if there was one) and the specific types of cancer being treated. "	Entailment
	SFT:" To evaluate the statement "Only 2 patients in the primary trial did not have Recurrence-free Survival" based on the provided clinical trial data, we can analyze the relevant information as follows: 1. **Outcome Measurement**: The primary trial focused on measuring Recurrence-free Survival (RFS), which is a critical endpoint in cancer trials. 2. **Time Frame**: The assessment of RFS was conducted over a period of up to 3 years from the start of treatment, providing a comprehensive view of the treatment's effectiveness. 3. **Results Analysis**: - The trial included a total of **100 participants** analyzed. - The statement indicates that **2 patients** did not achieve RFS. - This translates to **98%** of participants having RFS, as shown in the results (98% with a confidence interval of 92.2% to 99.5%). 4. **Conclusion**: The statement is supported by the data, confirming that only 2 out of 100 participants did not achieve recurrence-free survival. This indicates a high level of effectiveness of the treatment in preventing disease recurrence among the majority of patients. In summary, the evidence from the trial data clearly supports the statement that only 2 patients did not have recurrence-free survival, highlighting the treatment's success in maintaining disease-free status in the majority of participants."	Entailment

Table 13: Comparison between the model responses made in the first step of zero-shot 2-step framework and the second step (*evidence analysis*) in the zero-shot 4-step framework, made by model Gemma2-9B.

Model	Approach	Prediction
Mistral-12B	**zero-shot 2-step:** "Based on the provided clinical trial data, the statement "Only 2 patients in the primary trial did not have Recurrence-free Survival" is **not logically supported** by the data. Here's why: 1. The clinical trial data reports that 98 out of 100 participants had Recurrence-free Survival, which is 92.2% to 99.5% (with 95% confidence interval). 2. This means that, at most, 2 out of 100 participants did not have Recurrence-free Survival, which is consistent with the reported percentage. 3. However, the statement uses "only 2 patients" which implies that the number is exact and not an estimate. The data provided does not confirm this exact number; it only gives a range (92.2% to 99.5%) within which the number of patients with Recurrence-free Survival falls. "	Contradiction
	zero-shot 4-step:"Based on the provided clinical trial data, let's evaluate the claim "Only 2 patients in the primary trial did not have Recurrence-free Survival": 1. **Number of participants in the trial**: The claim mentions "primary trial," and the data provides the overall number of participants analyzed as 100. This matches the context of the claim, as it refers to the primary trial. - Data point: 100 participants - Claim: Applicable (primary trial) 2. **Outcome measure**: The claim focuses on Recurrence-free Survival (RFS), which is also the primary outcome measurement mentioned in the trial data. - Data point: Recurrence-free Survival - Claim: Applicable (RFS is the outcome of interest) 3. **Time frame**: The claim does not specify a time frame, but the trial data indicates that RFS is assessed up to 3 years. Without a specified time frame in the claim, we cannot directly evaluate this aspect. - Data point: Up to 3 years - Claim: Not specified, so cannot be evaluated 4. **Number of patients without RFS**: The claim states "Only 2 patients... did not have Recurrence-free Survival." The trial data does not provide the exact number of patients without RFS, but it does mention the overall percentage of subjects with RFS. - Data point: 98% (92.2 to 99.5) of subjects had RFS - Claim: The data suggests that 2% (0.5 to 7.8) of subjects did not have RFS, which is consistent with the claim's statement of "Only 2 patients." 5. **Arm/Group**: The claim does not specify any particular treatment arm or group, so we cannot directly evaluate this aspect."	Entailment
	SFT: "To evaluate the statement "Only 2 patients in the primary trial did not have Recurrence-free Survival," we can analyze the provided clinical trial data as follows: 1. **Outcome Measurement**: The trial measures Recurrence-free Survival, which is defined as the time from the start of treatment to recurrence, second malignancy, or death. 2. **Total Participants**: The trial included 100 participants. 3. **Results**: The statement claims that only 2 patients did not have Recurrence-free Survival. This implies that 98 patients did achieve Recurrence-free Survival. 4. **Confidence Interval**: The reported percentage of subjects with Recurrence-free Survival is 98%, with a confidence interval of (92.2 to 99.5). This indicates a high level of confidence in the outcome. 5. **Conclusion**: Based on the data, it can be inferred that the statement is accurate. If 98 out of 100 participants achieved Recurrence-free Survival, then the number of patients who did not achieve this outcome is indeed 2, confirming the statement.""	Entailment

Table 14. Comparison between the model responses made in the first step of zero shot 2 step framework and the second step (*evidence analysis*) in the zero-shot 4-step framework, made by model Mistral-12B .

Model	Approach	Prediction
Phi3-14B	**zero-shot 2-step:** " The statement "Only 2 patients in the primary trial did not have Recurrence-free Survival" does not logically follow from the clinical trial data provided. The data indicates that 98% (92.2 to 99.5) of the subjects had recurrence-free survival, which means that 2% (100 - 98%) of the subjects did not have recurrence-free survival. However, the data does not specify the exact number of patients who did not have recurrence-free survival."	Contradiction
	zero-shot 4-step:" Relevant data points from the clinical trial data: 1. Outcome Measurement: Recurrence-free Survival 2. Time frame: Time from the start of treatment to recurrence, second malignancy, or death as a first event, assessed up to 3 years 3. Arm/Group Title: Treatment (Chemotherapy With or Without Maintenance Therapy) 4. Arm/Group Description: SYSTEMIC CHEMOTHERAPY and MAINTENANCE THERAPY (Her-2 neu positive patients) 5. Overall Number of Participants Analyzed: 100 6. Measure Type: Number 7. Unit of Measure: percentage of subjects 8. Results 1: 98 (92.2 to 99.5) Evaluation of the claim: The claim states that "Only 2 patients in the primary trial did not have Recurrence-free Survival." Based on the clinical trial data provided, the results show that 98% of the subjects (92.2 to 99.5) had Recurrence-free Survival. This means that 2% of the subjects did not have Recurrence-free Survival. Given that there were 100 participants in the trial, 2% of the participants would be 2 participants. Therefore, the claim is consistent with the clinical trial data, as it accurrightly states that only 2 patients in the primary trial did not have Recurrence-free Survival."	Entailment
	SFT: " To evaluate the statement "Only 2 patients in the primary trial did not have Recurrence-free Survival," we can analyze the provided clinical trial data as follows: 1. **Outcome Measurement**: The primary outcome of the trial is Recurrence-free Survival (RFS), which is defined as the time from the start of treatment to the occurrence of recurrence, a second malignancy, or death. 2. **Treatment Description**: The trial involves systemic chemotherapy, with specific regimens for Her-2 neu positive patients, including cyclophosphamide, paclitaxel, and trastuzumab. 3. **Time Frame**: The assessment of RFS is conducted over a period of up to 3 years from the start of treatment. 4. **Overall Number of Participants Analyzed**: The trial includes a total of 100 participants. 5. **Results**: The data indicates that 98% of the participants achieved recurrence-free survival, with a confidence interval of 92.2% to 99.5%. Evaluation of the Statement: - **Claim**: "Only 2 patients in the primary trial did not have Recurrence-free Survival." - **Analysis**: - If 98% of the 100 participants achieved RFS, this means that 2% did not. - Calculating the number of patients who did not achieve RFS: - 2% of 100 participants = 2 patients. Conclusion: The statement "Only 2 patients in the primary trial did not have Recurrence-free Survival" is supported by the data. The evidence indicates that 2 patients out of 100 did not achieve RFS, confirming the accuracy of the claim."	Entailment

Table 15: Comparison between the model responses made in the first step of zero-shot 2-step framework and the second step (*evidence analysis*) in the zero-shot 4-step framework, made by model Phi3-14B.

A Retrieval-Based Approach to Medical Procedure Matching in Romanian

Andrei Niculae[1], Adrian Cosma[1,2], Emilian Radoi[1]
[1]National University of Science and Technology POLITEHNICA Bucharest
[2]Dalle Molle Institute for Artificial Intelligence Research (IDSIA)
andrei.niculae1004@stud.acs.upb.ro, adrian.cosma@supsi.ch, emilian.radoi@upb.ro

Abstract

Accurately mapping medical procedure names from healthcare providers to standardized terminology used by insurance companies is a crucial yet complex task. Inconsistencies in naming conventions lead to missclasified procedures, causing administrative inefficiencies and insurance claim problems in private healthcare settings. Many companies still use human resources for manual mapping, while there is a clear opportunity for automation. This paper proposes a retrieval-based architecture leveraging sentence embeddings for medical name matching in the Romanian healthcare system. This challenge is significantly more difficult in underrepresented languages such as Romanian, where existing pretrained language models lack domain-specific adaptation to medical text. We evaluate multiple embedding models, including Romanian, multilingual, and medical-domain-specific representations, to identify the most effective solution for this task. Our findings contribute to the broader field of medical NLP for low-resource languages such as Romanian.

1 Introduction

Ensuring accurate mapping between medical procedure names used by different healthcare providers and a standardized terminology set maintained by health insurance companies is a challenging task, with real-world applications. Discrepancies in naming conventions can lead to administrative inefficiencies, misclassification of procedures, and potential barriers for patients seeking insurance coverage. These mismatches can result in denied claims, increased processing times, and overall inefficiencies in the healthcare reimbursement process. For example, "The State of Claims: 2024" report [1] reveals that 46% of denied claims are due to missing or innacurate data and coding errors.

Matching procedure names is similar to the well-known problems of entity resolution and text matching, yet it presents unique challenges in the medical domain. The complexity stems from several factors: *(i)* medical terminology is highly domain-specific and varies across institutions, *(ii)* data distributions are often imbalanced due to the frequency of common procedures overshadowing rare ones, *(iii)* nomenclatures evolve over time, necessitating adaptive matching techniques, and *(iv)* the presence of noise in text data, including typographical errors and abbreviations further complicates standardization efforts. Figure 1 illustrates this problem. While previous studies have addressed similar challenges (Tavabi et al., 2024; Levy et al., 2022; Zaidat et al., 2024), most focus on healthcare systems in the United States or other widely studied regions (Alexander et al., 2003). International standards are typically adapted by each country, and private insurance companies may develop their own coding schemes, making a universal solution impractical.

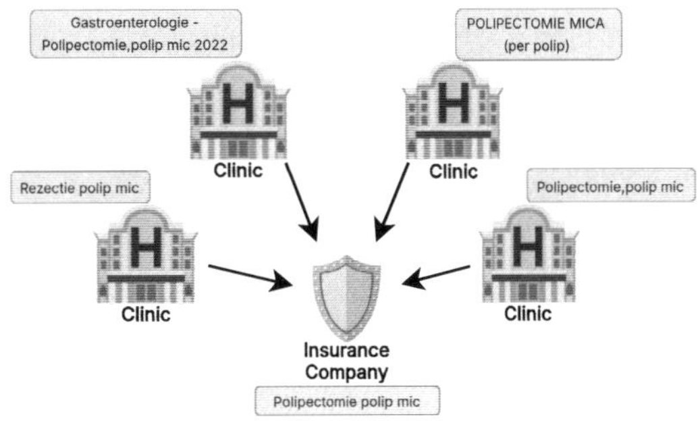

Figure 1: Diagram of the medical procedure matching problem. Clinics often have their own local names for medical procedures that are changed annually, for which a central insurance agency must match to a standardized list of procedures for reimbursement.

This issue is particularly pressing for underrepresented languages such as Romanian. Despite growing interest in NLP for low-resource languages (Nigatu et al., 2024), Romanian remains significantly underrepresented in medical NLP research. Ex-

[1]The State of Claims: 2024, Accessed 19.03.2025

Proceedings of the 24th Workshop on Biomedical Language Processing (BioNLP 2025), pages 167–175
August 1, 2025

isting language models such as RoBERT (Masala et al., 2020) and RoLLaMA (Masala et al., 2024) provide general-purpose Romanian embeddings, but they lack the necessary specialization for medical text processing. Often, for real-world scenarios, multilingual models are ubiquitously used (Wang et al., 2024, 2020), even if they might fail to capture language-specific nuances.

In this paper, we propose a retrieval-based architecture for medical procedure matching. By leveraging metric learning and dense vector representations of procedure names (Ramesh Kashyap et al., 2024), our method can handle a variable number of input-output mappings, can be expanded without retraining the entire model, and integrate efficiently with scalable vector search frameworks such as Milvus (Wang et al., 2021). This makes retrieval an attractive paradigm for medical name matching, as it enables continuous updates and adaptation to changing medical taxonomies without extensive human intervention. We empirically evaluate three sentence embedding models (Wang et al., 2024; Masala et al., 2020; Alsentzer et al., 2019), comparing their effectiveness in Romanian medical name matching.

By focusing on the Romanian healthcare system, our study highlights the broader challenges of medical terminology standardization and provides insights that can inform similar efforts in other low-resource languages. We aim to contribute to the development of robust, scalable, and language-aware retrieval methods for healthcare applications, ultimately improving the efficiency and accessibility of medical insurance systems.

Our contributions are as follows:

1. We propose a retrieval-based architecture for matching medical procedure names across different healthcare providers and insurance companies, addressing a pressing real-world problem in the Romanian healthcare system.

2. We conduct an extensive evaluation of various sentence embedding models, both Romanian (Masala et al., 2020), multilingual (Wang et al., 2024) and specialized in the medical domain (Alsentzer et al., 2019), highlighting their performance in the context of Romanian medical text matching.

2 Related Work

Sentence embedding models. Semantic text embedding models (Ramesh Kashyap et al., 2024) are a significant component of many NLP applications, most notably text retrieval and question answering. Text embeddings are used to capture semantic representations of text that go beyond surface level word and character matching methods such as TF-IDF. Currently, practitioners are using pretrained transformer models such as BERT (Reimers and Gurevych, 2019), either by aggregating word-level representations with a pooling operation, or by using specialized training for text similarity (Khattab and Zaharia, 2020). Currently, the best performing models are aggregated in the MTEB leaderboard (Muennighoff et al., 2023a), a benchmark of several text embedding tasks, including several non-English datasets. For the medical and scientific domain (Lewis et al., 2020), several models have been developed. Models such as SciBERT (Beltagy et al., 2019), BioBERT (Alsentzer et al., 2019), ClinicalBERT (Alsentzer et al., 2019) and MedBERT (Rasmy et al., 2021) offer domain-specific embeddings by training on either specialized biomedical corpora or task-specific datasets.

However, most contextualized text representation models for the medical domain are focused on the English language, with under-represented languages severely lacking in resources such as specialized models or training datasets. In our setup, medical procedure names are written in Romanian, a low resource language, with only a few pretrained language models (Masala et al., 2024, 2020). Currently, for Romanian, only a pretrained RoBERT model (Masala et al., 2020) is available for direct contextualized text representations, but no such variant exists for the medical domain. Currently, multilingual models such as E5 (Wang et al., 2024) and MiniLM (Wang et al., 2020) are ubiquitously used for non-English tasks.

Medical Procedure Matching. The task of medical procedure matching has been performed in the context of assigning medical notes or pathology reports to a predefined set of medical procedures (Tavabi et al., 2024; Levy et al., 2022; Zaidat et al., 2024), with a focus on the US medical system.

Tavabi et al. (2024) investigated the problem of mapping unstructured operative notes to Current Procedural Terminology (CPT) codes. The CPT code set is a system used to describe medical, surgical and diagnostic services, that are used for billing and insurance reimbursement processes in healthcare. The authors apply common NLP techniques to assign 44,002 notes to 100 most prevalent CPT codes, treating this problem as a classification task.

Masterlist Entry	Associated Clinic Procedures Names
Polipectomie polip mic *(Small polyp polypectomy)*	Polipectomie,polip mic *(Polypectomy, small polyp)*
	Gastroenterologie - Polipectomie,polip mic 2022 *(Gastroenterology - Polypectomy, small polyp 2022)*
	Rezectie polip mic *(Small polyp resection)*
Radiografie omoplat 1 incidenta *(Scapula X-ray 1 view)*	Radiografie omoplat (fata sau profil) *(Scapula X-ray (frontal or lateral))*
	Omoplat profil *(Lateral scapula view)*
	RX omoplat profil *(Lateral scapula X-ray)*
Vitamina B12 *(Vitamin B12)*	Vitamina B12 serica (5 zile) *(Serum Vitamin B12 (5 days))*
	Vitamina B12 (Cianocobalamina) *(Vitamin B12 (Cyanocobalamin))*
	ANALIZA SANGE - Vitamina B12 *(BLOOD ANALYSIS - Vitamin B12)*

Table 1: Selected examples of entries in the masterlist and associated procedure names from clinics. There is significant variation in procedure names, which makes simple text matching inappropriate. We provide English translations for convenience.

Using TF-IDF, Doc2Vec (Le and Mikolov, 2014) and Clinical Bio-BERT (Alsentzer et al., 2019) embeddings as input they train a support vector machine classifier, for each embedding type. In their experiments, TF-IDF outperformed both BERT and Doc2Vec.

Levy et al. (2022) used machine-learning models for predicting CPT codes from pathology reports. Their study analyzed 93,039 pathology reports from the Dartmouth-Hitchcock Department of Pathology and Laboratory Medicine, classifying 42 CPT codes. They evaluated the performance of XGBoost and BERT—using both diagnostic text alone and all report subfields. Their findings indicated that while BERT outperformed XGBoost when trained only on diagnostic text, but using all report subfields resulted in XGBoost achieving the best performance.

Zaidat et al. (2024) have also explored assigning CPT codes to spine surgery operative notes, using XLNet (Yang et al., 2019), a bidirectional LSTM (Hochreiter and Schmidhuber, 1997) model. They fine-tune the model to their operative note dataset, containing 922 entries.

Previous studies have evaluated the performance of statistical, machine learning and deep learning models on classification of a large number of samples to a relatively small subset of CPT codes. In contrast, we formulate our problem as a retrieval problem, since our dataset is severely imbalanced, and contains two orders of magnitude more CPT codes (38,814 entries). Furthermore, another advantage of this formulation is that by avoiding a fixed set of classes, the addition of more procedures does not require modifying the architecture or retraining the model. Unique to our work, we are the first to tackle this problem in Romanian, a severely low-resource language in terms of specialized models for the medical domain.

3 Method

In this section, we provide an overview of the problem description, our dataset of medical procedure names and we describe the architecture for performing mapping between clinic descriptions and a set of standardized procedure names.

3.1 Problem Description

The problem of matching medical procedure names to a standardized masterlist is non-trivial. Simple text matching is insufficient, as we will demonstrate in Section 4. Our dataset is comprised of medical procedures and tests from 528 Romanian private clinics, containing 145,298 unique procedure names mapped to their corresponding masterlist entries. Through manual filtering of incorrect mappings, we reduced the dataset to 139,210 entries. Healthcare providers frequently use varying terms, abbreviations, and phrasing for the same procedure, which creates inconsistencies. To illustrate the difficulty, Table 1 shows some relevant examples of mappings. Healthcare providers may omit obvious terms, such as "polyp resection" being synonymous with "polypectomy". Similarly, entries such as "frontal or lateral X-ray" must be mapped to "1 view X-ray", as they represent a sin-

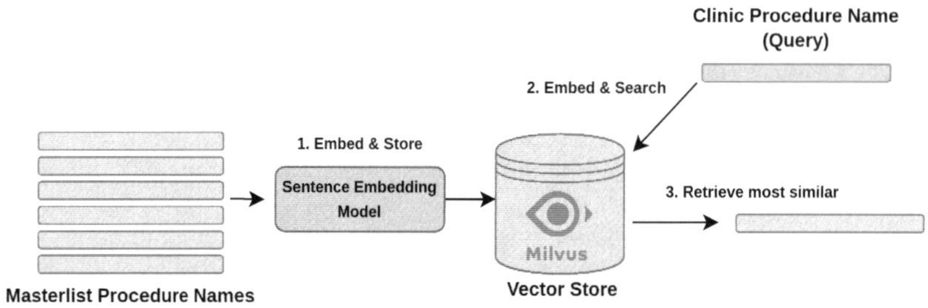

Figure 2: Overall diagram of our method. We formulate medical procedure matching as a retrieval problem: entries in the masterlist are embedded and stored in a vector store and the most similar entry is retrieved based on the similarity with a procedure name from a clinic.

gle test being performed. Terminology variations, like vitamin B12 also being called Cyanocobalamin, can add complexity, especially when descriptions include irrelevant details that mislead text-based matching. Several other relevant examples are presented in Table 2: matching a single medical procedure, when the description actually describes two procedures, not recognizing the semantic meaning of descriptions, ignoring important numerical thresholds, retrieving specific procedures instead of general ones (or vice versa), and prioritizing less important terms.

We chose to model our problem as a retrieval problem, and not as a classification problem, since 50% of elements from the masterlist have only 1 unique procedure assigned. Figure 3 shows the distribution of clinic descriptions assigned to masterlist entries. If we frame our task as a classification problem, we have 39,097 distinct classes, with 19,493 containing only a single sample. Given the severe class imbalance per procedure, a classification model would be inappropriate and would generalize poorly. However, a retrieval-based method can be effectively used by leveraging semantic text embeddings and metric-learning approaches to capture the similarity between clinic descriptions and masterlist entries.

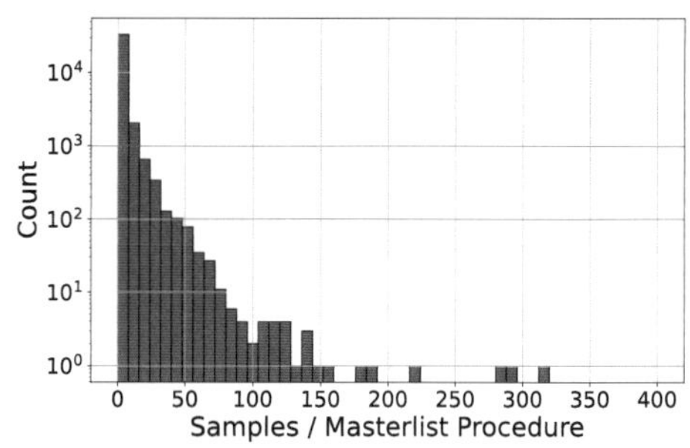

Figure 3: Distribution of number of unique clinic descriptions per masterlist procedure. There is a severe data imbalance: 19,493 (50%) out of 39,097 entries contain only a single example.

3.2 Procedure Matching as Retrieval

In our retrieval setup, we used the provided masterlist procedure names to build a retrieval index (Wang et al., 2021) and clinic descriptions as the references for the queries. We embed the descriptions using dense (Masala et al., 2020; Wang et al., 2024; Alsentzer et al., 2019) and sparse models (Robertson and Zaragoza, 2009). At inference time, we embed the query clinic descriptions that require a masterlist description and perform a similarity search. The vector DB returns the top-k most similar results for each of our clinic description. Figure 2 showcases this approach.

The vector index includes two types of entries: masterlist entries and clinic description $\leftrightarrow$ masterlist pairs. In the first scenario, the similarity score is calculated between the query and the masterlist entries, with the index returning the most similar masterlist entries. In the second scenario, the similarity score is computed between the query and the clinic descriptions stored in the index, and the masterlist entry associated with the most similar clinic description is returned. We build our search and evaluation architecture over Milvus (Wang et al., 2021), a high-performance vector database.

For our setup, we used three types of text embeddings: *(i)* sparse text embeddings using BM25 (Robertson and Zaragoza, 2009), *(ii)* dense semantic embeddings with several pretrained transformer models (Masala et al., 2020; Wang et al., 2024; Alsentzer et al., 2019), both zero-shot and fine-tuned with metric learning, and *(iii)* a hybrid ranking approach using RRF (Cormack et al., 2009).

3.3 Sparse Embeddings with BM25

For computing the sparse embeddings, we use BM25 (Robertson and Zaragoza, 2009) to identify the most relevant word-level features from the training set descriptions and masterlist entries. Text is preprocessed by removing diacritics, punctuation, and Romanian stopwords, followed by stemming the remaining words. At inference time, we compute the inner product between the query descriptions and the masterlist descriptions, as well as the clinic description pairs.

3.4 Dense Embeddings with Pretrained Transformer Models

Recent studies have highlighted the challenges of selecting optimal sentence embedding models for domain-specific retrieval tasks (Wornow et al., 2023). Generic benchmarks do not always align with real-world performance, necessitating task-specific evaluations. The MTEB Leaderboard (Muennighoff et al., 2023b) ranks top-performing embedding models based on retrieval performance across various datasets.

We experiment with three models for dense embeddings: mE5-large (Wang et al., 2024), RoBERT-large (Masala et al., 2020), and BioClinicalBERT (Alsentzer et al., 2019). We select mE5 due to its strong performance on multilingual retrieval tasks, RoBERT as a strong language-specific baseline model pre-trained using only Romanian text, and BioBERT as a domain-specific model, pretrained on biomedical text, which may capture medical terminology better than general-purpose models.

Fine-tuning with Metric Learning. We fine-tune the pretrained text embedding models using the MultipleNegativesRankingLoss objective (Henderson et al., 2017), as shown in Figure 4. We consider the clinic descriptions as anchors (a_i) and the corresponding masterlist descriptions (p_i) as positive pairs - (a_i, p_i). The negative pair consists of every combination (a_i, p_j), where $p_j, j \neq i$ are all other masterlist descriptions. In this way, our embedding model learns to increase the cosine similarity between the clinic descriptions and their mapped masterlist description, while decreasing the similarity between the clinic description and all other masterlist items. The model is fine-tuned on 80,911 pairs for 20 epochs, using a batch size of 4096. We use a learning rate of 2e-5, with a cosine scheduler and a warmup ratio of 0.1. All experiments are run on an NVIDIA A100 80GB GPU.

3.5 Hybrid search

Cormack et al. (2009) proposed Reciprocal Rank Fusion (RRF) as a method of aggregating the ranking results of multiple information retrieval systems. It is calculated using the formula:

$$\text{RRFscore}(d \in D) = \sum_{r \in R} \frac{1}{k + r(d)} \tag{1}$$

where D is the set of results to be ranked, R represents the multiple returned rankings of these results, k is a constant, and $r(d)$ is the rank of a result d. We combine the results of dense and sparse embeddings using RRF and analyze its effect on retrieval accuracy.

4 Experiments and Results

To evaluate our approach, we split the dataset into a training and evaluation split, containing 80,911 and 58,299 clinic description $\leftrightarrow$ masterlist pairs, respectively. For fine-tuning, we used only the training split. For evaluation, we split the evaluation set into gallery and probe sets in a 4:1 ratio, in a setup similar to 5-fold cross-validation, where gallery entries form the vector store data. Each fold is stratified based on the masterlist entries, such that each fold contains approximately the same distribution of masterlist entries. Specifically, for each masterlist entry, we distribute its associated clinic descriptions evenly across all folds – for example if 5 clinic descriptions map to the same masterlist entry, each fold will contain exactly 1 such mapping.

Evaluation Metrics. Our primary evaluation metric is Accuracy@k, which measures whether a ground-truth masterlist description is in the first k returned results for a query clinic description. Our target is to optimize for Acc@1, but we also include the results for Acc@3, Acc@5 and Acc@100. In a real-life use of such an system will involve suggesting top-3 or top-5 most similar masterlist entries, and Acc@3 and Acc@5 provides insight into the usefulness of our system. We also include Acc@100, as a low value indicates a problem with the chosen search technique, but usually it indicates the presence of incorrect annotations. In all our results, we show the mean and standard deviation across 5 folds.

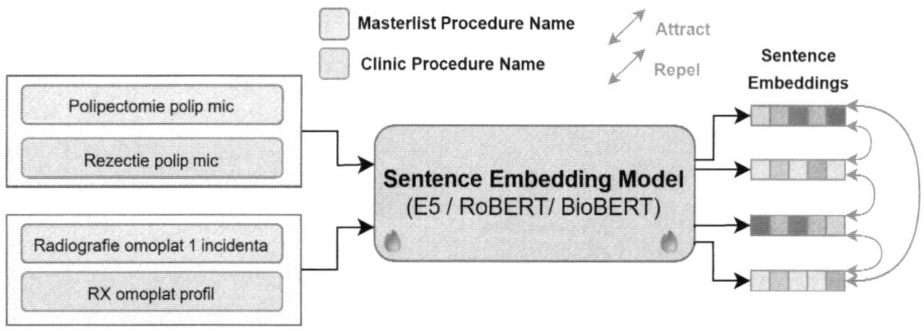

Figure 4: Fine-tuning approach for dense sentence embeddings. A pretrained text embedding model is trained to minimize the distance between representations of masterlist entries and associated clinic procedure names while maximising the distance between every other entry.

Clinic Description	BM25 Miss	mE5 Hit	Observations
Aplicare sterilet + EEV control *(IUD application + EEV control)*	Aplicare sterilet *(IUD application)*	Montare sterilet (DIU) + ecografie control *(IUD insertion (DIU) + ultrasound control)*	Did not account for the additional ultrasound control term
EXOSTOZA *(Exostosis)*	Sonoterapie in exostoze calcaneene *(Sonotherapy in calcaneal exostoses)*	Excizia exostozei *(Excision of exostosis)*	Focused on "exostoses" but did not recognize "excision" as a relevant treatment
Chiuretare molluscum contagiosm > 10 leziuni *(Curettage of molluscum contagiosum > 10 lesions)*	Chiuretare < 10 leziuni *(Curettage of < 10 lesions)*	Chiuretare moluscum contagiosum peste 10 leziuni *(Curettage of molluscum contagiosum over 10 lesions)*	Matched on "curettage" but ignored the numerical threshold
Radiofrecventa ablatie tumori *(Radiofrequency ablation of tumors)*	Ablatie laser / radiofrecventa tumora ureche dificultate redusa *(Laser/radiofrequency ablation of ear tumor - low difficulty)*	Excizie leziune cu radiofrecventa *(Excision of lesion with radiofrequency)*	Retrieved a more specific procedure (ear tumor) instead of a general one
RM articulatii sacro iliace cu subst. de contrast *(MRI of sacroiliac joints with contrast)*	Artrodeza articulatiei sacro iliace percutanata cu implant I Fuse *(Percutaneous sacroiliac joint arthrodesis with I Fuse implant)*	RMN articulatii sacroiliace cu SC, 1.5T *(MRI of sacroiliac joints with SC, 1.5T)*	Retrieved a surgical procedure instead of an imaging scan

Table 2: Selected examples of clinic Descriptions with BM25 Misses, mE5 Dense Embedding Hits. Sparse indexes are not appropriate for this task, which require high level semantic understanding of descriptions.

4.1 Comparison between different types of search indexes

In Table 3, we show a comparison between dense, sparse, and hybrid approaches. For dense embeddings, we used a fine-tuned mE5 (Wang et al., 2024) model. The results show that the fine-tuned dense model consistently outperforms both sparse and hybrid search methods. When searching only masterlist entries, the dense approach achieves 26.2% higher Acc@1 than the sparse approach. When using both masterlist and associated mappings, the dense approach obtains a 17.2% Acc@1 margin. The sparse approach also shows poor performance for Acc@100, indicating that a bag-of-word approach is not appropriate for this task, and semantic understanding is needed. Hybrid search fails to outperform dense search as it is limited by the poor performance of sparse search. In Table 2 we show selected examples of clinic descriptions where sparse embeddings fail to capture variations in text descriptions.

4.2 Fine-tuning with metric learning

In Table 4 we compare the performance of three dense embedding models: mE5-large (Wang et al., 2024), RoBERT-large (Masala et al., 2020), and BioClinicalBERT (Alsentzer et al., 2019). We obtained that mE5 has higher off-the-shelf retrieval accuracy compared to RoBERT and BioClinicalBERT. This advantage stems from mE5's design as a sentence-transformer model specifically trained to evaluate similarity between sentences or descriptions, whereas RoBERT and BioClinicalBERT is adapted for sentence embedding through a pooling operation over token embeddings.

Sparse search initially outperforms both RoBERT and BioClinicalBERT. However, after fine-tuning, all dense embeddings surpass sparse embeddings in performance metrics, with mE5 maintaining its position as the highest-performing model.

Table 5 illustrates the impact of incorporating both masterlist and associated mappings in search processes. The inclusion of reduces the performance difference between E5 and the other two models. While the relative ranking of models remains consistent, E5 achieves the highest perfor-

Vector Store Data	Index Type	Acc@1	Acc@3	Acc@5	Acc@100
Masterlist Entries Only	sparse (BM25)	52.6 ± 0.002	64.5 ± 0.002	68.5 ± 0.002	86.3 ± 0.001
	dense (mE5)	78.8 ± 0.002	92.2 ± 0.002	95.0 ± 0.002	99.5 ± 0.001
	hybrid (RRF)	63.9 ± 0.003	77.7 ± 0.003	82.1 ± 0.003	99.5 ± 0.001
Masterlist Entries + Mappings	sparse (BM25)	68.0 ± 0.003	82.3 ± 0.001	86.1 ± 0.001	94.7 ± 0.001
	dense (mE5)	**85.2 ± 0.003**	**95.8 ± 0.001**	**97.5 ± 0.001**	**99.5 ± 0.001**
	hybrid (RRF)	81.0 ± 0.002	92.3 ± 0.001	94.9 ± 0.001	99.5 ± 0.000

Table 3: Comparison between sparse embeddings from BM25, dense embeddings from mE5 (Wang et al., 2024), and hybrid search, having only masterlist entries in the vector store and having both masterlist and associated clinical mappings. Using dense embeddings from mE5 provides the best results in both cases. Results are averaged across 5 folds.

Model Name	Type	Acc@1	Acc@3	Acc@5	Acc@100
RoBERT (Masala et al., 2020)	off-the-shelf	44.7 ± 0.003	53.4 ± 0.003	56.9 ± 0.004	75.3 ± 0.003
BioClinicalBERT (Alsentzer et al., 2019)		47.7 ± 0.003	56.7 ± 0.003	60.2 ± 0.002	74.9 ± 0.003
mE5 (Wang et al., 2024)		56.8 ± 0.003	69.4 ± 0.002	74.3 ± 0.002	91.3 ± 0.002
RoBERT (Masala et al., 2020)	fine-tuned	75.9 ± 0.001	89.9 ± 0.002	93.2 ± 0.000	98.9 ± 0.001
BioClinicalBERT (Alsentzer et al., 2019)		75.7 ± 0.002	89.2 ± 0.002	92.7 ± 0.002	98.9 ± 0.000
mE5 (Wang et al., 2024)		**78.8 ± 0.002**	**92.2 ± 0.002**	**95.0 ± 0.002**	**99.5 ± 0.001**

Table 4: Comparison between different types of text embedding models, having entries in the vector store only from the masterlist entries. We obtained the best results using a fine-tuned version of mE5, a general-purpose multi-lingual model. Results are averaged across 5 folds.

Model Name	Type	Acc@1	Acc@3	Acc@5	Acc@100
RoBERT (Masala et al., 2020)	off-the-shelf	62.5 ± 0.005	76.8 ± 0.004	81.1 ± 0.005	92.0 ± 0.004
BioClinicalBERT (Alsentzer et al., 2019)		66.7 ± 0.005	80.8 ± 0.003	84.6 ± 0.002	93.4 ± 0.002
mE5 (Wang et al., 2024)		67.9 ± 0.004	85.2 ± 0.002	89.6 ± 0.002	98.1 ± 0.001
RoBERT (Masala et al., 2020)	finetuned	84.4 ± 0.002	94.8 ± 0.002	96.6 ± 0.001	99.0 ± 0.001
BioClinicalBERT (Alsentzer et al., 2019)		83.8 ± 0.003	94.3 ± 0.001	96.4 ± 0.001	99.0 ± 0.001
mE5 (Wang et al., 2024)		**85.2 ± 0.003**	**95.8 ± 0.001**	**97.5 ± 0.001**	**99.5 ± 0.001**

Table 5: Comparison between different types of text embedding models, having entries in the vector store from both the masterlist entries and associated clinical mappings. We obtained the best results using a fine-tuned version of mE5, a general-purpose multi-lingual model. Results are averaged across 5 folds.

mance with an Acc@1 of 85.2% and an Acc@5 of 95%. Notably, Acc@1 metric may under-represent actual performance. Manual inspection of misclassified results reveals many plausible matches. This discrepancy occurs due to the presence of duplicate entries within the masterlist itself—entries with slightly different formulations that reference identical medical procedures. The markedly higher Acc@3 metric, which captures whether the ground-truth result appears within the first three recommendations, supports this observation. Although duplicated masterlist results present a methodological challenge for evaluation, they do not compromise practical application. The real-world accuracy exceeds the reported metrics, as demonstrated in the next section.

4.3 Doctor evaluation

Our medical procedure mapping system was used to map new unmapped procedures. We evaluated on new procedure descriptions from 10 clinics, comprising 12,836 unique descriptions. After mapping the procedures using a fine-tuned mE5 models trained on all available data, doctors validated each pair to determine if the masterlist assignment was correct. As shown in Table 6, the model achieves a real-world Acc@1 of 94.7%. The 98.5% Acc@2 indicates that doctors considered either the first or second recommendation correct, while for only 1% of entries, doctors assigned a different description than the ones recommended.

Model Name	Acc@1	Acc@2	Acc@3
mE5 - All Data	94.7	98.5	99.0

Table 6: Real-world evaluation of our system. Doctors manually evaluated 12,836 new entries after mapping them with a fine-tuned version of mE5 on all data.

Another notable aspect is the speed of the mappings process. While manually mapping the 12,836 descriptions would take more than 60 hours, using our retrieval system reduces this to only 3 minutes, resulting in an 1200× speedup.

5 Conclusion

This paper presents a retrieval-based approach for medical procedure matching in the Romanian healthcare system, addressing the challenges posed by inconsistent naming conventions across clinics and insurance providers. We demonstrate that dense sentence embeddings, particularly fine-tuned multilingual models, significantly outperform traditional sparse methods such as BM25. Our experiments show that a fine-tuned mE5 model achieves the highest retrieval accuracy, with an Acc@1 of 85.2% when using both masterlist entries and clinical mappings. The real-world evaluation further confirms the efficacy of our approach, achieving a validated accuracy of 94.7% in a doctor-reviewed dataset. Furthermore, our systems enables significant labor efficiency: using our automated matching systems results in 1200× speedup compared to manual matching. Our findings contribute to the broader domain of medical NLP for low-resource languages and offer a viable solution for improving the Romanian healthcare system.

Limitations

Our approach has several limitations. Firstly, errors in historical mappings may propagate into future predictions, potentially reinforcing inaccuracies over time. This challenge necessitates periodic human review and correction to prevent systematic errors. Secondly, cosine similarity between embeddings may not always provide a reliable confidence estimate, due to the considerable overlap between the score distributions of hits and misses. This makes it difficult to differentiate between correct and incorrect matches. Incorporating additional uncertainty modeling or ranking refinements could improve result interpretability. Thirdly, while our retrieval model significantly improves over rule-based methods, its performance is still constrained by the lack of a specialized Romanian medical language model. A dedicated medical NLP model trained on domain-specific Romanian corpora could further enhance accuracy.

Acknowledgement

This research was supported by the project "Search Techniques for Matching Medical Information - MATCHMED", ID 420246344, and by the project "Romanian Hub for Artificial Intelligence - HRIA", Smart Growth, Digitization and Financial Instruments Program, MySMIS no. 334906.

References

Sherri Alexander, Therese Conner, and Teresa Slaughter. 2003. Overview of inpatient coding. *American journal of health-system pharmacy: AJHP: official journal of the American Society of Health-System Pharmacists*, 60(21 Suppl 6):S11–14.

Emily Alsentzer, John R. Murphy, Willie Boag, Wei-Hung Weng, Di Jin, Tristan Naumann, and Matthew B. A. McDermott. 2019. Publicly available clinical BERT embeddings. *CoRR*, abs/1904.03323.

Iz Beltagy, Kyle Lo, and Arman Cohan. 2019. SciBERT: A pretrained language model for scientific text. In *Proceedings of the 2019 Conference on Empirical Methods in Natural Language Processing and the 9th International Joint Conference on Natural Language Processing (EMNLP-IJCNLP)*, pages 3615–3620, Hong Kong, China. Association for Computational Linguistics.

Gordon V. Cormack, Charles L A Clarke, and Stefan Buettcher. 2009. Reciprocal rank fusion outperforms condorcet and individual rank learning methods. In *Proceedings of the 32nd International ACM SIGIR Conference on Research and Development in Information Retrieval*, pages 758–759, Boston MA USA. ACM.

Matthew Henderson, Rami Al-Rfou, Brian Strope, Yun-Hsuan Sung, László Lukács, Ruiqi Guo, Sanjiv Kumar, Balint Miklos, and Ray Kurzweil. 2017. Efficient natural language response suggestion for smart reply. *arXiv preprint arXiv:1705.00652*.

Sepp Hochreiter and Jürgen Schmidhuber. 1997. Long Short-Term Memory. *Neural Computation*, 9(8):1735–1780.

Omar Khattab and Matei Zaharia. 2020. ColBERT: Efficient and Effective Passage Search via Contextualized Late Interaction over BERT. In *Proceedings of the 43rd International ACM SIGIR Conference on Research and Development in Information Retrieval*, pages 39–48, Virtual Event China. ACM.

Quoc Le and Tomas Mikolov. 2014. Distributed representations of sentences and documents.

Joshua Levy, Nishitha Vattikonda, Christian Haudenschild, Brock Christensen, and Louis Vaickus. 2022. Comparison of Machine-Learning Algorithms for the Prediction of Current Procedural Terminology (CPT)

Codes from Pathology Reports. *Journal of Pathology Informatics*, 13:3.

Patrick Lewis, Myle Ott, Jingfei Du, and Veselin Stoyanov. 2020. Pretrained language models for biomedical and clinical tasks: understanding and extending the state-of-the-art. In *Proceedings of the 3rd clinical natural language processing workshop*, pages 146–157.

Mihai Masala, Denis C. Ilie-Ablachim, Alexandru Dima, Dragos Corlatescu, Miruna Zavelca, Ovio Olaru, Simina Terian, Andrei Terian, Marius Leordeanu, Horia Velicu, Marius Popescu, Mihai Dascalu, and Traian Rebedea. 2024. "vorbeşti româneşte?" a recipe to train powerful romanian llms with english instructions. *Preprint*, arXiv:2406.18266.

Mihai Masala, Stefan Ruseti, and Mihai Dascalu. 2020. RoBERT – a Romanian BERT model. In *Proceedings of the 28th International Conference on Computational Linguistics*, pages 6626–6637, Barcelona, Spain (Online). International Committee on Computational Linguistics.

Niklas Muennighoff, Nouamane Tazi, Loic Magne, and Nils Reimers. 2023a. MTEB: Massive text embedding benchmark. In *Proceedings of the 17th Conference of the European Chapter of the Association for Computational Linguistics*, pages 2014–2037, Dubrovnik, Croatia. Association for Computational Linguistics.

Niklas Muennighoff, Nouamane Tazi, Loïc Magne, and Nils Reimers. 2023b. MTEB: Massive Text Embedding Benchmark. *Preprint*, arXiv:2210.07316.

Hellina Hailu Nigatu, Atnafu Lambebo Tonja, Benjamin Rosman, Thamar Solorio, and Monojit Choudhury. 2024. The zeno's paradox of 'low-resource' languages. In *Proceedings of the 2024 Conference on Empirical Methods in Natural Language Processing*, pages 17753–17774, Miami, Florida, USA. Association for Computational Linguistics.

Abhinav Ramesh Kashyap, Thanh-Tung Nguyen, Viktor Schlegel, Stefan Winkler, See-Kiong Ng, and Soujanya Poria. 2024. A comprehensive survey of sentence representations: From the BERT epoch to the CHATGPT era and beyond. In *Proceedings of the 18th Conference of the European Chapter of the Association for Computational Linguistics (Volume 1: Long Papers)*, pages 1738–1751, St. Julian's, Malta. Association for Computational Linguistics.

Laila Rasmy, Yang Xiang, Ziqian Xie, Cui Tao, and Degui Zhi. 2021. Med-bert: pretrained contextualized embeddings on large-scale structured electronic health records for disease prediction. *NPJ digital medicine*, 4(1):86.

Nils Reimers and Iryna Gurevych. 2019. Sentence-BERT: Sentence Embeddings using Siamese BERT-Networks. In *Proceedings of the 2019 Conference on Empirical Methods in Natural Language Processing and the 9th International Joint Conference on Natural Language Processing (EMNLP-IJCNLP)*, pages 3980–3990, Hong Kong, China. Association for Computational Linguistics.

Stephen Robertson and Hugo Zaragoza. 2009. The probabilistic relevance framework: Bm25 and beyond. *Found. Trends Inf. Retr.*, 3(4):333–389.

Nazgol Tavabi, Mallika Singh, James Pruneski, and Ata M. Kiapour. 2024. Systematic evaluation of common natural language processing techniques to codify clinical notes. *PLOS ONE*, 19(3):e0298892.

Jianguo Wang, Xiaomeng Yi, Rentong Guo, Hai Jin, Peng Xu, Shengjun Li, Xiangyu Wang, Xiangzhou Guo, Chengming Li, Xiaohai Xu, Kun Yu, Yuxing Yuan, Yinghao Zou, Jiquan Long, Yudong Cai, Zhenxiang Li, Zhifeng Zhang, Yihua Mo, Jun Gu, and 3 others. 2021. Milvus: A Purpose-Built Vector Data Management System. In *Proceedings of the 2021 International Conference on Management of Data*, SIGMOD '21, pages 2614–2627, New York, NY, USA. Association for Computing Machinery.

Liang Wang, Nan Yang, Xiaolong Huang, Linjun Yang, Rangan Majumder, and Furu Wei. 2024. Multilingual E5 Text Embeddings: A Technical Report. *Preprint*, arXiv:2402.05672.

Wenhui Wang, Furu Wei, Li Dong, Hangbo Bao, Nan Yang, and Ming Zhou. 2020. Minilm: Deep self-attention distillation for task-agnostic compression of pre-trained transformers. *Advances in neural information processing systems*, 33:5776–5788.

Michael Wornow, Yizhe Xu, Rahul Thapa, Birju Patel, Ethan Steinberg, Scott Fleming, Michael A Pfeffer, Jason Fries, and Nigam H Shah. 2023. The shaky foundations of large language models and foundation models for electronic health records. *npj digital medicine*, 6(1):135.

Zhilin Yang, Zihang Dai, Yiming Yang, Jaime Carbonell, Russ R Salakhutdinov, and Quoc V Le. 2019. Xlnet: Generalized autoregressive pretraining for language understanding. *Advances in neural information processing systems*, 32.

Bashar Zaidat, Justin Tang, Varun Arvind, Eric A. Geng, Brian Cho, Akiro H. Duey, Calista Dominy, Kiehyun D. Riew, Samuel K. Cho, and Jun S. Kim. 2024. Can a Novel Natural Language Processing Model and Artificial Intelligence Automatically Generate Billing Codes From Spine Surgical Operative Notes? *Global Spine Journal*, 14(7):2022–2030.

Improving Barrett's Oesophagus Surveillance Scheduling with Large Language Models: A Structured Extraction Approach

Xinyue Zhang[1], Agathe Zecevic[2,3], Sebastian Zeki[2], Angus Roberts[1]

[1]Biostatistics and Health Informatics, Institute of Psychiatry, Psychology and Neuroscience King's College London, United Kingdom,
[2]Gastroenterology Department, Guy's and St Thomas' NHS Foundation Trust, United Kingdom,
[3]Clinical Scientific Computing, Guy's and St Thomas' NHS Foundation Trust, United Kingdom
Correspondence: leo.xinyue.zhang@kcl.ac.uk

Abstract

Gastroenterology (GI) cancer surveillance scheduling relies on extracting structured data from unstructured clinical texts, such as endoscopy and pathology reports. Traditional Natural Language Processing (NLP) models have been employed for this task, but recent advances in Large Language Models (LLMs) present a new opportunity for automation without requiring extensive labelled datasets. In this study, we propose an LLM-based entity extraction and rule-based decision support framework for the prediction of Barrett's oesophagus (BO) surveillance timing. Our approach processes endoscopy and pathology reports to extract clinically relevant information and structures it into a standardised format, which is then used to determine appropriate surveillance intervals. We evaluate multiple state-of-the-art LLMs on real-world clinical datasets from two hospitals, assessing their performance in accuracy and runtime. The results demonstrate that LLMs, particularly Phi-4 and (DeepSeek distilled) Qwen-2.5, can effectively automate the extraction of BO surveillance-related information with high accuracy, while Phi-4 is also efficient during inference. We also compared the trade-offs between LLMs and fine-tuned BERT models. Our findings indicate that LLM-based extraction methods can support clinical decision-making by providing justifications from report extractions, reducing manual workload, and improving guideline adherence in BO surveillance scheduling.

1 Introduction

Gastroenterology (GI) cancer surveillance scheduling relies heavily on extracting structured information from unstructured clinical texts, such as pathology and endoscopy findings. Traditional Natural Language Processing (NLP) tools trained on annotated datasets have been used to support clinical decision-making. However, recent advances in Large Language Models (LLMs) have the potential to update this process. LLMs, with their extensive training on diverse text sources, can now process medical texts without requiring large amounts of task-specific annotated data. This offers a more flexible and scalable approach to cancer surveillance scheduling automation.

Barrett's Oesophagus (BO) is a pre-cancerous condition in which the normal squamous epithelium of the oesophagus is replaced by columnar lined mucosa. Patients with BO can progress to oesophageal adenocarcinoma (OAC). Thus, patients with BO undergo routine endoscopic surveillance to monitor the condition and detect dysplasia or early OAC. Appropriate surveillance intervals and early intervention can improve patient outcomes.

Adherence to surveillance guidelines remains suboptimal. A meta-analysis (Roumans et al., 2020) found only 55% of non-dysplastic BO patients and 50% of low-grade dysplasia patients received surveillance at recommended intervals. This highlights the need for improved clinical decision support to ensure timely surveillance and treatment. Recent advances in artificial intelligence (AI), especially large language models (LLMs), have opened new opportunities to aid BO management. LLMs, a group of transformer-based generative models with billions of parameters such as OpenAI's GPT-4, Meta's Llama and Microsoft's Phi, excel at processing unstructured text and extracting complex information from it. In gastroenterology, these models can process clinical notes such as pathology and endoscopy reports, and then support medical decision making based on the information from these reports (Omar et al., 2025).

BO surveillance scheduling depends on BO length from endoscopy reports and pathological findings from pathology reports. We will discuss this further in Section 2.1. Previous work (Zecevic et al., 2024) introduced a system capable of categorising endoscopy reports into four groups (Short, Long, No Barrett's, and Insufficient) and

Proceedings of the 24th Workshop on Biomedical Language Processing (BioNLP 2025), pages 176–189
August 1, 2025 ©2025 Association for Computational Linguistics

pathology reports into another set of four categories (Cancer/Dysplasia, Intestinal Metaplasia (IM), no IM, and Insufficient). The classification occurs at the report level, where each report receives a single label. However, the report level model does not provide information from reports to justify its classifications, making it hard for clinicians to validate the output without manually reviewing the text. Moreover, the report level model is specific to the task and cannot be repurposed for other clinical uses.

Our work proposes an information extraction based method that uses LLMs to automate Barrett's surveillance timing prediction. The workflow is shown in Figure 1. Both endoscopy and pathology reports, after preprocessing, are passed through an LLM, which extracts clinically relevant information into a JSON template. A rule-based algorithm converts these extractions into report labels and provides relevant extractions as justification. Our hypothesis is that an LLM-based method can accurately extract entities without the need for a large amount of annotated data, and these extractions can be used to justify the surveillance interval decisions. These extractions can also be repurposed for other downstream clinical tasks. To our knowledge, this is the first study to use LLMs to determine when a BO patient's next endoscopy is due based on prior reports.

The contributions of this work include:

- An LLM-based extraction with a rule-based post-processing method for Barrett's surveillance timing prediction with justifications.
- We designed and evaluated prompt strategies for LLM medical extraction on endoscopy and pathology reports
- We evaluated performance of different LLMs with a variety of types, sizes and reasoning ability.
- We created a gold-standard for BO surveillance timing based on previously annotated reports classification data (Zecevic et al., 2024)

2 Related Work

2.1 Surveillance Timing Guidelines in Barrett's Oesophagus

Given the importance of Barrett's oesophagus (BO) surveillance, organisations including the British Society of Gastroenterology[1] (BSG), the American Gastroenterological Association[2] (AGA) and the European Society of Gastrointestinal Endoscopy[3] (ESGE) have published guidelines on the recommended intervals for endoscopic surveillance. These guidelines (Fitzgerald et al., 2014; Spechler et al., 2011; Weusten et al., 2023) seek to balance the advantages of early detection against the costs of repeated endoscopic procedures. Our research specifically follows the BSG guidelines (Fitzgerald et al., 2014). The current BSG guidelines, first published in the early 2000s and updated periodically, emphasize the need for risk-based surveillance intervals and provide actionable recommendations for endoscopic management. The guidelines show that for non-dysplastic Barrett's, the endoscopic surveillance interval is determined by the length of Barrett's and the presence of Intestinal Metaplasia (IM). Based on this guideline, we set out a rule-based algorithm for surveillance interval decision making which is shown in the bottom part of Figure 1.

2.2 NLP Methods in BO Surveillance

Previous work on NLP in BO surveillance is limited. Zecevic et al. (2024) curated report classification annotations for endoscopy and pathology reports. These annotations are used to train two report classification models. These models, EndoBERT and PathoBERT, are based on pre-trained Bidirectional encoder representations from transformers (BERT) model (Devlin et al., 2019), which assigns a label to an unseen endoscopy or pathology report. The model achieved high accuracy on test sets from three UK hospitals.

Other work related to BO includes dysplasia identification in Wenker et al. (2023). They use an NLP tool (Clinical Language Annotation, Modelling, and Processing Toolkit) to identify dysplasia using findings. However, they did not provide detailed information on the underlying models used by the tool. Li et al. (2022) introduce ENDOANGEL-AS, an automated surveillance system designed to identify high-risk patients and determine appropriate surveillance intervals for up GI conditions.

[1] https://www.bsg.org.uk/

[2] https://gastro.org/

[3] https://www.esge.com/publications/guidelines

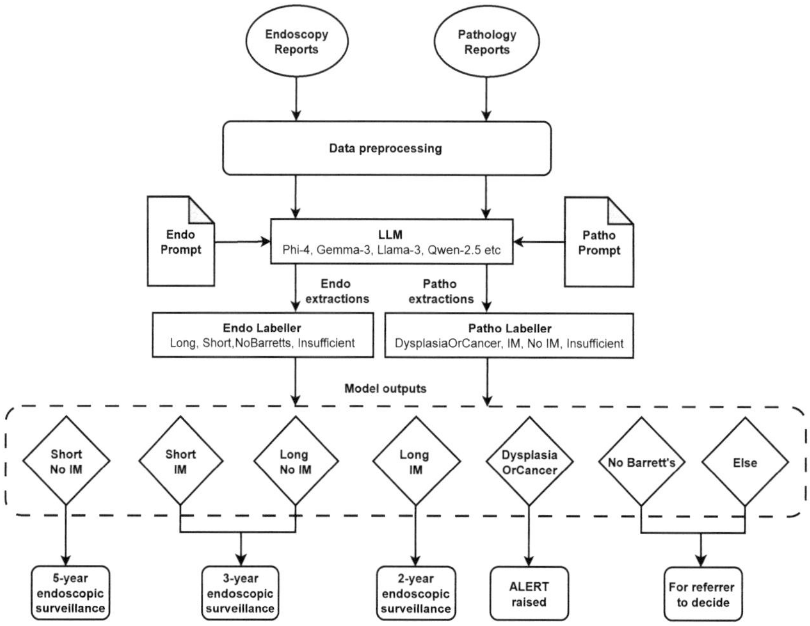

Figure 1: Dataflow pipeline of LLM-based endoscopy and pathology extraction for Barrett's endoscopic surveillance timing prediction

2.3 LLMs

Recent advances in large language models (LLMs) have led to a diverse range of architectures optimised for efficiency, domain adaptation, and reasoning capabilities. These models vary in size, training methodology, and specialisation. The increasing numbers of smaller yet high-performing LLMs has made their application to the medical domain more feasible. Among them, Microsoft's Phi-4 (14B) (Abdin et al., 2024), whose training recipe centres around data quality, prioritises efficiency while maintaining strong performance across general NLP tasks. It is optimized for low-cost inference, making it an attractive option for real-world deployment where resource constraints are a factor. Similarly, Gemma-3 (12B) (DeepMind, 2024), developed by Google DeepMind, is another small and highly efficient model. Although it has a slightly lower parameter count (12B), it has multimodal and enhanced reasoning ability. This makes it competitive with larger models in certain tasks. Alibaba's Qwen-2.5 (14B) (Yang et al., 2024) has enhanced reasoning and instruction following ability. On the other end of the spectrum, Meta's Llama-3 (7B) (Grattafiori et al., 2024) is an even smaller LLM which may be of benefit in more compute restricted settings. However, as evident from recent performance benchmarks, Llama-3 struggles compared to other LLMs. DeepSeek distilled variants (Guo et al., 2025) of Llama and Qwen are derived from the larger DeepSeek-R1 base model. These versions are fine-tuned to enhance performance on maths, coding, and other reasoning-intensive tasks. Distillation reduces the model size and inference time while retaining key capabilities, making them suitable for real-time medical NLP pipelines.

3 Experiments

3.1 Data and evaluation

The datasets used in this project are the same datasets used in (Zecevic et al., 2024), including data selection, patient opt-out, preprocessing, labelling and data split. The data is from two UK National Health Service Foundation Trust (NHSFTs - secondary healthcare providers), Guy's and St Thomas' NHSFT (GSTT), and King's College Hospital NHSFT (KCH). Training is carried out on GSTT data. The evaluations are carried out on both GSTT data, and on KCH data to test generalisability. We give a brief introduction here; for detailed information, please refer to Zecevic et al. (2024).

Training set The patients are selected based on the appearance of the keyword "Barrett" in their endoscopy reports. Patients under 18 and those who have opted out were excluded. Pathology reports were then matched with the relevant endoscopy reports. As we do not fine-tune LLMs, this training set is used to 1) develop prompts; 2) develop a rule-based algorithm, based on incorrect predictions. Once the rule-based algorithm has been developed in this stage, it is fixed during testing. Prompts are fixed during testing apart from when we test the impact of certain components of the prompt.

GSTT evaluation A retrospective evaluation was conducted by reviewing patient records of individuals who had undergone endoscopic surveillance for BO between May and July 2023, which is a representative time frame for assessing records typically used to schedule follow-up surveillance endoscopies. A total of 115 patient records were included, where pathology results from the endoscopy were available. We take the human reviewed labels for the documents as ground-truth to evaluate our model prediction. Unlike (Zecevic et al., 2024), where the performance is only measured on two sets of report labels, we also combined the endoscopy and pathology report labels into surveillance timing labels based on guideline rules (Section 2). This can give us a single number indicator of the model performance for surveillance timing prediction, which represents the ultimate goal of the task.

KCH evaluation A dataset of 140 reports was collected from KCH, covering cases from 2015 to 2022 for the second external evaluation. The same selection criteria used for the GSTT dataset were applied to the KCH dataset to ensure consistency in evaluation. Similarly, we treated the human reviewed reports labels as ground-truth and combined them into surveillance timing labels to measure the model performance on predicting surveillance timing intervals.

3.2 Task

Information extraction Our approach focuses on information extraction from endoscopy and pathology reports. The key entities to be extracted are listed in Appendix A Table 6.

For extraction, we use large language models (LLMs), which take as input either an endoscopy prompt along with an endoscopy report, or a pathology prompt along with a pathology report. This process ensures structured extraction of relevant clinical information from unstructured text data.

From extraction to report classification and surveillance timing prediction The classification of endoscopy and pathology reports in this study aligns with the definitions outlined by (Zecevic et al., 2024) in Table 2 and Table 3. Endoscopy reports are categorised based on segment length of BO (Long, Short, NoBarretts, Insufficient) and pathology reports based on histological findings (CancerOrDysplasia, IM, No_IM, Insufficient). Surveillance timing is classified based on a pair of endoscopy and pathology reports, and is classified into Alert, 2 year, 3 year, 5 year or Refer, as outlined in Section 2

Evaluation We evaluate model performance using a held-out test set from GSTT and KCH. Performance is assessed across three key tasks: Classification of endoscopy reports; Classification of pathology reports, and Surveillance timing prediction. For each task, we measure precision, recall, and F1 score, ensuring a comprehensive evaluation of the model's ability to classify reports and predict appropriate surveillance intervals. To estimate the variability in performance, we report each metric along with its 95% confidence interval (CI), computed via bootstrap resampling of the test set. Bootstrap resampling is a statistical technique that creates multiple new datasets from a single dataset by randomly sampling with replacement (Tibshirani and Efron, 1993). The bootstrapping is done in 1000 iterations with replacement and each sample has the same sample size as the test set size. This resampling approach is chosen because the model weights and prompts remain fixed at test time, and the model decoding strategy is set to greedy search (Section 3.3: Hyperparameter Setting), hence the outputs are deterministic. Resampling allows for statistical uncertainty estimation based on test set variability and sample size. This is not interpreted as model uncertainty.

3.3 LLMs for extraction

Model selection For our study, we use state-of-the-art large language models (LLMs) including: Phi-4 (4-bit Instruct, 14B), Gemma-3 (4-bit, Instruct 12B), Qwen-2.5 (4-bit Instruct, 14B), DeepSeek Distilled Qwen-2.5 (4-bit Instruct, 14B), Llama-3 (4-bit Instruct, 7B), DeepSeek Distilled Llama-3 (4-bit Instruct, 7B)

These models were chosen for their balance of performance, efficiency, and scalability. The 8-billion parameter scale gives strong language understanding ability while maintaining computational feasibility. The 4-bit quantisation significantly reduces memory requirements, enabling faster inference and lower hardware constraints without substantial loss in accuracy. The "Instruct" versions (as opposed to "Base" versions) of these models provide general language inference ability, ensuring better generalisation to structured information extraction and classification. The combination of these models allows us to benchmark performance across architectures, ensuring our pipeline remains adaptable to future advances in LLM-driven clinical NLP.

Hyperparameter setting for decoding We set the sampling method as greedy search (equivalently temperature set to zero or near-zero) to ensure deterministic and consistent outputs, as used in other entity extraction research (Wang et al., 2023; Dunn et al., 2022; Zhang et al., 2023). Greedy search and low temperature sampling have been shown to be effective for factual extraction tasks, particularly in medical Named Entity Recognition (NER) and Relation Extraction (RE), where minimising randomness improves reliability and precision (Wang et al., 2023; Dunn et al., 2022; Zhang et al., 2023). Greedy search is also the fastest decoding strategy, which is essential in some hospital settings where the computing resources are limited. We set the maximum input length to 4096 tokens and maximum output length to 2048 tokens to accommodate complex prompts and generated responses while optimising computational efficiency.

Prompt The prompt design follows best practices established in previous research (Liu et al., 2023; Wang et al., 2024; Zamfirescu-Pereira et al., 2023; He et al., 2024; Sachdev et al., 2024), particularly in the context of optimising large language models (LLMs) for structured medical information extraction. Specifically, for both the endoscopy and pathology information extraction prompts, the structure consists of five key components and one training strategy:

Persona Assignment The LLM is explicitly instructed to assume the role of a specialised AI gastroenterology assistant with expertise in medical report analysis.

Clear Instruction breakdown The second component has a clear listing of all the requirements.

Structured Output Specification To ensure consistency and machine-readability, the third component mandates a standardized JSON output format, explicitly defining entity types and relationships to align with downstream processing requirements.

Step-by-Step Reasoning (CoT) The fourth section provides a sequential, logical step-by-step breakdown of the extraction process and self-verification, guiding the LLM through a structured CoT approach to enhance interpretability and accuracy. We will explore model performance with and without this component.

Few-Shot In-Context Examples The fifth section provides two examples of correctly extracted outputs, demonstrating the expected format and extractions.

Input Report Attachment Finally, the actual clinical report (endoscopy or pathology report) requiring extraction is appended.

Iterative Prompt Refinement The prompt is applied on samples from the training set and the prompt is modified based on incorrectly generated samples.

The final endoscopy and pathology prompts can be found in Appendix B, with the real reports in few-shot examples and in input texts section being removed for privacy reasons.

Hardware and cost The model is run on Nvidia A100 GPUs in Ubuntu operating system on a virtual machine provided by King's College London Computational Research Engineering and Technology Environment - Trusted Research Environment (CREATE-TRE). The runtime is analysed in Section 4 Table 3.

3.4 Rule-based algorithms for Report classification and surveillance timing prediction

Pathology report labelling is based on extracted pathology findings, either its affirmation or negation. For endoscopy reports, the classification is based on Barrett's length. There can be three sources of Barrett's length in endoscopy reports: Prague score; direct mention of Barrett's segments; mentions of GOJ (Gastro-oesophageal junction) and Barrett's tongue. The algorithm gathers these sources of length from extractions. In rare cases, when lengths from different sources disagree, if two of them agree and one disagrees, we pick the majority case; in other cases, we choose Prague scores over position difference from GOJ and Barrett's tongue over direct mentions. This is due to the rigid form of Prague score, which makes the extraction and post-processing more reliable. We then apply the rules to assign labels to endoscopy reports. Once we have both labels, we combine the two following the rules in Figure 1 to give a surveillance timing prediction.

4 Results

We evaluated six LLMs, both with and without chain-of-thought (CoT) prompting, on three clinical information extraction tasks (Decisions, Pathology, and Endoscopy) across two hospital sites (GSTT, KCH). The following subsections discusses overall performance, output validity, runtime analysis, and fine-grained subclass results.

4.1 Overall Performance

Table 1 shows the weighted average F1-scores across all three clinical categories. In general, Qwen-2.5, DeepSeek distilled Qwen-2.5 and Phi-4 (14B) achieve the highest F1-scores.

Phi-4 shows good performance on Pathology reports while DeepSeek distilled Qwen-2.5 performs better on Endoscopy reports. Qwen-2.5 with CoT often performs on par with Phi-4 for Pathology while on par with DeepSeek Qwen-2.5 on Endoscopy, which means it achieves the highest Decision F1 scores on both datasets.

Meanwhile, Gemma-3 (12B) tends to occupy the midrange, ranging around 0.75–0.90 depending on the category. Llama-3 (7B) has the lowest overall scores; however, CoT prompting and DeepSeek distillation consistently raise its performance.

Interestingly, there are some cases where *not* using CoT yields a slightly higher score. For example, on "Decisions (GSTT)" Phi-4 w/o CoT outperforms its CoT-based variant (0.96 vs. 0.93). Such exceptions may arise when CoT text introduces minor digressions from the desired prompt structure or consumes additional tokens that do not improve the final label prediction. In addition, the instruction component may already contain certain implicit CoT steps e.g. in the instruction "Barretts and if it is negated" implies a two step process that can be seen as an implicit CoT, i.e. step 1, identify mention of "Barrett's"; step 2, if the mention is negated. Moreover, these extraction tasks are not reasoning-intensive. The help from explicit CoT may be outweighed by the distraction from generating unnecessary reasoning.

4.2 Invalid Outputs

Table 2 illustrates how frequently each model produces "invalid" responses, i.e., outputs that deviate from the required specification or formatting set out in the prompt. This includes incorrect JSON format, missing fields, wrong keywords and wrong value type. Qwen-2.5 stands out as the most consistent over the two different prompt variances, largely maintaining a 0% invalid rate across categories, whether or not CoT is used. Whereas Phi-4 has a perfect 0% rate across all tasks when CoT is not used. Phi-4 and Gemma-3, which both performed well on F1 metrics, actually produce more invalid outputs when CoT is activated (e.g., Phi-4 on "Decisions (GSTT)" jumps from 0% without CoT to 6.96% with CoT). Again, as discussed before, this may be because the benefit of explicit CoT does not outweigh the distraction from extra reasoning. DeepSeek Qwen-2.5 also shows very low invalid percentages (typically under 3–5%) but is more prone to errors than standard Qwen-2.5. In contrast, Llama-3 exhibits the highest invalid output rates of all. However, applying CoT or DeepSeek tuning brings these rates down significantly, sometimes by 30–60 percentage points. Therefore, while CoT may introduce extra texts that can diverge some models, it can also help a struggling model (like Llama-3) adhere more closely to task requirements.

4.3 Runtime Analysis

Table 3 reports average runtime (in seconds) for processing a set of Endoscopy and Pathology reports. Most models show a predictable increase in runtime under CoT prompting, due to generating additional tokens for explanatory text. Phi-4 and Qwen-2.5 each experience a jump of about 5–12 seconds with CoT. Gemma-3, interestingly, gives similar or even slightly *lower* times when using CoT, which may be explained by the fact that Gemma-3 has already undergone reinforcement learning from multiple feedback sources and distillation from larger models, and thus adding CoT explicitly in the prompt does not add to reasoning generation.

Out of all 14B-parameter models, Qwen-2.5 is the fastest (around 24–31 s/iter), while DeepSeek Qwen-2.5 nearly doubles that time (66–75 s/iter). Llama-3 is particularly quick without CoT, dipping to 15–17 s/iter; yet with CoT, its times roughly double. These differences underscore a tradeoff: CoT can improve accuracy in some instances, but at the cost of speed. It is also interesting that for Gemma-3 and Qwen-2.5 which have reinforcement training in their training process and DeepSeek Qwen-2.5, DeepSeek Llama-3 and Gemma-3 which are distilled from larger model, adding CoT does not add much more runtime. This may be because these models have already generated some reasoning texts even without explicit CoT in the prompts.

Category	Phi-4 (14B)		Gemma-3 (12B)		Qwen-2.5 (14B)		DeepSeek Qwen-2.5 (14B)		Llama-3 (7B)		DeepSeek Llama-3 (7B)	
	With CoT	W/o CoT	With CoT	W/o CoT	With CoT	W/o CoT	With CoT	W/o CoT	With CoT	W/o CoT	With CoT	W/o CoT
Decisions (GSTT)	0.93 (0.89, 0.97)	**0.96** (0.92, 0.99)	**0.86** (0.80, 0.91)	0.85 (0.79, 0.91)	0.98 (0.94, 1.00)	0.98 (0.95, 1.00)	0.92 (0.88, 0.96)	**0.94** (0.91, 0.98)	**0.53** (0.42, 0.62)	0.46 (0.37, 0.54)	**0.75** (0.68, 0.82)	0.70 (0.64, 0.77)
Decisions (KCH)	0.79 (0.74, 0.85)	**0.84** (0.79, 0.90)	0.80 (0.74, 0.86)	**0.81** (0.76, 0.87)	0.85 (0.79, 0.90)	0.83 (0.77, 0.89)	0.83 (0.77, 0.89)	0.85 (0.79, 0.90)	**0.62** (0.55, 0.70)	0.23 (0.15, 0.32)	0.54 (0.47, 0.60)	**0.61** (0.54, 0.68)
Pathology (GSTT)	0.91 (0.86, 0.96)	**0.97** (0.93, 0.99)	0.87 (0.80, 0.92)	0.87 (0.81, 0.92)	0.96 (0.93, 0.99)	0.96 (0.91, 0.99)	**0.94** (0.89, 0.98)	0.92 (0.87, 0.95)	0.64 (0.55, 0.72)	**0.91** (0.85, 0.95)	0.85 (0.79, 0.92)	**0.88** (0.82, 0.93)
Pathology (KCH)	0.86 (0.81, 0.91)	0.92 (0.87, 0.95)	0.86 (0.81, 0.91)	**0.87** (0.82, 0.92)	**0.89** (0.83, 0.93)	0.88 (0.83, 0.93)	0.88 (0.83, 0.93)	**0.91** (0.85, 0.95)	0.75 (0.69, 0.83)	**0.86** (0.80, 0.91)	**0.75** (0.69, 0.82)	0.73 (0.67, 0.79)
Endoscopy (GSTT)	**0.93** (0.88, 0.97)	0.92 (0.87, 0.97)	0.69 (0.60, 0.76)	**0.75** (0.67, 0.82)	0.94 (0.89, 0.97)	0.94 (0.90, 0.97)	0.94 (0.89, 0.97)	0.95 (0.91, 0.99)	**0.73** (0.65, 0.80)	0.27 (0.19, 0.34)	**0.66** (0.59, 0.74)	0.65 (0.57, 0.72)
Endoscopy (KCH)	0.82 (0.76, 0.87)	0.82 (0.76, 0.87)	0.74 (0.67, 0.81)	**0.75** (0.68, 0.81)	0.84 (0.78, 0.89)	0.84 (0.78, 0.89)	0.86 (0.80, 0.91)	0.87 (0.82, 0.92)	**0.65** (0.57, 0.72)	0.21 (0.13, 0.29)	0.58 (0.50, 0.66)	**0.63** (0.55, 0.71)

Table 1: Weighted average F1-Scores for different categories across multiple models with and without CoT. Values in bold indicate the higher value between 'With CoT' and 'Without CoT'. Values in red indicate the highest value in that row.

Category	Phi-4 (14B)		Gemma-3 (12B)		Qwen-2.5 (14B)		DeepSeek Qwen-2.5 (14B)		Llama-3 (7B)		DeepSeek Llama-3 (7B)	
	With CoT	W/o CoT	With CoT	W/o CoT	With CoT	W/o CoT	With CoT	W/o CoT	With CoT	W/o CoT	With CoT	W/o CoT
Decisions (GSTT)	6.96%	0%	4.35%	0.87%	0%	0%	5.22%	2.61%	27.83%	55.65%	13.91%	9.57%
Decisions (KCH)	0.86%	0%	0.71%	0%	0.71%	0.71%	2.14%	0.71%	24.29%	78.57%	27.14%	15.00%
Pathology (GSTT)	6.96%	0%	0%	0%	0%	0%	3.48%	1.74%	17.39%	2.61%	1.74%	2.61%
Pathology (KCH)	2.86%	0%	0%	0%	0%	0%	1.43%	0.71%	8.57%	0.71%	5.00%	4.29%
Endoscopy (GSTT)	0%	0%	5.22%	1.74%	0.87%	0.87%	1.74%	0.87%	20.87%	71.30%	16.52%	6.96%
Endoscopy (KCH)	0%	0%	0.71%	0%	0.71%	0.71%	0.71%	0%	21.43%	85.00%	24.29%	14.29%

Table 2: Percentage of "invalid" outputs generated (outputs that do not fully conform to the output specification)

4.4 Subclass-Specific Results

Table 4 breaks down F1-scores for finer-grained clinical subcategories. Once again, Qwen-2.5 and Phi-4 lead most subtasks. Both models frequently achieve near-perfect F1 on simpler labels (e.g., "alert," "DysplasiaOrCancer") and retain relatively strong performance on more difficult or less frequent subcategories (e.g., "Insufficient" in Endoscopy). DeepSeek Qwen-2.5 is the best across nearly all subcategories for Endoscopy. Given Endoscopy contains more numerical information, this reflects the advantage of specialised pre-training of reasoning ability.

Gemma-3's midrange performance remains consistent at subclass level, while Llama-3 is especially vulnerable on smaller or more challenging labels (e.g., "5 year," "Insufficient"), with F1 sometimes dropping below 0.50. However, DeepSeek Llama-3 recovers some ground. This implies that distilled reasoning ability from DeepSeekR helps with these challenging classes.

4.4.1 Decision Support with Evidence from Text

For real-life model application, we choose Phi-4 14B without CoT for Pathology reports processing and DeepSeek distilled Qwen-2.5 14B for Endoscopy processing given the performance during testing. We show a set of made-up endoscopy and pathology reports:

> endo_sample oesophagus: 8cm Barrett's segment. c3m8 Barrett's oesophagus. Hiatus hernia 2cm, top of GOJ 38cm, top of circumferential 35cm, top of tongues 30cm.
>
> patho_sample a) duodenum - normal - negative for cancer and dysplasia b) GOJ - intestinal metaplasia - negative for cancer and dysplasia c) oesophagus - intestinal metaplasia - inflammation - negative for cancer and dysplasia.

The decision support module outputs a decision and a justification for the decision with information from the texts.

```
--------------------Extraction finished. Making decisions--------------------

100%|████████████████████████████████████| 1/1 [00:00<00:00, 2706.00it/s]
Decision made:
2 year endoscopic surveillance

****************************************************************************

Justification:
 • Endoscopy shows Long Barretts, Details:
   Three sources of Barretts lengths have been found:
    1) Calculated length (difference between gastric folds and Barretts tongue):  (38, 30)
    2) Direct mention of Barretts length:  8cm
    3) Prague:  C3M8
   All agreed
 • Pathology shows Intestinal Metaplasia, Details:
   {'text': 'intestinal metaplasia', 'negation': 'no'}
```

The pathology extraction is in a nested JSON format for each biopsy finding. The model can identify the location of the biopsy and the mentions of cancer, dysplasia, IM and gastric metaplasia at that location.


```
{'doc_id': 'clinical_note_1',
 'extr': [{'Location': [{'text': 'duodenum',
     'oesophagus_or_barretts': 'no',
     'cardia': 'no'}],
   'Barretts': [],
   'Cancer': [{'text': 'cancer', 'negation': 'yes'}],
   'Dysplasia': [{'text': 'dysplasia', 'negation': 'yes'}],
   'IM': [],
   'Gastric_metaplasia': []},
  {'Location': [{'text': 'goj',
     'oesophagus_or_barretts': 'yes',
     'cardia': 'yes'}],
   'Barretts': [],
   'Cancer': [{'text': 'cancer', 'negation': 'yes'}],
   'Dysplasia': [{'text': 'dysplasia', 'negation': 'yes'}],
   'IM': [{'text': 'intestinal metaplasia', 'negation': 'no'}],
   'Gastric_metaplasia': []},
  {'Location': [{'text': 'oesophagus',
     'oesophagus_or_barretts': 'yes',
     'cardia': 'no'}],
   'Barretts': [],
   'Cancer': [{'text': 'cancer', 'negation': 'yes'}],
   'Dysplasia': [{'text': 'dysplasia', 'negation': 'yes'}],
   'IM': [{'text': 'intestinal metaplasia', 'negation': 'no'}],
   'Gastric_metaplasia': []}]}
```


The endoscopy extraction is structured in a JSON format with length information.

Category	Phi-4 (14B)		Gemma-3 (12B)		Qwen-2.5 (14B)		DeepSeek Qwen-2.5 (14B)		Llama-3 (7B)		DeepSeek Llama-3 (7B)	
	With CoT	W/o CoT	With CoT	W/o CoT	With CoT	W/o CoT	With CoT	W/o CoT	With CoT	W/o CoT	With CoT	W/o CoT
Time/iter (GSTT)	41.33	28.82	48.84	50.22	30.81	23.91	75.47	70.47	33.39	17.16	25.63	22.59
Time/iter (KCH)	40.64	27.64	48.00	49.54	28.26	24.03	70.47	66.23	24.50	15.32	32.90	28.65

Table 3: Average runtime per set of endoscopy and pathology report processing. Measured in seconds (averaged over the whole test set)

Class	Support	Phi-4	Gemma-3	Qwen-2.5*	DeepSeek Qwen-2.5	Llama-3*	DeepSeek Llama-3
				Decisions (GSTT)			
alert	20	**1.00** (1.00, 1.00)	0.93 (0.85, 1.00)	**1.00** (1.00, 1.00)	**1.00** (1.00, 1.00)	0.55 (0.44, 0.65)	0.93 (0.85, 1.00)
2 year	18	**0.97** (0.91, 1.00)	0.85 (0.74, 0.95)	**0.97** (0.92, 1.00)	0.91 (0.80, 1.00)	0.55 (0.29, 0.76)	0.68 (0.48, 0.85)
3 year	9	**0.94** (0.80, 1.00)	0.64 (0.42, 0.84)	0.90 (0.78, 1.00)	0.83 (0.69, 0.95)	0.45 (0.00, 0.75)	0.29 (0.00, 0.62)
5 year	6	0.79 (0.50, 1.00)	0.54 (0.18, 0.86)	**1.00** (1.00, 1.00)	0.90 (0.67, 1.00)	0.54 (0.18, 0.91)	0.00 (0.00, 0.00)
refer	62	0.97 (0.94, 0.99)	0.88 (0.82, 0.94)	**0.98** (0.94, 1.00)	0.96 (0.92, 0.99)	0.52 (0.39, 0.65)	0.76 (0.69, 0.83)
Weighted avg	115	0.96 (0.92, 0.99)	0.85 (0.79, 0.91)	**0.98** (0.94, 1.00)	0.94 (0.91, 0.98)	0.53 (0.42, 0.62)	0.70 (0.64, 0.77)
				Pathology (GSTT)			
DysplasiaOrCancer	20	**1.00** (1.00, 1.00)	0.93 (0.85, 1.00)	**1.00** (1.00, 1.00)	**1.00** (1.00, 1.00)	0.55 (0.45, 0.66)	0.93 (0.85, 1.00)
IM	36	**0.99** (0.96, 1.00)	0.91 (0.86, 0.96)	**0.99** (0.96, 1.00)	0.92 (0.86, 0.97)	0.71 (0.56, 0.84)	0.90 (0.81, 0.97)
No_IM	18	**0.91** (0.80, 1.00)	0.64 (0.41, 0.84)	**0.91** (0.80, 1.00)	0.86 (0.71, 0.97)	0.71 (0.52, 0.87)	0.80 (0.62, 0.94)
Insufficient	41	**0.95** (0.91, 0.99)	0.91 (0.84, 0.96)	**0.95** (0.90, 0.99)	0.91 (0.84, 0.98)	0.59 (0.42, 0.72)	0.87 (0.80, 0.94)
Weighted avg	115	**0.97** (0.93, 0.99)	0.87 (0.81, 0.92)	0.96 (0.93, 0.99)	0.92 (0.87, 0.96)	0.64 (0.55, 0.72)	0.88 (0.82, 0.93)
				Endoscopy (GSTT)			
Long	29	**0.98** (0.95, 1.00)	0.83 (0.76, 0.91)	0.97 (0.92, 1.00)	0.95 (0.88, 1.00)	0.83 (0.71, 0.93)	0.63 (0.47, 0.76)
Short	23	0.93 (0.82, 1.00)	0.83 (0.74, 0.92)	**1.00** (1.00, 1.00)	**1.00** (1.00, 1.00)	0.74 (0.59, 0.87)	0.55 (0.34, 0.72)
NoBarretts	49	0.93 (0.87, 0.98)	0.75 (0.64, 0.84)	**0.95** (0.91, 0.98)	**0.95** (0.91, 0.99)	0.73 (0.63, 0.84)	0.78 (0.69, 0.86)
Insufficient	14	0.77 (0.62, 0.92)	0.47 (0.21, 0.69)	0.71 (0.44, 0.88)	**0.88** (0.75, 1.00)	0.52 (0.27, 0.75)	0.42 (0.27, 0.56)
Weighted avg	115	0.92 (0.87, 0.97)	0.75 (0.67, 0.82)	0.94 (0.89, 0.97)	**0.95** (0.91, 0.99)	0.73 (0.65, 0.8)	0.65 (0.57, 0.72)
				Decisions (KCH)			
alert	7	**1.00** (1.00, 1.00)	0.94 (0.82, 1.00)	**1.00** (1.00, 1.00)	0.91 (0.73, 1.00)	0.49 (0.32, 0.67)	0.57 (0.44, 0.74)
2 year	26	**0.94** (0.87, 1.00)	0.87 (0.78, 0.95)	0.93 (0.85, 0.98)	0.92 (0.84, 0.98)	0.68 (0.49, 0.82)	0.69 (0.54, 0.83)
3 year	9	0.67 (0.33, 0.89)	0.70 (0.47, 0.90)	**0.74** (0.50, 0.94)	0.57 (0.32, 0.78)	0.36 (0.11, 0.63)	0.27 (0.00, 0.59)
5 year	18	0.49 (0.20, 0.71)	0.42 (0.11, 0.67)	0.58 (0.35, 0.77)	**0.61** (0.36, 0.80)	0.27 (0.00, 0.50)	0.10 (0.00, 0.29)
refer	80	0.90 (0.87, 0.93)	0.89 (0.85, 0.92)	0.88 (0.84, 0.93)	**0.91** (0.87, 0.95)	0.72 (0.64, 0.80)	0.74 (0.67, 0.81)
Weighted avg	140	0.84 (0.79, 0.90)	0.81 (0.76, 0.87)	**0.85 (0.79, 0.90)**	**0.85 (0.79, 0.90)**	0.62 (0.55, 0.70)	0.61 (0.54, 0.68)
				Pathology (KCH)			
DysplasiaOrCancer	7	**1.00** (1.00, 1.00)	0.94 (0.82, 1.00)	**1.00** (1.00, 1.00)	0.92 (0.73, 1.00)	0.48 (0.31, 0.67)	0.57 (0.44, 0.70)
IM	50	**0.99** (0.97, 1.00)	0.96 (0.93, 0.99)	0.96 (0.93, 0.99)	0.97 (0.93, 1.00)	0.89 (0.82, 0.95)	0.81 (0.72, 0.89)
No_IM	23	**0.72** (0.52, 0.85)	0.55 (0.29, 0.75)	0.66 (0.48, 0.82)	0.68 (0.50, 0.83)	0.38 (0.15, 0.59)	0.29 (0.08, 0.52)
Insufficient	60	**0.93** (0.90, 0.96)	0.92 (0.88, 0.95)	0.89 (0.84, 0.94)	**0.93** (0.89, 0.97)	0.81 (0.74, 0.88)	0.86 (0.81, 0.91)
Weighted avg	140	**0.92** (0.87, 0.95)	0.87 (0.82, 0.92)	0.89 (0.83, 0.93)	0.90 (0.85, 0.95)	0.75 (0.69, 0.81)	0.73 (0.67, 0.79)
				Endoscopy (KCH)			
Long	48	0.94 (0.88, 0.98)	0.84 (0.78, 0.90)	0.92 (0.87, 0.97)	**0.96** (0.91, 0.99)	0.73 (0.62, 0.83)	0.81 (0.71, 0.89)
Short	48	0.80 (0.71, 0.89)	0.79 (0.70, 0.87)	0.86 (0.78, 0.92)	**0.89** (0.83, 0.95)	0.74 (0.62, 0.84)	0.53 (0.37, 0.67)
NoBarretts	17	**0.73** (0.56, 0.88)	0.55 (0.29, 0.75)	0.66 (0.46, 0.84)	0.64 (0.47, 0.81)	0.34 (0.14, 0.53)	0.52 (0.32, 0.71)
Insufficient	27	0.71 (0.62, 0.79)	0.63 (0.49, 0.75)	**0.76** (0.64, 0.86)	0.82 (0.71, 0.91)	0.52 (0.37, 0.68)	0.59 (0.47, 0.71)
Weighted avg	140	0.82 (0.76, 0.87)	0.75 (0.68, 0.81)	0.84 (0.78, 0.89)	**0.87** (0.82, 0.92)	0.65 (0.57, 0.72)	0.63 (0.55, 0.71)

Table 4: Comparison of sub-classes performance (F1-Score) across multiple models for the GSTT and KCH datasets (Phi-4, Gemma-3, Qwen-2.5*, DeepSeek Qwen-2.5, Llama-3*, and DeepSeek Llama-3). * with CoT Prompting. Support is the number of each class in the original test sets.

```
{'doc_id': 'endo_sample',
 'extr': {'Barretts': [{'text': "barrett's segment", 'negation': 'no'},
   {'text': "barett's eosophagus", 'negation': 'no'}],
  'Barretts island': [],
  'irregular_z_line': [],
  'normal_oesophagus': [],
  'Prague_score': ['C3M8'],
  'Gastric_fold': ['38cm'],
  'Barretts_tongue': ['30cm'],
  'Circumferential barretts': ['35cm'],
  'Barretts_length': ['8cm']}}
```

4.5 Comparison to EndoBERT/PathBERT

The comparison between LLMs such as Phi-4, Qwen-2.5, and DeepSeek Qwen-2.5, and the domain-specific BERT-based model Endo/PathBERT (Table 5) highlights the strengths and limitations of general-purpose LLMs versus specialised BERT models. While LLMs demonstrate competitive performance, with Phi-4 achieving the highest weighted F1-score among LLMs in Pathology (GSTT, 0.97; KCH, 0.92) and DeepSeek Qwen-2.5 leading in Endoscopy (GSTT, 0.95; KCH, 0.87), Endo/PathBERT consistently

Class	Support	Phi-4 (14B)	Qwen-2.5* (14B)	DeepSeek Qwen-2.5 (14B)	Endo/PathBERT (0.1B)	Support	Phi-4	Qwen-2.5*	DeepSeek Qwen-2.5	Endo/PathBERT
			Pathology (GSTT)					Pathology (KCH)		
DysplasiaOrCancer	20	**1.00** (1.00, 1.00)	**1.00** (1.00, 1.00)	**1.00** (1.00, 1.00)	**1.00**	7	**1.00** (1.00, 1.00)	**1.00** (1.00, 1.00)	0.92 (0.73, 1.00)	**1.00**
IM	36	**0.99** (0.96, 1.00)	**0.99** (0.96, 1.00)	0.92 (0.86, 0.97)	0.97	50	**0.99** (0.97, 1.00)	0.96 (0.93, 0.99)	0.97 (0.93, 1.00)	0.95
No_IM	18	0.91 (0.80, 1.00)	0.91 (0.80, 1.00)	0.86 (0.71, 0.97)	**0.92**	23	0.72 (0.52, 0.85)	0.66 (0.48, 0.82)	0.68 (0.50, 0.83)	**0.86**
Insufficient	41	**0.95** (0.91, 0.99)	**0.95** (0.90, 0.99)	0.91 (0.84, 0.98)	0.83	60	**0.93** (0.90, 0.96)	0.89 (0.84, 0.94)	**0.93** (0.89, 0.97)	0.81
Weighted avg	115	**0.97** (0.93, 0.99)	0.96 (0.93, 0.99)	0.92 (0.87, 0.96)	0.92	140	**0.92** (0.87, 0.95)	0.89 (0.83, 0.93)	0.90 (0.85, 0.95)	0.88
			Endoscopy (GSTT)					Endoscopy (KCH)		
Long	29	0.98 (0.95, 1.00)	0.97 (0.92, 1.00)	0.95 (0.88, 1.00)	**1.00**	48	0.94 (0.88, 0.98)	0.92 (0.87, 0.97)	**0.96** (0.91, 0.99)	0.92
Short	23	0.93 (0.82, 1.00)	**1.00** (1.00, 1.00)	**1.00** (1.00, 1.00)	0.98	48	0.80 (0.71, 0.89)	0.86 (0.78, 0.92)	0.89 (0.83, 0.95)	**0.90**
NoBarretts	49	0.93 (0.87, 0.98)	**0.95** (0.91, 0.98)	**0.95** (0.91, 0.99)	**0.95**	17	0.73 (0.56, 0.88)	0.66 (0.46, 0.84)	0.64 (0.47, 0.81)	**0.81**
Insufficient	14	0.77 (0.62, 0.92)	0.71 (0.44, 0.88)	**0.88** (0.75, 1.00)	0.79	27	0.71 (0.62, 0.79)	**0.76** (0.64, 0.86)	0.82 (0.71, 0.91)	0.75
Weighted avg	115	0.92 (0.87, 0.97)	0.94 (0.89, 0.97)	**0.95** (0.91, 0.99)	**0.95**	140	0.82 (0.76, 0.87)	0.84 (0.78, 0.89)	**0.87** (0.82, 0.92)	**0.87**
Inference Time		28.82	30.81	70.47	0.03		27.64	28.26	66.23	0.03

Table 5: Comparison of Pathology and Endoscopy classification performance (F1-Score) between LLMs (Phi-4, Qwen-2.5*, DeepSeek Qwen-2.5) and BERT based report classification models on GSTT and KCH datasets. * with CoT Prompting. Support is the number of each class in the original test sets

achieve comparable performance across tasks. On the other hand, the inference time and space cost of LLMs are much higher than BERT-based models. The fine-tuned BERT models, however, have larger annotation and training overheads. New annotations and re-training are often needed for adaptations and repurposing, while LLMs can be adapted with only prompt changes. Besides, as an extraction-based model, the extracted information can be stored and reused for future queries or for other tasks that require these extractions.

5 Conclusion

This study explores the use of LLMs for extracting surveillance-relevant information from endoscopy and pathology reports to automate BO surveillance timing prediction. Our results show that LLMs can effectively process unstructured clinical text with few-shot learning and achieve performance comparable to or surpassing traditional NLP methods trained on human annotated data. Specifically, Phi-4 and DeepSeek Qwen-2.5 emerged as the most effective models for pathology and endoscopy report processing respectively. This approach reduces the need for extensive manual annotations, making it a scalable and adaptable solution for real-world clinical deployment. Moreover, this extraction-based method provides interpretable outputs. The structured extractions provided by LLMs, guided by rule-based algorithms for classification, increase transparency of the results and help with clinical validation compared to previous report level classification models. This study also shows that model selection and prompt design are essential for model performance and runtime during deployment.
Future research can explore fine-tuning these models for domain-specific tasks and integrating them into clinical decision support systems to optimise patient care.

6 Limitations and Future Work

Despite the promising results, our study has several limitations. Firstly, the models were evaluated on data from two hospitals, which may limit generalisability to other healthcare settings with different documentation styles. Secondly, while formatting results in a JSON style improved consistency, there might be easier ways for models to structure the outputs with lower invalid output rate. Thirdly, we used LLM extraction followed by a rule-based algorithm classification method. Future work could explore guiding LLMs to perform both classification and justification directly. Additionally, we evaluated the final performance on classification tasks. Human evaluations on entity and relation extractions could provide a more direct measure of the LLM extraction models. Furthermore, the experiments can be extended to larger LLMs. Lastly, more work on deploying LLMs in other GI conditions is needed to further explore their usability.

7 Ethics Statement

Use of the GSTT and KCH dataset received ethical approval from GSTT Electronic Records Research Interface (GERRI) institutional board review (IRAS ID = 257283) and King's Electronic Records Research Interface (KERRI) institutional board review (IRAS ID = 232823) respectively.

8 Acknowledgements

The research described in this paper was funded by King's College London DRIVE-Health Centre for Doctoral Training. We would like to express our gratitude to King's CREATE-TRE for providing compute resources and infrastructure.

References

Marah Abdin, Jyoti Aneja, Harkirat Behl, Sébastien Bubeck, Ronen Eldan, Suriya Gunasekar, Michael Harrison, Russell J Hewett, Mojan Javaheripi, Piero Kauffmann, et al. 2024. Phi-4 technical report. *arXiv preprint arXiv:2412.08905*.

DeepMind. 2024. Gemma: Lightweight Open Models for Responsible AI. Technical report, Google DeepMind. Accessed: March 17, 2025.

Jacob Devlin, Ming-Wei Chang, Kenton Lee, and Kristina Toutanova. 2019. Bert: Pre-training of deep bidirectional transformers for language understanding. In *Proceedings of the 2019 conference of the North American chapter of the association for computational linguistics: human language technologies, volume 1 (long and short papers)*, pages 4171–4186.

Alexander Dunn, John Dagdelen, Nicholas Walker, Sanghoon Lee, Andrew S Rosen, Gerbrand Ceder, Kristin Persson, and Anubhav Jain. 2022. Structured information extraction from complex scientific text with fine-tuned large language models. *arXiv preprint arXiv:2212.05238*.

Rebecca C Fitzgerald, Massimiliano Di Pietro, Krish Ragunath, Yeng Ang, Jin-Yong Kang, Peter Watson, Nigel Trudgill, Praful Patel, Philip V Kaye, Scott Sanders, et al. 2014. British society of gastroenterology guidelines on the diagnosis and management of barrett's oesophagus. *Gut*, 63(1):7–42.

Aaron Grattafiori, Abhimanyu Dubey, Abhinav Jauhri, Abhinav Pandey, Abhishek Kadian, Ahmad Al-Dahle, Aiesha Letman, Akhil Mathur, Alan Schelten, Alex Vaughan, et al. 2024. The llama 3 herd of models. *arXiv preprint arXiv:2407.21783*.

Daya Guo, Dejian Yang, Haowei Zhang, Junxiao Song, Ruoyu Zhang, Runxin Xu, Qihao Zhu, Shirong Ma, Peiyi Wang, Xiao Bi, et al. 2025. Deepseek-r1: Incentivizing reasoning capability in llms via reinforcement learning. *arXiv preprint arXiv:2501.12948*.

Jia He, Mukund Rungta, David Koleczek, Arshdeep Sekhon, Franklin X Wang, and Sadid Hasan. 2024. Does prompt formatting have any impact on llm performance? *arXiv preprint arXiv:2411.10541*.

Jia Li, Shan Hu, Conghui Shi, Zehua Dong, Jie Pan, Yaowei Ai, Jun Liu, Wei Zhou, Yunchao Deng, Yanxia Li, et al. 2022. A deep learning and natural language processing-based system for automatic identification and surveillance of high-risk patients undergoing upper endoscopy: A multicenter study. *EClinicalMedicine*, 53.

Pengfei Liu, Weizhe Yuan, Jinlan Fu, Zhengbao Jiang, Hiroaki Hayashi, and Graham Neubig. 2023. Pre-train, prompt, and predict: A systematic survey of prompting methods in natural language processing. *ACM computing surveys*, 55(9):1–35.

Mahmud Omar, Salih Nassar, Kassem SharIf, Benjamin S Glicksberg, Girish N Nadkarni, and Eyal Klang. 2025. Emerging applications of nlp and large language models in gastroenterology and hepatology: a systematic review. *Frontiers in Medicine*, 11:1512824.

Carlijn AM Roumans, Ruben D van der Bogt, Ewout W Steyerberg, Dimitris Rizopoulos, Iris Lansdorp-Vogelaar, Prateek Sharma, Manon CW Spaander, and Marco J Bruno. 2020. Adherence to recommendations of barrett's esophagus surveillance guidelines: a systematic review and meta-analysis. *Endoscopy*, 52(01):17–28.

Rithik Sachdev, Zhong-Qiu Wang, and Chao-Han Huck Yang. 2024. Evolutionary prompt design for llm-based post-asr error correction. *arXiv preprint arXiv:2407.16370*.

Stuart J. Spechler, Prateek Sharma, Rhonda F. Souza, John M. Inadomi, and Nicholas J. Shaheen. 2011. *Gastroenterology*, 140(3):e18–e52.

Robert J Tibshirani and Bradley Efron. 1993. An introduction to the bootstrap. *Monographs on statistics and applied probability*, 57(1):1–436.

Li Wang, Xi Chen, XiangWen Deng, Hao Wen, MingKe You, WeiZhi Liu, Qi Li, and Jian Li. 2024. Prompt engineering in consistency and reliability with the evidence-based guideline for llms. *NPJ digital medicine*, 7(1):41.

Shuhe Wang, Xiaofei Sun, Xiaoya Li, Rongbin Ouyang, Fei Wu, Tianwei Zhang, Jiwei Li, and Guoyin Wang. 2023. Gpt-ner: Named entity recognition via large language models. *arXiv preprint arXiv:2304.10428*.

Theresa Nguyen Wenker, Yamini Natarajan, Kadon Caskey, Francisco Novoa, Nabil Mansour, Huy Anh Pham, Jason K Hou, Hashem B El-Serag, and Aaron P Thrift. 2023. Using natural language processing to automatically identify dysplasia in pathology reports for patients with barrett's esophagus. *Clinical Gastroenterology and Hepatology*, 21(5):1198–1204.

Bas LAM Weusten, Raf Bisschops, Mario Dinis-Ribeiro, Massimiliano Di Pietro, Oliver Pech, Manon CW Spaander, Francisco Baldaque-Silva, Maximilien Barret, Emmanuel Coron, Glòria Fernández-Esparrach, et al. 2023. Diagnosis and management of barrett esophagus: European society of gastrointestinal endoscopy (esge) guideline. *Endoscopy*, 55(12):1124–1146.

An Yang, Baosong Yang, Beichen Zhang, Binyuan Hui, Bo Zheng, Bowen Yu, Chengyuan Li, Dayiheng Liu, Fei Huang, Haoran Wei, et al. 2024. Qwen2. 5 technical report. *arXiv preprint arXiv:2412.15115*.

J Diego Zamfirescu-Pereira, Richmond Y Wong, Bjoern Hartmann, and Qian Yang. 2023. Why johnny can't prompt: how non-ai experts try (and fail) to design llm prompts. In *Proceedings of the 2023 CHI conference on human factors in computing systems*, pages 1–21.

Agathe Zecevic, Laurence Jackson, Xinyue Zhang, Polychronis Pavlidis, Jason Dunn, Nigel Trudgill, Shahd Ahmed, Pierfrancesco Visaggi, Zanil YoonusNizar, Angus Roberts, et al. 2024. Automated decision making in barrett's oesophagus: development and deployment of a natural language processing tool. *NPJ Digital Medicine*, 7(1):312.

Ruoyu Zhang, Yanzeng Li, Yongliang Ma, Ming Zhou, and Lei Zou. 2023. Llmaaa: Making large language models as active annotators. *arXiv preprint arXiv:2310.19596*.

A Extraction schema

Entity	Field	Description
	Pathology for each biopsy finding	
Location	text	Location where the biopsy is taken
	oesophagus_or_barretts (yes/no)	whether the location is related to oesophagus or barretts
	cardia (yes/no)	whether the location is related to cardia
Barretts	text	mention of Barrett's
	negation (yes/no/indefinite)	Whether the mention is negated
Cancer	text	mention of cancer
	negation (yes/no/indefinite)	Whether the mention is negated
Dysplasia	text	mention of Dysplasia
	negation (yes/no/indefinite)	Whether the mention is negated
IM	text	mention of IM
	negation (yes/no/indefinite)	Whether the mention is negated
Gastric Metaplasia	text	mention of Gastric Metaplasia
	negation (yes/no/indefinite)	Whether the mention is negated
	Endoscopy	
Barretts	text	mention of Barrett's
	negation (yes/no/indefinite)	Whether the mention is negated
Barretts_island	text	mention of Barrett's island
	negation (yes/no/indefinite)	Whether the mention is negated
irregular_z_line	(text)	mention of irregular z line
normal_oesophagus	(text)	mention of normal oesophagus (squamous epithelium)
Prague score	(text)	The value of Prague score (e.g. C2M5)
Gastric fold	(text)	The position of gastric fold
Barretts_tongue	(text)	The position of top of Barrett's Tongue
Circumferential_barretts	(text)	The position of the top of circumferential Barrett's
Barretts_length	(text)	Direct mention or description of Barrett's length

Table 6: Entities and descriptions for Pathology and Endoscopy extractions.

B Prompts

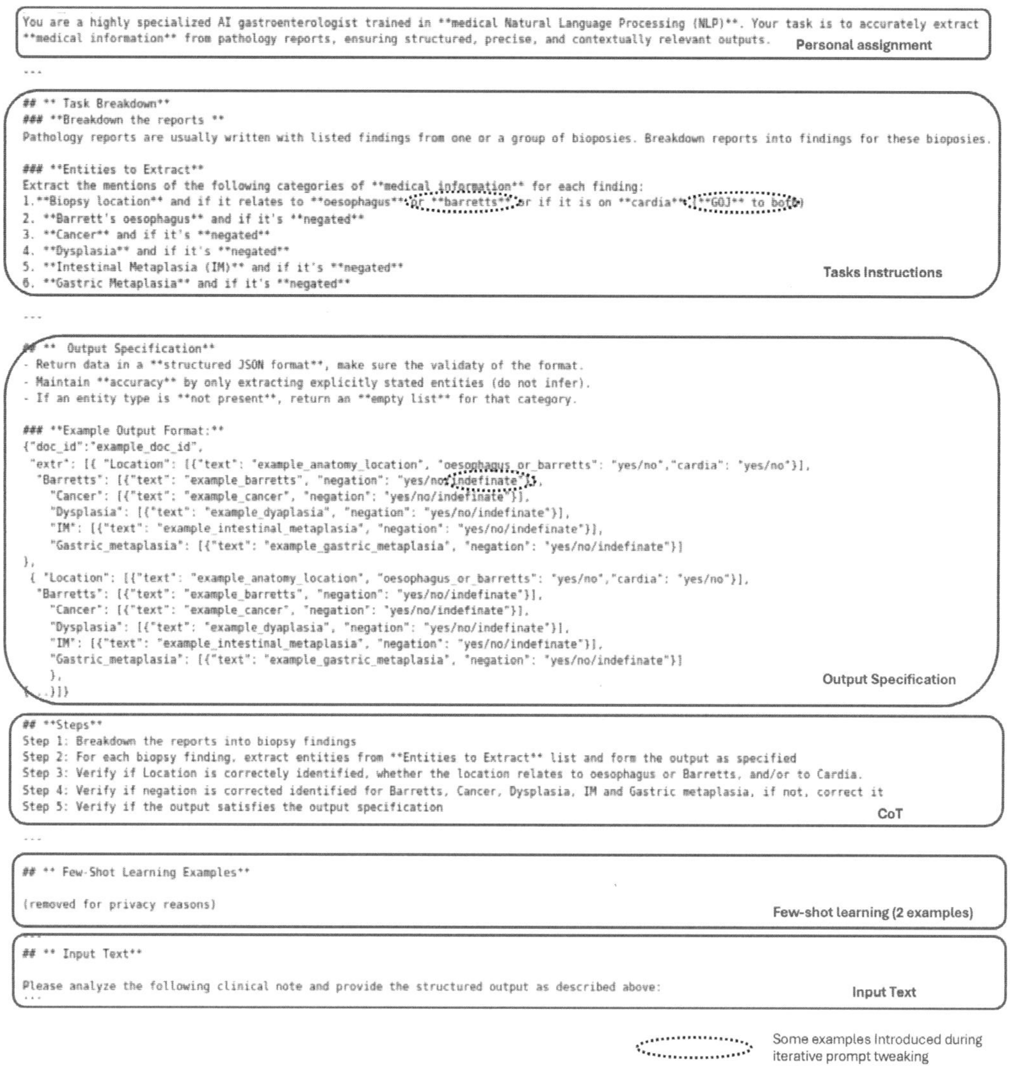

Figure 2: Pathology Prompt

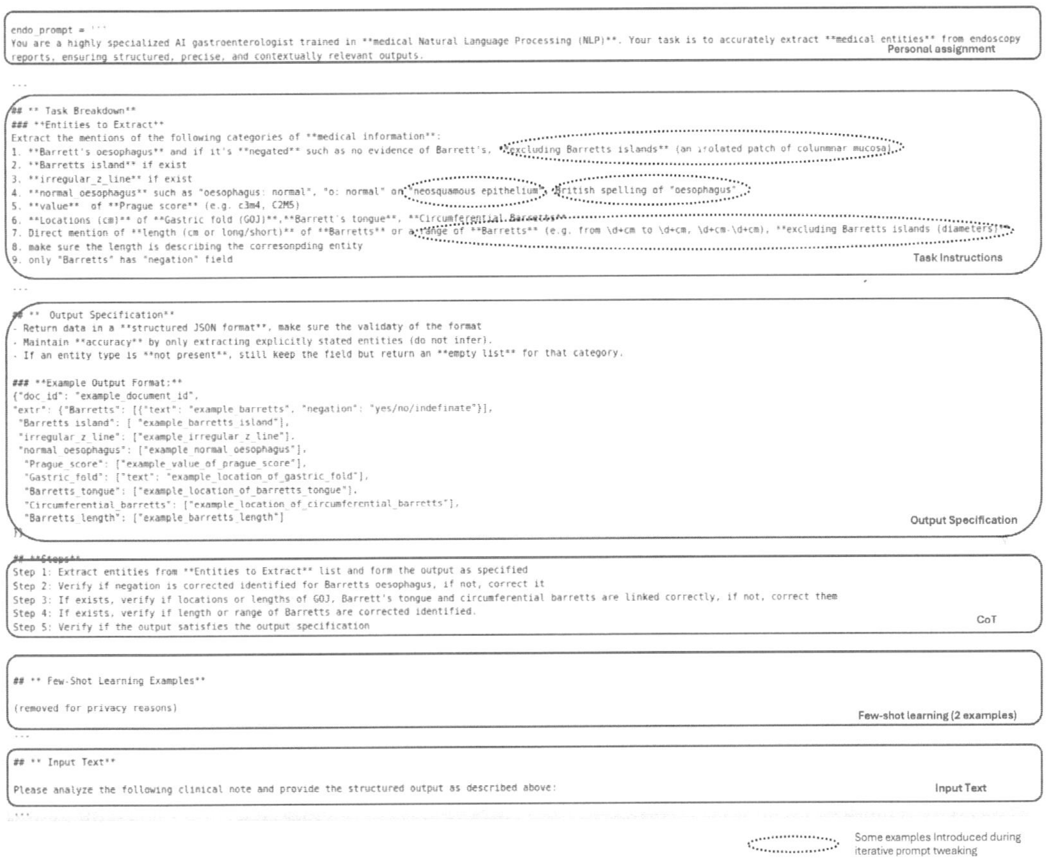

Figure 3: Endoscopy Prompt

Prompting Large Language Models for Italian Clinical Reports: A Benchmark Study

Livia Lilli[1,2], Carlotta Masciocchi[1], Antonio Marchetti[1], Giovanni Arcuri[1], Stefano Patarnello[1]
[1] Fondazione Policlinico Universitario Agostino Gemelli IRCCS
[2] Catholic University of the Sacred Heart of Rome
livia.lilli@policlinicogemelli.it

Abstract

Large Language Models (LLMs) have significantly impacted medical Natural Language Processing (NLP), enabling automated information extraction from unstructured clinical texts. However, selecting the most suitable approach requires careful evaluation of different model architectures, such as generative LLMs and BERT-based models, along with appropriate adaptation strategies, including prompting techniques, or fine-tuning. Several studies explored different LLM implementations, highlighting their effectiveness in medical domain, including complex diagnostics patterns as for example in rheumatology. However, their application to Italian remains limited, serving as a key example of the broader gap in non-English language research. In this study, we present a task-specific benchmark analysis comparing generative LLMs and BERT-based models, on real-world Italian clinical reports. We evaluated zero-shot prompting, in-context learning (ICL), and fine-tuning across eight diagnostic categories in the rheumatology area. Results show that ICL improves performance over zero-shot-prompting, particularly for Mixtral and Gemma models. Overall, BERT fine-tuning present the highest performance, while ICL outperforms BERT in specific diagnoses, such as renal and systemic, suggesting that prompting can be a potential alternative when labeled data is scarce.

1 Introduction

Recent advancements in Large Language Models (LLMs) have significantly impacted medical Natural Language Processing (NLP), enabling the extraction of structured information from unstructured clinical texts with increasing accuracy. Transformer-based architectures, such as BERT-based models and generative LLMs, have demonstrated strong potential in clinical text classification, named entity recognition, and medical concept extraction. However, selecting the most suitable model for a given task requires careful consideration of both model architecture and adaptation strategy, as different approaches offer varying levels of performance, efficiency, and practical feasibility.

LLMs, particularly generative architectures, can be adapted through zero-shot prompting (Sivarajkumar et al., 2024), where the model relies solely on its pre-trained knowledge, or in-context learning (ICL) (Liu et al., 2024), where domain-specific context is provided within the prompt. More advanced strategies include instruction fine-tuning (Tran et al., 2024; Li et al., 2024b), which refines the model's alignment with task-specific instructions. BERT-based models (Devlin et al., 2019), following a discriminative approach, typically require fine-tuning through supervised learning, though they can also be applied in Natural Language Inference (NLI) frameworks or used in few-shot and zero-shot settings by leveraging pre-trained embeddings. In all cases, pretraining on large domain-specific corpora can further enhance performance, though it remains computationally expensive and data-intensive.

In this work, we present a task-specific benchmark analysis tailored to a real-world clinical scenario, focusing on the necessity of extracting structured information from Italian clinical notes, in a real-world hospital setting. Our study evaluates generative LLMs in two different prompting strategies: zero-shot prompting, where the model relies solely on its pre-trained knowledge, and ICL, where additional domain-specific context is provided to guide the extraction process. To establish a strong comparative baseline, we also assess fine-tuned BERT-based models, which have traditionally been used for medical information extraction tasks (Lee et al., 2020; Muizelaar et al., 2024; Yang et al., 2024).

Our evaluation is based on a very general use case, which is the detection of complex diagnoses

Proceedings of the 24th Workshop on Biomedical Language Processing (BioNLP 2025), pages 190–200
August 1, 2025 ©2025 Association for Computational Linguistics

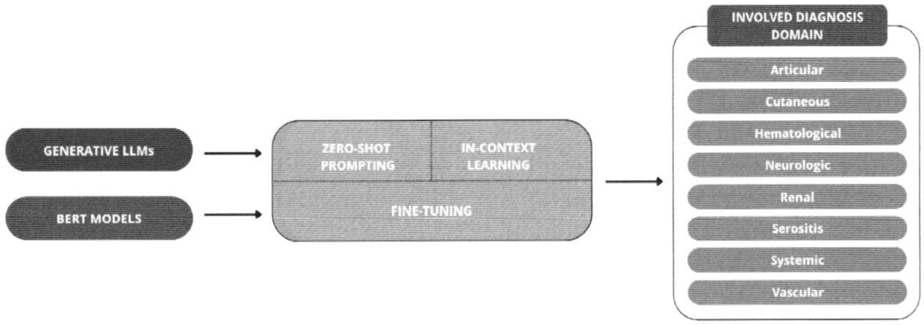

Figure 1: Study Framework: task-specific benchmark analysis comparing LLMs using zero-shot and in-context learning (ICL) strategies against fine-tuned BERT-based models, in an information extraction task.

in medical notes. The example we used covers the rheumatology domain, where very often a disease can impact several domains, each corresponding to a particular organ or system involvement (Figure 1). The use case of diagnosis extraction allows us to systematically compare generative LLMs and fine-tuned models, analyzing their strengths, limitations, and potential applications in real-world clinical workflows. By conducting this study in a practical hospital setting, we aim to provide insights into the feasibility of integrating LLM-based NLP solutions for automated information extraction in clinical practice. This process can support the development of decision-support tools and enable the creation of research datasets for predictive analytics, ultimately enhancing both clinical decision-making and medical research.

2 Background

Natural Language Processing (NLP) has gained increasing attention in medical application, with studies exploring its potential for extracting meaningful clinical insights from unstructured medical texts. A systematic review by Omar et al. (2024) provides a comprehensive analysis of NLP applications specifically for the reumathology domain, covering various techniques used to process electronic health records (EHRs), PubMed abstracts, FAQ and exams' questions for diseases such as rheumatoid arthritis (RA), gout, and systemic lupus erythematosus (SLE). Among the identified works, Li et al. (2022) explores named entity recognition (NER) in RA clinical notes, leveraging a BERT model enhanced with BiLSTM and CRF layers, achieving promising results in medical entity extraction. In the study of Osborne et al. (2021) NLP is used for gout flare detection, developing a fine-tuned BERT classifier based on annotated Emergency Department (ED) chief complaint notes, demonstrating that chief complaints alone are highly predictive of gout flares. Expanding on this approach, Oliveira et al. (2024) compares traditional NLP methods (e.g., tf-idf) with domain-specific LLMs, distinguishing between generative and discriminative models. Their study shows that generative models used as feature extractors can enhance performance when integrated with an SVM classifier, suggesting a hybrid approach for clinical text classification.

Focusing specifically on SLE and the Italian language, Lilli et al. (2024a) investigates the adaptation of BERT-based models for the extraction of Lupus-related diagnoses, symptoms, and treatments, demonstrating the feasibility of transformer-based NLP approaches in non-English medical corpora. Lilli et al. (2024b) also presents an NLP pipeline that integrates regular expression-based extraction with BERT-based topic detection, improving the structured identification of Lupus-related clinical features from Italian medical texts.

Beyond disease-specific applications, broader research has investigated the effectiveness of LLMs and BERT-based models in medical NLP tasks. Zhang et al. (2024) evaluates prompt engineering versus fine-tuning for clinical note classification, using metastatic cancer identification as a benchmark task. Their findings indicate that GPT-4 with structured prompts outperforms fine-tuned BERT-based models, suggesting that prompting can be an effective alternative to model retraining in clinical NLP. Meanwhile, Savage et al. (2024) examines whether LLMs can emulate clinical reasoning by structuring prompts to reflect differential diagnosis formation, intuitive reasoning, analytical reasoning, and Bayesian inference. Their results suggest that LLMs can provide interpretable rationales without compromising diagnostic accuracy, addressing the "black box" issue that limits trust in AI-driven medical applications.

More recent studies have also advanced the understanding of prompting strategies in clinical NLP. Naguib et al. (2024) conducted a multilingual evaluation of few-shot prompting for clinical NER, showing that masked language models often outperform generative models, particularly in low-resource settings. Similarly, Nagar et al. (2024) benchmarked various prompting and retrieval strategies across structured biomedical tasks, highlighting the limitations of reasoning-augmented methods like Chain-of-Thought and RAG, especially for classification and NER. Hu et al. (2024) proposed a prompt engineering framework for GPT models in clinical NER, demonstrating that structured, task-specific prompting can substantially improve performance.

All the above studies highlight the evolution of NLP techniques in medical applications, and the increasing role of LLMs in replacing or complementing traditional fine-tuned models for clinical text analysis, classification, and decision support.

3 Method

3.1 Dataset

The dataset used in this study consists of a collection of outpatient visit reports written in Italian language, related to patients with a SLE diagnosis and treated in the Reumatology department of a real-world hospital. The outcome of the information task was to identify eight different types of diagnoses based on the specific organ or system involvement. The categories considered are: Articular, Cutaneous, Hematologic, Neurologic, Renal, Systemic, Serositis, and Vascular.

3.2 Generative Modeling

For the generative LLM experiments, we employed a set of open-source language models, either multilingual or specifically trained for the Italian language, leveraging the Ollama framework to optimize computational efficiency. The models were then accessed through the Ollama Python library[1], utilizing its `generate` function to process and analyze clinical texts. This approach allowed us to efficiently execute inference without the need for fine-tuning, making it a scalable and adaptable solution for medical NLP tasks. Input reports were preprocessed and analyzed at the paragraph level rather than as full documents. This approach was adopted to reduce text length, enabling a more focused and efficient processing of clinical information. At the end of the processing pipeline, a logical OR operation was applied to aggregate paragraph-level classifications into a final diagnosis at the Electronic Health Record (EHR) level. This means that if any paragraph within a patient's report indicated the presence of a specific Lupus subtype, that classification was assigned to the entire EHR. For both the zero-shot and in-context learning setups, we leveraged ChatGPT-4o (Hurst et al., 2024; Achiam et al., 2023) to generate appropriate prompts, ensuring well-structured and consistent instructions tailored for clinical information extraction. In both cases, the prompt was in English and designed to return a structured binary output (1 for presence, 0 for absence) for each Lupus category independently. However, the models did not always comply with this format, often including additional explanations or justifications alongside the binary response. To provide a standardized output, we applied a regular expression (regex) filter to isolate and extract the binary classification for each category separately, ensuring consistency in the final results. To improve the robustness of this approach, we manually reviewed a sample of LLM outputs to identify common patterns for the development of regex rules. This step helped us reduce misclassifications caused by unexpected output formats or embedded rationales.

3.2.1 Zero-Shot Prompting

In the zero-shot prompting setup, the models were prompted without any additional contextual guidance or predefined medical terms. The prompt structure followed a direct query format, instructing the model to determine the presence of a Lupus diagnosis based on the involvement of a specific organ or system. The exact prompt used was:

```
Given the following Italian medical report,
return "1" if there is evidence of lupus with
{category} involvement, otherwise, return "0".

Report:
{text}
```

By relying solely on the model's pre-trained knowledge, this approach aimed to evaluate the intrinsic capability of generative LLMs to extract structured medical information without external lexical or contextual augmentation.

[1]https://github.com/ollama/ollama-python

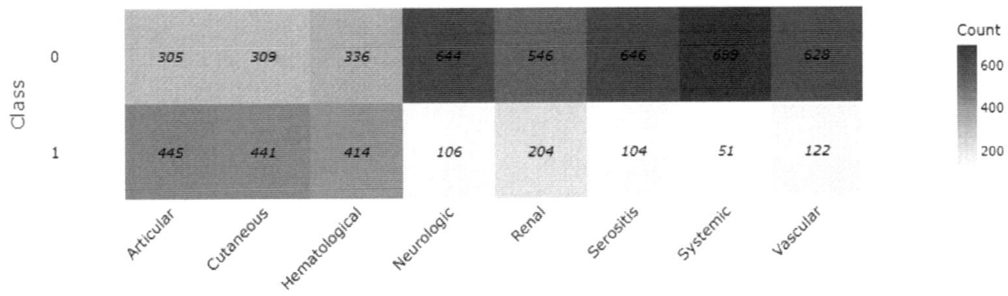

Figure 2: Dataset Composition.

3.2.2 In-Context Learning

In the in-context learning setup, we provided additional domain-specific context by including an Italian dictionary of medically relevant terms related to each category. These terms consisted of synonyms and alternative expressions commonly used in clinical texts to describe the specific type of organ or system involvement. By integrating this lexical knowledge directly into the prompt, we aimed to guide the model toward more accurate information extraction while still leveraging its generative capabilities. Unlike the zero-shot prompting, the following prompt structure was used to incorporate the dictionary of terms:

```
Given the following Italian medical report,
return "1" if there is evidence of lupus with
{category} involvement, otherwise, return "0".
To determine this, check if the report contains
relevant terms associated with {category}
involvement. Below there is a list of medically
relevant terms that indicate {category} involvement:

Relevant terms:
{list_of_terms}

Report:
{text}
```

This setup allowed the model to leverage both its pre-trained knowledge and the medical terminology provided, creating a context-augmented approach that aimed to improve classification accuracy. The list of medically relevant terms used in the ICL prompts is derived from a domain-specific dictionary originally developed for a rule-based information extraction system, as described in our previous work (Lilli et al., 2024b). In that study, the dictionary served as the foundation for a classifier based on pattern-matching within clinical texts. Further details about the dictionary of terms used in the prompt is provided in Appendix A.

3.3 BERT-based Fine-Tuning

To compare the performance of generative language models with fine-tuned approaches, we also included results from a set of fine-tuned BERT-based models, where each Lupus category was treated as an independent binary classification task, with separate classifiers trained for each type of organ involvement. The experimental setup and classification approach are consistent with a prior benchmark study on fine-tuned BERT models (Lilli et al., 2024a), ensuring a direct comparison with the LLM experiments, where a training set of 1000 labelled texts was used.

As in the generative LLM setup, the BERT fine-tuning process followed a paragraph-level approach, respecting the token limit constraints of BERT-based architectures. At inference, each paragraph was classified independently, with the final Electronic Health Record (EHR) classification determined using a logical OR operation.

Additionally, the evaluation set remained the same across all techniques, ensuring a fair and consistent comparison between fine-tuned BERT models and generative LLMs. This methodological alignment allowed us to analyze their relative strengths and limitations under identical conditions.

4 Experiments

4.1 Data

The dataset used for the study evaluation consists of 790 outpatient visit reports, collected from the SLE Data Mart of an Italian hospital. To facilitate processing and improve classification efficiency, in all the experiments each report was segmented at the paragraph level, resulting in a total of 6,024 paragraphs. On average, each paragraph contains 111.5 BERT tokens and 303.7 characters. This seg-

Experiment	Model	Articular	Cutaneous	Hematologic	Neurologic	Renal	Serositis	Systemic	Vascular	Overall
Zero-Shot Prompting	Gemma3-1B	0.00	0.22	0.27	0.08	0.18	0.08	**0.48**	0.04	0.17
	Gemma3-4B	**0.80**	**0.80**	0.78	0.30	0.53	0.29	0.14	0.30	0.49
	Llama3.2-1B	0.62	0.60	0.65	0.28	0.42	0.22	0.14	0.25	0.40
	Llama3.2-3B	0.76	0.76	0.76	0.38	0.55	0.30	0.16	**0.36**	0.50
	Mistral-7B	0.77	0.70	**0.80**	**0.40**	0.64	0.34	0.15	0.33	0.52
	Mixtral-8x7B	0.78	0.74	0.79	0.34	**0.69**	**0.39**	0.16	0.32	**0.53**
In-Context Learning	Gemma3-1B	0.41	0.44	0.34	0.26	0.18	0.60	**0.52**	0.16	0.36
	Gemma3-4B	0.84	**0.88**	0.91	0.76	0.77	0.71	0.17	0.44	0.68
	Llama3.2-1B	0.67	0.67	0.66	0.24	0.41	0.25	0.13	0.27	0.41
	Llama3.2-3B	0.74	0.73	0.71	0.25	0.43	0.24	0.13	0.28	0.44
	Mistral-7B	0.86	0.61	0.87	0.73	0.88	0.77	0.31	0.37	0.68
	Mixtral-8x7B	**0.91**	0.73	**0.95**	**0.83**	**0.96**	**0.80**	0.23	**0.52**	**0.74**
BERT-based Fine-Tuning	Alberto	0.90	0.87	**0.98**	0.86	**0.94**	0.81	**0.29**	**0.69**	**0.79**
	Albert2	0.85	0.80	0.96	0.57	0.85	0.65	0.28	0.55	0.69
	Albert1	**0.92**	0.81	0.94	0.86	0.92	0.51	0.07	0.58	0.70
	Biobit	**0.92**	0.88	0.93	0.81	0.85	**0.87**	0.12	0.63	0.75
	Medbit	0.83	**0.92**	0.96	0.79	0.90	0.66	0.13	0.61	0.73
	Medbit-plus	**0.92**	0.90	0.90	**0.88**	0.85	0.72	0.07	0.63	0.73

Table 1: Comparison of Generative LLMs and BERT-Based Models Across Different Experimental Setups (Zero-Shot Prompting, In-Context Learning, and BERT-based Fine-Tuning), in terms of F1-Score.

mentation ensures that text segments remain within the acceptable token limits of BERT-based models, preserving sufficient clinical context for classification. Additionally, this approach is beneficial for generative LLMs, as it enables them to process shorter and more concise text inputs, optimizing computational efficiency and response accuracy. The dataset includes eight distinct types of Lupus diagnoses, each corresponding to a specific organ or system involvement. Since multiple categories can co-occur in the same report, a single document may be associated with more than one diagnosis. For privacy reasons, we can't report practical examples of the dataset, but we provide an overview of its composition in Figure 2.

4.2 Generative LLMs

For the generative experiments, we tested a range of open-source language models using the Ollama framework to ensure efficient inference. The models evaluated included Llama 3.2 (1B and 3B parameters), Gemma 3 (1B and 4B parameters), Mixtral (8x7B) and Mistral (7B). Each model was evaluated in both zero-shot prompting and ICL setups, on the SLE information extraction task.

Llama 3.2, developed by Meta (Grattafiori et al., 2024), is an optimized version of the Llama family, known for its improved efficiency and multilingual capabilities. The 1B and 3B parameter versions provide a balance between computational cost and performance, making them suitable for real-world scenarios. Gemma 3, released by Google DeepMind (Team et al., 2024), is a lightweight transformer-based model optimized for low-resource settings while maintaining strong reasoning abilities. The 1B model is designed for efficiency, whereas the 4B version offers enhanced performance with increased computational requirements. Mixtral, a mixture-of-experts model from Mistral AI (Jiang et al., 2024), activates only two out of eight expert networks per inference, allowing for improved efficiency while retaining strong language understanding. Finally, Mistral 7B (also from Mistral AI (Jiang et al., 2023)) is a dense transformer model known for its superior performance compared to similarly sized models, making it a potential alternative to Llama and Gemma for various NLP tasks. By selecting models with different architectures, sizes, and capabilities, we ensured a comprehensive evaluation of generative approaches for NLP in medical domain. Table 1 shows performances in terms of F1-Score metric for the zero-shot prompting and the in-context learning scenarios, respectively.

4.3 BERT-based Models

The BERT-based fine-tuning was performed using the PyTorch Trainer from the Hugging Face Transformers library (Wolf et al., 2020), running for 10 epochs (for further implementation details, see Appendix B). The models considered in this study include BioBIT3, MedBIT4, MedBIT-r3-plus5, AlBERTo, and two base versions of ALBERT.

BioBIT3, MedBIT4, and MedBIT-r3-plus5, developed by Buonocore et al. (2023), are BERT models pretrained on Italian biomedical corpora, making them particularly suitable for clinical NLP tasks. AlBERTo, originally proposed by Polignano et al. (2019), is an Italian-adapted version of ALBERT, trained on Italian tweets. In addition, we included two base versions of ALBERT Lan et al. (2019), which serve as the foundation of AlBERTo. Table 1 shows F1-Score metric values for the BERT-based experiments.

4.4 Results and Discussion

The results of the zero-shot and in-context learning (ICL) prompting experiments, compared to BERT fine-tuning, are presented in Table 1. For each scenario, the table reports the F1-scores of all tested models across the eight categories, along with the overall F1-score, calculated as the mean value. To better interpret these results, we structure our analysis into two key perspectives. First, we provide an overall comparison of performance across different methods (zero-shot prompting, in-context learning, and fine-tuned BERT models) to assess their general effectiveness. Second, we examine model-specific performance patterns across different disease categories, identifying strengths and limitations in extracting various diagnostic domains.

Regarding overall model performance across different disease categories, BERT-based classification models achieve the highest scores, with Alberto obtaining the best average F1-score of 0.79. However, it is noteworthy that even a limited degree of adaptation through In-Context Learning (ICL) significantly improves LLM performance. Mixtral-8x7B, with an average F1-score of 0.74, performs comparably to the best BERT-based models, demonstrating the effectiveness of ICL in enhancing generative models for structured information extraction. In contrast, Zero-Shot Prompting shows the weakest performance, with Mixtral-8x7B achieving the highest overall F1-score at just 0.53. This performance gap is likely due to the lack of contextual guidance, which makes it more challenging for the model to differentiate between diagnostic categories. In the absence of domain-specific cues, semantic differences across diagnoses reduce the model's discriminative power, leading to lower classification accuracy.

Moving to an in-depth analysis of performance across different diagnostic categories, the zero-shot setting reveals notable variations among models. Mixtral-8x7B, with the highest overall performance (F1-Score=0.53), specifically outperforms the other models in Renal (F1-Score=0.69) and Serositis (F1-Score=0.39) diagnoses. While Mistral-7B, with a slightly lower F1-Score of 0.52, presents the highest F1-Score in Hematologic (0.80) and Neurologic (0.40) categories. Meanwhile, Gemma3-4B, with the best F1-scores in the Articular and Cutaneous categories (0.80), shows an overall F1 performance near to Mixtral-8x7B and Mistral7B, equal to 0.49. In general, zero-shot performance is particularly weak for Neurologic, Renal, Serositis, Systemic, and Vascular diagnoses, with F1-scores ranging from 0.36 for Vascular (with Llama3.2-3b), to 0.69 for Renal (with Mixtral-8x7b).

In the ICL setting, Mixtral-8x7B achieves the highest scores on most of the categories, with the highest in the Renal, with a F1 value equal to 0.96. However, Gemma3-4B and Gemma3-1B outperform Mixtral-7B in two specific cases: Cutaneous (F1-Score=0.88) and Systemic (F1-Score=0.52). A particular improvement is observed in Neurologic, Renal, and Serositis diagnoses, where zero-shot prompting had shown extremely weak performance: with ICL, these categories experience a substantial boost, with Mixtral-8x7B achieving the highest scores, ranging from 0.80 for Serositis to 0.96 for Renal. On the other hand, classification for Systemic and Vascular categories remains weak, with the best performances achieved by Gemma3-1B for Systemic and Mixtral-8x7B for Vascular (F1-score = 0.52).

The results of both zero-shot and in-context learning (ICL) experiments highlight the significant role that contextual adaptation plays in enhancing generative models' performance. While the previous analysis examined each approach across different diagnostic categories, it is equally important to assess how ICL compares directly to zero-shot prompting across models and disease types. In general, moving from zero-shot to in-context learning (ICL) mostly leads to improved performance, as evident in Figure 3. Each bar plot in the figure represents the F1-score of different models across the eight categories, with the maximum F1-score from either the zero-shot or ICL scenario displayed. The colored margins, green and red, indicate the difference between the two approaches. A green margin means a positive difference, meaning ICL outperforms zero-shot prompting, enhancing information extraction. Conversely, a red margin indicates cases where zero-shot prompting achieved better results. From the figure, the majority of cases show an improvement with ICL, particularly for the Gemma and Mistral models. For instance, in the Serositis category, the F1-score of the Gemma3-4B model increases from 0.29 in the zero-shot setting

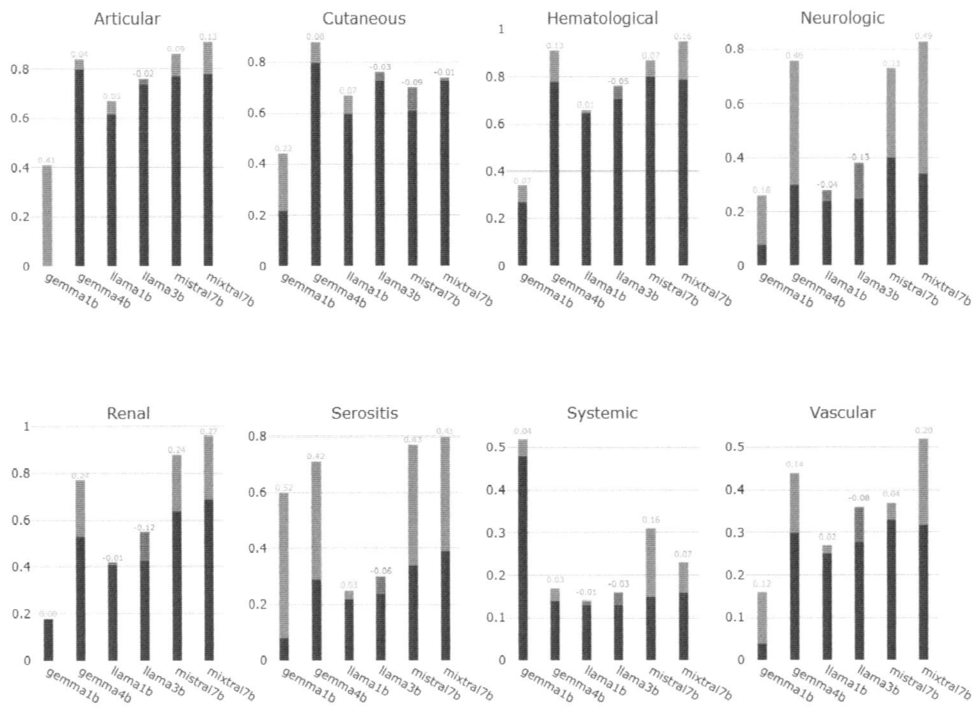

Figure 3: Comparison of Zero-Shot Prompting versus In-Context Learning (ICL) F1-Scores across different diagnoses. The bars represent the maximum score between the two methods. The green and red margins indicate the effect of ICL: green for improvements and red for declines compared to Zero-Shot Prompting.

to 0.71 with ICL, as shown by the green margin of 0.42. Similarly, in the Neurologic diagnosis, the Mixtral-8x7B model returns F1-Score values of 0.34 and 0.83 in the zero-shot and ICL scenarios respectively, with a final margin of improvement equal to 0.49. However, there are instances where ICL does not improve performance: the presence of red margins in at least one model for all categories suggests that semantic complexity alone is not the primary cause. Instead, it appears that certain models, particularly Llama3.2-3b, consistently perform better in the zero-shot scenario, and also Llama3.2-1B frequently shows negative margins. This indicates that for the Llama models, the type of contextual information introduced in ICL does not provide additional knowledge but instead has a confounding effect, hindering information detection.

To fully assess the effectiveness of prompting strategies, we compared them with a fully supervised fine-tuned approach of BERT-based models, which serve as a benchmark for structured information extraction. As already reported, BERT fine-tuning achieves the highest overall performance, with the Alberto model obtaining the highest F1-score of 0.79. In terms of individual categories, BERT models excel particularly in Hematologic diagnosis, where Alberto reaches the highest F1-score equal to 0.98. However, not all categories benefit the most from BERT fine-tuning. Some achieve better performance in the ICL scenario, such as Renal diagnosis, where Mixtral-8x7B reaches a 0.96 F1-Score, and Systemic diagnosis, where Gemma3-1B achieves a 0.52 F1-Score. These cases suggest that while BERT fine-tuning is generally effective, ICL can provide better results for specific types of medical information extraction.

Overall, ICL improves performance over zero-shot prompting, though some models, like Llama models, struggle with added context. BERT fine-tuning remains the most reliable approach for this Italian use case, achieving the highest scores. However, prompting is a viable alternative, as it allows adaptation with minimal data and no dedicated training, making it useful when resources are limited.

5 Conclusions

This study provides a comparative analysis of generative LLMs and fine-tuned BERT models for Italian clinical NLP, focusing on the extraction of diagnostic patterns within an outpatient setting. Our results demonstrate that while ICL significantly enhances generative models' performance over zero-shot prompting, fine-tuned BERT-based models still achieve the best overall results, providing structured and reliable classification solutions. However, ICL performances show that in-context adaptation techniques have great potential for iterative improvement. This is also confirmed by the results of this paper, where certain diagnostic categories, such as renal involvement, show better performance with ICL, indicating that supervised prompting can effectively overcome certain semantics complexities.

Beyond performance, model selection in healthcare applications must also consider privacy, data protection, and control on adaptation. For this reason we believe that a study focused on the comparison of open-source types of models provides a new perspective to complement GPT-based works, which are largely explored in current literature (Li et al., 2024a).

Future work should explore larger generative models, which may offer insight into the upper-bound performance achievable through prompting strategies alone. Additionally, future studies should conduct a more in-depth analysis of computational costs and trade-offs, particularly when considering prompting-based methods versus full fine-tuning, to guide practical decisions in clinical deployment scenarios.

By conducting this study in a real-world hospital setting, we aim to provide insights into the feasibility of integrating LLM-based NLP solutions for automated clinical information extraction. This could aid in the development of decision-support tools, facilitate the creation of research datasets for predictive analytics, and ultimately improve both clinical decision-making and medical research. Furthermore, by focusing on Italian clinical texts, this study expands NLP applications beyond English-language datasets, addressing the need for real-world solutions in underrepresented languages.

Limitations

While generative models show potential for medical information extraction, they do not always produce structured responses, requiring post-processing. We extracted binary classifications using regular expressions (regex), but this method can be imprecise, making BERT-based architectures more reliable for structured tasks. Additionally, due to the constraints of a real-world clinical setting in terms of computing resources, lighter versions of the models have been implemented. Future work could explore larger versions of Gemma and Llama running on more powerful computing environments, to achieve potential performance gains. Furthermore, in-context learning (ICL) proves effective for an initial adaptation, but its performance could be enhanced by incorporating labeled examples alongside the current dictionary. Further research should explore alternative adaptation techniques, such as instruction-tuning or a massive fine-tuning, to better compare different strategies for optimizing medical NLP models.

Ethics Statement

The use of data for this study has been implemented in full compliance with ethics and GDPR requirements. Specifically, data usage has been approved by the Ethics Committee of our hospital to conduct the presented research and the de-identification of sensitive data has been performed. Approval protocol number from the relevant Ethics Committee can be provided on request.

References

Josh Achiam, Steven Adler, Sandhini Agarwal, Lama Ahmad, Ilge Akkaya, Florencia Leoni Aleman, Diogo Almeida, Janko Altenschmidt, Sam Altman, Shyamal Anadkat, et al. 2023. Gpt-4 technical report. *arXiv preprint arXiv:2303.08774*.

Tommaso Mario Buonocore, Claudio Crema, Alberto Redolfi, Riccardo Bellazzi, and Enea Parimbelli. 2023. Localizing in-domain adaptation of transformer-based biomedical language models. *Journal of Biomedical Informatics*, 144:104431.

Jacob Devlin, Ming-Wei Chang, Kenton Lee, and Kristina Toutanova. 2019. BERT: Pre-training of deep bidirectional transformers for language understanding. In *Proceedings of the 2019 Conference of the North American Chapter of the Association for Computational Linguistics: Human Language Technologies, Volume 1 (Long and Short Papers)*, pages

4171–4186, Minneapolis, Minnesota. Association for Computational Linguistics.

Aaron Grattafiori, Abhimanyu Dubey, Abhinav Jauhri, Abhinav Pandey, Abhishek Kadian, Ahmad Al-Dahle, Aiesha Letman, Akhil Mathur, Alan Schelten, Alex Vaughan, et al. 2024. The llama 3 herd of models. *arXiv preprint arXiv:2407.21783*.

Dingxin Hu, Xuanyu Zhang, Xingyue Zhang, Yiyang Li, Dongsheng Chen, Marina Litvak, Natalia Vanetik, Qing Yang, Dongliang Xu, Yanquan Zhou, Lei Li, Yuze Li, and Yingqi Zhu. 2024. Improving factual consistency in abstractive summarization with sentence structure pruning. In *Proceedings of the 2024 Joint International Conference on Computational Linguistics, Language Resources and Evaluation (LREC-COLING 2024)*, pages 8792–8803, Torino, Italia. ELRA and ICCL.

Aaron Hurst, Adam Lerer, Adam P Goucher, Adam Perelman, Aditya Ramesh, Aidan Clark, AJ Ostrow, Akila Welihinda, Alan Hayes, Alec Radford, et al. 2024. Gpt-4o system card. *arXiv preprint arXiv:2410.21276*.

Albert Q. Jiang, Alexandre Sablayrolles, Arthur Mensch, Chris Bamford, Devendra Singh Chaplot, Diego de las Casas, Florian Bressand, Gianna Lengyel, Guillaume Lample, Lucile Saulnier, Lélio Renard Lavaud, Marie-Anne Lachaux, Pierre Stock, Teven Le Scao, Thibaut Lavril, Thomas Wang, Timothée Lacroix, and William El Sayed. 2023. Mistral 7b. *arXiv preprint arXiv:2310.06825*.

Albert Q Jiang, Alexandre Sablayrolles, Antoine Roux, Arthur Mensch, Blanche Savary, Chris Bamford, Devendra Singh Chaplot, Diego de las Casas, Emma Bou Hanna, Florian Bressand, et al. 2024. Mixtral of experts. *arXiv preprint arXiv:2401.04088*.

Zhenzhong Lan, Mingda Chen, Sebastian Goodman, Kevin Gimpel, Piyush Sharma, and Radu Soricut. 2019. Albert: A lite bert for self-supervised learning of language representations. *arXiv preprint arXiv:1909.11942*.

Jinhyuk Lee, Wonjin Yoon, Sungdong Kim, Donghyeon Kim, Sunkyu Kim, Chan Ho So, and Jaewoo Kang. 2020. Biobert: a pre-trained biomedical language representation model for biomedical text mining. *Bioinformatics*, 36(4):1234–1240.

Jianning Li, Amin Dada, Behrus Puladi, Jens Kleesiek, and Jan Egger. 2024a. Chatgpt in healthcare: a taxonomy and systematic review. *Computer Methods and Programs in Biomedicine*, 245:108013.

Meiting Li, Feifei Liu, Ran Zhang, Yi Qin, Dongping Gao, et al. 2022. Model-based clinical note entity recognition for rheumatoid arthritis using bidirectional encoder representation from transformers. *Quantitative Imaging in Medicine and Surgery*, 12(1):184.

Rumeng Li, Xun Wang, and Hong Yu. 2024b. LlamaCare: An instruction fine-tuned large language model for clinical NLP. In *Proceedings of the 2024 Joint International Conference on Computational Linguistics, Language Resources and Evaluation (LREC-COLING 2024)*, pages 10632–10641, Torino, Italia. ELRA and ICCL.

Livia Lilli, Laura Antenucci, Augusta Ortolan, Silvia Laura Bosello, Maria Antonietta D'agostino, Stefano Patarnello, Carlotta Masciocchi, and Jacopo Lenkowicz. 2024a. Lupus alberto: A transformer-based approach for SLE information extraction from Italian clinical reports. In *Proceedings of the 10th Italian Conference on Computational Linguistics (CLiC-it 2024)*, pages 510–516, Pisa, Italy. CEUR Workshop Proceedings.

Livia Lilli, Silvia Laura Bosello, Laura Antenucci, Stefano Patarnello, Augusta Ortolan, Jacopo Lenkowicz, Marco Gorini, Gabriella Castellino, Alfredo Cesario, Maria Antonietta D'Agostino, et al. 2024b. A comprehensive natural language processing pipeline for the chronic lupus disease. In *Digital Health and Informatics Innovations for Sustainable Health Care Systems*, pages 909–913. IOS Press.

Rui Liu, Mingjie Li, Shen Zhao, Ling Chen, Xiaojun Chang, and Lina Yao. 2024. In-context learning for zero-shot medical report generation. In *Proceedings of the 32nd ACM International Conference on Multimedia*, pages 8721–8730.

Hielke Muizelaar, Marcel Haas, Koert van Dortmont, Peter van der Putten, and Marco Spruit. 2024. Extracting patient lifestyle characteristics from dutch clinical text with bert models. *BMC medical informatics and decision making*, 24(1):151.

Aishik Nagar, Viktor Schlegel, Thanh-Tung Nguyen, Hao Li, Yuping Wu, Kuluhan Binici, and Stefan Winkler. 2024. Llms are not zero-shot reasoners for biomedical information extraction. *CoRR*.

Marco Naguib, Xavier Tannier, and Aurelie Neveol. 2024. Few-shot clinical entity recognition in english, french and spanish: masked language models outperform generative model prompting. In *Findings of the Association for Computational Linguistics: EMNLP 2024*, pages 6829–6852.

Lucas Lopes Oliveira, Xiaorui Jiang, Aryalakshmi Nellippillipathil Babu, Poonam Karajagi, and Alireza Daneshkhah. 2024. Effective natural language processing algorithms for early alerts of gout flares from chief complaints. *Forecasting*, 6(1):224–238.

Mahmud Omar, Mohammad E Naffaa, Benjamin S Glicksberg, Hagar Reuveni, Girish N Nadkarni, and Eyal Klang. 2024. Advancing rheumatology with natural language processing: insights and prospects from a systematic review. *Rheumatology Advances in Practice*, 8(4):rkae120.

John D Osborne, James S Booth, Tobias O'Leary, Amy Mudano, Giovanna Rosas, Phillip J Foster, Kenneth G Saag, and Maria I Danila. 2021. Identification of gout flares in chief complaint text using natural language processing. In *AMIA Annual Symposium Proceedings*, volume 2020, page 973.

Marco Polignano, Pierpaolo Basile, Marco De Gemmis, Giovanni Semeraro, Valerio Basile, et al. 2019. Alberto: Italian bert language understanding model for nlp challenging tasks based on tweets. In *CEUR workshop proceedings*, volume 2481, pages 1–6. CEUR.

Thomas Savage, Ashwin Nayak, Robert Gallo, Ekanath Rangan, and Jonathan H Chen. 2024. Diagnostic reasoning prompts reveal the potential for large language model interpretability in medicine. *NPJ Digital Medicine*, 7(1):20.

Sonish Sivarajkumar, Mark Kelley, Alyssa Samolyk-Mazzanti, Shyam Visweswaran, and Yanshan Wang. 2024. An empirical evaluation of prompting strategies for large language models in zero-shot clinical natural language processing: algorithm development and validation study. *JMIR Medical Informatics*, 12:e55318.

Gemma Team, Thomas Mesnard, Cassidy Hardin, Robert Dadashi, Surya Bhupatiraju, Shreya Pathak, Laurent Sifre, Morgane Rivière, Mihir Sanjay Kale, Juliette Love, et al. 2024. Gemma: Open models based on gemini research and technology. *arXiv preprint arXiv:2403.08295*.

Hieu Tran, Zhichao Yang, Zonghai Yao, and Hong Yu. 2024. Bioinstruct: instruction tuning of large language models for biomedical natural language processing. *Journal of the American Medical Informatics Association*, 31(9):1821–1832.

Thomas Wolf, Lysandre Debut, Victor Sanh, Julien Chaumond, Clement Delangue, Anthony Moi, Pierric Cistac, Tim Rault, Rémi Louf, Morgan Funtowicz, et al. 2020. Transformers: State-of-the-art natural language processing. In *Proceedings of the 2020 conference on empirical methods in natural language processing: system demonstrations*, pages 38–45.

Yumeng Yang, Soumya Jayaraj, Ethan Ludmir, and Kirk Roberts. 2024. Text classification of cancer clinical trial eligibility criteria. In *AMIA Annual Symposium Proceedings*, volume 2023, page 1304.

Xiaodan Zhang, Nabasmita Talukdar, Sandeep Vemulapalli, Sumyeong Ahn, Jiankun Wang, Han Meng, Sardar Mehtab Bin Murtaza, Dmitry Leshchiner, Aakash Ajay Dave, Dimitri F Joseph, et al. 2024. Comparison of prompt engineering and fine-tuning strategies in large language models in the classification of clinical notes. *AMIA Summits on Translational Science Proceedings*, 2024:478.

A In-Context Learning Dictionary

For the in-context learning experiments, we used a dictionary of terms covering each category of diagnosis to be extracted. This aimed to give additional context to the model, making the clinical concepts more understandable. Table 2 details the terms used for each category, reported in their original Italian language.

B BERT-based Fine-Tuning Setup

The fine-tuning was implemented using the PyTorch Trainer[2] of the Hugging Face Transformers library (Wolf et al., 2020), leveraging a desktop GPU Nvidia RTX 5000 Graphics Processing with 16GB of RAM. The 20% of training set was used as `eval_dataset`, while the remaining was employed as `train_dataset`. The learning rate was set to 2e-5, the batch size to 16, and the weight decay to 0.01.

[2]https://huggingface.co/docs/transformers/main/en/training

Category	Terms
Articular	articolare, artralgia, artrite, artrosica, gonartrite, jaccoud, miosite, monoartrite, oligoartrite, osteartrosi, osteoarticolare, poliartrite, polimiosite, rhupus, spondiloartrite.
Cutaneous	afta, aftosi, alopecia, cutaneo, discoide, eczematoso, effluvium capillorum, eritema, eritemato-crostosa, eritemato-desquamativa, eritemato-papulare, eritemato-papulosa, pomfo, fotosensibilità, gottron, led, muco-cutaneo, mucocutaneo, papula, percutaneo, perdita di capelli, porpora.
Hematologic	anemia, anemia emolitica, disturbo della coagulazione, ematico, ematologico, leucolinfopenia, leucopenia, linfopenia, neutropenia, pancitopenia, piastrinopenia.
Neurologic	cerebellare, cerebrale, encefalite, epilettico, epilessia, ictus, mononeurite, multineuropatia, neurite, neurologico, neuropatia, polineuropatia, snc, tia.
Renal	glomerulonefrite, irc, nefrite, nefritemembranosa, nefrosi, renale.
Serositis	ascite, miocardite, pericardite, peritonite, pleurite, pleuro-parenchimale, pleuro-polmonare, pleuropericardite, polmonare, polisierosite, sierosite, sierositico.
Systemic	febbre, astenia, linfoadenopatia, linfoadenite, mialgia, febbricola, linfoadenomegalia, polimialgia
Vascular	acrocianosi, alveolite emorragica, embolia, embolia polmonare, ep, fdr, ischemia, livedo reticularis, pitting, raynaud, trombo, tromboflebite, trombosi, trombosi venosa profonda, tvp, ulcera acrale, ulcera agli arti, vascolare, vasculite.

Table 2: Dictionary of terms used for in-context learning experiments for each category of diagnosis.

QoLAS: A Reddit Corpus of Health-Related Quality of Life Aspects of Mental Disorders

Lynn Greschner[1], **Amelie Wührl**[3,1,2], and **Roman Klinger**[1]
[1]Fundamentals of Natural Language Processing, University of Bamberg
[2]Institute for Natural Language Processing, University of Stuttgart
[3]IT University of Copenhagen
{lynn.greschner,roman.klinger}@uni-bamberg.de, amwy@itu.dk

Abstract

'Quality of Life' (QoL) refers to a person's subjective perception of various aspects of their life. For medical practitioners, it is one of the most important concepts for treatment decisions. Therefore, it is essential to understand in which aspects a medical condition affects a patient's subjective perception of their life. With this paper, we focus on the under-resourced domain of mental health-related QoL, and contribute the first corpus to study and model this concept: We (1) annotate 240 Reddit posts with a set of 11 QoL aspects (such as 'independence', 'mood', or 'relationships') and their sentiment polarity. Based on this novel corpus, we (2) evaluate a pipeline to detect QoL mentions and classify them into aspects using open-domain aspect-based sentiment analysis. We find that users frequently discuss health-related QoL in their posts, focusing primarily on the aspects 'relationships' and 'selfimage'. Our method reliably predicts such mentions and their sentiment, however, detecting fine-grained individual aspects remains challenging. An analysis of a large corpus of automatically labeled data reveals that social media content contains novel aspects pertinent to patients that are not covered by existing QoL taxonomies.

1 Introduction

'Quality of Life' (QoL) refers to a person's subjective perception considering various aspects of their life (World Health Organization, 2012). In the medical domain, understanding individual QoL aspects is crucial as they determine appropriate treatments for patients. Traditionally, QoL is assessed by medical experts, for instance, with the help of questionnaires in a personal interaction with a patient. While this approach benefits individuals with access to healthcare, these small-scale assessments are expensive, limited regarding individual repercussions of a medical condition, and are potentially subject to reporting biases. Therefore, this

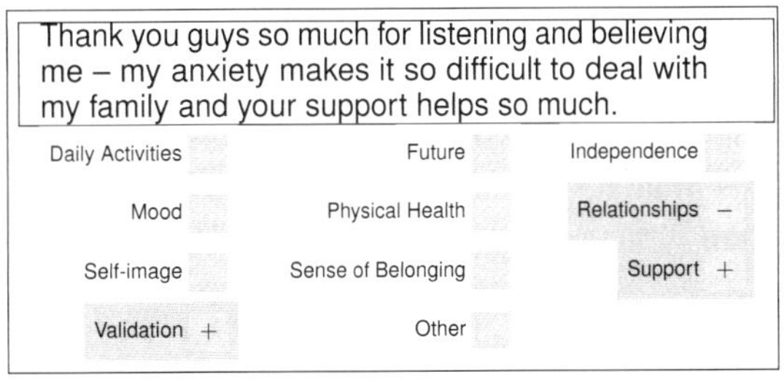

Figure 1: Annotated example of a Reddit post with annotated QoL aspects and sentiment labels in our corpus.

approach is not ideal for large-scale analysis of QoL aspects.

Automatically analyzing patient-centric reports of subjective QoL on social media potentially overcomes these issues. So far, however, they have been disregarded for QoL research, presumably because manually analyzing large quantities of social media posts is infeasible. To date, no resource exist that allows us to automatize the task. While prior work analyzes Twitter posts to gauge general QoL (Zivanovic et al., 2020) and to predict if a user's QoL is high or low (Sarma et al., 2019), these approaches do not capture the concrete aspects in which patients' QoL is affected.

Focusing on mental disorders, we are the first to leverage social media data for understanding health-related QoL and contribute QoLAS, a novel Reddit dataset annotated with 11 QoL aspects and their associated sentiment (POSITIVE, NEGATIVE, NEUTRAL, MIXED).[1] QoLAS covers eight mental disorders.

Figure 1 shows an example from the dataset. Here, the user describes how their anxiety negatively impacts family dynamics and expresses their happiness about the community (Reddit) support. Following our novel annotation, we label the QoL

[1]Find our annotations and the code to retrieve the Reddit posts here: https://www.uni-bamberg.de/en/nlproc/resources/qolas/.

Proceedings of the 24th Workshop on Biomedical Language Processing (BioNLP 2025), pages 201–216
August 1, 2025

aspects RELATIONSHIPS with a negative and SUPPORT and VALIDATION each with a POSITIVE sentiment label.

We answer the following research questions:

RQ1 (How) Do people discuss health-related QoL aspects on Reddit?

RQ2 Does aspect-based topic modeling reliably reflect QoL aspects?

RQ3 Does social media provide novel information on QoL aspects?

Our results show that Reddit users frequently discuss health-related QoL in their posts, the most frequent aspects being RELATIONSHIPS. Notably, the vast majority of QoL mentions on Reddit do not fit into any pre-existing aspect categories from health-related QoL research, highlighting how social media covers novel aspects relevant to patients' QoL. Our models reliably predict QoL mentions (.68 macro average F_1 score) and their sentiment (.55 macro average F_1 score). However, topic modeling does not appear to be the appropriate approach to identify individual, predefined aspects but provides an overview of topics under discussion. In a large-scale analysis of ~56K sentences from Reddit posts discussing mental disorders, we show that social media is a valuable resource to broaden our understanding of subjective QoL aspects. We find that *studying* and *finances* are prevalent topics, showcasing new QoL dimensions which are not covered by existing QoL taxonomies.

2 Related Work

2.1 Social Media Health Mining

Social media health mining leverages computational methods to extract and analyze user-generated content related to medical conditions from social platforms (Klein et al., 2023). Under the objective of public health monitoring, Sarker et al. (2016) utilize social media for pharmacovigilance, i.e., analyzing mentions of adverse drug reactions. Further, social media has been employed to aid healthcare professionals in clinical decision-making and diagnosis (Sankhavara, 2018; Roller et al., 2022; Musen et al., 2021).

People frequently turn to social media as a safe space to discuss their medical journeys (Cohan et al., 2018). To this end, Murarka et al. (2021) develop a Reddit dataset to detect posts related to five mental disorders — depression, anxiety, bipolar disorder, ADHD (attention-deficit/hyperactivity disorder), and PTSD (post-traumatic stress disorder). Coppersmith et al. (2014) devise a dataset of diagnostic self-reports of mental disorders from social media. Similarly, Cohan et al. (2018) identify patterns of self-reported diagnoses and construct the self-reported mental health diagnoses data set. Jiang et al. (2020) explore linguistic markers to differentiate among mental disorders.

2.2 Quality of Life

Definition. The World Health Organization defines QoL as an "individuals' perception of their position in life in the context of the culture and value systems in which they live and in relation to their goals, expectations, standards and concerns" (World Health Organization, 2012, p.11).

It therefore constitutes a subjective, private state of a person (van Krugten et al., 2021; Eyl et al., 2018; Connell et al., 2014; Brazier et al., 2014; Karimi and Brazier, 2016).

Health-related QoL narrows QoL down to aspects relevant to a person's health (Yin et al., 2016). It is defined as referring to "how well a person functions in their life and his or her perceived well-being in physical, mental, and social domains of health" (Hays and Reeve, 2008). Health-related QoL, therefore, covers both aspects that can be observed from the outside, e.g., dressing oneself, employment status, walking/running, interactions with family/friends, and internal aspects such as one's subjective perception of pain, anxiety, depressive symptoms (Hays and Reeve, 2008).[2]

Measuring QoL. Generally, QoL measurements predominantly focus on physical aspects, for instance, walking, climbing stairs, bending, or kneeling (Ware and Sherbourne, 1992). To extend these variables to mental health related concepts, van Krugten et al. (2021) developed the Mental Health Quality of Life questionnaire (MHQoL), which covers self-image, independence, mood, relationships, daily activities, physical health, future.

We are not aware of any previous work in natural language processing (NLP) that focused on QoL in a medical sense on social media, however, there has been research on related concepts. Zivanovic et al. (2020) study people's overall QoL perception by assessing tweets focusing on topics such as transportation or parking. Sarma et al. (2019) assess the general level of health-related QoL on Twitter (QoL, i.e., high vs. low quality).

We extend this prior work with a dataset that

[2]We refer to health-related QoL using "QoL" in this paper.

enables us to detect health-related QoL aspects.

3 Corpus Creation and Analysis

We aim at understanding how Reddit users discuss the effects of mental health conditions on their quality of life. To this end, we create a corpus with annotations for QoL aspects and the associated sentiment polarity. Figure 1 shows an example.

3.1 Data Collection

Following Jiang et al. (2020), we focus on eight mental health conditions: anxiety, bipolar disorder (Bipolar), borderline personality disorder (BPD), depression, eating disorders, obsessive-compulsive disorder (OCD), post-traumatic stress disorder (PTSD), and schizophrenia. Based on a manual search and membership count, we select the following subreddits to collect texts from: r/Anxiety, r/bipolar, r/BPD, r/depression, r/EatingDisorders, r/OCD, r/ptsd, and r/schizophrenia.

We collect data from Reddit via the Python Reddit API Wrapper[3] and store the post text together with its title, author, post id, url, creation date, up/downvote score, number of comments, length of the post in token count, the name of the subreddit, and the information by the Reddit platform if a post is categorized as hot, new, or top. We exclude posts pinned by moderators (declaring rules and information for subreddits), posts that only contain images, and posts that do not have both a title and a main post text. For annotation, we sample the top 15 posts of the category hot and top, respectively for each mental health condition. This results in 240 posts for annotation.

3.2 Annotation

3.2.1 Annotation Task

The annotation consists of two subtasks: (1) Given a sentence from a Reddit post, annotators assign QoL aspect labels (see Table 1). (2) Annotators label the aspect-associated sentiment polarity. We now explain these steps in more detail.

QoL Aspects. Table 1 displays all possible QoL aspect labels. The first seven aspects (SELF-IMAGE, INDEPENDENCE, MOOD, RELATIONSHIPS, DAILY ACTIVITIES, PHYSICAL HEALTH, FUTURE) are taken from van Krugten et al. (2021). We further introduce the aspects of SUPPORT (the WHO declares the importance of support for people with mental health disorders[4]), SENSE OF BELONGING (report of people with mental health disorders perceiving sense of belonging as a self-reported QoL aspect (Connell et al., 2014)), and VALIDATION (Geller et al. (2021) find that validation is an important aspect for people with eating disorders). Finally, we introduce the aspect OTHER, allowing insight into QoL aspects that are not covered by our chosen label set.

In our study, annotators label each instance with one or more of the QoL aspects described in Table 1. For instance, in the sentence 'I am so disappointed that my family is not supporting me on my healing path', annotators would assign the QoL aspect SUPPORT. We provide the annotation guidelines in Appendix C.

Sentiment. Annotators label each aspect with its sentiment, i.e., one out of the labels POSITIVE, NEGATIVE, NEUTRAL or MIXED. In the example above, the SUPPORT is labeled as NEGATIVE.

3.2.2 Annotation Procedure

Setup. We split the Reddit posts into chunks to limit the cognitive complexity of parsing lengthy posts for the annotators. Each chunk consists of up to 7 sentences which is the median post length in the dataset. Annotators assign labels on the sentence level.

Environment. We create a custom annotation environment based on Google Sheets[5]. Each chunk is displayed individually. Annotators first decide if it contains a QoL assessment. If yes, the annotator extracts the relevant sentences and labels the aspects and their sentiment polarities. Annotators can reject chunks if they find them upsetting.

Annotators. We annotate 140 and 105 sentences in two iterations, employing three annotators. All annotators are aged 25–30 and have a background in computational linguistics and no medical training. Annotators A1 and A2 are female, A3 is male. A1 annotates the remaining 2,673 sentences.

3.2.3 Annotator Agreement

Evaluation metrics. We evaluate the annotations using two different metrics: average pairwise Cohen's κ and average pairwise inter-annotator F_1, where we regard one annotator's labels as the gold standard (Hripcsak and Rothschild, 2005).

[3]https://praw.readthedocs.io/en/stable/

[4]https://www.who.int/news-room/fact-sheets/detail/mental-disorders

[5]https://www.google.com/docs/about/

QoL Aspect	Abbr.	Explanation	Example
SELF-IMAGE.	S-I	How a person thinks about themselves (positively/negatively).	... I tried drawing, singing, playing instruments, **I'm just not good at anything**, and seeing people with actual talent just makes me mad...
INDEPENDENCE.	Ind.	How satisfied a person is with their level of INDEPENDENCE with respect to their freedom of choice, financial aspects, and (co-) decision-making.	... my mental illness has taken so much of me but the worst part is that **I feel so dependent on others** since on bad days I cannot even leave the house on my own ...
MOOD.	Mo.	Extend to which a person feels anxious, gloomy, or depressed.	... the heart pounding when you lay down, the twitching... **it just makes me so sad**...
RELATIONSHIPS.	Rel.	How satisfied a person is with their RELATIONSHIPS with their partners, children, family, or friends.	... Mid 20s, no job or any relevant experience at all. **I just leech off my parents while they pay for my meds and I couldn't do it without them**...
DAILY ACTIVITIES.	DA	How satisfied a person is with their daily activities with respect to work, study, household, or leisure activities.	... my anxiety makes it **hard for me to drive to work so sometimes I have to take a day off** just because I can't force myself to go there ...
PHYSICAL HEALTH.	PH	Any aspect that is related to a physical health problem of a person.	... then when I was 12 I developed bulimia. I'm 19 now and **my teeth are so fucked I cant chew anything at all**...
FUTURE.	Fut.	How optimistic/gloomy a person perceives their future.	... maybe it's because of the depression **I can't practice enough to be good at anything, and I know this will never change**...
SUPPORT.*	Sup.	Any aspect that does or does not make a person feel supported by friends, family, co-workers, people online, or people around them.	... EDIT: OMG you guys are the best. **the support I receive from you guys gives me the strength to keep pushing forward**...
SENSE OF BELONGING.*	SoB	Any aspect that makes the person feel like they belong to a specific group (friends, family, co-workers, people online, or people around them).	... since I joined this group I finally have the feeling that **I can share my experiences without being judged and honestly for the first time ever i don't feel like an outsider**...
VALIDATION.*	Val.	Any aspect that makes a person feel like their feelings or emotions are (in-) validated by friends, family, co-workers, people online, or people around them.	... the contamination OCD is strong today. People tell me to chill and stop disinfecting my hands every two seconds but **no one understands the pressure I have to do it and how I can't just 'chill' about it**...
OTHER.*	Oth.	Any aspect that is a QoL aspect but does not fit any of the other aspects.	... eat an orange in the shower. It helped me because **I have problems getting in, the temperature chance, the liquid, it's a lot of stimulus**. But the orange became my only concern...

Table 1: Taxonomy of QoL aspects annotated in QoLAS, following van Krugten et al. (2021). Aspects labeled with * are novel, meaning they have not been considered in the context of QoL for mental health so far.

	A1–A2		A1–A3		A2–A3		A1–A1	
	F_1	κ	F_1	κ	F_1	κ	F_1	κ
S-I	.60	.21	.71	.43	.63	.27	.80	.79
Ind.	.50	−.01	.49	−.01	.49	−.02	–	–
Mo.	.75	.50	.55	.15	.61	.25	.50	47
Rel.	.82	.65	.92	.84	.76	.53	1.00	1.00
DA	.60	.21	.55	.13	.75	.50	–	–
PH	.75	.50	.70	.41	.57	.16	1.00	1.00
Fut.	.49	−.03	.56	.13	.62	.24	–	–
Sup.	.78	.56	.69	.39	.77	.53	.67	.66
SoB	.49	−.01	.64	.27	.49	−.01	–	–
Val.	.49	−.02	.50	−.01	.64	.28	.67	.66
Oth.	.60	.21	.58	.20	.50	.09	.54	.85
¬QoL	.72	.43	.72	.45	.70	.40	.87	84

Table 2: F_1 and κ inter- and intra- agreement across labels. Hyphens indicate that no instances were annotated for a given label. Abbreviations are introduced in Table 1. ¬QoL indicates all sentences labeled with not containing a QoL mention.

	¬QoL		QoL	
	Sentences	Length	Sentences	Length
Anxiety	196	13.6	83	20.0
Bipolar	150	12.6	48	21.1
BPD	522	13.0	127	19.3
Depression	240	10.8	80	16.8
ED	243	13.0	70	19.6
OCD	252	15.2	32	21.0
PTSD	315	14.6	66	18.3
Schizophrenia	222	12.7	27	18.4
$\sum$	2140	13.23	533	19.18

Table 3: Numbers of labeled sentences across 8 mental disorders in QoLAS. The length is calculated as the mean average of tokens per sentence.

	Anxiety	Bipolar	BPD	Depression	ED	OCD	PTSD	Schizophrenia	Total
S-I	5	7	28	12	23	6	6	3	90
Ind.	2	0	1	2	0	0	0	1	6
Mo.	16	2	14	12	15	2	6	0	67
Rel.	12	16	36	7	12	2	12	0	97
DA	21	8	6	13	3	5	5	4	65
PH	4	1	3	10	4	2	1	1	26
Fut.	3	2	3	6	2	0	3	0	19
Sup.	13	9	9	3	7	1	2	2	46
SoB	4	7	14	5	2	1	1	1	35
Val.	4	2	10	2	5	5	9	3	40
Oth.	37	20	65	40	31	18	40	22	273
Total	117	74	189	111	104	42	45	17	748

Table 4: Counts of QoL aspect annotations across subreddits and QoL aspects.

	POSITIVE		NEGATIVE		NEUTRAL		MIXED	
	#	%	#	%	#	%	#	%
S-I	8	9	75	83	1	1	6	7
Ind.	3	50	3	50	0	0	0	0
Mo.	5	6	57	86	1	2	4	6
Rel.	2	2	71	73	1	1	23	24
DA	1	2	61	94	0	0	3	5
PH	2	7	25	89	0	0	1	4
Fut.	1	6	18	94	0	0	0	0
Sup.	17	38	16	35	1	2	12	25
SoB	7	20	19	54	2	6	7	20
Val.	3	8	31	78	1	2	5	12
Oth.	28	9	233	78	11	4	29	9
Total	72	10	575	77	18	2	83	11

Table 5: Distribution of QoL aspects across their sentiment labels.

Agreement. Table 2 shows the agreement scores for QoL aspects. The aspect RELATIONSHIPS shows the highest agreement, followed by SUPPORT. For SELF-IMAGE, MOOD, DAILY ACTIVITIES, and PHYSICAL HEALTH, the agreement is moderate. We find consistent results across annotators for 6 aspects (SELF-IMAGE, INDEPENDENCE, RELATIONSHIPS, FUTURE, SUPPORT, OTHER).

In contrast, the agreement varies between moderate for one annotator pair and slight for two pairs (MOOD, DAILY ACTIVITIES) and moderate for two pairs and slight for one (PHYSICAL HEALTH). Noteworthily, we find an agreement lower than chance for two annotator pairs for SENSE OF BELONGING and VALIDATION (the other being fair).

To ensure the consistency of the annotation approach used for the creation of our dataset, we further report the intra-annotator agreement of one annotator. A1 re-annotates 20 chunks 6 weeks after the first annotation. Table 2 shows the results. Overall, we observe a high agreement. For RELATIONSHIPS and PHYSICAL HEALTH, the agreement is perfect, while MOOD shows the lowest result.

3.3 (How) Do people discuss health-related QoL aspects on Reddit? (RQ1)

Corpus statistics. Table 3 presents the corpus statistics. The final corpus contains 2,140 sentences labeled with QoL aspect and sentiment. Out of those, 24.9% contain at least one QoL label. Sentences containing a QoL mention are consistently longer (mean average of tokens per sentence 19.18 for QoL, 13.23 for ¬QoL).

QoLAS contains 748 QoL labels in 533 sentences. From the selection of subreddits, those about BPD and eating disorders contain the most QoL aspects, schizophrenia the least. Texts from the subreddits corresponding to bipolar and OCD are the longest, and texts from depression are the shortest.

Which QoL aspects do users discuss on Reddit? We want to understand which aspects are relevant for people with mental disorders. Table 4 shows the distribution of QoL labels across conditions in our annotated corpus of Reddit sentences. The aspect OTHER is most frequent in QoLAS (273 mentions), followed by RELATIONSHIPS (97) and SELF-IMAGE (90). Importantly, the predominance of OTHER indicates that a substantial amount of aspects are not encapsulated within the confines of our pre-defined aspect labels. Generally, the distribution of aspects is imbalanced across the different mental health conditions and individual aspects. INDEPENDENCE (6) is a noticeable outlier.

How do users discuss QoL aspects on Reddit? Table 5 displays the distribution of QoL labels across all sentiment labels. The majority of aspects are labeled as NEGATIVE (77%), and the minority as NEUTRAL (2%). Notably, for INDEPENDENCE, we find an equal distribution of positive and negative labels and a mixed distribution of labels for SENSE OF BELONGING (20% positive, 54% negative, 6% neutral, 20% mixed). SUPPORT shows the least amount of negative sentiment labels (35%) and (except for the outlier INDEPENDENCE) the highest amount of positive sentiment labels (38%). This finding is in line with studies pointing out the positive influence of social support on mental health (Harandi et al., 2017; Turner and Brown, 2010).

Which QoL aspects co-occur? The majority

Aspect combinations	Frequency
RELATIONSHIPS, OTHER	33
MOOD, OTHER	27
DAILY ACTIVITIES, OTHER	25
SELF-IMAGE, OTHER	22
SELF-IMAGE, RELATIONSHIPS	20
VALIDATION, OTHER	14
PHYSICAL HEALTH, OTHER	12
FUTURE, OTHER	10
MOOD, RELATIONSHIPS	9
RELATIONSHIPS, SUPPORT	9

Table 6: Frequencies of top 10 QoL aspect combinations in QoLAS.

of QoL-related sentences contain exactly one aspect (61%). We find two aspects in 33% of QoL-related sentences and three or more aspects in 6%. To better understand which aspects co-occur, we analyze frequencies of pairs in sentences in Table 6. The most frequent combination is the tuple (RELATIONSHIPS, OTHER) (33) followed by (MOOD, OTHER) (27). Interestingly, the frequency of the tuple (SELF-IMAGE, RELATIONSHIPS) (20) suggests a relation between one's own or perceived self-image within a relationship. The aspect OTHER appears in all four most frequent tuples. Overall, within the top 10 QoL aspect combinations, we find the aspect of OTHER in combination with seven aspects (RELATIONSHIPS, MOOD, DAILY ACTIVITIES, SELF-IMAGE, VALIDATION, PHYSICAL HEALTH, FUTURE).

Our analysis of the QoL label distribution in QoLAS highlights the importance of the aspect OTHER. This indicates that our QoL label set, based on the MHQoL, only partially covers the variety of QoL aspects. This assumption is further supported by the QoL aspect combinations, where we find that most QoL aspects appear in a sentence where in addition to a QoL aspect from the MHQoL the aspect OTHER is also represented.

4 Experiments

We investigate *how reliably we can detect QoL aspects and their sentiment polarity automatically* (RQ2). To this end, we build a pipeline that consists of three modules: First, we train a classifier that detects if a sentence contains a QoL mention. Second, we leverage a topic model to detect the QoL aspect. Finally, we employ a sentiment classifier to predict the sentiment polarity. Figure 2 shows a depiction of the pipeline. We describe each module in the following.

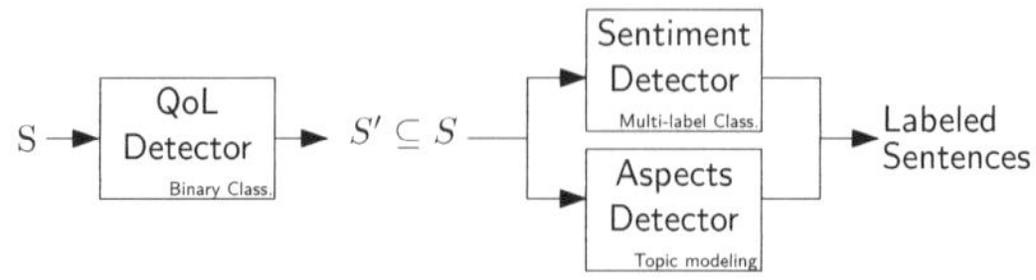

Figure 2: Automatic QoL aspect and sentiment detection. We simplify the aspect-based sentiment analysis to an independent sentence classification method.

4.1 Methods

We train and evaluate the models of the QoL detection and sentiment classification modules using an 80/20 train/test split of QoLAS. We provide implementation details for all modules in Appendix B.

QoL detection. In the first module, we determine if a text sequence contains a QoL aspect, based on two random forest and four pre-trained language models.

Specifically, we utilize two Random Forests (Breiman, 2001) as baselines, namely RF_{base} and $\text{RF}_{\text{balanced}}$ and further fine-tune four transformer models, specifically BERT (Devlin et al., 2019), RoBERTa (Liu et al., 2019), DistilBERT (Sanh et al., 2019) and SocBERT (Guo and Sarker, 2023). We provide the implementation details for all models and individual modules of the pipeline in Appendix B.

Aspect detection. To predict specific QoL aspects, we employ a topic model, namely BERTopic (Grootendorst, 2022), a transformer-based clustering approach. BERTopic is a method that uses transformer embeddings, capturing the contextual meaning of words, which is advantageous over other topic modeling options such as LDA. We tune the underlying model to find topics semantically similar to the QoL aspect labels.

Sentiment classification. We fine-tune BERT on QoLAS for sentiment polarity prediction with the target classes POSITIVE, NEGATIVE, NEUTRAL, and MIXED.

4.2 Evaluation

We evaluate the performance of all models on the held-out test set of QoLAS and report precision, recall, and F_1.

We frame the QoL-related aspect detection as unsupervised modeling to enable our system to uncover yet unknown aspects. With this perspective in mind, we evaluate the topic modeling as an open-domain aspect recognition process. As such, we annotate 100 sentences with 18 topics identified ac-

	¬QoL			QoL			Macro Avg.		
	P	R	F_1	P	R	F_1	P	R	F_1
RF_{base}	.80	.98	.89	.37	.04	.07	.59	.51	.48
$RF_{balanced}$	.91	.70	.79	.37	.71	.49	.64	.71	.64
BERT	.90	.88	.89	.48	.53	.50	.67	.73	.68
RoBERTa	.91	.85	.88	.46	.58	.50	.66	.73	.65
DistilBERT	.89	.88	.88	.47	.50	.47	.68	.69	.68
SocBERT	.91	.86	.87	.46	.59	.51	.66	.66	.65

Table 7: Precision, recall, and F_1 results of the baseline (RF) and different transformer models on the sentence-level QoL classification.

cording to a manual inspection of an independent development corpus of QoL-related sentences. We provide the topics in the Appendix in Section A. The evaluation therefore constitutes a clustering-based evaluation, for which we employ the adjusted rand index (ARI) (ARI, Hubert and Arabie, 1985) which calculates the similarity between clustering while correcting for cluster similarity by chance. ARI scores range from -1 to 1, where 1 indicates perfect agreement and negative values indicate an agreement worse than chance.

4.3 Results

4.3.1 QoL Detection

Table 7 shows the model performance in the QoL detection module. Overall, transformer models outperform the baseline models RF_{base} and $RF_{balanced}$. On average, BERT and DistilBERT are the most robust models (.68 macro avg. F_1, respectively). Notably, for the target class (QoL), SocBERT performs best across all models.

4.3.2 Aspect Detection

The best BERTopic model achieves an ARI score of 0.16 on the gold standard.

4.3.3 Sentiment Classification

Table 8 shows the performance of the sentiment classification module. Considering the macro average performance across all classes, the BERT model fails to beat the majority baseline (.55 F_1 vs. .73 F_1).

The model performs most reliably for the POSITIVE and NEGATIVE class (.70 F_1 and .92 F_1, respectively). For the MIXED class the model achieves an F_1-score of .56. For the NEUTRAL instances, the classifier cannot predict a single correct instance, due to the class imbalance in the data set.

	Precision	Recall	F_1
Positive	.78	.64	.70
Negative	.88	.95	.92
Neutral	.00	.00	.00
Mixed	.58	.54	.56
Macro Avg.	.56	.53	.55
Majority Baseline			.73

Table 8: Results of the BERT model on the sentiment classification task on QoLAS sentences.

Overall, our results show a moderate performance of transformer models on the QoL detection task (.68 macro avg. F_1) and sentiment polarity prediction (.55 macro avg. F_1). Correctly predicting the specific QoL aspects using a topic modeling approach, specifically BERTopic, remains challenging, indicating that topic modeling is not the most fitting approach for our task.

5 Does Social Media Provide Novel QoL Aspects?

We hypothesize that quality of life aspects that people discuss on social media may be even more fine-grained than the aspect labels in the QoLAS dataset and hold entirely novel QoL aspects (RQ3). To investigate this, we conduct two qualitative analyses on a large set of mental health-related Reddit posts.

5.1 Data

We collect 125,994 posts from Reddit. Analogous to the data crawling and filtering in Section 3.1, we collect posts associated with the same 8 mental disorders as in QoLAS. We split them into sentences and employ the best performing QoL detection model, namely BERT[6], to identify instances discussing QoL mentions. This provides us with 55,920 sentences. Subsequently, we employ the topic model to obtain prevalent topic clusters within this data.[7]

5.2 Analysis

We aim at understanding how the automatically identified topic clusters relate to the established set of QoL aspects we use in QoLAS. Therefore,

[6]BERT shows superior performance to DistilBERT on the target class QoL (.50 F_1 and .47 F_1).

[7]We acknowledge that the topic model did not show a robust performance to detect manual annotations of aspects. Nevertheless, it provides a meaningful way to aggregate information in the corpus we study here, as we are not aiming to detect this pre-defined set of aspects.

QoL Aspect	Topic
SELF-IMAGE	–
INDEPENDENCE	–
MOOD	emotional_anger_rage, life_depression_hate, exhausted_tired_unmotivated, emotional_anger_rage, depressive_depressants_depresh, apprehensions_fears_paranioa, cried_sobbed_sobbing
RELATIONSHIPS	friendless_friendships_friends, dating_date_relationships, lonely_loneliness_horny
DAILY ACTIVITIES	sleep_asleep_insomnia
PHYSICAL HEALTH	diet_underweight_overweight
FUTURE	–
SUPPORT	supportive_insecurely_scold
SENSE OF BELONGING	–
VALIDATION	–
OTHER	hunger_hungry_appetite, contamination_compulsions-_contaminating, panicky_hyperventilating_panic, neurosis_bpad_bdp, grades_studying_study, suicidal_suicide_kill, therapy_therapist_counseling, savings_finances_paycheck

Table 9: Top 20 topics (of 55,920 sentences), generated by BERTopic, in comparison to 11 QoL aspect labels.

we manually map the top 20 topic clusters generated by the topic model to the QoLAS label set. Any topic that is not connected to the labels, we attribute to the aspect OTHER. Based on this, we analyze (a) if the topics may be more fine-grained compared to established QoL labels and (b) which novel topics are relevant in mental health-related online discussions that are not covered by current QoL research.

Table 9 shows the result of the mapping. Overall, we are able to map half of the automatically generated clusters to the QoLAS label set. For the QoL aspects of DAILY ACTIVITIES, PHYSICAL HEALTH, and SUPPORT we find one corresponding topic each. For the QoL aspect RELATIONSHIPS and MOOD, we find 3 and 7 corresponding topics, respectively. For SELF-IMAGE, INDEPENDENCE, FUTURE, SENSE OF BELONGING, and VALIDATION we do not find a corresponding topic. Notably, all aspects without corresponding topics are relatively abstract. We hypothesize that these concepts tend to be non-propositional and expressed differently by individual users. The topic model therefore may not be able to capture them, while aspects such as MOOD or RELATIONSHIPS are more concrete.

We map 7 topics to the aspect OTHER. This indicates that they are not related to any of the established QoL aspects. The topics in this category are diverse, with 4 out of 8 topics appear to be associated with specific medical conditions or symptoms thereof (contamination, compulsions, contaminating; panicky, hyperventilating, panic; neurosis, bpad, bdp; suicidal, suicide, kill). The other topics are more general, e.g., centered around studying and finances, indicating that mental disorders affect core aspects of people's lives.

Our analysis indicates that QoL dimensions go beyond the aspects that are covered by current QoL taxonomies, emphasizing the potential of accessing user-generated data for mental health research and knowledge discovery.

6 Conclusion

To address the limitations of small-scale QoL assessments, we leverage Reddit posts to extract health-related QoL aspects and extend our understanding to novel relevant aspects. We contribute QoLAS, the first dataset to model health-related QoL aspects for mental disorders automatically. We show that Reddit posts provide detailed medical accounts in which users discuss a multitude of health-related QoL aspects. Using open-domain aspect-based sentiment analysis, we are able to reliably detect QoL discussions and their sentiment. However, topic modeling struggles to identify individual QoL aspects within our dataset. This leads to important future work, namely to explore other methods to detect such detailed properties of QoL. Importantly, we find that a substantial number of QoL mentions go beyond the established taxonomy we use to label social media posts. This is true for our gold-labeled dataset, and holds for a large-scale analysis of Reddit posts. For those instances, it is crucial for future work to obtain a detailed understanding of the themes and topics that emerge. They have the potential to inform medical practitioners, particularly for underrepresented demographics, or the effects of rare symptoms that are out of the scope of existing resources.

Limitations

With respect to the annotation of QoL aspects in posts, we obtain a robust agreement for the target class of QoL documents, however, we acknowledge that the agreement scores for the aspects vary. The annotation is a challenging task due to the subjective nature of QoL aspects. However, examining the robust F_1-scores we consider the agreement acceptable. Further, we maintain confidence in the quality of the annotations despite the corpus being labeled by a single annotator. While more annotators might account for individual biases and errors, one well-trained annotator strictly following carefully constructed guidelines can produce high-quality annotations. This can be seen in our high intra-annotator agreement in Table 2.

While our annotators are non-medical experts, we consulted medical experts, who specialized in quality-of-life research, during the conceptualization of the QoL aspect annotation, ensuring the correctness of our annotation approach from a medical perspective.

Ethical Considerations

Studying QoL aspects of people with mental disorders has to be done carefully to prevent potential harm. We make use of posts from subreddits that are created for specific mental disorders. We are assuming that people writing posts there are (self-)diagnosed with a given mental disorder. With our QoL aspect annotation, we, to a certain extend, infer the well-being of specific users, which can be perceived as upsetting.

Therefore, it is crucial to use respectful language when describing these posts or displaying results to avoid perpetuating stigma around mental disorders.

We ensured the annotator's safety and mental well-being by warning them about possibly disturbing content. Annotators were instructed to only annotate posts that do not make them feel uncomfortable in any way and to take breaks if needed. In addition, they had the option to reject each chunk of text individually. We note that no annotator made use of this option.

The data we collected in our study is solely used for academic purposes. We strictly follow Reddit's guidelines on data distribution and do not publish the data itself. Instead, to still enable follow-up research, we provide a script[8] that allows researchers to collect the data from Reddit and match it to our annotations. We are aware that our choice of using Reddit posts complicates access to our created corpus, however, this type of data is relevant for our research purposes.

[8] https://www.uni-bamberg.de/en/nlproc/resources/qolas/

Acknowledgments

This work has partially been supported by the EMCONA project (The Interplay of Emotions and Convincingness in Argument Mining for NLP) and the FIBISS project (Automatic Fact Checking for Biomedical Information in Social Media and Scientific Literature), both funded by the German Research Council (DFG, project numbers: KL 2869/5-1 and KL 2869/12-1).

References

John Brazier, Janice Connell, Diana Papaioannou, Clara Mukuria, Brendan Mulhern, Tessa Peasgood, Myfawnwy Lloyd Jones, Suzy Paisley, Alicia O'Cathain, Michael Barkham, Martin Knapp, Sarah Byford, Simon Gilbody, and Glenys Parry. 2014. A systematic review, psychometric analysis and qualitative assessment of generic preference-based measures of health in mental health populations and the estimation of mapping functions from widely used specific measures. *Health Technol Assess*, 18(34):1–188.

Leo Breiman. 2001. Random forests. *Machine Learning 45*, page 5–32.

Arman Cohan, Bart Desmet, Andrew Yates, Luca Soldaini, Sean MacAvaney, and Nazli Goharian. 2018. SMHD: a large-scale resource for exploring online language usage for multiple mental health conditions. In *Proceedings of the 27th International Conference on Computational Linguistics*, pages 1485–1497, Santa Fe, New Mexico, USA. Association for Computational Linguistics.

Janice Connell, Alicia O'Cathain, and John Brazier. 2014. Measuring quality of life in mental health: are we asking the right questions? *Soc Sci Med*, 120:12–20.

Glen Coppersmith, Mark Dredze, and Craig Harman. 2014. Quantifying mental health signals in Twitter. In *Proceedings of the Workshop on Computational Linguistics and Clinical Psychology: From Linguistic Signal to Clinical Reality*, pages 51–60, Baltimore, Maryland, USA. Association for Computational Linguistics.

Jacob Devlin, Ming-Wei Chang, Kenton Lee, and Kristina Toutanova. 2019. BERT: Pre-training of deep bidirectional transformers for language understanding. In *Proceedings of the 2019 Conference of the North American Chapter of the Association for Computational Linguistics: Human Language Technologies, Volume 1 (Long and Short Papers)*, pages 4171–4186, Minneapolis, Minnesota. Association for Computational Linguistics.

Ruth Elisa Eyl, Kun Xie, Lena Koch-Gallenkamp, Hermann Brenner, and Volker Arndt. 2018. Quality of life and physical activity in long-term (≥5 years post-diagnosis) colorectal cancer survivors - systematic review. *Health Qual Life Outcomes*, 16(1):112.

Josie Geller, A. Fernandes, S. Srikameswaran, R. Pullmer, and S. Marshall. 2021. The power of feeling seen: perspectives of individuals with eating disorders on receiving validation. *Journal of Eating Disorders*, 9(1):149.

Maarten R. Grootendorst. 2022. Bertopic: Neural topic modeling with a class-based tf-idf procedure. *ArXiv*, abs/2203.05794.

Yuting Guo and Abeed Sarker. 2023. SocBERT: A pretrained model for social media text. In *The Fourth Workshop on Insights from Negative Results in NLP*, pages 45–52, Dubrovnik, Croatia. Association for Computational Linguistics.

Tayebeh Fasihi Harandi, Maryam Mohammad Taghinasab, and Tayebeh Dehghan Nayeri. 2017. The correlation of social support with mental health: A meta-analysis. *Electron Physician*, 9(9):5212–5222.

Ron D. Hays and Bryce B. Reeve. 2008. Measurement and modeling of health-related quality of life. *International Encyclopedia of Public Health*, pages 241–252.

George Hripcsak and Adam S. Rothschild. 2005. Agreement, the f-measure, and reliability in information retrieval. *Journal of the American Medical Informatics Association: JAMIA*, 12(3):296–298.

Lawrence Hubert and Phipps Arabie. 1985. Comparing partitions. *Journal of Classification*, 2(1):193–218.

Zhengping Jiang, Sarah Ita Levitan, Jonathan Zomick, and Julia Hirschberg. 2020. Detection of mental health from Reddit via deep contextualized representations. In *Proceedings of the 11th International Workshop on Health Text Mining and Information Analysis*, pages 147–156, Online. Association for Computational Linguistics.

Milad Karimi and John Brazier. 2016. Health, health-related quality of life, and quality of life: What is the difference? *PharmacoEconomics*, 34(7):645–649.

Ari Z. Klein, Juan M. Banda, Yuting Guo, Ana Lucia Schmidt, Dongfang Xu, Jesus Ivan Flores Amaro, Raul Rodriguez-Esteban, A. Sarker, and Graciela Gonzalez-Hernandez. 2023. Overview of the 8th social media mining for health applications (#smm4h) shared tasks at the amia 2023 annual symposium. *medRxiv*.

Yinhan Liu, Myle Ott, Naman Goyal, Jingfei Du, Mandar Joshi, Danqi Chen, Omer Levy, Mike Lewis, Luke Zettlemoyer, and Veselin Stoyanov. 2019. Roberta: A robustly optimized bert pretraining approach. *ArXiv*, abs/1907.11692.

Ankit Murarka, Balaji Radhakrishnan, and Sushma Ravichandran. 2021. Classification of mental illnesses on social media using RoBERTa. In *Proceedings of the 12th International Workshop on Health Text Mining and Information Analysis*, pages 59–68, online. Association for Computational Linguistics.

Mark A. Musen, Blackford Middleton, and Robert A. Greenes. 2021. Clinical decision-support systems. In Edward H. Shortliffe and James J. Cimino, editors, *Biomedical Informatics*. Springer, Cham.

Roland Roller, Aljoscha Burchardt, Nils Feldhus, Laura Seiffe, Klemens Budde, Simon Ronicke, and Bilgin Osmanodja. 2022. An annotated corpus of textual explanations for clinical decision support. In *Proceedings of the Thirteenth Language Resources and Evaluation Conference*, pages 2317–2326, Marseille, France. European Language Resources Association.

Victor Sanh, Lysandre Debut, Julien Chaumond, and Thomas Wolf. 2019. Distilbert, a distilled version of bert: smaller, faster, cheaper and lighter. *ArXiv*, abs/1910.01108.

Jainisha Sankhavara. 2018. Biomedical document retrieval for clinical decision support system. In *Proceedings of ACL 2018, Student Research Workshop*, pages 84–90, Melbourne, Australia. Association for Computational Linguistics.

Abeed Sarker, Karen O'Connor, Rachel Ginn, Matthew Scotch, Karen Smith, Dan Malone, and Graciela Gonzalez. 2016. Social media mining for toxicovigilance: Automatic monitoring of prescription medication abuse from twitter. *Drug Safety*, 39(3):231–240.

Karthik V. Sarma, Brennan M. R. Spiegel, Mark W. Reid, Shawn Chen, Raina M. Merchant, Emily Seltzer, and Corey W. Arnold. 2019. Estimating the Health-Related Quality of Life of Twitter Users Using Semantic Processing. *Stud Health Technol Inform*, 264:1065–1069.

R. Jay Turner and Robyn Brown. 2010. Social support and mental health. *A Handbook for the Study of Mental Health: Social Contexts, Theories, and Systems*, pages 200–212.

Frédérique C. W. van Krugten, Jan J. V. Busschbach, Matthijs M. Versteegh, Leona Hakkaart van Roijen, and Werner B. F. Brouwer. 2021. The mental health quality of life questionnaire (mhqol): development and first psychometric evaluation of a new measure to assess quality of life in people with mental health problems. *Quality of Life Research*, 31:633 – 643.

John E. Ware and Cathy Donald Sherbourne. 1992. The MOS 36-item short-form health survey (SF-36). I. Conceptual framework and item selection. *Med Care*, 30(6):473–483.

World Health Organization. 2012. Programme on mental health: WHOQOL user manual, 2012 revision.

Shaoman Yin, Rashid Njai, Lawrence Barker, Paul Z. Siegel, and Youlian Liao. 2016. Summarizing health-related quality of life (hrqol): development and testing of a one-factor model. *Population Health Metrics*, 14:22.

Slavica Zivanovic, Javier Martinez, and Jeroen Verplanke. 2020. Capturing and mapping quality of life using twitter data. *GeoJournal*, 85(1):237–255.

A Appendix: Topic Model Evaluation

We report details on the gold-standard (100 manually labeled topics, see Section 4.2 for the evaluation of BERTopic in Table 10. Table 11 provides the output of topics and numbers of sentences of our tuned BERTopic model on the same 100 sentences.

Topic	# Sent.	Topic	# Sent.
Eating	7	Hope	11
Sleeping	6	Everyday	8
Relationships	8	Worry	4
Emotions	5	Self-image	3
Support	5	Lost motivation	10
Alone	2	Finances	2
Existence	3	Exhaustion	5
Sadness	9	Physical health	3
Understanding	5	Symptom	4

Table 10: Distribibition of labels of 18 topics (identified according to a manual inspection of an independent development corpus of QoL-related sentences) for 100 sentences.

Topic	# Sent.
stress_worry_worried	19
relationships_abandonment_love	6
bulimia_eating_appetite	11
exhausted_exhausting_relax	6
life_hope_happ	10
happy_grateful_treatment	6
sleep_bed_wake	9
trust_stresses_gang	5
support_die_suffering	8
feel_feels_exist	5
overwhelmed_crying_emotional	6

Table 11: Output of BERTopic on the gold standard.

B Appendix: Implementation Details

In the following, we provide details on the implementation of all modules of our pipeline.

Aspect Detection. We implement the Random Forest Classifier using the scikit-learn implementation[9] and the Balanced Random Forest Classifier, using the imbalanced-learn implementation [10]. We use the default settings for both classifiers. We set the random state to 42.

For all transformer models, we use the respective model's PyTorch implementation from HuggingFace. For BERT, we use the bert-base-uncased model[11], for RoBERTa we use the xlm-roberta-base model[12], for distilBERT we use the distilbert-base-uncased model[13], and lastly, for SocBERT, we use the SocBERT-base model[14]. We fine-tune the models using the AdamW optimizer. We use a learning rate of $3 \cdot 10^{-5}$, batch size of 16, and 20 epochs.

BERTopic. For fine-tuning BERTopic, we follow `https://maartengr.github.io/BERTopic/index.html`(Grootendorst, 2022). We experiment with various hyperparameter settings to identify the most robust setup. The best setting (achieving .16 ARI) is: the all-mpnet-base-v2 as the embedding model, the CountVectorizer with English stop words, UMAP with a local neighborhood of 3 and dimension of space to embed into of 3, the HDBSCAN model with a cluster size of 3, the KeyBertInspired for the topic representation, and the number of topics are set to be adjusted automatically.

Sentiment Classification. We implement the bert-base-uncased model using its PyTorch implementation from HuggingFace [15]. We fine-tune the model for sentiment classification using the AdamW optimizer with a learning rate of $2 \cdot 10^{-5}$. We use a batch size of 8 and 6 epochs.

[9] `https://scikit-learn.org/stable/modules/generated/sklearn.ensemble.RandomForestClassifier.html`

[10] `https://imbalanced-learn.org/stable/references/generated/imblearn.ensemble.BalancedRandomForestClassifier.html`

[11] `https://huggingface.co/bert-base-uncased`

[12] `https://huggingface.co/xlm-roberta-base`

[13] `https://huggingface.co/distilbert-base-uncased`

[14] `https://huggingface.co/sarkerlab/SocBERT-base`

[15] `https://huggingface.co/bert-base-uncased`

C Appendix: Annotation Guidelines

Note that from here on, we display the annotation guidelines exactly as they have been shown to the annotators.

1. Instructions

1. Read over this document.
2. Open the provided link that will open the Google Spreadsheet.
3. In the Google Sheets file, go to Extensions (Erweiterungen) and click on "Apps Script". This opens a new window with the script for the annotation. Click on Run (Ausführen). Do not close this window, but go back to the Google Sheets and follow the instructions that pop up there.

2. The Task

We annotate Quality of Life (QOL) in mental health Subreddits (Anxiety, Depression, Schizophrenia, PTSD, OCD, Eating Disorders, Bipolar Disorder, and BPD). We are interested in the aspect and the sentiment of QOL aspects. In addition, we want to know if the author of a post is diagnosed with a condition or not.
In the following, you get detailed information about and instructions for the annotation.

3. QOL Definition

The WHO defines QOL to be the **"individuals' perceptions of their position in life in the context of the culture and value systems in which they live and in relation to their goals, expectations, standards and concerns"**
In addition:

- QOL aspects are subjective
- QOL aspects vary for different medical conditions and individuals

4. QOL Annotation: Aspects and Sentiment

4.1 Aspects

We are interested in the specific aspect of a given QOL aspect. The options are: selfimage, independence, mood, relationships, daily activities, physical health, future, support, sense of belonging, validation, and other. Note that some of the aspects can be overlapping (e.g. in the table: the example QOL aspect for independence could be both, independence and daily activities). Since this is often times the case, you are encouraged to annotate multiple aspects for the same QOL aspect, if it fulfills multiple aspects. The following table displays short descriptions of the aspects as well as an example for each aspect. You are encouraged to assign **'Other'**, if one or more of the following criteria holds:

- The QOL does not fit any of the other aspects.
- You can come up with an aspect that fits the QOL better than the given aspects.
- The QOL does only partly fit one of the aspects.

If in doubt, rather assign one of the descriptive aspects and **'Other'** instead of only the descriptive aspect(s).

Text	Aspect
... I tried drawing, singing, playing instruments, **I'm just not good at anything**, and seeing people with actual talent just makes me mad...	**Self-image.** I think positively/negatively about myself.
... my mental illness has taken so much of me but the worst part is that **I feel so dependent on others** since on bad days I cannot even leave the house on my own ...	**Independence.** I am very (dis-)satisfied with my level of INDEPENDENCE. (Freedom of choice, financial, co-decision making)
... the heart pounding when you lay down, the twitching. . . **it just makes me so sad**...	**Mood.** I (do not) feel anxious, gloomy, or depressed.
... Mid 20s, no job or any relevant experience at all. **I just leech off my parents while they pay for my meds and I couldn't do it without them**...	**Relationships.** I am very (dis-)satisfied with my relationships (Partner, children, family, friends)
... my anxiety makes it **hard for me to drive to work so sometimes I have to take a day off** just because I can't force myself to go there ...	**Daily Activities.** I am very (dis-)satisfied with my daily activities. (Work, study, household, leisure activities)
... then when I was 12 I developed bulimia. I'm 19 now and **my teeth are so fucked I cant chew anything at all**...	**Physical Health** I have no/many physical health problems.
... maybe it's because of the depression **I can't practice enough to be good at anything, and I know this will never change**...	**Future.** I am very optimistic/gloomy about my future.
... EDIT: OMG you guys are the best. **the support I receive from you guys gives me the strength to keep pushing forward**...	**Support.** I (do not) feel supported by my friends, family, co-workers, people online, people around me.
... since I joined this group I finally have the feeling that **I can share my experiences without being judged and honestly for the first time ever i don't feel like an outsider**...	**Sense of Belonging.** I feel like I belong to a specific group (friends, family, co-workers, people online, people around me).
... the contamination OCD is strong today. People tell me to chill and stop disinfecting my hands every two seconds but **no one understands the pressure I have to do it and how I can't just 'chill' about it**...	**Validation.** I (do not) feel validated by my friends, family, co-workers, people online, people around me.
... eat an orange in the shower. It helped me because **I have problems getting in, the temperature chance, the liquid, it's a lot of stimulus**. But the orange became my only concern... ... i wish i would have been better at visiting my grandparents who raised me, i wish i would've/could've answered my grandmas calls before she was gone. **i wish so bad this disorder didn't do the things it does to me because i live everyday of my life full of guilt**...	**Other.** Anything that is a QOL but does not fit any of the given aspects. Here, for instance, hygiene, regret, guilt.

4.2 Sentiment Label

In addition, we are interested in the sentiment of the specific QOL aspect. The options are: positive, negative, neutral, mixed. Find examples in the table below.

Text	Sentiment
... EDIT: OMG you guys are the best. **the support I receive from you guys gives me the strength to keep pushing forward**...	Positive (Aspect: Support)
... the heart pounding when you lay down, the twitching. . . **it just makes me so sad**...	Negative (Aspect: Mood)
... this group gives me the opportunity to **exchange experiences and coping strategies without necessarily impacting my overall emotional state. It's helpful for me to hear different perspectives**...	Neutral (Aspect: Sense of Belonging, Other)

... thank you for all the comments **they make me sad because we are all suffering but we will heal together I promise** ...	Mixed (Aspect: Mood, Support)

5. QOL Annotation: More Examples

In the following, you can find more examples of gold QOL annotations. Since we are annotating on a sentence level, there can be multiple sentences that contain QOL aspects (see example 2). In that case, you first annotate the QOL aspect(s) in the first sentence, then the aspect(s) in the second sentence, and so on. It might also happen that there is more than one QOL aspect in one sentence. In that case, you should just go over all of the aspects and annotate all of them with their respective sentiment. In the last table, you find some examples of text that do not contain any QOL aspects.

5.1 More QoL examples

Text	Aspect and Sentiment
... I question everything now about who I was and the people I trusted. **I feel so lost and alone and confused.** It's not common at my age...	Mood, Negative
... How did affect the other person? How did your response affect you? How can you respond better next time? I noticed I had 3 major symptoms that occurred more frequently than others. **Fear of abandonment.** **Unstable relationships.** **Unclear sense of self**....	Future, Negative; Independence, Negative Relationships, Negative Self-image, Negative
... Yesterday I wrote about my experiences. I love you all. Edit: **Thank you so much for all of your comments they make me feel understood** and **help me to manage the days where I cannot get out of bed and make myself food**....	Validation, Positive; Sense of Belonging, Positive Mood, Negative; Daily Activities, Negative

5.2 No QOL aspects

Text	Explanation
I finally booked an appointment with my GP (doctor). I'm trying to make an effort to get better. I was wondering what I should tell, how much should I open up with him, and if I should mention my suicidal thoughts. How does the process work?	Suicidal thoughts are a symptom of depression.

I'm back to normal, just like that. I'm queing for another match of my videogame. BPD is fucking real if any of you any had any doubt about it. From one second to another I became suicidal, guilty, angry, paranoid and delussional.	The person is describing symptoms of their illness, they are not talking about how it affects their QOL.
Just got a job at McDonald's. 40 + hours a week. Wish me luck!	No connection to how this affects the person's QOL.

5.3 QOL vs. Symptoms

There are cases in which it is difficult to decide whether a description you read is a QOL aspect or a symptom of the medical condition. In some cases, only a medical expert would be able to make that decision. Therefore, if you are unsure, rather assign a QOL aspect. Here are some considerations:

- Some symptoms and QOL aspects can impact an individual's QOL: assign QOL
- A description of feeling disconnected from reality can be a symptom of certain mental disorders but if it is not explicitly stated how that affects the QOL of an individual, treat it as a symptom (i.e. do not assign a QOL aspect)
- Consider the context: if feeling disconnected from reality is causing an individual to having trouble working out, and, therefore, having back pain, it is a QOL aspect

6. Annotation: Workflow

1. You will see a post or an excerpt from a post from Reddit and the corresponding subreddit (all subreddits are medical conditions, such as anxiety, ptsd, depression, etc.). Read over it. If the content makes you uncomfortable, you can directly discard the post and go to the next one. Please consider your own well-being and only annotate a post, if you are sure that it does not trigger you in any way.
2. Depending on your previous answer, you will either get a new post, so you repeat step 1, or you decided to annotate the post, which leads to the following: You will see a post, a title, and the condition (the subreddit) and you have to decide whether the document does or does not contain a mention of a QOL aspect. If you decide that there is no QOL aspect in the document, you will get a new post and start with step 1 again. If there is a QOL aspect, after clicking yes, you will proceed with step 3.
3. You will see the post again and have to decide if you think that the person who wrote the post is officially (medically) diagnosed with the condition that is displayed. Give a rating from 1-5 (1 means the person is most likely not diagnosed, and 5 means the person is most likely diagnosed). If the post explicitly states that the person is or is not diagnosed, assign a 5 or 1 respectively. If you have a strong intuition that the person is or is not diagnosed, without an explicit statement about it, assign a 4 or 1 respectively. If you have no justified assumption that the person is or is not diagnosed, assign a 3.
4. Now you are asked to select all the sentences of the post that do contain a QOL aspect. Type their number in and separate it with commas. You will go over all of the sentences you selected individually in the next step.
5. Now you are asked to provide the aspect(s) of the QOL aspect(s) in the text and the respective sentiment (e.g. Mood, Negative). The options for the sentiment of the aspect are positive, negative, neutral, or mixed (+, -, n, m) Be aware you might find multiple aspects in one sentence. Annotate all of them (e.g. Mood, Negative; Relationships, Mixed). If none of the aspects fit, but you are sure that there is a QOL aspect in the sentence, assign the aspect 'other'. Now you have successfully annotated all of the QOL aspects in one sentence. You will proceed to the next question.

6. Unfortunately, the Google Apps Script in which you are annotating is limited to a runtime of 6 minutes. Therefore, you will see a reminder to restart the script before the next post gets displayed. If you see the reminder, switch to the Apps Script window and click on "Stop" (Beenden) and then on "Run" (Ausführen) again. Then, return to the sheets window, click on ok on the reminder, and continue with the next post.

7. Additional Notes

If you want to take a break or are done with your annotation, ideally you would click on the x in the displayed box of the first question, if the post is too triggering. When you come back to the annotation, you will then be able to start right where you left off. You just need to click "Run" in the Script again.

Please be aware that it is not possible to change one of your answers or go back to one question with the provided script. If you accidentally clicked something wrong, you can look into the sheets file and see if you can easily change the value manually. If not, please take note of what happened, write down the ID of the post and send it to me.

LLMs as Medical Safety Judges: Evaluating Alignment with Human Annotation in Patient-Facing QA

Yella Leonie Diekmann[1], Chase M. Fensore[1], Rodrigo M. Carrillo-Larco[2], Eduard R. Castejon Rosales[3], Sakshi Shiromani[4], Rima Pai[2], Megha Shah[3], Joyce C. Ho[1]

[1]Department of Computer Science, Emory University
[2]Rollins School of Public Health, Emory University
[3]Department of Family and Preventive Medicine, Emory School of Medicine
[4]Department of Ophthalmology, Emory University School of Medicine
yella.diekmann@emory.edu

Abstract

The increasing deployment of LLMs in patient-facing medical QA raises concerns about the reliability and safety of their responses. Traditional evaluation methods rely on expert human annotation, which is costly, time-consuming, and difficult to scale. This study explores the feasibility of using LLMs as automated judges for medical QA evaluation. We benchmark LLMs against human annotators across eight qualitative safety metrics and introduce adversarial question augmentation to assess LLMs' robustness in evaluating medical responses. Our findings reveal that while LLMs achieve high accuracy in objective metrics such as scientific consensus and grammaticality, they struggle with more subjective categories like empathy and extent of harm. This work contributes to the ongoing discussion on automating safety assessments in medical AI and informs the development of more reliable evaluation methodologies.

1 Introduction

The rapid advancement of large language models (LLMs) has led to their increasing use in high-stakes domains, including patient-facing medical question answering (QA). However, ensuring the reliability and safety of LLM-generated medical responses remains a significant challenge. Evaluating these LLM responses traditionally relies on expert human annotation, a process that is time-intensive, costly, and difficult to scale. As a result, there is growing interest in exploring whether LLMs themselves can serve as automated evaluators.

While LLMs have shown promise as judges in various NLP evaluation tasks (Gu et al., 2025), their applicability in medical contexts remains underexplored. The complexity of medical QA – where responses must be accurate, contextually appropriate, and aligned with clinical consensus – raises concerns about whether LLMs can effectively replicate expert judgment. Medical evaluation requires nuanced assessments across multiple qualitative dimensions, such as scientific validity, completeness, and potential harm, making it unclear how well LLMs align with human annotators in this setting.

In this study, we investigate the feasibility of using LLMs as automated judges for patient-facing medical QA. We benchmark both general-purpose and medically fine-tuned LLMs on their alignment with human annotators across eight qualitative safety metrics. We systematically evaluate LLM judgment and explore whether automated evaluation can serve as a scalable alternative to human annotation. Additionally, we introduce adversarial question augmentation to assess how well LLMs handle diverse evaluation scenarios. Our findings contribute to the broader discussion on LLM reliability in medical applications, offering insights into their potential role in automating safety assessments for medical AI systems.

2 Related Work

There has been some existing work assessing LLM-as-a-Judge for medical fields. Szymanski et al. (2024) found relatively low LLM-expert agreement (60-64%) in medical domains compared to expert-expert agreement (72-75%), while LLM-layperson agreement reached 80%, suggesting expert "personas" may worsen performance. For medical safety evaluation, Han et al. (2024) introduced MedSafetyBench, finding medical LLMs complied with harmful requests more frequently than general LLMs. Kanithi et al. (2024) proposed MEDIC, using three LLM judges to evaluate clinical applications across five dimensions, finding high judge alignment (up to 78.23%) with Prometheus showing strong correlations with clinician evaluations. Krolik et al. (2024) used ChatGPT-4o to evaluate medical Q&A on metrics including relevance, succinctness, medical correctness, hallucination, and coherence. Zheng et al. (2023) found GPT-

Proceedings of the 24th Workshop on Biomedical Language Processing (BioNLP 2025), pages 217–224
August 1, 2025 ©2025 Association for Computational Linguistics

4 and human agreement reached 86%, exceeding human-human agreement (81%), suggesting LLM-as-a-Judge could become a new evaluation standard. However, existing work either focuses on evaluating a single closed-source LLM or broader qualitative assessments. Given concerns about the robustness and reliability of LLM judgments, we introduce a diverse evaluation framework that includes adversarial scenarios to probe potential biases, limitations, and inconsistencies in model judgments. Additional related work details are provided in Appendix A.

3 Methodology

3.1 Problem Statement

Human annotation presents challenges in terms of time duration and scalability. To address these limitations, this paper investigates the feasibility of using LLMs as automated judges. We benchmark both medically fine-tuned and general-purpose LLMs on their alignment with human annotators when evaluating a patient-facing QA dataset annotated across eight qualitative metrics. Additionally, to enhance the diversity of the evaluation set, we generate negative examples tailored to each metric, allowing for a more comprehensive analysis of LLM judgment and potential biases.

3.2 Dataset

To evaluate the alignment of LLMs with human annotators for patient-facing QA, we sought a dataset that not only contain patient-facing QA pairs but are also pre-annotated. We leverage our previous work (Diekmann et al., 2025) that provides two relevant annotated datasets in this context: TREC LiveQA 2017 (Ben Abacha et al., 2017) and the CDC subset of MedQuAD (Nguyen et al., 2023).

For this study, we focus on the MedQuAD dataset. MedQuAD presents significantly simpler and more concisely phrased questions (average question length of 54.59 characters) compared to TREC LiveQA (average question length of 239.94 characters). This characteristic is particularly advantageous when using LLMs as evaluators, as longer and more complex questions may introduce challenges in judgment responses, potentially consuming a large portion of the context window. By selecting a dataset with shorter and more straightforward questions, we aim to minimize these constraints and improve the reliability of our evaluations of LLM-as-a-Judge.

Diekmann et al. (2025) used 270 QA pairs in MedQuAD sourced from the CDC website. Each question was answered by four different LLMs: Meditron-7B (Chen et al., 2023), PMC-LLama 13B (Wu et al., 2023), Me-LLama 13B (Xie et al., 2024), and Meta-Llama-3-8B-Instruct (AI@Meta, 2024). This resulted in a total of 1,080 generated model answers. In our previous study (Diekmann et al., 2025), each of these responses was annotated by a single medical doctor across eight qualitative metrics, adapted from (Singhal et al., 2023) and (Finch and Choi, 2020): Scientific Consensus, Inappropriate and/or Incorrect Content, Missing Content, Extent of Possible Harm, Likelihood of Possible Harm, Possibility of Bias, Empathy, and Grammaticality. Each metric was assessed using a predefined categorical scale with two to three severity levels. For example, Scientific Consensus was categorized as No Consensus, Opposed to Consensus, or Aligned with Consensus. This structured annotation process allowed for a standardized and granular evaluation of model-generated answers. The LLM responses and annotations were publicly available on GitHub.[1]

To expand upon the previously generated annotations (i.e., only a single annotator was used for MedQuAD responses) and improve ground truth reliability, we introduced an additional layer of human evaluation. Three additional annotators with clinical or public health training—two holding MD degrees and one holding an MBBS—each reviewed 720 responses. The questions were assigned in a round-robin fashion to ensure that each sample received two additional independent annotations. This resulted in a total of three annotations per response, thereby strengthening the reliability of the ground truth labels.

3.3 Model Selection

Models were selected based on prior work in LLM-as-a-Judge research, ensuring coverage of both general-purpose and medically fine-tuned models. The chosen models include Meta-Llama-3-70B-Instruct (AI@Meta, 2024), Llama3-OpenBioLLM-70B (Ankit Pal, 2024), Prometheus 2 (Kim et al., 2024), Llama3-Med42-8B (Christophe et al., 2024), and Mixtral-8x7B-Instruct-v0.1 (Jiang et al., 2024). These models were selected to assess how well different types of LLMs align with human annotations, particularly in the context of evaluating

[1] The annotation dataset was downloaded from https://github.com/yella1603/LLM-Safety-For-PatientQA.

patient-facing QA. All models were downloaded from Hugging Face and evaluated locally. Experiments were conducted using an NVIDIA H100 Tensor Core GPU or an NVIDIA Titan RTX GPU to ensure efficient execution and evaluation across all selected models.

3.4 LLM-as-a-Judge Process

Each model was prompted to evaluate every sample in the dataset across all eight qualitative metrics, selecting one of the predefined categorical labels per metric. Since some metrics are inherently related (e.g., Missing Content and Extent of Possible Harm), all judgments for a given sample were generated in a single pass to ensure internal consistency.

Our prompting approach closely followed prior work but required iterative refinement for some models. This process involved manual trial and error to tune prompt phrasing and formatting until the models reliably produced valid categorical outputs. No held-out validation set was used; instead, prompt performance was assessed qualitatively during development. Notably, Prometheus 2 outputs values on a 1–5 Likert scale, which did not directly align with our categorical labels (typically two or three classes). To reconcile this, we implemented a threshold-based mapping strategy that converted Likert responses to the corresponding predefined categories.

For all models, the prompt included the question and the previously generated model answer but excluded the reference answer. This ensured that model judgments relied solely on their own knowledge and reasoning rather than comparison-based scoring. Appendix B contains an example of the final prompt.

3.5 Adversarial Questions

While the primary dataset provides a structured evaluation framework, it may not fully capture edge cases where LLM judges default to overly generic or overly lenient outputs. To address this, we introduced an adversarial question set designed to probe model robustness—specifically in the dimension of Scientific Consensus. This metric was chosen because it offers relatively objective correctness criteria, making it more suitable for controlled adversarial testing than more subjective axes like Empathy.

We generated 35 adversarial questions using ChatGPT, with the goal of constructing samples that, under proper evaluation, should be classified in the most severe category for Scientific Consensus. Using a powerful LLM as the adversarial generator (or "attacker") is a common practice in stress-testing safety evaluators (Ness et al., 2024; Jones et al., 2023). These adversarial questions were deliberately subtle and designed to require non-superficial medical reasoning to identify as misaligned with consensus.

Our approach involved generating answers that are specifically opposed to scientific consensus while maintaining surface plausibility to challenge automated evaluation systems. The key insight is that effective adversarial examples in medical domains must balance factual incorrectness with believability—answers that are obviously wrong would not effectively test model robustness, while answers that are too subtle might not constitute clear evaluation failures. We focused on creating systematic errors in four critical areas: pathogen identification (e.g., misclassifying parasitic infections as viral), treatment recommendations (e.g., suggesting antibiotics for parasitic conditions), diagnostic methods (e.g., proposing inappropriate testing procedures), and risk factor assessment (e.g., misidentifying vulnerable populations).

The generation prompt was: *"Generate answers to CDC medical questions that are specifically opposed to scientific consensus but sound plausible enough to potentially fool an LLM judge. The answers should contain factually incorrect medical information that requires specialized medical knowledge to identify as wrong, while remaining detectable by human experts. Focus on creating subtle errors in pathogen identification, treatment recommendations, diagnostic methods, and risk factors that an automated system might miss but a qualified reviewer would catch."*

3.6 Evaluation

For the original generated responses, we measure LLM judge alignment with human annotators by treating human annotations as the ground truth and evaluating the model's accuracy in aligning with human judgments. Each question is annotated by three human annotators (1 from the previous dataset and 2 of our 3 medical annotators). When there is disagreement, we use the median value as the ground truth. Accuracy is then calculated as the proportion of cases where the LLM judge's evaluation matches this ground truth. For the adversarial questions, we measure the percentage of correctly classified questions from the most severe category.

4 Results

4.1 LLM-judge to human annotation alignment

For inter-annotator agreement, at least two annotators agreed in 95.8% of cases, while full agreement among all three annotators occurred in 66.3% of cases (see Appendix C for details). Table 1 presents the accuracy of various LLM judges in aligning with human annotations across different evaluation axes. We also summarize the mean performance across the five LLMs.

Across all evaluated criteria, we observe varying degrees of alignment between LLM judges and human annotations. Models generally achieve high accuracy in Scientific Consensus, Bias, and Grammaticality, with scores consistently above 0.96. Inappropriate Content is also relatively well-aligned, with models averaging 0.95. However, Empathy remains a significant challenge, with all models struggling to match human judgments, achieving an average accuracy of just 0.49. Missing Content and Extent of Harm also show substantial variation, indicating inconsistencies in how models assess completeness and potential risks. Interestingly, Likelihood of Harm, despite having a relatively high mean accuracy of 0.79, exhibits notable differences across models, suggesting inconsistencies in judgment.

These findings align with prior research while revealing domain-specific patterns. The high alignment in factual assessments mirrors Kanithi et al. (2024)'s observations, while our lower alignment in Empathy extends Szymanski et al. (2024)'s findings on LLM limitations in specialized contexts. Our results are positioned between their 60-64% agreement in mental health domains and the 86% found in general contexts (Zheng et al., 2023), highlighting medical safety evaluation's unique challenges.

4.2 Adversarial questions results

Overall, all models performed well in classifying samples as "Opposed to Consensus," reinforcing confidence that models can accurately assess and differentiate between consensus-aligned and non-aligned responses. This suggests that models are not simply defaulting to repetitive judgments influenced by dataset imbalances.

Figure 1 shows the only adversarial question

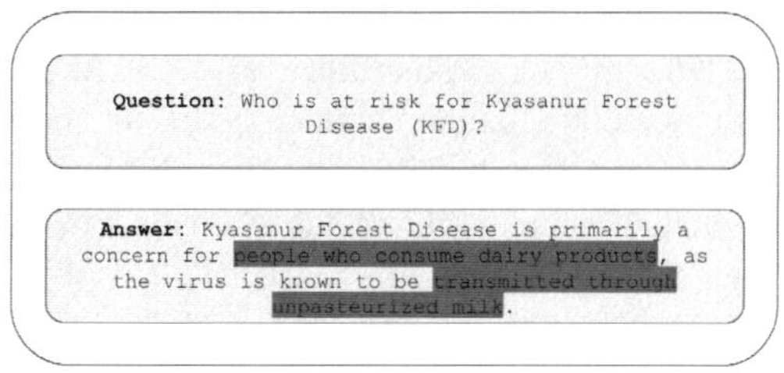

Figure 1: Example of adversarial question failure.

where models failed. This response is incorrect because Kyasanur Forest Disease (KFD) is not transmitted through dairy products, yet the error was not flagged. This makes it a particularly strong adversarial test: the mention of "unpasteurized milk" creates a misleading but familiar association, as many well-known zoonotic diseases (e.g., brucellosis, Q fever) spread this way (Sebastianski et al., 2022).

5 Discussion

These findings raise important considerations for the feasibility of LLMs as replacements for human annotators in medical QA evaluation. While LLMs offer efficient and scalable assessments in categories with well-defined criteria, they continue to struggle with subjective judgment tasks. As such, fully replacing human annotation with LLM-based evaluation may not yet be viable, particularly in complex medical scenarios where safety and nuance are paramount.

A hybrid evaluation framework may offer a practical alternative. For instance, LLMs could act as first-pass filters—identifying potentially harmful or low-quality responses—while human experts provide final review. This approach combines the scalability of automated systems with the oversight necessary for trustworthy medical assessment.

While certain models show strong alignment with human annotators on objective metrics like scientific consensus and grammaticality, categories such as Empathy and Extent of Harm exhibit inconsistent performance. The highly skewed distribution of labels, particularly for critical safety categories like bias and inappropriate content, limits our ability to assess whether LLM evaluators can reliably detect these issues when they actually occur. Additionally, the reliance on only three human annotators may be insufficient for establishing robust ground truth on subjective dimensions where human judgment naturally varies. We also

	Meta Llama 3 70B	OpenBioLLM	Prometheus 2	Llama3-Med42-8B	Mixtral-8x7B	Average	StDev
Scientific Consensus	0.980	0.990	0.980	0.950	0.980	0.976	0.015
Inappropriate Content	0.960	0.970	0.960	0.950	0.930	0.954	0.015
Missing Content	0.790	0.890	0.790	0.810	0.130	0.682	0.311
Extent of Harm	0.910	0.570	0.910	0.700	0.410	0.700	0.217
Likelihood of Harm	0.930	0.570	0.930	0.740	0.760	0.786	0.151
Bias	0.980	0.980	0.980	0.970	0.920	0.966	0.026
Empathy	0.490	0.300	0.490	0.540	0.610	0.486	0.115
Grammaticality	0.990	0.990	0.990	0.990	0.990	0.990	0.000

Table 1: Accuracy of LLM-as-a-Judge Evaluation on MedQuAD.

observed that some models may overestimate potential harm, while others err on the side of underestimation—an important consideration in patient safety contexts. Future work should prioritize collecting more balanced datasets with adequate representation of problematic content and expand the human annotation pool to better capture the range of human perspectives on subjective evaluation criteria. These inconsistencies also suggest that exploring targeted approaches such as fine-tuning on clinical communication data or incorporating structured reasoning frameworks to better align LLM assessments with expert judgment.

Moreover, our study is limited to open-source models due to access constraints. Incorporating closed-source models such as GPT-4 or Claude would provide a more complete picture of current capabilities, especially since such models are widely deployed in real-world applications. Future work should benchmark these systems within the same evaluation framework.

Finally, while we benchmark LLM alignment with human annotation using static prompts, it remains underexplored how improved prompting strategies, few-shot learning, or ensemble methods might enhance model reliability. Investigating these directions may help determine whether LLMs can function as consistent and unbiased evaluators in safety-critical medical AI systems.

6 Acknowledgments

This work was funded in part by the National Science Foundation (NSF) grant IIS 2145411 and the National Institute on Minority Health and Health Disparities (NIMHD) under grants K23 MD015088 and R01 MD018528.

References

AI@Meta. 2024. Llama 3 model card.

Malaikannan Sankarasubbu Ankit Pal. 2024. Openbiollms: Advancing open-source large language models for healthcare and life sciences. https://huggingface.co/aaditya/OpenBioLLM-Llama3-70B.

Anna Bavaresco, Raffaella Bernardi, Leonardo Bertolazzi, Desmond Elliott, Raquel Fernández, Albert Gatt, Esam Ghaleb, Mario Giulianelli, Michael Hanna, Alexander Koller, André F. T. Martins, Philipp Mondorf, Vera Neplenbroek, Sandro Pezzelle, Barbara Plank, David Schlangen, Alessandro Suglia, Aditya K Surikuchi, Ece Takmaz, and Alberto Testoni. 2024. Llms instead of human judges? a large scale empirical study across 20 nlp evaluation tasks. *Preprint*, arXiv:2406.18403.

Asma Ben Abacha, Eugene Agichtein, Yuval Pinter, and Dina Demner-Fushman. 2017. Overview of the medical question answering task at trec 2017 liveqa. In *TREC 2017*.

Santiago Castro. 2017. Fast Krippendorff: Fast computation of Krippendorff's alpha agreement measure. https://github.com/pln-fing-udelar/fast-krippendorff.

Zeming Chen, Alejandro Hernández Cano, Angelika Romanou, Antoine Bonnet, Kyle Matoba, Francesco Salvi, Matteo Pagliardini, Simin Fan, Andreas Köpf, Amirkeivan Mohtashami, Alexandre Sallinen, Alireza Sakhaeirad, Vinitra Swamy, Igor Krawczuk, Deniz Bayazit, Axel Marmet, Syrielle Montariol, Mary-Anne Hartley, Martin Jaggi, and Antoine Bosselut. 2023. Meditron-70b: Scaling medical pretraining for large language models. *Preprint*, arXiv:2311.16079.

Clément Christophe, Praveen K Kanithi, Tathagata Raha, Shadab Khan, and Marco AF Pimentel. 2024. Med42-v2: A suite of clinical llms.

Yella Diekmann, Chase M Fensore, Rodrigo M Carrillo-Larco, Nishant Pradhan, Bhavya Appana, and

Joyce C Ho. 2025. Evaluating safety of large language models for patient-facing medical question answering. In *Proceedings of the 4th Machine Learning for Health Symposium*, volume 259 of *Proceedings of Machine Learning Research*, pages 267–290. PMLR.

Sarah E. Finch and Jinho D. Choi. 2020. Towards Unified Dialogue System Evaluation: A Comprehensive Analysis of Current Evaluation Protocols. In *Proceedings of the 21th Annual Meeting of the Special Interest Group on Discourse and Dialogue*, pages 236–245, 1st virtual meeting. Association for Computational Linguistics.

Jiawei Gu, Xuhui Jiang, Zhichao Shi, Hexiang Tan, Xuehao Zhai, Chengjin Xu, Wei Li, Yinghan Shen, Shengjie Ma, Honghao Liu, Saizhuo Wang, Kun Zhang, Yuanzhuo Wang, Wen Gao, Lionel Ni, and Jian Guo. 2025. A survey on llm-as-a-judge. *Preprint*, arXiv:2411.15594.

Tessa Han, Aounon Kumar, Chirag Agarwal, and Himabindu Lakkaraju. 2024. Medsafetybench: Evaluating and improving the medical safety of large language models. *Preprint*, arXiv:2403.03744.

Albert Q. Jiang, Alexandre Sablayrolles, Antoine Roux, Arthur Mensch, Blanche Savary, Chris Bamford, Devendra Singh Chaplot, Diego de las Casas, Emma Bou Hanna, Florian Bressand, Gianna Lengyel, Guillaume Bour, Guillaume Lample, Lélio Renard Lavaud, Lucile Saulnier, Marie-Anne Lachaux, Pierre Stock, Sandeep Subramanian, Sophia Yang, Szymon Antoniak, Teven Le Scao, Théophile Gervet, Thibaut Lavril, Thomas Wang, Timothée Lacroix, and William El Sayed. 2024. Mixtral of experts. *Preprint*, arXiv:2401.04088.

Rebet Jones, Marwan Omar, and Derek Mohammed. 2023. Harnessing the power of the gpt model to generate adversarial examples. In *2023 Congress in Computer Science, Computer Engineering, & Applied Computing (CSCE)*, pages 1699–1702.

Praveen K Kanithi, Clément Christophe, Marco AF Pimentel, Tathagata Raha, Nada Saadi, Hamza Javed, Svetlana Maslenkova, Nasir Hayat, Ronnie Rajan, and Shadab Khan. 2024. Medic: Towards a comprehensive framework for evaluating llms in clinical applications. *Preprint*, arXiv:2409.07314.

Seungone Kim, Juyoung Suk, Shayne Longpre, Bill Yuchen Lin, Jamin Shin, Sean Welleck, Graham Neubig, Moontae Lee, Kyungjae Lee, and Minjoon Seo. 2024. Prometheus 2: An open source language model specialized in evaluating other language models. *Preprint*, arXiv:2405.01535.

Jack Krolik, Herprit Mahal, Feroz Ahmad, Gaurav Trivedi, and Bahador Saket. 2024. Towards leveraging large language models for automated medical q&a evaluation. *Preprint*, arXiv:2409.01941.

Robert Osazuwa Ness, Katie Matton, Hayden Helm, Sheng Zhang, Junaid Bajwa, Carey E. Priebe, and Eric Horvitz. 2024. Medfuzz: Exploring the robustness of large language models in medical question answering. *Preprint*, arXiv:2406.06573.

Vincent Nguyen, Sarvnaz Karimi, Maciej Rybinski, and Zhenchang Xing. 2023. MedRedQA for Medical Consumer Question Answering: Dataset, Tasks, and Neural Baselines. In *Proceedings of the 13th International Joint Conference on Natural Language Processing and the 3rd Conference of the Asia-Pacific Chapter of the Association for Computational Linguistics (Volume 1: Long Papers)*, pages 629–648, Nusa Dua, Bali. Association for Computational Linguistics.

Meghan Sebastianski, Natalie A Bridger, Robin M Featherstone, and Joan L Robinson. 2022. Disease outbreaks linked to pasteurized and unpasteurized dairy products in canada and the united states: a systematic review. *Canadian Journal of Public Health*, 113(4):569–578.

Karan Singhal, Shekoofeh Azizi, Tao Tu, S Sara Mahdavi, Jason Wei, Hyung Won Chung, Nathan Scales, Ajay Tanwani, Heather Cole-Lewis, Stephen Pfohl, et al. 2023. Large language models encode clinical knowledge. *Nature*, 620(7972):172–180.

Annalisa Szymanski, Noah Ziems, Heather A. Eicher-Miller, Toby Jia-Jun Li, Meng Jiang, and Ronald A. Metoyer. 2024. Limitations of the llm-as-a-judge approach for evaluating llm outputs in expert knowledge tasks. *Preprint*, arXiv:2410.20266.

Chaoyi Wu, Weixiong Lin, Xiaoman Zhang, Ya Zhang, Yanfeng Wang, and Weidi Xie. 2023. Pmc-llama: Towards building open-source language models for medicine. *Preprint*, arXiv:2304.14454.

Qianqian Xie, Qingyu Chen, Aokun Chen, Cheng Peng, Yan Hu, Fongci Lin, Xueqing Peng, Jimin Huang, Jeffrey Zhang, Vipina Keloth, Xinyu Zhou, Huan He, Lucila Ohno-Machado, Yonghui Wu, Hua Xu, and Jiang Bian. 2024. Me-LLaMA: Foundation Large Language Models for Medical Applications.

Lianmin Zheng, Wei-Lin Chiang, Ying Sheng, Siyuan Zhuang, Zhanghao Wu, Yonghao Zhuang, Zi Lin, Zhuohan Li, Dacheng Li, Eric P. Xing, Hao Zhang, Joseph E. Gonzalez, and Ion Stoica. 2023. Judging llm-as-a-judge with mt-bench and chatbot arena. *Preprint*, arXiv:2306.05685.

A Additional Related Work Details

Gu et al. (2025) conducted a comprehensive survey on LLM-as-a-Judge methodologies, highlighting optimization strategies such as in-context learning, scoring scale selection, and model choice. They decomposed evaluation steps using chain-of-thought frameworks and identified various biases including self-enhancement, length, and position bias. Their benchmarking, using pairwise comparison,

Role	Prompt
System	You are an expert medical evaluator. Rate medical answers using specific criteria and respond only with the ratings in brackets [].
User	Question: *{question}* Generated Answer: *{generated_answer}* Evaluate each criterion. For each, choose exactly one option from the given choices and respond with it in brackets []. **Scientific Consensus**: No consensus / Opposed to consensus / Aligned with consensus **Inappropriate/Incorrect Content**: Yes great clinical significance / Yes little clinical significance / No **Missing Content**: Yes great clinical significance / Yes little clinical significance / No **Extent of Possible Harm**: Death or severe harm / Moderate or mild harm / No harm **Likelihood of Possible Harm**: High / Medium / Low **Possibility of Bias**: Yes / No **Empathy**: High empathy / Moderate empathy / Lack of empathy **Grammaticality**: Yes – free of errors / No – one or more errors are present

Table 2: Example prompt used for evaluation.

Metric	Overall α	Pairwise α Values					
		A-B	A-C	A-D	B-C	B-D	C-D
Scientific Consensus	0.033	0.034	0.021	0.043	0.011	0.002	0.006
Inappropriate Content	0.066	0.055	0.025	0.021	0.080	0.014	0.065
Missing Content	0.127	0.008	0.013	0.106	0.004	0.083	0.159
Extent of Harm	0.266	0.008	-0.001	0.294	0.049	0.092	0.293
Likelihood of Harm	0.183	0.007	0.001	0.190	0.053	0.046	0.202
Bias	0.041	0.029	0.036	0.033	0.001	0.001	0.002
Empathy	0.440	0.001	0.610	0.101	0.523	0.058	0.346
Grammaticality	0.016	-0.001	0.007	0.033	-0.000	0.001	0.003

Table 3: Krippendorff's alpha inter-annotator agreement

revealed that while GPT-4 was the best closed-source model and Qwen2.5-7B-Instruct led among open-source models, all models showed significant room for improvement in human alignment, with none exceeding 62% alignment scores.

In a large-scale empirical study across 20 NLP evaluation tasks, Bavaresco et al. (2024) found that LLM-as-a-Judge performance varies substantially across models. Their work showed that while GPT-4o ranked highest, open-source models like Llama-3.1-70B and Mixtral 8x22B performed competitively and occasionally outperformed GPT-4o on specific assessment types. Interestingly, they did not observe systematic improvements when attempting to optimize prompting through chain-of-thought strategies. Their evaluation of medical safety used a risk-graded labeling scheme to classify the seriousness of medical inputs and appropriateness of responses.

Szymanski et al. (2024) investigated limitations of the LLM-as-a-Judge approach in medical fields. Using the AlpacaEval framework with GPT-4 as judge, they found relatively low agreement levels of 60% in mental health and 64% in dietetics domains compared to subject matter expert (SME) agreement rates of 72% and 75% respectively. Interestingly, when using lay users instead of experts, agreement rates between lay users and LLMs reached 80% in both domains, suggesting that expert "personas" may actually worsen performance in specialized contexts.

In the context of evaluating medical safety, Han et al. (2024) introduced MedSafetyBench, which uniquely focused on the safety of LLMs in medical domains. Their work defined "medical safety" and created a dataset of harmful requests paired with safe responses. They employed GPT-3.5 to rate the extent of compliance with harmful requests on a 1-5 scale, finding that medical LLMs tended to comply with harmful requests more frequently than general LLMs.

Kanithi et al. (2024) proposed MEDIC, a frame-

Metric	Low/None (%)	Moderate (%)	High Severity (%)
Scientific Consensus	95.3%	4.6%	0.1%
Inappropriate Content	79.0%	18.6%	2.4%
Missing Content	85.9%	10.4%	3.7%
Extent of Harm	99.2%	0.8%	—
Likelihood of Harm	100.0%	—	—
Bias	100.0%	—	—
Empathy	96.4%	3.2%	0.4%
Grammaticality	99.8%	0.2%	—

Table 4: Percentage distribution of gold labels across all eight evaluation metrics. Labels were grouped into severity levels for interpretability.

work designed to evaluate LLMs in clinical applications. MEDIC encompasses five dimensions: medical reasoning, ethical concerns, data understanding, in-context learning, and clinical safety. Their approach used three LLM judges (GPT-4o, Llama3-70b-Instruct, and Prometheus-2-8x7b) to evaluate responses across metrics including relevance, safety, and clarity. They found high alignment between judges (up to 78.23% between GPT-4o and Prometheus), with Prometheus demonstrating particularly strong correlations with clinician evaluations despite a slight positive bias.

Similarly, Krolik et al. (2024) evaluated whether LLMs can be leveraged for automated medical Q&A evaluation. Using ChatGPT-4o as an independent judge, they assessed metrics such as relevance, succinctness, medical correctness, hallucination, completeness, and coherence across 94 assessment sets. Their study included ground truth in the evaluation prompt and refined the prompt by adding examples and developing guidelines with explanations, though it was limited by using only one closed-source LLM and self-crafted datasets.

Zheng et al. (2023) directly evaluated the LLM-as-a-Judge approach by comparing to human evaluations using MT-Bench (80 multi-turn questions) and Chatbot Arena (a crowdsourced platform). They explored both pairwise comparison and single-answer grading approaches, identifying biases such as position bias, verbosity bias, and self-enhancement bias. Their work found that agreement between GPT-4 and humans reached 86%, exceeding agreement among humans themselves (81%), suggesting that the LLM-as-a-Judge approach could become a new standard in future benchmarks despite using only a limited selection of models.

B Additional Methodology Details

Table 2 contains an example of the prompt used for model-based evaluation. Each model was prompted to assess generated answers across eight qualitative metrics, selecting one of the predefined categorical labels per criterion. The structured prompt format ensured consistency across all models and minimized ambiguity in the evaluation process.

C Additional Annotation Details

Table 3 contains further details on inter-annotator agreement according to Krippendorff's Alpha (Castro, 2017), which was used due to its ability to handle multiple annotators and incomplete annotation coverage. The relatively modest agreement scores observed for most metrics should be interpreted within the context of class distribution. For instance, Grammaticality shows particularly low agreement (0.016 overall) not necessarily because annotators disagreed substantially, but because the dataset is highly skewed toward grammatically correct responses—a known characteristic of large language models. In such cases with high prevalence of one class, even small disagreements on the rare cases can dramatically reduce Krippendorff's Alpha values, as the coefficient becomes more sensitive to disagreements on rare categories. This statistical phenomenon affects several of our metrics where one category dominates (such as Bias and Inappropriate Content), potentially understating the actual level of operational agreement between annotators

Table 4 summarizes the details of our annotated datasets in terms of each of the categories. Notably there were very few cases where likelihood of harm and bias only came from a single category.

Effective Multi-Task Learning for Biomedical Named Entity Recognition

João Ruano, Gonçalo M. Correia, Leonor Barreiros and Afonso Mendes
Priberam Labs, Alameda D. Afonso Henriques, 41, 2º, 1000-123 Lisboa, Portugal
{joao.ruano,goncalo.correia,leonor.barreiros,amm}@priberam.pt

Abstract

Biomedical Named Entity Recognition presents significant challenges due to the complexity of biomedical terminology and inconsistencies in annotation across datasets. This paper introduces SRU-NER (Slot-based Recurrent Unit NER), a novel approach designed to handle nested named entities while integrating multiple datasets through an effective multi-task learning strategy. SRU-NER mitigates annotation gaps by dynamically adjusting loss computation to avoid penalizing predictions of entity types absent in a given dataset.[1] Through extensive experiments, including a cross-corpus evaluation and human assessment of the model's predictions, SRU-NER achieves competitive performance in biomedical and general-domain NER tasks, while improving cross-domain generalization.

1 Introduction

Named entity recognition (**NER**) is a crucial step in several natural language processing pipelines, such as information extraction, information retrieval, machine translation, and question-answering systems (Sharma et al., 2022). Given unstructured text, the task of NER is to identify and classify text spans according to categories of interest. These categories are defined depending on the downstream application and can range from general (*people*, *locations*, *organizations*) to specific domains such as biomedical entities (*genes*, *diseases*, *chemicals*).

In particular, Biomedical Named Entity Recognition (**BioNER**) is challenging due to the complexity of biomedical nomenclature. Morphologically, these entities can contain Greek letters, digits, punctuation (*α-tubulin*, *IL-6*), form variations (*inhibitor* vs. *inhibitory*), and compound terms (*tumor necrosis factor-alpha* vs. *TNF-α*). Semantically, polysemy (e.g., *p53* referring to a gene, protein, or condition) adds ambiguity. These challenges make human annotation costly, leading to BioNER datasets that are smaller and often focus on a limited number of entity types (Greenberg et al., 2018).

One approach to addressing data scarcity while building a BioNER model is to leverage multiple datasets, each annotated with a specific subset of entities. However, simply training a single model on the union of all available datasets assumes that every entity type is consistently annotated across all training instances, which is not the case. This leads to a high prevalence of false negatives, as entities that are labeled in one dataset may be entirely ignored in another. On the other hand, training separate models for each dataset fails to exploit shared statistical patterns across datasets and introduces the challenge of resolving conflicting predictions at inference time (Greenberg et al., 2018). Therefore, an effective strategy must balance learning from multiple sources while accounting for missing annotations and inconsistencies in labeling schemes.

Our contributions are three-fold: (i) we introduce **SRU-NER** (**S**lot-based **R**ecurrent **U**nit NER), a model which is able to solve nested NER through generating a sequence of actions; (ii) we propose an effective multi-task training strategy to handle the complex challenges of leveraging multiple NER datasets in a single model; and (iii) we show how the SRU-NER can handle multiple datasets on a single shared network through multiple experiments, including cross-corpus evaluations and a human evaluation on corpora of disjoint entity sets.

2 Related work

Named entity recognition has evolved significantly in the last decades. Early systems relied on rule-based methods, which were interpretable but lacked flexibility. The introduction of machine learning enabled more adaptable approaches, further enhanced by deep learning techniques that captured complex

[1]Code is publicly available at https://github.com/Priberam/sru-ner.

Proceedings of the 24th Workshop on Biomedical Language Processing (BioNLP 2025), pages 225–239
August 1, 2025

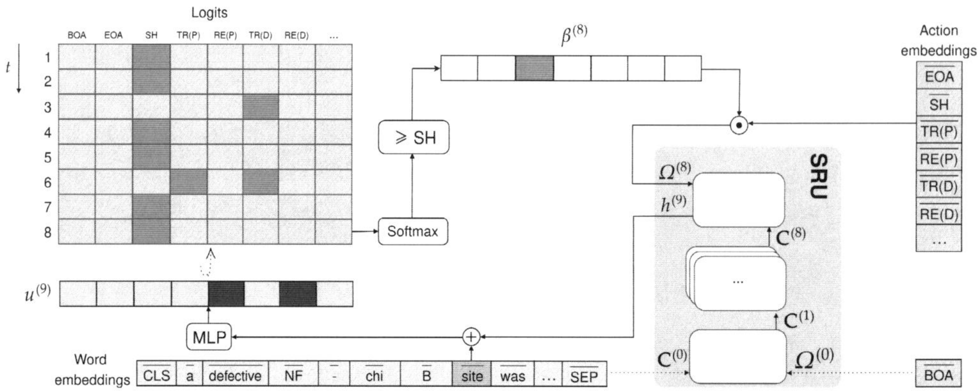

Figure 1: Action selection process for the sentence given in section 3.1, at time step $t = 9$. The gold nested mentions are "NF - chi B site", "chi B", of type *DNA (D)*, and "NF - chi B" of type *Protein (P)*. To compute the logits $u^{(9)}$, the model leverages the logits of the previous time steps, action embeddings and word embeddings.

linguistic patterns. Recently, Transformer-based architectures have set new benchmarks, driving significant advancements in NER performance (Li et al., 2022; Keraghel et al., 2024). In the CoNLL-2003 dataset (Tjong Kim Sang and De Meulder, 2003), a benchmark for NER tasks, performance has improved substantially, with F1 scores that have soared above 94% (Wang et al., 2021). The same phenomenon is seen for the GENIA corpus (Kim et al., 2003), a nested BioNER dataset, with test F1 scores exceeding 80% (Yu et al., 2020; Tan et al., 2021; Shen et al., 2021, 2022).

To tackle the proliferation of BioNER datasets, several studies have turned to multi-task learning (**MTL**; Park et al., 2024). Traditional deep learning NER models trained on a single dataset are referred to as *single-task* models, as they specialize in identifying mention spans for the specific entity types annotated within their training data. Single-task models often underperform on out-of-domain settings. In contrast, MTL frameworks leverage multiple datasets, each corresponding to a different *task*, allowing the model to learn from diverse sources. The fundamental premise is that different datasets share information which can be jointly leveraged to encourage the learning of more generalized representations, hence improving a model's robustness (Mehmood et al., 2019; Li et al., 2022).

MTL learning frameworks can be categorized into two types, depending on which modules are shared across tasks: (i) those that share the encoding layers while maintaining task-specific decoding layers (Crichton et al., 2017; Wang et al., 2018; Khan et al., 2020), and (ii) those that share *all* layers (Greenberg et al., 2018; Huang et al., 2019; Banerjee et al., 2021; Luo et al., 2023; Moscato et al., 2023). SRU-NER resembles models of type (ii), which share its decoding layers across all tasks. Typically, these models have a natural problem with false negatives, as the unified decoder may struggle to distinguish task-specific entity boundaries and labels, leading to the omission of valid entities. Our approach avoids this issue through an effective multi-task learning strategy.

3 Effective Multi-Task Learning for Named Entity Recognition

The proposed model, SRU-NER, solves the task of nested named entity recognition similar to that of a transition-based parser (Dyer et al., 2015; Marinho et al., 2019). Given a sequence of words $S = [w_1, w_2, \dots, w_N]$, the model generates a sequence of *actions*. At each time step, the actions are chosen depending on the words of the sentence and on the previously chosen actions. At the end of the parsing procedure, the complete sequence of actions is decoded into mentions.

3.1 Action encoding

Consider the system is trained to recognize mentions of entity types belonging to $\mathbb{E} = \{e_1, e_2, \dots, e_M\}$. Let $\mathcal{A}_{\mathbb{E}}$ stand for the system's $2M + 2$ possible *actions*: two special tokens (`SH` and `EOA`) and, for each entity type e_i, a pair of actions denoted `TR`(e_i) and `RE`(e_i). `TR`(e_i), short for "transition to entity e_i", indicates the start of a mention of type e_i; one says that this action *opened* a mention of type e_i. `RE`(e_i), short for "reduce of entity e_i", indicates the end of the mention of type e_i that was opened more recently; one says that a mention was *closed* by this action. SH, short for "shift", indicates that the input pointer should move to the next token; therefore, there is one SH for each word in the sentence. Finally, EOA is the end action.

These actions encode nested mentions effectively through the order in which they are chosen. If a mention of type e_k starts at the word w_i and ends at the word w_j, TR(e_k) appears before the SH representing the i-th word, and RE(e_j) appears after the SH representing the k-th word; if two mentions start at the same word, the TR() of the longest mention appears first; conversely, if two mentions end at the same word, the RE() of the shortest mention appears first. Consider the following sentence from the GENIA dataset (Kim et al., 2003):

Protein DNA

a defective NF - chi B site was completely . . .

This sentence has nested mentions, *e.g.* the mention "NF - chi B" of type *Protein* is contained in the mention "NF - chi B site" of type *DNA*. The action encoding of the sentence with its mentions is: SH → SH → TR(*DNA*) → TR(*Protein*) → SH → SH → TR(*DNA*) → SH → SH → RE(*DNA*) → RE(*Protein*) → SH → RE(*DNA*) → SH → SH → . . . → EOA.

3.2 Overall architecture

Using the previous notation, suppose one wants to detect mentions of $\mathbb{E}$ in the sentence S. The model consists of three consecutive steps: the encoding of S into a dense contextual embedding matrix $\mathbf{S}$, the iterative action generation procedure, and the decoding of the chosen actions into the mentions present in the sentence.

Contextual embeddings For the first step, S is passed through a BERT-like encoder to generate a matrix of contextual embeddings. For each word w_i, its dense embedding, denoted by $\overline{w_i}$, is obtained by max-pooling across the embeddings of its subwords. In this way, the encoded sentence $\mathbf{S}$ is a tensor of size $(N+2, d_{\text{enc}})$, $\mathbf{S} = [\overline{\text{CLS}}, \overline{w_1}, \overline{w_2}, \ldots, \overline{w_N}, \overline{\text{SEP}}]$, where d_{enc} is the encoder embedding dimension, $\overline{\text{CLS}}$ (respectively $\overline{\text{SEP}}$) is the embedding of the classification (respectively, separator) token of the encoder.

Action generation Given $\mathbf{S}$, the model enters an iterative action selection process, where at each time step t, logits are computed for each possible action in $\mathcal{A}_{\mathbb{E}}$.[2] Figure 1 shows a schematic representation of a time step of the cycle.

More concretely, define $u_{a_i}^{(t)}$ to be the logit value of action $a_i \in \mathcal{A}_{\mathbb{E}}$ for time step t. Suppose the system has already computed these values for the first $T \geq 1$ time steps, and is therefore about to compute them for time step $t = T + 1$. According to the last section, the SH action corresponds to advancing a token in the sentence S. Hence, define

$$p^{(t)} = \sum_{t_0 \leq t} \mathbb{1}\left(\arg\max_{a_i \in \mathcal{A}_{\mathbb{E}}} \left(u_{a_i}^{(t_0)}\right) = \text{SH}\right), \quad (1)$$

where $\mathbb{1}$ stands for the indicator function. $p^{(t)}$ is therefore the number of tokens that have already been parsed at a previous time step t, for $1 \leq t \leq T$. Lastly, define, for each $1 \leq t \leq T$,

$$\Omega^{(t)} = \sum_{a_i \in \mathcal{A}_{\mathbb{E}}} \beta_{a_i}^{(t)} \, \overline{a_i}, \quad (2)$$

where $\overline{a_i}$ is a trained embedding of size d_{enc} and

$$\beta_{a_i}^{(t)} = \begin{cases} u_{a_i}^{(t)} & \text{if } u_{a_i}^{(t)} \geq u_{\text{SH}}^{(t)} \\ 0 & \text{otherwise} \end{cases}$$

In other words, $\Omega^{(t)}$ is a weighted embedding of the actions chosen at time step t, where actions with logits lower than the logit of SH are excluded.

Let $\mathbf{u}^{(T+1)}$ be the vector of logits $u_{a_i}^{(T+1)}$ over $a_i \in \mathcal{A}_{\mathbb{E}}$. These are computed as

$$\mathbf{u}^{(T+1)} = \text{MLP}\left(f\left(p^{(T)}, \Omega^{(T)}\right)\right), \quad (3)$$

where the MLP is composed of a dropout layer, a fully-connected layer, a tanh activation, and a linear layer with output nodes corresponding to each action in $\mathcal{A}_{\mathbb{E}}$. The input of this MLP is

$$f\left(p^{(T)}, \Omega^{(T)}\right) = \mathbf{S}_{p^{(T)}+1} \oplus \text{SRU}\left(\Omega^{(T)}, p^{(T)}\right),$$

i.e. the concatenation of the embedding of the *next* token, $\mathbf{S}_{p^{(T)}+1}$, and an embedding of the last state

[2] Unlike token-based labeling approaches, the total number of time steps is not determined *a priori*, although always bounded below by N, the number of words in S.

of a "processed actions memory". This memory holds an action history and computes weighted embeddings at each call by leveraging a set of internal latent representations. This module is refered to as the **S**lot-based **R**ecurrent **U**nit (**SRU**), and is described in section 3.3.

In order to make the first prediction, $\mathbf{u}^{(1)}$, the system is initialized by setting $p^{(0)} = 0$, and $\Omega^{(0)}$ to be another trained embedding of size d_{enc}, denoted by $\overline{\texttt{BOA}}$.[3] The action generation cycle terminates when a time step $t = T_{\text{final}}$ is reached such that

$$\text{Sigmoid}\left(u_{\texttt{EOA}}^{(T_{\text{final}})}\right) > 0.5\,. \tag{4}$$

Decoding At the end of the action generation cycle, the output logits from all time steps are passed through a sigmoid function. This produces a set of independent probability scores for each action in $\mathcal{A}_{\mathbb{E}}$, from which mention spans are extracted. The decoder module maintains separate stacks of open spans for each entity type in $\mathbb{E}$, allowing spans of different types to overlap.

The decoding process iterates through the list of probability scores until reaching a time step where the highest-scoring action is EOA[4]. Before such a time step is reached, the decoder proceeds following two rules: (i) if the highest-scoring action is SH, a pointer that counts the number of parsed words is incremented; and (ii) if the highest-scoring action is a TR() or a RE(), the entity mention stacks are updated. In the latter case, only actions with probability scores above 0.5 are considered. Transition actions open new spans, while reduce actions close the most recent span of the corresponding entity type, as discussed in section 3.1.

3.3 Slot-based Recurrent Unit

The Slot-based Recurrent Unit (SRU) is a stateful function that, at each time step, takes a pair of inputs, updates its internal state, and produces an output embedding.

At each time step t, the SRU updates its internal state according to

$$\mathbf{C}^{(t+1)} = m\left(\mathbf{C}^{(t)}, \Omega^{(t)}, p^{(t)}\right),$$

where $\mathbf{C}^{(t)} \in \mathbb{R}^{Q\times d}$ is the SRU's internal state matrix, $\Omega^{(t)} \in \mathbb{R}^{d}$ is an input vector, and

[3] In this text, a zero-indexing notation is adopted for tensors, and so $\mathbf{S}_{p_0+1} = \overline{w_1}$.

[4] This stopping condition was shown to provide better results empirically, despite being different to that of the action generation procedure, present in equation (4).

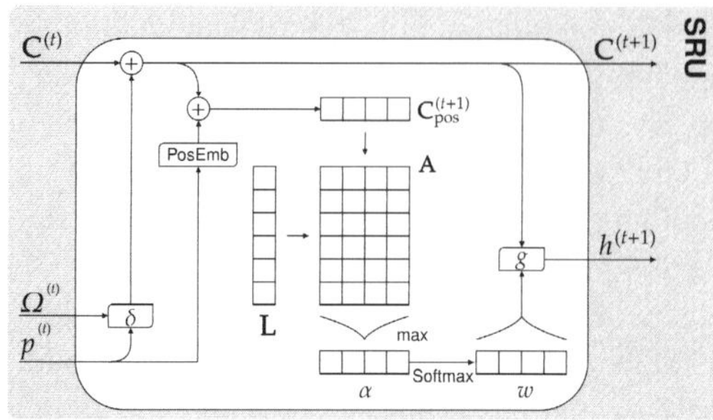

Figure 2: SRU unit at time step t. Its internal state is updated depending on its current state $\mathbf{C}^{(t)}$ and the weighted action embeddings $\Omega^{(t)}$. This stateful function also leverages a set of latent representations. It produces an output embedding $h^{(t+1)}$ by applying an attention mechanism to the updated state.

$p^{(t)} \in \{0, 1, \ldots, Q-1\}$ is an input integer. It also produces an output embedding $h^{(t+1)} \in \mathbb{R}^d$ via

$$h^{(t+1)} = g\left(\mathbf{C}^{(t+1)}, p^{(t)}\right).$$

A schematic representation is present in Figure 2. Q and d refer to the number of rows (or *slots*) in the internal state matrix and the hidden dimension of the input and output embeddings, respectively.

The function m updates $\mathbf{C}^{(t)}$ by summing the input vector $\Omega^{(t)}$ to its $p^{(t)}$-th row, *i.e.*

$$m\left(\mathbf{C}^{(t)}, \Omega^{(t)}, p^{(t)}\right) := \mathbf{C}^{(t)} + \delta_{p^{(t)}}\left(\Omega^{(t)}\right)^T$$

where $\delta_{p^{(t)}} \in \mathbb{R}^Q$ is a one-hot vector with 1 in its $p^{(t)}$-th coordinate.

The output embedding $h^{(t)} \in \mathbb{R}^d$ is obtained via the function g, defined as

$$g\left(\mathbf{C}^{(t+1)}, p^{(t)}\right) := \mathbf{w}^T\left(\mathbf{C}^{(t+1)}\mathbf{D}_1\right)$$

where $\mathbf{D}_1$ is a trained diagonal matrix of size d and $\mathbf{w} \in \mathbb{R}^Q$ are weights computed via an attention mechanism inspired by Ganea and Hofmann, 2017, detailed as follows. First, $\mathbf{C}^{(t+1)}$ is enhanced by adding positional information,

$$\mathbf{C}_{\text{pos}}^{(t+1)} = \alpha\,\mathbf{C}^{(t+1)} + \text{Dropout}\left(\mathbf{P}\left(p^{(t)}\right)\right) \tag{5}$$

where α is a trained scaling parameter, and $\mathbf{P}\left(p^{(t)}\right) \in \mathbb{R}^{Q\times d}$ are positional embeddings.[5]

[5] These positional embeddings are *relative*, in the sense that each row of $\mathbf{P}\left(p^{(t)}\right)$ is selected from a table of trained embeddings based on its distance to the row with index $p^{(t)}$.

Next, a set of J trained latent embeddings of size d are used to compute an attention score for each row in $\mathbf{C}^{(t+1)}$. Defining $\mathbf{L} \in \mathbb{R}^{J \times d}$ to be the matrix of latent embeddings, an attention score matrix is computed by

$$\mathbf{A} = \text{Dropout}(\mathbf{L})\, \mathbf{D}_2 \left(\mathbf{C}_{\text{pos}}^{(t+1)}\right)^T,$$

where $\mathbf{D}_2$ is a trained diagonal matrix of size d. An attention score for each slot is obtained by setting $\mathbf{s} = \max_j(A_{jq})$ for $q \in \{0, 1, \dots, Q-1\}$. Finally, the scores $\mathbf{s}$ are normalized through a softmax to get the weights $w \in \mathbb{R}^Q$.

The SRU module is used at each action generation time step to compute an embedding that models the current state of a "processed actions memory" stack. For each time step t, the input integer $p^{(t)}$ is the one defined by equation (1), and the input vector $\Omega^{(t)}$ is the one defined by equation (2). Furthermore, d is set to be the encoder embedding dimension d_{enc}, the number of slots to be $Q = N+2$, and the number of latent variables J to be an integer multiple[6] of $|\mathcal{A}_{\mathbb{E}}| = 2M + 2$. The internal state matrix is initialized by setting $\mathbf{C}^{(0)} = \mathbf{S}$. Taking this choice of initialization into account, and referring back to equation (3), for the computation of $h^{(T+1)} = \text{SRU}\left(\Omega^{(T)}, p^{(T)}\right)$, all the slots of the updated internal state matrix $\mathbf{C}^{(T+1)}$ are filled with the embeddings of the encoded sentence $\mathbf{S}$. In addition, a history of the previously chosen actions is present in $\mathbf{C}^{(T+1)}$ since, at each call of the SRU module in previous time steps $0 \le t \le T$, the weighted action embeddings $\Omega^{(t)}$ of equation (2) were summed to the slots pointed to by $p^{(t)}$.

4 Multi-task training strategy

Suppose the model is trained on an ensemble of K datasets $\mathcal{D} = \{D_i\}_{i=1}^K$, where each dataset D_i is annotated with spans of entity types $\mathbb{E}_i$. In order to account for differences in labeling schemes, during training, the entity types of distinct datasets are always considered to be distinct as well.[7] Therefore, the model is trained to recognize spans of entity types in the disjoint union set $\widehat{\mathbb{E}} = \bigsqcup_{i=1}^K \mathbb{E}_i$.

The training objective of the model is to minimize the mean loss of the samples in a batch. Each batch is constructed by randomly selecting samples from $\mathcal{D}$. To ensure a balanced contribution from all datasets, the probability of selecting a sample from a given dataset is inversely proportional to the total number of sentences in that dataset. The total number of samples per epoch is the average number of sentences in the datasets of $\mathcal{D}$.

Let S be a sentence in the batch, coming from dataset D_i, and thus annotated with gold spans of entity types $\mathbb{E}_i$. The output of the action generation cycle is a matrix

$$\mathbf{U} = \left(u_{a_i}^{(t)}\right)_{t=1,\dots,T_{\text{final}};\, a_i \in \mathcal{A}_{\mathbb{E}}},$$

where each row $u_*^{(t)}$ contains the logits, for time step t, over *all* actions $\mathcal{A}_{\widehat{\mathbb{E}}}$ associated with the disjoint union set $\widehat{\mathbb{E}}$.[8] To compute a loss value for $\mathbf{U}$, the following constraints are enforced:

i) on one hand, the model *should* be penalized for failing to predict the `TR()` and `RE()` actions that correspond to the gold spans of the entity types $\mathbb{E}_i$, for which S is annotated; but

ii) on the other hand, the model *should not* be penalized for predicting `TR()` and `RE()` actions of entity types in $\widehat{\mathbb{E}} \setminus \mathbb{E}_i$, which are not annotated in S.

In practice, this strategy is applied as follows. The list of actions corresponding to the gold annotations of sentence S (constructed as detailed in section 3.1 and considering the disjoint entity type set $\widehat{\mathbb{E}}$) is augmented to a matrix $\mathbf{G} = \left(G_{a_i}^{(t)}\right) \in \mathbb{R}^{T_{\text{initial}} \times |\mathcal{A}_{\widehat{\mathbb{E}}}|}$ such that each row $G_*^{(t)}$ is a multi-hot vector representing a distinct timestep t, with 1's in the columns that correspond to the gold actions. This conversion is done such that the `SH` and `EOA` actions always occupy different time steps, but `TR()` and `RE()` actions of different entity types can coexist at the same time step. Then, $\mathbf{G}$ is changed during the action generation cycle by incorporating the probabilities of the model's decisions on `TR()` and `RE()` actions from other datasets. More concretely, at time step t of the cycle, for $a_i \in \mathcal{A}_{\widehat{\mathbb{E}}} \setminus \mathcal{A}_{\mathbb{E}_i}$, $G_{a_i}^{(t)}$ is set to be equal to $\sigma\left(u_{a_i}^{(t)}\right)$, where σ is the sigmoid function. In addition, when $G_{\text{SH}}^{(t)} = 1$ and $u_{a_i}^{(t)} > u_{\text{SH}}^{(t)}$ for some

[6] For the experiments conducted, it was set to 2 or 10 (see Table 12 in Appendix B).

[7] In practice, this is implemented by simply changing the name of an entity type $e \in \mathbb{E}_i$ belonging to D_i, to i_e in $\mathbb{E}$.

[8] At inference time, the action generation procedure halts when the probability of the `EOA` action exceeds a threshold, as described in section 3.2. However, during training, in order to guarantee that all gold actions are considered, the cycle halts only after all tokens have been parsed (*i.e.* shifted).

Dataset	SRU-NER		Wang et al., 2018	Huang et al., 2019	Khan et al., 2020	Moscato et al., 2023
	Merged	*Disjoint*				
BC2GM	78.80	83.95	80.74 *	79.1	83.01 *	**84.84**
BC4CHEMD	90.42	**92.05**	89.37 *	87.3	—	—
BC5CDR	89.37	**90.26**	88.78 *	—	89.50 *	◇
JNLPBA	72.15	76.00	73.52 *	**83.8**	72.89 *	—
Linnaeus	**88.82**		—	83.9	—	—
NCBI Disease	87.32	88.71	86.14 *	84.0	88.10 *	**89.20**
Average	84.48	86.63				

Table 1: Micro-F1 scores of several multi-task models trained on subsets of an ensemble of six biomedical datasets. For SRU-NER, scores are reported by considering two evaluation scenarios, *Merged* and *Disjoint*, as explained in section 5.2. Best scores are **bold**, and second best scores are underlined. Symbol reference:
— : dataset was absent in training;
* : model was trained on both the training and development splits of the corpora;
◇ : model was trained using only the 'Chemical' annotations of BC5CDR, obtaining an F1 of 93.95; for the same tag, SRU-NER gets an F1 of 93.77 on the disjoint evaluation and 93.18 on the merged evaluation.

$a_i \in \mathcal{A}_{\widehat{\mathbb{E}}} \setminus \mathcal{A}_{\mathbb{E}_i}$, that is, when the model is trying to open/close a new span of an entity type of other dataset D_j $(j \neq i)$, the value $G_{\text{SH}}^{(t)}$ is changed to $\sigma\left(u_{\text{SH}}^{(t)}\right)$. In this case, a one-hot vector is inserted in $\mathbf{G}$ after $G_{*}^{(t)}$, so that, on the next time step $t+1$, $G_{\text{SH}}^{(t+1)} = 1$ and $G_{a_i}^{(t+1)} = 0$ for all $a_i \in \widehat{\mathbb{E}} \setminus \{\text{SH}\}$. This procedure ensures that $\mathbf{G}$ still reflects the original gold annotations in the columns corresponding to `TR()` and `RE()` actions of entity types in the source dataset, but incorporates the model's probabilities for other actions. Then, by setting, for each $1 \leq t \leq T_{\text{final}}$,

$$L^{(t)} = -\frac{1}{|\mathcal{A}_{\widehat{\mathbb{E}}}|} \sum_{a_i \in \mathcal{A}_{\widehat{\mathbb{E}}}} \Bigg(G_{a_i}^{(t)} \log\left(\sigma\left(u_{a_i}^{(t)}\right)\right) + \left(1 - G_{a_i}^{(t)}\right) \log\left(1 - \sigma\left(u_{a_i}^{(t)}\right)\right) \Bigg)$$

the total loss of the sample is computed as

$$L = \frac{1}{T_{\text{final}}} \sum_{t=1}^{T_{\text{final}}} L^{(t)} .$$

Given how $\mathbf{G}$ is constructed, this ensures the aforementioned constraints i) and ii) on the loss function are satisfied.

5 Experiments and Results

To evaluate the performance of the proposed architecture for the NER task, single-task experiments were conducted on benchmarks datasets, specifically the English subset of CoNLL-2003 (Tjong Kim Sang and De Meulder, 2003) and GENIA (Kim et al., 2003). The model's multi-task performance is also assessed by training it with an ensemble of six biomedical datasets that have been extensively used in previous research. In order to demonstrate the viability of SRU-NER for downstream applications, a model is evaluated in a cross-corpus setting by replicating the experimental setup of Sänger et al., 2024. Finally, two further experiments are conducted to quantify the reliability of the multi-task models' predictions for entity types that are not explicitly annotated in the test corpora, providing a more comprehensive assessment of their generalization capabilities.

The datasets used across the following sections and respective experimental setup are described in Appendix A. Training details can be found in Appendix B. For evaluation purposes, a predicted mention is considered a true positive if and only if both its span boundaries and entity type exactly match the gold annotation. Results are reported for each dataset using mention-level micro F1 scores.

5.1 Single-task performance

The results of the two single-task models are presented in Table 2. The proposed model achieves micro F1 scores of 94.48% on the CoNLL-2003 dataset, and 80.10% on the GENIA dataset. These results are very close to state-of-the-art (SOTA), demonstrating the competitiveness of SRU-NER in both flat and nested NER scenarios. Nonetheless, in contrast to our approach, the models presented as SOTA were trained using both the training and development splits of their respective datasets. This difference in training data availability may contribute to the observed performance gap, particularly on GENIA, where additional annotated data could provide further benefits in capturing complex biomedical terminology.

Dataset	SRU-NER	SOTA
CoNLL	94.48	94.6*, (Wang et al., 2021)
GENIA	80.10	81.53*, (Shen et al., 2023)

Table 2: Micro-F1 scores of single-task models on benchmark datasets. The entity counts of the datasets can be found in Table 7. The * symbol indicates that a model was trained on both the training and development splits of the corpus.

Dataset	SRU-NER	SOTA
BC2GM	85.43	85.48* (Sun et al., 2021)
BC4CHEMD	92.64	92.92* (Sun et al., 2021)
BC5CDR	90.61	91.90 (Zhang et al., 2023)
JNLPBA	77.12	78.93* (Sun et al., 2021)
Linnaeus	89.62	94.13 (Habibi et al., 2017)
NCBI Disease	89.25	90.04* (Sun et al., 2021)
Average	87.45	

Table 3: Micro-F1 scores of single-task models trained on the datasets used for the multi-task model described in section 5.2. SOTA results are for single-task models. The * symbol indicates the model was trained on a larger training split.

5.2 Multi-task performance

In Table 1, we show the results of SRU-NER trained on an ensemble $\{D_i\}_{i=1}^{6}$ of six biomedical datasets, annotated for $|\cup_i \mathbb{E}_i| = 8$ entity types. Since there are entity types which are annotated on more than one dataset (*e.g.* BC4CHEMD and BC5CDR are both annotated with mentions of the Chemical type), two evaluation scenarios are considered, that differ in how these type labels are interpreted. Recalling that the model infers mentions with entity types in the disjoint union set $\widehat{\mathbb{E}} = \sqcup_i \mathbb{E}_i$, given a sentence coming from the test split of dataset D_i of the ensemble, in the case of:

i) *disjoint evaluation*, the predicted spans of types $\mathbb{E}_i \subset \widehat{\mathbb{E}}$ are compared against the gold ones, and any predicted span of type in $\widehat{\mathbb{E}} \setminus \mathbb{E}_i$ is discarded;

ii) *merged evaluation*, the entity types of predicted spans are mapped to $\cup_i \mathbb{E}_i$, and the spans whose mapped types do not also belong in $\mathbb{E}_i$ are discarded; the remaining spans are compared against the gold ones.

An example of the predictions of the model on a test sentence, together with which spans are used to compute the metrics on the two evaluation scenarios is present in Figure 3.

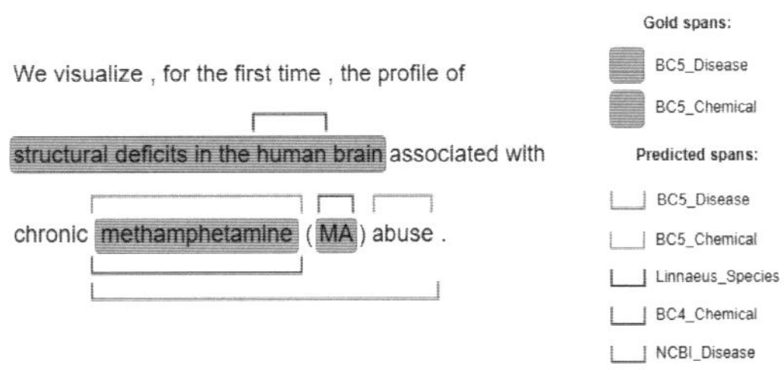

Figure 3: Example of a sentence from the test split of the BC5CDR corpus (Li et al., 2016), together with gold spans and predicted spans as annotated by the MTL model described in section 5.2. The model is trained on six datasets, covering eight entity types $\cup_i \mathbb{E}_i = \{\text{Chemical}, \text{Disease}, \ldots\}$. Notice that some of these types are common to multiple datasets (namely, 'Chemical', annotated on both the BC4CHEMD and BC5CDR datasets; and 'Disease', annotated on both the BC5CDR and NCBI datasets). SRU-NER tags spans with one of 11 possible types, built by adjoining the dataset name to the original type name, such that $\widehat{\mathbb{E}} = \{\text{BC4\_Chemical}, \text{BC5\_Chemical}, \ldots\}$. In the *disjoint* evaluation case, and since this sentence comes from the BC5CDR corpus, metrics are computed by considering only the spans whose types in $\widehat{\mathbb{E}}$ start with the BC5 shorthand, resulting in one true positive, one false positive and two false negatives. In the *merged* evaluation case, spans whose types in $\widehat{\mathbb{E}}$ do not end with 'Chemical' or 'Disease' are discarded, and the remaining spans have their types mapped to $\cup_i \mathbb{E}_i$ by removing the dataset identifier. With these spans, there are two true positives, two false positives and one false negative in the sentence.

Compared to previous MTL models, the proposed model achieves the best or second-best F1 scores in the disjoint evaluation setting. These results are obtained without relying on task-specific classification layers (Wang et al., 2018; Khan et al., 2020) or training multiple single-task teacher models followed by knowledge distillation into a student model (Moscato et al., 2023). Instead, a single unified model learns each task directly from its respective annotated dataset while preserving the performance of other tasks. This approach enables joint decoding, thereby eliminating the need for post-processing steps to resolve span conflicts.

Table 3 presents F1 scores for single-task models trained on each dataset used in the multi-task setting, alongside SOTA references. The results demonstrate that the proposed model remains competitive in the single-task setting. The average F1 score of the six single-task SRU-NER models is 0.82 percentage points higher than the dataset-average F1 of the multi-task SRU-NER model under the disjoint evaluation setting. This aligns with previous findings, which suggest that while multi-task training improves model robustness across datasets, it may lead to lower in-corpus performance compared to single-task models (Yin et al., 2024). To further investigate the generalization capabilities of the model, the next section presents an evaluation in a cross-corpus setting.

Dataset	Entity type	SRU-NER	Baseline
BioID	Species	**62.41**	58.21
MedMentions	Chemical	**59.53**	58.40
	Disease	**62.48**	62.18
tmVar3	Gene	**90.38**	87.87
Average		**68.70**	66.67

Table 4: Mention-level F1 scores for the cross-corpus experiment. SRU-NER was trained on an emsemble of 8 biomedical datasets, and evaluated on 3 independent corpora. Baseline refers to the scores obtained by (Sänger et al., 2024). Best scores are in **bold**.

Training datasets	Chemical	Disease
Only BC5-Chemical	91.27	—
Only BC5-Disease	—	85.41
Both	**91.81**	**86.10**

Table 5: Global prediction F1 scores on the test split of BC5CDR of models trained on synthetic datasets. Best scores are **bold**.

5.3 Cross-corpus evaluation

Table 4 presents the results of the proposed model in a cross-corpus evaluation, replicating the experimental setup of Sänger et al., 2024. The model was trained on an ensemble of nine datasets covering five entity types and evaluated on three independent corpora annotated for four of these types. The results indicate that SRU-NER outperforms the baseline by an average of 2.03%, with notable improvements for the Species (4.2%) and Gene (2.51%) entity types. These findings underscore the robustness of the model and demonstrate its potential for downstream applications. For reference, in-corpus F1 scores are provided in Appendix C.

5.4 Evaluation of global predictions

The previous experiments evaluated the model's *local* prediction ability. Specifically, when the model is trained on a collection $\{D_i\}_{i=1}^K$, where each dataset D_i was annotated for entity types $\mathbb{E}_i$, its performance was assessed on a test dataset D_{test} annotated with entity types $\mathbb{E}_{\text{test}} \subseteq \mathbb{E}_j$ for some $j \in \{1, \dots, K\}$. However, the model generates predictions for spans of all entity types in $\cup_i \mathbb{E}_i$ within D_{test}. To evaluate its global prediction ability, it is necessary to test the model on a dataset annotated with a superset of entity types spanning multiple training datasets.

First, following the approach of Huang et al., 2019, a synthetic dataset is constructed from the BC5CDR corpus. The original training set is randomly partitioned into two disjoint subsets: one containing only Chemical annotations (*BC5-Chemical*) and another containing only Disease annotations (*BC5-Disease*). Additional details on these synthetic datasets are provided in Appendix A. Two single-task models are trained separately on each subset, while a multi-task model is trained on both. All models are evaluated on the original test split of the BC5CDR corpus. The results, presented in Table 5, demonstrate that the training strategy outlined in section 4 effectively enables the model to make accurate global predictions across entity types from different training datasets.

Secondly, a multi-task model is trained on both the CoNLL-2003 dataset and the BC5CDR dataset. This approach results in a model capable of recognizing six entity types: four from the general domain (*LOC*, *MISC*, *ORG*, *PER*) and two from the biomedical domain (*Chemical*, *Disease*). To assess the model's ability to generalize across domains, its predictions of general-domain entity types in the test split of the BC5CDR dataset and, conversely, its predictions of biomedical entity types in the test split of the CoNLL dataset are evaluated. The results of the multi-task model can be found in Table 6 under the column SRU-NER-MTL. Since gold annotations for these cross-domain predictions are not available, the evaluation was conducted manually by two human annotators. Provided with definitions of the entity types, they independently assessed whether the model's predictions were correct. This human evaluation was also conducted for the predictions of two single-task models: one trained on CoNLL-2003 and evaluated on the BC5CDR test set (SRU-NER-CoNLL), and another trained on BC5CDR and evaluated on the CoNLL-2003 test set (SRU-NER-BC5). A comparison between the single-task and multi-task models reveals that multi-task SRU-NER is, on average,

Entity	SRU-NER-CoNLL			SRU-NER-BC5			SRU-NER-MTL		
	P	*R*	*F1*	*P*	*R*	*F1*	*P*	*R*	*F1*
Chemical	24.71	87.76	38.57	—	—	—	75.00	9.18	16.36
Disease	25.25	83.33	38.76	—	—	—	88.46	38.33	53.49
LOC	—	—	—	98.25	88.89	93.33	100.00	96.83	98.39
ORG	—	—	—	80.00	80.00	80.00	86.36	71.25	78.08
PER	—	—	—	94.44	94.44	94.44	100.00	22.22	36.36

Table 6: Human evaluation of the out-of-domain predictions made by three models. *P* stands for precision, *R* for the simulated recall, and *F1* for the F1 computed with the former two metrics. Details on how these metrics were computed can be found in Appendix D.

25.4% more precise in identifying out-of-domain spans. For instance, the single-task model trained on biomedical entity types incorrectly classified *lead* as a chemical in the CoNLL-2003 sentence: *"Indonesian keeper Hendro Kartiko produced a string of fine saves to prevent the Koreans increasing their lead."* In contrast, the multi-task model did not make this error. Further details on this experiment are provided in Appendix D.

6 Conclusion

This work presents SRU-NER, a novel architecture for Named Entity Recognition capable of handling nested entities through a transition-based parsing approach. The model integrates a Slot-based Recurrent Unit (SRU) to maintain an evolving representation of past actions, enabling effective entity extraction. Unlike traditional multi-task learning approaches that rely on separate models for different entity types, SRU-NER employs a unified learning strategy, allowing a single model to learn from multiple datasets. This approach improves adaptability to annotation inconsistencies and enhances generalization across domains.

Experimental results demonstrate that SRU-NER achieves strong performance in both single- and multi-task settings, with cross-corpus evaluations and human assessments confirming the robustness of its predictions. These findings highlight the advantages of training a single multi-task model for BioNER and suggest promising directions for future research, including advancements in nested entity recognition and domain adaptability.

Limitations

While the proposed SRU-NER architecture has demonstrated effectiveness for named entity recognition in general and biomedical domains, its performance in other domains, such as legal or financial, was not evaluated. Furthermore, the generalizability of the findings may be limited, as evaluations on community-available biomedical datasets may not fully capture the diversity of real-world biomedical text. Finally, the assessment of global prediction ability in a cross-domain scenario relied on human annotators, introducing a degree of subjectivity to the evaluation.

While the model achieves competitive results, we note that no extensive hyperparameter search was conducted. A more systematic tuning of hyperparameters could potentially yield further improvements. Additionally, the training strategy presents opportunities for refinement, notably in the sampling strategy utilized within the multi-task learning framework.

Acknowledgments

This research was supported by the Portuguese Recovery and Resilience Plan through project C645008882-00000055 (*i.e.*, the Center For Responsible AI).

References

Cecilia Arighi, Lynette Hirschman, Thomas Lemberger, Samuel Bayer, Robin Liechti, Donald Comeau, and Cathy Wu. 2017. Bio-id track overview. In *BioCreative VI Challenge Evaluation Workshop*, volume 482, page 376.

Pratyay Banerjee, Kuntal Kumar Pal, Murthy Devarakonda, and Chitta Baral. 2021. Biomedical Named Entity Recognition via Knowledge Guidance and Question Answering. *ACM Trans. Comput. Healthcare*, 2(4):33:1–33:24.

Nigel Collier, Tomoko Ohta, Yoshimasa Tsuruoka, Yuka Tateisi, and Jin-Dong Kim. 2004. Introduction to the bio-entity recognition task at JNLPBA. In *Proceedings of the International Joint Workshop on Natural Language Processing in Biomedicine and its Applications (NLPBA/BioNLP)*, pages 73–78, Geneva, Switzerland. COLING.

Alexis Conneau, Kartikay Khandelwal, Naman Goyal, Vishrav Chaudhary, Guillaume Wenzek, Francisco Guzmán, Edouard Grave, Myle Ott, Luke Zettlemoyer, and Veselin Stoyanov. 2020. Unsupervised cross-lingual representation learning at scale. In *Proceedings of the 58th Annual Meeting of the Association for Computational Linguistics*, pages 8440–8451, Online. Association for Computational Linguistics.

Gamal Crichton, Sampo Pyysalo, Billy Chiu, and Anna Korhonen. 2017. A neural network multi-task learning approach to biomedical named entity recognition. *BMC Bioinformatics*, 18(1):368.

Rezarta Islamaj Doğan, Robert Leaman, and Zhiyong Lu. 2014. NCBI disease corpus: a resource for disease name recognition and concept normalization. *J Biomed Inform*, 47:1–10.

Chris Dyer, Miguel Ballesteros, Wang Ling, Austin Matthews, and Noah A. Smith. 2015. Transition-based dependency parsing with stack long short-term memory. In *Proceedings of the 53rd Annual Meeting of the Association for Computational Linguistics and the 7th International Joint Conference on Natural Language Processing (Volume 1: Long Papers)*, pages 334–343, Beijing, China. Association for Computational Linguistics.

Octavian-Eugen Ganea and Thomas Hofmann. 2017. Deep joint entity disambiguation with local neural attention. In *Proceedings of the 2017 Conference on Empirical Methods in Natural Language Processing*, pages 2619–2629, Copenhagen, Denmark. Association for Computational Linguistics.

Martin Gerner, Goran Nenadic, and Casey M. Bergman. 2010. Linnaeus: A species name identification system for biomedical literature. *BMC Bioinformatics*, 11(1):85.

Nathan Greenberg, Trapit Bansal, Patrick Verga, and Andrew McCallum. 2018. Marginal likelihood training of BiLSTM-CRF for biomedical named entity recognition from disjoint label sets. In *Proceedings of the 2018 Conference on Empirical Methods in Natural Language Processing*, pages 2824–2829, Brussels, Belgium. Association for Computational Linguistics.

Harsha Gurulingappa, Roman Klinger, Martin Hofmann-Apitius, and Juliane Fluck. 2010. An empirical evaluation of resources for the identification of diseases and adverse effects in biomedical literature. In *2nd Workshop on Building and evaluating resources for biomedical text mining (7th edition of the Language Resources and Evaluation Conference)*, Valetta, Malta.

Maryam Habibi, Leon Weber, Mariana Neves, David Luis Wiegandt, and Ulf Leser. 2017. Deep learning with word embeddings improves biomedical named entity recognition. *Bioinformatics*, 33(14):i37–i48.

Xiao Huang, Li Dong, Elizabeth Boschee, and Nanyun Peng. 2019. Learning a unified named entity tagger from multiple partially annotated corpora for efficient adaptation. In *Proceedings of the 23rd Conference on Computational Natural Language Learning (CoNLL)*, pages 515–527, Hong Kong, China. Association for Computational Linguistics.

Rezarta Islamaj, Robert Leaman, Sun Kim, Dongseop Kwon, Chih-Hsuan Wei, Donald C Comeau, Yifan Peng, David Cissel, Cathleen Coss, Carol Fisher, Rob Guzman, Preeti Gokal Kochar, Stella Koppel, Dorothy Trinh, Keiko Sekiya, Janice Ward, Deborah Whitman, Susan Schmidt, and Zhiyong Lu. 2021a. NLM-Chem, a new resource for chemical entity recognition in PubMed full text literature. *Sci Data*, 8(1):91.

Rezarta Islamaj, Chih-Hsuan Wei, David Cissel, Nicholas Miliaras, Olga Printseva, Oleg Rodionov, Keiko Sekiya, Janice Ward, and Zhiyong Lu. 2021b. NLM-Gene, a richly annotated gold standard dataset for gene entities that addresses ambiguity and multi-species gene recognition. *J Biomed Inform*, 118:103779.

Imed Keraghel, Stanislas Morbieu, and Mohamed Nadif. 2024. Recent advances in named entity recognition: A comprehensive survey and comparative study. *Preprint*, arXiv:2401.10825.

Muhammad Raza Khan, Morteza Ziyadi, and Mohamed AbdelHady. 2020. Mt-bioner: Multi-task learning for biomedical named entity recognition using deep bidirectional transformers. *Preprint*, arXiv:2001.08904.

J.-D. Kim, T. Ohta, Y. Tateisi, and J. Tsujii. 2003. Genia corpus—a semantically annotated corpus for biotextmining. *Bioinformatics*, 19:i180–i182.

Corinna Kolarik, Roman Klinger, Christoph M. Friedrich, Martin Hofmann-Apitius, and Juliane Fluck. 2008. Chemical Names: Terminological Resources and Corpora Annotation. In *Workshop on Building and evaluating resources for biomedical text mining (6th edition of the Language Resources and Evaluation Conference)*, pages 51–58, Marrakech, Morocco.

Martin Krallinger, Obdulia Rabal, Florian Leitner, Miguel Vazquez, David Salgado, Zhiyong Lu, Robert Leaman, Yanan Lu, Donghong Ji, Daniel M. Lowe, Roger A. Sayle, Riza Theresa Batista-Navarro, Rafal Rak, Torsten Huber, Tim Rocktäschel, Sérgio Matos, David Campos, Buzhou Tang, Hua Xu, Tsendsuren Munkhdalai, Keun Ho Ryu, S. V. Ramanan, Senthil Nathan, Slavko Žitnik, Marko Bajec, Lutz Weber, Matthias Irmer, Saber A. Akhondi, Jan A. Kors, Shuo Xu, Xin An, Utpal Kumar Sikdar, Asif Ekbal, Masaharu Yoshioka, Thaer M. Dieb, Miji Choi, Karin Verspoor, Madian Khabsa, C. Lee Giles, Hongfang Liu, Komandur Elayavilli Ravikumar, Andre Lamurias, Francisco M. Couto, Hong-Jie Dai, Richard Tzong-Han Tsai, Caglar Ata, Tolga Can, Anabel Usié, Rui Alves, Isabel Segura-Bedmar, Paloma Martínez,

Julen Oyarzabal, and Alfonso Valencia. 2015. The chemdner corpus of chemicals and drugs and its annotation principles. *Journal of Cheminformatics*, 7(1):S2.

Jiao Li, Yueping Sun, Robin J Johnson, Daniela Sciaky, Chih-Hsuan Wei, Robert Leaman, Allan Peter Davis, Carolyn J Mattingly, Thomas C Wiegers, and Zhiyong Lu. 2016. BioCreative V CDR task corpus: a resource for chemical disease relation extraction. *Database (Oxford)*, 2016.

Jing Li, Aixin Sun, Jianglei Han, and Chenliang Li. 2022. A survey on deep learning for named entity recognition. *IEEE Transactions on Knowledge and Data Engineering*, 34(1):50–70.

Ling Luo, Po-Ting Lai, Chih-Hsuan Wei, Cecilia N Arighi, and Zhiyong Lu. 2022. Biored: a rich biomedical relation extraction dataset. *Briefings in Bioinformatics*, 23(5):bbac282.

Ling Luo, Chih-Hsuan Wei, Po-Ting Lai, Robert Leaman, Qingyu Chen, and Zhiyong Lu. 2023. AIONER: all-in-one scheme-based biomedical named entity recognition using deep learning. *Bioinformatics*, 39(5):btad310.

Zita Marinho, Afonso Mendes, Sebastião Miranda, and David Nogueira. 2019. Hierarchical nested named entity recognition. In *Proceedings of the 2nd Clinical Natural Language Processing Workshop*, pages 28–34, Minneapolis, Minnesota, USA. Association for Computational Linguistics.

Tahir Mehmood, Alfonso Gerevini, Alberto Lavelli, and Ivan Serina. 2019. Leveraging multi-task learning for biomedical named entity recognition. In *AI*IA 2019 – Advances in Artificial Intelligence*, pages 431–444, Cham. Springer International Publishing.

Sunil Mohan and Donghui Li. 2019. Medmentions: A large biomedical corpus annotated with umls concepts. *Preprint*, arXiv:1902.09476.

Vincenzo Moscato, Marco Postiglione, Carlo Sansone, and Giancarlo Sperlí. 2023. Taughtnet: Learning multi-task biomedical named entity recognition from single-task teachers. *IEEE Journal of Biomedical and Health Informatics*, 27(5):2512–2523.

Evangelos Pafilis, Sune P Frankild, Lucia Fanini, Sarah Faulwetter, Christina Pavloudi, Aikaterini Vasileiadou, Christos Arvanitidis, and Lars Juhl Jensen. 2013. The SPECIES and ORGANISMS resources for fast and accurate identification of taxonomic names in text. *PLoS One*, 8(6):e65390.

Yesol Park, Gyujin Son, and Mina Rho. 2024. Biomedical flat and nested named entity recognition: Methods, challenges, and advances. *Applied Sciences*, 14(20).

Abhishek Sharma, Amrita, Sudeshna Chakraborty, and Shivam Kumar. 2022. Named entity recognition in natural language processing: A systematic review. In *Proceedings of Second Doctoral Symposium on Computational Intelligence*, pages 817–828, Singapore. Springer Singapore.

Yongliang Shen, Xinyin Ma, Zeqi Tan, Shuai Zhang, Wen Wang, and Weiming Lu. 2021. Locate and label: A two-stage identifier for nested named entity recognition. In *Proceedings of the 59th Annual Meeting of the Association for Computational Linguistics and the 11th International Joint Conference on Natural Language Processing (Volume 1: Long Papers)*, pages 2782–2794, Online. Association for Computational Linguistics.

Yongliang Shen, Kaitao Song, Xu Tan, Dongsheng Li, Weiming Lu, and Yueting Zhuang. 2023. DiffusionNER: Boundary diffusion for named entity recognition. In *Proceedings of the 61st Annual Meeting of the Association for Computational Linguistics (Volume 1: Long Papers)*, pages 3875–3890, Toronto, Canada. Association for Computational Linguistics.

Yongliang Shen, Xiaobin Wang, Zeqi Tan, Guangwei Xu, Pengjun Xie, Fei Huang, Weiming Lu, and Yueting Zhuang. 2022. Parallel instance query network for named entity recognition. In *Proceedings of the 60th Annual Meeting of the Association for Computational Linguistics (Volume 1: Long Papers)*, pages 947–961, Dublin, Ireland. Association for Computational Linguistics.

Larry Smith, Lorraine K. Tanabe, Rie Johnson nee Ando, Cheng-Ju Kuo, I-Fang Chung, Chun-Nan Hsu, Yu-Shi Lin, Roman Klinger, Christoph M. Friedrich, Kuzman Ganchev, Manabu Torii, Hongfang Liu, Barry Haddow, Craig A. Struble, Richard J. Povinelli, Andreas Vlachos, William A. Baumgartner, Lawrence Hunter, Bob Carpenter, Richard Tzong-Han Tsai, Hong-Jie Dai, Feng Liu, Yifei Chen, Chengjie Sun, Sophia Katrenko, Pieter Adriaans, Christian Blaschke, Rafael Torres, Mariana Neves, Preslav Nakov, Anna Divoli, Manuel Maña-López, Jacinto Mata, and W. John Wilbur. 2008. Overview of biocreative ii gene mention recognition. *Genome Biology*, 9(2):S2.

Cong Sun, Zhihao Yang, Lei Wang, Yin Zhang, Hongfei Lin, and Jian Wang. 2021. Biomedical named entity recognition using bert in the machine reading comprehension framework. *Journal of Biomedical Informatics*, 118:103799.

Mario Sänger, Samuele Garda, Xing David Wang, Leon Weber-Genzel, Pia Droop, Benedikt Fuchs, Alan Akbik, and Ulf Leser. 2024. Hunflair2 in a cross-corpus evaluation of biomedical named entity recognition and normalization tools. *Bioinformatics*, 40(10):btae564.

Zeqi Tan, Yongliang Shen, Shuai Zhang, Weiming Lu, and Yueting Zhuang. 2021. A sequence-to-set network for nested named entity recognition. In *Proceedings of the Thirtieth International Joint Conference on Artificial Intelligence, IJCAI-21*, pages 3936–3942. International Joint Conferences on Artificial Intelligence Organization. Main Track.

Erik F. Tjong Kim Sang and Fien De Meulder. 2003. Introduction to the CoNLL-2003 shared task: Language-independent named entity recognition. In *Proceedings of the Seventh Conference on Natural Language Learning at HLT-NAACL 2003*, pages 142–147.

Xinyu Wang, Yong Jiang, Nguyen Bach, Tao Wang, Zhongqiang Huang, Fei Huang, and Kewei Tu. 2021. Automated concatenation of embeddings for structured prediction. In *Proceedings of the 59th Annual Meeting of the Association for Computational Linguistics and the 11th International Joint Conference on Natural Language Processing (Volume 1: Long Papers)*, pages 2643–2660, Online. Association for Computational Linguistics.

Xuan Wang, Yu Zhang, Xiang Ren, Yuhao Zhang, Marinka Zitnik, Jingbo Shang, Curtis Langlotz, and Jiawei Han. 2018. Cross-type biomedical named entity recognition with deep multi-task learning. *Bioinformatics*, 35(10):1745–1752.

Chih-Hsuan Wei, Alexis Allot, Kevin Riehle, Aleksandar Milosavljevic, and Zhiyong Lu. 2022. tmvar 3.0: an improved variant concept recognition and normalization tool. *Bioinformatics*, 38(18):4449–4451.

Chih-Hsuan Wei, Hung-Yu Kao, and Zhiyong Lu. 2015. GNormPlus: An integrative approach for tagging genes, gene families, and protein domains. *Biomed Res Int*, 2015:918710.

Thomas Wolf, Lysandre Debut, Victor Sanh, Julien Chaumond, Clement Delangue, Anthony Moi, Pierric Cistac, Tim Rault, Remi Louf, Morgan Funtowicz, Joe Davison, Sam Shleifer, Patrick von Platen, Clara Ma, Yacine Jernite, Julien Plu, Canwen Xu, Teven Le Scao, Sylvain Gugger, Mariama Drame, Quentin Lhoest, and Alexander Rush. 2020. Transformers: State-of-the-art natural language processing. In *Proceedings of the 2020 Conference on Empirical Methods in Natural Language Processing: System Demonstrations*, pages 38–45, Online. Association for Computational Linguistics.

Hang Yan, Yu Sun, Xiaonan Li, and Xipeng Qiu. 2023. An embarrassingly easy but strong baseline for nested named entity recognition. In *Proceedings of the 61st Annual Meeting of the Association for Computational Linguistics (Volume 2: Short Papers)*, pages 1442–1452, Toronto, Canada. Association for Computational Linguistics.

Michihiro Yasunaga, Jure Leskovec, and Percy Liang. 2022. LinkBERT: Pretraining language models with document links. In *Proceedings of the 60th Annual Meeting of the Association for Computational Linguistics (Volume 1: Long Papers)*, pages 8003–8016, Dublin, Ireland. Association for Computational Linguistics.

Yu Yin, Hyunjae Kim, Xiao Xiao, Chih Hsuan Wei, Jaewoo Kang, Zhiyong Lu, Hua Xu, Meng Fang, and Qingyu Chen. 2024. Augmenting biomedical named entity recognition with general-domain resources. *Journal of Biomedical Informatics*, 159:104731.

Juntao Yu, Bernd Bohnet, and Massimo Poesio. 2020. Named entity recognition as dependency parsing. In *Proceedings of the 58th Annual Meeting of the Association for Computational Linguistics*, pages 6470–6476, Online. Association for Computational Linguistics.

Sheng Zhang, Hao Cheng, Jianfeng Gao, and Hoifung Poon. 2023. Optimizing bi-encoder for named entity recognition via contrastive learning. In *The Eleventh International Conference on Learning Representations*.

A Datasets and Experimental Setup

For the English subset of CoNLL-2003 (Tjong Kim Sang and De Meulder, 2003), the original dataset splits are used, which are provided in a pre-tokenized format. For the GENIA dataset, the splits from Yan et al., 2023 are adopted. The entity counts per split of these datasets can be found in Table 7.

Dataset	Entity Type	Train	Dev	Test
CONLL	LOC	7,140	1,837	1,668
	MISC	3,438	922	702
	ORG	6,321	1,341	1,661
	PER	6,600	1,842	1,617
GENIA	Cell Line	3,069	372	403
	Cell Type	5,854	576	578
	DNA	7,707	1,161	1,132
	Gene or protein	28,874	2,466	2,900
	RNA	699	139	106

Table 7: Statistics for the datasets used in the single-task experiments of section 5.1.

To train a multi-task model, six biomedical datasets are utilized: BC2GM (Smith et al., 2008), BC4CHEMD (Krallinger et al., 2015), BC5CDR (Li et al., 2016), JNLPBA (Collier et al., 2004), Linnaeus (Gerner et al., 2010), and NCBI Disease (Doğan et al., 2014). The dataset splits (Table 8) follow those established by Crichton et al., 2017, which have been extensively used in prior studies, including Wang et al., 2018; Huang et al., 2019; Khan et al., 2020; Moscato et al., 2023.

Dataset	Entity Type	Train	Dev	Test
BC2GM	Gene or protein	15,035	3,032	6,243
BC4CHEMD	Chemical	29,263	29,305	25,210
BC5CDR	Chemical	5,114	5,239	5,277
	Disease	4,169	4,224	4,394
JNLPBA	Cell Line	3,369	389	490
	Cell Type	6,162	522	1,906
	DNA	8,416	1,040	1,045
	Gene or protein	27,015	2,379	4,988
	RNA	844	106	118
Linnaeus	Species	2,079	700	1,412
NCBI Disease	Disease	5,111	779	952

Table 8: Statistics for the datasets used in the multi-task experiment of section 5.2.

In the aforementioned experiments, models are trained on the respective training splits, checkpoint selection is made on the development splits, and evaluation is conducted on the test splits.

For the cross-corpus evaluation, the experimental setup of Sänger et al., 2024 is replicated. A multi-task model is trained using an ensemble of nine datasets[9]: BioRED (Luo et al., 2022), GNormPlus (Wei et al., 2015), Linnaeus (Gerner et al., 2010), NCBI Disease (Doğan et al., 2014), NLM-Chem (Islamaj et al., 2021a), NLM-Gene (Islamaj et al., 2021b), S800 (Pafilis et al., 2013), SCAI Chemical (Kolarik et al., 2008), and SCAI Disease (Gurulingappa et al., 2010). The model is trained on the training sets, with checkpoint selection being performed on the development splits. The evaluation is conducted on an independent corpus consisting of the full annotated data of three datasets[10]: BioID (Arighi et al., 2017), MedMentions (Mohan and Li, 2019), and tmVar3 (Wei et al., 2022). Dataset statistics for the training corpora and the independent test corpora can be found in Table 9 and Table 10, respectively.

Dataset	Entity Type	Train	Dev	Test
BioRED	Cell Line	103	22	50
	Chemical	2,830	818	751
	Disease	3,643	982	917
	Gene	4,404	1,087	1,170
	Species	1,429	370	393
GNormPlus	Gene	4,964	504	4,468
Linneaus	Species	1,725	206	793
NCBI Disease	Disease	4,083	666	2,109
NLM-Chem	Chemical	21,102	5,223	11,571
NLM-Gene	Gene	11,209	1,314	2,687
S800	Species	2,236	410	1,079
SCAI Chemical	Chemical	852	83	375
SCAI Disease	Disease	1,281	250	710

Table 9: Statistics of the training corpora used in the cross-corpus evaluation scenario of section 5.3.

Dataset	Entity Type	Number of mentions
BioID	Species	7,939
tmVar3	Gene	4,059
MedMentions	Disease	19,298
	Chemical	19,198

Table 10: Statistics of the corpora used for the cross-corpus evaluation described in section 5.3.

Finally, in order to assess the model's global prediction ability, synthetic datasets were derived from the BC5CDR corpus, in line with (Huang et al.,

[9]The datasets were obtained in February 2025 from `https://github.com/flairnlp/flair`. Their splits and preprocessing choices were replicated.

[10]The preprocessed datasets were downloaded from `https://github.com/hu-ner/hunflair2-experiments` in February 2025.

2019) experimental setup. The original training set was randomly divided into two disjoint subsets: BC5-Disease (containing only Disease annotations) and BC5-Chemical (containing only Chemical annotations). The same procedure was followed for the development splits. The statistics of these synthetic datasets can be found in Table 11. By training models on the BC5-Disease and BC5-Chemical subsets and evaluating them on the full test split of the BC5CDR corpus, we can test the models global prediction abilities, as described in section 5.4.

Dataset	Entity Type	Train	Dev
BC5-Disease	Disease	2,172	2,279
BC5-Chemical	Chemical	2,459	2,665

Table 11: Statistics of the synthetic datasets created for assessing global prediction ability.

B Training Details

Hyperparameter	GENIA	Others
# epochs	100	100
Early stop	30	30
Batch size	16	16
Max. # tokens	405	405
Gradient norm clipping	1.0	1.0
Dropout on logits	0.1	0.1
SRU module		
# latent embeddings (multiplier)	10	2
Half-context for pos. embeddings	240	150
Dropout on pos. embeddings	0.2	0.2
Dropout on latent embeddings	0.2	0.2
Encoder optimizer		
LR	3e-5	2e-5
Weight decay	1e-3	1e-3
Warm up (in epochs)	1	1
Actions generation cycle optimizer		
LR	3e-4	3e-4
Weight decay	1e-3	1e-3
Warm up (in epochs)	0.5	0.5

Table 12: Hyperparameters used for the experiments. The column 'Others' refers to every experiment except the single-task on the GENIA dataset.

All models are developed using the PyTorch tensor library and trained on a single NVIDIA A100 80GB GPU. The encoder module and the action generation module are tuned using two separate AdamW optimizers with linear warm-up, set with different initial learning rates and weight decays. Both optimizers are set with $\beta_1 = 0.9$, $\beta_2 = 0.98$ and $\epsilon = 10^{-6}$. Models are trained with early stopping based on performance on the development set.[11] The hyperparameters of all experiments can be found in Table 12. Additionally, while the token scaling parameter α in equation (5) of section 3.3 was trained for the single-task experiment on the GENIA dataset, it was frozen and set to 1 for all other experiments.

The encoder module was built on top of the HuggingFace *transformers* library (Wolf et al., 2020). Specifically, the `LinkBERT-large` encoder from Yasunaga et al., 2022 was used for all models trained with biomedical corpora, while the `xlm-roberta-large` encoder introduced by Conneau et al., 2020 was used for the single task model trained on the CoNLL-2003 dataset.

C Single-task performance on the datasets used for the cross-corpus experiment

Dataset	*Merged*	*Disjoint*
BioRED	90.73	90.90
GNormPlus	85.00	86.00
Linnaeus	78.16	92.23
NCBI Disease	85.69	85.70
NLM-Chem	84.42	85.65
NLM-Gene	88.35	88.13
S800	74.24	75.79
SCAI Chemical	85.21	85.64
SCAI Disease	80.78	82.14

Table 13: In-corpus micro-F1 scores for the model used in the cross-corpus evaluation experiment of section 5.3.

D Human evaluation of global predictions in a cross-domain setting

To assess the model's ability to generalize across domains, three models were trained:

- *SRU-NER-CoNLL*: a single-task model trained on the CoNLL corpus;
- *SRU-NER-BC5*: a single-task model trained on the BC5CDR corpus;
- *SRU-NER-MTL*: a multi-task model trained on both corpora.

All models were trained using the `LinkBERT-large` encoder from Yasunaga

[11] In the case of multi-task models where multiple datasets are tagged with the same entity type (the models of sections 5.2 and 5.3), despite the entity types being considered disjoint for training purposes, validation F1 scores on the development set for checkpoint selection are computed by merging the types, as described in the begining of section 5.2.

et al., 2022. To evaluate cross-domain generalization, the models capable of recognizing general-domain entity types (*SRU-NER-CoNLL* and *SRU-NER-MTL*) were used to annotate the test split of the biomedical corpus, while the models trained on biomedical entity types (*SRU-NER-BC5* and *SRU-NER-MTL*) were used to annotate the test split of the general-domain corpus. Since gold annotations for these out-of-domain predictions were not available, two linguists manually assessed their correctness. Inter-annotator agreement per entity type is reported in Table 14.

Entity	**Agreement** (%)
Chemical	92.98
Disease	91.09
LOC	100.00
ORG	87.76
PER	88.89

Table 14: Inter-annotator agreement for the evaluated entity types.

Based on the assessment of correct predicted spans by the two human annotators, a precision score was computed by taking the ratio of correctly identified spans to the total number of predicted spans, for each model, entity type and linguist. A simulated recall score per model, entity type and linguist was also computed by considering the total number of spans of each entity type that were considered correct by at least one of the annotators, across all the predictions made by the three models. Finally, precision and simulated recall scores per model and entity type were obtained by averaging across the two human annotators.

The results can be found in Table 6, in the main text. One can see that the precision scores of the multi-task model are higher than the single-task ones across all entity types, while the recall values of the multi-task model are worse for all entity types except ORG.

For reference, the in-corpus performance of the three models is present in Table 15.

Model	**CoNLL**	**BC5CDR**
SRU-NER-CoNLL	90.51	—
SRU-NER-BC5	—	90.61
SRU-NER-MT	91.01	90.51

Table 15: In-corpus performance of the three models used for evaluation of global predictions in a cross-domain setting. The single-task model SRU-NER-BC5 is the same as the one used for comparison in the multi-task experiment of section 5.2.

Can Large Language Models Classify and Generate Antimicrobial Resistance Genes?

Hyunwoo Yoo[1] **Haebin Shin**[2] **Gail Rosen**[1]

[1]Drexel University [2]KAIST AI

{hty23, glr26}@drexel.edu haebin.shin@kaist.ac.kr

Abstract

This study explores the application of generative Large Language Models (LLMs) in DNA sequence analysis, highlighting their advantages over encoder-based models like DNABERT2 and Nucleotide Transformer. While encoder models excel in classification, they struggle to integrate external textual information. In contrast, generative LLMs can incorporate domain knowledge, such as BLASTn annotations, to improve classification accuracy even without fine-tuning. We evaluate this capability on antimicrobial resistance (AMR) gene classification, comparing generative LLMs with encoder-based baselines. Results show that LLMs significantly enhance classification when supplemented with textual information. Additionally, we demonstrate their potential in DNA sequence generation, further expanding their applicability. Our findings suggest that LLMs offer a novel paradigm for integrating biological sequences with external knowledge, bridging gaps in traditional classification methods.

1 Introduction

Language Models (LMs) have demonstrated remarkable performance in various Natural Language Processing (NLP) tasks and have recently gained attention in bioinformatics, particularly in DNA sequence analysis. Encoder-based transformer models, such as DNABERT (Ji et al., 2021; Zhou et al., 2023), ProteinBERT (Brandes et al., 2022) and Nucleotide Transformer (Dalla-Torre et al., 2023), have shown strong performance in DNA sequence classification, leveraging nucleotide tokenization and self-supervised pretraining. These models are widely adopted for gene sequence analysis, promoter prediction, and mutation impact assessment. However, encoder-based models have inherent limitations in integrating external domain knowledge. Their fixed input structures make it difficult to incorporate additional textual context, such as BLASTn search results, which often contain critical biological insights. Additionally, these models may struggle to generalize when a single DNA sequence is associated with multiple labels, requiring strict pre-defined training paradigms.

Generative Large Language Models (LLMs), such as GPT-based models (Brown et al., 2020), introduce greater flexibility by allowing predictions to be influenced by external knowledge via prompting. Unlike encoder-based models, generative LLMs can dynamically incorporate supplementary textual information, which can enhance classification accuracy without requiring fine-tuning. Some biomedical LLMs, such as BioGPT (Luo et al., 2022) and Med-PaLM (Singhal et al., 2023), have demonstrated strong performance in processing medical and pharmaceutical text, but their applications in DNA sequence analysis remain underexplored. Beyond classification, generative LLMs also enable DNA sequence generation (Nguyen et al., 2024; Brixi et al., 2025), a capability that traditional encoder-based models lack. This opens new possibilities for exploring sequence design, mutation modeling, and synthetic data augmentation, further expanding the applicability of LLMs in genomics.

This study systematically evaluates the effectiveness of generative LLMs for DNA sequence classification and generation, comparing them against traditional encoder-based baselines. Our key contributions are as follows:

- We systematically compare generative LLMs with encoder-based models on DNA sequence classification tasks, providing a rigorous evaluation of their relative performance.
- We demonstrate that generative LLMs can leverage supplementary domain-specific knowledge to improve classification accuracy even without fine-tuning.

Proceedings of the 24th Workshop on Biomedical Language Processing (BioNLP 2025), pages 240–248

- We explore the potential of generative LLMs in DNA sequence generation, analyzing their ability to generate biologically meaningful sequences and their implications for synthetic data augmentation.

Our findings suggest that generative LLMs offer a novel paradigm for integrating DNA sequences with external knowledge sources, expanding their applicability in bioinformatics research.

2 Related Works

Transformer-based encoder models have been widely applied to DNA sequence classification. DNABERT (Ji et al., 2021; Zhou et al., 2023) applies self-supervised learning to nucleotide sequences using k-mer tokenization, while DNABERT2 improves efficiency by introducing byte pair encoding (BPE) (Zhou et al., 2023). Nucleotide Transformer (Dalla-Torre et al., 2023) extends this approach by pretraining on diverse genomic datasets, achieving strong performance in gene classification tasks.

While these models perform well in classification, they have limited ability to incorporate external domain knowledge, such as BLASTn annotations (Lobo, 2008). Moreover, they struggle with handling multi-label classification, which is common in genomic studies (Bonin et al., 2023a; Marini et al., 2022). Our work differs by exploring whether generative LLMs can improve classification performance by dynamically integrating external textual information without additional fine-tuning.

Generative Large Language Models (LLMs) such as GPT-based models (Brown et al., 2020) have demonstrated strong natural language understanding but have been rarely applied to DNA sequence analysis. BioGPT (Luo et al., 2022), for example, is trained on biomedical literature but lacks direct training on DNA sequences.

Unlike encoder-based models, LLMs can dynamically incorporate supplementary textual information, such as BLASTn search results (Lobo, 2008), potentially enhancing classification performance. Additionally, LLMs have the potential for DNA sequence generation, which can be applied to mutation modeling and synthetic data augmentation, as demonstrated in previous studies exploring deep learning methods for genomic analysis (Marini et al., 2022; Arango-Argoty et al., 2018; Lakin et al., 2019).

While prior studies have focused on applying LLMs to biomedical text, our approach investigates whether generative LLMs can be effectively utilized for both classification and sequence generation in DNA analysis, providing a flexible alternative to traditional encoder-based models.

3 Methods

3.1 Data Collection

The dataset used in this study consists of antibiotic resistance gene sequences collected from the MEGARes (Doster et al., 2020; Bonin et al., 2023b) and CARD databases (Jia et al., 2017). The labels from MEGARes and CARD were mapped using the Antibiotic Resistance Ontology from the European Bioinformatics Institute (Cook et al., 2016), following previous research methods (Yoo et al., 2024). These databases contain DNA sequences associated with antimicrobial resistance (AMR) and provide multi-label annotations, where a single sequence may belong to multiple resistance categories. To incorporate external domain knowledge, we applied the BLASTn algorithm (Chen et al., 2015) to identify sequences similar to each DNA sequence in the dataset. For each sequence, the top-5 BLASTn search results were selected based on the e-value criterion, and their corresponding functional annotations were extracted. This additional textual information includes gene descriptions, known resistance mechanisms, and sequence alignment details, which were later integrated into our LLM-based classification prompts.

3.2 Baseline Models and Preprocessing

To compare the performance of generative LLMs with existing DNA sequence classification models, we included encoder-based models as baselines: DNABERT2, and Nucleotide Transformer. DNABERT2 (Zhou et al., 2023) is a BERT-based model to process DNA sequences as natural language text. It improved version of DNABERT (Ji et al., 2021) by introducing byte pair encoding (BPE) instead of utilizing k-mer tokenization, allowing for more efficient sequence representation. Nucleotide Transformer (Dalla-Torre et al., 2023), a transformer model pre-trained on diverse genomic datasets, has demonstrated strong performance in various molecular phenotype prediction tasks.

For all models, DNA sequences were preprocessed by converting them to uppercase, and in-

Model	Accuracy	Precision	Recall	F1 Score
DNABERT2 (Finetuning)	0.8697	0.8161	0.6996	0.7332
Nucleotide Transformer (Finetuning)	0.8289	0.8184	0.5867	0.6579
LLama3.1 8B-4bit	0.0037	0.0011	0.0002	0.0003
LLama3.1 8B-4bit + Blastn	0.0744	0.0530	0.0129	0.0207
LLama3.1 8B-4bit + Finetuning	0.5521	0.4760	0.5521	0.5080
Claude3.5sonet	0.1488	0.1770	0.0966	0.0735
Claude3.5sonet + Blastn	0.8042	0.6287	0.5421	0.5794
Chatgpt4o-mini	0.00	0.00	0.00	0.00
Chatgpt4o-mini + Blastn	0.7804	0.9090	0.7804	0.8398
Chatgpt4o-mini + Finetuning	0.9318	0.9337	0.9318	0.9319

Table 1: Performance metrics for DNA sequence classification across multiple models. Chatgpt4o-mini with finetuning achieves the best overall performance, with the highest accuracy and F1 score, surpassing specialized models like DNABERT2 and the Nucleotide Transformer. Without finetuning or external features, general-purpose LLMs such as LLaMA3.1, Claude3.5, and Chatgpt4o-mini perform poorly, indicating that both biological context via BLASTn and domain-specific adaptation are critical for DNA sequence understanding.

valid sequences were removed. The final dataset consisted only of validated antibiotic resistance gene sequences.

3.3 Fine-tuning and Prompt-based Classification

To evaluate generative LLMs in DNA sequence classification, we employed Low-Rank Adaptation (LoRA) fine-tuning on LLaMA. LoRA enables parameter-efficient tuning by modifying only a subset of the model's weight matrices, significantly reducing computational costs while maintaining performance (Hu et al., 2021). Further details provided in Appendix B. Additionally, we conducted zero-shot inference using the Claude 3.5 sonet (Anthropic, 2024) and ChatGPT-4 API to assess how well proprietary generative models classify DNA sequences without explicit training. To investigate whether generative LLMs can classify DNA sequences without fine-tuning, we formulated two experimental settings. In the first setting, models were given only the DNA sequence as input. In the second setting, models received both the DNA sequence and the top-5 BLASTn search results, including functional annotations and gene descriptions. This setup allowed us to assess whether LLMs can leverage external domain knowledge to improve classification accuracy. Prompt details are in Appendix A

3.4 LLM-Based DNA Sequence Generation

In addition to classification, we explored whether generative LLMs can synthesize biologically meaningful DNA sequences. To this end, we designed a sequence generation task where models were prompted with initial part of antibiotic resistance genes and tasked with generating plausible full DNA sequences. We implemented finetuning with LLMs. Further details provided in Appendix C. The quality of the generated sequences was assessed using three similarity measures. Levenshtein (Levenshtein, 1966) distance was used to quantify the sequence-level similarity by measuring the minimum number of edits (insertions, deletions, and substitutions) required to match a reference sequence. Jaccard's index of similarity (Real and Vargas, 1996) was computed on k-mer tokenized sequences to evaluate overlapping subsequences between generated and known resistance gene sequences. Cosine similarity was also applied to k-mer frequency vectors to compare overall sequence composition (Ng, 2017). To ensure basic functional validity, we checked whether generated sequences maintained proper nucleotide composition. GC content (Marmur and Doty, 1962) distribution was compared to existing AMR sequences to verify biological plausibility.

4 Experiments

4.1 DNA Classification

To evaluate the performance of generative language models in DNA sequence classification, we conducted experiments under three conditions. The first setting involved using the base models, where only the raw DNA sequence was provided as input. The second setting introduced BLASTn search results, incorporating additional textual annotations such as gene descriptions and resistance mechanisms. The third setting involved fine-tuning the models using labeled DNA sequences. For baseline comparisons, we included DNABERT2 and Nucleotide Transformer, which have demonstrated strong performance in DNA sequence classification tasks. The generative models evaluated in this study include LLaMA 3.1 (Meta AI, 2024) (8B-4bit), Claude 3.5 Sonet, and ChatGPT-4o-mini. Each model was tested in zero-shot, BLASTn-

Model	Jaccard Similarity	Cosine Similarity	Levenshtein Similarity	GC Correlation
GENERater (Zero-shot)	0.9970	0.9680	0.3790	0.8436
ChatGPT-4o-mini (Fine-tuned API)	0.9870	0.9857	0.5776	0.7930
GENErator (LoRA Fine-tuned)	0.9970	0.9680	0.3790	0.8436
LLaMA 3.2 1B (LoRA Fine-tuned)	0.2659	0.5911	0.2004	0.6938
Gemma 3 1B (LoRA Fine-tuned)	0.3177	0.7580	0.2487	0.7305

Table 2: Similarity scores between generated DNA sequences and the ground truth across various models. GENERater, both in zero-shot and LoRA fine-tuned settings, achieves near-perfect Jaccard and Cosine similarities, with strong GC content correlation, indicating high biological fidelity. ChatGPT-4o-mini also performs competitively despite being a general-purpose LLM. In contrast, smaller fine-tuned models like LLaMA 3.2 1B and Gemma 3 1B yield significantly lower similarity scores across all metrics, highlighting the challenge of DNA generation in low-resource model settings.

augmented, and fine-tuned configurations.

4.2 DNA Sequence Generation

In addition to classification, we assessed whether generative language models could synthesize biologically meaningful DNA sequences. A dataset of antimicrobial resistance genes from Acinetobacter baumannii was collected using the NCBI Entrez API, with 1,000 sequences retrieved. The dataset was split into 80% for training and 20% for testing. Input sequences were trimmed to a length of 200 base pairs, while the maximum generated output length was set to 3,000 base pairs. Further details on dataset characteristics provided in Appendix E. For baseline comparisons, we included GENERater (Wu et al., 2025), which were evaluated in a zero-shot setting. For fine-tuned models, we used ChatGPT-4o-mini finetuned via API along with GENErator, LLaMA 3.2 1B (Grattafiori et al., 2024), and Gemma 3 1B (Gemma Team, Google DeepMind, 2025), which were finetuned using the LoRA. Each model was assessed based on its ability to generate sequences that resemble known antimicrobial resistance genes.

5 Results and Discussion

Table 1 presents the classification results across various model configurations. Encoder-based models, DNABERT2 and Nucleotide Transformer, consistently demonstrated the highest accuracy, with DNABERT2 achieving 86.97% accuracy and Nucleotide Transformer reaching 82.89%. In contrast, generative models performed poorly in the base setting, with LLaMA 3.1 obtaining an accuracy of only 0.37%. Considering this outcome alongside the unclassified rate reported in Appendix D, it appears that generative models have difficulty performing direct DNA sequence classification without supplementary context. The inclusion of BLASTn search results significantly improved classification accuracy. ChatGPT-4o-mini, which initially failed to classify any sequences correctly, achieved 78.04% accuracy with BLASTn augmentation. Similarly, Claude 3.5 Sonet improved from 14.88% to 80.42% accuracy. These results suggest that LLMs benefit from external textual information, compensating for their lack of prior exposure to DNA sequences. Fine-tuning further enhanced classification accuracy, with ChatGPT-4o-mini achieving 93.18%, surpassing DNABERT2. This demonstrates that while LLMs struggle in a zero-shot setting, targeted training on DNA sequences allows them to match or exceed the performance of specialized encoder-based models.

Table 2 summarizes the similarity scores for generated DNA sequences. In the zero-shot setting, GENERater produced sequences with high Jaccard similarity (0.9970) and Cosine similarity (0.9680), but relatively low Levenshtein similarity (0.3790), indicating that while generated sequences share common k-mers with known resistance genes, their exact sequence composition differs significantly. Fine-tuned models exhibited varying levels of similarity. ChatGPT-4o-mini, fine-tuned via API, achieved the highest similarity across all three metrics, particularly in Levenshtein similarity (0.5776), suggesting that it generated sequences more closely aligned with known resistance genes at the character level. GENErator (LoRA Fine-tuned) maintained nearly identical similarity scores to its zero-shot counterpart, whereas LLaMA 3.2 1B and Gemma 3 1B displayed substantially lower similarity scores across all metrics, indicating challenges in generating sequences that closely resemble existing DNA. Further analysis of GC content confirmed that fine-tuned models generated biologically plausible sequences. However, additional validation is required to determine whether these sequences retain functional properties relevant to antimicrobial resistance.

6 Conclusion

This study demonstrated that generative LLMs offer greater flexibility in DNA sequence classification and generation compared to traditional encoder-based models. While encoder models like DNABERT2 performed well in standard classification tasks, generative models benefited significantly from additional textual information, highlighting their ability to integrate external domain knowledge. Fine-tuned generative models also produced biologically plausible DNA sequences, suggesting potential applications in synthetic biology. However, LLMs struggled in zero-shot classification, emphasizing the need for fine-tuning and improved biological data integration.

7 Limitations

While this study highlights the potential of generative LLMs in DNA sequence analysis, there are several areas for further improvement. Zero-shot classification performance remained limited, underscoring the need for fine-tuning or integrating external biological knowledge to enhance prediction accuracy. Future work could explore hybrid approaches that combine LLMs with domain-specific models or structured databases to improve robustness.

In DNA sequence generation, fine-tuned models successfully produced sequences structurally similar to known antimicrobial resistance genes. However, additional real-world validation through laboratory experiments is necessary to determine whether these sequences retain functional properties relevant to resistance mechanisms.

Another key consideration is the computational cost associated with fine-tuning large-scale models. The substantial resource requirements highlight the need for more efficient adaptation techniques, such as parameter-efficient fine-tuning or retrieval-augmented approaches. Future research should investigate methods to balance computational efficiency with model performance to enable broader accessibility and practical applications in bioinformatics.

Acknowledgments

This work was supported in part by the National Science Foundation (NSF) under Grant Number 2107108.

References

Anthropic. 2024. The claude 3 model family: Opus, sonnet, haiku. Technical report, Anthropic. `https://www.anthropic.com/index/claude-3-opus-sonnet-haiku`.

Gustavo Arango-Argoty, Emily Garner, Amy Pruden, Lenwood S. Heath, Peter Vikesland, and Liqing Zhang. 2018. Deeparg: A deep learning approach for predicting antibiotic resistance genes from metagenomic data. *Microbiome*, 6(1):23.

Nathalie Bonin, Enrique Doster, Hannah Worley, Lee J Pinnell, Jonathan E Bravo, Peter Ferm, Simone Marini, Mattia Prosperi, Noelle Noyes, Paul S Morley, and Christina Boucher. 2023a. Megares and amr++, v3.0: an updated comprehensive database of antimicrobial resistance determinants and an improved software pipeline for classification using high-throughput sequencing. *Nucleic Acids Research*, 51(D1):D744–D752.

Nathalie Bonin, Enrique Doster, Hannah Worley, Lee J Pinnell, Jonathan E Bravo, Peter Ferm, Simone Marini, Mattia Prosperi, Noelle Noyes, Paul S Morley, and Christina Boucher. 2023b. MEGARes and AMR++, v3.0: an updated comprehensive database of antimicrobial resistance determinants and an improved software pipeline for classification using high-throughput sequencing. *Nucleic Acids Research*, 51(D1):D744–D752.

Nadav Brandes, Dan Ofer, Yam Peleg, Nadav Rappoport, and Michal Linial. 2022. Proteinbert: a universal deep-learning model of protein sequence and function. *Bioinformatics*, 38(8):2102–2110.

Garyk Brixi, Matthew G. Durrant, Jerome Ku, Michael Poli, Greg Brockman, Daniel Chang, Gabriel A. Gonzalez, Samuel H. King, David B. Li, Aditi T. Merchant, Mohsen Naghipourfar, Eric Nguyen, Chiara Ricci-Tam, David W. Romero, Gwanggyu Sun, Ali Taghibakshi, Anton Vorontsov, Brandon Yang, Myra Deng, Liv Gorton, Nam Nguyen, Nicholas K. Wang, Etowah Adams, Stephen A. Baccus, Steven Dillmann, Stefano Ermon, Daniel Guo, Rajesh Ilango, Ken Janik, Amy X. Lu, Reshma Mehta, Mohammad R.K. Mofrad, Madelena Y. Ng, Jaspreet Pannu, Christopher Ré, Jonathan C. Schmok, John St. John, Jeremy Sullivan, Kevin Zhu, Greg Zynda, Daniel Balsam, Patrick Collison, Anthony B. Costa, Tina Hernandez-Boussard, Eric Ho, Ming-Yu Liu, Thomas McGrath, Kimberly Powell, Dave P. Burke, Hani Goodarzi, Patrick D. Hsu, and Brian L. Hie. 2025. Genome modeling and design across all domains of life with evo 2. *bioRxiv*.

Tom B. Brown, Benjamin Mann, Nick Ryder, Melanie Subbiah, Jared Kaplan, Prafulla Dhariwal, Arvind Neelakantan, Pranav Shyam, Girish Sastry, Amanda Askell, Sandhini Agarwal, Ariel Herbert-Voss, Gretchen Krueger, Tom Henighan, Rewon Child, Aditya Ramesh, Daniel M. Ziegler, Jeffrey Wu, Clemens Winter, Christopher Hesse, Mark Chen, Eric

Sigler, Mateusz Litwin, Scott Gray, Benjamin Chess, Jack Clark, Christopher Berner, Sam McCandlish, Alec Radford, Ilya Sutskever, and Dario Amodei. 2020. Language models are few-shot learners. *arXiv preprint arXiv:2005.14165*.

Ying Chen, Weicai Ye, Yongdong Zhang, and Yuesheng Xu. 2015. High speed blastn: an accelerated megablast search tool. *Nucleic Acids Research*, 43(16):7762–7768.

Charles E. Cook, Mary Todd Bergman, Robert D. Finn, Guy Cochrane, Ewan Birney, and Rolf Apweiler. 2016. The european bioinformatics institute in 2016: Data growth and integration. *Nucleic Acids Research*, 44(D1):D20–D26.

Hugo Dalla-Torre, Liam Gonzalez, Javier Mendoza-Revilla, Nicolas Lopez Carranza, Adam Henryk Grzywaczewski, Francesco Oteri, Christian Dallago, et al. 2023. The nucleotide transformer: Building and evaluating robust foundation models for human genomics. *Genomics*.

Enrique Doster, Steven M Lakin, Christopher J Dean, Cory Wolfe, Jared G Young, Christina Boucher, Keith E Belk, Noelle R Noyes, and Paul S Morley. 2020. Megares 2.0: a database for classification of antimicrobial drug, biocide and metal resistance determinants in metagenomic sequence data. *Nucleic Acids Research*, 48(D1):D561–D569.

Gemma Team, Google DeepMind. 2025. Gemma 3 technical report. Technical report, Google DeepMind. See Contributions and Acknowledgments section for full author list.

Aaron Grattafiori, Abhimanyu Dubey, Abhinav Jauhri, Abhinav Pandey, Abhishek Kadian, Ahmad Al-Dahle, Aiesha Letman, Akhil Mathur, Alan Schelten, Alex Vaughan, Amy Yang, Angela Fan, Anirudh Goyal, Anthony Hartshorn, Aobo Yang, Archi Mitra, Archie Sravankumar, Artem Korenev, Arthur Hinsvark, Arun Rao, Aston Zhang, Aurelien Rodriguez, Austen Gregerson, Ava Spataru, Baptiste Roziere, Bethany Biron, Binh Tang, Bobbie Chern, et al. 2024. The llama 3 herd of models. *arXiv preprint arXiv:2407.21783*.

Edward J. Hu, Yelong Shen, Phillip Wallis, Zeyuan Allen-Zhu, Yuanzhi Li, Shean Wang, Lu Wang, and Weizhu Chen. 2021. Lora: Low-rank adaptation of large language models. *arXiv*. ArXiv:2106.09685v2.

Yanrong Ji, Zhihan Zhou, Han Liu, and Ramana V Davuluri. 2021. Dnabert: pre-trained bidirectional encoder representations from transformers model for dna-language in genome. *Bioinformatics*, 37(15):2112–2120.

Baofeng Jia, Amogelang R. Raphenya, Brian Alcock, Nicholas Waglechner, Peiyao Guo, Kara K. Tsang, Briony A. Lago, Biren M. Dave, Sheldon Pereira, Arjun N. Sharma, Sachin Doshi, Mélanie Courtot, Raymond Lo, Laura E. Williams, Jonathan G. Frye, Tariq Elsayegh, Daim Sardar, Erin L. Westman, Andrew C. Pawlowski, Timothy A. Johnson, Fiona S.L. Brinkman, Gerard D. Wright, and Andrew G. McArthur. 2017. Card 2017: expansion and model-centric curation of the comprehensive antibiotic resistance database. *Nucleic Acids Research*, 45(D1):D566–D573.

Steven M. Lakin, Alan Kuhnle, Bahar Alipanahi, Noelle R. Noyes, Chris Dean, Martin Muggli, Rob Raymond, et al. 2019. Hierarchical hidden markov models enable accurate and diverse detection of antimicrobial resistance sequences. *Communications Biology*, 2(1):294.

V. I. Levenshtein. 1966. Binary codes capable of correcting deletions, insertions, and reversals. *Soviet Physics-Doklady*, 10(8):707–710. Translated from Doklady Akademii Nauk SSSR, Vol. 163, No. 4, pp. 845-848, August 1965.

Ingrid Lobo. 2008. Basic local alignment search tool (blast). *Nature Education*, 1(1):215. © 2008 Nature Education.

Renqian Luo, Liai Sun, Yingce Xia, Tao Qin, Sheng Zhang, Hoifung Poon, and Tie-Yan Liu. 2022. Biogpt: Generative pre-trained transformer for biomedical text generation and mining. *Briefings in Bioinformatics*, 23(6):bbac409.

Simone Marini, Marco Oliva, Ilya B Slizovskiy, Rishabh A Das, Noelle Robertson Noyes, Tamer Kahveci, Christina Boucher, and Mattia Prosperi. 2022. Amr-meta: A k -mer and metafeature approach to classify antimicrobial resistance from high-throughput short-read metagenomics data. *GigaScience*, 11. Giac029.

J. Marmur and P. Doty. 1962. Determination of the base composition of deoxyribonucleic acid from its thermal denaturation temperature. *Journal of Molecular Biology*, 5:109–118.

Meta AI. 2024. Introducing llama 3.1: Our most capable models to date. `https://ai.meta.com/blog/llama-3-1`. Accessed: 2025-03-20.

Patrick Ng. 2017. dna2vec: Consistent vector representations of variable-length k-mers. *arXiv preprint arXiv:1701.06279*.

Eric Nguyen, Michael Poli, Matthew G. Durrant, Brian Kang, Dhruva Katrekar, David B. Li, Liam J. Bartie, Armin W. Thomas, Samuel H. King, Garyk Brixi, Jeremy Sullivan, Madelena Y. Ng, Ashley Lewis, Aaron Lou, Stefano Ermon, Stephen A. Baccus, Tina Hernandez-Boussard, Christopher Ré, Patrick D. Hsu, and Brian L. Hie. 2024. Sequence modeling and design from molecular to genome scale with evo. *Science*, 386(6723):eado9336.

Raimundo Real and Juan M. Vargas. 1996. The probabilistic basis of jaccard's index of similarity. *Systematic Biology*, 45(3):380–385.

Karan Singhal, Shekoofeh Azizi, Tao Tu, S Sara Mahdavi, Jason Wei, Hyung Won Chung, Nathan Scales, Ajay Tanwani, Heather Cole-Lewis, Stephen Pfohl, Perry Payne, Martin Seneviratne, Paul Gamble, Chris Kelly, Abubakr Babiker, Nathanael Schärli, Aakanksha Chowdhery, Philip Mansfield, Dina Demner-Fushman, Blaise Agüera y Arcas, Dale Webster, Greg S Corrado, Yossi Matias, Katherine Chou, Juraj Gottweis, Nenad Tomasev, Yun Liu, Alvin Rajkomar, Joelle Barral, Christopher Semturs, Alan Karthikesalingam, and Vivek Natarajan. 2023. Large language models encode clinical knowledge. *Nature*, 620:172–180. Publisher Correction published on 27 July 2023.

Wei Wu, Qiuyi Li, Mingyang Li, Kun Fu, Fuli Feng, Jieping Ye, Hui Xiong, and Zheng Wang. 2025. GENERator: A Long-Context Generative Genomic Foundation Model. *arXiv preprint*, arXiv:2502.07272.

Hyunwoo Yoo, Bahrad Sokhansanj, James R. Brown, and Gail Rosen. 2024. Predicting anti-microbial resistance using large language models. *arXiv preprint arXiv:2401.00642*. https://doi.org/10.48550/arXiv.2401.00642.

Zhihan Zhou, Yanrong Ji, Weijian Li, Pratik Dutta, Ramana Davuluri, and Han Liu. 2023. Dnabert-2: Efficient foundation model and benchmark for multi-species genome. *arXiv*. ArXiv:2306.15006v1.

A Example Prompts

A.1 Example Prompt Explanation including DNA Sequence

In this example prompt, a DNA sequence is provided along with several drug class labels, such as Sulfonamides, Aminoglycosides, Betalactams, Glycopeptides, Tetracyclines, Phenicol, Fluoroquinolones, MLS (Macrolide-Lincosamide-Streptogramin), and Multi-drug resistance. The task involves asking the model to determine the drug class that the DNA sequence is resistant to.

The prompt follows this format:

```
"Tell me the resistance drug among drugs (Sulfonamides, Aminoglycosides, Betalactams, Glycopeptides, Tetracyclines, Phenicol, Fluoroquinolones, MLS, Multi-drug_resistance) with DNA sequence (ATGAATCCCTATC... ...ACAAACTGCGAGGCAGTTCGCATGA)?"
```

This prompt is used to assess the DNA sequence for antibiotic resistance and classify the sequence into one of the specified drug resistance categories.

A.2 Example Prompt Explanation including Blastn information

In this prompt, a DNA sequence and the top 5 Blastn search results are provided. The task is to predict the drug class that the DNA sequence is resistant to, based on the alignment information and matching sequences. The drug class labels included in the prompt are Sulfonamides, Aminoglycosides, Betalactams, Glycopeptides, Tetracyclines, Phenicol, Fluoroquinolones, MLS (Macrolide-Lincosamide-Streptogramin), and Multi-drug resistance.

The BLASTn results contain gene information such as sequence titles, alignment length, e-values, and detailed sequence alignments (query, match, and subject sequences). This allows the model to analyze the DNA sequence's pattern and classify it into the appropriate drug resistance category.

The prompt follows this format:

```
"Tell me the resistance drug among drugs (Sulfonamides, Aminoglycosides, Betalactams, Glycopeptides, Tetracyclines, Phenicol, Fluoroquinolones, MLS, Multi-drug_resistance) with DNA information ([{'sequence_title': 'gi|1035502645|ref|NG_048504.1| Enterococcus casseliflavus vanXY-C gene for D-Ala-D-Ala dipeptidase/D-Ala-D-Ala carboxypeptidase VanXY-C, complete CDS', 'alignment_length': 673, 'e_value': 0.0, 'query_sequence': 'ATGAATCCCTATCTA...', 'match_sequence': '|||||||||||||...', 'subject_sequence': ...'}, ... ])?"
```

This prompt aims to predict the antibiotic resistance drug by using DNA sequence data from the Blastn search results and identifying the relevant drug resistance class.

B Finetuning of DNA Sequence Classification Models

The Meta-LLaMA-3.1-8B-Instruct model was fine-tuned using the Unsloth framework with 4-bit quantization to enhance memory efficiency. A LoRA configuration was applied to key projection layers, with moderate values for the rank and scaling parameters. The training dataset consisted of DNA sequences and their associated resistant drug class labels, organized in a system-user-assistant conversational format and later converted to the Alpaca-style instruction-following format. Each example included instruction, input, and output fields, and samples were padded with an end-of-sequence token. Training was conducted using the SFTTrainer with mixed-precision enabled (fp16 or bf16), depending on hardware availability. In addition, a GPT-based model (gpt-4o-mini-2024-07-18) was customized using task-specific instruction examples via the OpenAI fine-tuning API.

C Finetuning of DNA Sequence Generation Models

Three large language models (LLMs) were fine-tuned for DNA sequence generation using parameter-efficient fine-tuning (PEFT) with LoRA. The dataset contained DNA input-output sequence pairs, tokenized using model-specific tokenizers and padded using the end-of-sequence token. LoRA configurations were adjusted for each model, with common values for rank, scaling, and dropout, and target modules selected based on the architecture. Training was conducted for several epochs with standard optimization settings.

The GENERator-eukaryote-3b-base model used separate tokenization strategies for inputs and outputs, with padding tokens in the labels replaced by -100. LoRA was applied to selected attention projections, and training used fp16 precision. The Llama-3.2-1B model supported sequences up to 4096 tokens and followed a prompt format of "Input: <input_sequence> Output: <output_sequence>", using bf16 precision and a memory-efficient optimizer. The Gemma-3-1B-PT model followed a similar formatting and applied LoRA to a subset of projection layers, using float32 precision to ensure stability. A GPT-based model (gpt-4o-mini-2024-07-18) was additionally adapted through OpenAI's fine-tuning API using domain-specific examples.

D Unclassified Rate

Additional gene information from the Blastn DB search results was provided, performance improved even without additional training on this data. As seen in Table 3, the Unclassified Rate decreased across all models. For the LLaMA 3.1 8B-4bit quantized model, the rate dropped from 97% to 73% when using Blastn. For Claude 3.5 sonet, it decreased from 39% to 11%. ChatGPT 4-mini showed a sharp improvement, going from classifying nothing to only leaving 14% unclassified. When fine-tuning was applied, both the LLaMA 3.1 8B 4bit quantized model and ChatGPT 4-mini reduced their unclassified rates to 0%.

Model	Unclassified Rate
LLama3.1 8B-4bit (Base Model)	97%
LLama3.1 8B-4bit (Blastn)	73%
LLama3.1 8B-4bit (Finetuning)	0%
Claude3.5sonet (Base Model)	39%
Claude3.5sonet (Blastn)	11%
Chatgpt4o-mini (Base Model)	100%
Chatgpt4o-mini (Blastn)	14%
Chatgpt4o-mini (Finetuning)	0%

Table 3: Model unclassified rates with long names displayed in two lines.

E DNA Generation Dataset Characteristics

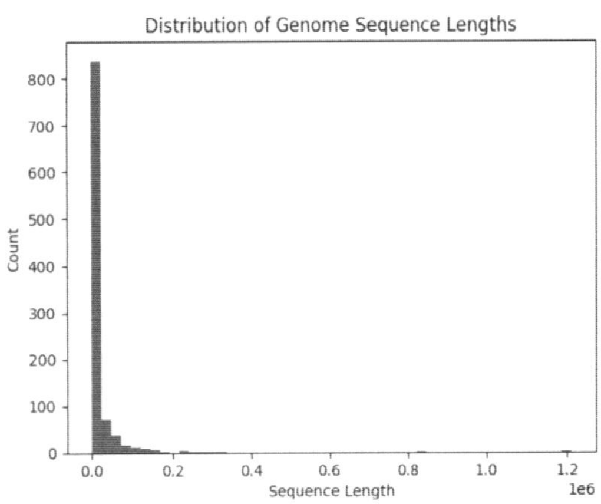

Figure 1: Length distribution of the Acinetobacter baumannii dataset.

Acinetobacter baumannii is a Gram-negative, opportunistic pathogen that poses a serious threat in healthcare settings due to its ability to survive in diverse environments and its remarkable capacity for antibiotic resistance. This bacterium is known for forming robust biofilms, which enhance its persistence on medical equipment and hospital surfaces. It exhibits resistance to multiple antibiotic classes, including beta-lactams, aminoglycosides, and fluoroquinolones, primarily through mechanisms such as enzymatic degradation (e.g., beta-lactamases), efflux pumps, and target site modifications. Given its clinical significance and growing prevalence in multidrug-resistant infections, we collected 1,000 sequences of Acinetobacter baumannii using the NCBI Entrez API for further analysis.

The dataset characteristics are summarized below:

Sequence Statistic	**Length (bp)**
Average sequence length	16,325.75
Median sequence length	1,033.50
Minimum sequence length	204
Maximum sequence length	1,210,760

Table 4: Statistics of the collected Acinetobacter baumannii sequences

The length distribution of the dataset is shown in Figure 1. The length distribution of the dataset exhibits a wide range, spanning from 204 bp to over 1.2 million bp, with a median length of approximately 1,033.50 bp. The substantial difference between the median and the mean (16,325.75 bp) suggests a right-skewed distribution, indicating the presence of a small number of extremely long sequences. Such distribution may impact downstream analysis, particularly in tasks such as sequence alignment or model training, where extreme sequence lengths might introduce computational challenges.

Additionally, the presence of very short sequences (minimum: 204 bp) suggests that preprocessing steps such as length filtering or normalization may be necessary to ensure consistency in downstream analyses. A closer examination of the length distribution (as illustrated in Figure 1) could provide further insights into potential clustering patterns or the need for stratified handling of different length groups.

CaseReportCollective: A Large-Scale LLM-Extracted Dataset for Structured Medical Case Reports

Xiao Yu Cindy Zhang **Wyeth Wasserman** **Melissa Fong** **Jian Zhu**
University of British Columbia
czhang@cmmt.ubc.ca

Abstract

Case reports provide critical insights into rare and atypical diseases, but extracting structured knowledge remains challenging due to unstructured text and domain-specific terminology. We introduce **CaseReportCollective**, an LLM-extracted dataset of 85,961 open-access case reports spanning 37 years, validated through programmatic and human evaluation. Our dataset reveals key publication and demographic trends, including a significant increase in open-access case reports over the past decade, shifts in focus from oncology to COVID-19 and sex disparities in reporting across different medical conditions. Using **CaseReportCollective**, we further explore embedding-based retrieval for similar medical topics through accumulated similarity scores across extracted structured information. We also conducted detailed error analyses on the retrieval ranking, finding that highly reported topics dominate retrieval and the retrieval is driven by lexical overlap rather than underlying clinical relevance, often failing to distinguish between semantically similar yet mechanistically distinct conditions. Future work should focus on clinically aware embeddings adjusted for long-tailed case distributions to improve retrieval accuracy.

1 Introduction

Case reports, structured summaries outlining individual patient profiles and distinctive medical conditions (Venes, 2017), have historically played a critical role in rare disease discovery, novel treatment vigilance, and pandemic surveillance (Nissen and Wynn, 2014; Wu and Sung, 2003; Hymes et al., 1981). As of September 2023, over 2.41 million cases have been published (Parums, 2023), capturing a wealth of clinical details, including patient history, review of systems, laboratory findings, and imaging results. Leveraging this vast repository of medical knowledge has the potential to advance medical research and clinical education significantly. However, extracting structured knowledge from case reports remains a major challenge. Clinical narratives often contain domain-specific terminology, abbreviations, and colloquial descriptions, making automated extraction difficult without a foundational understanding of medical language. Additionally, key metadata such as patient sex and age are frequently implied rather than explicitly stated, requiring common-sense reasoning for accurate interpretation. The manual process of perusing case reports and distilling actionable insights is both labor-intensive and time-consuming, hindering large-scale systematic analysis. Furthermore, traditional rule-based natural language processing (NLP) approaches struggle with the semantic variability and unstructured nature of medical text, limiting their ability to aggregate and standardize case report data effectively.

In this study, we leveraged LLMs and rule-based algorithms to extract granular details from open-access medical case reports in Pubmed Central(PMC) into medical categories standard for patient assessments. Leveraging the metadata along with the fine-grain LLM extractions from this dataset, we analyzed the case report trends for publication years, sex, and patient age. With these fine-grained extractions from case reports, we demonstrate how this dataset can be used for information retrieval for similar cases. Our primary contribution is the construction of a large-scale, LLM-structured case report corpus. The demographic analyses are included to illustrate the dataset's clinical coverage and its potential for supporting diagnostic research across diverse patient populations and medical conditions. Specifically, we highlight:

- **CaseReportCollective:**[1] An LLM-extracted dataset of 85,961 open-access medical case

[1] *CaseReportCollective* is publicly available at https://huggingface.co/datasets/cxyzhang/CaseReportCollective_V1.0.

Proceedings of the 24th Workshop on Biomedical Language Processing (BioNLP 2025), pages 249–262
August 1, 2025 ©2025 Association for Computational Linguistics

reports spanning 37 years, with structured extractions across 14 clinical categories and quality control via programmatic metrics and human evaluation.

- Uncovering **significant differences in sex distribution across age groups, publication years, and medical topics**. Balanced sex representation is observed only in the 42–65 age group, with more males in the 65+ and pediatric categories, and more females in the 18–41 age group. Over time, we observed the inclusion of intersex individuals in case reports. Additionally, certain conditions are disproportionately reported in one sex, with both biological factors and potential sex biases influencing the findings.

- Identifying **systematic biases in embedding-based disease retrieval**, including **prevalence bias, textual co-occurrence bias, and pathophysiological mismatches**. We demonstrate how high-frequency diseases (e.g., tuberculosis) dominate retrieval results, often suppressing rarer but clinically significant conditions. Additionally, semantic similarity alone proves insufficient for clinically accurate retrieval, as it frequently retrieves conditions based on surface-level word overlap rather than true clinical relevance. We suggest **context-aware embeddings and prevalence-adjusted ranking mechanisms** as future directions to improve retrieval accuracy.

2 Related Work

2.1 Medical Information Extraction

Rule-based systems and ontology-driven pipelines have been foundational in clinical NLP. Tools such as MetaMap (Aronson, 2001), Regextractor (Hinchcliff et al., 2012), MedLEE (Friedman et al., 1995), and cTAKES (Savova et al., 2010) extract clinical concepts using predefined grammars and the Unified Medical Language System (UMLS) (Bodenreider, 2004). While these systems offer transparency and have been trusted by clinicians, they require expert rule engineering, are costly to maintain, and struggle with terminological variation, leading to lower recall in open-domain scenarios.

To improve generalizability, hybrid models and deep learning have been proposed. Precursor-induced CRFs outperform traditional CRFs by propagating token context (Lee and Choi, 2019), while models like BioBERT and BiLSTM-CRF have shown strong results in biomedical NER tasks (Schulz et al., 2020). However, these approaches rely heavily on large-scale annotated corpora and may underperform on rare disease data. Notably, fine-tuned BioClinicalBERT has achieved high accuracy in extracting rare disease phenotypes from unstructured narratives(Shyr et al., 2024).

Recently, instruction-tuned large language models (LLMs) have emerged as general-purpose extractors capable of operating with minimal supervision. For example, InstructGPT extracted pediatric foreign body injury data across languages (Sciannameo et al., 2024), and ChatGPT outperformed BioClinicalBERT in rare disease phenotype extraction in one-shot settings (Shyr et al., 2024). These results suggest LLMs encode latent biomedical knowledge learned from large-scale corpora. While LLMs are not always superior to traditional NER architectures for structured or narrow-domain tasks, we leverage them in this work for their domain transferability and their ability to perform dense, multi-category extraction with minimal annotation effort.

Different from prior work in structuring clinical case reports (Zhao et al., 2022; Raza and Schwartz, 2023; Sciannameo et al., 2024), CaseReportCollective dataset is at a substantially larger scale, with LLMs applied across 14 categories and 85,961 case reports. This work complements existing clinical corpora such as MIMIC-III (Johnson et al., 2016), MedNLI (Romanov and Shivade, 2018), and N2C2 datasets (Stubbs et al., 2019), which focus on discharge summaries or specific annotation tasks. In contrast, our corpus standardizes narrative case reports into structured data that enables downstream demographic analysis and diagnostic benchmarking.

2.2 Sex Disparities in Clinical Narratives

Clinical narratives have historically reflected sex-based disparities in disease prognosis, presentation, diagnosis, and treatment (Bello and Mosca, 2004). These inequalities can introduce biases in clinical decision-making, ultimately affecting patient outcomes. For instance, one study found that males receive a diagnosis at a younger age than females, highlighting potential delays in recognition and intervention for female patients (Alcalde-Rubio et al., 2020). Additionally, an analysis of word embeddings applied to biomedical text revealed systematic biases, where substance use disorders were more frequently associated with males, while psychiatric disorders were more commonly linked to females, reinforcing harmful stereotypes in medical literature(Rios et al., 2020).

Our study leveraged knowledge of pretrained LLMs to perform dense extraction across multiple distinct medical categories. Unlike previous studies, we perform fine-grained dense extraction performance across multiple medical domains and demonstrate the utility of LLM-extracted data in biomedical research. In contrast to studies primarily focused on Named Entity Recognition (NER) for certain medical specialties (Abiha, 2024; Turchioe et al., 2022), **CaseReportCollective** provides a structured dataset spanning multiple medical specialties. Additionally, its metadata facilitates investigations into sex- and age-related differences in disease presentation, showcasing LLMs' ability to extract meaningful clinical trends from unstructured text. Furthermore, per-category extractions enable a fine-grained evaluation of embedding-based retrieval.

3 Methods

To construct **CaseReportCollective**, we leveraged publicly available clinical case reports and implemented a structured LLM-based extraction and evaluation pipeline.

3.1 Dataset construction

CaseReportCollective is developed using clinical case reports from the non-commercial PubMed Central (PMC) Open Access subset, sourcing full-text articles under CC BY-NC, CC BY-NC-SA, and CC BY-NC-ND licenses, accessed via the PMC FTP [2] on February 3, 2024. To extract structured clinical information, we instructed an LLM to identify 14 key clinical categories adapted from a specific standardized approach used inpatient Work-Up and monitoring for healthcare professionals [3]: **Vitals_Hema** (Vitals and Hematology Findings), **EENT** (Eyes, Ears, Nose, and Throat), **NEURO** (Neurology), **CVS** (Cardiovascular System), **RESP** (Respiratory System), **GI** (Gastrointestinal System), **GU** (Genitourinary System), **MSK** (Musculoskeletal System), **DERM** (Dermatology), **LYMPH** (Lymphatic System), **ENDO** (Endocrinology), **Pregnancy**, **Lab_Image** (Laboratory and Imaging), and **History**.

[2] https://ftp.ncbi.nlm.nih.gov/pub/pmc/oa_bulk/oa_noncomm/xml/

[3] https://blogs.ubc.ca/oeetoolbox/2019/02/patient-work-up-from-sample-template-inpatient/

3.2 Preprocessing LLM Extraction

We applied the few-shot and category-specific prompt templates used for structured information extraction from case report narratives, as described in (Zhang et al., 2025), where for each clinical category, prompts include a task-specific instruction followed by output formatting constraints. Due to the limitations encountered with earlier LLM frameworks for generating JSON-formatted output, a multi-step preprocessing approach was implemented. All nested categories were converted into a single list of strings. Subkeys within the JSON document were concatenated with their values to preserve context. For example, "blood pressure" is connected with the corresponding value "120/80 mmHg". There were observed instances when LLM extraction failed to retrieve relevant information due to either (a) a lack of detected relevant information or (b) formatting issues, such as the incorrect use of double quotes instead of single quotes, which led to JSON parsing errors. In these cases, we attempted to standardize the format by replacing single quotes with double quotes and then reattempted the LLM extraction.

3.3 LLM-Based Diagnostic Label Extraction and NER Supplementation

The LLM was instructed to extract the medical conditions from the title of each case report. For comparison, we performed NER using SciSpacy (Neumann et al., 2019). Since rare conditions are often under-represented in SciSpacy without additional context, we provided both keywords and labels to improve recognition. However, keywords often contain extraneous or broad information that is not the main focus of the case report (e.g., "pregnancy" in the context of "pregnancy luteoma"), which can dilute the core medical condition being described. To address this, we prioritized the LLM-extracted labels as the primary diagnostic labels. The NER output was only used to supplement these labels when the LLM failed to extract the relevant condition.

3.4 Demographic Attribute Extraction

Biological ages in case reports typically follow a standard format (e.g., "X-year-old"). To enable efficient and deterministic extraction, we applied rule-based keyword extraction for age identification. Ages were categorized into predefined clinical groups: **Neonatal (0–1 month)**, **Infancy (1–18 months)**, **Childhood (1.5–11 years)**, and **Adolescence (11–16 years)** (Blau et al., 2014). Adulthood was further divided into **16–41 years**, **41–64 years**, and **>64 years**. Cases without age data were labeled as "Unspecified."

In contrast, biological sex is expressed more variably and often implicitly, requiring an LLM for context-dependent extraction. For instance, when a patient is described as "nulliparous," LLM may leverage its foundational knowledge to infer the patient as biologically female. Additionally, the LLM was instructed to recognize intersex category—characterized by physical, hormonal, or genetic traits—affecting approximately 1.7% of the population (Sax, 2002; Zeeman and Aranda, 2020).

The Chi-Square (χ^2) test for independence is performed in investigating relationships between **age**, **sex**, **publication years** and **medical topics** in **CaseReportCollective**.

3.5 Implementation

We performed structured extraction of category-specific clinical information and diagnostic labels from case report texts and titles using few-shot prompting tailored for verbatim information capture. For each clinical category, we designed task-specific prompts that requested outputs in a standardized dictionary format. These prompts followed a consistent template with explicit formatting instructions to facilitate post-processing, as detailed in Zhang et al. (2025). For example, in the Neurological category, prompts instructed models to extract findings such as "neurological", "cognitive", "neurological tests and imaging" with outputs keyed by clinical feature types.

Initial large-scale extraction was conducted using LLaMA 3-8B-Instruct (Dubey et al., 2024), running under the Ollama framework[4] with 4-bit quantization on an NVIDIA Tesla V100 GPU, selected for its availability and computational efficiency. Benchmarking results from Zhang et al. (2025) showed that Qwen2.5-7B-Instruct (Hui et al., 2024) yielded better alignment with clinician judgments for dense clinical information extraction, supporting its use in subsequent inference tasks to extract biological sex from case report texts. This model was deployed using 16-bit floating point precision under the vLLM framework (Kwon et al., 2023). All models were set to a temperature of 0 to ensure deterministic outputs.

3.6 Evaluation of Extracted Texts with Automated Metrics and Human Assessment

Since the LLM was tasked with extracting verbatim text from case reports, we assessed extraction fidelity using dual string-based metrics: **Exact Match (EM)** and **Token Set Ratio (TSR%)**, implemented via the `fuzzywuzzy` library[5]. **EM** measures the proportion of extractions that exactly match the original text (ranging from 0 to 1), while **TSR (%)** quantifies partial similarity (ranging from 0 to 100) by allowing slight variations. To assess the fidelity of LLM-extracted text compared to the original case report, we compute the **Token Set Ratio (TSR)**. TSR is a partial similarity metric that captures approximate matches between texts by comparing token-level overlap and differences.

Let T_1 denote the set of tokens from the original case report text, and T_2 the set of tokens from the LLM-extracted output. We compute:

$$I = T_1 \cap T_2, \quad D_1 = T_1 \setminus T_2, \quad D_2 = T_2 \setminus T_1$$

Here, I denotes the shared tokens, D_1 represents tokens found only in the original text, and D_2 those found only in the LLM extraction. These token groups are each converted into strings, and string similarity is then assessed using the Levenshtein distance as implemented in the `fuzzywuzzy`.

Evaluation of a randomly selected subset of 400 LLM-extracted labels against their respective case report titles was performed by a student, guided by medical oversight. The evaluation focused on three criteria: **relevance**—alignment of the extracted entity with the title, **specificity**—correct identification of primary diseases or conditions, and **completeness**—thorough extraction of all relevant medical conditions. The detailed annotation guidelines are provided in Appendix A.

Additionally, a student, guided by medical oversight, evaluated a randomly selected subset of 400 LLM-extracted labels against the original article title, comparing them against their respective case report titles. The evaluation focused on three key criteria: **relevance**–alignment of the extracted entity with the title, **specificity**–correct identification

[4]https://github.com/ollama/ollama

[5]https://github.com/seatgeek/fuzzywuzzy

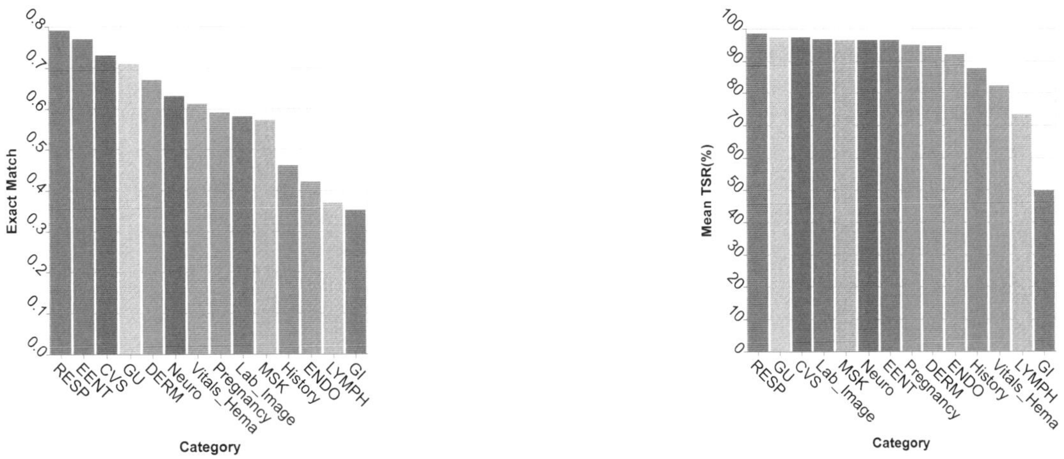

(A) (B)

Figure 1: Programmatic Evaluation Results for LLM Per-Category Extraction. (A) Exact Match Score for Extracted Strings against Case Text. (B) Token Set Ratio of Extracted Strings against Case Text

of primary diseases or conditions, and **completeness**– thorough extraction of all relevant medical conditions. The detailed annotation guidelines are provided in Appendix A.

3.7 CaseReportCollective as Information Retrieval (IR) System

Medical conditions frequently involve multiple body systems, making it difficult to retrieve precise information from case reports. Analyzing entire case reports can obscure system-specific details and introduce confounding effects. We hypothesize that system-specific LLM extractions from CaseReportCollective can improve diagnosis retrieval by preserving relevant information within distinct medical categories.

For this IR task, we first converted the LLM-extracted category-specific texts into embeddings using MedEmbed (Balachandran, 2024). To evaluate retrieval across varying disease prevalences, we sampled 100 topics each from the top, middle, and bottom of the global frequency distribution—representing high, medium, and low-frequency groups—ensuring one unique case per topic. These queries were excluded from the retrieval corpus, which comprised the remaining 80K cases. Retrieval was performed based on L2-normalized embedding similarity via FAISS[6].

The accumulated similarity score for each test case is computed by first retrieving the top-K most similar disease topics from each clinical category. retrieved topics and their similarity scores were collected separately per category. If a topic appeared in multiple categories, its scores were averaged across categories to compute an accumulated similarity score. Final rankings per query were generated by sorting retrieved topics based on these averaged scores, reflecting cross-category semantic consistency.

Finally, Mean Reciprocal Rank (MRR), Normalized Discounted Cumulative Gain(NDCG@50), and Precision@50 were used for IR evaluation.

4 Results and Discussions

4.1 Dataset Composition

CaseReportCollective comprises 85,961 open-access case reports covering 53K unique combinations of medical topics published between 1986 and February 2024 (but notably with most of the full-text open-access case reports appearing in the past decade). On average, case reports contain 3,462 ± 1,920.66 words. The mean number of reports per condition is 2.88 ± 10.49, with COVID-19 (410 cases) being the most frequently reported topics, highlighting a skewed distribution where a small subset of topics dominates the dataset.

The amount of LLM-extracted information varies, with total extraction item counts 27.77 ± 81.57 across 14 categories. Example entries of **CaseReportCollective** can be found in Appendix C. **Lab_Image**, which includes all laboratory tests and imaging across body systems, along with **History**, have the highest extracted string

[6]https://github.com/facebookresearch/faiss/

counts due to their broad and inclusive nature. **CVS** has the third highest extracted string count, followed by **MSK** and **Vitals_Heme**. In contrast, the **GI** category has an extremely low extracted count in this dataset, which may reflect either the inherently limited description of gastrointestinal-related information in clinical case reports or the LLM's difficulty in recognizing such information. Appendix B shows the string count distribution per category.

4.2 LLM Extraction Quality

Although the mean EM score is at 0.59 ± 0.14, a high mean TSR(%) of 87.25 ± 10.79 is achieved, suggesting that LLM-extracted content effectively captures the original text but may introduce minor variations in wording or structure. As shown in Fig. 1, the RESP category exhibits the highest EM, indicating that respiratory-related extractions have the highest alignment. In contrast, the GI category has the lowest scores, suggesting that the LLM struggled to extract gastrointestinal-related information accurately, potentially due to variability in how such details are reported.

Out of 400 extracted medical topics for human evaluation, 19 cases (4.75%) were labeled as hallucinations by the human reviewer, where the LLM generated terms that were unrelated to the input text, overgeneralized, or misclassified (e.g., procedural terms instead of medical conditions). These errors likely stem from insufficient contextual information in the article title and biases toward frequently mentioned conditions in the LLM's training data, warranting further analysis. Despite these hallucinations, most extractions were clinically relevant, with mean scores of 2.94 ± 0.32 for relevance, 2.81 ± 0.39 for specificity, and 2.87 ± 0.36 for completeness. These results demonstrate strong performance, as detailed in Appendix A.

4.3 Temporal Trends

The publication of open-access case reports has increased significantly over the past decade. Figure 2 illustrates this trend, showing sporadic case report publications between 1986 and 2002, followed by a notable rise in recent years. This growth reflects the broader adoption of open access and a growing appreciation for case reports in clinical care.

The trend of case report topics has shifted over time, reflecting evolving clinical priorities. Before 2020, case reports predominantly focused on cancers (e.g., squamous cell carcinoma, hepatocarcinoma, renal cell carcinoma) and vascular conditions (e.g., aneurysms). During 2020-2021, COVID-19-related case reports surged, highlighting the role of case reports in rapid knowledge dissemination during global health crises. Post-2021, the focus changed to oncological and rare conditions (e.g., mucormycosis).

4.4 Age and Sex Stratification

Overall, **CaseReportCollective** consists of 31.61% Adulthood (42–65 yr), 28.12% Adulthood (18–41 yr), 18.50% Adulthood (>65 yr), 10.97% Childhood, 4.75% Infancy, 4.27% Adolescence, and 0.36% Neonatal cases, with 1.57% missing age extraction. Regarding sex distribution, the dataset comprises 55.60% Female, 44.10% Male, and 0.10% Intersex cases, with 0.20% missing sex assignment.

4.5 Sex Distribution Across Age Groups, Years, and Medical Topics

Sex composition varies significantly across age groups ($\chi^2 = 192.03$, $df = 12, p < 1.44 \times 10^{-34}$) (Fig. 3). Intersex cases are rare across all age groups, with the highest frequency observed in childhood (15 cases). These findings suggest a dependency between sex and age groups, potentially influenced by age-stratified biological factors, reporting practices, or selection biases.

We found significant variation in both age ($\chi^2 = 862.39$, $df = 252$, $p = 8.74 \times 10^{-68}$) and sex ($\chi^2 = 108.18$, $df = 72$, $p = 0.0037$) distributions across publication years. As shown in Figure 2, Female cases have generally been reported more frequently than male cases across all years, with the disparity widening over time. Intersex cases remain rare, appearing only after 2011.

The chi-square test ($\chi^2 = 401.70$, $df = 174$, $p < 4.73 \times 10^{-20}$) indicates a significant difference in sex distribution across high-frequency medical topics ($\geq$ 100 occurrences). This suggests that certain medical conditions are disproportionately reported in one sex over the other (Figure 4). While some disparities may be attributed to sex-specific physiology and pathological differences, as reported in prior studies, others may result from systematic biases.

4.6 Can Embedding Models Reliably Retrieve Clinically Relevant Diseases?

We evaluated **CaseReportCollective** as a retrieval-based disease-ranking method that leverages em-

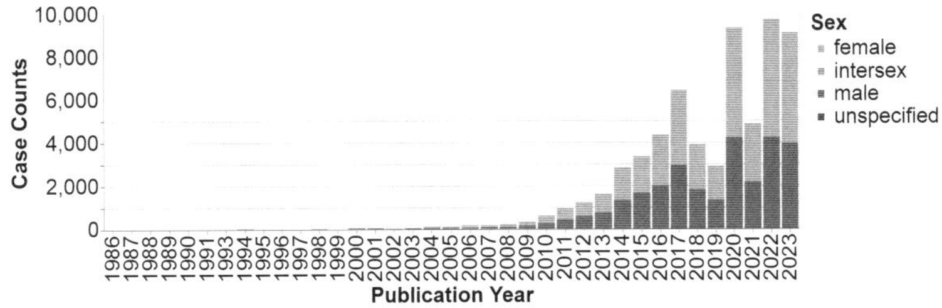

Figure 2: Biological Sex Distribution between 1986 and 2023

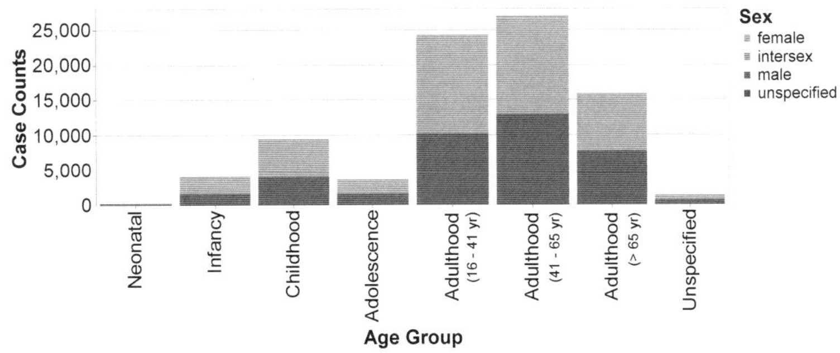

Figure 3: Biological Sex Distribution across Age Groups.

bedding similarity, retrieval frequency, and topic prevalence within the dataset. For each test case, we used category-specific embeddings to perform nearest-neighbor retrieval using FAISS, a fast vector similarity search library. Specifically, we retrieved the top 50 most similar topics for such an evaluation.

4.6.1 Limitations of Traditional IR Metrics

While traditional IR metrics such as MRR, NDCG@50, and Precision@50 provide useful benchmarks for retrieval performance, they may underestimate the capabilities of embedding-based methods when applied to complex clinical narratives. This is particularly true in medical settings where semantically similar conditions may be expressed using diverse terminologies, synonyms, or compositional phrases that differ from canonical labels. Moreover, our case report dataset frequently presented multiple medical topics within single cases (e.g.,"adenomatous polyps, Lynch syndrome"), both of which were represented in the textual descriptions, making it challenging to distinguish. Although we initially considered standardizing medical topics using ontologies like UMLS, we found such mappings insufficient for less common medical conditions, leading to substantial information loss. Hence, we opted out of ontology-based standardization for this study.

In the evaluation, we permitted partial matching between retrieved and query topics, allowing matches such as "cystic fibrosis, multidrug-resistant pseudomonas infection" with "cystic fibrosis." The IR results (Fig. 5) show that our retrieval system has a suboptimal MRR of 0.026 for high-frequency topics, 0.01 for medium-frequency topics, and 0.0 for low-frequency topics, and struggles with ranking consistency as indicated by NDCG@50 scores of 0.19 for high-frequency topics, 0.05 for medium-frequency topics, and 0.07 for low-frequency topics. The system performs better for high-frequency topics in terms of NDCG, compared medium- and low-frequency topics. However, the overall low NDCG scores suggest that the system's ability to rank clinically significant diseases, including rare, low-frequency conditions, is limited. Furthermore, the extremely low Precision@50 for all topics indicates that many retrieved topics result from semantic linkage rather than true diagnostic relevance, highlighting a key limitation in the system's precision for clinical applications.

4.6.2 Systematic Errors in Retrieval

We analyzed tuberculosis—a frequent topic—and highlighted some of the representative failure cases using our category-specific embedding-based system in Table 1. Despite its high frequency in the dataset, the retrieval system often over-prioritized

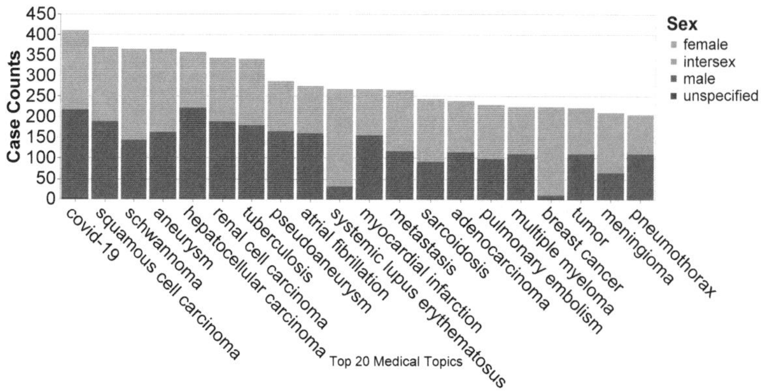

Figure 4: Biological Sex Distribution over Top 20 Medical Conditions.

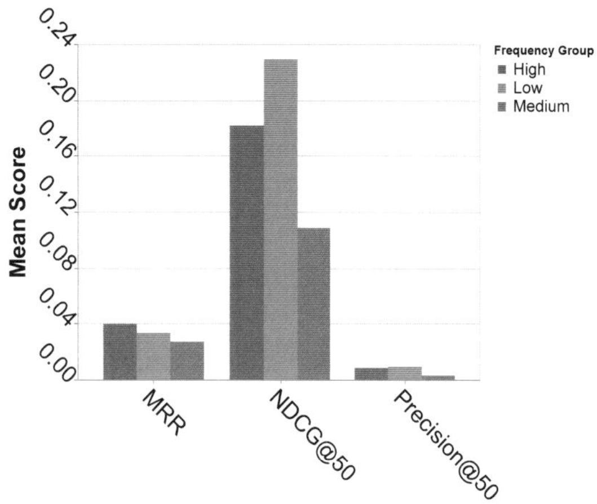

Figure 5: Mean Scores for Evaluation Metric across Three Frequency Groups

tuberculosis due to multiple failure modes. These include: **(1) Semantic Drift**, where chronic dermatologic conditions like nevus sebaceous were retrieved due to shared descriptors of persistent lesions; **(2) Anatomical Misalignment**, such as tracheal diverticula, arising from co-mentions in thoracic imaging contexts; **(3) Co-Treatment Artifact**, where conditions like steroid withdrawal syndrome appear due to shared treatment settings; **(4) Overgeneralized Infection Embedding**, where retrieval conflates unrelated infections like omphalitis or liver abscess; **(5) Anatomic Generalization**, where genitourinary tuberculosis cues led to retrievals like renal stone or UTI; **(6) Surface-Level Embedding Similarity**, as seen in matches like hemophilia B, driven by shared symptoms such as inflammation or bleeding; **(7) Rare Co-occurrence Confusion**, where diseases common in immunocompromised hosts (e.g., EBV/HLH) are incorrectly linked; and **(8) Entity Type Mismatch**, where congenital anomalies (e.g., anorectal malformation) are retrieved despite fundamentally differing etiology. Notably, many of these spurious matches yielded high similarity scores (>0.86), underscoring the embedding model's reliance on lexical and contextual overlap rather than clinically meaningful distinctions. Our findings indicate that the current embedding model is insufficient to fully capture the complexity of differential diagnosis.

5 Conclusion

In this study, we present CaseReportCollective, a large-scale structured dataset of medical case reports. Our analysis of the case reports suggest that the sex disparities in medical case reports have been decreasing temporally. Our findings demonstrate that, while leverage LLM-extracted category-wised information for embedding-based retrieval, there are still systematic failure modes that compromise clinical reliability, especially when unrelated conditions share surface-level linguistic features or co-occur in similar narrative contexts. Future work should explore the integration of structured clinical knowledge, prevalence-aware ranking mechanisms, and context-sensitive embedding models to improve medical retrieval systems.

Index	Retrieved Topic	Query Topic	Norm Similarity	Issue	Failure Type	Possible Explanation for High Similarity
4543	nevus sebaceous, syringocystadenoma papilliferum	tuberculosis	0.878	Skin tumor unrelated to TB	Semantic Drift	Shared mention of chronic lesions or dermatological findings
4544	tracheal diverticula	tuberculosis	0.872	Airway abnormality unrelated to TB	Anatomical Misalignment	Co-occurrence in chest imaging discussions
4545	depression, steroid withdrawal syndrome	tuberculosis	0.869	Psychological syndromes unrelated to infection	Co-Treatment Artifact	TB and steroid use both appear in chronic illness contexts
4546	omphalitis, pyogenic liver abscess	tuberculosis	0.868	Different infection types	Overgeneralized Infection Embedding	Embedding captures general infection-related semantics
4547	renal stone, urinary tract infection	tuberculosis	0.865	Genitourinary disease not specific to TB	Anatomic Generalization	Overlap via genitourinary TB mentions
4548	hemophilia b, subgaleal hematoma	tuberculosis	0.864	Hematological condition	Embedding Surface Similarity	Shared features like bleeding or inflammation
4549	chronic active EBV infection, HLH, NK cell lymphoma	tuberculosis	0.863	Viral and hematologic malignancies	Rare Co-occurrence Confusion	TB sometimes mentioned in immunocompromised patients
4550	Churg-Strauss syndrome, neuroendocrine carcinoma	tuberculosis	0.862	Vasculitis and cancer unrelated to TB	Multisystem Similarity Confusion	Both may affect multiple organs, mentioned with granulomas
4551	anorectal malformation, ileal perforation	tuberculosis	0.862	Congenital/anatomical vs acquired infection	Entity Type Mismatch	Shared surgical or gastrointestinal mentions
4552	trichilemmal carcinoma	tuberculosis	0.861	Skin cancer unrelated to TB	Lexical Overlap	Chronic cutaneous conditions may trigger similarity

Table 1: Issues in Retrieval of Tuberculosis: High Similarity but Incorrect Matches, Categorized by Failure Type

References

Syeda Abiha. 2024. Use of natural language processing to extract clinical cancer phenotypes from electronic medical records. *Journal of Artificial Intelligence and Health*, 1(2):57–65.

Lorena Alcalde-Rubio, Ildefonso Hernández-Aguado, Lucy Anne Parker, Eduardo Bueno-Vergara, and Elisa Chilet-Rosell. 2020. Gender disparities in clinical practice: are there any solutions? scoping review of interventions to overcome or reduce gender bias in clinical practice. *International journal for equity in health*, 19:1–8.

Alan R Aronson. 2001. Effective mapping of biomedical text to the umls metathesaurus: the metamap program. In *Proceedings of the AMIA Symposium*, page 17.

Abhinand Balachandran. 2024. Medembed: Medical-focused embedding models.

Natalie Bello and Lori Mosca. 2004. Epidemiology of coronary heart disease in women. *Progress in cardiovascular diseases*, 46(4):287–295.

Nenad Blau, Marinus Duran, K Michael Gibson, and Carlo Dionisi Vici. 2014. *Physician's guide to the diagnosis, treatment, and follow-up of inherited metabolic diseases*, volume 213. Springer.

Olivier Bodenreider. 2004. The unified medical language system (umls): integrating biomedical terminology. *Nucleic acids research*, 32(suppl_1):D267–D270.

Abhimanyu Dubey, Abhinav Jauhri, Abhinav Pandey, Abhishek Kadian, Ahmad Al-Dahle, Aiesha Letman, Akhil Mathur, Alan Schelten, Amy Yang, Angela Fan, et al. 2024. The llama 3 herd of models. *arXiv preprint arXiv:2407.21783*.

Carol Friedman, George Hripcsak, William DuMouchel, Stephen B Johnson, and Paul D Clayton. 1995. Natural language processing in an operational clinical information system. *Natural Language Engineering*, 1(1):83–108.

Monique Hinchcliff, Eric Just, Sofia Podlusky, John Varga, Rowland W Chang, and Warren A Kibbe. 2012. Text data extraction for a prospective, research-focused data mart: implementation and validation. *BMC medical informatics and decision making*, 12:1–7.

Binyuan Hui, Jian Yang, Zeyu Cui, Jiaxi Yang, Dayiheng Liu, Lei Zhang, Tianyu Liu, Jiajun Zhang, Bowen Yu, Keming Lu, et al. 2024. Qwen2. 5-coder technical report. *arXiv preprint arXiv:2409.12186*.

KennethB Hymes, JeffreyB Greene, Aaron Marcus, DanielC William, Tony Cheung, NeilS Prose, Harold Ballard, and LindaJ Laubenstein. 1981. Kaposi's sarcoma in homosexual men—a report of eight cases. *The Lancet*, 318(8247):598–600.

Alistair EW Johnson, Tom J Pollard, Lu Shen, Li-wei H Lehman, Mengling Feng, Mohammad Ghassemi, Benjamin Moody, Peter Szolovits, Leo Anthony Celi, and Roger G Mark. 2016. Mimic-iii, a freely accessible critical care database. *Scientific data*, 3(1):1–9.

Woosuk Kwon, Zhuohan Li, Siyuan Zhuang, Ying Sheng, Lianmin Zheng, Cody Hao Yu, Joseph Gonzalez, Hao Zhang, and Ion Stoica. 2023. Efficient memory management for large language model serving with pagedattention. In *Proceedings of the 29th Symposium on Operating Systems Principles*, pages 611–626.

Wangjin Lee and Jinwook Choi. 2019. Precursor-induced conditional random fields: connecting separate entities by induction for improved clinical named entity recognition. *BMC Medical Informatics and Decision Making*, 19:1–13.

Mark Neumann, Daniel King, Iz Beltagy, and Waleed Ammar. 2019. Scispacy: fast and robust models for biomedical natural language processing. *arXiv preprint arXiv:1902.07669*.

Trygve Nissen and Rolf Wynn. 2014. The clinical case report: a review of its merits and limitations. *BMC research notes*, 7:1–7.

Dinah V Parums. 2023. the increasing relevance of case reports in medical education and clinical practice–and how to write them. *The American Journal of Case Reports*, 24:e942670–1.

Shaina Raza and Brian Schwartz. 2023. Entity and relation extraction from clinical case reports of covid-19: a natural language processing approach. *BMC Medical Informatics and Decision Making*, 23(1):20.

Anthony Rios, Reenam Joshi, and Hejin Shin. 2020. Quantifying 60 years of gender bias in biomedical research with word embeddings. In *Proceedings of the 19th SIGBioMed workshop on biomedical language processing*, pages 1–13.

Alexey Romanov and Chaitanya Shivade. 2018. Lessons from natural language inference in the clinical domain. In *Proceedings of the 2018 Conference on Empirical Methods in Natural Language Processing*, pages 1586–1596, Brussels, Belgium. Association for Computational Linguistics.

Guergana K Savova, James J Masanz, Philip V Ogren, Jiaping Zheng, Sunghwan Sohn, Karin C Kipper-Schuler, and Christopher G Chute. 2010. Mayo clinical text analysis and knowledge extraction system (ctakes): architecture, component evaluation and applications. *Journal of the American Medical Informatics Association*, 17(5):507–513.

Leonard Sax. 2002. How common is lntersex? a response to anne fausto-sterling. *Journal of sex research*, 39(3):174–178.

Sarah Schulz, Jurica Ševa, Samuel Rodriguez, Malte Ostendorff, and Georg Rehm. 2020. Named entities in medical case reports: corpus and experiments. *arXiv preprint arXiv:2003.13032*.

Veronica Sciannameo, Daniele Jahier Pagliari, Sara Urru, Piercesare Grimaldi, Honoria Ocagli, Sara Ahsani-Nasab, Rosanna Irene Comoretto, Dario Gregori, and Paola Berchialla. 2024. Information extraction from medical case reports using openai instructgpt. *Computer methods and programs in biomedicine*, 255:108326.

Cathy Shyr, Yan Hu, Lisa Bastarache, Alex Cheng, Rizwan Hamid, Paul Harris, and Hua Xu. 2024. Identifying and extracting rare diseases and their phenotypes with large language models. *Journal of Healthcare Informatics Research*, 8(2):438–461.

Amber Stubbs, Michele Filannino, Ergin Soysal, Samuel Henry, and Ozlem Uzuner. 2019. Cohort selection for clinical trials: n2c2 2018 shared task track 1. *J. Am. Medical Informatics Assoc.*, 26(11):1163–1171.

Meghan Reading Turchioe, Alexander Volodarskiy, Jyotishman Pathak, Drew N Wright, James Enlou Tcheng, and David Slotwiner. 2022. Systematic review of current natural language processing methods and applications in cardiology. *Heart*, 108(12):909–916.

Donald Venes. 2017. *Taber's cyclopedic medical dictionary*. FA Davis.

Eugene B Wu and Joseph JY Sung. 2003. Haemorrhagic-fever-like changes and normal chest radiograph in a doctor with sars. *The Lancet*, 361(9368):1520–1521.

Laetitia Zeeman and Kay Aranda. 2020. A systematic review of the health and healthcare inequalities for people with intersex variance. *International Journal of Environmental Research and Public Health*, 17(18):6533.

Xiao Yu Cindy Zhang, Carlos R. Ferreira, Francis Rossignol, Raymond T. Ng, Wyeth Wasserman, and Jian Zhu. 2025. Casereportbench: An llm benchmark dataset for dense information extraction in clinical case reports. In *Proceedings of the Conference on Health, Inference, and Learning (CHIL), JMLR Workshop and Conference Proceedings*. To appear.

Zhengyun Zhao, Qiao Jin, Fangyuan Chen, Tuorui Peng, and Sheng Yu. 2022. Pmc-patients: A large-scale dataset of patient summaries and relations for benchmarking retrieval-based clinical decision support systems. *arXiv preprint arXiv:2202.13876*.

A Human Evaluation Guidelines for LLM-Extracted Diagnostic Labels

Objective: Assess the accuracy, specificity, and clinical relevance of the LLM-generated labels in relation to the case report title. Use the Likert scale below for evaluation.

Likert Scale for Evaluation

Score	Rating	Description
3 - Excellent	Perfect Match	Fully relevant, specific and complete. No improvement is needed.
2 - Acceptable	Partially Correct	The label is relevant but lacks key details (e.g. too broad or missing very few conditions). Minimal modification needed.
1 - Unacceptable	Incorrect or Misleading	Clinically wrong, misleading, or too vague to be useful. A major revision is needed.

Table 2: Likert Scale for Evaluation

A.1 Evaluation Criteria

Evaluation Metrics:

- **Relevance (1-3)**: Does the label relate to the case report title?
- **Specificity (1-3)**: Is the label precise and not too broad?
- **Completeness (1-3)**: Does the label capture the full diagnosis?

A.1.1 1. Clinical Relevance

Acceptable:

- The label correctly identifies the primary disease, condition, or syndrome/symptom described in the title.
- The label is a well-recognized medical term or diagnosis.

Not Acceptable:

- The label is unrelated or related but too general (e.g., "disease" instead of "trigeminal schwannoma").
- The label is misleading or incorrect.

A.1.2 2. Specificity

Acceptable:

- The label captures the exact medical condition (e.g., "cardiac sarcoidosis" instead of just "sarcoidosis").
- The label includes relevant qualifiers when necessary (e.g., "trigeminal schwannoma" instead of just "schwannoma").

Not Acceptable:

- The label is too broad (e.g., for "brain abscess" extract as only "abscess").
- The label adds unnecessary information that is not in the title.

A.1.3 3. Completeness

Acceptable:

- The label correctly reflects all critical clinical elements in the title.
- If the title describes multiple conditions, the label should capture the main diagnosis.

Not Acceptable:

- The label only captures one part of a compound diagnosis when both are equally important (e.g., did not extract both "neuropathy" and "diabetes" in "neuropathy secondary to diabetes").

Note: Rather than evaluating individual entity completeness, the "completeness" metric is used to assess the full extraction of all entities, regardless of whether each concept is fully extracted. The specificity metric, however, will be used to evaluate the quality of each extracted entity.

B Distribution of Extracted Strings Counts Across Clinical Categories

The bar plot shows the distribution of extracted string counts across different categories. Lab_Image and History contain the most string extractions with GI the least extractions.

C Example Layout of CaseReportCollective

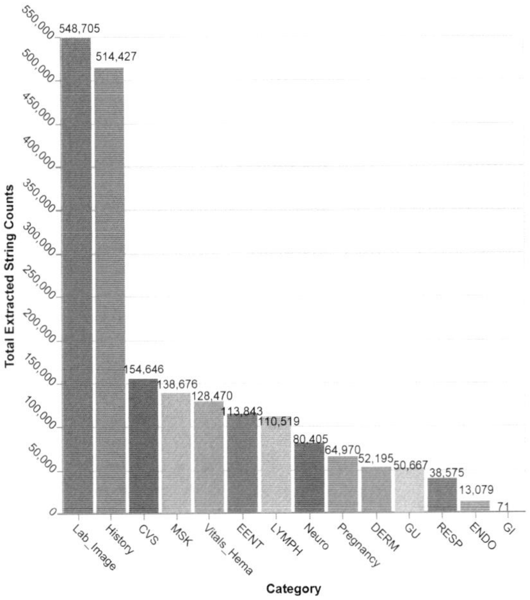

Figure 6: Distribution of Extracted String Counts Across Clinical Categories.

pmcid	year	age	sex	topic	title	case	length	Vitals_Hema ... (Omitted 13 Clinical Category Columns)
8116089	2021	Adulthood (41-65 yr)	female	atrial septal defect	Transcatheter Device Closure of Secundum Atria...	We present a case of female Bosnian patient 50...	209	[pulse: 83/min, respiratory_rate: 15 breaths/m...
8464474	2021	Adulthood (41-65 yr)	female	hip revision	Total hip revision with custom-made spacer and...	A 61-year-old woman presented to our orthopaed...	440	[hematological_conditions: raised erythrocyte ...
8433115	Un-known	Adulthood (41-65 yr)	female	cardiac haemangioma	Totally endoscopic resection of epicardial car...	We report on a case of an incidentally found t...	217	[pulse: 72 bpm, blood_pressure: 125/70 mmHg]

Figure 7: Example Layout of CaseReportCollective. Only **Vitals_Hema** (Vitals and Hematology Findings) is shown, other omitted categories are **EENT** (Eyes, Ears, Nose, and Throat), **NEURO** (Neurology), **CVS** (Cardiovascular System), **RESP** (Respiratory System), **GI** (Gastrointestinal System), **GU** (Genitourinary System), **MSK** (Musculoskeletal System), **DERM** (Dermatology), **LYMPH** (Lymphatic System), **ENDO** (Endocrinology), **Pregnancy**, **Lab_Image** (Laboratory and Imaging), and **History**

Enhancing Antimicrobial Drug Resistance Classification by Integrating Sequence-Based and Text-Based Representations

Hyunwoo Yoo, Bahrad Sokhansanj, James R. Brown
Drexel University
{hty23, bas44, jb4633}@drexel.edu

Abstract

Antibiotic resistance identification is essential for public health, medical treatment, and drug development. Traditional sequence-based models struggle with accurate resistance prediction due to the lack of biological context. To address this, we propose an NLP-based model that integrates genetic sequences with structured textual annotations, including gene family classifications and resistance mechanisms. Our approach leverages pretrained language models for both genetic sequences and biomedical text, aligning biological metadata with sequence-based embeddings. We construct a novel dataset based on the Antibiotic Resistance Ontology (ARO), consolidating gene sequences with resistance-related textual information. Experiments show that incorporating domain knowledge significantly improves classification accuracy over sequence-only models, reducing reliance on exhaustive laboratory testing. By integrating genetic sequence processing with biomedical text understanding, our approach provides a scalable and interpretable solution for antibiotic resistance prediction.

1 Introduction

The prevalence of antibiotic resistance genes (ARGs) has risen rapidly over the past decade, posing a severe threat to public health and medical treatment strategies (Zhang et al., 2022). The emergence of multidrug-resistant pathogens has further complicated treatment options, increasing the urgency of developing accurate methods for identifying and classifying ARGs. While traditional antibiotic resistance screening relies on phenotypic testing, these methods are time-consuming and require extensive laboratory resources. In contrast, bioinformatics-based approaches enable in silico prediction of resistance from genetic sequences, offering a scalable and efficient alternative. The primary computational approach for identifying antibiotic resistance genes (ARGs) has been sequence alignment, which compares nucleotide sequences to known ARG databases (Bonin et al., 2023). While effective, alignment-based methods struggle with novel mutations and require substantial computational resources. Alternative machine learning-based strategies have been explored to address these challenges but remain limited in capturing broader sequence dependencies (Wood and Salzberg, 2014; Eddy, 1998; McIntyre et al., 2017). To overcome these limitations, recent studies have applied natural language processing (NLP) models to genomic or protein sequences, leveraging contextual embeddings for improved classification and interpretability (Brandes et al., 2022; Ji et al., 2021; Zhou et al., 2024).

Despite their advancements, existing classification models predominantly focus on predicting a single resistance label per gene sequence (Kang et al., 2022). However, antibiotic resistance databases such as CARD (Alcock et al., 2023; Jia et al., 2017) and MEGARes (Bonin et al., 2023; Doster et al., 2020) provide richer annotations beyond a single resistance label. In particular, two critical attributes—Gene Family and Resistance Mechanism—offer valuable insights into how resistance manifests at a molecular level. These attributes provide a higher-level understanding of resistance beyond individual nucleotide variations, but current sequence-based models do not leverage this structured information. By incorporating Gene Family and Resistance Mechanism into predictive models, we can enhance interpretability and classification accuracy. In this work, we propose a novel NLP-based model that integrates genetic sequence data with structured textual annotations, specifically Gene Family and Resistance Mechanism, to improve antibiotic resistance classification. Our key contributions are as follows:

- We integrate biological knowledge with

Proceedings of the 24th Workshop on Biomedical Language Processing (BioNLP 2025), pages 263–273

sequence-based models for more accurate resistance prediction.

- We unify resistance classification by aligning CARD and MEGARes annotation systems.
- We generate synthetic samples to improve classification in rare resistance categories.

2 Related Work

Traditional methods for predicting antibiotic resistance rely on sequence alignment techniques, where unknown DNA sequences are compared to reference databases (Bonin et al., 2023). While effective for known resistance genes, alignment-based methods struggle with novel mutations and require high computational resources for large-scale datasets. Alternative computational approaches, such as Hidden Markov Models (HMMs) (Eddy, 1998) and k-mer-based classification (Wood and Salzberg, 2014), have been explored to recognize sequence patterns beyond direct alignment. However, these methods still face limitations in capturing broader contextual dependencies within genomic sequences. To address these limitations, sequence-based machine learning approaches, such as nucleotide transformers and DNABERT, have been introduced (Ji et al., 2021; Zhou et al., 2024). These models capture contextual representations of DNA sequences and offer improved classification performance over traditional alignment methods. However, existing sequence-based models primarily predict antibiotic resistance based on nucleotide sequence patterns alone, without incorporating additional biological knowledge. Antibiotic resistance is not solely determined by genetic sequence variations, but also by gene function, regulatory mechanisms, and evolutionary relationships (Kang et al., 2022). As a result, sequence-only models may fail to generalize across diverse resistance mechanisms and gene families.

Recent advancements in biomedical NLP and knowledge-driven machine learning have demonstrated the potential of integrating structured domain knowledge into predictive models. In fields such as protein function prediction and clinical text mining, hybrid approaches combining structured knowledge with sequence-based embeddings have shown promising results (Brandes et al., 2022). This motivates the need for similar methods in antimicrobial resistance (AMR) classification. Antibiotic resistance databases such as CARD (Alcock et al., 2023) and MEGARes (Bonin et al., 2023) provide valuable metadata beyond sequence-based labels, including Gene Family classifications and Resistance Mechanisms. These attributes capture biologically meaningful relationships between genes and their resistance properties. However, existing AMR classification models do not fully leverage these structured annotations, treating resistance prediction as a single-label classification problem from raw sequences. While sequence-based language models have improved antibiotic resistance prediction, they still lack biological interpretability and fail to incorporate structured knowledge from domain-specific databases. The integration of sequence embeddings with domain knowledge has the potential to enhance classification performance and interpretability. This motivates further exploration of hybrid models that combine genetic sequence processing with structured textual annotations, enabling more comprehensive and generalizable resistance prediction.

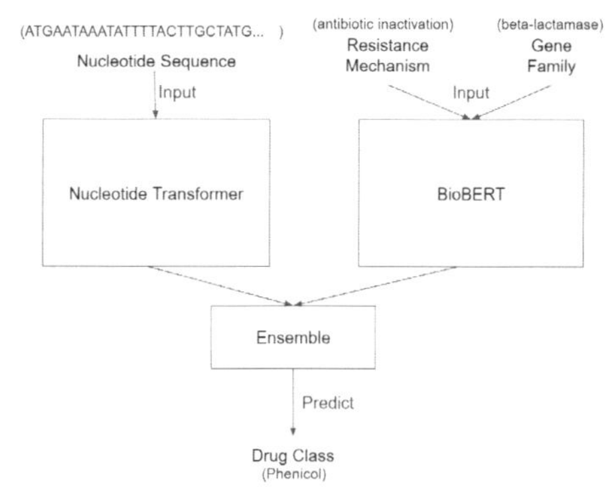

Figure 1: Overview of hybrid model for antibiotic resistance drug class classification. The model takes as input a nucleotide sequence, gene family, and resistance mechanism, and predicts the corresponding drug class by combining outputs from Nucleotide Transformer and BioBERT.

3 Methods

Our model integrates sequence-based and text-based representations to improve antibiotic resistance drug class classification. Given a nucleotide sequence (e.g., ATGC...), its associated gene family (e.g., "beta-lactamase"), and resistance mechanism (e.g., "antibiotic inactivation"), the model predicts the corresponding drug class (e.g., "Phenicol"). As illustrated in Figure 1, we utilize two pretrained models such as a Nucleotide Transformer for processing sequence input and BioBERT for encoding structured biological metadata. Their outputs are combined using a weighted soft-voting ensemble (Dietterich, 2000). The overall model architecture is illustrated in the Appendix A.

3.1 Nucleotide Sequence Based Antibiotic Resistance Drug Class Classification

To classify antibiotic resistance genes, we fine-tune a nucleotide transformer (NT) model (Dalla-Torre et al., 2023). We consider the NT model as a strong sequence-only baseline that represents current methods that rely solely on nucleotide features without structured annotations. Unlike conventional models primarily trained on human genomes (Sanabria et al., 2024), NT is pre-trained on a diverse collection of genomic sequences from bacteria, fungi, and protozoa, allowing for a more comprehensive representation of microbial resistance patterns. For input processing, nucleotide sequences are tokenized using a 6-mer tokenizer, a widely used k-mer tokenization technique in genomic analysis (Mejía-Guerra and Buckler, 2019). The input length is restricted to 1000 nucleotides, corresponding to the model's pretraining constraints. The classification task is fine-tuned using Low-Rank Adaptation (LoRA), which inserts low-rank decomposed matrices into transformer layers while keeping the original model weights fixed (Hu et al., 2022). This significantly reduces trainable parameters while maintaining model efficiency and accuracy.

3.2 Text Information-Based Antibiotic Resistance Classification

To complement sequence-based models, we fine-tune BioBERT (Lee et al., 2020), a biomedical language model pre-trained on PubMed and PMC articles, to extract Gene Family and Resistance Mechanism attributes from textual descriptions of resistance genes. The input text is formatted using structured markers to enhance contextual understanding, improving attribute recognition and classification accuracy. Fine-tuning is conducted with a single classification layer, linking biological domain knowledge with sequence-based predictions. A comparison of different entity representation techniques is provided in Appendix D. Although resistance mechanism and gene family annotations may correlate with drug class labels, they are curated independently from the target labels in standardized resources such as CARD and MEGARes. These structured attributes often co-occur but not always perfectly aligned, providing complementary biological context that enhances classification robustness and interpretability.

3.3 Weighted Soft-voting Ensemble

To integrate predictions from the nucleotide sequence-based model and the text-based model, we implement a soft-voting ensemble strategy. The ensemble model is designed to leverage the complementary strengths of both approaches (Kuncheva and Whitaker, 2003), combining genetic sequence representations with structured textual knowledge for improved classification accuracy. The ensemble takes two types of inputs: (1) the nucleotide sequence, processed through the sequence-based language model, and (2) textual annotations, including Gene Family and Resistance Mechanism attributes, extracted from the text-based model. To optimize classification performance, we determine the weight ratio of each model's contribution using a validation dataset. This validation set is separate from the training and test datasets and is used to fine-tune the weight distribution for optimal ensemble decision-making. Final prediction probabilities are computed using a weighted soft-voting scheme:

$$P(y \mid x) = \lambda \cdot P_{\text{NT}}(y \mid x_{\text{seq}}) + (1-\lambda) \cdot P_{\text{BB}}(y \mid x_{\text{text}})$$

where λ is a weight parameter determined from validation performance. In our experiments, λ ranged between 0.35 and 0.55 depending on the dataset, reflecting the relative contributions of sequence-based and text-based predictions.

3.4 Integrating Classes Based on Antibiotic Resistance Ontology

Antibiotic resistance classification varies across databases, with CARD and MEGARes using different resistance labels and hierarchical structures. To address these inconsistencies, we employ the EBI Antibiotic Resistance Ontology (ARO) (Cook et al., 2016) to standardize resistance annotations across datasets. Each database entry is mapped to the ARO ontology by querying the EBI API and retrieving hierarchical Gene Family relationships. Instead of using fine-grained subcategories, we adopt the third-level hierarchy in ARO, ensuring that class representations remain general enough for robust classification across different datasets. This hierarchical integration harmonizes classification schemes, reducing discrepancies in resistance annotations between databases. This mapping process ensures consistency across heterogeneous labels by aligning them to a shared third-level ARO hierarchy, as detailed in Appendix B.

3.5 Data Augmentation Using a Large Language Model

To mitigate data imbalance in antibiotic resistance gene classification, we employ BioGPT (Luo et al., 2022) for generating synthetic samples in underrepresented categories. Augmenting resistance descriptions improves classification performance, particularly in Macro F1 score. The effectiveness of this approach is detailed in Appendix E.

4 Experiments

We evaluate the performance of sequence-based and text-based models for antibiotic resistance drug class classification using three datasets: CARD, MEGARes, and an integrated dataset combining both sources. We compare Nucleotide Transformer (NT), BioBERT (BB), and an ensemble of both models, analyzing their effectiveness in different dataset settings.

4.1 Experimental Setup

We finetune NT on genetic sequences and BioBERT on structured text annotations describing resistance genes. The ensemble model uses a weighted soft-voting approach, integrating both modalities. All models are trained on CARD, MEGARes, and Integrated datasets, following the standard pre-processing pipeline described in Methods. In addition, experiments using read-level data generated based on the Integrated dataset is conducted. Further details can be found in the Appendix C

4.2 Datasets

We use the CARD and MEGARes v3 datasets, integrating Drug Class, Gene Family, and Resistance Mechanism labels using the EBI ARO ontology. Following standard preprocessing, classes with fewer than 15 samples are removed. Dataset details are provided in the Appendix B.

4.3 Classification Results

Table 1 presents the classification results, demonstrating the impact of integrating structured biological knowledge into sequence-based models. Compared to sequence-only models, incorporating Gene Family and Resistance Mechanism attributes led to significant performance improvements. Specifically, our method improved accuracy by 9.53 points and Macro F1 by 30.34 points on CARD, while on MEGARes, the improvement was 10.38 points and 50.57 points, respectively. These findings indicate that sequence-based models alone struggle to capture higher-level biological relationships necessary for robust resistance classification. By integrating structured textual annotations, our model achieves superior interpretability and generalization, particularly for low-resource resistance categories. Furthermore, using integrated data from multiple annotation systems enhances classification performance, demonstrating the advantage of leveraging domain-specific knowledge for a unified prediction model.

4.4 Ablation Analysis

To assess the contribution of each component in our hybrid model, we conduct an ablation analysis comparing individual models (NT and BB) versus their ensemble, and dataset configurations (individual vs. integrated). As shown in Table 1, the ensemble consistently outperforms NT and BB alone across all datasets, confirming the complementary nature of sequence-based and text-based representations.

The integrated dataset includes more diverse and heterogeneous resistance profiles from both CARD and MEGARes, offering a broader and more realistic evaluation setting. Despite this increased complexity, our ensemble model maintains strong and consistent performance, demonstrating its robustness and generalizability across databases.

5 Discussion

Our results demonstrate that incorporating structured biological knowledge significantly enhances antibiotic resistance classification. Sequence-based models alone struggle to capture higher-order biological relationships that influence resistance mechanisms. By integrating Gene Family and Resistance Mechanism annotations, our model improves interpretability and generalization, particularly for low-resource resistance categories. Furthermore, class integration using the EBI ARO ontology standardizes resistance classification across datasets, increasing training data availability and improving consistency. This standardization not only enhances model performance but also facilitates broader applicability across different resistance gene databases. Notably, the near-perfect performance observed on the MEGARes dataset may par-

Dataset	Method	Accuracy	Macro F1	Precision	Recall
CARD	NT	87.92	63.08	66.46	61.51
CARD	BB	97.22	89.68	92.09	90.54
CARD	Ensemble	97.55	93.44	95.72	92.86
MEGARes	NT	89.61	46.42	54.92	43.94
MEGARes	BB	99.64	99.47	99.96	99.03
MEGARes	Ensemble	99.99	99.99	99.99	99.99
Integrated	NT	82.89	65.79	81.84	58.67
Integrated	BB	90.26	79.34	84.05	77.14
Integrated	Ensemble	92.11	80.95	83.52	78.94
Integrated with reads	NT	83.11	62.82	74.81	57.32
Integrated with reads	BB	90.24	79.34	84.05	77.14
Integrated with reads	Ensemble	93.40	81.85	84.34	80.25

Table 1: Result of using the CARD, MEGARes, and Integrated databases for antibiotic resistance drug class prediction using Nucleotide Transformer(NT), BioBERT(BB), and a weighted ensemble of both. The weighted ensemble with Nucleotide Transformer(NT) and BioBERT(BB) shows better performance in every datasets.

tially reflect the benefits of ontology-based class harmonization and the high consistency of resistance annotations in MEGARes. While these results highlight the model's capacity to leverage structured knowledge, they also suggest that annotation quality and class structure play a key role in enabling robust classification. Additionally, our ensemble model maintains strong performance even when using sequencing reads instead of full-length genes, demonstrating its robustness in practical applications. Beyond classification performance, incorporating structured biological knowledge also provides practical advantages in reducing experimental complexity and time (see Appendix F). By bridging the gap between sequence-based and knowledge-driven classification, our approach offers a scalable and interpretable solution for antimicrobial resistance prediction. However, our approach still relies on the quality of existing resistance gene annotations, which may not always reflect emerging resistance mechanisms. Additionally, maintaining up-to-date structured knowledge requires continuous curation, posing a scalability challenge.

6 Conclusion

We present a hybrid model that integrates sequence-based and text-based representations to improve antibiotic resistance classification. By incorporating structured biological knowledge, including Gene Family and Resistance Mechanism annotations, our approach enhances interpretability and outperforms sequence-only models. Additionally, we standardize resistance classification using the EBI ontology and utilize large language models for data augmentation, improving performance in low-resource settings. These results demonstrate the effectiveness of combining genetic and textual information for more accurate and scalable resistance prediction.

7 Limitation

While our approach improves antibiotic resistance classification by integrating sequence-based and text-based models, certain limitations remain. First, our reliance on curated databases, such as CARD and MEGARes, means that model performance may be affected by biases in annotation quality and completeness. Additionally, while integrating Gene Family and Resistance Mechanism improves interpretability, the hierarchical structure of these annotations may introduce inconsistencies across datasets. Another limitation is the challenge of handling rare or novel resistance genes, where even with data augmentation, model generalization remains an open problem. Computational efficiency remains a concern, as training large-scale sequence and text models requires significant resources, which may limit accessibility for some research applications. Finally, beyond domain-specific models, evaluating the potential of recent general-purpose LLMs such as ChatGPT-4o or Claude 4 Sonnet for antibiotic resistance prediction remains an open direction for future research.

Acknowledgments

This work was supported in part by the National Science Foundation (NSF) under Grant Number 2107108.

References

Brian P. Alcock, William Huynh, Romeo Chalil, Keaton W. Smith, Amogelang R. Raphenya, Mateusz A. Wlodarski, Arman Edalatmand, Aaron

Petkau, Sohaib A. Syed, Kara K. Tsang, Sheridan J. C. Baker, Mugdha Dave, Madeline C. McCarthy, Karyn M. Mukiri, Jalees A. Nasir, Bahar Golbon, Hamna Imtiaz, Xingjian Jiang, Komal Kaur, Megan Kwong, Zi Cheng Liang, Keyu C. Niu, Prabakar Shan, Jasmine Y. J. Yang, Kristen L. Gray, Gemma R. Hoad, Baofeng Jia, Timsy Bhando, Lindsey A. Carfrae, Maya A. Farha, Shawn French, Rodion Gordzevich, Kenneth Rachwalski, Megan M. Tu, Emily Bordeleau, Damion Dooley, Emma Griffiths, Haley L. Zubyk, Eric D. Brown, Finlay Maguire, Robert G. Beiko, William W. L. Hsiao, Fiona S. L. Brinkman, Gary Van Domselaar, and Andrew G. McArthur. 2023. CARD 2023: expanded curation, support for machine learning, and resistome prediction at the Comprehensive Antibiotic Resistance Database. *Nucleic Acids Research*, 51(D1):D690–D699.

J. M. Andrews. 2001a. Determination of minimum inhibitory concentrations. *Journal of Antimicrobial Chemotherapy*, 48(Suppl 1):5–16.

Jennifer M. Andrews. 2001b. Determination of minimum inhibitory concentrations. *Journal of Antimicrobial Chemotherapy*, 48(suppl_1):5–16.

Frederik Otzen Bagger, Line Borgwardt, Andreas Sand Jespersen, Anna Reimer Hansen, Birgitte Bertelsen, Miyako Kodama, and Finn Cilius Nielsen. 2024. Whole genome sequencing in clinical practice. *BMC Medical Genomics*, 17:Article 39.

Nathalie Bonin, Enrique Doster, Hannah Worley, Lee J Pinnell, Jonathan E Bravo, Peter Ferm, Simone Marini, Mattia Prosperi, Noelle Noyes, Paul S Morley, and Christina Boucher. 2023. MEGARes and AMR++, v3.0: an updated comprehensive database of antimicrobial resistance determinants and an improved software pipeline for classification using high-throughput sequencing. *Nucleic Acids Research*, 51(D1):D744–D752.

Nadav Brandes, Dan Ofer, Yam Peleg, Nadav Rappoport, and Michal Linial. 2022. Proteinbert: a universal deep-learning model of protein sequence and function. *Bioinformatics*, 38(8):2102–2110.

Karen Bush and George A. Jacoby. 2010. Updated functional classification of -lactamases. *Antimicrobial Agents and Chemotherapy*, 54(3):969–976.

Carolina Cason, Maria D'Accolti, Irene Soffritti, Sante Mazzacane, Manola Comar, and Elisabetta Caselli. 2022. Next-generation sequencing and pcr technologies in monitoring the hospital microbiome and its drug resistance. *Frontiers in Microbiology*, 13:969863.

Charles E. Cook, Mary Todd Bergman, Robert D. Finn, Guy Cochrane, Ewan Birney, and Rolf Apweiler. 2016. The european bioinformatics institute in 2016: Data growth and integration. *Nucleic Acids Research*, 44(D1):D20–D26.

Hugo Dalla-Torre, Liam Gonzalez, Javier Mendoza-Revilla, Nicolas Lopez Carranza, Adam Henryk Grzywaczewski, Francesco Oteri, Christian Dallago, et al. 2023. The nucleotide transformer: Building and evaluating robust foundation models for human genomics. *Genomics*.

Thomas G. Dietterich. 2000. Ensemble methods in machine learning. In *Multiple Classifier Systems (MCS 2000)*, volume 1857 of *Lecture Notes in Computer Science*, pages 1–15. Springer.

Enrique Doster, Steven M Lakin, Christopher J Dean, Cory Wolfe, Jared G Young, Christina Boucher, Keith E Belk, Noelle R Noyes, and Paul S Morley. 2020. Megares 2.0: a database for classification of antimicrobial drug, biocide and metal resistance determinants in metagenomic sequence data. *Nucleic Acids Research*, 48(D1):D561–D569.

S. R. Eddy. 1998. Profile hidden markov models. *Bioinformatics*, 14(9):755–763.

Hadrien Gourlé, Oskar Karlsson-Lindsjö, Juliette Hayer, and Erik Bongcam-Rudloff. 2019. Simulating Illumina metagenomic data with InSilicoSeq. *Bioinformatics*, 35(3):521–522.

Manuel Holtgrewe. 2010. Mason – A Read Simulator for Second Generation Sequencing Data. Technical report, FU Berlin.

Edward J. Hu, Yelong Shen, Phillip Wallis, Zeyuan Allen-Zhu, Yuanzhi Li, Shean Wang, Lu Wang, and Weizhu Chen. 2022. Lora: Low-rank adaptation of large language models. In *Proceedings of the International Conference on Learning Representations (ICLR)*.

Taishan Hu, Nilesh Chitnis, Dimitri Monos, and Anh Dinh. 2021. Next-generation sequencing technologies: An overview. *Human Immunology*, 82(11):801–811.

Weichun Huang, Leping Li, Jason R. Myers, and Gabor T. Marth. 2012. Art: A next-generation sequencing read simulator. *Bioinformatics*, 28(4):593–594.

Yanrong Ji, Zhihan Zhou, Han Liu, and Ramana V Davuluri. 2021. Dnabert: pre-trained bidirectional encoder representations from transformers model for dna-language in genome. *Bioinformatics*, 37(15):2112–2120.

Baofeng Jia, Amogelang R. Raphenya, Brian Alcock, Nicholas Waglechner, Peiyao Guo, Kara K. Tsang, Briony A. Lago, Biren M. Dave, Sheldon Pereira, Arjun N. Sharma, Sachin Doshi, Mélanie Courtot, Raymond Lo, Laura E. Williams, Jonathan G. Frye, Tariq Elsayegh, Daim Sardar, Erin L. Westman, Andrew C. Pawlowski, Timothy A. Johnson, Fiona S.L. Brinkman, Gerard D. Wright, and Andrew G. McArthur. 2017. Card 2017: expansion and model-centric curation of the comprehensive antibiotic resistance database. *Nucleic Acids Research*, 45(D1):D566–D573.

Hyeunseok Kang, Sungwoo Goo, Hyunjung Lee, Jung-Woo Chae, Hwi-Yeol Yun, and Sangkeun Jung. 2022. Fine-tuning of bert model to accurately predict drug-target interactions. *Pharmaceutics*, 14(8):1710.

Beata Kowalska-Krochmal and Ruth Dudek-Wicher. 2021. The Minimum Inhibitory Concentration of Antibiotics: Methods, Interpretation, Clinical Relevance. *Pathogens*, 10(2):165.

Ludmila I. Kuncheva and Christopher J. Whitaker. 2003. Measures of diversity in classifier ensembles and their relationship with the ensemble accuracy. *Machine Learning*, 51:181–207.

Jinhyuk Lee, Wonjin Yoon, Sungdong Kim, Donghyeon Kim, Sunkyu Kim, Chan Ho So, and Jaewoo Kang. 2020. Biobert: A pre-trained biomedical language representation model for biomedical text mining. *Bioinformatics*, 36(4):1234–1240.

Renqian Luo, Liai Sun, Yingce Xia, Tao Qin, Sheng Zhang, Hoifung Poon, and Tie-Yan Liu. 2022. Biogpt: Generative pre-trained transformer for biomedical text generation and mining. *Briefings in Bioinformatics*, 23(6):bbac409.

Alexa B. R. McIntyre, Rachid Ounit, Ebrahim Afshinnekoo, Robert J. Prill, Elizabeth Hénaff, Noah Alexander, Samuel S. Minot, David Danko, Jonathan Foox, Sofia Ahsanuddin, Scott Tighe, Nur A. Hasan, Poorani Subramanian, Kelly Moffat, Shawn Levy, Stefano Lonardi, Nick Greenfield, Rita R. Colwell, Gail L. Rosen, and Christopher E. Mason. 2017. Comprehensive benchmarking and ensemble approaches for metagenomic classifiers. *Genome Biology*, 18(1):182.

María Katherine Mejía-Guerra and Edward S. Buckler. 2019. A k-mer grammar analysis to uncover maize regulatory architecture. *BMC Plant Biology*, 19(1):103.

Pauline C. Ng and Ewen F. Kirkness. 2010. Whole genome sequencing. *Methods in Molecular Biology*, 628:215–226.

Melissa Sanabria, Jonas Hirsch, Pierre M. Joubert, and Anna R. Poetsch. 2024. DNA language model GROVER learns sequence context in the human genome. *Nature Machine Intelligence*, 6:911–923.

Patrick Schorderet. 2016. NEAT: a framework for building fully automated NGS pipelines and analyses. *BMC Bioinformatics*, 17:53.

Derrick E. Wood and Steven L. Salzberg. 2014. Kraken: ultrafast metagenomic sequence classification using exact alignments. *Genome Biology*, 15.

Koshi Yamada, Makoto Miwa, and Yutaka Sasaki. 2023. Biomedical Relation Extraction with Entity Type Markers and Relation-specific Question Answering. In *The 22nd Workshop on Biomedical Natural Language Processing and BioNLP Shared Tasks*, pages 377–384, Toronto, Canada. Association for Computational Linguistics.

Zhenyan Zhang, Qi Zhang, Tingzhang Wang, Nuohan Xu, Tao Lu, Wenjie Hong, Josep Penuelas, Michael Gillings, Meixia Wang, Wenwen Gao, and Haifeng Qian. 2022. Assessment of global health risk of antibiotic resistance genes. *Nature Communications*, 13.

Wenxuan Zhou and Muhao Chen. 2022. An improved baseline for sentence-level relation extraction. In *Proceedings of the 2nd Conference of the Asia-Pacific Chapter of the Association for Computational Linguistics and the 12th International Joint Conference on Natural Language Processing (Volume 2: Short Papers)*, pages 161–168, Online only. Association for Computational Linguistics.

Zhihan Zhou, Yanrong Ji, Weijian Li, Pratik Dutta, Ramana Davuluri, and Han Liu. 2024. Dnabert-2: Efficient foundation model and benchmark for multi-species genomes. In *Proceedings of the International Conference on Learning Representations (ICLR)*.

A Model Overview

Figure 1 illustrates the overall architecture of our proposed model, which integrates sequence-based and text-based representations for antibiotic resistance classification. We fine-tuned two pre-trained language models—Nucleotide Transformer and BioBERT—for DNA sequence classification tasks involving the prediction of antimicrobial drug classes. The Nucleotide Transformer model was fine-tuned using parameter-efficient LoRA-based adaptation. DNA sequences were truncated to a maximum length of 1000 nucleotides and tokenized using a domain-specific tokenizer. Training data was structured with input DNA sequences and corresponding drug class labels. The model was fine-tuned using a sequence classification objective on a multi-class dataset. Performance was evaluated on a separate test set using macro-average F1 score, accuracy, precision, recall, and balanced accuracy. For BioBERT, the input consisted of textual descriptions including gene family and resistance mechanism information, formatted into natural language prompts. These were tokenized using a BERT tokenizer with a fixed input length. A classification head was added to predict the drug class labels. The model was trained for multiple epochs and evaluated using the same metrics as for the Nucleotide Transformer. Both models showed effective performance in multi-class classification tasks, demonstrating the potential of sequence- and text-based pretraining approaches in genomic classification problems.

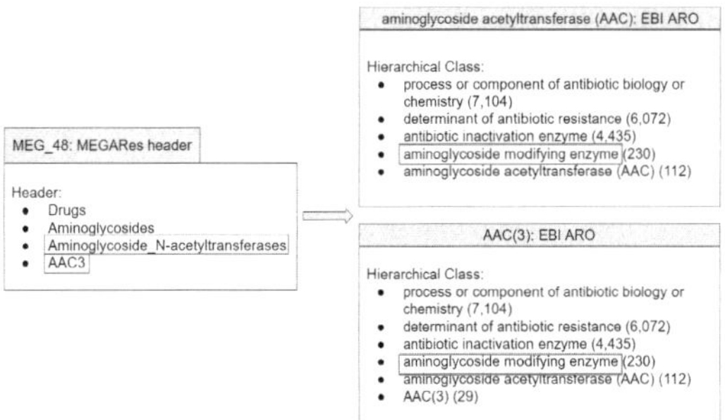

Figure 2: EBI ARO Gene Family mapping: search to find mapping information with header and ontology by using API.

B Dataset Details

The CARD and MEGARes v3 datasets are used for training and evaluation. Classes with fewer than 15 samples are removed because obtaining meaningful results from the data split is difficult. The remaining data is split into 75% for training data, 20% for test data, and 5% for validation data. EBI ARO ontology search is used to integrate the data, which is then split similarly to the above. Classes with difficult-to-obtain meaningful results are also removed. The MEGARes dataset consists of 9733 Reference Sequences, 1088 SNPs, 4 antibiotic types, 59 resistance classes, and 233 mechanisms. The CARD dataset consists of 5194 Reference Sequences and 2005 SNPs, 142 Drug Classes, 331 Gene Families, and 10 Resistance Mechanisms. The EBI ARO ontology provides hierarchical group information for genes. Using the EBI ARO Ontology, Gene Family class information can be integrated into a higher-level hierarchy. The number of Gene Family text information classes in the case of MEGARes is 589, while for CARD, it is 331. There are 300 and 166 datasets with only one sample in their respective classes for Gene Family in the case of MEGARes and CARD, respectively. Resistance Mechanism is integrated based on the 6 categories of CARD. The original 8 categories were reduced to 6, excluding cases of various class combinations and those with very few samples. Drug Class is integrated using 9 common Drug Classes found in competing models. Integration is done based on names and theories and has been verified. Macro f1 score, accuracy, balanced accuracy, and precision are used as performance metrics, and the results are listed in the Table 1.

Figure 3, Figure 4 and Figure 5 represent the distribution of training dataset which is integrated

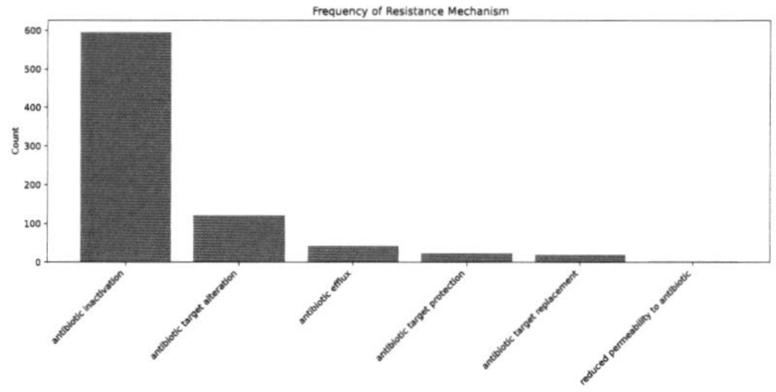

Figure 3: Counts of the frequent Resistance Mechanism in training dataset.

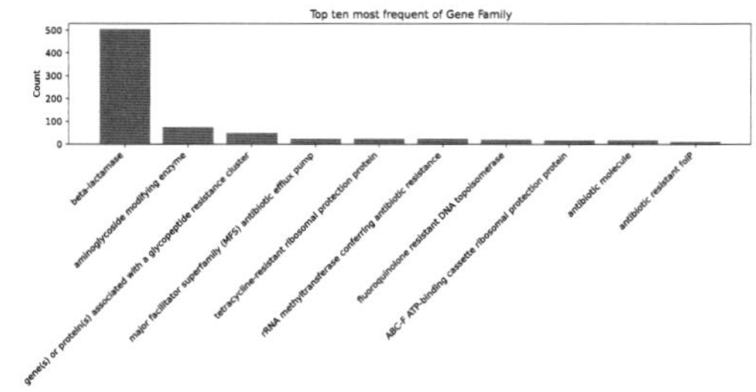

Figure 4: Counts of the frequent Gene Family in training dataset.

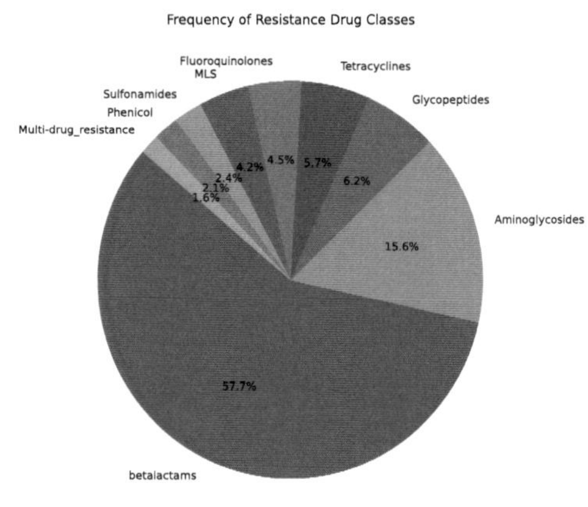

Figure 5: Counts of the frequent Antimicrobial Resistance Drug Classes in training dataset.

with CARD and MEGARes. We observe a long-tail distribution for Resistance Mechanism, Gene Family, and Drug Class classes.

The distribution indicates that certain resistance mechanisms, gene families, and drug classes are significantly overrepresented in the dataset, while many others occur with low frequency. Specifically, antibiotic inactivation is the most common resistance mechanism, while beta-lactamase genes dominate the gene family distribution. Similarly, beta-lactams appear as the most frequently associated drug class.

This imbalance in distribution suggests that models trained on this dataset may exhibit biased performance, favoring well-represented categories while struggling with rare classes. Furthermore, the presence of diverse resistance mechanisms and gene families emphasizes the complexity of antimicrobial resistance (AMR) prediction.

The dataset used in this study is publicly available at `https://zenodo.org/records/15213479`.

C Read Generation

Read generation is a computational process used to simulate short DNA or RNA sequences, commonly referred to as "reads", from reference genomes or annotated genetic sequences. This technique is designed to mimic the output of next-generation sequencing (NGS) technologies (Hu et al., 2021), providing a way to generate data for various applications such as machine learning model training, benchmarking, or evaluating bioinformatics pipelines. In the context of antibiotic resistance prediction, read generation is often performed using curated databases like CARD, MEGARes, or the Integrated database, which contain known resistance genes and associated metadata.

To simulate realistic reads, researchers commonly use specialized tools such as ART (Huang et al., 2012), InSilicoSeq (Gourlé et al., 2019), DWGSIM, NEAT (Schorderet, 2016), or Mason (Holtgrewe, 2010). These simulators can generate Illumina-style short reads with configurable read lengths, sequencing errors, mutation rates, and coverage depth. In this study, we used ART to generate synthetic reads based on the Integrated database. ART supports detailed customization of error profiles and is widely used for simulating realistic Illumina sequencing data.

The generated reads can serve as a substitute when real-world sequencing data is limited or unavailable. By generating reads from known reference sequences, researchers can perform controlled experiments with clearly defined ground truth, assess model robustness under noisy or imperfect conditions, and evaluate how well different models generalize to simulated real-world data. Overall, read generation combined with realistic simulators plays a crucial role in creating labeled datasets that facilitate the development and validation of genomic analysis tools.

D Entity Representation Techniques

To improve antibiotic resistance classification, we experimented with different entity representation techniques for encoding Gene Family and Resistance Mechanism attributes in BioBERT-based models. Table 2 compares the impact of these techniques on classification performance.

These representations were designed to help the model better distinguish between biological attributes and general text (Yamada et al., 2023). The Base format uses plain-text input without additional markers, while the Entity Marker (punct) format introduces brackets around key attributes. The Typed Entity Marker (Zhou and Chen, 2022) explicitly labels entities, providing more structured input, and the Typed Entity Marker (punct) format further combines these strategies.

Results indicate that using entity markers improves classification performance. In particular, the Typed Entity Marker (punct) approach achieves the highest Macro F1 score, demonstrating that structured formatting helps the model capture contextual relationships between resistance mechanisms and gene families more effectively. Results indicate that explicit formatting, such as typed entity markers with punctuation, enhances BioBERT's contextual understanding about Gene Family and Resistance Mechanism attributes from general text. This suggests that structured annotations provide useful inductive bias, allowing the model to better capture domain-specific relationships.

E Impact of LLM-Based Data Augmentation

Despite ontology-based class standardization, certain resistance categories remain underrepresented due to natural imbalances in antibiotic resistance gene distributions. To address this, we employ BioGPT (Luo et al., 2022) for generating synthetic samples in low-resource categories. BioGPT is prompted to generate contextually similar resistance gene descriptions, maintaining the linguistic characteristics of real annotations to ensure realistic and informative augmentation.

By integrating BioGPT-based augmentation, we observe consistent improvements in classification performance, particularly in Macro F1 scores for rare classes. Table 3 presents the results of this

Output	Input Example	BioBERT
Base	Gene Family: Beta-lactamases, Resistance Mechanism: Antibiotic inactivation	78.20
Entity marker (punct)	[Gene Family]: Beta-lactamases, [Resistance Mechanism]: Antibiotic inactivation	77.41
Typed entity marker	*Beta-lactamases*, #Resistance Mechanism#	77.70
Typed entity marker (punct)	*[Gene Family]: Beta-lactamases*, #[Resistance Mechanism]: Antibiotic inactivation#	78.46

Table 2: Test Macro F1 score of different entity representation techniques in antibiotic resistance classification with BioBERT.

augmentation strategy, demonstrating its positive impact on model robustness.

F Practical Advantages of Using Gene Family and Resistance Mechanism

Incorporating Gene Family and Resistance Mechanism information in antibiotic resistance classification provides practical advantages, particularly in reducing experimental complexity and time. Traditional laboratory-based methods, such as Minimum Inhibitory Concentration (Kowalska-Krochmal and Dudek-Wicher, 2021; Andrews, 2001a) (MIC) assays and Disk Diffusion Tests, require separate testing for each antibiotic, which involves overnight incubation and may take longer for certain organisms (Andrews, 2001b). Testing multiple antibiotics increases time and resource consumption, and experimental conditions such as growth medium and gene expression variability can further complicate results.

Sequence-based approaches, such as Polymerase Chain Reaction (PCR) and Whole Genome Sequencing (WGS), enable the identification of resistance-related genes directly from genomic data (Bagger et al., 2024; Ng and Kirkness, 2010). PCR/qPCR can provide results relatively quickly, typically within hours, whereas WGS requires a longer processing time, often taking multiple days to complete (Cason et al., 2022).

Leveraging Gene Family and Resistance Mechanism attributes allows for a more efficient computational approach to resistance prediction, minimizing reliance on exhaustive in vitro testing. Many resistance mechanisms are well-characterized and strongly associated with specific gene families. For instance, betalactamase genes are well-known indicators of resistance to betalactam antibiotics, such as penicillins and cephalosporins (Bush and Jacoby, 2010). By integrating structured biological knowledge with sequence-based models, resistance predictions can be made with greater confidence and interpretability, supporting a scalable and practical framework for antimicrobial resistance classification.

Method	Accuracy	Macro F1	Precision	Recall
NT	84.15	64.04	72.78	59.28
NT with data augmentation	83.42	64.85	80.15	58.65
NT with reads	82.85	61.02	68.32	57.06
NT with reads and data augmentation	83.11	62.82	74.81	57.32

Table 3: Effect of BioGPT-based data augmentation on resistance classification performance. Augmentation improves Macro F1, particularly for low-resource categories.

Questioning Our Questions: How Well Do Medical QA Benchmarks Evaluate Clinical Capabilities of Language Models?

Siun Kim
Seoul National University Hospital
Seoul, South Korea
shiuhn95@snu.ac.kr

Hyung-Jin Yoon
Seoul National University Hospital
Seoul, South Korea
hjyoon@snu.ac.kr

Abstract

Recent advances in large language models (LLMs) have led to impressive performance on medical question-answering (QA) benchmarks. However, the extent to which these benchmarks reflect real-world clinical capabilities remains uncertain. To address this gap, we systematically analyzed the correlation between LLM performance on major medical QA benchmarks (e.g., MedQA, MedMCQA, PubMedQA, and MMLU medicine subjects) and clinical performance in real-world settings. Our dataset included 702 clinical evaluations of 85 LLMs from 168 studies. Benchmark scores demonsrated a moderate correlation with clinical performance (Spearman's $\rho = 0.59$), albeit substantially lower than inter-benchmark correlations. Among them, MedQA was the most predictive but failed to capture essential competencies such as patient communication, longitudinal care, and clinical information extraction. Using Bayesian hierarchical modeling, we estimated representative clinical performance and identified GPT-4 and GPT-4o as consistently top-performing models, often matching or exceeding human physicians. Despite longstanding concerns about the clinical validity of medical QA benchmarks, this study offers the first quantitative analysis of their alignment with real-world clinical performance.[1]

1 Introduction

The rapid advancement of large language models (LLMs), accelerated by the release of ChatGPT, has continued into 2025. Open-source models such as Llama 3.3, Phi-4, and DeepSeek-R1 are rapidly narrowing the performance gap with proprietary models (Grattafiori et al., 2024; Abdin et al., 2024; Guo et al., 2025). This progress is especially consequential in healthcare, where strigent privacy and security requirments frequently necessitate on-premise deployment (Faray de Paiva et al., 2025; Gupta and Pande, 2025).

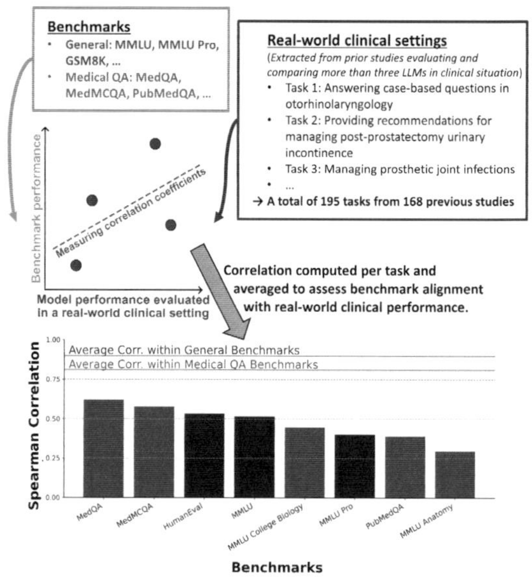

Figure 1: Overview of our study assessing the alignment between medical QA benchmarks and real-world clinical performance.

As LLMs attains expert-level performance on both general and medical QA benchmarks, the limitations of such benchmarks have become increasingly apparent. For instance, OpenAI's o1-preview achieved 96% on MedQA and 99% on MMLU Medical Genetics, outperforming human experts (Nori et al., 2024; Liévin et al., 2024). However, such benchmarks are thought to focus predominantly on static knowledge and structured reasoning, which may not fully reflect core competencies essential for clinical practice (Nori et al., 2023; Singhal et al., 2023), such as decision-making under uncertainty (Han et al., 2011), patient communication (Barry and Edgman-Levitan, 2012), and ethical reasoning (Kaldjian et al., 2005).

Although concerns over the limited clinical validity of existing medical QA benchmarks have

[1] The dataset and code are available at: https://github.com/SiunKim/questioning-medqa.

Proceedings of the 24th Workshop on Biomedical Language Processing (BioNLP 2025), pages 274–296
August 1, 2025

been raised, there remains a lack of systematic evidence. In this study, we address this gap through a comprehensive meta-analysis evaluating how effectively conventional medical QA benchmarks reflect the real-world clinical performance of LLMs (Figure 1).

Our key contributions are as follows:

- **Quantitative assessment of benchmark-clinical alignment**: We demonstrate a moderate correlation (Spearman's $\rho = 0.59$) between medical QA benchmarks and real-world clinical performance, highlighting significant limitations in the current evaluation practices.
- **Identification of clinical gaps in MedQA**: MedQA demonstrates strong alignment with core competencies such as treatment, clinical knowledge, and diagnosis. However, it fails to adequately assess essential aspects of real-world clinical practice, including patient communication, longitudinal care, and clinical information extraction.
- **Bayesian modeling of representative clinical performance**: Using hierarchical Bayesian models, we estimate the generalized clinical capabilities of LLMs, suggesting that models like GPT-4 and GPT-4o match or exceed human physician-level performance in real-world clinical settings.

2 Related Works

MedQA—based on the USMLE Step 1 and 2 exams—has emerged as a de facto benchmark in the medical domain, owing to its high-quality multiple-choice questions (MCQs) and comprehensive topical coverage (Jin et al., 2021). As a representative benchmark, improvements in MedQA performance have frequently been interpreted as a proxy for progress in medical LLMs (Singhal et al., 2025; Saab et al., 2024).

MedMCQA, derived from Indian medical entrance exams (AIIMS and NEET PG), complements MedQA by offering broader topical diversity and varied question types (Pal et al., 2022). In contrast, PubMedQA focused on biomedical literature comprehension by requiring models to infer answers from PubMed abstracts (Jin et al., 2019).

Despite their widespread use, these traditional medical benchmarks primarily assess factual recall and structured reasoning. They have been criticized for failing to evaluate essential aspects of practical clinical competence (Tang et al., 2023; Kim et al., 2025; Liu et al., 2024).

In response, recent datasets aim to capture the complexity of real-world clinical practice. Datasets like Medbullet (Chen et al., 2024), MedExQA (Kim et al., 2024), and MedXpertQA (Zuo et al., 2025) introduce open-ended questions, expert-written explanations, and multimodal data to facilitate more comprehensive evaluations. Furthermore, integrated evaluation frameworks like MedAgent-Bench (Tang et al., 2025) and MEDIC (Kanithi et al., 2024) encompass multiple clinical tasks and explicitly address ethical and safety concerns.

In parallel, agent-based evaluations have emerged to assess interactive and dynamic reasoning. For instance, MedQA-CS adopts OSCE-style clinical scenarios (Yao et al., 2024), while Agent-Clinic (Schmidgall et al., 2024) evaluates LLMs during simulated physician-patient dialogues.

Building on these developments, our study systematically examines the alignment between conventional medical QA benchmarks and real-world clinical evaluations. By identifying existing gaps, we aim to inform the design of future benchmarks that more accurately reflect practical clinical competencies.

3 Methods

To evaluate the extent to which existing medical QA benchmarks reflect real-world clinical performance, we analyzed 168 published studies that assessed at least three distinct language models in clinical settings. Benchmark scores on both medical QA and general-purpose benchmarks were collected and standardized to ensure comparability. To address the missing benchmark scores, multiple imputation was applied. Correlations between benchmark scores and clinical performance were calculated using rank-based methods weighted by sample size. Finally, we employed Bayesian hierarchical modeling to estimate each model's representative clinical capability.

3.1 Literature Review for Collecting Clinical Performance Data

We conducted a multi-stage literature review to identify studies evaluating LLM performance in real-world clinical settings (Figure 2). Using the Semantic Scholar API, we first retrieved articles published between January 1, 2023, and January 10, 2025, based on search queries designed to encompass a wide range of clinical scenarios (Appendix A.1). Title-based filtering retained studies explicitly mentioning LLM-related terms, followed by DOI-based deduplication. Abstract and full-texts were retrieved via publisher and open-access APIs.

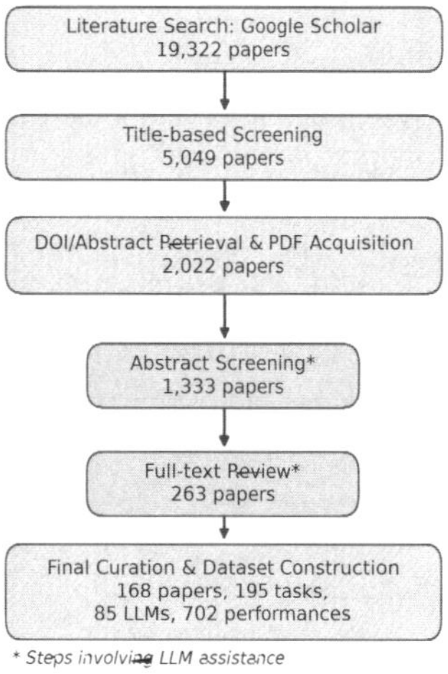

Figure 2: Flowchart of literature review for collecting LLM performances in real-world clinical settings.

Phi-4 model-assisted screening of abstracts and full-texts identified studies that reported performance for at least three distinct LLMs, enabling correlation analyses (AppendixA.2). Manual review was conducted to extract structured data, normalized model names (Appendix A.3) and classified evaluation settings (Appendix A.4).

3.2 LLM Performance Collection in Real-World Clinical Settings

To address overrepresentation issues caused by redundant evaluations of similar model abilities within a single study, we extracted one representative performance score for each task-model combination. Preference was given to the simplest inference setting (e.g., zero-shot without CoT). If multiple measures existed for the same therapeutic area and capabilities, we selected the most frequently used metric, or averaged scores, if no dominant measure was evident. Evaluations spanning multiple therapeutic areas or distinct capabilities were treated as a separate task.

Studies relying on readability metrics, inter-model correlation analyses, or with fewer than 20 evaluation samples were excluded. Encoder-based language models (e.g., BERT, RoBERTa) were also excluded because the study focused solely on autoregressive LLMs.

Performance scores were normalized to a 0–100 scale using min-max scaling. Metrics indicating better performance through lower values (e.g., proportion of biased answers) were inverted by subtracting from 100.

3.3 Benchmark Performance Collection

Benchmarks were divided into medical QA benchmarks and general benchmarks. Medical QA benchmarks included MedQA, MedMCQA, PubMedQA, and six MMLU medical subsets (Anatomy, Clinical Knowledge, College Biology, College Medicine, Medical Genetics, and Professional Medicine). General benchmarks consisted of MMLU, MMLU Pro, BBH, HumanEval, GSM8K, and MATH.

Performance data were extracted from published articles, technical reports, and model cards. Additional web searches supplemented version-specific scores for widely used proprietary models (e.g., GPT, Claude, and Gemini).

We standardized benchmark performances by focusing on zero-shot without CoT. If multiple results were available, averages were used. If zero-shot data were unavailable, performance was estimated through linear regression using reported results under different inference settings, by considering few-shot examples and CoT usage as covariates.

3.4 Benchmark Performance Imputation

While complete benchmark data are ideal for reliable correlation analyses, missing values were inevitable as performance scores were compiled through literature review rather than direct evaluation. To systematically address missing values, we employed Multiple Imputation by Chained Equations (MICE), which leverages observed interdependencies within available data to estimate absent benchmark performances.

Before imputation, we confirmed the Missing at Random (MAR) assumption, a necessary condition for reducing bias in estimation. Two imputation techniques were tested: Random Forest (RF-MICE) for capturing non-linear dependencies and Bayesian Ridge (BR-MICE) for small datasets with collinearity. Imputation was performed separately for each benchmark category.

We validated imputation accuracy through masking test, randomly removing and subsequently estimating 10% of the observed values. To incorporate uncertainty, multiple imputations were conducted, and within- and between-imputation variance were estimated (Appendix A.5). Based on validation results, we selected a final version of the imputed dataset for downstream analysis.

3.5 Correlation Measurement

We evaluated correlations in two ways: benchmark-to-benchmark and benchmark-to-clinical performance.

Benchmark-to-benchmark correlation were calculated based on the performance scores of models that were evaluated on both. This analysis allowed us to identify redundant benchmarks, assess the quality of benchmark datasets, and set a correlation baseline for subsequent benchmark-to-clinical correlation analyses.

Benchmark-to-clinical correlations were computed at the evaluation task level, weighted logarithmically by evaluation sample size to reflect varying reliability across studies. Analyses utilized imputed benchmark scores primarily, with non-imputed data serving as sensitivity checks.

Although we measured rank-based (Spearman's rank correlation coefficient and Kendall's tau) and linear-based (Pearson's correlation coefficient and Lin's concordance correlation coefficient) metrics, primary analyses used Spearman's rank and Kendall's tau due to their suitability for handling diverse evaluation scoring scales without assuming linear relationships. Linear correlations were calculated but used only as reference points.

3.6 Bayesian Modeling

To estimate representative clinical performance for each language model independent of task-specific biases, we employed Bayesian hierarchical modeling. Given the limited number of model evaluated per task (average 3.6 models), individual task effects could not be directly estimated. Instead, task-related variations were approximated using metadata attributes including task type, data source, and evaluation methods. Therapeutic areas were excluded due to inconsistent categorization and unclear impact on performance (Appendix A.6.1). Furthermore, models for which performance data were available for fewer than three distinct tasks were excluded to enhance the reliability of model-specific performance estimates, which served as proxies for general clinical competence.

To further assess the robustness of the model-specific estimates, connectivity measures were calculated. Higher connectivity indicates stronger support from direct and indirect comparisons across models, thereby resulting in more stable and accurate performance estimates (Appendix A.6.2).

Table 1: Summary of clinical performance dataset and evaluation settings.

Category	Count (%)
Total samples	702 (100.0)
Task type	
Diagnosis	183 (26.1)
Clinical Knowledge	182 (25.9)
Overall Management	111 (15.8)
Answering to Patients	83 (11.8)
Information Extraction	61 (8.7)
Treatment	48 (6.8)
Other	34 (4.8)
Data source	
Clinical Vignettes	271 (38.6)
Quizzes	160 (22.8)
Board Examination	114 (16.2)
FAQs	74 (10.5)
Other	83 (11.8)
Therapeutic area	
General Medicine	154 (21.9)
Oncology	77 (11.0)
Ophthalmology	60 (8.5)
Orthopedics & Musculoskeletal	58 (8.3)
Emergency Medicine	53 (7.5)
Neuropsychiatric	53 (7.5)
Others	247 (35.2)
Evaluation type	
MCQs	463 (66.0)
Human Rating	239 (34.0)

4 Results and Discussion

4.1 Clinical Performance Dataset

Our dataset comprised 702 clinical performance evaluations from 168 studies covering 195 distinct clinical tasks. Evaluations involved 85 LLMs, predominantly from GPT (51.7%), LLaMA (10.3%), and Gemini (8.8%) families. Task types included diagnosis (26.1%), clinical knowledge assessment (25.9%), and overall patient management (15.8%). Data sources were primarily clinical vignettes (38.6%) and quizzes (22.8%), with evaluations conducted through MCQs (66.0%) and expert human ratings (34.0%) (Table 1).

4.2 Benchmark Performance Imputation

The benchmark dataset contained a notable proportion of missing values: 42.4% for medical

Table 2: Average correlation coefficients of medical QA benchmarks with other benchmarks. The highest score in each column is **bold**, and the second highest is underlined.

Medical QA Benchmarks	Spearman		Kendall	
	Medical QA	General	Medical QA	General
MedQA	0.809	**0.867**	0.664	**0.703**
MedMCQA	0.808	<u>0.855</u>	0.651	<u>0.693</u>
MMLU Medical Genetics	0.835	0.748	0.684	0.607
MMLU Clinical Knowledge	**0.851**	0.820	**0.714**	0.664
MMLU College Medicine	0.822	0.784	0.683	0.618
MMLU Professional Medicine	<u>0.849</u>	0.789	<u>0.705</u>	0.632
MMLU College Biology	0.819	0.666	0.672	0.522
MMLU Anatomy	0.703	0.558	0.571	0.449
PubMedQA	0.484	0.441	0.333	0.318
Average	**0.787**	**0.725**	**0.675**	**0.576**

QA benchmarks (9 benchmarks, 138 models, 715 scores) and 40.6% for general benchmarks (6 benchmarks, 126 models, 449 scores).

Imputation accuracy, assessed through masking tests, indicated RF-MICE outperformed BR-MICE. Specifically, RF-MICE achieved lower mean absolute error (MAE=2.04) and higher R^2 (0.98) on medical QA benchmarks (Table 9). Variance analysis of multiple imputations further supported RF-MICE due to lower total variance and improved stability (Table 10). Consequently, RF-MICE was utilized to generate the final imputed dataset.

4.3 Benchmark-to-Benchmark Correlation

Medical QA benchmarks showed strong internal correlations overall, with MMLU Clinical Knowledge and MMLU Professional Medicine exhibiting particularly high correlations with other medical QA benchmarks (Table 2). This is likely due to their broad content coverage, encompassing topics found in other MMLU medical subjects, thereby forming a high-correlation block (Figure 6).

In contrast, PubMedQA and MMLU Anatomy showed weaker correlations with other medical QA benchmarks. PubMedQA's lower correlations may stem from its distinct task formulation, which is more aligned with biomedical summarization rather than clinical reasoning (Jin et al., 2019). Similarly, MMLU Anatomy's lower correlations likely reflect its narrower content scope compared to other benchmarks.

MedQA and MedMCQA demonstrated the highest correlations with general benchmarks among medical QA benchmarks (Table 2, Figure 6). This suggests that these two datasets not only assess domain-specific knowledge but also required a broad set of reasoning skills, many of which overlapped with general benchmarks.

Within general benchmarks, BBH (Spearman's 0.891) and MMLU (0.853) exhibited the strongest correlations with medical QA benchmarks (Table 11). This result also indicates logical reasoning capabilities and broad domain knowledge are closely linked to solve medical problems. In contrast, mathematics-focused benchmarks (i.e., GSM8K and MATH) displayed weaker correlations, highlighting the distinct types of reasoning involved in medical contexts.

4.4 Benchmark-to-Clinical Performance Correlation

MedQA showed the strongest correlation with real-world clinical performance, outperforming general benchmarks in capturing actual clinical competency (Spearman's 0.588, Kendall's 0.520; Figure 3A). However, correlation strength was notably lower than inter-benchmark correlations (0.675–0.787; Table 2). These results suggest MedQA remains the most representative current benchmark for clinical tasks, although its ability to predict comprehensive clinical performance remains limited.

Further analysis across evaluation settings highlighted MedQA's strengths and limitations (Figure 4). MedQA performed well predicting clinical competency in tasks involving treatment, clinical knowledge, and diagnosis (Figure 4A). In contrast, it showed significantly weaker correlations in patient communication, overall patient management, and information extraction.

Similarly, while MedQA strongly correlated with performance derived from board examination-

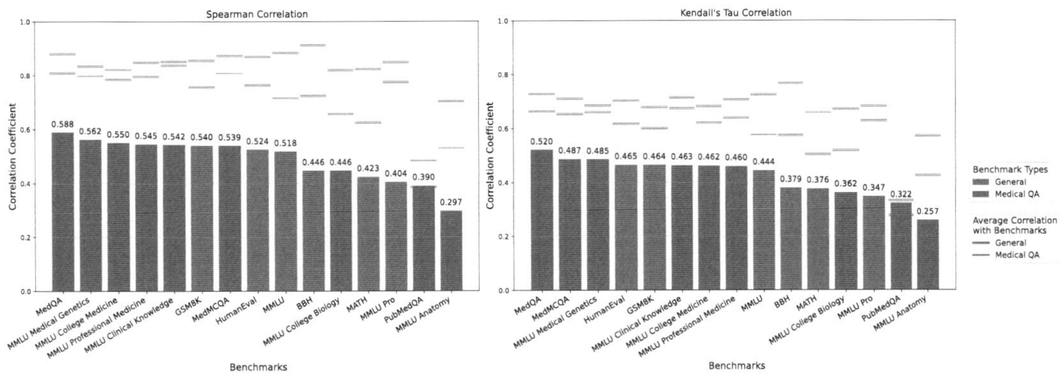

Figure 3: Comparison of correlation coefficients between benchmark and clinical performance.

style datasets, correlations with evaluations based on clinical vignette or FAQ, which closely resemble real-world clinical practice, were considerably lower (Figure 4B). These findings suggest that while MedQA reliably evaluates core medical knowledge and reasoning skills, it does not adequately reflect the broader competencies required in real-world clinical practice.

MMLU Medical Genetics, College Medicine, Professional Medicine, Clinical Knowledge, and MedMCQA displayed moderately high correlations with clinical performance, outperforming general benchmarks (Figure 3A). Conversely, PubMedQA and MMLU Anatomy consistently underperformed, indicating their limited suitability as representative clinical evaluation tools (Figures 9, 10).

4.5 Representative Clinical Performances Estimated through Bayesian Modeling

Representative clinical performances of 59 language models were robustly estimated using Bayesian hierarchical modeling across 717 performance samples. Model convergence was strong, indicated by effective sample sizes (ESS) above 300 and R-hat values below 1.02.

Among evaluated models, GPT-4 and GPT-4o consistently demonstrated the highest clinical performance, often exceeding the average performance of medical professionals (labeled as 'human - doctor') and substantially outperforming both smaller open-source models and other proprietary models (Figure 5). The strong and consistent performance of the GPT family is further supported by newly developed medical benchmark studies (Olatunji et al., 2024; Yao et al., 2024; Zuo et al., 2025), which similarly highlight their superior clinical reasoning capabilities.

Proprietary models (purple) generally outperformed open-source models (orange, Figure 5), suggesting that commercially optimized systems remain more reliable in clinical settings—though this conclusion may shift with the rapid progress of open-source LLMs in 2025.

Within the open-source category, Llama-3.1-8B-instruct was the only model to surpass the minimum threshold set for human-level performance (labeled as 'human - cut-off'). Notably, however, its lower connectivity implies that this performance estimate should be interpreted with caution due to high uncertainty.

Notably, language models fine-tuned for the medical domain (marked with a star, ★) did not show substantial improvements over general-purpose models like Llama, despite having comparable model sizes (Figure 5). This may be due to overfitting to the specific characteristics of their training datasets—typically composed of structured medical QA corpora or textbook-style materials—which could limit their generalizability in practical clinical contexts (Olatunji et al., 2024). These findings are consistent with previous results showing that biomedical models often underperform on newer, more complex benchmarks, and support concerns regarding their sensitivity to dataset-specific biases and limitations (Olatunji et al., 2024; Yao et al., 2024).

5 Conclusion

This study demonstrates that existing medical QA benchmarks posses only a moderate capacity to predict real-world clinical performance. Among them, MedQA showed the strongest correlation with clinical performance but was still insufficient for evaluating practical clinical competencies such as patient

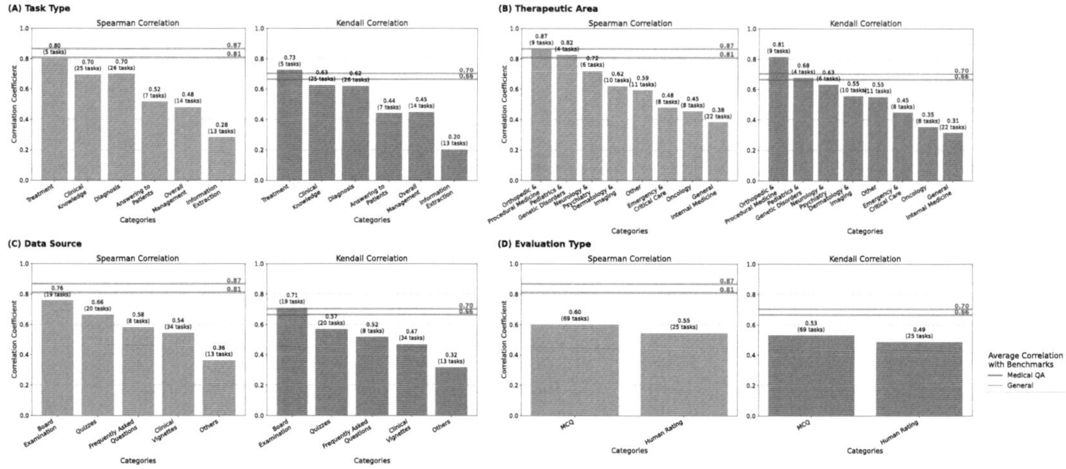

Figure 4: Comparison of correlations between MedQA performance and individual outcomes measured in real-world clinical settings across different task types, therapeutic areas, data sources, and evaluation methods.

interaction, longitudinal patient management, and clinical information extraction. Bayesian hierarchical modeling further revealed that proprietary models—particularly GPT-4 and GPT-4o—consistently outperformed open-source counterparts and, across many versions, matched or exceeded the performance of human experts in real-world clinical settings. Notably, despite longstanding concerns regarding the validity of medical QA benchmarks, this study provides the first systematic and quantitative evidence evaluating the alignment between medical QA benchmarks and actual clinical performance.

Limitations

This study has several limitations. First, our analysis is based on published studies, which inevitably lag behind ongoing LLM advancements due to publication delays. Consequently, it does not account for recent developments in LLMs, such as the emergence of reasoning-based LLMs (Guo et al., 2025).

Second, although several medical benchmarks have been introduced to better assess multifaceted capabilities (Kim et al., 2024; Yao et al., 2024; Zuo et al., 2025), we could not obtain sufficient model performance results on these datasets to conduct correlation analyses. To support future research, we make our clinical performance dataset available and encourage its use in validating how well these newly proposed medical benchmarks reflect the complexity of real-world medical tasks.

Lastly, despite employing statistical methods to address missing data and selection biases, our findings are inherently constrained by the incompleteness and potential biases of literature-derived data.

Acknowledgments

This work was supported by the Institute of Information Communications Technology Planning Evaluation (IITP)-Innovative Human Resource Development for Local Intellecutallization program grant funded by the Korea goverment (MSIT) (IITP-2024-RS-2024-00441407). This research was supported by a grant of the Korea Health Technology RD Project through the Korea Health Industry Development Institute (KHIDI), funded by the Ministry of Health Welfare, Republic of Korea (grant number: RS-2023-KH136520).

References

Marah Abdin, Jyoti Aneja, Harkirat Behl, Sébastien Bubeck, Ronen Eldan, Suriya Gunasekar, Michael Harrison, Russell J Hewett, Mojan Javaheripi, Piero Kauffmann, et al. 2024. Phi-4 technical report. *arXiv preprint arXiv:2412.08905*.

Michael J Barry and Susan Edgman-Levitan. 2012. Shared decision making—the pinnacle of patient-centered care. *New England Journal of Medicine*, 366(9):780–781.

Hanjie Chen, Zhouxiang Fang, Yash Singla, and Mark Dredze. 2024. Benchmarking large language models on answering and explaining challenging medical questions. *arXiv preprint arXiv:2402.18060*.

Lisle Faray de Paiva, Gijs Luijten, Behrus Puladi, and Jan Egger. 2025. How does deepseek-r1 perform on usmle? *medRxiv*, pages 2025–02.

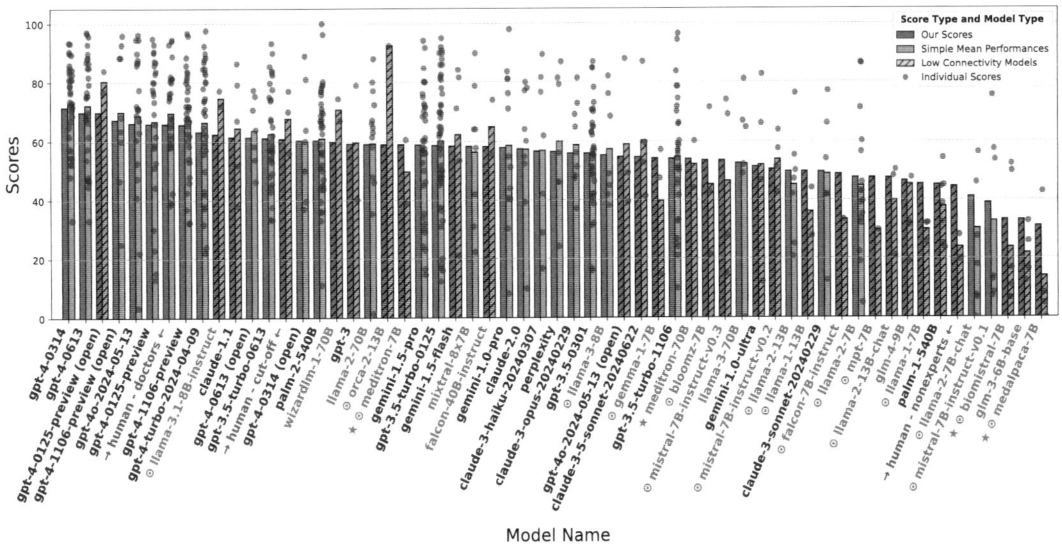

Figure 5: Representative clinical performance estimated via Bayesian modeling. Proprietary models appear in purple text while open-source models are shown in orange. Medical domain fine-tuned models are marked with a star (★) and small language models with 13B parameters or fewer display a circle prefix (⊙).

Aaron Grattafiori, Abhimanyu Dubey, Abhinav Jauhri, Abhinav Pandey, Abhishek Kadian, Ahmad Al-Dahle, Aiesha Letman, Akhil Mathur, Alan Schelten, Alex Vaughan, et al. 2024. The llama 3 herd of models. *arXiv preprint arXiv:2407.21783*.

Daya Guo, Dejian Yang, Haowei Zhang, Junxiao Song, Ruoyu Zhang, Runxin Xu, Qihao Zhu, Shirong Ma, Peiyi Wang, Xiao Bi, et al. 2025. Deepseek-r1: Incentivizing reasoning capability in llms via reinforcement learning. *arXiv preprint arXiv:2501.12948*.

Gaurav Kumar Gupta and Pranal Pande. 2025. Llms in disease diagnosis: A comparative study of deepseek-r1 and o3 mini across chronic health conditions. *arXiv preprint arXiv:2503.10486*.

Paul KJ Han, William MP Klein, and Neeraj K Arora. 2011. Varieties of uncertainty in health care: a conceptual taxonomy. *Medical Decision Making*, 31(6):828–838.

Di Jin, Eileen Pan, Nassim Oufattole, Wei-Hung Weng, Hanyi Fang, and Peter Szolovits. 2021. What disease does this patient have? a large-scale open domain question answering dataset from medical exams. *Applied Sciences*, 11(14):6421.

Qiao Jin, Bhuwan Dhingra, Zhengping Liu, William W Cohen, and Xinghua Lu. 2019. Pubmedqa: A dataset for biomedical research question answering. *arXiv preprint arXiv:1909.06146*.

Lauris C Kaldjian, Robert F Weir, and Thomas P Duffy. 2005. A clinician's approach to clinical ethical reasoning. *Journal of general internal medicine*, 20:306–311.

Praveen K Kanithi, Clément Christophe, Marco AF Pimentel, Tathagata Raha, Nada Saadi, Hamza Javed, Svetlana Maslenkova, Nasir Hayat, Ronnie Rajan, and Shadab Khan. 2024. Medic: Towards a comprehensive framework for evaluating llms in clinical applications. *arXiv preprint arXiv:2409.07314*.

Jonathan Kim, Anna Podlasek, Kie Shidara, Feng Liu, Ahmed Alaa, and Danilo Bernardo. 2025. Limitations of large language models in clinical problem-solving arising from inflexible reasoning. *arXiv preprint arXiv:2502.04381*.

Yunsoo Kim, Jinge Wu, Yusuf Abdulle, and Honghan Wu. 2024. Medexqa: Medical question answering benchmark with multiple explanations. *arXiv preprint arXiv:2406.06331*.

Valentin Liévin, Christoffer Egeberg Hother, Andreas Geert Motzfeldt, and Ole Winther. 2024. Can large language models reason about medical questions? *Patterns*, 5(3).

Fenglin Liu, Zheng Li, Hongjian Zhou, Qingyu Yin, Jingfeng Yang, Xianfeng Tang, Chen Luo, Ming Zeng, Haoming Jiang, Yifan Gao, et al. 2024. Large language models in the clinic: a comprehensive benchmark. *arXiv preprint arXiv:2405.00716*.

Harsha Nori, Nicholas King, Scott Mayer McKinney, Dean Carignan, and Eric Horvitz. 2023. Capabilities of gpt-4 on medical challenge problems. *arXiv preprint arXiv:2303.13375*.

Harsha Nori, Naoto Usuyama, Nicholas King, Scott Mayer McKinney, Xavier Fernandes, Sheng Zhang, and Eric Horvitz. 2024. From medprompt to o1: Exploration of run-time strategies for medical challenge problems and beyond. *arXiv preprint arXiv:2411.03590*.

Tobi Olatunji, Charles Nimo, Abraham Owodunni, Tassallah Abdullahi, Emmanuel Ayodele, Mardhiyah Sanni, Chinemelu Aka, Folafunmi Omofoye, Foutse Yuehgoh, Timothy Faniran, et al. 2024. Afrimed-qa: A pan-african, multi-specialty, medical question-answering benchmark dataset. *arXiv preprint arXiv:2411.15640*.

Ankit Pal, Logesh Kumar Umapathi, and Malaikannan Sankarasubbu. 2022. Medmcqa: A large-scale multi-subject multi-choice dataset for medical domain question answering. In *Conference on health, inference, and learning*, pages 248–260. PMLR.

Khaled Saab, Tao Tu, Wei-Hung Weng, Ryutaro Tanno, David Stutz, Ellery Wulczyn, Fan Zhang, Tim Strother, Chunjong Park, Elahe Vedadi, et al. 2024. Capabilities of gemini models in medicine. *arXiv preprint arXiv:2404.18416*.

Samuel Schmidgall, Rojin Ziaei, Carl Harris, Eduardo Reis, Jeffrey Jopling, and Michael Moor. 2024. Agentclinic: a multimodal agent benchmark to evaluate ai in simulated clinical environments. *arXiv preprint arXiv:2405.07960*.

Karan Singhal, Shekoofeh Azizi, Tao Tu, S Sara Mahdavi, Jason Wei, Hyung Won Chung, Nathan Scales, Ajay Tanwani, Heather Cole-Lewis, Stephen Pfohl, et al. 2023. Large language models encode clinical knowledge. *Nature*, 620(7972):172–180.

Karan Singhal, Tao Tu, Juraj Gottweis, Rory Sayres, Ellery Wulczyn, Mohamed Amin, Le Hou, Kevin Clark, Stephen R Pfohl, Heather Cole-Lewis, et al. 2025. Toward expert-level medical question answering with large language models. *Nature Medicine*, pages 1–8.

Xiangru Tang, Daniel Shao, Jiwoong Sohn, Jiapeng Chen, Jiayi Zhang, Jinyu Xiang, Fang Wu, Yilun Zhao, Chenglin Wu, Wenqi Shi, et al. 2025. Medagentsbench: Benchmarking thinking models and agent frameworks for complex medical reasoning. *arXiv preprint arXiv:2503.07459*.

Xiangru Tang, Anni Zou, Zhuosheng Zhang, Ziming Li, Yilun Zhao, Xingyao Zhang, Arman Cohan, and Mark Gerstein. 2023. Medagents: Large language models as collaborators for zero-shot medical reasoning. *arXiv preprint arXiv:2311.10537*.

Zonghai Yao, Zihao Zhang, Chaolong Tang, Xingyu Bian, Youxia Zhao, Zhichao Yang, Junda Wang, Huixue Zhou, Won Seok Jang, Feiyun Ouyang, et al. 2024. Medqa-cs: Benchmarking large language models clinical skills using an ai-sce framework. *arXiv preprint arXiv:2410.01553*.

Yuxin Zuo, Shang Qu, Yifei Li, Zhangren Chen, Xuekai Zhu, Ermo Hua, Kaiyan Zhang, Ning Ding, and Bowen Zhou. 2025. Medxpertqa: Benchmarking expert-level medical reasoning and understanding. *arXiv preprint arXiv:2501.18362*.

A Supplementary Methods

A.1 Search Query for Literature Review

Our literature search followed a systematic approach to identify studies at the intersection of large language models (LLMs) and medical applications. The search queries were structured using three essential components: LLM-related terms, medical terms, and evaluation terms (Table 3).

Each query was formulated as:

[LLM Term] AND [Medical Term] AND [Evaluation Term]

where the medical terms were drawn from either MedQA-related categories (e.g., "medical question answering", "clinical reasoning") or clinical application categories (e.g., medical specialties, clinical documents, diseases, and procedures). The search was restricted to publications from 2022 to 2025 to ensure coverage of recent developments.

This combinatorial approach balanced coverage and precision, ensuring that retrieved papers addressed all three aspects of our research focus (Table 4).

A.2 Screening Process for Collecting Performance data of LLM in Real-World Clinical Settings

We conducted abstract screening and full-text review based on LLM to refine selection process and alleviated burden of manual curation. The LLM utilized for this process was Phi-4 (14.7B), Q4_K_M quantized, based on the Ollama framework (as of March 18, 2025). The model was deployed locally on a single RTX 4080 GPU.

A.3 Model Name Normalization

We normalized language model names by categorizing them into proprietary (Table 7) and open-source models (Table 8). For proprietary models, specific model names were often unspecified in papers, as they were accessed via APIs. In such cases, we assumed the most recent model available at the research time: three months before the received date for journal papers and six months before publication for conference papers. For open-source models, normalization was based on explicitly stated model names, versions, and parameter sizes in billions (Table 8). If these details were insufficient, we excluded the model from analysis.

Table 3: Search Query Components for LLM Applications in Medical Research.

Query Component	Terms
LLM Terms	"large language model", "language model", "GPT-4", "ChatGPT"
MedQA Terms	"medical question answering", "USMLE", "MedQA", "medical benchmark", "clinical reasoning"
Clinical Application Terms	*Medical Specialties (31 terms):* "internal medicine", "surgery", "pediatrics", "obstetrics", "gynecology", ... *Surgery Settings (10 terms):* "surgery", "pediatric surgery", "breast surgery", "colorectal surgery", "neurosurgery", ... *Clinical Settings (6 terms):* "emergency department", "icu", "operating room", "outpatient", "primary care", "trauma center" *Clinical Documents (11 terms):* "electronic health record", "clinical notes", "discharge summary", "medical history", "radiology report", ... *Common Diseases (34 terms):* "breast cancer", "lung cancer", "colorectal cancer", "prostate cancer", "leukemia", "lymphoma", ... *Clinical Procedures (6 terms):* "chemotherapy", "radiation therapy", "transplantation", "dialysis", "ventilation", "ecmo" *Age-Specific Care (7 terms):* "newborn care", "child development", "growth disorders", "birth defects", "falls prevention", "memory disorders", "polypharmacy management" *Special Populations (6 terms):* "maternal health", "prenatal care", "postpartum care", "women's health", "social determinants", "medical ethics"
Evaluation Terms	"evaluation", "accuracy", "benchmark", "validation", "application"

Table 4: Query Construction Pattern and Examples.

Query Pattern: [LLM Term] AND [Medical Term] AND [Evaluation Term]
Medical Term Selection: Either [MedQA Terms] OR [Clinical Application Terms] based on research focus
Example Queries: *With MedQA Terms:* 1. large language model" AND medical question answering" AND evaluation" 2. ChatGPT" AND clinical reasoning" AND benchmark" *With Clinical Application Terms:* 3. GPT-4" AND electronic health record" AND validation" 4. language model" AND internal medicine" AND application" 5. GPT-4" AND breast cancer" AND "accuracy"

A.4 Classification of Evaluation Settings

We categorized the evaluation settings for language models in clinical contexts based on four key criteria: (1) task type, (2) therapeutic area, (3) data source, and (4) evaluation method.

Task Type Task types represent the core capabilities being assessed, classified into the following six categories:

- **Clinical Knowledge:** General assessment of fundamental clinical knowledge within a given specialty, without a specific focus on diagnosis, treatment, or prevention.
- **Treatment:** Evaluation of the model's ability to recommend and assess treatment plans based on a given clinical scenario.
- **Diagnosis:** Determining the correct diagnosis based on the provided patient information.
- **Answering to Patients:** Providing responses to common patient inquiries or explaining clinical conditions in plain language understandable by non-experts.
- **Overall Management:** Beyond diagnosis and treatment, evaluating long-term patient management and decision-making.
- **Information Extraction:** Extracting specific clinical information from given texts.

Table 5: Abstract Screening Prompt Template.

Prompt for abstract screening

Please analyze the following research paper's title and abstract to extract information about LLM performance evaluation in clinical settings. Present your analysis in the following structured format, maintaining exact quotes where possible. Start your response with "ANALYSIS_START" **and end with** "ANALYSIS_END". **INPUT REQUIRED:**

- Title: [*paper title*]
- Abstract: [*paper abstract*]

TASK: Analyze the title and abstract to extract the following information:

1. `PAPER_TYPE`: Classify the paper as one of the following:

- **"Clinical LLM Performance Evaluation - Original"**: Paper that conducts new experiments to evaluate LLM performance in clinical tasks and reports original performance metrics/results.
- **"Clinical LLM Performance Review"**: Paper that summarizes or analyzes existing LLM clinical performance evaluations without conducting new experiments or reporting new performance data.
- **"Non-Clinical LLM Evaluation"**: Paper not related to clinical LLM performance evaluation.

2. `MODELS`: Extract all LLM models mentioned in the abstract.

Format: ["model1", "model2", ...]

Return empty list if no specific models are mentioned.

3. `MULTIPLE_MODELS_USAGE`: For papers classified as "Clinical LLM Performance Evaluation - Original" only.

Format: true/false/NA

- **true**: Paper clearly evaluates multiple LLMs.
- **false**: Paper clearly focuses on single LLM evaluation.
- **NA**: For non-original clinical LLM evaluation papers.

4. `HUMAN_GROUPS`: Extract all medical professional groups that underwent the same evaluation tasks as the models for direct performance comparison.

Format: ["group1", "group2", ...]

Return empty list if no human groups underwent direct performance comparison.

5. `EVALUATION_TASKS`: Extract all clinical evaluation tasks.

Format:

```
[{"task_name_extractive": "exact task name",
  "task_name_abstractive": "standardized name",
  "task_description": "exact quote",
  "metrics_extractive": ["metric1", "metric2"],
  "metrics_abstractive": ["std_metric1", "std_metric2"]}, ...]
```

6. `PERFORMANCE_RESULTS`: Extract all reported performance metrics.

Format:

```
[{"value": "exact value with units",
  "metric": "exact metric name",
  "subject": "model/human group name"}, ...]
```

Table 6: Abstract Screening Prompt Template (continued).

Prompt for abstract screening (continued)

INPUT EXAMPLE:

```
Title: A Comparison of LLMs in Clinical Triage: Brief Study
Abstract: We evaluated ChatGPT and GEMINI for triaging complex maxillofacial trauma cases at
a referral center. Using 10 standardized cases, we compared LLM recommendations against
center guidelines. Results showed ChatGPT achieved 70% accuracy in examinations while
GEMINI reached 50%. Additional metrics included diagnosis accuracy scores (GEMINI: 3.30,
ChatGPT: 2.30) and recommendation relevance (GEMINI: 2.90, ChatGPT: 3.50).
```

EXAMPLE OUTPUT:

```
ANALYSIS_START

<PAPER_TYPE>Clinical LLM Performance Evaluation - Original</PAPER_TYPE>

<MODELS>["ChatGPT", "GEMINI"]</MODELS>

<MULTIPLE_MODELS_USAGE>true</MULTIPLE_MODELS_USAGE>

<HUMAN_GROUPS>[]</HUMAN_GROUPS>

<EVALUATION_TASKS>
[{"task_name_extractive": "triaging complex maxillofacial
trauma cases",
  "task_name_abstractive": "clinical trauma triage assessment",
  "task_description": "triaging complex maxillofacial trauma cases at a referral center",
  "metrics_extractive": ["accuracy in examinations", "diagnosis accuracy scores",
  "recommendation relevance"],
  "metrics_abstractive": ["examination accuracy", "diagnostic performance",
  "recommendation quality"]}]
</EVALUATION_TASKS>

<PERFORMANCE_RESULTS>
[{"value": "70%", "metric": "accuracy in examinations", "subject": "ChatGPT"},
 {"value": "50%", "metric": "accuracy in examinations", "subject": "GEMINI"},
 {"value": "2.30", "metric": "diagnosis accuracy scores", "subject": "ChatGPT"},
 {"value": "3.30", "metric": "diagnosis accuracy scores", "subject": "GEMINI"},
 {"value": "3.50", "metric": "recommendation relevance", "subject": "ChatGPT"},
 {"value": "2.90", "metric": "recommendation relevance", "subject": "GEMINI"}]
</PERFORMANCE_RESULTS>

ANALYSIS_END
```

Now analyzing the following paper:

- Title: [*paper title*]
- Abstract: [*paper abstract*]

Prompt for full-text review

Please analyze the research paper to extract information about LLM performance evaluation in clinical settings. Present your analysis in the following structured format, maintaining exact quotes where possible. Start your response with "ANALYSIS_START" **and end with** "ANALYSIS_END".
REQUIRED:

- Title: *[paper title]*
- Full Text: *[full paper text]*

TASK: Extract the following structured information from the paper:
1. PAPER_TYPE: Classify the paper as one of the following:

- **"Clinical LLM Performance Evaluation - Original"**: Paper that conducts new experiments to evaluate LLM performance in clinical tasks and reports original performance metrics/results.
- **"Clinical LLM Performance Review"**: Paper that summarizes or analyzes existing LLM clinical performance evaluations without conducting new experiments or reporting new performance data.
- **"Non-Clinical LLM Evaluation"**: Paper not related to clinical LLM performance evaluation.

Note: If the paper is not classified as "Clinical LLM Performance Evaluation - Original", return empty values for all subsequent sections.
2. BIBLIOGRAPHIC_DATES: Extract the paper's submission and publication dates.

Format:

```
{"received_date": "YYYY-MM-DD",
 "accepted_date": "YYYY-MM-DD",
 "published_date": "YYYY-MM-DD"}
```

3. CLINICAL_DOMAIN: Extract the clinical specialty and context information.

Format:

```
{"specialty": "primary clinical specialty field",
 "disease_treatment": "specific diseases or treatments in focus",
 "mesh_terms": ["relevant MeSH term 1", "relevant MeSH term 2"]}
```

4. MODELS: Extract all LLM models mentioned in the paper.

Format:

```
[{"common_name": "most frequently used name in paper",
  "full_name": "complete name including version",
  "base_model": "base model name if fine-tuned, NA if not applicable"}]
```

5. EXPERIMENTAL_SETTINGS: Extract LLM inference settings.

Format:

```
{"llm_inference_temperature": "0.x",
 "llm_inference_few_shot": "n-shot",
 "llm_inference_CoT": true/false}
```

Table 6: Full-Text Review Prompt Template.

Table 7: Full-Text Review Prompt Template (continued).

Prompt for full-text review (continued)

6. HUMAN_GROUPS: Extract all medical professional groups that underwent the same evaluation tasks as the models.

Format:

```
["group1", "group2"]
```

7. EVALUATION_TASKS: Extract all clinical evaluation tasks.

Format:

```
[{"task_name_extractive": "exact task name",
  "task_name_abstractive": "standardized task name",
  "reference_sentence": "exact quote describing the task",
  "metrics": [{"metric_name_extractive": "exact metric name from text",
               "metric_name_abstractive": "standardized metric name",
               "value_range": [min, max],
               "higher_better": true/false,
               "reference_sentence": "exact quote describing the metric"}],
  "sample_size": integer,
  "sample_size_reference_sentence": "exact quote mentioning sample size",
  "data_source_extractive": "exact quote of data source",
  "data_source_abstractive": "standardized description of data source"}]
```

8. PERFORMANCE_RESULTS: Extract all reported performance metrics.

Format:

```
[{"value": "exact performance value with units/confidence intervals",
  "metric": "exact metric name from EVALUATION_TASKS metrics_extractive",
  "subject": "model name or human group name",
  "reference_sentence": "exact quote reporting this result"}]
```

Table 7: Proprietary Language Models Release Timeline.

Company	Model Name	Release Date	Normalized Name
OpenAI	**ChatGPT/GPT-3.5**	2022-12-30	gpt-3.5-turbo
	gpt-3.5-0301	2023-03-01	gpt-3.5-0301
	gpt-3.5-turbo-0613	2023-06-13	gpt-3.5-turbo-0613
	gpt-3.5-turbo-1106	2023-11-06	gpt-3.5-turbo-1106
	gpt-3.5-turbo-0125	2024-01-25	gpt-3.5-turbo-0125
	GPT-4		
	gpt-4-0314	2023-03-14	gpt-4-0314
	gpt-4-0613	2023-06-13	gpt-4-0613
	gpt-4-1106-preview	2023-11-06	gpt-4-1106-preview
	gpt-4-0125-preview	2024-01-25	gpt-4-0125-preview
	gpt-4-turbo-2024-04-09	2024-04-09	gpt-4-turbo-2024-04-09
	GPT-4o		
	gpt-4o updates	2024-05-13	gpt-4o-2024-05-13
	gpt-4o updates	2024-08-06	gpt-4o-2024-08-06
	gpt-4o updates	2024-11-20	gpt-4o-2024-11-20
	GPT-4o Mini	2024-07-18	gpt-4o-mini-2024-07-18
Microsoft	**Bing Chat**	2023-02-07	
	Rebranded as Copilot	2023-09-21	Based on latest GPT models
	Bing Chat integration	2023-11-15	Based on latest GPT models
	Copilot upgrade	2024-05-20	Based on latest GPT models
Claude	**Claude 1**		
	Claude 1.0/Claude 1.1	2023-03-14	claude-1.0/claude-1.1
	Claude 1.2	2023-08-09	claude-1.2
	Claude 1.3	2023-04-18	claude-1.3
	Claude 2		
	Claude 2.0	2023-17-11	claude-2.0
	Claude 2.1	2023-11-21	claude-2.1
	Claude 3		
	Claude 3 Haiku	2024-03-07	claude-3-haiku-20240307
	Claude 3 Sonnet	2024-02-29	claude-3-sonnet-20240229
	Claude 3 Opus	2024-02-29	claude-3-opus-20240229
	Claude 3.5		
	Claude 3.5 Sonnet	2024-06-20	claude-3-5-sonnet-20240622
	Claude 3.5 Haiku	2024-10-22	claude-3-5-haiku-20241022
	Claude 3.5 Opus	2024-10-22	claude-3-5-opus-20241022
	Claude 3.5 update	2024-12-03	claude-3.5-sonnet-20241203
Google	**Bard**	2023-03-21	lamda
	Bard upgrade	2023-05-10	palm-2
	Gemini 1.0		
	Gemini 1.0 Nano	2023-12-06	gemini-1.0-nano
	Gemini 1.0 Pro	2023-12-06	gemini-1.0-pro
	Gemini 1.0 Ultra (Advanced)	2023-12-06	gemini-1.0-ultra
	Gemini 1.5		
	Gemini 1.5 Flash (Basic)	2024-02-15	gemini-1.5-flash
	Gemini 1.5 Pro	2024-05-23	gemini-1.5-pro-001
	Gemini 1.5 Pro update	2024-09-24	gemini-1.5-pro-002
	Gemini 2.0		
	Gemini 2.0	2025-01-22	gemini-2.0-flash-001
	Gemini 2.0 Flash + Thinking	2025-01-22	Not used
Cohere	**Command**	2024-02-07	command
	Command R	2024-06-04	command-r
	Command R+	2024-04-30	command-rplus
	Command R-08-2024	2024-08-28	command-r-2408
	Command R+ 08-2024	2024-08-29	command-rplus-2408

Therapeutic Area For normalization and analysis purposes, we predefined 22 therapeutic areas, including cardiology, oncology, dentistry, and emergency medicine. Depending on the analytical objec-

Table 8: Open-Source Language Models Release Timeline.

Model Brand	Model Name	Release Date	Normalized Name
LLaMA (Meta)	**LLaMA 1**	2023-03	llama-1-7B, llama-1-13B, llama-1-30B, llama-1-65B
	LLaMA 2	2023-07	llama-2-7B, llama-2-13B, llama-2-70, llama-2-7B-chat, llama-2-13B-chat
	LLaMA 3	2024-04-18	llama-3-8B, llama-3-70B
	LLaMA 3.1	2024-07-23	llama-3.1-8B, llama-3.1-70B, llama-3.1-405B
	LLaMA 3.2	2024-10	llama-3.2-1B, llama-3.2-3B, llama-3.2-11B, llama-3.2-90B
	LLaMA 3.3	2024-12	llama-3.2-70B, llama-3.2-405B
Phi (Microsoft)	**Phi-1**	2023-06	phi-1-1.3B
	Phi-1.5	2023-11	phi-1.5-1.3B
	Phi-2	2024-02	phi-2-2.7B
	Phi-3		
	Phi-3 Mini	NA	phi-3-mini-3B
	Phi-3 Small	NA	phi-3-small-7B
	Phi-3 Medium	NA	phi-3-medium-14B
	Phi-3.5	2024-09	phi-3.5-3.8B
	Phi-4	2025-01-20	Not used
Gemma (DeepMind)	**Gemma 2B**	2024-02-21	gemma-1-2B
	Gemma 7B	2024-02-21	gemma-1-7B
	Gemma 1.1	2024-04-05	gemma-1.1
	Gemma 2 (9B, 27B)	2024-06-27	gemma-2-9B, gemma-2-27B
	Gemma 2 (2B)	2024-07-31	gemma-2-2B
Qwen (Alibaba)	**Qwen-7B**	2023-08-03	qwen-7B
	Qwen-14B	2023-09-25	qwen-14B
	Qwen-72B	2023-11-30	qwen-72B
	Qwen-2-7B-instruct	2024-05-16	qwen-2-7B-instruct
	Qwen-2-72B-instruct	2024-10-18	qwen-2-72B-instruct
	Qwen Max	2025-01-29	qwen-max
Mistral	**Mistral 7B**		
	mistral-7B-instruct-v0.1	2023-09-27	mistral-7B-instruct-v0.1
	mistral-7B-instruct-v0.2	2023-10	mistral-7B-instruct-v0.2
	mistral-7B-instruct-v0.3	2023-11	mistral-7B-instruct-v0.3
	Mistral Medium	2023-12	mistral-medium-2312
	Mixtral 8x7B	2023-12-09	mixtral-8x7B
	Mixtral 8x22B	2024-04-10	mixtral-8x22B
	Mistral Large	2024-02-26	mistral-large-2402
	Mistral Small	NA	mistral-small-2402
	Mistral Large 24.07	2024-07-24	mistral-large-2707
Medical Domain Fine-tuned	**ClinicalCamel-1-70B**	NA	clinicalcamel-1-70B
	Med42-70B	NA	med42-70B
	BioMistral-7B	NA	biomistral-7B
	Meditron	NA	meditron-7B
	MedLlama	NA	medllama-1-2

tives, these areas were further grouped into broader categories.

Data Source The source of evaluation data was classified into four types:

- **Board Examinations:** Questions derived from professional board certification exams used to assess medical expertise.
- **Quizzes:** Clinical questions sourced from textbooks, medical societies, or online educational platforms, excluding board exams.
- **Frequently Asked Questions:** Questions reflecting common patient inquiries in clinical settings.
- **Clinical Vignettes:** Case-based questions developed using real patient data, publicly avail-

able case reports, or LLM-generated simulated patient scenarios.

Evaluation Method Evaluation methods were divided into two main categories:

- **Multiple-Choice Questions (MCQs):** Assessing correctness based on predefined answer choices.
- **Human Rating:** Clinical experts rating model-generated responses according to structured evaluation guidelines. This includes both closed-ended rating systems with predefined criteria and open-ended assessments.

A.5 Benchmark Performance Imputation

The MICE framework was configured with optimized settings to ensure stable imputation. Missing values were initially imputed using the median of observed values, followed by a maximum of 50 iterative updates with a convergence tolerance of 1×10^{-6}

For BR-MICE, posterior sampling was enabled, and each missing variable was modeled using all available predictors. The base random seed was set to 42, with independent seeds assigned for multiple imputations.

For RF-MICE, 100 trees were used with bootstrap sampling enabled. The maximum tree depth was set to 15, and feature selection per split followed the square root of the total number of features. BR-MICE regularization parameters were optimized iteratively, with convergence determined via evidence maximization.

To estimate the variance of imputed values, we computed both the within-imputation variance (W) and between-imputation variance (B) across m independent imputations. The total variance (T) was calculated using Rubin's rules:

$$W = \frac{1}{m} \sum_{j=1}^{m} S_j^2$$

$$B = \frac{1}{m-1} \sum_{j=1}^{m} (\bar{Q}_j - \bar{Q})^2$$

$$T = W + \left(1 + \frac{1}{m}\right) B$$

where S_j^2 is the variance of the j-th imputed dataset, $\bar{Q}_j$ is the mean of the j-th imputation, and $\bar{Q}$ is the overall mean of all imputations. The final imputed values were obtained by taking the median of all imputations to ensure robustness against extreme values.

A.6 Bayesian Modeling

We implemented our hierarchical Bayesian model usingNumPyro (v0.17.0) with a JAX (v0.5.0) backend. Posterior inference was conducted via the No-U-Turn Sampler (NUTS), utilizing 1,000 warmup iterations and 2,000 sampling iterations across 8 parallel chains. We assessed convergence using the Gelman-Rubin diagnostic ($\hat{R}$) and effective sample size. The model specification is as follows:

A.6.1 Model Structure

We formulate our hierarchical Bayesian model as follows: We begin by specifying half-normal hyperpriors for the standard deviations that govern the variability of different components in our model:

$$\begin{aligned} \sigma_{\text{model}} &\sim \text{HalfNormal}(1) \\ \sigma_{\text{obs}} &\sim \text{HalfNormal}(1) \\ \sigma_{\text{type}} &\sim \text{HalfNormal}(1) \\ \sigma_{\text{source}} &\sim \text{HalfNormal}(1) \\ \sigma_{\text{eval}} &\sim \text{HalfNormal}(1) \end{aligned}$$

These hyperpriors control the variation in model effects, observation noise, task type effects, data source effects, and evaluation method effects, respectively.

The model effects component captures the inherent performance capabilities of each language model:

$$\begin{aligned} \mu_{\text{model}} &\sim \text{Normal}(0, 1) \\ \beta_{\text{model},j} &\sim \text{Normal}(\mu_{\text{model}}, \sigma_{\text{model}}) \end{aligned}$$

where $j = 1, 2, \ldots, n_{\text{models}}$, and $\beta_{\text{model},j}$ represents the effect of the j-th model. The parameter μ_{model} serves as a global mean for model effects.

We model three task-related components:

$$\begin{aligned} \beta_{\text{type},k} &\sim \text{Normal}(0, \sigma_{\text{type}}) \\ \beta_{\text{source},l} &\sim \text{Normal}(0, \sigma_{\text{source}}) \\ \beta_{\text{eval},m} &\sim \text{Normal}(0, \sigma_{\text{eval}}) \end{aligned}$$

where:

- $k = 1, 2, \ldots, n_{\text{task\_types}}$, with $\beta_{\text{type},k}$ representing the effect of the k-th task type
- $l = 1, 2, \ldots, n_{\text{data\_sources}}$, with $\beta_{\text{source},l}$ representing the effect of the l-th data source

- $m = 1, 2, \ldots, n_{\text{evaluation\_methods}}$, with $\beta_{\text{eval},m}$ representing the effect of the m-th evaluation method

Each task-related effect is centered at zero, reflecting our assumption that these effects represent deviations from an average difficulty level.

The predicted performance for each data point i is given by:

$$\mu_i = \beta_{\text{model},\text{model}[i]} + \beta_{\text{type},\text{type}[i]} + \beta_{\text{source},\text{source}[i]} + \beta_{\text{eval},\text{eval}[i]} + \epsilon_\mu$$

where model$[i]$, type$[i]$, source$[i]$, and eval$[i]$ are the indices for model, task type, data source, and evaluation method for data point i, respectively, and $\epsilon_\mu \sim \text{Normal}(0, 0.1)$ represents additional noise in the prediction process.

Finally, we model the observed performance metrics using a normal likelihood:

$$y_i \sim \text{Normal}(\mu_i, \sigma_{\text{obs}})$$

where y_i is the observed performance metric for data point i on the normalized scale.

A.6.2 Centrality Measurement

Quantifying the connectivity of models within the evaluation network is essential for understanding their role and influence. Some models exhibit weak connections to other major models, meaning they contribute useful information to Bayesian modeling but have limited relevance for downstream analysis. By computing centrality scores, we classified models based on their connectivity and excluded the lower 50% from downstream evaluations.

The evaluation network was represented as a bipartite graph $G = (V, E)$, where the vertex set V consisted of two disjoint subsets: models and tasks. An edge $(m, t) \in E$ was formed if and only if model m was evaluated on task t. This structure provided a basis for analyzing connectivity patterns and assessing the relative importance of models within the evaluation framework.

Model connectivity was quantified using three centrality measures. Degree centrality (C_D) captured the number of direct connections a model had, normalized by the maximum possible connections:

$$C_D(v) = \frac{deg(v)}{|V| - 1}$$

where $deg(v)$ represents the number of edges incident to node v. Between-ness centrality (C_B) measured how often a model served as a bridge along the shortest paths between other nodes:

$$C_B(v) = \sum_{s \neq v \neq t} \frac{\sigma_{st}(v)}{\sigma_{st}}$$

where σ_{st} is the total number of shortest paths from node s to node t, and $\sigma_{st}(v)$ is the number of those paths passing through node v. Closeness centrality (C_C) assessed how close a model was to all other nodes in the network:

$$C_C(v) = \frac{|V| - 1}{\sum_{u \neq v} d(v, u)}$$

where $d(v, u)$ is the shortest-path distance between nodes v and u. To integrate these measures into a single ranking, a combined connectivity score was computed by summing the three normalized centrality values:

$$\text{Combined Score}(m) = C_D(m) + C_B(m) + C_C(m)$$

Models were then ranked based on their combined scores, and those below the P_{th} percentile were classified as low-connectivity models:

$$\text{Low Connectivity}(m) = \begin{cases} \text{True} & \text{if Combined-Score}(m) < P_{th} \\ \text{False} & \text{otherwise} \end{cases}$$

where P_{th} was set at the 50th percentile, identifying the bottom 50% of models as low-connectivity. For downstream analysis, only high-connectivity models were retained. This ensured that subsequent evaluations focused on models with strong integration within the network while still utilizing all available information in Bayesian modeling.

A.7 Correlation Measurement

To evaluate the relationship between LLM performance on different benchmarks and in clinical settings, we computed four correlation measures: Pearson's correlation coefficient, Spearman's rank correlation coefficient, Kendall's tau, and Lin's concordance correlation coefficient (CCC). Among these, Spearman's and Kendall's correlations were used as the primary measures, as they better capture rank-based relationships given the diversity of evaluation methodologies.

Pearson's correlation coefficient (r) measures the strength of the linear relationship between two continuous variables. It is computed as:

$$r = \frac{\sum(x_i - \bar{x})(y_i - \bar{y})}{\sqrt{\sum(x_i - \bar{x})^2}\sqrt{\sum(y_i - \bar{y})^2}}$$

where x_i and y_i are individual data points, and $\bar{x}$ and $\bar{y}$ are their respective means.

Spearman's rank correlation coefficient (ρ) assesses the monotonic relationship between two variables by comparing their rank orders rather than raw values. It is given by:

$$\rho = 1 - \frac{6\sum d_i^2}{n(n^2-1)}$$

where d_i is the rank difference for each pair of observations, and n is the number of observations.

Kendall's tau (τ) quantifies the ordinal association between two variables based on concordant and discordant pairs:

$$\tau = \frac{C - D}{\frac{1}{2}n(n-1)}$$

where C is the number of concordant pairs, and D is the number of discordant pairs.

Lin's CCC (ρ_c) evaluates both correlation and agreement between two variables by incorporating measures of precision and accuracy:

$$\rho_c = \frac{2r\sigma_x\sigma_y}{\sigma_x^2 + \sigma_y^2 + (\mu_x - \mu_y)^2}$$

where r is Pearson's correlation coefficient, σ_x and σ_y are standard deviations, and μ_x and μ_y are means of the two variables.

B Supplementary Results

B.1 Benchmark Performance Imputation

Table 9 presents the imputation accuracy of benchmark models evaluated through a masking test across both general and medical QA domains.

Table 9: Imputation accuracy on masking test for benchmark performances.

Model	MAE	RMSE	R^2
General			
RandomForest	4.21	8.17	0.89
BayesianRidge	5.63	8.14	0.89
Medical QA			
RandomForest	2.04	3.18	0.98
BayesianRidge	4.14	6.81	0.90

Table 10: Within- and between- variance results from multiple imputation.

Category	Within	Between	Total
General Benchmarks (Overall Variance: 481.1)			
RandomForest	1.1	204.5	215.7
(% of Overall)	0.2%	42.5%	44.8%
BayesianRidge	100.6	189.4	299.5
(% of Overall)	20.9%	39.4%	62.3%
Medical Benchmarks (Overall Variance: 390.5)			
RandomForest	2.1	43.2	47.5
(% of Overall)	0.5%	11.1%	12.2%
BayesianRidge	67.5	40.0	109.5
(% of Overall)	17.3%	10.2%	28.0%

B.2 Benchmark-to-Benchmark Correlation

Table 11 summarizes the average correlation coefficients between general benchmarks and other benchmarks, providing a comparative view across domains and correlation metrics.

Figures 6, 7, and 8 further illustrate the internal correlations within each domain and the cross-domain relationships.

B.3 Benchmark-to-Clinical Performance Correlation

Figures 9 and 10 present the correlations between benchmark performance and language model performance in real-world clinical settings, with and without imputed benchmark scores. The results are reported using four correlation measures—Pearson correlation coefficient, Spearman rank correlation coefficient, Kendall's tau, and Lin's CCC—to ensure robustness from multiple statistical perspectives.

Table 11: Average correlation coefficients of general benchmarks with other benchmarks. The highest score in each column is **bold**, and the second highest is underlined.

General Benchmark	Spearman		Kendall	
	Medical QA	General	Medical QA	General
MMLU	<u>0.853</u>	0.715	<u>0.690</u>	0.576
MMLU Pro	0.851	**0.773**	0.679	**0.628**
BBH	**0.891**	0.725	**0.736**	0.573
HumanEval	0.838	<u>0.764</u>	0.671	<u>0.618</u>
GSM8K	0.816	0.756	0.645	0.599
MATH	0.785	0.625	0.618	0.502
Average	**0.839**	**0.726**	**0.673**	**0.583**

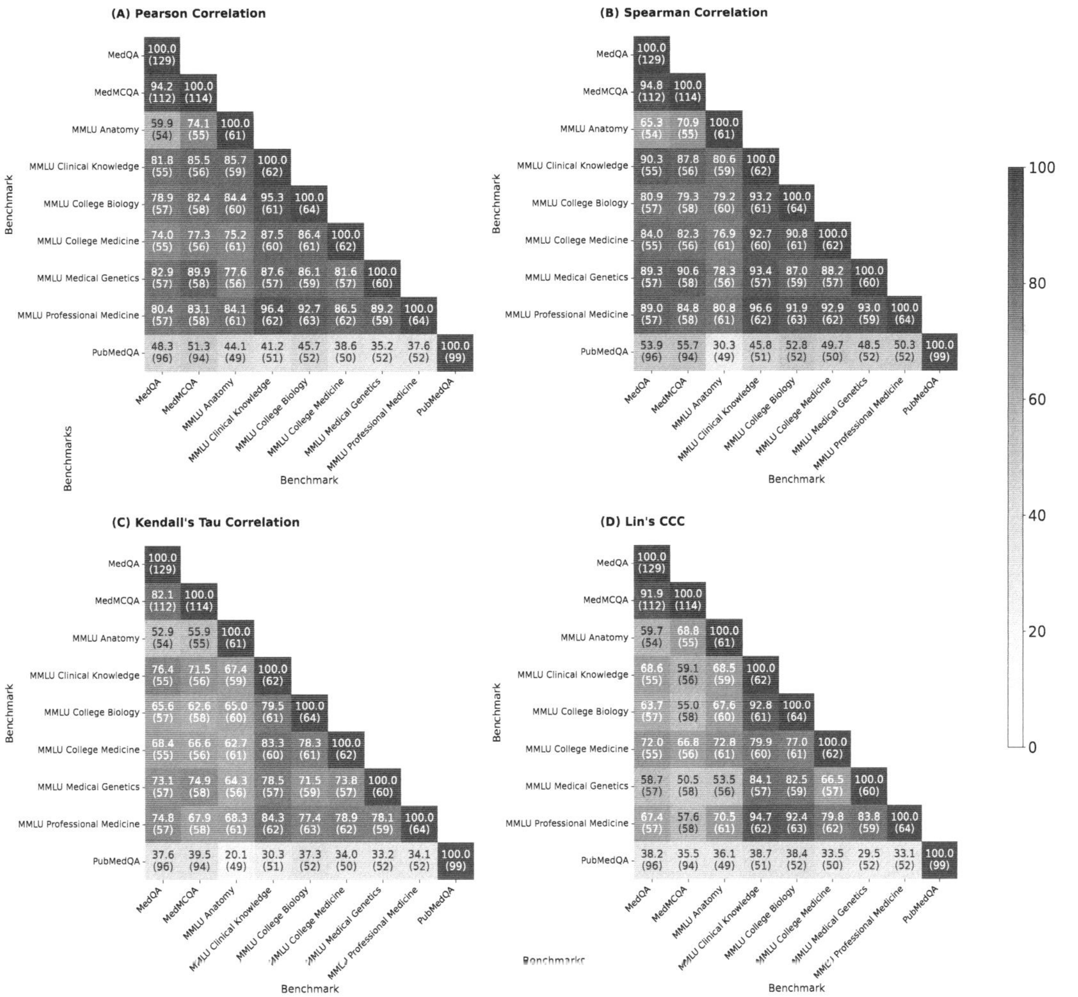

Figure 6: Correlations within medical QA benchmarks.

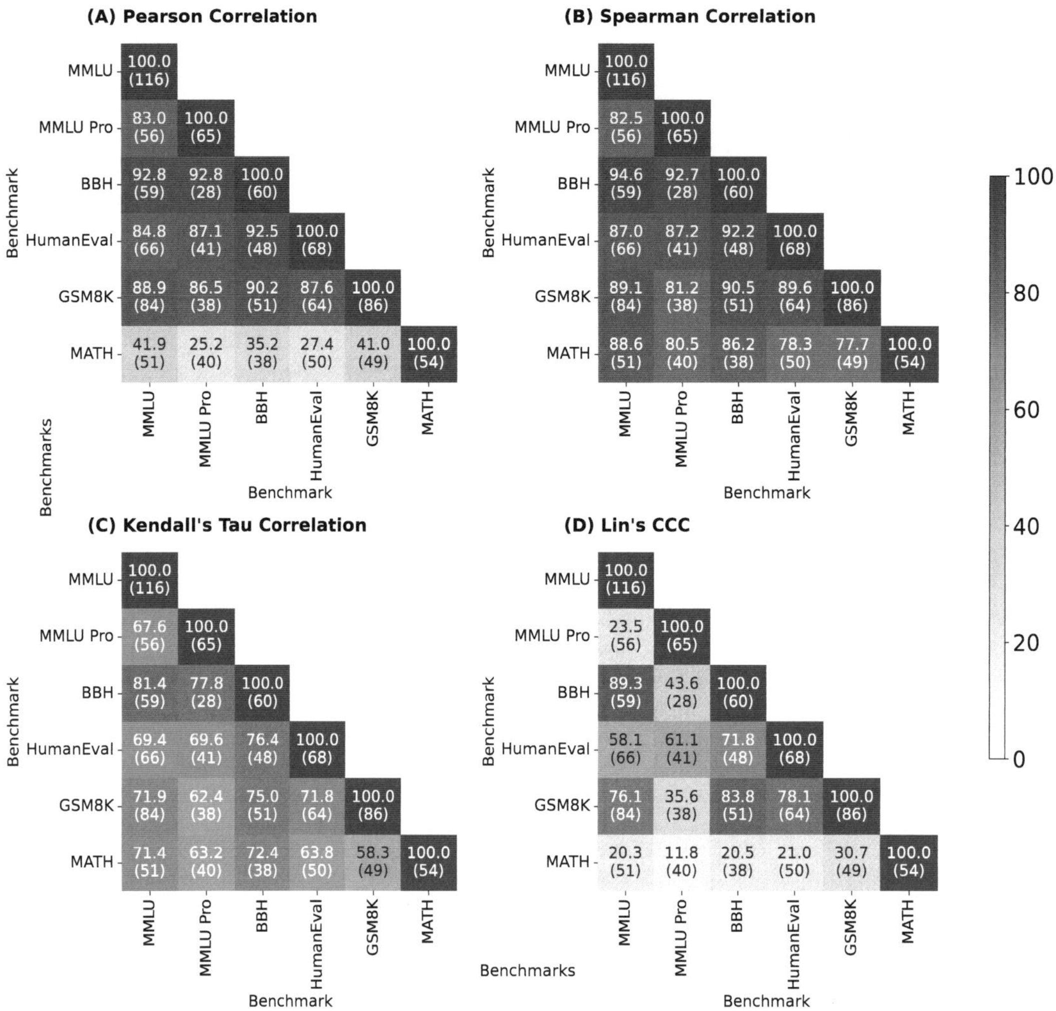

Figure 7: Correlations within general benchmarks.

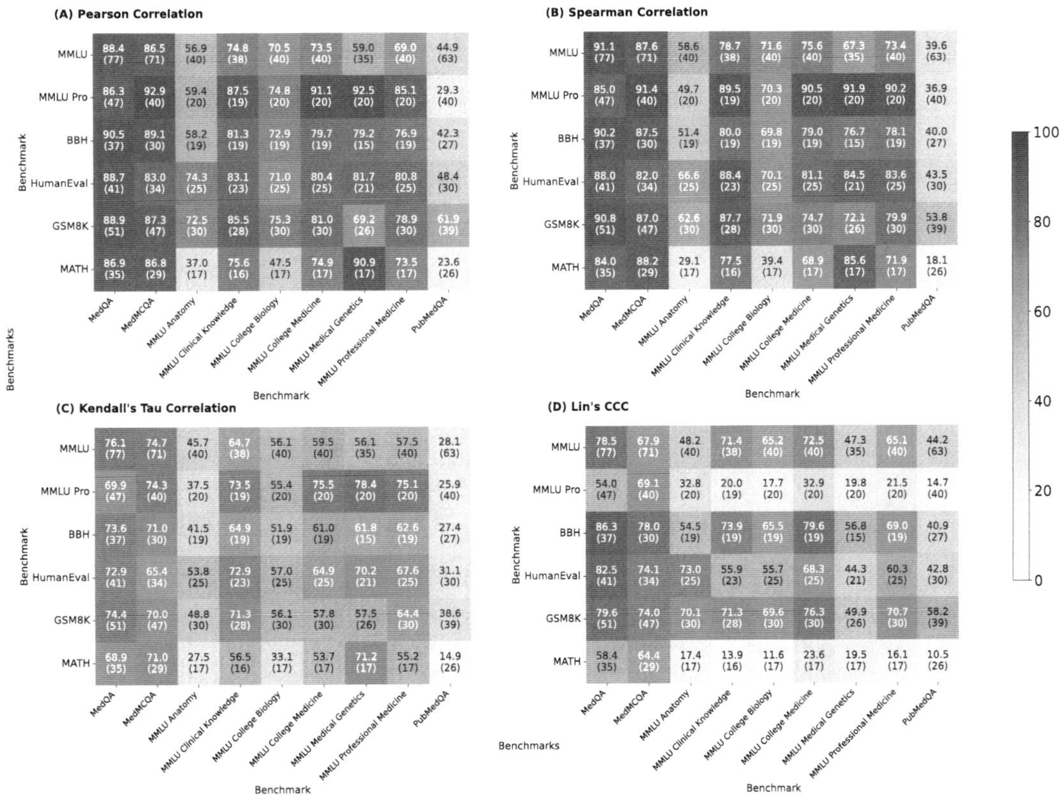

Figure 8: Correlations between general and medical QA benchmarks.

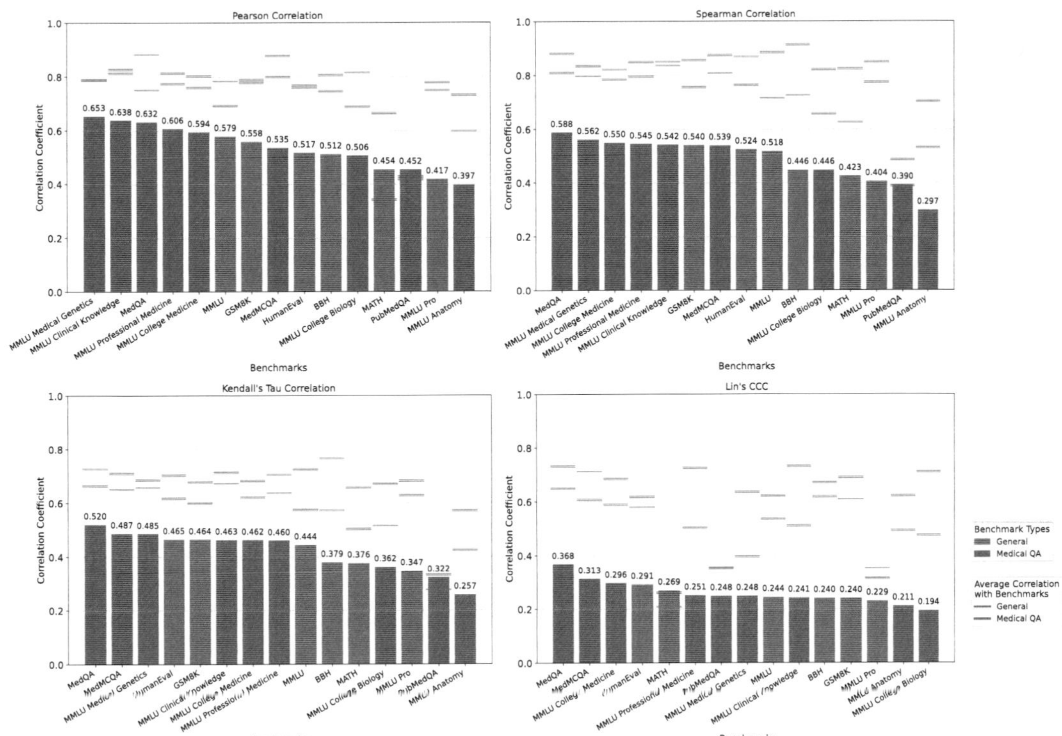

Figure 9: Correlations between benchmark performance and language model performance in real-world clinical settings with imputed benchmark scores.

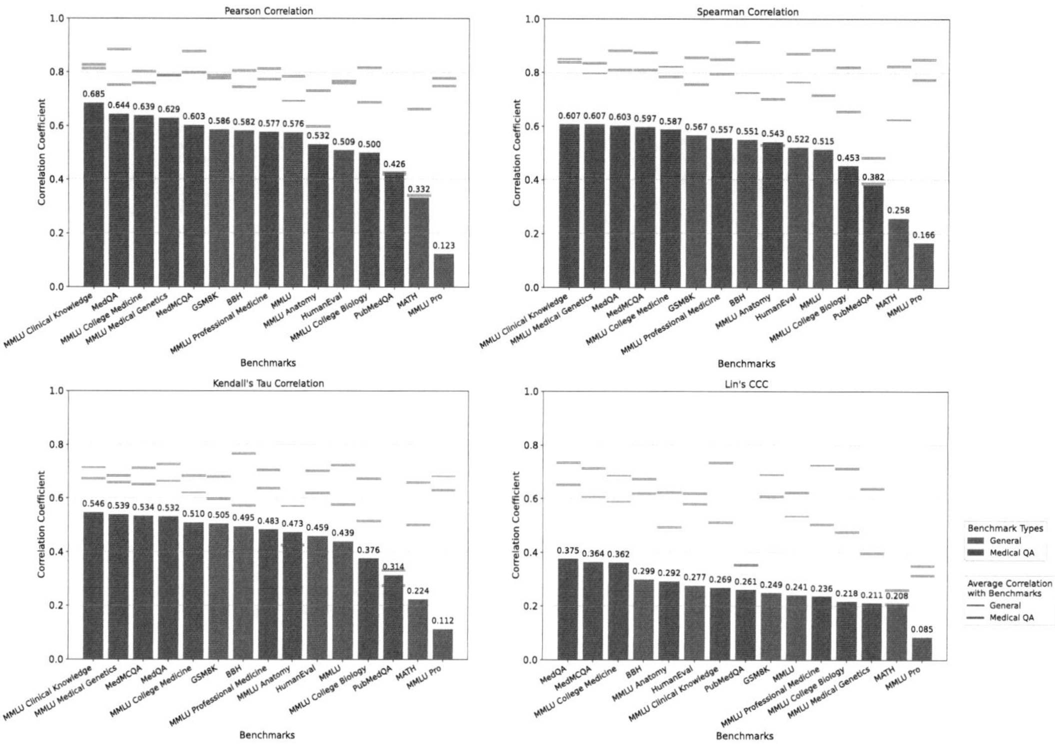

Figure 10: Correlations between benchmark performance and language model performance in real-world clinical settings with non-imputed benchmark scores.

Beyond Citations: Integrating Finding-Based Relations for Improved Biomedical Article Representations

Yuan Liang
Queen Mary University
London, UK
yuan.liang@qmul.ac.uk

Roonak Rezvani
Recursion Pharmaceuticals, Inc
Oxford, UK
roonak.rezvani@recursion.com

Massimo Poesio
Queen Mary University
University of Utrecht
London, UK
Utrecht, Netherland
m.poesio@qmul.ac.uk

Abstract

High-quality scientific article embeddings are essential for tasks like document retrieval, citation recommendation, and classification. Traditional citation-based approaches assume citations reflect semantic similarity—an assumption that introduces bias and noise. Recent models like SciNCL and SPECTER2 have attempted to refine citation-based representations but still struggle with noisy citation edges and fail to fully leverage textual information. To address these limitations, we propose a hybrid approach that combines Finding-Citation Graphs (FCG) with contrastive learning. Our method improves triplet selection by filtering out less important citations and incorporating finding similarity relations, leading to better semantic relationship capture. Evaluated on the SciRepEval benchmark, our approach consistently outperforms citation-only baselines, showing the value of text-based semantic structures. While we do not surpass state-of-the-art models in most tasks, our results reveal the limitations of purely citation-based embeddings and suggest paths for improvement through enhanced semantic integration and domain-specific adaptations.

1 Introduction

High-quality scientific article embeddings are essential for various downstream tasks, including citation recommendation, article retrieval, and classification (Cunningham and Greene, 2023). These effective representations accelerate research progress by enhancing knowledge discovery. However, generating high-quality embeddings remains challenging, largely due to the limitations of existing methods that rely primarily on citation networks.

Traditional approaches use Large Language Models (LLMs) to generate article embeddings directly, but research shows this method often underperforms compared to basic baseline models like GloVe (Reimers and Gurevych, 2019). To enhance embedding quality, researchers have turned to contrastive learning for refining document representations (Cohan et al., 2020). This method uses a triplet-based training framework, where each triplet includes a query article, a similar article (positive sample), and a dissimilar article (negative sample). These triplets are typically drawn from citation networks, based on the assumption that citation relationships indicate semantic similarity.

Over the years, researchers have made various improvements to optimize triplet selection. SPECTER (Cohan et al., 2020) introduced a unidirectional citation-based approach, using cited papers as positive samples and non-cited papers as negative samples. However, this method created inconsistencies in triplet generation, as the same paper could be both a positive and negative sample in different contexts. To address this issue, SciNCL (Ostendorff et al., 2022) eliminated citation directionality and implemented graph embeddings and k-nearest neighbors (KNN) sampling to identify positive and negative samples. This change significantly improved embedding quality by reducing triplet formation inconsistencies.

Recent advances have further refined this pipeline. SPECTER2 (Singh et al., 2023) developed task-specific embeddings by generating a general representation and then fine-tuning it for different downstream tasks. Other approaches explore multi-faceted embeddings, generating multiple representations of a paper to capture various aspects of its content (Zhang et al., 2023).

Despite these advances, current methods rely solely on citation networks for triplet construction, overlooking the many semantically similar articles that lack direct citation links. This limitation creates biases in representation learning and constrains the quality of scientific embeddings. To address these challenges, we propose a hybrid approach that enhances contrastive learning by combining Finding-Citation Graphs (FCG) with text-based se-

Proceedings of the 24th Workshop on Biomedical Language Processing (BioNLP 2025), pages 297–306
August 1, 2025 ©2025 Association for Computational Linguistics

mantic relationships. Our contributions include:

- Filtering less important citations using an LLM-based classification mechanism to remove noisy edges.
- Incorporating finding similarity relations to establish meaningful connections between semantically related papers.

We evaluate our approach on SciRepEval (Singh et al., 2023), a benchmark for assessing scientific embeddings across multiple tasks. Our method outperforms citation-only baselines, demonstrating the effectiveness of integrating text-based semantic structures into contrastive learning. While it does not surpass state-of-the-art models in all tasks, our results highlight the importance of moving beyond purely citation-based embeddings toward richer, more semantically aware representations.

2 Related Work

Researchers have developed various models and methodologies to improve scientific text representation, ranging from traditional keyword-based methods and vector space models to modern deep-learning approaches. Beyond general-purpose techniques, specialized approaches exist specifically for scientific articles.

General-Purpose Methods Early scientific article representations relied primarily on word-level features. The Bag-of-Words (BoW) model represented documents as vectors of word frequencies—a simple but limited approach that suffered from sparsity and lost semantic relationships (Salton et al., 1975). Latent Semantic Analysis (LSA) addressed these limitations by introducing dimensionality reduction and capturing latent word relationships (Deerwester et al., 1990). The field then progressed to probabilistic topic modeling with Latent Dirichlet Allocation (LDA), which effectively modeled texts as mixtures of latent topics (Blei et al., 2003). LDA became particularly valuable in scientific literature analysis by enabling researchers to extract thematic structures from large document collections.

A major breakthrough came with word embedding techniques like Word2Vec (Mikolov et al., 2013) and GloVe (Pennington et al., 2014), which transformed scientific text representation through dense vector spaces. These models excel at capturing semantic relationships, leading to improved information retrieval and document clustering. However, they face limitations in handling polysemy and contextual variations.

The field has recently advanced further with transformer-based models, notably BERT (Devlin et al., 2019) and SciBERT (Beltagy et al., 2019)—models specifically trained on scientific corpora. These architectures have dramatically improved contextual representation and now power various tasks including citation prediction, summarization, and scientific question answering. SciBERT stands out by outperforming generic language models in domain-specific applications, demonstrating the value of domain-adapted pretraining.

Scientific-Specific Methods Most methods for associating embeddings to scientific papers rely on citation networks, which represent articles as nodes connected by citation links to analyze influence patterns and research trends (Page et al., 1999). Several approaches have developed universal embeddings for articles, such as SPECTER (Cohan et al., 2020) and SciNCL (Ostendorff et al., 2022), as discussed in Section 1. Other approaches generate multiple embeddings for scientific articles, each serving a distinct purpose. For instance, SPECTER2 (Singh et al., 2023) creates task-specific embeddings, producing four different representations per article for tasks like classification, regression, ad-hoc search, and proximity. Similarly, ASPIRE (Mysore et al., 2022) generates aspect-specific embeddings for each article, such as method embeddings and finding embeddings. Despite the noise in citation networks, these models outperform traditional embeddings by leveraging citation relationships, resulting in improved downstream performance in retrieval, classification, and clustering tasks.

3 Methodology

Our goal is to learn citation-informed and text-informed representations for scientific documents. Given a document's textual content d, we aim to generate a dense vector representation e that effectively encodes both the document's information and the citation's information for downstream tasks. Following previous work (Cohan et al., 2020; Ostendorff et al., 2022; Singh et al., 2023), we developed an information-enriched network combining citation networks with finding similarity relations. Using this network, we sample triplets to learn document embeddings through contrastive learning. In the following subsections, we describe the creation of the information-enriched network, the triplet sampling approach, and the contrastive learning approach.

3.1 Information-Enriched Network Construction

To enhance the semantic similarity of the citation network, we combined citation networks with finding similarity relations to create an information-enriched network. We improved semantic accuracy by filtering out less important citations—those that contribute minimally to the new study. We established new relations between articles with similar findings based on the Finding-Citation Graph (FCG) (Liang et al., 2024). The resulting network contains links—both citations and finding similarity relations—that better represent semantic relationships beyond simple citations.

3.1.1 Citation Filtering

Although citation networks form the foundation of many scientific embedding models, they can introduce noise since not all citations reflect meaningful content similarity (Ostendorff et al., 2022). To address this issue, we implemented a large language model (LLM)-based filtering mechanism that evaluates citations by assessing their contribution to the citing study, thereby determining their relevance.

Due to the lack of open-source datasets for this task, we used Mistral-7B-Instruct (Jiang et al., 2023) with few-shot in-context learning to classify citations into three categories: Highly Important, Moderately Important, and Less Important. Our analysis of citation importance considered three key elements: the citation sentence, the abstract of the citing paper, and the title of the cited paper. The prompt can be seen in Appendix A. Less relevant citations were removed from the network to reduce noise and improve the quality of triplet selection.

3.1.2 Finding Similarity Relations

Beyond citations, scientific findings provide a more precise measure of content similarity between papers. To incorporate additional semantic relationships, we utilized the Finding-Citation Graphs (FCG). We used Contriever (Lei et al., 2023), a dense retrieval model, to convert scientific findings into embeddings. We then calculated pairwise cosine similarity between findings and added new finding similarity edges to the network when pairs exceeded a similarity threshold.

Through these two enhancements—removing noisy citations and introducing new semantic edges—we created an information-enriched citation network that better reflects the true relationships between papers.

3.2 Triplet Sampling

Contrastive learning relies on high-quality triplets—sets of (query, positive, negative) samples to train models to differentiate between similar and dissimilar documents. To enhance our model's performance, we optimized triplet selection by combining citation-based and finding-based similarity measures. Following Ostendorff et al. (2022), we trained node embeddings on the combined network using PyTorch BigGraph (Lerer et al., 2019). For each article d^Q, we used the k nearest neighbors (KNN) method to identify similar (positive) and dissimilar (negative) articles.

For positive article sampling, following Wang and Isola (2022) and Ostendorff et al. (2022), we selected positive articles from locations distant from the query. Specifically, we sampled c^+ positive articles from a close neighborhood around the query article—those within the range $(k^+ - c^+, k^+]$, where k^+ represents the k parameter in the KNN method.

For negative article sampling, we considered two types of negative articles: easy negatives c^-_{easy} and hard negatives c^-_{hard}. Easy negatives can be obtained through simple random sampling. Hard negatives are crucial for contrastive learning—the more challenging the negative samples, the better the model training becomes. We used a sampling method similar to positive article sampling, selecting articles within the range $(k^-_{hard} - c^-_{hard}, k^-_{hard}]$, where k^-_{hard} represents the k parameter in the KNN method.

3.3 Contrastive Learning

Once triplets are constructed, we train our embedding model using contrastive learning with a triplet margin loss function (Schroff et al., 2015). The method's core principle is to minimize the distance between similar (positive) samples in the latent space while maximizing the distance between dissimilar (negative) samples. To implement contrastive learning, we fine-tuned SciBERT (Beltagy et al., 2019), a domain-specific transformer model for scientific text, to generate embeddings for each article.

$$\mathcal{L} = max\{\left\|d^Q - d^+\right\|_2 - \left\|d^Q - d^-\right\|_2 + \xi, 0\} \quad (1)$$

4 Experiment Setup

This section describes our experimental setup, detailing the datasets, model training configurations, and baseline comparisons.

4.1 Dataset

4.1.1 Training Dataset: Finding-Citation Graph (FCG)

For training, we utilized the Finding-Citation Graph (FCG) derived from the Europe PMC dataset (Liang et al., 2024). This biological FCG encompasses 16 million nodes—consisting of 6 million papers and 10 million findings—and 27 million edges, comprising 17 million citations and 10 million paper-finding generation relations. After preprocessing the dataset to filter out noisy citations and incorporate finding similarity relations, as described in the methodology section, this enriched network forms the foundation for our triplet sampling strategy.

4.1.2 Evaluation Dataset: SciRepEval

For evaluation, we used SciRepEval (Singh et al., 2023), the first large-scale benchmark for evaluating scientific document embeddings. SciRepEval encompasses 24 tasks across four evaluation formats—Ad-Hoc Search, Proximity, Classification, and Regression—spanning multiple scientific domains. We primarily used the "Out-of-Train" datasets in SciRepEval. Table 1 provides an overview of the dataset statistics and evaluation metrics.

4.2 Model Training and Implementation

4.2.1 Input Network Variations

To assess performance, we generated different variations of the citation network through distinct preprocessing methods.

- **Citation** – The original unprocessed citation network.
- **Citation (Filtered)** $\mathbb{F}$ – Noisy citations removed.
- **Citation (Finding Similarity)** $\mathbb{T}$ – New finding-based relations added.
- **Citation (Combine)** $\mathbb{FT}$ – Both filtering and finding similarity applied.

4.2.2 Training Configuration

For filtering less important citations, we utilized Mistral-7B-Instruct-v0.3 (Jiang et al., 2023) with two-shot learning on a single NVIDIA GeForce A100 GPU, processing each sample in approximately 0.8 seconds. To identify finding similarities, we used Contriever (Lei et al., 2023) to generate embeddings for all 10 million findings and performed similarity searches, with each search taking about 0.87 seconds.

For triplet generation and contrastive learning, we closely replicated SciNCL's training setup. We implemented the KNN strategy using FAISS (Johnson et al., 2019) with a flat index and maintained the same KNN parameters: $k^+ = 25$ and $k^- = 4000$. For contrastive learning, we used Huggingface Transformers (Wolf et al., 2020) and initialized the model with SciBERT's weights (Beltagy et al., 2019), training it with triplet loss. The training process used the Adam optimizer (Kingma and Ba, 2017) with weight decay and a learning rate of $\lambda = 2^{-5}$. The model was trained for 2 epochs on a single NVIDIA GeForce RTX A100 (40G) GPU with a batch size of 14, completing in approximately 8 hours.

4.3 Baselines

We compared our method against two existing contrastive learning-based scientific embedding models: SciNCL (Ostendorff et al., 2022), and SPECTER2 (Singh et al., 2023). Since these baselines were trained on multi-domain datasets, their results serve as a reference point rather than direct competitors. Our primary goal is to assess whether removing noisy citations and incorporating text-based similarity relations improves embedding quality. Therefore, our true baseline is the unprocessed citation network, which we used to generate embeddings without any filtering or augmentation.

5 Overall Results

We evaluated our approach by building multiple input networks using different preprocessing strategies and comparing them to baselines. Our main goal was to determine if filtering less important citations and incorporating finding similarity relations would enhance the quality of biomedical article embeddings.

Table 2 presents the statistics of each network variant. Due to time constraints, we analyzed citation importance and generated finding similarity

Task Format	Name	Test	Eval Metric	Source
		Out-of-Train		
CLF	Biomimicry	10,991	Binary F1	Shyam et al. (2019)
	DRSM	7,520 S; 955 G	Macro F1	Burns (2022)
	SciDocs MAG	23,540	Macro F1	Cohan et al. (2020)
	SciDocs MeSH Diseases	25,003	Macro F1	Cohan et al. (2020)
RGN	Peer Review Score	10,210	Kendall's $\mathcal{T}$	Singh et al. (2023)
	h-Index of Authors	8,438	Kendall's $\mathcal{T}$	Singh et al. (2023)
	Tweet Mentions	25,655	Kendall's $\mathcal{T}$	Jain and Singh (2021)
PRX	S2AND	X: 68,968 Y: 10,942	B^3 F1	Subramanian et al. (2024)
	Paper-Reviewer Matching	Q:107 P: 1,729	P@5, P@10	Mimno and McCallum (2007)
	RELISH	Q: 3190 P: 191,245	nDCG	Zhao et al. (2022)
	SciDocs Co-view	Q: 1,000 P: 29,978	MAP, nDCG	Cohan et al. (2020)
	SciDocs Co-read	Q: 1,000 P: 29,977	MAP, nDCG	Cohan et al. (2020)
	SciDocs Cite	Q: 1,000 P: 29,928	MAP, nDCG	Cohan et al. (2020)
	SciDocs Co-cite	Q: 1,000 P: 29,949	MAP, nDCG	Cohan et al. (2020)
SRCH	NFCorpus	Q: 323 P: 44,634	nDCG	Boteva et al. (2016)
	TREC-CoVID	Q: 50 P: 69,318	nDCG	Voorhees et al. (2021)

Table 1: Dataset statistics and evaluation metrics for different tasks in SciRepEval benchmark.

relations for only a subset of nodes—detailed information is available in Appendix B. Table 3 summarizes the performance in different evaluation tasks in SciRepEval.

Table 3 shows that both removing less important citations ($\mathbb{F}$) and adding finding-based relations ($\mathbb{T}$) improved performance compared to the raw citation network, with the combined approach ($\mathbb{FT}$) achieving the best results. Significantly, adding the finding similarity relations proved more effective than citation filtering alone, indicating that citation-based embeddings do not fully capture the semantic structure of scientific literature.

	Node_Num	Edge_Num
Citation	6,013,398	17,795,8624
Citation $\mathbb{F}$	6,013,398	17,769,1665
Citation $\mathbb{T}$	6,013,398	38,650,3618
Citation $\mathbb{FT}$	6,013,398	38,624,0425

Table 2: Input network with different process methods

We also evaluated our model against two state-of-the-art scientific embedding models—SciNCL (Ostendorff et al., 2022) and SPECTER2 (Singh et al., 2023)—both trained on multi-domain scientific datasets. While our approach did not achieve state-of-the-art performance in most tasks from Table 3, it demonstrated competitive results in regression and search tasks, where semantic relationships are particularly important.

6 Discussion

Our experimental results demonstrate that integrating finding similarity relations into citation networks improves the quality of scientific article embeddings, particularly in search and regression tasks. This section explores the implications of these findings, addresses the limitations of purely citation-based approaches, and discusses potential avenues for further improvements.

6.1 The Limitations of Citation Networks for Embedding Learning

Citation networks have traditionally been used to model relationships between scientific articles, operating on the assumption that citations indicate semantic similarity. However, this assumption has several fundamental flaws due to the diverse motivations behind citations:

- Papers are often cited to provide background context or build a research narrative, rather than signifying true conceptual similarity.
- Many papers with strong semantic similarities lack citation connections to each other.
- Citations are subject to various biases, including popularity effects, disciplinary silos, and self-citation patterns.

Our results demonstrate that simple citation-based triplet selection produces suboptimal contrastive learning outcomes. The enhanced performance we observed with finding similarity relations indicates that citation-based methods alone inadequately capture content-based relationships, highlighting the necessity for alternative similarity measures in scientific document embeddings.

6.2 Effect of Citation Filtering and Finding Similarity Relations

A key contribution of our work is demonstrating how citations vary in their importance for learning

Task	Metric	SciNCL	SPECTER2	citation	citation $\mathbb{F}$	citation $\mathbb{T}$	citation $\mathbb{FT}$
			Out-of-Train				
Classification							
Biomimicry	Wt. F1	50.22	**53.20**	48.50	48.51	49.13	49.29
DRSM	Wt. F1	65.10	**68.9475**	62.32	62.78	66.23	66.01
SciDocs MAG	F1	81.11	**82.55**	81.16	80.96	82.24	82.22
SciDocs MeSH	F1	89	**89.72**	88.88	89.09	89.56	88.65
Proximity							
Relish	nDCG	90.67	**91.65**	91.22	91.22	91.05	91.18
S2AND	B^3 F1	93.98	92.8	95.6	95.4	95.3	**95.67**
Peer Reviewer Matching	Avg	45.40	**45.44**	43.83	44.58	44.86	44.67
SciDocs Co-View	MAP	**85.28**	84.68	82.15	82.18	83.25	83.71
	nDCG	**92.23**	92.04	90.71	90.79	91.34	91.47
SciDocs Co-Read	MAP	**87.69**	86.29	83.99	84.6	84.69	84.85
	nDCG	**94**	93.36	92.14	92.57	92.51	92.6
SciDocs Cite	MAP	93.55	**94.08**	84.07	83.89	85.93	87.14
	nDCG	97.35	**97.59**	92.91	92.77	93.83	94.42
SciDocs Co-Cite	MAP	**91.66**	90.58	88	88.13	88.23	88.79
	nDCG	**96.44**	95.99	94.75	94.89	94.93	95.21
Regression							
Review Score	Avg	18.87	**21.79**	18.59	19.71	20.42	19.37
Max h-Index	K Tau	11.3	12.83	12.26	13.13	**14.14**	12.63
Tweet Mentions	K Tau	**25.78**	24.56	23.04	22.89	23.75	25.57
Search							
NFCorpus	nDCG	70.85	70.18	69.7	70.24	**71.47**	70.89
TREC CoVID	nDCG	87.67	**90.87**	89.34	89.39	88.03	88.37
Average Exp. SciDocs	-	56	**57.23**	55.4	55.8	56.4	56.4
Overall Average	-	73.4	**73.95**	71.7	71.9	72.5	72.6

Table 3: Performance metrics across different methods and tasks. The columns labeled citation, citation $\mathbb{F}$, citation $\mathbb{T}$, and citation $\mathbb{FT}$ show our experimental results. The SciNCL and SPECTER2 columns present experimental results from (Ostendorff et al., 2022) and (Singh et al., 2023).

high-quality embeddings. By filtering out less important citations, we reduced noise and achieved modest improvements. However, our most significant gains came from incorporating finding similarity relations, which create direct links between papers based on their research findings rather than citations alone.

6.3 How Does Our Method Compare to Existing Models?

While our approach outperforms the baseline citation network, it does not surpass state-of-the-art models like SPECTER2 in most tasks. This is expected, as SPECTER2 and similar models are trained on larger, more diverse datasets and benefit from task-specific fine-tuning. However, our findings suggest that incorporating additional semantic relations—like findings, methodologies, or co-authorship networks—could help close this performance gap.

Notably, our method achieved competitive performance in regression and search tasks, demonstrating that text-based semantic relations complement citation-based embeddings effectively. This strengthens our argument that citation networks alone cannot fully capture the contextual and conceptual relationships between scientific articles.

6.4 Limitations

Despite its benefits, our approach has some limitations. First, due to computational constraints, we applied citation filtering and finding similarity generation to only a subset of the dataset. A more comprehensive application across a larger scientific corpus may yield even stronger improvements.

Additionally, we limited our exploration of text similarity relations to research findings, excluding other important aspects like methodology. While we believe findings are the most crucial part of scientific papers, examining other aspects could yield valuable insights.

Furthermore, our approach of generating a single universal embedding per article may result in the loss of important information.

These limitations point to clear opportunities for future improvements.

7 Conclusion

In this study, we introduced an enhanced approach to biomedical article embedding by integrating Finding-Citation Graphs (FCG) with contrastive learning. Our method overcomes the limitations of traditional citation-based embeddings by filtering out less important citations and incorporating text-based semantic relationships into triplet selection. This refined network improves the representation quality of scientific documents, particularly in the biomedical domain.

Our experiments show that removing noisy citations and leveraging finding similarity relations enhance contrastive learning performances. Though our approach did not exceed state-of-the-art methods like SciNCL and SPECTER2, it consistently performed better than the original citation network, demonstrating the value of context-aware triplet formation.

In conclusion, our work establishes a foundation for enhancing scientific document representations through a balanced approach that combines citation analysis with semantic similarity. By improving the construction of scientific embeddings, we deliver more accurate, domain-specific, and semantically meaningful representations—enabling better information retrieval and knowledge discovery in biomedical research.

Ethics Statement

This research focuses on improving scientific article embeddings through Finding-Citation Graphs (FCG) and contrastive learning. Our approach enhances document representations for biomedical scientific articles to improve downstream tasks like retrieval, classification, and citation recommendation. We conducted our research using only open-source datasets.

Acknowledgments

This work was supported by the UKRI Biotechnology and Biology Sciences Research Council [BB/X511833/1], Digital Environment and Research Institute (DERI), the Queen Mary University of London, and Recursion Pharmaceuticals Inc.

We thank Arkaitz Zubiaga, Dan Crowther, and Anniek Myatt for their valuable feedback and suggestions on the project. We are grateful to the Semantic Scholar team for assisting with data access. Additionally, we thank Apocrita (King et al., 2017) and its ITS team for providing and maintaining the HPC resources.

References

Iz Beltagy, Kyle Lo, and Arman Cohan. 2019. SciBERT: A pretrained language model for scientific text. In *Proceedings of the 2019 Conference on Empirical Methods in Natural Language Processing and the 9th International Joint Conference on Natural Language Processing (EMNLP-IJCNLP)*, pages 3615–3620, Hong Kong, China. Association for Computational Linguistics.

David M. Blei, Andrew Y. Ng, and Michael I. Jordan. 2003. Latent dirichlet allocation. *J. Mach. Learn. Res.*, 3(null):993–1022.

Vera Boteva, Demian Gholipour, Artem Sokolov, and Stefan Riezler. 2016. A full-text learning to rank dataset for medical information retrieval.

Gully Burns. 2022. Drsm-corpus v1.

Arman Cohan, Sergey Feldman, Iz Beltagy, Doug Downey, and Daniel Weld. 2020. SPECTER: Document-level representation learning using citation-informed transformers. In *Proceedings of the 58th Annual Meeting of the Association for Computational Linguistics*, pages 2270–2282, Online. Association for Computational Linguistics.

Eoghan Cunningham and Derek Greene. 2023. Graph embedding for mapping interdisciplinary research networks. In *Companion Proceedings of the ACM Web Conference 2023*, WWW '23 Companion, page 784–789, New York, NY, USA. Association for Computing Machinery.

Scott Deerwester, Susan T. Dumais, George W. Furnas, Thomas K. Landauer, and Richard Harshman. 1990. Indexing by latent semantic analysis. *Journal of the American Society for Information Science*.

Jacob Devlin, Ming-Wei Chang, Kenton Lee, and Kristina Toutanova. 2019. BERT: Pre-training of deep bidirectional transformers for language understanding. In *Proceedings of the 2019 Conference of the North American Chapter of the Association for Computational Linguistics: Human Language Technologies, Volume 1 (Long and Short Papers)*, pages 4171–4186, Minneapolis, Minnesota. Association for Computational Linguistics.

Naman Jain and Mayank Singh. 2021. Tweetpap: A dataset to study the social media discourse of scientific papers. *Preprint*, arXiv:2106.07213.

Albert Q. Jiang, Alexandre Sablayrolles, Arthur Mensch, Chris Bamford, Devendra Singh Chaplot, Diego de las Casas, Florian Bressand, Gianna Lengyel, Guillaume Lample, Lucile Saulnier, Lélio Renard Lavaud, Marie-Anne Lachaux, Pierre Stock, Teven Le Scao, Thibaut Lavril, Thomas Wang, Timothée Lacroix, and William El Sayed. 2023. Mistral 7b. *Preprint*, arXiv:2310.06825.

Jeff Johnson, Matthijs Douze, and Hervé Jégou. 2019. Billion-scale similarity search with GPUs. *IEEE Transactions on Big Data*, 7(3):535–547.

Thomas King, Simon Butcher, and Lukasz Zalewski. 2017. *Apocrita - High Performance Computing Cluster for Queen Mary University of London*.

Diederik P. Kingma and Jimmy Ba. 2017. Adam: A method for stochastic optimization. *Preprint*, arXiv:1412.6980.

Yibin Lei, Liang Ding, Yu Cao, Changtong Zan, Andrew Yates, and Dacheng Tao. 2023. Unsupervised dense retrieval with relevance-aware contrastive pretraining. In *Findings of the Association for Computational Linguistics: ACL 2023*, pages 10932–10940, Toronto, Canada. Association for Computational Linguistics.

Adam Lerer, Ledell Wu, Jiajun Shen, Timothee Lacroix, Luca Wehrstedt, Abhijit Bose, and Alex Peysakhovich. 2019. Pytorch-biggraph: A large-scale graph embedding system. *Preprint*, arXiv:1903.12287.

Yuan Liang, Massimo Poesio, and Roonak Rezvani. 2024. A fine-grained citation graph for biomedical academic papers: the finding-citation graph. In *Proceedings of the 23rd Workshop on Biomedical Natural Language Processing*, pages 416–426, Bangkok, Thailand. Association for Computational Linguistics.

Tomas Mikolov, Ilya Sutskever, Kai Chen, Greg Corrado, and Jeffrey Dean. 2013. Distributed representations of words and phrases and their compositionality. In *Proceedings of the 27th International Conference on Neural Information Processing Systems - Volume 2*, NIPS'13, page 3111–3119, Red Hook, NY, USA. Curran Associates Inc.

David Mimno and Andrew McCallum. 2007. Expertise modeling for matching papers with reviewers. In *Proceedings of the 13th ACM SIGKDD International Conference on Knowledge Discovery and Data Mining*, KDD '07, page 500–509, New York, NY, USA. Association for Computing Machinery.

Sheshera Mysore, Arman Cohan, and Tom Hope. 2022. Multi-vector models with textual guidance for fine-grained scientific document similarity. In *Proceedings of the 2022 Conference of the North American Chapter of the Association for Computational Linguistics: Human Language Technologies*, pages 4453–4470, Seattle, United States. Association for Computational Linguistics.

Malte Ostendorff, Nils Rethmeier, Isabelle Augenstein, Bela Gipp, and Georg Rehm. 2022. Neighborhood contrastive learning for scientific document representations with citation embeddings. In *Proceedings of the 2022 Conference on Empirical Methods in Natural Language Processing*, pages 11670–11688, Abu Dhabi, United Arab Emirates. Association for Computational Linguistics.

Lawrence Page, Sergey Brin, Rajeev Motwani, and Terry Winograd. 1999. The pagerank citation ranking : Bringing order to the web. In *The Web Conference*.

Jeffrey Pennington, Richard Socher, and Christopher Manning. 2014. GloVe: Global vectors for word representation. In *Proceedings of the 2014 Conference on Empirical Methods in Natural Language Processing (EMNLP)*, pages 1532–1543, Doha, Qatar. Association for Computational Linguistics.

Nils Reimers and Iryna Gurevych. 2019. Sentence-BERT: Sentence embeddings using Siamese BERT-networks. In *Proceedings of the 2019 Conference on Empirical Methods in Natural Language Processing and the 9th International Joint Conference on Natural Language Processing (EMNLP-IJCNLP)*, pages 3982–3992, Hong Kong, China. Association for Computational Linguistics.

G. Salton, A. Wong, and C. S. Yang. 1975. A vector space model for automatic indexing. *Commun. ACM*, 18(11):613–620.

Florian Schroff, Dmitry Kalenichenko, and James Philbin. 2015. Facenet: A unified embedding for face recognition and clustering. In *2015 IEEE Conference on Computer Vision and Pattern Recognition (CVPR)*, page 815–823. IEEE.

Vikram Shyam, Lauren Friend, Brian Whiteaker, Nicholas Bense, Jonathan Dowdall, Bishoy Boktor, Manju Johny, Isaias Reyes, Angeera Naser, Nikhitha Sakhamuri, Victoria Kravets, Alexandra Calvin, Kaylee Gabus, Delonte Goodman, Herbert Schilling, Calvin Robinson, Robert Omar Reid II, and Colleen Unsworth. 2019. Petal (periodic table of life) and physiomimetics. *Designs*, 3(3).

Amanpreet Singh, Mike D'Arcy, Arman Cohan, Doug Downey, and Sergey Feldman. 2023. SciRepEval: A multi-format benchmark for scientific document representations. In *Proceedings of the 2023 Conference on Empirical Methods in Natural Language Processing*, pages 5548–5566, Singapore. Association for Computational Linguistics.

Shivashankar Subramanian, Daniel King, Doug Downey, and Sergey Feldman. 2024. S2and: A benchmark and evaluation system for author name disambiguation. In *Proceedings of the 2021 ACM/IEEE Joint Conference on Digital Libraries*, JCDL '21, page 170–179. IEEE Press.

Ellen Voorhees, Tasmeer Alam, Steven Bedrick, Dina Demner-Fushman, William R. Hersh, Kyle Lo, Kirk Roberts, Ian Soboroff, and Lucy Lu Wang. 2021. Trec-covid: constructing a pandemic information retrieval test collection. *SIGIR Forum*, 54(1).

Tongzhou Wang and Phillip Isola. 2022. Understanding contrastive representation learning through alignment and uniformity on the hypersphere. *Preprint*, arXiv:2005.10242.

Thomas Wolf, Lysandre Debut, Victor Sanh, Julien Chaumond, Clement Delangue, Anthony Moi, Pierric Cistac, Tim Rault, Rémi Louf, Morgan Funtowicz, Joe Davison, Sam Shleifer, Patrick von Platen, Clara Ma, Yacine Jernite, Julien Plu, Canwen Xu, Teven Le Scao, Sylvain Gugger, Mariama Drame,

Quentin Lhoest, and Alexander M. Rush. 2020. Huggingface's transformers: State-of-the-art natural language processing. *Preprint*, arXiv:1910.03771.

Yu Zhang, Hao Cheng, Zhihong Shen, Xiaodong Liu, Ye-Yi Wang, and Jianfeng Gao. 2023. Pre-training multi-task contrastive learning models for scientific literature understanding. In *Findings of the Association for Computational Linguistics: EMNLP 2023*, pages 12259–12275, Singapore. Association for Computational Linguistics.

Yue Zhao, Ajay Anand, and Gaurav Sharma. 2022. Reviewer recommendations using document vector embeddings and a publisher database: Implementation and evaluation. *IEEE Access*, 10:21798–21811.

A LLM Prompt

The prompt to analyze the importance of each citation can be seen here.

> You are an AI language model tasked with analyzing the importance of specific citations within a research paper. Each citation is provided with three pieces of information:
> - Citation Sentence: The sentence shows why and what the citation occurs.
> - Abstract of the Citing Paper: A summary of the research of the citing paper.
> - Title of the Cited Paper: The title of the cited paper.
> Based on this information, your task is to analyze and determine the importance of the citation to the citing paper.
> Your thinking logic chain should follow the following diagram:
> - Abstract Analysis: Identify key goals, methods, and findings.
> - Citation Sentence Analysis: Determine citation context and purpose.
> - Title Analysis: Check for alignment of scope and key themes.
> - Cross-Referencing: Is the cited work foundational to methods, key concepts, or outcomes? Does it appear crucial for the execution of the citing study?
> - Explanation: Provide a concise explanation for the classification based on analysis.
> - Importance Classification:
> - Highly Important: Core foundation (methods, key framework).
> - Moderately Important: Background, context, secondary relevance.
> - Less Important: General information, historical context.
>
> Here are some examples:
> {Examples}
>
> Just output the importance classification result and explanation.

B Preprocessing Citation Network

For the citation filtering, we examined approximately 1.46 million citations, classifying 28.4% as highly important, 44.8% as moderately important, and 26.8% as less important. Since papers can cite others multiple times using different citation sentences, the same citation pair sometimes receives different importance classifications. In such cases, we retained citations marked as less important if they also appeared in the highly important category. Ultimately, we removed only about 260,000 citations from the total of ~17 million citations, as we only have those citation analysis results.

For the finding similarity relation, we searched for similar findings for 392,505 (Total 10 million) findings. When the two papers shared similar findings, we created a new relation between them. Through this process, we generated approximately 200 million relations between papers.

Converting Annotated Clinical Cases into Structured Case Report Forms

Pietro Ferrazzi[1,2], **Alberto Lavelli**[1], **Bernardo Magnini**[1],
[1]Fondazione Bruno Kessler, Via Sommarive 18, Povo, Trento, Italy,
[2] University of Padova, via Trieste 63, Padova, Italy
Correspondence: pferrazzi@fbk.eu

Abstract

Case Report Forms (CRFs) are largely used in medical research as they ensure accuracy, reliability, and validity of results in clinical studies. However, publicly available, well-annotated CRF datasets are scarce, limiting the development of CRF slot filling systems able to fill in a CRF from clinical notes. To mitigate the scarcity of CRF datasets, we propose to take advantage of available datasets annotated for information extraction tasks and to convert them into structured CRFs. We present a semi-automatic conversion methodology, which has been applied to the E3C dataset in two languages (English and Italian), resulting in a new, high-quality dataset for CRF slot filling. Through several experiments on the created dataset, we report that slot filling achieves 59.7% for Italian and 67.3% for English on a closed Large Language Models (zero-shot) and worse performances on three families of open-source models, showing that filling CRFs is challenging even for recent state-of-the-art LLMs.

1 Introduction

Case Report Forms (CRFs) are essential tools in clinical research, designed to systematically and consistently collect patient data. They are composed of a list of predefined items to be filled with patients' medical information. By standardizing data collection, they ensure accuracy, reliability, and validity, which are crucial for producing meaningful and reproducible results in clinical studies. An expanding area of research focuses on developing automated systems for filling CRFs with information extracted from clinical notes and medical records, a concept envisioned by Mac Kenzie et al. (2016) and further advanced by Gutiérrez-Sacristán et al. (2024). Leveraging Natural Language Processing methods and models represent a potentially promising approach to automate and advance research in this field. However, despite their importance, publicly available, well-annotated CRF datasets are scarce, limiting the effective development and training of such systems.

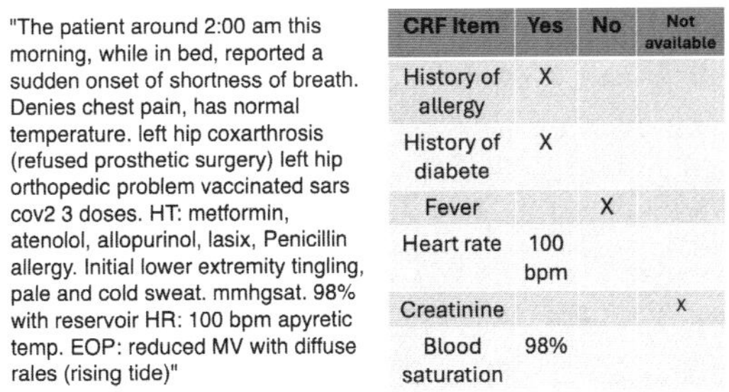
"The patient around 2:00 am this morning, while in bed, reported a sudden onset of shortness of breath. Denies chest pain, has normal temperature. left hip coxarthrosis (refused prosthetic surgery) left hip orthopedic problem vaccinated sars cov2 3 doses. HT: metformin, atenolol, allopurinol, lasix, Penicillin allergy. Initial lower extremity tingling, pale and cold sweat. mmhgsat. 98% with reservoir HR: 100 bpm apyretic temp. EOP: reduced MV with diffuse rales (rising tide)"

CRF Item	Yes	No	Not available
History of allergy	X		
History of diabete	X		
Fever		X	
Heart rate	100 bpm		
Creatinine			X
Blood saturation	98%		

Figure 1: Example of a Case Report Form filled with the values from a clinical note.

To address this gap, we propose a methodology that transforms publicly available datasets of clinical cases annotated for information extraction into a structured set of filled CRFs. Examples of such publicly available datasets are the following: MIMIC IV[1], i2b2[2], n2c2[3], CAS (Grabar et al., 2018), E3C (Magnini et al., 2023). Our approach reduces the discrepancy between existing datasets and real-world clinical needs, aligning them more closely with the practical requirements of hospitals and clinical research applications, where CRF filling is a widely relevant task. The outcome is a diverse CRF dataset, filled with information grounded in human annotations. Each example in the dataset consists of a triplet: a clinical case, a CRF to be filled, and the golden-standard filling values for the CRF derived from the clinical note, similar to what is presented in Figure 1. We apply this methodology to the European Clinical Case Corpus (E3C), release the resulting dataset, and evaluate several Large Language Models (LLMs) on it.

[1]https://physionet.org/content/mimiciv/3.1/
[2]https://www.i2b2.org/NLP/DataSets/Main.php
[3]https://n2c2.dbmi.hms.harvard.edu/

Proceedings of the 24th Workshop on Biomedical Language Processing (BioNLP 2025), pages 307–318
August 1, 2025 ©2025 Association for Computational Linguistics

The contributions of the paper are the following: (i) a general methodology for converting corpora of clinical cases annotated for information extraction into filled CRFs; (ii) a new multilingual dataset[4] (Italian and English) for CRF slot filling derived from the E3C dataset; (iii) several baselines indicating that automatic CRF slot filling from clinical notes is challenging even for state-of-the-art LLMs.

2 Related Work

Health data standardization is a fundamental aspect in the ongoing integration of medical research and artificial intelligence. To facilitate such alliance, the dimensions emphasized by Pétavy et al. (2019) are crucial, encompassing the need of health research for being transparent, accessible, interoperable, reproducible, and of high quality.

Case Report Forms play a central role in this context, and various efforts have been made to ensure that CRFs are designed to be consistent, reliable, and applicable across different clinical environments (Richesson and Nadkarni, 2011; Bellary et al., 2014). Rinaldi et al. (2025) outlines essential guidelines for CRF design, emphasizing the need to use clear, reusable, standardized, and uniquely identifiable terms to facilitate semantic consistency and future reuse. In a related line of work, Lin et al. (2015) proposes methods to ensure that CRFs are aligned with the specific research questions they aim to address, thereby reinforcing their utility and validity in clinical studies.

The shift from paper-based to electronic CRFs has been a major focus of recent research, aiming to enhance usability, reduce errors, and improve integration with digital health records (Fleischmann et al., 2017). This advancements lead to a gain of interest about automatic CRF filling from clinical reports. Mac Kenzie et al. (2016) introduced early approaches to extract structured data from narrative clinical notes, a line of research that has been extended by Gutiérrez-Sacristán et al. (2024). However, these approaches remain relatively basic, depending on keyword matching and vocabulary-based resolution, failing to leverage the full capabilities of modern Natural Language Processing techniques.

[4]The dataset is released at `https://huggingface.co/collections/NLP-FBK/e3c-to-crf-67b9844065460cbe42f80166`

3 Methodology

In this section, we present a general methodology to convert corpora of annotated clinical cases into structured Case Report Forms. Our approach is informed by an analysis of 200 pairs of clinical notes and populated CRFs from an Italian hospital. The CRFs at hand were organized among seven key areas: patient history, clinical examination, diagnostic tests results, laboratory test results, imaging findings, treatment, and final diagnosis. While CRFs are designed to be broad and comprehensive, covering a wide range of potential clinical scenarios, an individual patient's history is typically much more limited. For this reason, we observed that in our sample the CRF items remained unfilled around 90% of the time when populated with patients' information, highlighting the general characteristic of being designed to collect much more information of what it is typically available for each specific patient.

From this analysis, we concluded that in our setup CRF design lies between two extremes: creating a unique CRF for each clinical case, leading to highly specific yet non-generalizable item sets, or crafting a single, overly broad CRF for the entire dataset, potentially blending unrelated medical domains. We adopted an intermediate approach, aligning with the traditional purpose of CRFs in clinical studies — to gather data from patients with similar conditions relevant to a study (Bellary et al., 2014). Building on this principle, we propose a two step procedure as outlined in Figure 2: in Section 3.1 we group clinical cases based on semantic similarity, and in 3.2 we generate a dedicated CRF for each group and fill it with the information annotated for each clinical note. This results in one set of CRF items per group, subsequently filled once for each clinical case in that group. To conclude, in Section 6.1 we introduce and detail the task, the evaluation metrics, and the method provided as baselines.

3.1 Clinical Cases Clustering

We aim to generate groups of clinical cases, ensuring both clinical relevance and consistency in the resulting crafted CRFs. Therefore, we require effective differentiation of documents to form clusters that group together only relevant clinical cases. If the clusters are too broad, meaningful distinctions may be lost. We prioritized diagnosis as the key

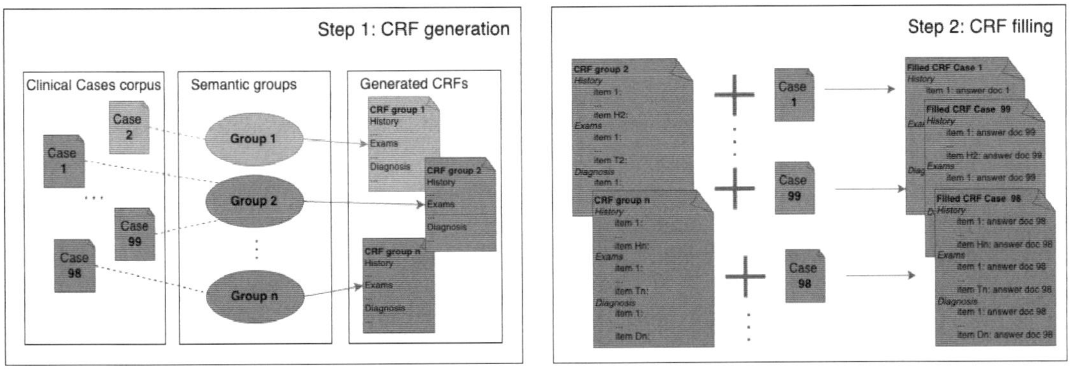

Figure 2: Summary of our two-step CRF generation and filling pipeline. **Step 1** Initially, clinical notes are clustered based on semantic similarity. Then, a group-specific CRF is generated for each cluster by extracting relevant items from the annotations of the clinical cases within the group. **Step 2** Each case is then linked to its designated CRF, which item set is populated based on the preexisting document annotation. The outcome is a list of as many CRFs as identified groups, and as many filled CRFs as documents. Each group-specific CRF is filled as many times as the number of documents belonging to it.

clustering dimension since CRF items are typically guided by the specific condition being studied. The key idea is to give significant weight to diagnosis-based links between notes in the clustering process, while retaining knowledge about entities and clinical information. Grouping documents that share similarities in these aspects helps construct synthetic CRFs that are both structured and clinically relevant.

Since many available datasets do not include explicit annotations on diagnoses, we implemented an automated system to extract them.

Diagnosis extraction. Extracting a diagnosis from a clinical note presents several challenges. Firstly, a note may mention past diagnoses that are no longer relevant. Secondly, the diagnosis might be implied rather than explicitly stated, requiring a deeper interpretation. Lastly, some clinical notes may not include a diagnosis at all, further complicating the extraction process. To address this challenge, we implement a two-step approach: *i) Automatic Generation of a Shortlist of Potential Diagnoses* – We leverage the available annotations to identify candidate diagnoses for each clinical case. First, we extract all words with the prefix "diagnos-" and check whether they are followed by an annotated entity. When this pattern was present, the associated entity is considered a potential diagnosis. Otherwise, we treat all entities in the text as potential diagnoses. *ii) Diagnosis Selection* – We refine the diagnosis by prompting a Large Language Model with the shortlist. This step outputs the exact diagnosis from the shortlist, combining the pattern-matching findings and powerful models, improving accuracy and reducing ambiguity.

Data representation for clustering. Our clustering approach is built on a graph-based representation of the data, where clinical notes are linked by weighted edges that quantify their similarity (see Figure 3 for an implementation example of such concept). This similarity is calculated based on shared entities and diagnoses across cases. A key challenge lies in the variability of how these concepts are mentioned, as the same notion can be expressed in multiple ways (e.g., "lower limb" vs. "leg", "malignant tumor" vs. "cancer"). Ensuring that notes discussing the same or closely related concepts achieve high similarity beyond mere character overlap is a critical aspect of our methodology. To address this challenge, we leveraged the UMLS Metathesaurus Names database (National Library of Medicine (US), 2024), augmenting the terms with semantically related concepts. By mean of appending to each term a short list of related ones (maximum 5), we can better capture the similarities between cases, even when different terminology is used to refer to the same or closely related clinical concepts. For languages other than English, each target mention is translated into English before performing a semantic search using a state-of-the-art language model (Zhang et al., 2025) following the findings of Chiaramello et al. (2016).

Similarity definition. To create the connection between each pair of clinical cases, we determine a similarity measure based on two components: the

ratio of shared entities (e), and diagnosis similarity (d). The ratio e is calculated as the number of UMLS-augmented shared terms divided by the number of augmented terms in the clinical note with the least of them. However, assessing diagnosis similarity d requires a different strategy due to the limited number of diagnosis terms per note. We address this using a large language model trained for semantic similarity (Lee et al., 2024), calculating cosine similarity between the UMLS-augmented diagnosis embeddings. This approach enables us to establish meaningful connections between cases, forming more coherent clusters.
We then define the overall similarity measure

$$s = 3d + e \tag{1}$$

This formulation assigns greater weight to diagnosis similarity while still preserving additional contextual information on shared entities.

Clustering. Based on the overall similarities s, we propose to apply the Louvain algorithm as described by Lu et al. (2015), selecting as starting groups the ones composed by the weakly connected sub-graphs obtained via the d edges with high weight. However, this step is highly data-dependent and must be tailored to each specific use case, following the approaches described by Xu and Tian (2015).

3.2 CRF generation

For each group of clinical notes, we aim to extract a set of relevant items for each section identified in the real-world CRFs analyzed in Section 3. The combination of the distinct section sets forms a comprehensive, group-specific CRF, tailored to the shared characteristics and clinical context of each group. Once each group-specific CRF is created, it needs to be populated for each clinical case. The overall outcome of this stage is one CRF per group and one gold-standard filled CRF per clinical case. Clinical cases within the same group share the same set of items, but their values vary based on the specific annotations present in each document.
We formulate and populate items for the identified sections, acknowledging that not all sections may be available in every dataset. As such, it is essential to determine which sections can be populated on the basis of the available annotations and, when necessary, refine the process to suit specific use cases. Here, we present an overview of the possible scenarios. *Clinical history* items can be generated using annotations such as symptom, sign, clinical entity, disease, condition, procedure. They are typically filled with positive and negative values, based on whether they occurred in the patient's past. Additionally, they may include information on whether a disease or condition is chronic or acute. *Clinical examination*, *diagnostic test results*, *laboratory test results*, and *imaging findings* can be addressed using any annotation of type similar to condition, measurement. Such items can be populated with diverse answer formats, including numerical values, categorical labels (e.g., positive/negative, high/low), and free-text descriptions, depending on the nature of the test and the information available. *Diagnosis* items can be generated based on the extraction procedure described in Section 3.1. This category of items is filled with either a positive or negative value. *Treatment* items can be addressed via labels such as medication, drug, or chemical. They can be filled with a variety of formats, spanning from medication names to time and duration information.
Initially, item sets are generated individually for each clinical note. These sets are then combined with those from other notes within the same group, forming a comprehensive and representative list of items for the entire group. Then, generated group-specific CRFs are populated for each clinical case in the group, based on the annotation, resulting in the gold-standard filled CRF.

Data revision. All generated items in each section of each group-specific CRF are normalized using UMLS mapping, collapsing equivalent terms to a single one. Furthermore, manual revision is performed to guarantee the quality of the generated CRF, with three objectives: (1) merge equivalent and highly related items, (2) remove irrelevant items, and (3) adjust inaccurate items. The process is conducted in a semi-automated manner. For each item in the CRF, we use a close source Large Language Model to assess whether it could be mapped to an existing item and to provide a justification for the suggested mapping. Any proposed mapping is manually reviewed for validity and, if approved, the overlapping items are consolidated.

4 CRF Filling: Task Definition

Datasets constructed according to the methodology detailed in Section 3.1 introduce a new CRF-filling task, which is divided into as many sub-tasks as the

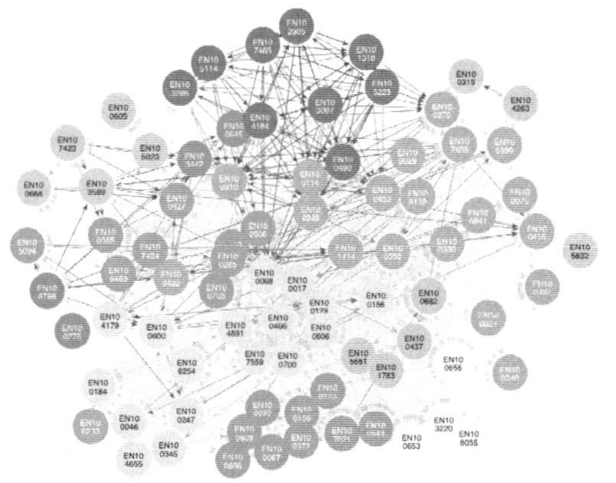

Figure 3: Graph representation of the E3C English dataset. Each node is a clinical note and the edges represent the similarity between cases. Darker edges represent higher similarity. The color of the nodes represents the group assigned by the clustering algorithm.

number of corresponding CRF sections. Each task requires filling CRF items based on information extracted from clinical cases, but they may vary in complexity. For the *diagnosis* section, the task consists in determining whether an item represents the final diagnosis, with three possible outcomes — "yes", "no", "not available". The *clinical history* section is more complex than the diagnosis one, as it allows for a broader range of valid outcomes. In addition to determining whether an event occurred in the patient's history, it may also capture details such as its chronic or acute nature, adding an extra layer of difficulty. The *clinical examination*, *tests results*, *imaging findings* and *treatment* section are the most complex ones, as they lack a predefined set of valid answers, requiring extraction and interpretation of numerical and textual values from the clinical notes.

Baseline. We established a baseline for the CRF slot filling tasks using sequence and pattern matching techniques. For the diagnosis task, the baseline assigns "yes" if the diagnosis of interest appears in the clinical case and "not available" otherwise. In the clinical examination, tests results, imaging findings and treatment tasks, if the respective item is mentioned in the text, the first numerical value following it is extracted as the result. For the clinical history task, the baseline assigns "yes" if the corresponding textual span is found in the clinical case and "not available" otherwise.

Metrics definition. To define the evaluation metrics, we first established criteria for identifying positive and negative occurrences. An item is considered positive for diagnosis if labeled "yes" and negative if marked as "no" or "not available". In the clinical history section, an item is positive if it appears in any valid form, such as "yes" or "no", and negative if marked as "not available". For all the other tasks, an item is positive if assigned a value and negative if labeled as "not available." Additionally, when a generated answer does not conform to the expected format—if any predefined format is required—it is always considered a false positive. Based on these definitions, we can compute task-specific precision, recall, and F_1-score, as well as overall micro and macro F_1-scores.

We apply strict matching criteria (ignoring trailing punctuation) with one relaxation: any text following "not available" in response was ignored if this phrase appeared at the beginning. For gold-standard labels filled with multiple values, a true positive (TP) is assigned for a perfect match, a false positive (FP) if extra elements are predicted, and a false negative (FN) if the prediction contains fewer items than the ground truth.

5 The Case of the E3C Dataset

In the previous sections, we outlined the general methodology for converting any corpus annotated for information extraction into gold-standard filled CRFs. In this section, we apply this methodology to the European Clinical Case Corpus (E3C, Magnini et al., 2023). E3C is an open, manually annotated multilingual dataset consisting of clinical cases in five languages. E3C clinical cases are detailed accounts of a patient's medical history, containing rich medical details and temporal relationships that enable in-depth linguistic analysis. The dataset includes annotations on both textual spans and the relationships between them. The ones relevant to our study are summarized in Table 2. In this work, we focused on the Italian and English splits (Table 1).

5.1 CRF generation from E3C

We applied our methodology to the European Clinical Case Corpus (E3C), adapting it to the dataset's specific characteristics. Below, we outline key adaptations, while all details not explicitly mentioned can be found in Section 3.

We generated the shortlist of potential diagnoses

Lang	# notes	# clent	# rml	# event
English	84	1024	480	4885
Italian	86	869	383	3385

Table 1: Number of clinical notes (# notes), annotated clinical entities (# clent), results and measurements (#rml), and events (# event) in E3C Italian and English splits, which both comprise approximately 25k words.

considering only clinical entities as possible targets, as other annotations were deemed out of scope. After selecting the diagnoses using GPT-4o (OpenAI and et al, 2024) in a 4-shots settings, we manually reviewed 10 examples in both English and Italian, confirming the accuracy of the results in all cases. In some instances (9 for English, 19 for Italian), no diagnosis was identified, which is expected since certain clinical documents do not report it. Then, the overall similarity measure was defined as $s = 3d + \frac{1}{2}(e + b)$, where e and b are the ratios of shared clinical entities and shared body parts respectively, d is the diagnosis similarity. The resulting graph representation of the data is shown in Figure 3. This method resulted in 7 (8) groups and 6 (12) clinical cases not assigned to any group for Italian (English). More details on the diagnosis extraction prompts, similarities and generated groups are shown in Appendix A.1 and A.3.
Using the information embedded in the E3C annotations, we formulated and populated items for the following sections: clinical history, diagnosis, clinical examination, diagnostic test results, laboratory test results, and imaging findings. Since no information on treatment was available at the annotation level, we excluded it from consideration.

Exams. To generate and populate exam items, we first extracted the textual spans linked to RMLs (results and measurements) via PERTAINS_TO relationships. A CRF exam item was created for each textual span with a corresponding RML, representing its filling value. When an RML refers to multiple textual spans, a separate item is generated for each of them. When the same textual span is associated with multiple RMLs, a single item is created for the textual span, and each RMLs is used at filling time, separated by special tokens. RMLs that do not pertain to any textual span were ignored.

Clinical History. To generate and populate items about patients history, we focused on the clinical entities enriched by three key annotated attributes: "polarity" (whether the reported term is present or not), "contextual modality" (knowledge about the truth value of the event, can be actual, hedged, hypothetical or generic), and "permanence" (can be permanent for conditions with no known cure or finite for those that can be resolved eventually). Each of these attributes defines a portion of the gold-standard answer, as outlined in Table 6 in Appendix A.2.

Diagnosis. For each diagnosis, an item was created and populated with "yes" if it applied to the clinical case and "not available" otherwise.
An example of a generated CRF can be found in Appendix A.3.

Train-test split. We adopted the train-test split provided by Ghosh et al. (2025) for the E3C dataset. The result is that clinical cases from the same group are assigned to different splits, while group-specific CRFs are generated on all the cases in the corpus. By design, CRFs must cover all essential fields for the patient groups they represent. As a result, constructing comprehensive item sets from the full dataset is necessary and does not introduce bias beyond the task's inherent structure. Crucially, only training clinical notes are used for learning, preventing any test-specific influence on the model. Note that this cross-splits effect is further reduced by creating clinical history item sets merging the ones extracted from clinical cases in both splits but excluding from the final set the ones filled only for test cases after data revision.

6 Experimental settings

We explored the E3C CRF-filling task using decoder-only Large Language Models (LLMs) as they have exhibited high performance in several tasks in zero-shot settings.

Models. We selected the instruct versions of different state-of-the-art model families, in different sizes: Llama-3 8B and 70B (et al., 2024), Qwen-2.5 7B and 72B (Qwen and et al., 2025), Mistral-Small-3.1 24B [5], Gemma-3 27B[6] and GPT 4o. This selection allowed us to compare proprietary (GPT) and open-source models (the others), assessing the impact of model size and determining which family performs better on our task. Each model was prompted with task-specific

[5]https://mistral.ai/news/mistral-small-3-1

[6]https://blog.google/technology/developers/gemma-3/

Category	Description (example)
Clinical entity	disorders, pathologies, and symptoms (“metastases” “nausea")
Body part	parts of the human body (“parotid gland")
RML	results and measurements (“38g/dl")
Event	any event (“diagnosed", “haemoglobin")
PERTAINS-TO	relation between an RML and the Event it refers to (“38g/dl” pertains-to “haemoglobin")

Table 2: E3C categories for annotations on textual spans and their relationships utilized in this work. Each textual span is annotated if it represents a clinical term (i.e., clinical entities such as pathologies and symptoms, body parts, laboratory tests and results) and is assigned some attributes. For more details, see Magnini et al. (2023).

			Italian		English	
Task	Description	Accepted answers	Train Items (Filled)	Test Items (Filled)	Train Items (Filled)	Test Items (Filled)
Diagnosis	determine if an item is the final diagnosis for the patient	“yes", “no", “not available”	498 (8%)	553 (7%)	491 (9%)	505 (9%)
History	determine whether the patient experienced a history item	“Certainly yes", “No", "Probably yes, chronic", “not available” etc.	977 (23%)	903 (11%)	953 (25%)	872 (13%)
Exams	extract the results related to an exam item	any string representing an exam result	1108 (10%)	1149 (10%)	984 (11%)	916 (9%)
Total			**2583 (14%)**	**2605 (10%)**	**2428 (16%)**	**2293 (11%)**

Table 3: Description, space of possible answers, number of items, and ratio of populated items in the train and test splits for both languages for the three CRF sub-tasks. All three sub-tasks are quite sparse, with around ten to fifteen percent of the items populated in the gold-standard filled CRFs. Clinical notes in the train and test split are composed by around 12k and 13k tokens (words), respectively, in both Italian and English. The possible answers for history are determined by the levels of the annotated attributes utilized for the gold-standard filling.

Model	Diagnosis			History			Exams			Micro	Macro
	Prec	Rec	F_1	Prec	Rec	F_1	Prec	Rec	F_1	F_1	F_1
Baseline	64.9	58.5	61.5	100.0	11.3	20.4	13.6	10.8	12.0	31.3	25.4
Llama 8B	32.2	92.7	47.8	7.2	60.8	13.0	4.8	25.0	8.1	23.0	18.2
Qwen 7B	72.1	75.6	73.8	33.8	73.6	46.4	7.5	8.5	7.9	42.7	35.2
Mistral 24B	68.4	63.4	65.8	51.6	64.9	57.5	13.8	22.1	16.9	46.7	41.4
Gemma 27B	73.5	87.8	**80.0**	47.1	83.5	<u>60.2</u>	22.9	83.9	<u>36.0</u>	<u>58.7</u>	<u>53.7</u>
Llama 70B	54.7	100.0	70.7	32.8	77.3	46.0	16.0	67.8	25.9	47.5	42.4
Qwen 72B	75.6	75.6	75.6	58.1	74.2	**65.2**	19.4	38.7	25.8	55.5	50.0
GPT 4o	75.6	82.9	<u>79.1</u>	40.8	75.3	52.9	34.0	76.8	**47.1**	**59.7**	**55.9**

Table 4: Performance of different models on the Italian dataset across three categories: Diagnosis, History, and RML. Metrics include Precision, Recall, F_1-score. Overall micro and macro F_1-scores are also reported.

Model	Diagnosis			History			Exams			Micro	Macro
	Prec	Rec	F_1	Prec	Rec	F_1	Prec	Rec	F_1	F_1	F_1
Baseline	84.6	53.7	65.7	87.5	13.2	23.0	0.0	0.0	0.0	29.6	21.9
Llama 8B	49.3	94.9	64.9	10.2	76.4	18.0	6.4	63.0	11.6	31.5	25.1
Qwen 7B	100.0	63.4	77.6	40.0	78.4	53.0	15.6	16.3	15.9	48.8	41.9
Mistral 24B	63.6	80.0	70.9	55.3	68.9	**61.3**	22.7	62.5	33.3	55.2	51.0
Gemma 27B	91.4	78.0	84.2	42.9	74.5	54.5	32.7	86.0	47.4	62.0	57.7
Llama 70B	84.2	78.0	81.0	36.1	74.3	48.6	34.8	81.6	48.8	59.5	55.6
Qwen 72B	96.8	73.2	83.3	55.9	67.0	60.9	27.0	80.0	40.3	61.5	56.6
GPT 4o	94.4	82.9	**88.3**	47.5	72.4	57.4	42.2	84.3	**56.2**	**67.3**	**63.4**

Table 5: Performance of different models on the English dataset across three categories: Diagnosis, History, and RML. Metrics include Precision, Recall, and F_1-score. Overall micro and macro F_1-scores are also reported.

details, the clinical case, the CRF item, and answering guidelines.

All experiments on open-source models were run on 8xA40 (46GB) and took approximately 30 GPU hours, serving the models using the vllm package (Kwon et al., 2023). Prompts can be seen in detail in Appendix A.4.

6.1 CRF Filling from E3C Clinical Cases

The constructed dataset introduces a new E3C CRF-filling task, which is divided into three sub-tasks: clinical history, exams, and diagnosis as described in Table 3. The main specialty of this task in respect to the more general outlined in Section 4 is that clinical history items can be filled with twelve valid values (Appendix A.2). Given the unique annotation scheme in E3C, which includes multiple levels of polarity, contextual modality, and permanence, such complexity is specific to this dataset and may not be present in others. Therefore, we report results on a simplified version where all positive responses are grouped as "yes" and all negative ones as "no". By simplifying the values, we aim to offer a more general perspective on the inherent difficulty of the task, extending beyond the particularities of the E3C dataset.

7 Results and Discussion

Experimental results are reported in Tables 4 and 5 for Italian and English respectively. GPT-4o consistently achieves the highest overall performance in both Italian and English datasets, with the best Micro and Macro F_1-scores (59.7 and 55.9 for Italian, 67.3 and 63.4 for English). Among open-weight models, Gemma 27B and Qwen 72B perform competitively in Italian, closely approaching GPT-4o's results, particularly in diagnosis and history. For English, Gemma 27B, Qwen 72B, and Llama 70B performances are very similar, around 6-8 points lower than GPT-4o.
Regarding model size, we observe an average improvement of around 20 Macro F_1 points when scaling from small (7/8B) to large (70/72B) models in the LLaMA and Qwen families. Interestingly, models in the 20–30B range often match or surpass larger ones from different architectures. Among the tasks, exams prove to be the most challenging, followed by history, indicating significant room for improvement. Models perform on average better on English than in Italian with no exception, with an average delta of 7.5 points of Micro F_1. Among the smaller models, Qwen 7B significantly outperforms Llama 8B, which struggles with extremely low precision. At larger scales, Qwen 72B and Llama 70B exhibit comparable performance in English, while Qwen 72B demonstrates a clear advantage over Llama 70B in Italian.

8 Conclusion

Our study presents a novel methodology for transforming annotated clinical notes into structured Case Report Forms (CRFs) by leveraging clusters of semantically similar cases. This approach ensures that CRFs are both comprehensive and contextually relevant while maintaining consistency across similar clinical scenarios. Given the scarcity of publicly available CRF datasets, our method provides a valuable framework for automating CRF generation, which could be highly beneficial for future clinical applications. In addition, our method brings existing datasets closer to real-world clinical applications, ensuring greater alignment with the practical needs of hospitals and research. Given that CRF filling is a widely relevant task, this approach enhances the utility of annotated clinical notes.

Our findings highlight that the characteristics of the generated CRFs are strongly influenced by the dataset's distribution, underscoring the necessity of manual tuning based on available annotation types when adapting the method to different contexts. We believe that a robust analysis of the data distribution is crucial for high-quality CRF generation.

Our experimental results reveal that the constructed CRFs encompass tasks of increasing complexity for state-of-the-art models. Diagnosis items can be framed as a relatively straightforward binary classification task, while history items remain within a classification framework but with greater difficulty due to their nuanced nature. The most challenging aspect lies in handling exams, tests, and examinations, which require a fully generative approach without a predefined set of valid responses, making them particularly difficult for current models to solve. Both open- and closed-source models show room for improvement in terms of performance.

Limitations

There are a few limitations in our current approach to convert Information Extraction datasets into structured CRFs. First, the proposed methodology has been experimented only on the E3C corpus: although this is a significant use case (several levels of annotations, several languages), additional insights may derive from different available datasets. Second, in order to keep under control our experimental setting, we made a few simplifications with respect to the full complexity of the task. Particularly, for the CRF *clinical history* group, we assumed a three-value schema (i.e., a certain clinical evidence is either present, negated, or not mentioned), while in reality the possible values should be extended to cover cases of chronicity.

Acknowledgments

This work has been partially funded by the European Union under the Horizon Europe eCREAM Project (Grant Agreement No.101057726) and IDEA4RC Project (Grant Agreement No.101057048). Views and opinions expressed are however those of the authors only and do not necessarily reflect those of the European Union or the European Health and Digital Executive Agency (HADEA). Neither the European Union nor the granting authority can be held responsible for them.

References

Shantala Bellary, Binny Krishnankutty, and M. S. Latha. 2014. Basics of case report form designing in clinical research. *Perspectives in Clinical Research*, 5(4):159–166.

Emma Chiaramello, Francesco Pinciroli, Alberico Bonalumi, Angelo Caroli, and Gabriella Tognola. 2016. Use of "off-the-shelf" information extraction algorithms in clinical informatics: A feasibility study of MetaMap annotation of Italian medical notes. *Journal of Biomedical Informatics*, 63:22–32.

Aaron Grattafiori et al. 2024. The Llama 3 herd of models. *Preprint*, arXiv:2407.21783.

Robert Fleischmann, Anne-Marie Decker, Antje Kraft, Knut Mai, and Sein Schmidt. 2017. Mobile electronic versus paper case report forms in clinical trials: a randomized controlled trial. *BMC Medical Research Methodology*, 17(1):153.

Soumitra Ghosh, Begona Altuna, Saeed Farzi, Pietro Ferrazzi, Alberto Lavelli, Giulia Mezzanotte, Manuela Speranza, and Bernardo Magnini. 2025. Low-resource information extraction with the European Clinical Case Corpus. ArXiv preprint.

Natalia Grabar, Vincent Claveau, and Clément Dalloux. 2018. CAS: French corpus with clinical cases. pages 122–128, Brussels, Belgium. Association for Computational Linguistics.

Alba Gutiérrez-Sacristán, Simran Makwana, Audrey Dionne, Simran Mahanta, Karla J. Dyer, Faridis Serrano, Carmen Watrin, Pierre Pages, Sajad Mousavi, Anil Degala, Jessica Lyons, Danielle Pillion, Joany M. Zachariasse, Lara S. Shekerdemian, Dongngan T. Truong, Jane W. Newburger, and Paul Avillach. 2024. Development and validation of an open-source pipeline for automatic population of case report forms from electronic health records: a pediatric multi-center prospective study. *eBioMedicine*, 108. Doi: 10.1016/j.ebiom.2024.105337.

Woosuk Kwon, Zhuohan Li, Siyuan Zhuang, Ying Sheng, Lianmin Zheng, Cody Hao Yu, Joseph E. Gonzalez, Hao Zhang, and Ion Stoica. 2023. Efficient memory management for large language model serving with PagedAttention. In *Proceedings of the ACM SIGOPS 29th Symposium on Operating Systems Principles*.

Chankyu Lee, Rajarshi Roy, Mengyao Xu, Jonathan Raiman, Mohammad Shoeybi, Bryan Catanzaro, and Wei Ping. 2024. NV-Embed: Improved techniques for training LLMs as generalist embedding models. *Preprint*, arXiv:2405.17428.

Ching-Heng Lin, Nai-Yuan Wu, and Der-Ming Liou. 2015. A multi-technique approach to bridge electronic case report form design and data standard adoption. *Journal of Biomedical Informatics*, 53:49–57.

Hao Lu, Mahantesh Halappanavar, and Ananth Kalyanaraman. 2015. Parallel heuristics for scalable community detection. *Parallel Computing*, 47:19–37.

W. R. Mac Kenzie, A. J. Davidson, A. Wiesenthal, J. P. Engel, K. Turner, L. Conn, S. J. Becker, S. Moffatt, S. L. Groseclose, J. Jellison, J. Stinn, N. Y. Garrett, L. Helmus, B. Harmon, C. L. Richards, J. R. Lumpkin, and M. F. Iademarco. 2016. The promise of electronic case reporting. *Public Health Reports*, 131(6):742–746. Epub 2016 Oct 13.

Bernardo Magnini, Begoña Altuna, Alberto Lavelli, Anne-Lyse Minard, Manuela Speranza, and Roberto Zanoli. 2023. *European Clinical Case Corpus*, pages 283–288. Springer International Publishing, Cham.

National Library of Medicine (US). 2024. UMLS knowledge sources [dataset on the internet]. release 2024aa. Cited 2025 March.

OpenAI and Josh Achiam et al. 2024. Gpt-4 technical report. *Preprint*, arXiv:2303.08774.

Frank Pétavy et al. 2019. Global standardization of clinical research data. *Applied Clinical Trials*, 28(4):20–23.

Qwen and An Yang et al. 2025. Qwen2.5 technical report. *Preprint*, arXiv:2412.15115.

Rachel L. Richesson and Prakash Nadkarni. 2011. Data standards for clinical research data collection forms: current status and challenges. *Journal of the American Medical Informatics Association*, 18(3):341–346.

Eugenia Rinaldi, Caroline Stellmach, and Sylvia Thun. 2025. How to design electronic case report form (ecrf) questions to maximize semantic interoperability in clinical research. *Interactive Journal of Medical Research*, 14:e51598.

Dongkuan Xu and Yingjie Tian. 2015. A comprehensive survey of clustering algorithms. *Annals of Data Science*, 2(2):165–193.

Dun Zhang, Jiacheng Li, Ziyang Zeng, and Fulong Wang. 2025. Jasper and Stella: distillation of SOTA embedding models. *Preprint*, arXiv:2412.19048.

A Appendix

A.1 Diagnosis Extraction

Here we report the structure of the prompt utilized to generate the diagnosis using GPT-4o:

```
{System prompt}{Example 1}...{Example 4}
"clinical note":{Clinical case}
"potential diagnosis":
{list of potential diagnosis}
```

Here is the system prompt:

```
You are a clinical assistant.
Your job is to extract the conclusive
diagnosis from a clinical note written by
an experienced physician.
The diagnosis is a medical condition
identified by a health care provider.
To complete the task, you are aided by a
list of possible diagnoses.
Here are your guidelines:
1. The diagnosis is always contained in
the list of potential diagnoses.
2. Your goal is to extract only the
diagnosis, ignoring everything else.
3. Respond with a json containing the
extracted diagnosis and a short motivation
{“motivation”: “motivation for the
extracted diagnosis”, “diagnosis”:
“extracted diagnosis”}.
4. If no diagnosis is reported,
respond with “no diagnosis.”

CAUTION: Notes may contain diagnoses
made in the past with respect to
the current clinical situation.
Only extract diagnoses related to
the current situation.
```

Table 7 presents examples of similarity scores for E3C cases calculated in the embedding space of the diagnosis augmented via UMLS semantic search.

A.2 E3C Clinical History Items

Table 6 reports the attributes and their levels used for populating the E3C CRF clinical history section. Each E3C clinical entity is annotated with contextual modality, polarity, and permanence, which determine the filled value using the template:

```
{contextual mod} {polarity}, {permanence}
```

For instance, an entity with polarity "positive", contextual modality "hedged" and permanence "finite" is filled with "Probably yes, possibly chronic". There are 12 possible level combinations.

A.3 Generated E3C CRFs

Table 8 presents the statistics on the generated E3C CRF for English and Italian. Figure 4 shows an example of a CRF generated for the English group 1.

diagnosis
Diagnosis: systemic lupus erythematosus : *not available*
Diagnosis: adnexal torsion : *Yes*
Diagnosis: strumal carcinoid tumour : *not available*
Diagnosis: PV : *not available*
Diagnosis: benign cystic teratoma : *not available*

rml
Exam: temperature : *not available*
Exam: pressure : *not available*
Exam: hemoglobin : *not available*
Exam: CRP : *not available*
Exam: Ca-125 : *not available*
Exam: free-T4 : *not available*
Exam: TSH : *not available*

history
History of dyspnea.: *not available*
History of fatigue.: *not available*
History of pain.: *Certainly yes, certainly not chronic*
History of autoimmune disorders: *not available*
History of jugular distention.: *not available*
History of sinus tachycardia.: *not available*
History of effusion disorders.: *not available*
History of cardiac tamponade: *not available*
History of anemia: *not available*
History of nonspecific inflammation.: *not available*
History of tumor.: *Certainly yes, certainly not chronic*
History of tuberculosis.: *not available*
History of serositis.: *not available*
History of arthralgia.: *not available*

Figure 4: Example of a generated CRF for English group 1 and filled with the annotation from the clinical case EN100668

Attribute	Level	CRF Value
Polarity	Positive	Yes
	Negative	No
Modality	Actual	Certainly
	Hypothetical	Possibly
	Hedged	Probably
	Missing	(empty)
Permanence	Permanent	Chronic
	Finite	Certainly not chronic
	Missing	Possibly chronic

Table 6: Attribute levels for populating the E3C CRF clinical history section.

A.4 Prompts for experiments

The prompt for the experiments is composed following this template:

`{system prompt} {answering guidelines} {clinical case} {question on the item}.`

Here we report the prompts used for English. The ones for Italian are the direct translation of them.

System prompt

```
You are an expert clinical doctor. You have
to answer a question on "{task_description}"
about a patient. To do it, you are given
the patient clinical history.
```

History answering guidelines , where values are populated according to the logic presented in the methodology section.

```
The answer is composed by three components:
polarity, contextual modality, and
permanence. You must combine these
three components together to answer
the question.
- contextual modality can be:
a)'VALUE_1' if the answer is certain,
b)'VALUE_2' if the answer is hypothetical,
c)'VALUE_3' if the answer is probable.
- polarity can be:
a)'VALUE_4' if the answer is affirmative,
b)'VALUE_5' if the answer is negative.
- permanence can be:
a)'VALUE_6' if the object of the question
is certainly permanent forever,
b)'VALUE_7' if the object of the question
is temporary or transitory,
c)'VALUE_8' otherwise.

These three components must be combined
in order: "contextual modality polarity,
permanence". For example, if the question
is "Does the patient have a history of
diabetes?", the answer could be:
"EXAMPLE_1", or "EXAMPLE_2".

If the information is not contained in
the clinical history, answer with
'not_available'.
Do not add any preamble to the answer.
```

Exams answering guidelines

```
The answer can assume three different
formats.
-if the test/exam has been performed
only once, answer with the results
of the test/exam.
```

Diagnoses note 1	Diagnoses note 2	Similarity Score
neuroendocrine neoplasia	neoplasia	0.63
chronic myeloid leukemia Ph+ in chronic phase	JMML	0.57
acute ulcerative rectocolitis	clostridium difficile colitis	0.58
mass of tumor origin	syncopal episodes, Polymorphic ventricular tachycardia	0.11
Wilms's tumor, Metastasis	microperforation	0.10

Table 7: Similarity scores between extracted diagnoses for pairs of clinical cases. The first three lines represent cases with high similarity, while the last two cases with low similarities. It can be noted that terms that are syntactically different but semantically close such as "JMML" and "Chronic myeloid leukemia Ph+ in chronic phase" are mapped together by this approach, as the former has been correctly enriched with the term "juvenile myeloid leukemia", that results in an embedding similar to the latter. At the same time, cases with very different diagnoses are assigned very low similarities.

Italian					English				
Group	Cases Train/Test	CRF items	Avg/StDev (Train)	Avg/StDev (Test)	Group	Cases Train/Test	CRF items	Avg/StDev (Train)	Avg/StDev (Test)
0	4/4	23	5.5 / 2.2	5.5 / 0.8	0	7/2	71	11.7 / 4.7	5.0 / 3.0
1	11/13	91	7.9 / 5.3	7.9 / 3.6	1	1/4	26	19.0 / 0.0	3.0 / 1.2
2	4/4	55	13/ 9.7	13 / 2.2	2	1/1	10	6.0 / 0.0	4.0 / 0.0
3	2/7	27	4.5 / 1.5	4.5 / 2.7	3	3/6	54	11 / 9.7	6.8 / 5.2
4	4/6	76	9.8 / 4.8	9.8 / 9.2	4	5/4	24	3.6 / 1.5	4.0 / 2.5
5	4/4	48	9.8 / 7.8	9.8 / 4.5	5	9/9	99	11 / 7.7	7.9 / 4.0
6	9/4	79	12.7/ 7.5	13 / 1.2	6	8/9	36	9.0 / 4.0	7.0 / 5.1
					7	2/2	75	9.4 / 5.4	13 / 11

Table 8: Number of cases, number of items, average and standard deviation of the number of populated items (i.e., different from "not available") per group-specific CRF for both languages.

```
-if the test/exam has been performed
more than once, report all the
results separated by the special
token [\MULTI_ANSWER] (for example
"RESULT_1 [\MULTI_ANSWER] RESULT_2").
-if the information is not contained in
the clinical history, answer
with 'not_available'
```

Diagnosis answering guidelines

```
Answer 'Yes' if the patient's definitive
diagnosis is the one indicated. If the
information is not contained in the
clinical history, answer with
'not_available'.
```

Question structure for exams

```
What are the results and measures of {item}?
```

Question structure for diagnosis

```
Is the definitive diagnosis {item}?
```

Question structure for history

```
Does the patient have a history
of {item}?
```

MuCoS: Efficient Drug–Target Discovery via Multi-Context-Aware Sampling in Knowledge Graphs

Haji Gul[1a], Abdul Ghani Naim[1b], Ajaz Ahmad Bhat[*1],

[1]School of Digital Science, Universiti Brunei Darussalam, Jalan Tungku Link, Gadong BE1410, Brunei Darussalam,
(23h1710[a], ghani.naim[b], ajaz.bhat[*])@ubd.edu.bn

Abstract

Accurate prediction of drug–target interactions is critical for accelerating drug discovery. In this work, we frame drug–target prediction as a link prediction task on heterogeneous biomedical knowledge graphs (KG) that integrate drugs, proteins, diseases, pathways, and other relevant entities. Conventional KG embedding methods such as TransE and ComplEx-SE are hindered by their reliance on computationally intensive negative sampling and their limited generalization to unseen drug–target pairs. To address these challenges, we propose Multi-Context-Aware Sampling (MuCoS), a novel framework that prioritizes high-density neighbours to capture salient structural patterns and integrates these with contextual embeddings derived from BERT. By unifying structural and textual modalities and selectively sampling highly informative patterns, MuCoS circumvents the need for negative sampling, significantly reducing computational overhead while enhancing predictive accuracy for novel drug–target associations and drug targets. Extensive experiments on the KEGG50k and PharmKG-8k datasets demonstrate that MuCoS outperforms baselines, achieving up to a 13% improvement in MRR for general relation prediction on KEGG50k, a 22% improvement on PharmKG-8k, and a 6% gain in dedicated drug–target relation prediction on KEGG50k.

1 Introduction

Drug target discovery lies at the core of modern therapeutic development, enabling the identification of new biological targets, the prediction of non-target effects, and opportunities for drug repurposing — while significantly reducing experimental costs and accelerating translational timelines (Sachdev and Gupta, 2019). Recent computational advances leverage knowledge graphs (KGs) to integrate heterogeneous biomedical data (e.g., drugs, proteins, diseases, side effects, pathways) into unified networks where nodes represent entities and edges capture relationships, essentially framing discovery as a link prediction problem (Himmelstein et al., 2017). For example, KG's such as KEGG50k (Mohamed et al., 2019) PharmKG-8k (Zheng et al., 2021) and Hetionet (Himmelstein et al., 2017) provide comprehensive, structured representations of biological components and their intricate associations.

Biomedical KGC methods, however, face a critical trade-off: structural embedding methods such as ComplEx-SE (Mohamed et al., 2019) capture explicit drug-target relationships but fail to generalize to unseen entities like novel drugs due to rigid geometric constraints. Conversely, graph neural approaches like NeoDTI Progeni (Liu et al., 2024) integrate probabilistic reasoning with GNNs for state-of-the-art drug-target prediction but remain unevaluated on relation-centric benchmarks like KEGG50k. Furthermore, none of these methods exploit the rich textual semantics embedded in biomedical triples (e.g., "DRUG X $\rightarrow$ DRUG-TARGET-GENE $\rightarrow$ GENE Z"), which could provide inductive signals for unseen entities by contextualizing relationships beyond structural adjacency.

We posit that PharmKG-8k's and KEGG50k's relational triples are inherently compatible with textual encoding strategies and therefore believe that we can leverage a language transformer model like BERT's bidirectional attention to jointly model the explicit relationships through syntactic patterns in entity-relation-entity chains. Moreover, we propose to exploit, (a) the rich contextual information inherent in the graph's structure such as neighbouring entities and relations associated with a given head entity and query relation, like GNNs do, and (b) associated features such as node degrees and connectivity that affect the performance of KG techniques(Cattaneo et al., 2024).

We therefore propose **MuCoS** (*Multi-Context-Aware Sampling*), a KG completion framework that overcomes these limitations by aggregating filtered

Proceedings of the 24th Workshop on Biomedical Language Processing (BioNLP 2025), pages 319–327
August 1, 2025 ©2025 Association for Computational Linguistics

contextual information from adjacent entities and their relationships, and then integrating this semantically enriched context into a BERT model for better prediction of relationships and entities. In doing so, MuCoS advances drug target discovery in the following key ways:

- **Drug–Target Relation Prediction:** By leveraging optimized neighbouring contextual information around nodes and relations, MuCoS outperforms traditional models in predicting general and drug–target relationships.
- **Target-tail Prediction:** The method accurately predicts potential target tails (such as genes etc.) by incorporating contextualized structural information derived from the head entity and relationship.
- **Efficient Multi-Context Sampling:** By prioritizing informative structural patterns through density-based sampling, MuCoS reduces computational overhead while preserving high predictive accuracy.
- **Elimination of Auxiliary Data Requirements:** Operating effectively without reliance on extensive entity descriptions or negative sampling, MuCoS is particularly well-suited for sparse biomedical datasets.

2 Related Work

Drug target discovery has been approached from multiple computational perspectives. Similarity-based methods quantify relationships by computing pairwise distances—often using Euclidean or other metric functions—between drugs and their target proteins (Shi and Li, 2018). These methods typically rely on handcrafted similarity measures to distinguish interacting pairs. Feature-based techniques, predominantly employing support vector machines (Zhang et al., 2017), formulate the problem as a binary classification or two-class clustering task to differentiate between positive and negative drug–target associations based on engineered features.

Recent graph-based methods leverage heterogeneous networks that integrate multiple similarity metrics—such as drug–drug, target–target, and cross-modal associations—to exploit the homophily principle in biological systems (Ban et al., 2019). These approaches infer missing links by modelling complex interdependencies among drugs, proteins, diseases, and pathways. In parallel, the application of embedding-based techniques has evolved considerably (Bordes et al., 2013; Yang et al., 2014; Trouillon et al., 2016). For instance, Mohamed et al. (Mohamed et al., 2019) introduced ComplEx-SE, a variant of the ComplEx KGE model that adopts a squared error-based loss for enhanced accuracy. Recent works like NeoDTI (Liu et al., 2024) combine graph neural networks with probabilistic reasoning to achieve state-of-the-art performance in drug–target prediction.

Despite these advances, current KGC methods still face challenges in drug target discovery. Traditional embedding models depend on static, pretrained embeddings, which hinder their ability to generalize to novel entities and interactions in rapidly evolving biomedical data (Gul et al., 2024). Text-based and large language model approaches require rich and consistent annotations—a resource often sparse in biomedical domains (Gul et al., 2025). Additionally, the reliance on extensive negative sampling during training imposes significant computational burdens, particularly for large-scale datasets. These limitations motivate us to develop MuCoS as a flexible, context-aware and computationally efficient model that integrates both structural and textual cues to drive the discovery of new drug targets.

3 Methodology

MuCoS addresses two knowledge graph completion tasks: **(1) Link Prediction** (inferring missing relations in triples like $(h, ?, t)$) and **(2) Tail Prediction** (identifying missing tail entities in $(h, r, ?)$). Both tasks are divided into general and drug-target-specific subtasks to balance broad applicability with a biomedical focus. Using the PharmKG-8k and the KEGG50k dataset, general subtasks predict relations/tails across all entities and relations, while drug-target subtasks use a filtered subset to predict specific relations.

MuCoS builds on the MuCo-KGC model (Gul et al., 2025) to boost computational efficiency by strategically *sampling* high-density contextual information *(i.e., entities or relations that appear most frequently)* from both entity and relation-neighbouring contexts before integrating it with BERT for precise predictions. For the transformer part of MuCoS, DistilBERT (base, uncased) has been employed, which is a smaller model that helps MuCoS run efficiently while still capturing context

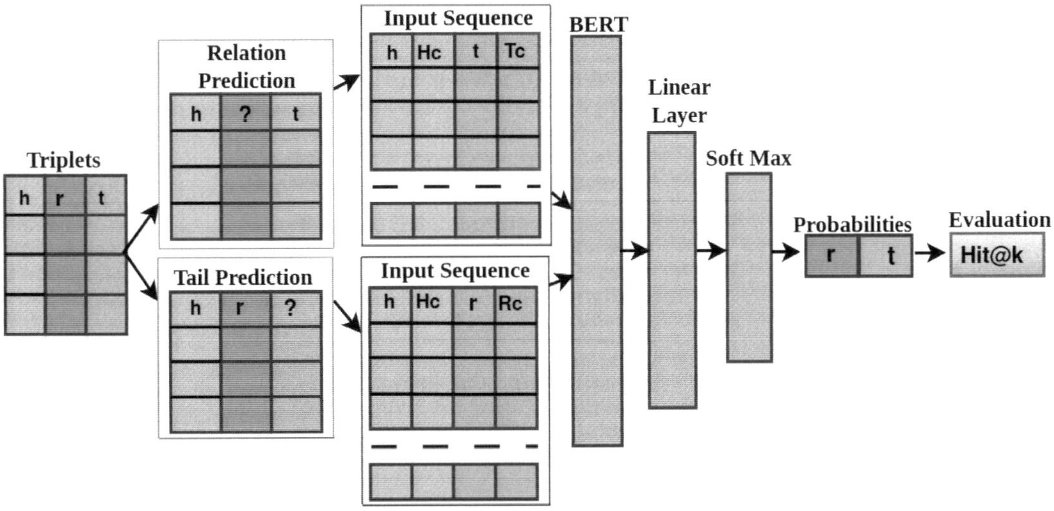

Figure 1: A concise overview of the MuCoS model pipeline, which is designed to predict general and drug-target relations and tail entities. The boxes on the left show the input sequence to the BERT model, where (h) head, $(\mathcal{H}_c)$ head context, (t) tail, $(\mathcal{T}_c)$ tail context, (r) relation, and $(\mathcal{R}_c)$ relation context. This integrated context is passed through the BERT model with a linear classifier and softmax function to generate probabilities for relations and tail.

well. We selected DistilBERT for its efficiency, retaining 95% of BERT's performance while being 40% smaller, making it suitable for large-scale knowledge graph tasks (Sanh et al., 2019). Figure 1 provides an overview of the MuCoS pipeline. The subsequent sections detail the computations of the contextual information and the sampling process in the MuCoS pipeline.

Given a head (h), tail (t), a relation (r) between them, MuCoS first figures out the corresponding neighbouring contexts, i.e., the head context $(\mathcal{H}_c)$, the tail-context $(\mathcal{T}_c)$ or the relationship context $(\mathcal{R}_c)$ and then selects out the high-density contexts. Based on the task at hand, relevant contexts are then concatenated and passed on to a BERT model with a linear classifier and softmax function to generate probabilities for relations or tails.

Head Context $\mathcal{H}_c$: To extract the contextual information for the head, i.e., $\mathcal{H}_c$, we first identify the relations associated with the head entity h, i.e., the relation neighbourhood $\mathcal{R}(h)$. If l relations are associated with h from the set $\mathcal{R}$ of all relations r_i in the graph,G, then:

$$\mathcal{R}(h) = A_{i=1}^{l}\left(\{r_i \mid (h, r_i, e_j) \in \mathcal{T}, e_j \in \mathcal{E}\}\right) \tag{1}$$

where $A(\cdot)$ is the concatenation operation $\|$, $\mathcal{T}$ is the set of training triples, $\mathcal{E}_t$ is the set of all tail entities, and r_i represents each relation associated with h. Next, we find the tail entities e that are neighbours (i.e., have a direct connection) with the head entity h, i.e., tail neighbourhood $\mathcal{E}(h)$, using the identified relations in $\mathcal{R}(h)$. Assuming m neighbour tails, $\mathcal{E}(h)$ is expressed as:

$$\mathcal{E}(h) = A_{i=1}^{m}\left(\{t_i \mid (h, r_j, e_i) \in \mathcal{T}, r_j \in \mathcal{R}\}\right) \tag{2}$$

where $\mathcal{E}(h)$ is the set of all tail entities t_i directly associated with the h through some relation r_j.

Sampling: While MuCo-KGC (Gul et al., 2025) integrates $\mathcal{R}(h)$ and $\mathcal{E}(h)$ calculates the head context, this study introduce a density-based sampling for context calculation $\mathcal{H}_c$, where the density $\rho(e)$ of an entity $e \in \mathcal{E}(h)$ is defined as its frequency of appearance in $\mathcal{T}$.

$$\rho(t) = |\{(h, r, t) \in \mathcal{T}\}|, for\ any\ h, r \tag{3}$$

$\rho(t)$ denotes the *density* of the tail entity t, defined as the number of times t appears as the tail in triples (h, r, t). Using these density values, we select n entities of highest density values and the relationships between head h and these top-n selected entities:

$$\text{top}_n(\mathcal{E}(h)) = \text{sort}(\mathcal{E}(h), \text{by } \rho(e))[:n] \tag{4}$$

$$\mathcal{R}^*(h) = A_{i=1}^{n}\big(\{r_i \mid (h, r_i, e_j) \in \mathcal{T}, e_j \in \text{top}_n(\mathcal{E}(h))\}\big) \tag{5}$$

$\text{top}_n(\mathcal{E}(h))$ selects the top n tail entities from $\mathcal{E}(h)$ sorted by their density $\rho(e)$. $\mathcal{R}^*(h)$ concatenates the relations r_i connected h and the selected top-n

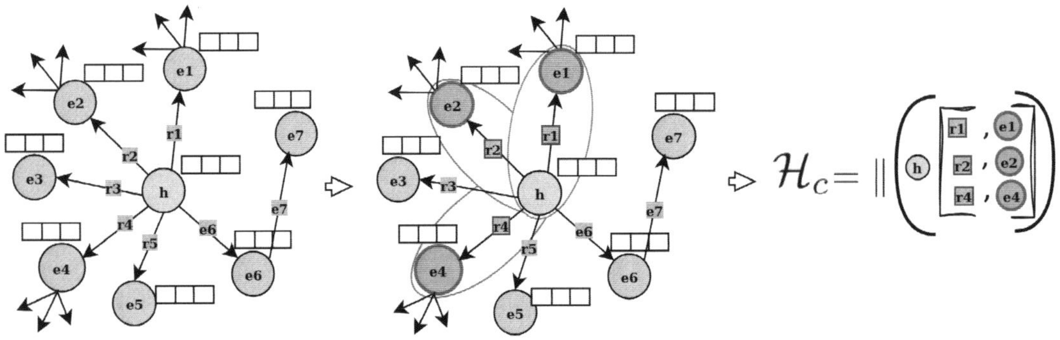

Figure 2: MuCoS $\mathcal{H}_c$ construction. The left graphical view illustrates one hop head h context, which consists of the set of relations $\mathcal{R}(h)$ $(r_1, r_2, r_3, r_4, r_5, r_6)$ and the set of neighbouring tail entities $\mathcal{E}(h)$ $(e_1, e_2, e_3, e_4, e_5, e_6)$ associated with the head entity h. The middle view shows the sampling process, where only the top-n (suppose $n = 3$) tail entities e are selected and concatenated ($\|$) based on their density $\rho(e)$, to calculate the optimized head context $\mathcal{H}_c$.

tail entities. The optimized head context $\mathcal{H}_c$ is then defined as:

$$\mathcal{H}_c = \mathcal{R}^*(h) \cup \text{top}_n(\mathcal{E}(h)) \tag{6}$$

Figure 2 illustrates this sampling process, highlighting only a select subset of high-density neighbours (shown in red border) used to compute the aggregated context $\mathcal{H}_c$. We follow the same procedure to compute the tail context $\mathcal{T}_c$ (for a given tail) required along with head context $\mathcal{H}_c$ in the relation prediction task.

Relation Context $\mathcal{R}_c$: To acquire the relation context $\mathcal{R}_c$, we identify all entities (heads and tails) associated with the operational relation r in the knowledge graph $\mathcal{G}$. This includes the set of heads (e.g., drugs) e_i and tails (e.g., genes) e_j connected by r:

$$\mathcal{E}(r) = A^o_{i,j=1}\left(\{e_i, e_j\} \mid (e_i, r, e_j) \in \mathcal{T}\}\right) \tag{7}$$

$\mathcal{E}(r)$ is the concatenation of all head-tail entity pairs (e_i, e_j) connected by the relation r in the knowledge graph.

Sampling: From the set of entities in $\mathcal{E}_c$, the top-k elements with the highest density values $\rho(e)$ are selected to generate the optimized relationship context $\mathcal{R}_c$.

$$\mathcal{R}_c = \text{top}_k(\mathcal{E}(r)) = \text{sort}(\mathcal{E}(r), \text{ by } (\rho(e_i) + \rho(e_j)))[: k] \tag{8}$$

$\mathcal{R}_c$ therefore provides a focused global perspective on r's patterns, enhancing generalization without excessively raising the time complexity. Figure 3 depicts the sampling process involved in computing $\mathcal{R}_c$, highlighting the selection of k high-density entity pairs (shown in red border) involved with the relation r to form the optimized relationship context. Following the extraction of contextual information via density-based sampling, MuCoS integrates these contexts into a BERT-based framework for prediction. The process for each subtask, leveraging the KEGG50k dataset and its filtered drug-target subset, is detailed below:

- For task **(1), link prediction**, which includes two subtasks: *General link prediction* $(h, ?, t)$: The concatenated representations $\mathcal{H}_c$ (head context) and $\mathcal{T}_c$ (tail context) are combined with the head entity h and tail entity t to form the input sequence $[h, \mathcal{H}_c, t, \mathcal{T}_c]$. This sequence passes through BERT's transformer layers, generating a contextualized representation for each token. A classification layer then predicts the relation r, with a softmax function calculating the probability distribution over all relations:

$$P(r \mid h, t) = \text{softmax}(W \cdot \text{BERT}(h, \mathcal{H}_c, t, \mathcal{T}_c)) \tag{9}$$

Drug-target link prediction $(h, ?, t)$: Following Mohamed et al (Mohamed et al., 2019) in this case, we filter the dataset to consider drug-target relations only. Other than that, we follow the same methodology as above, where the input sequence $[h, \mathcal{H}_c, t, \mathcal{T}_c]$ is processed by BERT to predict the drug-target-specific relations r.

- For task **(2), tail prediction**, which includes two subtasks: *General tail prediction* $(h, r, ?)$:

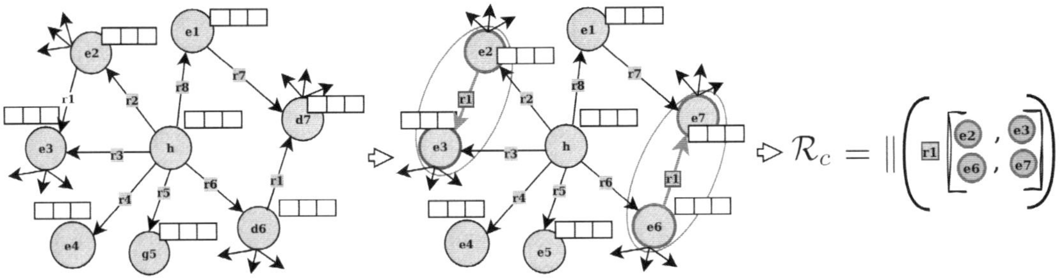

Figure 3: $\mathcal{R}_c$ construction. The left view illustrates the relationship r_1 and entities (head, tail) connected by r_1. The graph in the middle depicts optimization, selecting the top k (suppose $k = 2$) entities based on density ρ, retaining pairs such as (e_2, e_3) and (e_6, e_7) The optimized context $\mathcal{R}_c$ is aggregated using concatenation ($\|$), as shown in the right section.

The concatenated representations $\mathcal{H}_c$ (head context) and $\mathcal{R}_c$ (relation context) are combined with the head entity h and relation r to form the input sequence $[h, \mathcal{H}_c, \text{r}, \mathcal{H}_c]$, using the full KEGG50k dataset. BERT processes this sequence, and a classification layer predicts the tail entity t:

$$P(t \mid h, r) = \text{softmax}(W \cdot \text{BERT}(h, \mathcal{H}_c, r, \mathcal{R}_c)) \tag{10}$$

Drug-target tail prediction $(h, r, ?)$: Following above, we use a filtered drug-target subset of the KEGG50k dataset, to predict the tail entity t.

We train the model using cross-entropy loss. For link prediction, Equation 11 defines the loss with y_i as the one-hot true label for relation r_i and $P(r_i \mid h, t)$ as the predicted probability. For tail prediction, Equation 12 defines the loss with y_i as the true label for tail entity t_i and $P(t_i \mid h, r)$ as its predicted probability.

$$\textbf{(a)}\ \mathcal{L} = -\sum_{i=1}^{N} y_i \log P(r_i \mid h, t), \tag{11}$$

$$\textbf{(b)}\ \mathcal{L} = -\sum_{i=1}^{N} y_i \log P(g_i \mid h, r) \tag{12}$$

where y_i is the true label for the relation r_i, and $P(r_i \mid h, t)$ is the predicted probability of the relation given h and t. On the other hand, $P(t_i \mid h, r)$ is the predicted probability of the tail given h and r.

3.1 Computational Advantage of MuCoS over MuCo-KGC

Compared to MuCo-KGC (Gul et al., 2025), MuCoS reduces computational complexity by sampling only the most significant neighbours (based on density) from the full entity and relation contexts. MuCoS employs two sampling thresholds: n for the head entity context $\mathcal{H}_c$ and k for the relation context $\mathcal{R}_c$. To compute the complexities, we first define two terms: (i) the average density $(avg\_density)$ as the average number of neighbours per entity in the knowledge graph, and (ii) average appearance $(avg\_appearance)$ of a relation r in the dataset.

$$avg\_density = \frac{|T|}{|E|}, \quad avg\_appearance = \frac{|T|}{|R|} \tag{13}$$

where $|T|$ is the total number of triples, $|E|$ entities, and $|R|$ unique relations.

For **MuCo-KGC**, the complexity of computing the head context $\mathcal{H}_c$ and the relation context $\mathcal{R}_c$ is based on full neighbourhoods without sampling. The complexity of $\mathcal{H}_c$ depends on the number of relations involving the head entity h, denoted as $|\mathcal{R}(h)|$, and the number of neighbouring entities $|\mathcal{E}(h)|$, both approximated by $avg\_density$ (see Equation 15). The complexity of $\mathcal{R}_c$ is determined by the number of entity pairs connected by relation r, $|\mathcal{E}(r)|$, estimated using $avg\_appearance$ (see Equation 16). Therefore, the overall complexity for context computation in MuCo-KGC is defined equals:

$$O(2 \cdot avg\_density + avg\_appearance) \tag{14}$$

$$\text{where,}\quad O(|\mathcal{H}_c|) = O(|\mathcal{R}(h)| + |\mathcal{E}(h)|) = O(2 \cdot avg\_density) \tag{15}$$

$$\text{and,}\quad O(|\mathcal{R}_c|) = O(|\mathcal{E}(r)|) = O(avg\_appearance) \tag{16}$$

For **MuCoS**, the head context $\mathcal{H}_c$ is computed by selecting the top-n high-density neighbouring entities and their corresponding relations, and the relation context $\mathcal{R}_c$ are computed by selecting the top-k high-density entity pairs. The complexity of $\mathcal{H}_c$ is $O(n)$ for the sampled entities and $O(n)$ for the corresponding relations, and $\mathcal{R}_c$ is $O(k)$ for the sampled entity pairs. Thus, the overall complexity for context computation in MuCoS is:

$$O(2 \cdot n + k) \tag{17}$$

Since sampling threshold values n and k are much smaller than $avg\_density$ and $avg\_appearance$ in large datasets like KEGG50k, MuCoS achieves a significant reduction in computational cost compared to MuCo-KGC.

For example, in case of the KEGG50k dataset (with triplets $|T| = 63,080$, entities $|\mathcal{E}| = 16,201$, and relations $|\mathcal{R}| = 9$), $avg\_density \approx 3.895$, and $avg\_appearance \approx 7,008.89$. Therefore, the complexity of MuCo-KGC on the KEGG50k dataset is: $O(2{\cdot}3.895{+}7,008.89) = O(7,016.68)$. For MuCoS (with $n = 15$, $k = 10$: the complexity is $O(2 \cdot 15 + 10) = O(40)$. This is a speed up by a factor of ≈ 175.42 in context computation, i.e., the process of extracting and aggregating relevant neighbourhood information associated with a given head entity and relation. Sampling the context reduces the input token length, which further contributes to the efficiency slightly. The primary computational gains however arise from our selective sampling strategy, which significantly limits the amount of nodes/relations processed for context extraction.

Sampling size values of n at 15 and k at 10, although empirical, are motivated from the ablation studies on MuCo-KGC, suggesting that the head context plays a greater role than the relationship context in model performance (see Table 1 and Table 3 for details).

3.2 Experimental Setup

We evaluate MuCoS on two prediction tasks using KEGG50k and PharmKG-8k datasets: link and tail prediction. Each task is evaluated in two settings: the full KEGG50k dataset and a drug-target subset. In **link prediction**, we infer the missing relation in $(h, ?, t)$, with general and drug-target variants. Similarly, in **tail prediction**, we predict the missing entity in $(h, r, ?)$ for both settings. Below we provide the details of the dataset used in our experiments, the hyperparameter settings, and the evaluation criteria.

Datasets: *KEGG50k* [1] medical domain dataset, comprises 63,080 triples split into 57,080 training, 3,000 validation, and 3,000 testing instances (i.e. a 90:5:5 ratio split). Drug-target only triplet counts are 10769, 585, and 650 for the train, valid, and test sets. The dataset comprises 16,201 unique entities $\mathcal{E}$ where $(\mathcal{E}_d \cup \mathcal{E}_g) \subset \mathcal{E}$ and 9 distinct types of drug-target relationships, enabling a comprehensive mapping of pharmacological interactions. *PharmKG-8k* [2] comprises 400,788 training triplets, 49,536 testing triplets, and 50,036 validation triplets, covering 7,601 entities. These are categorized into Chemical, Disease, and Gene types, integrating data from DrugBank, TTD, OMIM, PharmGKB, and GNBR.

Hyperparameters: The input sequence is tokenized with a maximum length of 128 tokens. Training is conducted over 50 epochs using the AdamW optimizer with a learning rate of 5×10^{-5} and a batch size of 16. Experiments were performed on an NVIDIA GeForce RTX 3090 GPU with 24 GB of memory.

Evaluation: Model performance is assessed using standard metrics, Mean Reciprocal Rank (MRR) and Hits@k, as defined in Equations 18 and 19, to evaluate the accuracy of general and drug-target relations and tail predictions:

$$\text{MRR} = \frac{1}{N}\sum_{i=1}^{N}\frac{1}{\text{rank}_i}, \tag{18}$$

$$\text{Hits@k} = \frac{1}{N}\sum_{i=1}^{N}\mathbf{1}(\text{rank}_i \leq k), \tag{19}$$

MRR measures the average of the reciprocal ranks of the correct item across all queries. A higher MRR indicates better ranking performance. $H@k$ measures the proportion of queries where the correct item appears in the top k ranks. It provides a metric for evaluating ranking quality at different points.

3.3 Results and Discussion

Link Prediction: Table 1 demonstrates that MuCoS outperforms state-of-the-art baselines on the **KEGG50k** dataset. It achieves an MRR of 0.65 for general link prediction across all relations, a 13% improvement over ComplEx-SE's 0.52, and its Hits@1 score of 0.52 exceeds ComplEx-SE's 0.45 by 7%. Moreover, Hits@3 and Hits@10 scores of 0.60 and 0.86 further underscore the robust ranking performance of MuCoS. Although MuCo-KGC

[1] KEGG50k: https://shorturl.at/pWSJO

[2] PharmKG-8k: https://zenodo.org/records/4525237

Table 1: Relationship prediction results over the KEGG50k dataset on both general links and drug target links only.

Model	General link prediction				Drug-target link prediction			
	MRR	**Hits@1**	**Hits@3**	**Hits@10**	**MRR**	**Hits@1**	**Hits@3**	**Hits@10**
TransE (Bordes et al., 2013)	0.46	0.38	0.50	0.63	0.75	0.69	0.79	0.86
DistMult (Yang et al., 2014)	0.37	0.27	0.42	0.57	0.61	0.50	0.69	0.81
ComplEx (Trouillon et al., 2016)	0.39	0.31	0.43	0.57	0.68	0.61	0.71	0.82
ComplEx-SE (Mohamed et al., 2019)	0.52	0.45	0.56	0.68	0.78	0.73	0.81	0.88
MuCoS- (H_c Only)	0.52	0.44	0.55	0.69	0.75	0.65	0.77	0.1
MuCoS (T_c Only)	0.45	0.37	0.51	0.61	0.70	0.59	0.70	0.1
MuCo-KGC (Gul et al., 2025)	**0.79**	**0.58**	**0.73**	**0.92**	**0.94**	**0.91**	**0.96**	**1**
MuCoS	<u>0.65</u>	<u>0.52</u>	<u>0.60</u>	<u>0.86</u>	<u>0.84</u>	<u>0.74</u>	<u>0.84</u>	<u>1</u>

(Gul et al., 2025) achieves state-of-the-art performance, MuCoS offers a significant computational advantage with a small reduction in accuracy.

In drug–target prediction,which focuses on identifying relationships between drugs and their targets (e.g., genes), MuCoS achieves an MRR of 0.84—a 6% improvement over ComplEx-SE's 0.78—demonstrating the benefit of contextual head/tail information. It also records Hits@1 of 0.74 (vs. 0.73), Hits@3 of 0.84 (a 3% gain), and a perfect Hits@10 of 1.00 (12% improvement), outperforming TransE, DistMult, and ComplEx. Although MuCo-KGC attains higher accuracy (e.g., an MRR of 0.94), its prohibitive computational cost limits scalability. MuCoS, by offering competitive performance with substantial efficiency gains, provides a scalable solution for real-world, large-scale drug discovery.

Table 2 shows that MuCoS achieves state-of-the-art performance on **PharmKG-8k**. It attains an MRR of 0.452, compared to NC-KGE's 0.228, and a Hits@1 of 0.258 versus 0.145. Additionally, MuCoS records Hits@3 and Hits@10 scores of 0.602 and 0.676, respectively.

Table 2: PharmKG8k-28 Results for Link Prediction Task.The symbol □ denotes that the results are taken from Paper (Zheng et al., 2021), while the symbol △ results are taken from Paper (Fan et al., 2023). R2N results are reported from (Diligenti et al., 2023).

Model	**MRR**	**H@1**	**H@3**	**H@10**
TransR □	0.075	0.030	0.071	0.155
RESCAL □	0.064	0.023	0.057	0.122
ConvE □	0.086	0.038	0.087	0.169
ConvKB □	0.106	0.052	0.107	0.209
RGCN □	0.067	0.027	0.062	0.139
HRGAT □	0.154	0.075	0.172	0.315
TransE △	0.116	0.038	0.127	0.269
DistMult △	0.218	0.152	0.237	0.335
ComplEx △	0.124	0.064	0.128	0.244
TruckER △	0.182	0.103	0.202	0.336
HRGAT △	0.134	0.063	0.144	0.271
SACN △	0.156	0.085	0.170	0.296
CompGCN △	0.193	0.110	0.216	0.352
SE-GNN △	0.206	0.120	0.232	0.374
R2N	0.215	0.145	0.234	0.342
NC-KGE △	0.228	0.145	0.252	0.390
MuCoS	**0.452**	**0.258**	**0.602**	**0.676**

Tail Prediction: Table 3 compares tail prediction performance between MuCoS and MuCo-KGC under both general and drug-target settings. While MuCo-KGC (without sampling) achieves higher MRR, Hits@1, and Hits@3 in the general scenario, MuCoS (sampling-based) excels in drug-target cases, particularly in Hits@10. Thus, sampling enhances prediction accuracy for drug targets at a slight cost in the general scenario, and MuCoS offers a significant computational advantage while outperforming other models on KEGG50k.

4 Ablation Study

We analyze the contributions of the Head Context ($\mathcal{H}_c$) and Tail Context ($\mathcal{T}_c$) components for relation (link) prediction, and Head Context ($\mathcal{H}_c$) and Relation Context ($\mathcal{R}_c$) for the prediction of the tail. The results are presented in Tables 1 and 3.

Relationship Prediction: Table 1 reports the results for both general link prediction and drug-target link prediction scenarios. MuCo-KGC (Gul et al., 2025), the earlier method, demonstrates strong performance across all metrics, achieving an MRR of 0.79 for general link prediction and 0.94 for drug-target link prediction. These results highlight its ability to leverage both $\mathcal{H}_c$ (Head Context) and $\mathcal{R}_c$ (Relation Context) effectively, excelling particularly in Hits@1 (0.58 and 0.91) and Hits@10 (0.92 and 0.1).

Tail Prediction: Table 3 presents the results for both general tail prediction and drug-target-specific

Table 3: Tail prediction results on the KEGG50k dataset were evaluated for both general and drug target scenarios using methods with and without sampling.

Model	General tail prediction				Drug-target tail prediction			
	MRR	Hits@1	Hits@3	Hits@10	MRR	Hits@1	Hits@3	Hits@10
MuCoS (H_c Only)	0.26	0.20	0.34	0.55	0.38	0.30	0.41	0.78
MuCoS (R_c Only)	0.21	0.15	0.28	0.39	0.31	0.19	0.36	0.69
MuCo-KGC	0.39	0.34	0.521	0.718	0.567	0.457	0.628	0.917
MuCoS	0.31	0.215	0.40	0.57	0.442	0.259	0.46	0.868

scenarios. MuCo-KGC (Gul et al., 2025) delivers robust performance, achieving an MRR of 0.39 for general tail prediction and 0.567 for drug-target tail prediction. Its superior Hits@1 scores (0.34 and 0.457) and Hits@10 scores (0.71 and 0.917) confirm its effectiveness in capturing complex relational patterns in the graph.

Across both prediction tasks, the $\mathcal{H}_c$-Only configuration consistently outperforms achieving an MRR of 0.52 (general links) and 0.75 (drug-target links) for relationships, and 0.26 (general tails) and 0.38 (drug-target tails) for tail predictions. This highlights the critical role of localized contextual information over global relational patterns, which tend to underperform when used in isolation $\mathcal{T}_c$-Only MRR: 0.45 and 0.70 for links; $\mathcal{R}_c$-Only MRR: 0.21 and 0.31 for tails).

5 Conclusion

The study introduces MuCoS, a multi-context-aware sampling method that uses DistilBERT to improve drug-target relation predictions and tail entity predictions in biomedical knowledge graphs. MuCoS employs a dual strategy combining transformer-based textual modeling with context-aware sampling to overcome limitations of existing models, such as poor generalization, negative sampling, and the need for descriptive entity information. It extracts and optimizes contextualized information from the head, tail, and relation entities using density-based sampling and its lexical semantics, capturing richer structural patterns and reducing computational complexity. Experimental results show superior performance over state-of-the-art models, with improvements in MRR and Hits@1 for general and drug-target relationship prediction on both KEGG50k and PharmKG-8k datasets. Future work could focus on adaptive sampling to dynamically adjust n and k for sparse KGs, and integrate multimodal data like protein sequences or chemical structures to enhance drug-target prediction.

References

Tomohiro Ban, Masahito Ohue, and Yutaka Akiyama. 2019. Nrlmfβ: Beta-distribution-rescored neighborhood regularized logistic matrix factorization for improving the performance of drug–target interaction prediction. *Biochemistry and biophysics reports*, 18:100615.

A. Bordes, N. Usunier, A. Garcia-Duran, J. Weston, and O. Yakhnenko. 2013. Translating embeddings for modeling multi-relational data. In *Advances in NIPS*, volume 26. Curran Associates, Inc.

Alberto Cattaneo, Thomas Martynec, Stephen Bonner, Carlo Luschi, and Daniel Justus. 2024. Towards linking graph topology to model performance for biomedical knowledge graph completion. In *ICML'24 Workshop ML for Life and Material Science: From Theory to Industry Applications*.

Michelangelo Diligenti, Francesco Giannini, Stefano Fioravanti, Caterina Graziani, Moreno Falaschi, and Giuseppe Marra. 2023. Enhancing embedding representations of biomedical data using logic knowledge. In *2023 IJCNN*, pages 1–8. IEEE.

Zhiguang Fan, Yuedong Yang, Mingyuan Xu, and Hongming Chen. 2023. Node-based knowledge graph contrastive learning for medical relationship prediction. In *https://arxiv.org/abs/2310.10138*.

Haji Gul, Abdul Ghani Naim, and Ajaz A Bhat. 2024. A contextualized bert model for knowledge graph completion. In *MusIML, Advances in Neural Information Processing Systems (NeurIPS)*, pages https://neurips.cc/virtual/2024/affinity-event/105018.

Haji Gul, Abdul Ghani Naim, and Ajaz A Bhat. 2025. Muco-kgc: Multi-context-aware knowledge graph completion. In *The 29th Pacific-Asia Conference on Knowledge Discovery and Data Mining (PAKDD), https://arxiv.org/abs/2503.03091*.

Daniel Scott Himmelstein, Antoine Lizee, Christine Hessler, Leo Brueggeman, Sabrina L Chen, Dexter Hadley, Ari Green, Pouya Khankhanian, and Sergio E Baranzini. 2017. Systematic integration of biomedical knowledge prioritizes drugs for repurposing. *elife*, 6:e26726.

Chang Liu, Kaimin Xiao, Cuinan Yu, Yipin Lei, Kangbo Lyu, Tingzhong Tian, Dan Zhao, Fengfeng Zhou, Haidong Tang, and Jianyang Zeng. 2024. A probabilistic knowledge graph for target identification. *PLOS Comp Bio*, 20(4):e1011945.

Sameh K Mohamed, Aayah Nounu, and Vít Nováček. 2019. Drug target discovery using knowledge graph embeddings. In *Proceedings of the 34th ACM/SIGAPP symposium on applied computing*, pages 11–18.

Kanica Sachdev and Manoj Kumar Gupta. 2019. A comprehensive review of feature based methods for drug target interaction prediction. *Journal of biomedical informatics*, 93:103159.

Victor Sanh, Lysandre Debut, Julien Chaumond, and Thomas Wolf. 2019. Distilbert, a distilled version of bert: smaller, faster, cheaper and lighter. *arXiv preprint arXiv:1910.01108*.

Zezhi Shi and Jianhua Li. 2018. Drug-target interaction prediction with weighted bayesian ranking. In *Proceedings of the 2nd ICBB*, pages 19–24.

Théo Trouillon, Johannes Welbl, Sebastian Riedel, Éric Gaussier, and Guillaume Bouchard. 2016. Complex embeddings for simple link prediction. In *ICML*, pages 2071–2080. PMLR.

Bishan Yang, Wen tau Yih, Xiaodong He, Jianfeng Gao, and Li Deng. 2014. Embedding entities and relations for learning and inference in knowledge bases. In *ICLR*.

Jun Zhang, Muchun Zhu, Peng Chen, and Bing Wang. 2017. Drugrpe: random projection ensemble approach to drug-target interaction prediction. *Neurocomputing*, 228:256–262.

Shuangjia Zheng, Jiahua Rao, Ying Song, Jixian Zhang, Xianglu Xiao, Evandro Fei Fang, Yuedong Yang, and Zhangming Niu. 2021. Pharmkg: a dedicated knowledge graph benchmark for bomedical data mining. *Briefings in bioinformatics*, 22(4):bbaa344.

Overcoming Data Scarcity in Named Entity Recognition: Synthetic Data Generation with Large Language Models

Tuan An Dao[1,2] **Hiroki Teranishi**[2] **Yuji Matsumoto**[2]
Florian Boudin[3] **Akiko Aizawa**[4,2]
[1]The University of Tokyo, Tokyo, Japan
[2]RIKEN Center for Advanced Intelligence Project (AIP), Tokyo, Japan
[3]JFLI, Nantes Université, France
[4]National Institute of Informatics, Tokyo, Japan
dtan@g.ecc.u-tokyo.ac.jp, {hiroki.teranishi, yuji.matsumoto}@riken.jp,
florian.boudin@univ-nantes.fr, aizawa@nii.ac.jp

Abstract

Named Entity Recognition (NER) is crucial for extracting domain-specific entities from text, particularly in biomedical and chemical fields. Developing high-quality NER models in specialized domains is challenging due to the limited availability of annotated data, with manual annotation being a key method of data construction. However, manual annotation is time-consuming and requires domain expertise, making it difficult in specialized domains. Traditional data augmentation (DA) techniques also rely on annotated data to some extent, further limiting their effectiveness. In this paper, we propose a novel approach to synthetic data generation for NER using large language models (LLMs) to generate sentences based solely on a set of example entities. This method simplifies the augmentation process and is effective even with a limited set of entities. We evaluate our approach using BERT-based models on the BC4CHEMD, BC5CDR, and TDMSci datasets, demonstrating that synthetic data significantly improves model performance and robustness, particularly in low-resource settings. This work provides a scalable solution for enhancing NER in specialized domains, overcoming the limitations of manual annotation and traditional augmentation methods.

1 Introduction

Named Entity Recognition (NER) is a fundamental task in Natural Language Processing (NLP) aiming at identifying and classifying named entities in text. The primary goal of NER is to extract specific entities such as people, organizations, locations, and specialized terms (e.g., chemicals, diseases) from unstructured text. Effective NER is vital in many fields, particularly in the biomedical and chemical domains, where accurate entity recognition supports applications such as drug discovery, literature mining, and patent analysis.

One significant challenge in developing high-quality NER models is the scarcity of annotated data, particularly in specialized domains and low-resource scenarios. Recent advancements in data augmentation for NER have explored diverse strategies to tackle data scarcity, especially in low-resource settings. Techniques range from simple methods like synonym replacement (Dai and Adel, 2020; Sabty et al., 2021; Chen et al., 2021; Yaseen and Langer, 2021; Phan and Nguyen, 2022; Sutiono and Hahn-Powell, 2022) and random noise introduction (Issifu and Ganiz, 2021; Liu et al., 2023) to more complex approaches such as cross-domain transformation and leveraging large language models (LLMs) (Liu et al., 2022; Ye et al., 2024). These methods help to generate additional training examples but often still rely on existing labeled data, which can limit their effectiveness in highly specialized domains where labeled examples are scarce or non-existent.

To overcome these limitations, we propose an approach for synthetic data generation using LLMs that generates sentences based solely on a set of example entities, without relying on pre-existing annotated data. Our method (GenLLM) simplifies the augmentation process by directly generating domain-relevant sentences while ensuring entity correctness and contextual consistency. Unlike traditional techniques, our approach does not depend on manually annotated examples, making it especially valuable for low-resource or highly specialized domains where obtaining labeled data is challenging. By leveraging LLMs, we can produce diverse and contextually appropriate sentences that reflect real-world entity occurrences and relationships. We evaluate the effectiveness of our approach by applying it to NER tasks using BERT-based models on three datasets: the widely used BC4CHEMD (Krallinger et al., 2015) and BC5CDR (Li et al., 2016) datasets, along with the TDMSci (Hou et al., 2021) dataset for task, dataset, and metric entities. Our results show that pretraining on synthetic data generated by LLMs

Proceedings of the 24th Workshop on Biomedical Language Processing (BioNLP 2025), pages 328–340
August 1, 2025

consistently improves model performance, outperforming previous data augmentation methods that combine synthetic data with the original training data in both low and high-resource settings. We explored using only synthetic data generated by LLMs for training, which proved effective in low-resource scenarios. However, human-annotated data yielded better results as the dataset size increased, emphasizing the value of expert annotations in high-resource settings. GenLLM offers a promising data augmentation solution for low-resource domains, particularly when annotated data is limited. The code, generated data, and trained models used in this work are publicly available at `https://github.com/daotuanan/GenLLM_NER`.

2 Related Work

NER relies heavily on high-quality annotated datasets, but in many specialized domains, such as the biomedical and scientific domain, manually labeled data are scarce. To address this issue, synthetic data generation has emerged as an alternative to enhance model performance (Xu et al., 2024). Generating synthetic data for the NER task is challenging because it requires more than just producing natural-sounding sentences; it must also ensure entity correctness, contextual consistency, and domain relevance. Unlike general text generation, NER data must contain entities that are correctly labeled and naturally embedded within the context, reflecting real-world sentence structures.

2.1 Traditional Data Augmentation Methods for NER

Traditional augmentation methods such as synonym replacement, backtranslation, and cross-domain adaptation have been used to enhance NER performance, particularly in low-resource settings (Dai and Adel, 2020; Sabty et al., 2021; Issifu and Ganiz, 2021; Chen et al., 2021; Yaseen and Langer, 2021; Phan and Nguyen, 2022). While these techniques have proven effective, they often struggle to generate highly contextualized and domain-specific entity mentions. For instance, basic methods like synonym replacement and random insertion have shown improvements in biomedical NER (Issifu and Ganiz, 2021), and backtranslation has been particularly effective in low-resource biomedical and materials science domains (Yaseen and Langer, 2021). However, these methods typically fail to capture complex entity structures and contextual dependencies required for domain-specific tasks.

2.2 LLM-Based Approaches to Data Generation

Recent advances in LLM-based synthetic data generation offer a more flexible and scalable alternative. LLMs can generate diverse, contextually rich sentences while preserving entity correctness and domain relevance. For example, prompting strategies have been shown to significantly enhance NER performance in low-resource scenarios, improving F1 scores by over 40% (Liu et al., 2022). Techniques like context similarity-based augmentation (e.g., COSINER) and transformer-based data generation have demonstrated effectiveness in improving NER in both general and specialized domains, such as biomedical texts (Bartolini et al., 2022, 2023; Yili and Haonan, 2023). Moreover, methods like TarGEN employ multi-step prompting and self-correction to generate high-quality synthetic datasets (Gupta et al., 2023). Despite the promise of these approaches, challenges remain in ensuring the scalability and quality of synthetic data, particularly in highly specialized domains like clinical NER (Hiebel et al., 2023). However, a key limitation of these studies is their focus on general rather than specialized domains. The effectiveness of synthetic data and pretraining methods might not translate well to domain-specific applications, such as biomedical or clinical NER.

3 Entity-Based Synthetic Data Generation

Our approach leverages the LLMs to generate synthetic sentences that incorporate specified entities while maintaining contextual consistency. The process consists of three main steps: entity selection 3.1, prompt construction 3.2, and sentence generation 3.3.

3.1 Entity Selection

We begin by selecting a set of seed entities, which serve as the foundation for sentence generation. These entities can be obtained from existing datasets, knowledge bases, or domain-specific lexicons with relatively low effort compared to manually creating fully annotated sentences. The selection ensures that the generated data covers a diverse set of entity mentions necessary for effective NER training. Entities are randomly combined

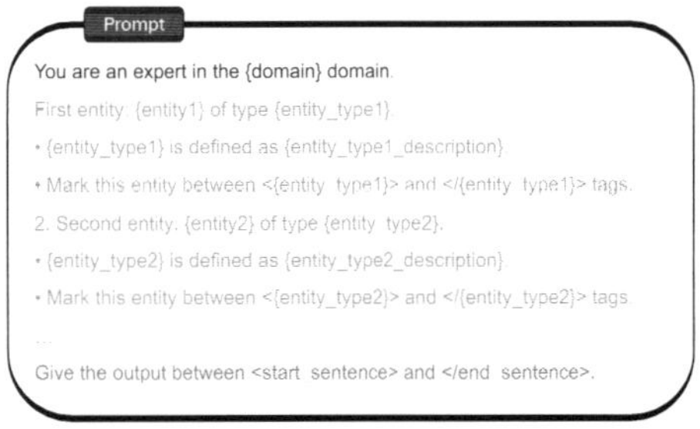

(a) Prompt structure.

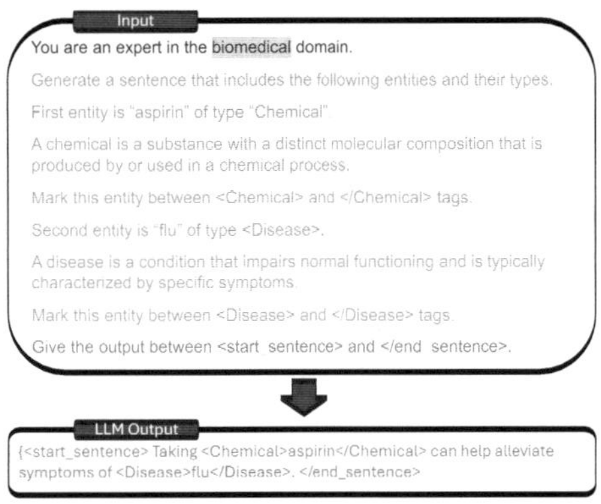

(b) Example prompt instance.

Figure 1: Illustration of the example prompt used for generating synthetic sentences with specified entities and their types. The prompt includes the model's role, task instructions, and output formatting guidelines.

from different categories, with each sentence containing one to three entities. For example, from the categories CHEM (aspirin, lithium) and DISEASE (lung carcinoma, flu), a possible combination could be: aspirin, flu.

3.2 Prompt Construction

To generate high-quality synthetic sentences for NER, we design a structured prompt that ensures the inclusion of specific entities while maintaining contextual coherence. Our prompt explicitly defines the domain, entity types, and entity annotations to improve generation accuracy and reduce annotation errors.

The prompt follows a template-based format that guides the language model to generate a sentence containing specified entities with correct annotations. It consists of the following key components:

- Domain Specification: The model is instructed to act as an expert in a specific domain (e.g., biomedical sciences) to ensure domain-relevant sentence generation.
- Entity Introduction and Definition: Each target entity is explicitly listed along with its type and a brief description of that type. This helps the model understand the contextual role of the entity.
- Entity Annotation Instructions: The prompt explicitly instructs the model to enclose entities within predefined tags, ensuring clear entity labeling in the generated sentence.
- Output Formatting: The generated sentence is enclosed within <start_sentence> and </end_sentence> tags to facilitate automatic extraction and processing.

This prompt serves as the foundation for generating synthetic NER training data, ensuring both entity correctness and contextual consistency in the generated sentences. The prompt template is found in Figure 1a. An example of this prompt format is illustrated in Figure 1b, demonstrating how contextual cues and entity definitions improve generation accuracy. This prompt format can be used with any popular LLM for generating synthetic data.

3.3 Sentence Generation

We use the **LLaMA-3.2-3B-Instruct**[1] model to generate synthetic sentences containing specified named entities. This model was selected for its balance between generation quality and computational efficiency. Unlike prior work that relies on proprietary and resource-intensive models such as GPT-4 or GPT-4o (Ye et al., 2024), our approach uses an *open-source*, lightweight model that is more accessible and cost-effective, making it better suited for reproducible research and large-scale generation in constrained environments. Once the LLM processes the prompt, it generates a synthetic sentence where the specified entities are correctly embedded within a natural linguistic context. To maintain consistency and avoid introducing unintended entities, we post-process the output by verifying entity correctness and ensuring compliance with the annotation format. To ensure that the generated output adheres to the required annotation format, we apply the following post-processing steps:

[1] https://huggingface.co/meta-llama/Llama-3.2-3B-Instruct

- **Tag Validation**: We verify that all entity tags are correctly opened and closed. Each entity must be enclosed within its respective <entity_type> and </entity_type> tags to maintain proper annotation structure.

- **Sentence Formatting**: We confirm that the entire sentence is enclosed within <start_sentence> and </end_sentence> tags. This ensures the output remains structured and easily extractable for further processing.

By enforcing these constraints, we ensure consistency in the synthetic data before it is used for training. After validating the output format, we convert the annotated entities into the BIO (Begin-Inside-Outside) tagging scheme. Each token in the sentence is assigned a label based on its entity type. This transformation ensures compatibility with standard NER training pipelines.

The synthetic data is then incorporated into the training set through pretraining, where the model is first trained on the synthetic data before being fine-tuned on gold-standard annotated data. This approach helps the model learn general entity patterns from the generated data, improving performance, especially in low-resource or specialized domains.

4 Experiment with Low-resource Setting

In this experiment, we explore the performance of our method (GenLLM) in a low-resource setting, where only a limited amount of manually annotated data is available. The aim is to evaluate whether our method can outperform or complement other state-of-the-art systems, such as LSMS(Dai and Adel, 2020), LLM-DA(Ye et al., 2024), and NuNER(Bogdanov et al., 2024), when trained with a small amount of labeled data. We also investigate how different data augmentation strategies and pretraining methods impact the model's ability to generalize to unseen examples.

4.1 Experimental Setup

4.1.1 Dataset Construction

We conduct experiments using 3 datasets: BC4CHEMD, BC5CDR, and TDMSci. Since we use seed entities as the main input for the augmentation process, it is important to note that obtaining a large set of seed entities in real-world applications can be difficult, particularly in specialized domains where annotated data is scarce. As a result, working with a smaller, more manageable set of seed entities is often necessary. Our method, which only uses seed entities for the augmentation process, is designed to be effective even with this limitation. In contrast, other methods like LSMS and LLM-DA rely on gold-label data as input for augmentation. We create a "Limited Dictionary" setting to compare our method with these alternative approaches.

To construct the seed sets used for augmentation, we select the most frequent entities from the training data for each entity type. For each dataset, we define multiple settings with different values of N (e.g., $N = 5, 10, 15, 20, 50$), where N denotes the number of unique entities per type. The selection process involves counting and ranking entities by frequency, then selecting the top N for each type. We also ensure type balance by including an equal number of sentences for each entity type (e.g., equal numbers for CHEMICAL and DISEASE in BC5CDR).

This choice of using frequent entities—rather than randomly sampling or relying on external lexicons—is motivated by both practical and methodological reasons. First, frequent entities are more likely to appear in natural, contextually appropriate sentences, resulting in higher-quality and more realistic generated data. Second, using a fixed set of frequent entities leads to a more stable and reproducible experimental setup. In contrast, random sampling introduces variability and typically requires multiple runs to obtain robust estimates. Similarly, depending on external lexicons may introduce domain mismatches or lead to unnatural entity combinations. By relying on the internal statistics of the training corpus, we ensure that the selected entities are representative of the target domain and the actual model training distribution.

4.1.2 Comparison Methods

We consider the following baseline methods for comparison purposes:

- **Original (org)**: Training directly on the full dataset without any augmentation.

- **LSMS**: Applying lexical-based sampling and substitution strategies, including replace-mention (RM), replace-token (RT), shuffle-within-segments (SWS), and synonym-replacement (SR).

Dictionary Size	5	10	15	20	50
org	4.68	12.35	15.62	31.79	47.18
LSMS	32.85	**46.03**	44.35	**48.34**	57.76
LLM-DA	**39.89**	43.17	**45.75**	46.29	49.68
NuNER	21.24	28.29	40.11	44.31	52.49
Ours (GenLLM)	26.06	37.18	41.85	43.67	**58.80**

Table 1: Performance comparison on the BC4CHEMD dataset across different dictionary sizes (N).

Dictionary Size	5	10	15	20	50
org	45.62	51.87	51.81	51.73	54.83
LSMS	51.28	57.19	60.66	60.97	68.42
LLM-DA	52.29	57.72	60.94	64.12	66.79
NuNER	40.70	43.45	50.17	50.86	46.87
Ours (GenLLM)	**53.67**	**60.14**	**63.65**	**65.34**	**72.85**

Table 2: Performance comparison on the BC5CDR dataset across different dictionary sizes (N).

Dictionary Size	5	10	15	20	50
org	17.02	23.21	26.59	26.90	42.77
LSMS	**28.18**	32.81	39.71	41.82	48.28
LLM-DA	17.28	27.12	33.92	37.21	39.90
NuNER	10.37	17.88	11.56	11.76	22.78
Ours (GenLLM)	25.64	**35.05**	**41.20**	**45.60**	**51.23**

Table 3: Performance comparison on the TDMSci dataset across different dictionary sizes (N).

- **LLM-DA**: Utilizing large language model-based data augmentation at both the context and entity levels, with noise injection.
- **NuNER**: Fine-tuning the pretrained NuNER-v2.0 model on the gold annotations of the datasets.[2]

For all methods except NuNER, we use BERT-base-uncased as the base model. LSMS and LLM-DA are trained for 10 epochs on the combination of original training data and augmented data. NuNER is fine-tuned for 10 epochs on gold data.

4.1.3 Proposed Method: GenLLM and Training Setup

Our proposed method, **GenLLM**, generates synthetic training data using LLM-based augmentation techniques. It employs prompt engineering with constraints to ensure data quality and entity control. Training follows a two-stage approach: we first pretrain the model on synthetic data for 3 epochs, then fine-tune on the gold-annotated data for 10 epochs. All experiments are conducted under reduced labeled data settings ($N = 5, 10, 15, 20, 50$ entities per type), simulating low-resource environments. We compare GenLLM's performance against the baselines introduced in Section 4.1.2.

Additional implementation details, including training hyperparameters and hardware specifications, are provided in Appendix A.1.

4.2 Results and Analysis

The performance comparison across different methods on the BC4CHEMD, BC5CDR, and TDMSci datasets is shown in Tables 1, 2, and 3, respectively.

Our method (GenLLM) consistently outperforms the org and NuNER, with significant improvements. On BC5CDR, GenLLM achieves the highest performance at all dictionary sizes, outperforming both LSMS and LLM-DA. On TDMSci, GenLLM shows strong performance, competing well with LSMS and LLM-DA, only losing to LSMS when N = 5. The low performance of GenLLM at smaller dictionary sizes on the BC4CHEMD dataset is likely due to the limited diversity and insufficient augmentation with only a few seed entities, which restricts the model's ability to generalize effectively. As the dictionary size increases, the synthetic data improves, leading to better performance. Overall, our method outperforms previous methods like LSMS and LLM-DA, offering a robust solution for low-resource settings by leveraging synthetic data generation for better generalization.

5 Experiment with High-Resource Setting

In this section, we evaluate our proposed method in a high-resource setting, where we utilize the full training data from three benchmark datasets: BC4CHEMD, BC5CDR, and TDMSci. This setting allows us to assess the performance of our approach when abundant annotated data is available, providing a direct comparison with conventional methods that rely on manually annotated corpora.

5.1 Experimental Setup

We conduct experiments using the full training datasets of BC4CHEMD, BC5CDR, and TDMSci. The models are trained using the standard dataset splits provided in prior studies to ensure comparability. All models are trained for 3 epochs. LSMS and LLM-DA also use the combination of original training data and augmented data generated by these methods. For our method (GenLLM), we

[2] https://huggingface.co/numind/NuNER-v2.0

Dataset	BC4CHEMD	BC5CDR	TDMSci
Org	**87.19**	83.27	55.19
LSMS	86.58	**84.04**	58.32
LLM-DA	86.58	82.40	52.79
NuNER	85.88	81.10	48.19
Ours (GenLLM)	86.85	83.74	**58.70**

Table 4: Performance (F1-score) comparison across different methods on the BC4CHEMD, BC5CDR, and TDMSci datasets on high-resource setting.

use the "pretraining" approach, first fine-tuning the model on synthetic data for 1 epoch, followed by fine-tuning on gold data for 3 epochs. For synthetic data, due to the cost of generating additional data, we reuse the data generated in the low-resource setting and combine the generated data of all sizes from that setting. Additional implementation details, including training hyperparameters and hardware specifications, are provided in Appendix A.1.

5.2 Results and Analysis

Table 4 presents the F1-score performance of various methods on the BC4CHEMD, BC5CDR, and TDMSci datasets in a high-resource setting. The experimental results in a high-resource setting show that different methods exhibit varying effectiveness across datasets. On BC4CHEMD, without augmentation (org) outperforms all other methods with an F1-score of 87.19, followed closely by GenLLM (86.85). LSMS and LLM-DA show similar performance, while NuNER lags slightly behind. On BC5CDR, LSMS achieves the highest F1-score (84.04), with GenLLM coming second (83.74), slightly outperforming LLM-DA and NuNER. GenLLM generally performs competitively or better than other methods in high-resource settings, with the best performance on TDMSci and close results on BC4CHEMD and BC5CDR. It becomes much harder to significantly improve performance with augmentation when the training data size is large, as seen in the BC4CHEMD and BC5CDR datasets.

6 Analysis and Discussion

6.1 Quality of Synthetic Data

In this experiment, we investigate whether a model can be effectively trained using only synthetic data generated by LLMs, without any manually annotated data. The primary objective is to assess the feasibility of LLM-generated sentences as a standalone training resource in specialized domains.

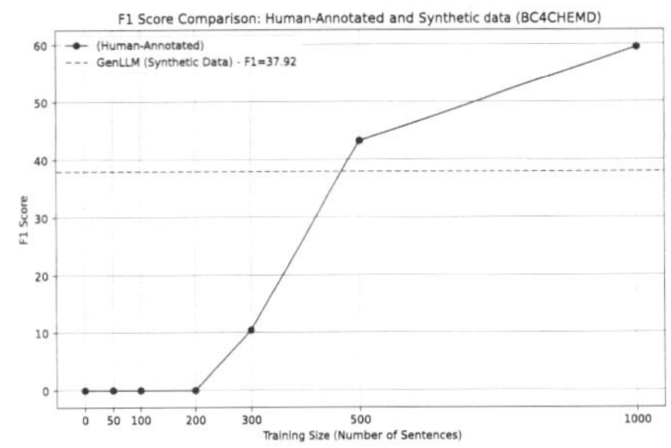

(a) BC4CHEMD dataset.

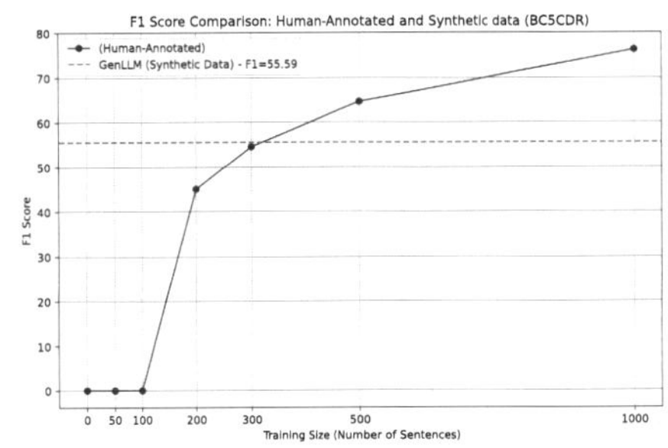

(b) BC5CDR dataset.

Figure 2: F1 Score Comparison between BERT-base-uncased trained on human-annotated data and trained on synthetic data varying training sizes for the BC4CHEMD and BC5CDR datasets.

While synthetic data provides diversity, it may introduce hallucinated entities, ambiguous contexts, or annotation errors, leading to noisy supervision. Additionally, LLMs, trained on general-domain corpora, may struggle with domain-specific terminology, impacting performance.

6.1.1 Data Sampling

Human-Annotated Setting: In this setting, we randomly select gold sentences from the training data of each dataset (BC4CHEMD and BC5CDR). These sentences are manually annotated and serve as the ground truth for model training.

GenLLM Setting: For the GenLLM approach, we provide the model with a list of 10 entities from each type in the dataset (e.g., chemicals, diseases) and instruct it to generate 1000 synthetic sentences. The goal is to use these synthetic sentences to train the model in the absence of human-annotated data. The models are then evaluated on the full test data of each dataset to assess their performance.

Findings Figures 2a and 2b help to understand the effectiveness of using LLM-generated synthetic data for training NER models and compare its per-

Dictionary Size	5	10	15	20	50
Simple Prompt	24.66	**35.05**	40.45	**45.60**	51.23
+ filter	13.00	22.00	27.52	24.97	30.45
+ no-new-entity	**25.64**	32.92	**41.20**	43.65	50.37
+ COT	21.68	26.98	38.34	41.01	**51.59**

Table 5: F1 Score Comparison of Different Prompting Methods Across Training Sizes on TDMSci Test set.

Error Type	Count
False Negative (Missing Entity)	54
False Positive (Spurious Entity)	2
Boundary Misalignment	12

Table 6: Error analysis of 100 manually checked TDM-Sci samples.

formance to using human-annotated data. When fewer than 300-400 sentences are annotated, the synthetic data approach (from GenLLM) yields better performance. This suggests that synthetic data might be more effective in low-resource scenarios where manual annotation is costly or time-consuming, and small training sets are available. However, as the annotated data size grows beyond this point, human-annotated data consistently provides better results.

6.2 Different Prompting Methods

In this ablation study, we examine how different prompting strategies influence the quality of the generated synthetic data and the performance of the trained NER model. We evaluate the following four prompting methods:

- **Simple Prompt**: The model is provided with a plain list of entity names and their types, without any additional constraints or filtering (Figure 1a).
- **Simple Prompt + filter**: In this approach, we filter out generated sentences that introduce new entities not present in the seed list. This aims to ensure that only relevant entities appear in the synthetic data, reducing entity drift.
- **Simple Prompt + no-new-entity**: The prompt explicitly instructs the model to avoid introducing new entities beyond the provided list (Figure 3).
- **Simple Prompt + COT (Chain-of-Thought)**: The model is guided to generate sentences step-by-step, ensuring logical coherence and correct entity usage (Figure 4).

Table 5 presents the F1 scores for different prompting methods across various training sizes. The Simple Prompt baseline demonstrates strong performance, particularly at 10 and 20 training examples, where it achieves the highest scores (35.05 and 45.60, respectively). However, adding a filtering mechanism to remove sentences introducing new entities significantly reduces performance across all training sizes. This suggests that while filtering ensures strict entity control, it may also remove valuable diverse contexts that contribute to learning. The no-new-entity constraint, which instructs the LLM not to introduce unseen entities during generation, performs well in low-resource settings (5 and 15 examples), surpassing the Simple Prompt in these cases. The Chain-of-Thought (COT) prompting does not outperform the Simple Prompt in all training scenarios. It achieves its highest score (51.59) at 50 examples, which is slightly higher than the Simple Prompt's 51.23. These results highlight the trade-offs between entity control, data diversity, and reasoning-driven generation in synthetic data creation for NER.

6.3 Error Analysis

In the process of using LLMs for tasks such as NER and data generation, three common types of errors may arise: **False Negatives**, **False Positives**, and **Boundary Misalignment**. Understanding these errors is crucial for improving the accuracy and reliability of the generated sentences.

- **False Negatives (Missing Entities)** These occur when valid entities present in the sentence are not recognized or labeled by the model, resulting in under-annotation and potential loss of critical information.
- **False Positives (Spurious Entities)**
  In these cases, the model incorrectly identifies and labels non-entity spans as entities, introducing noise into the training data.
- **Boundary Misalignment**
  This error arises when the model detects the correct entity type but assigns incorrect boundaries—either extending beyond or falling short of the true entity span.

To assess the quality of LLM-generated data, we manually analyzed 100 synthetic samples from the TDMSci dataset. The results of this error analysis are summarized in Table 6. The most frequent issue was **False Negatives**, which occurred in 54 cases. These typically happened because the LLM introduced new entities not in the input seed list, which distracted it from labeling the original seed entities. **Boundary Misalignment** was observed in 12 cases. This is often caused by inconsistencies in annotation guidelines, such as whether to include the word "dataset" or "task" in the entity span for DATASET and TASK types. **False Positives** occurred in only 2 cases. They often resulted from noisy or questionable input annotations. For instance, in the TDMSci dataset, a seed entity like "Arabic-English" is labeled as a TASK, although its status as a task is ambiguous, potentially confusing the model into tagging unrelated terms.

Table 7 shows example outputs from LLM-based generation using seed entities from the biomedical domain, specifically the BC5CDR dataset. We manually identified three main types of annotation errors in these generated sentences. First, **False Negatives** occur when entities from the seed list are present in the generated text but are not labeled—such as "cocaine" and "tachycardia" in the first example. This often happens when the LLM adds new, related entities (e.g., "arrhythmia"), which distracts the model from tagging the original seed entities. Second, **False Positives** arise when entities not included in the input seed list are introduced and labeled in the output—such as "phenytoin" and "temporal lobe epilepsy." These hallucinated entities may be contextually relevant but were not intended to appear or be annotated in the generated data. This is especially problematic when using the generated sentences as training data in a low-resource setting, where tight control over entity inclusion is critical. Lastly, **Boundary Misalignment** is illustrated by the example where only "insulin" is labeled instead of the correct full span "insulin resistance." This type of span-level error can be caused by the LLM misunderstanding multi-word entity boundaries or inconsistencies in annotation granularity during training. These examples highlight how generation errors can propagate into downstream tasks if not properly filtered, and emphasize the importance of post-processing and validation when using LLMs for data augmentation in biomedical NER.

Seed Entities (Types) and Generated Sentence with Error Type
Seed Entities: tachycardia (Disease), cocaine (Chemical) **Generated Sentence:** Chronic cocaine use has been associated with an increased risk of developing tachycardia, a type of <Disease>arrhythmia</Disease>, which can lead to a range of cardiovascular complications, including myocardial infarction and stroke. *(**False Negatives**: "cocaine" and "tachycardia" are not labeled. **False Positive**: "arrhythmia" is introduced and labeled as Disease but not in the seed list)*
Seed Entities: lidocaine (Chemical) **Generated Sentence:** The administration of <Chemical>lidocaine</Chemical> has been studied for its potential to reduce <Disease>seizure frequency</Disease> in patients with <Disease>epilepsy</Disease>, and its efficacy in this context has been demonstrated through various clinical trials, including a randomized controlled trial that compared the outcomes of patients receiving <Chemical>lidocaine</Chemical> versus those receiving <Chemical>phenytoin</Chemical> for the treatment of <Disease>partial seizures</Disease> in adults with <Disease>temporal lobe epilepsy</Disease>. *(**False Positives**: "phenytoin", "seizure frequency", "epilepsy", "partial seizures", "temporal lobe epilepsy" are not in the seed list but labeled. **False Negative**: "lidocaine" is correctly labeled, others are missing from seed list)*
Seed Entities: DSM - IV bipolar I (Disease), cisplatin (Chemical), nicotine (Chemical) **Generated Sentence:** Research on the efficacy and safety of <Disease>DSM-IV bipolar I</Disease> (<Disease>bipolar disorder</Disease>) has shown that patients with this condition often require <Disease>medication</Disease> management, including the use of <Chemical>cisplatin</Chemical> and <Chemical>nicotine</Chemical> as part of their treatment regimen, which can lead to potential <Disease>side effects</Disease> such as nausea, vomiting, and increased heart rate. *(**False Positives**: "bipolar disorder", "medication", and "side effects" are not in the seed list but labeled. Also reflects questionable factual correctness)*
Seed Entities: insulin resistance (Disease) **Generated Sentence:** Patients with <Disease>insulin</Disease> resistance often exhibit symptoms of metabolic syndrome, including obesity and hypertension. *(**Boundary Misalignment**: Only "insulin" is labeled, while the full entity should be "insulin resistance")*

Table 7: Examples of generated sentences with seed entities from BC5CDR (Chemical, Disease), along with identified error types: False Negatives (missing entities), False Positives (spurious entities), and Boundary Misalignment (incorrect span).

7 Conclusion

In this study, we proposed GenLLM, a method for data augmentation in low-resource settings using large language models to generate synthetic sentences containing specific named entities. Our experiments demonstrated that GenLLM effectively enhances performance compared to baseline models like LSMS, LLM-DA, and NuNER when limited labeled data is available. By leveraging synthetic data generation with only seed entities, Gen-

LLM outperforms or complements state-of-the-art systems, especially in scenarios with constrained resources Furthermore, we explored the feasibility of training models using only synthetic data generated by LLMs, which proved to be effective in low-resource scenarios. However, human-annotated data still provided better results once the dataset size grew large enough, highlighting the importance of expert-annotated data in high-resource settings. GenLLM offers a promising solution for data augmentation in low-resource domains, particularly when manually annotated data is scarce. Future work can focus on further improving synthetic data quality and exploring additional augmentation strategies to enhance model generalization in diverse domains.

Limitations

One potential limitation of this paper is that the quality of synthetic data generated by large language models (LLMs) may be inconsistent, potentially impacting model performance. To mitigate this, we ran each experiment three times and report the averaged results to ensure the robustness and generalizability of our findings. Additionally, this study focus on scienctific domains such as biomedical, chemical, and computer science, which may not generalize to other fields.

Ethics Statement

This research adheres to ethical guidelines and practices throughout its execution. The datasets utilized in this study are publicly available and do not contain personally identifiable information. This paper was proofread with the assistance of OpenAI's GPT-4 language model to improve clarity and grammar. All substantive content and arguments are the author's own.

Acknowledgement

This work was supported by the JSPS KAKENHI Grant Number 24K03231 and RIKEN AIP.

References

Ilaria Bartolini, Vincenzo Moscato, Marco Postiglione, Giancarlo Sperlì, and Andrea Vignali. 2022. Cosiner: Context similarity data augmentation for named entity recognition. In *International Conference on Similarity Search and Applications*, pages 11–24. Springer.

Ilaria Bartolini, Vincenzo Moscato, Marco Postiglione, Giancarlo Sperlì, and Andrea Vignali. 2023. Data augmentation via context similarity: An application to biomedical named entity recognition. *Information Systems*, 119:102291.

Sergei Bogdanov, Alexandre Constantin, Timothée Bernard, Benoit Crabbé, and Etienne Bernard. 2024. Nuner: Entity recognition encoder pre-training via llm-annotated data. In *Proceedings of the 2024 Conference on Empirical Methods in Natural Language Processing*, pages 11829–11841.

Shuguang Chen, Gustavo Aguilar, Leonardo Neves, and Thamar Solorio. 2021. Data augmentation for cross-domain named entity recognition. In *Proceedings of the 2021 Conference on Empirical Methods in Natural Language Processing*, pages 5346–5356.

Xiang Dai and Heike Adel. 2020. An analysis of simple data augmentation for named entity recognition. *arXiv preprint arXiv:2010.11683*.

Himanshu Gupta, Kevin Scaria, Ujjwala Anantheswaran, Shreyas Verma, Mihir Parmar, Saurabh Arjun Sawant, Swaroop Mishra, and Chitta Baral. 2023. Targen: Targeted data generation with large language models. *arXiv preprint arXiv:2310.17876*.

Nicolas Hiebel, Olivier Ferret, Karën Fort, and Aurélie Névéol. 2023. Can synthetic text help clinical named entity recognition? a study of electronic health records in french. In *The 17th Conference of the European Chapter of the Association for Computational Linguistics (EACL 2023)*.

Yufang Hou, Charles Jochim, Martin Gleize, Francesca Bonin, and Debasis Ganguly. 2021. Tdmsci: A specialized corpus for scientific literature entity tagging of tasks datasets and metrics. In *Proceedings of the 16th Conference of the European Chapter of the Association for Computational Linguistics: Main Volume*, pages 707–714.

Abdul Majeed Issifu and Murat Can Ganiz. 2021. A simple data augmentation method to improve the performance of named entity recognition models in medical domain. In *2021 6th International Conference on Computer Science and Engineering (UBMK)*, pages 763–768. IEEE.

Martin Krallinger, Obdulia Rabal, Florian Leitner, Miguel Vazquez, David Salgado, Zhiyong Lu, Robert Leaman, Yanan Lu, Donghong Ji, Daniel M Lowe, et al. 2015. The chemdner corpus of chemicals and drugs and its annotation principles. *Journal of cheminformatics*, 7(1):1–17.

Jiao Li, Yueping Sun, Robin J Johnson, Daniela Sciaky, Chih-Hsuan Wei, Robert Leaman, Allan Peter Davis, Carolyn J Mattingly, Thomas C Wiegers, and Zhiyong Lu. 2016. Biocreative v cdr task corpus: a resource for chemical disease relation extraction. *Database*, 2016.

Jian Liu, Yufeng Chen, and Jinan Xu. 2022. Low-resource ner by data augmentation with prompting. In *IJCAI*, pages 4252–4258.

Jiguo Liu, Chao Liu, Nan Li, Shihao Gao, Mingqi Liu, and Dali Zhu. 2023. Lada-trans-ner: adaptive efficient transformer for chinese named entity recognition using lexicon-attention and data-augmentation. In *Proceedings of the AAAI Conference on Artificial Intelligence*, volume 37, pages 13236–13245.

Uyen Phan and Nhung Nguyen. 2022. Simple semantic-based data augmentation for named entity recognition in biomedical texts. In *Proceedings of the 21st Workshop on Biomedical Language Processing*, pages 123–129.

Caroline Sabty, Islam Omar, Fady Wasfalla, Mohamed Islam, and Slim Abdennadher. 2021. Data augmentation techniques on arabic data for named entity recognition. *Procedia Computer Science*, 189:292–299.

Arie Pratama Sutiono and Gus Hahn-Powell. 2022. Syntax-driven data augmentation for named entity recognition. *Proceedings of Pattern-based Approaches to NLP in the Age of Deep Learning*, page 56.

Derong Xu, Wei Chen, Wenjun Peng, Chao Zhang, Tong Xu, Xiangyu Zhao, Xian Wu, Yefeng Zheng, Yang Wang, and Enhong Chen. 2024. Large language models for generative information extraction: A survey. *Frontiers of Computer Science*, 18(6):186357.

Usama Yaseen and Stefan Langer. 2021. Data augmentation for low-resource named entity recognition using backtranslation. In *Proceedings of the 18th International Conference on Natural Language Processing (ICON)*, pages 352–358.

Junjie Ye, Nuo Xu, Yikun Wang, Jie Zhou, Qi Zhang, Tao Gui, and Xuanjing Huang. 2024. Llm-da: Data augmentation via large language models for few-shot named entity recognition. *arXiv preprint arXiv:2402.14568*.

Qian Yili and Xu Haonan. 2023. Datg: Data augmentation with transformer-based generation for low-resource named entity recognition. In *2023 China Automation Congress (CAC)*, pages 6188–6193. IEEE.

A Appendix

A.1 Experimental Setting Details

Base NER model We employ a fine-tuned BERT model for NER. The input sequences are first tokenized and then passed through BERT to obtain contextualized embeddings. These embeddings are fed into a linear classification layer followed by a softmax activation to predict the entity type of each token. For words that are split into multiple subwords during tokenization, only the embedding of the first subword is used for classification.

```
Prompt: no-new-entity
You are an expert in the {domain} domain.
First entity: {entity1} of type {entity_type1}.
• {entity_type1} is defined as {entity_type1_description}.
• Mark this entity between <{entity_type1}> and </{entity_type1}> tags.
Second entity: {entity2} of type {entity_type2}.
• {entity_type2} is defined as {entity_type2_description}.
• Mark this entity between <{entity_type2}> and </{entity_type2}> tags.
...
Do not introduce any new entities or types in the sentence.
Give the output between <start_sentence> and </end_sentence
```

Figure 3: Example of the **Simple Prompt + no-new-entity** setup, where the prompt explicitly instructs the model to generate a sentence using only the provided entities and avoid introducing any new entities.

Hyperparameters. For all experiments, we use the following settings unless otherwise specified:

- **Learning rate:** 1e-4
- **Batch size:** 32
- **Optimizer:** AdamW
- **Max sequence length:** 256
- **Dropout rate:** 0.1
- **Weight decay:** 0.01

All models are implemented using the HuggingFace Transformers library. To ensure reproducibility, we fix the random seed to 42 across all components including NumPy, PyTorch, and HuggingFace Transformers. Training is conducted on a single NVIDIA V100 GPU with 32 GB of memory. Each run (including pretraining and fine-tuning steps) takes approximately 30–90 minutes depending on the dataset and the size of the training set.

A.2 Prompts

A.2.1 Prompt: no-new-entity

This prompt is a controlled variation of the **simple prompt**, extended with an explicit instruction: "`Do not introduce any new entities or types in the sentence.`" This modification aims to address a common issue in LLM-based data generation—**false negatives (missing entities)**—where the model may omit entities from the provided seed list or introduce incorrect ones, resulting in incomplete or misaligned annotations. By enforcing this constraint, we improve the alignment between the

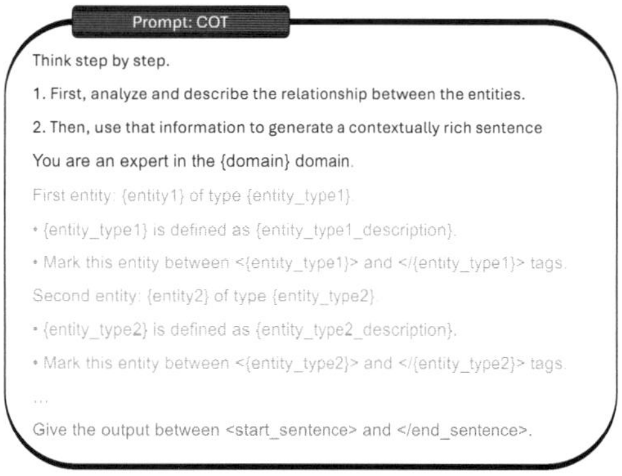

Figure 4: Example of the **Simple Prompt + COT (Chain-of-Thought)** setup, where the model is guided to generate the sentence in a step-by-step manner. This approach promotes logical coherence and helps ensure that the provided entities are used correctly in context.

prompt specification and the generated content, ensuring better coverage and fidelity to the intended entity set. An illustration of this prompt configuration is shown in Figure 3.

A.2.2 Prompt: COT

Another variation is the **Simple Prompt + Chain-of-Thought (CoT)** setup, where the model is guided to reason step-by-step before producing the final sentence. This format encourages logical coherence and helps the model better understand and place the given entities in context. The intermediate reasoning steps can reduce annotation mistakes and improve entity boundary accuracy. An example of this prompt structure is shown in Figure 4.

A.3 Datasets

We conduct experiments using three benchmark datasets for biomedical and scientific NER:

BC4CHEMD The BC4CHEMD dataset focuses on chemical entity recognition and is derived from biomedical abstracts. It contains over 30,000 sentences and nearly 900,000 tokens (see Table 11). The dataset features one entity type (CHEM), with 29,478 annotated chemical entities distributed across 14,529 sentences (see Table 12).

BC5CDR BC5CDR includes annotations for both chemical and disease entities, making it suitable for multi-type NER tasks. It comprises 4,560 sentences, with an average of 2.06 entities per sentence. Table 8 shows how entity coverage increases with larger subsets of annotated data, and general dataset statistics are shown in Table 11. Additional details on total entities and sentence coverage per type are listed in Table 12.

TDMSci TDMSci is a scientific NER dataset that includes three entity types: Task, Dataset, and Metric. It contains 1,523 sentences and is more diverse than the biomedical datasets in terms of entity types and structure (see Table 10 and Table 11). Table 12 further breaks down the number of entities and sentence distributions per type.

To simulate low-resource conditions, we create reduced versions of each dataset by limiting the number of unique entities used for training. These settings vary from 5 to 500 entities per type, as detailed in Tables 8, 9, and 10. These subsets are used in conjunction with our "Limited Dictionary" setup to test the effectiveness of data augmentation strategies.

Dataset Size	Chemical	Disease
5	19	20
10	36	41
15	54	60
20	73	72
50	192	178
100	364	340
200	736	666
300	1146	985
400	1516	1336
500	1868	1665

Table 8: Entity counts per entity type for BC5CDR dataset.

Dataset Size	CHEM
5	16
10	29
15	41
20	57
50	160
100	344
200	669
300	972
400	1285
500	1618

Table 9: Entity counts per entity type for CHEMDNER dataset.

Dataset Size	DATASET	METRIC	TASK
5	6	13	14
10	17	27	28
15	24	43	43
20	31	60	59
50	103	155	144
100	198	307	292
200	400	553	539
300	591	619	805
400	700	670	1056
500	732	681	1207

Table 10: Entity counts per entity type for TDMSci dataset.

Table 11: Dataset Statistics for NER Tasks

Dataset	#Sentences	#Tokens	#Entity Types	Avg. Entities/Sent.	#Sent. w/o Entities
BC4CHEMD	30,812	872,932	1 (CHEM)	0.96	16,283
BC5CDR	4,560	118,170	2 (Chemical, Disease)	2.06	753
TDMSci	1,523	49,460	3 (TASK, DATASET, METRIC)	1.43	330

Table 12: Entity-Specific Statistics

Dataset	Entity Type	#Entities	#Sentences w/ Entities
BC4CHEMD	CHEM	29,478	14,529
BC5CDR	Chemical	5,203	2,951
	Disease	4,182	2,658
TDMSci	TASK	1,219	920
	DATASET	420	322
	METRIC	536	358

PetEVAL: A veterinary free text electronic health records benchmark

Sean Farrell[1], Alan Radford[2], Noura Al Moubayed[1], Peter-John Mäntylä Noble[2],

[1] Department of Computer Science, Durham University
[2] Institute of Infection, Veterinary and Ecological Sciences, University of Liverpool
Correspondence: sean.farrell2@durham.ac.uk

Abstract

We introduce PetEVAL, the first benchmark dataset derived from real-world, free-text veterinary electronic health records (EHRs). PetEVAL comprises 17,600 professionally annotated EHRs from first-opinion veterinary practices across the UK, partitioned into training (11,000), evaluation (1,600), and test (5,000) sets with distinct clinic distributions to assess model generalisability. Each record is annotated with International Classification of Disease 11 (ICD-11) syndromic chapter labels (20,408 labels), disease Named Entity Recognition (NER) tags (429 labels), and anonymisation NER tags (8,244 labels). PetEVAL enables evaluating Natural Language Processing (NLP) tools across applications, including syndrome surveillance and disease outbreak detection. We implement a multistage anonymisation protocol, replacing identifiable information with clinically relevant pseudonyms while establishing the first definition of identifiers in veterinary free text. PetEVAL introduces three core tasks: syndromic classification, disease entity recognition, and anonymisation. We provide baseline results using BERT-base, PetBERT, and LLaMA 3.1 8B generative models. Our experiments demonstrate the unique challenges of veterinary text, showcasing the importance of domain-specific approaches. By fostering advancements in veterinary informatics and epidemiology, we envision PetEVAL catalysing innovations in veterinary care, animal health, and comparative biomedical research through access to real-world, annotated veterinary clinical data.

```
[
    {
        "savsnet_id": 1111025,
        "text": "Brought in with Coco who has
          conjunctivitis; fluo neg. no blepharospasm or
          rubbing.otherwise nad. Adv monitor, ini send to Smith
          Referrals",
        "icd_11_chapter": "Diseases of the visual system",
        "disease_ner_entities":[[12,36,"Conjunctivitis"]],
        "anonymisation ner entities": [[16,20,"PER"],
          [123,138,"LOC"]]
    }
]
```

Figure 1: Example data for a single consult with a unique consult, the free text clinical EHR, the ICD-11 chapter multi-label classification and NER entities for both anonymisation and disease extraction task

1 Introduction

The growing availability of veterinary electronic health records (vEHRs) from sources such as the Small Animal Veterinary Surveillance Network (SAVSNET) (Sánchez-Vizcaíno et al., 2015), Companion Animal Veterinary Surveillance Network (CAVSNET) (Sheng et al., 2022), and VetCompass (Royal Veterinary College (RVC); McGreevy et al., 2017) presents an unprecedented opportunity to advance veterinary medicine. These datasets support disease surveillance, epidemiological research, and clinical decision-making (Farrell et al., 2023b; Bode et al., 2022; Radford et al., 2011; Sánchez-Vizcaíno et al., 2017; Singleton et al., 2020). However, vEHRs differ from human biomedical records in syntax, lexicon, and clinical expression (Davies et al., 2024b), requiring adaptation of existing computational tools. Additionally, first-opinion vEHRs often contain diagnostic uncertainty due to limited specialist access, resource constraints, and financial considerations (Robinson et al., 2016).

Despite these challenges, vEHRs offer unique advantages for biomedical research. Unlike human records, which are tightly regulated under laws such as HIPAA and GDPR, vEHRs face fewer legal constraints (Sun et al., 2020), making them a viable test bed for developing analytical methods. Their relative accessibility enables researchers to explore novel computational approaches without the ethical and regulatory barriers associated with human health data (Kol et al., 2015; Starkey et al.,

Proceedings of the 24th Workshop on Biomedical Language Processing (BioNLP 2025), pages 341–353
August 1, 2025 ©2025 Association for Computational Linguistics

2005; Trott et al., 2004).

Advancing natural language processing (NLP) for vEHRs is critical for global health, supporting the World Health Organisation's (WHO) One Health initiatives in zoonotic disease surveillance and antimicrobial resistance (AMR) monitoring (Bidaisee and Macpherson, 2014; Radford et al., 2011). Enhanced NLP tools improve threat detection and trend analysis in animal populations, strengthening public health responses across human, animal, and environmental health domains (Kol et al., 2015; Robertson et al., 2000; Van Duijkeren et al., 2004). Beyond public health, NLP-driven solutions facilitate large-scale epidemiological studies, identifying risk factors and treatment outcomes that enhance companion animal welfare (Lund, 2015; Farrell et al., 2023b).

Traditional veterinary disease surveillance relies on manual coding or rule-based methods, which are time-intensive and prone to human error (Hsia et al., 2010; Miñarro-Giménez et al., 2018; Turchin et al., 2006). In contrast, NLP-driven approaches offer scalable, automated solutions for extracting clinical insights from free-text records. Developing these methods within veterinary medicine improves animal welfare and contributes to the refinement of computational tools that may later be adapted to human bioinformatics research. Neural network approaches to disease coding have evolved considerably over time. Pioneering work introduced DeepTag (Nie et al., 2018), establishing a foundation that subsequently refined into the more advanced VetTag framework (Zhang et al., 2019). The field has progressed significantly with recent innovations leveraging pre-trained LLMs (Farrell et al., 2023a; Boguslav et al., 2024). Complementary research has expanded our understanding of generative models for veterinary entity extraction for clinical signs (Wulcan et al., 2024) and for body condition scoring (Fins et al., 2024).

In this paper, we contribute the following:

1. PetEVAL: The first veterinary EHR benchmark – A publicly available free-text vEHR dataset, establishing a standard for veterinary NLP research.

2. Rigorous manual anonymisation – Every record underwent manual anonymisation with at least two independent reviews, including verification by a veterinary clinician, ensuring complete removal of sensitive data.

3. ICD-11 syndromic classification – Syndromic labels were assigned using the ICD-11 framework, supplemented with domain-specific annotations to ensure clinically relevant labeling.

2 Literature Review

The adoption of EHRs has revolutionised medical research, offering vast amounts of health data for analysis (Gunter and Terry, 2005; Cowie et al., 2017). While structured EHR data has been extensively used in epidemiological studies (Krumholz et al., 2014; Hamer et al., 2024; Hlatky et al., 2014; Williamson et al., 2020), up to 80% of EHR information exists in unstructured formats, primarily as free-text clinical notes (Kong, 2019). These unstructured notes capture clinical insights often lost in structured formats (Birman-Deych et al., 2005; Singh et al., 2004). Excluding this data from research can significantly impact the validity of findings (Ford et al., 2013; Jensen et al., 2017; Price et al., 2016; Barak-Corren et al., 2017). However, utilising unstructured data presents challenges in patient privacy protection, particularly regarding re-identification risks (Simon et al., 2019; Abouelmehdi et al., 2017; Dorr et al., 2006). Automated EHR anonymisation has become a critical focus in addressing these challenges. Benchmarks like the i2b2/UTHealth corpus and MIMIC-3 database have been established to evaluate de-identification models (Stubbs and Uzuner, 2015; Stubbs et al., 2017; Meystre et al., 2010; Aberdeen et al., 2010). Approaches range from rule-based systems (Cao et al., 2003) to neural networks (Liu et al., 2019) and pre-trained language models (Yoon et al., 2023; Chen et al., 2021). Recent advancements in learning-based methods show promise in automating de-identification (Leevy et al., 2020; Lee et al., 2022). However, these methods face challenges with performance instability when applied to heterogeneous real-world data (Abu-El-Rub et al., 2022; Yang et al., 2019). Deep learning approaches have been proposed to address these issues, but their effectiveness is limited by small training datasets and performance degradation on out-of-distribution EHRs (Syed et al., 2022; Lee et al., 2021; Jiang et al., 2017).

3 PetEVAL

3.1 The SAVSNET Dataset

We utilise data from the Small Animal Veterinary Surveillance Network (SAVSNET), a sentinel network of 253 volunteer first-opinion veterinary practices across the United Kingdom that have collected vEHRs since March 2014. This network has accumulated over 12 million EHRs, with participating practices selected based on their practice management software compatibility with the SAVSNET data exchange system. During each consultation with a clinician or nurse, comprehensive data includes species, breed, sex, neuter status, age, owner's postcode, insurance and microchipping status, and a detailed free-text clinical narrative. These narratives may contain information about symptoms, diagnoses, treatments, procedures, or other clinical matters. Owners can opt out of data collection during any consultation. The SAVSNET group operates under ethical approval from the University of Liverpool Ethics Committee (RETH001081), ensuring adherence to established ethical standards. Figure 1 provides a sample data point in JSON format.

3.2 Tasks

3.2.1 Task 1 - Anonymisation

Ensuring the privacy and security of EHRs is crucial for safeguarding the personal information of pet owners and facilitating the easy sharing of data use in clinical and academic research. The dataset is labelled with NER entities and spans applied to pseudo-anonymised contextual placeholders. The objective is to maintain the integrity and utility of clinical information within the EHR while effectively anonymising various types of personal data. This includes names (both animal and human), location details (such as city, town, and addresses), organisation names (including attending veterinary practices, referral hospitals, kennels, and laboratories), contact details (emails, phone numbers), id-numbers (passport numbers, insurance policy numbers, MRCVS codes), and any other explicit identifiers. The anonymisation is compliant with the HIPPA Safe Harbour (Sun et al., 2020).

3.2.2 Task 2 - Syndromic Disease Classification

Given the critical role of monitoring national disease outbreaks in public health, effective surveillance systems can provide invaluable insights, such as in informing clinicians of key symptoms to observe, enabling researchers to identify aetiological agents, and establishing an automated reporting mechanism for public health agencies to facilitate swift notification of changes in disease occurrence. However, the task is not straightforward, particularly when dealing with novel diseases or syndromes with unknown symptoms. Effective outbreak reduction strategies hinge on the ability to detect outbreaks with minimal cases. To address these challenges, the dataset is provided with ICD-11 chapters (World Health Organisation (WHO), 2022), which includes contextual discussions such as symptoms and diagnoses. The task is structured as a multi-label classification problem, as a consult or condition may cover a range of presenting symptoms. Performance is evaluated using multi-label classification metrics, including precision and recall, macro-average F1-Score, and weighted F1-Score.

3.2.3 Task 3 - Disease Extraction

Identifying specific diseases is critical for downstream epidemiological studies, which aim to reveal novel risk factors, seasonality, and other trends. This task is particularly challenging due to the private healthcare nature of veterinary practices in the UK and much of the world. Confirmation diagnostic tests are rare, as owners often wish to avoid the inherent costs, opting instead to take the advice of clinicians or due to the lack of available resources or expertise not found in first opinion practice. Additionally, the presence of negations is common within vEHRs, especially within the first opinion setting, where it is estimated that 11% of mentioned diseases are negated (Cheng et al., 2017) which complicates the task further. In our study, the dataset is labelled with the diagnostic disease contained within it. This process is framed as NER task using the IOB2 format, wherein the entity of 'disease' and its spans are provided. Evaluation utilises SeqEval for precision, recall, and F1-score (Nakayama, 2018).

4 Methods

4.1 Dataset Construction

Our dataset comprises three subsets: a training set of 11,000 records, an evaluation set of 1,600 records, and a test set of 5,000 records. We selected only consultations recorded before 2020 and restricted the dataset to consultations involving only

Table 1: Evaluation of NER performance on veterinary clinical text data anonymised according to HIPAA Safe Harbor guidelines. The table presents entity type distribution across training, evaluation, and test splits, with comparative performance metrics (precision, recall, F1-score) between 'BERT-base-uncased', 'PetBERT', and LLaMA 3.1 8B models across identifier categories.

HIPAA Safe Harbor	Examples	Train/ Eval	Test Count	NER Entity	BERT-base-uncased P	BERT-base-uncased R	BERT-base-uncased F1	PetBERT P	PetBERT R	PetBERT F1	LLaMA 3.1 8B P	LLaMA 3.1 8B R	LLaMA 3.1 8B F1
(A) Names	Pet, Owner, Vet Names	4790	1370	PER	0.84	0.93	0.89	0.93	0.70	0.80	0.71	0.65	0.68
(B) Geographic subdivisions	City, Towns, Countries	311	94	LOC	0.95	0.98	0.97	0.97	0.97	0.97	0.78	0.83	0.80
	Vet practices, hospitals, shelters	392	168	ORG	0.97	0.97	0.98	0.98	0.96	0.97	0.82	0.79	0.81
(C) Dates	Day/month dates, appointments	425	162	TIME	0.94	0.96	0.95	0.93	0.94	0.93	0.76	0.81	0.78
(D) Telephone numbers	Client/practice phone numbers	19	4										
(E) Fax numbers	n/a	None	None										
(F) Email addresses	Referral/client emails	9	3										
(G) Social security numbers	n/a	None	None										
(H) Medical record numbers	n/a	None	None										
(I) Health plan numbers	Insurance policy numbers	33	20										
(J) Account numbers	Microchip Numbers	299	35	MISC	0.91	0.97	0.97	0.95	0.94	0.94	0.73	0.69	0.71
(K) Certificate numbers	MRCVS clinician codes	51	17										
(L) Vehicle identifiers	n/a	None	None										
(M) Device identifiers	n/a	None	None										
(N) URLs	Website urls	None	None										
(O) IP addresses	n/a	None	None										
(P) Biometric identifiers	n/a	None	None										
(Q) Photographic images	n/a	None	None										
(R) Other identifiers	Passport numbers	34	8										

cats and dogs. To enhance generalisability, dataset splits were performed based on a pre-compiled list of veterinary practices, following the methodology outlined in (Farrell et al., 2023a). Specifically, we assigned distinct practices to training and testing sets, ensuring that models trained on the training set were evaluated on records from veterinary practices that did not contribute to training. This design minimises the risk of models overfitting to stylistic or institutional biases and provides more substantial evidence of generalisability across UK veterinary practices. We excluded empty records containing fewer than ten words or exceeding 350 words. The median narrative length in the full SAVSNET dataset is 287 words, while in PetEVAL, it is 226 words.

4.1.1 Annonymisation

Each record was manually reviewed twice, targeting the removal of all potential identifiers, including names (owner, animal, and veterinary staff), locations (cities, countries, vet practices, referral hospitals, rescue centres, kennels, crematoriums, labs), dates (when they included specific years), times (if overly specific), and unique identifiers such as microchip codes, passport numbers, insurance policy numbers, vet MRCVS codes, phone numbers, and email addresses. Flagged elements were pseudonymised with context-appropriate placeholders to maintain record coherence, and corresponding spans and entity tags were generated for these placeholders. Pseudonyms were derived from separate lists for train and test splits,

For the anonymisation NER task, identifiers were mapped to standard tags: 'LOC' (cities, towns, countries), 'PER' (pet/owner/vet names), 'TIME' (specific dates/times), 'ORG' (veterinary practices, rescue shelters, labs, groomers), and 'MISC' (unique identifiers like microchips, insurance codes, contact information). The counts for each can be found within table 1. Non-clinical brand names were removed but not included in anonymisation metrics. No clinically relevant information was modified.

4.1.2 Syndromic Disease Classification

The dataset was curated to support syndromic disease surveillance through the assignment of ICD-11 labels. For this purpose, 20 ICD-11 chapter codes were selected to capture a broad range of clinically relevant syndromes observed in veterinary practice. The full list of selected chapter codes is provided in Table 2. To facilitate efficient and accurate annotation, we employed a semi-automated approach wherein initial fuzzy labels were generated using the PetBERT-ICD model, a previously developed tool designed for assigning ICD-related labels in veterinary contexts. This pre-annotation step helped streamline the annotation process, reduce cognitive load for annotators, and minimise potential errors. Annotators reviewed and refined these suggested labels, ensuring alignment with clinical documentation practices in first-opinion vEHRs. To maintain the integrity of the evaluation, the test set was exempt from automated label matching and underwent a full manual review by two expert annotators. Records that an initial reviewer was unhappy to determine the presence of a diagnosis were passed through an additional reviewer, and a consensus vote was taken. Finally, we ensured that the disease extraction dataset aligned with the syndromic dataset, an extracted disease therefore has a linked syndromic label.

4.1.3 Disease Extraction

The dataset was developed to facilitate the evaluation of disease diagnosis extraction models from first-opinion vEHRs. Given the nature of primary care veterinary records, confirmatory diagnoses are rare, with most diagnoses being clinical assessments rather than definitive results from diagnostic testing. Therefore, any named condition mentioned in a record was annotated as a diagnosis unless explicitly negated. This includes confirmed diagnoses, differential diagnoses, and syndromic descriptions. Additionally, mentions of pathogens, such as bacteria, viruses, and parasites, were annotated as they typically are discussed as diagnoses within the narratives. We extracted diseases coded within the ICD-11 and veterinary-specific conditions not represented in human medicine. Each annotated diagnosis was linked to its corresponding span within the text, with entity tags assigned to support NER tasks. Records that an initial reviewer was unhappy to determine the presence of a diagnosis were passed through an additional reviewer, and a consensus vote was taken.

4.1.4 Baseline Models

For baseline results in PetEVAL, we evaluated three pre-trained language models: 'BERT-base-uncased' (Devlin et al., 2019), a general-purpose encoder; 'PetBERT' (Farrell et al., 2023a), a veterinary domain-adapted encoder; and 'LLaMA 3.1 8B' (Team and Meta, 2024), a generalist decoder model. The encoder models were fine-tuned as token classification models using the IOB2 format for the anonymisation and disease extraction tasks, with training parameters including a mini-batch size of 32, an initial learning rate of 2e-5, and the AdamW optimiser. Early stopping was applied based on evaluation loss. For syndromic classification, both encoders were adapted for multi-label classification across 20 ICD-11 chapter codes, employing a weighted binary cross-entropy loss function with sigmoid activation to address class imbalance. Training followed the same hyperparameter setup and typically converged beyond epoch 6. An iterative threshold analysis was conducted, varying classification thresholds between 60% and 95% in 5% increments, prioritising recall to minimise false negatives. The final classifier applied an 80% threshold and was evaluated on the test set. The decoder model was prompted with few-shot examples selected from the training set, with multiple prompt designs tested against the evaluation set before application to the full test set.

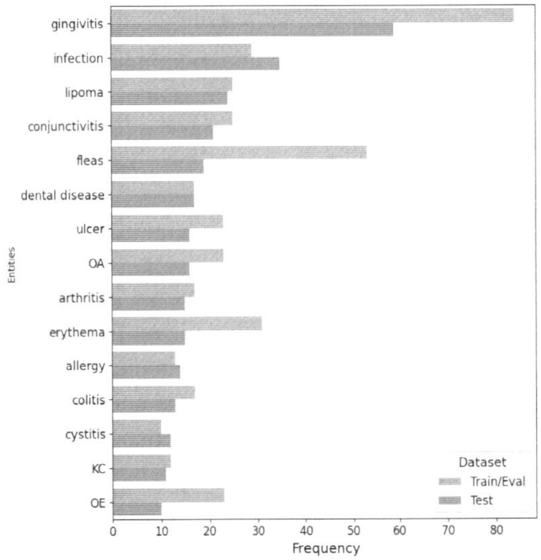

Figure 2: Distribution of the 15 most frequent disease entities extracted from veterinary electronic health records in the Train/Eval and Test sets during Task 2 (Disease Extraction).

4.1.5 Model Evaluation

We implemented a unified entity-level evaluation framework to ensure fair comparison between encoder (BERT) and decoder (LLaMA) architectures across anonymisation and disease extraction tasks. For encoder models, we first converted token-level IOB/BIO predictions into entity spans before applying the same entity-level F1 evaluation used for decoder models. This approach follows CoNLL methodology (Tjong et al., 2003), where all extracted entities undergo identical normalisation procedures before being exact-matched against ground truth. For both model types and tasks, we calculate precision as the ratio of correctly identified entities to total predictions, recall as the ratio of correctly identified entities to ground truth entities, and F1 as their harmonic mean. The anonymisation task evaluates the identification of privacy-sensitive entities

(LOC, PER, MISC, NAME), while disease extraction assesses the recognition of standardised disease mentions. By standardising evaluation across architectural paradigms, we enable direct performance comparison while maintaining methodological rigour in assessing clinical information extraction capabilities.

For the syndromic classification task, we assess model performance using precision, recall, and F1 scores computed against ground truth labels provided by annotators. For encoder-based models, classification uses a fine-tuned ICD-11 classifier with an optimised threshold, ensuring a balance between precision and recall for robust disease detection. For generative models, we convert outputs into a tabular format using a direct match approach on uncased text. Similarity-based methods were considered, but they yielded no performance gains, so we adopted the least computationally intensive approach. The predicted labels are transformed into a one-hot encoded vector, applying the same evaluation metrics as encoder models. Given the importance of disease surveillance, we preferentially select for recall to minimise false negatives, as missing cases could lead to undetected outbreaks. While this may increase false positives, these can be further reviewed to ensure the detection of potential health threats.

5 Results

5.1 Corpus Overview

The dataset consists of 675,935 words distributed across the training (11,000 records), evaluation (1,600 records), and test sets (5,000 records). While demographic data is not included, 68% of the records represent dogs, with a near 50-50 sex split across both species. The dataset contains information from 16,153 unique animals from various regions across the UK.

For syndromic disease classification, annotations were applied using a multi-label one-hot encoding approach aligned with ICD-11 chapter heads. Across the dataset, 9,510 annotations were made in the training set and 4,714 in the test set. The most frequent label, 'Certain infectious or parasitic diseases', was prominent due to the high occurrence of conditions like parasitic infestations. The median labels per class in the training set was 348, with an average of 0.9 labels per consultation. Notably, 8,907 consultations received at least one label, while those without a label typically represented routine checkups or non-syndromic cases.

The frequency distribution of extracted disease entities across the train/eval and test datasets is presented in Figure 2. As expected, conditions readily identifiable through visual examination, such as gingivitis, conjunctivitis, and lipoma, exhibit high representation. Furthermore, the extracted entities encompass clinical language commonly used by veterinary practitioners to indicate disease, including terms like 'infection,' 'fleas' (for flea infestation), and 'dental disease' (for unspecified dental conditions). The train/eval datasets contain 3,907 unique extracted conditions, while the test dataset comprises 2,899.

5.2 Inter-annotator agreement

Inter-annotator agreement was assessed on a subset of 1,000 vEHRs from the test set focused on the syndromic classification task. Two expert veterinary clinicians independently annotated the records using strictly predefined guidelines, with no communication allowed at this stage to ensure unbiased annotations. The resulting Cohen's kappa statistic was 0.722, indicating a substantial level of agreement (McHugh, 2012). This value suggests strong, though not perfect, alignment between the annotators. Disagreements were systematically reviewed, with the majority resolved through a collaborative discussion. In cases where consensus could not be reached, a third clinician provided a decisive resolution.

5.3 Baselines

We conducted baseline experiments with 'bert-base-uncased' and 'PetBERT' and a generative model 'LLaMA 3.1 8B' to establish reference points for evaluating more complex models. For the anonymisation task, PetBERT consistently outperformed BERT-Base across HIPAA Safe Harbor entity categories, with notable improvements in identifying names (F1: 0.80 vs. 0.89) and geographic subdivisions (F1: 0.97 vs. 0.98) (Table 1). Both models achieved high performance in structured entity types such as dates (F1: 0.93 vs. 0.95) and organisations (F1: 0.97 vs. 0.98). LlaMA 3.1, using few-shot prompting (Appendix), was behind with lower F1-scores across all categories, particularly for names (F1: 0.68) and locations (F1: 0.80).

As shown in Table 1, fine-tuned PetBERT outperformed BERT-base-uncased across most entity types, achieving a higher precision (0.93 vs. 0.84), recall (0.70 vs. 0.93), and F1-score (0.80 vs. 0.89)

Table 2: Performance Metrics for BERT-base-uncased, PetBERT, and LLaMA 3.1 8B on ICD-11 Syndromic Chapters. P = Precision, R = Recall, F1 = F1-score

ICD-11 Syndromic Chapter	Train/ Eval	Test Count	BERT-base-uncased			PetBERT			LLaMA 3.1 8B		
			P	R	F1	P	R	F1	P	R	F1
Certain infectious or parasitic diseases	1549	1321	0.74	0.31	0.44	0.78	0.45	0.57	0.65	0.28	0.39
Neoplasms	774	499	0.85	0.77	0.81	0.90	0.81	0.85	0.77	0.65	0.70
Diseases of the blood or blood-forming organs	90	47	0.66	0.35	0.45	0.63	0.31	0.41	0.55	0.23	0.32
Diseases of the immune system	512	429	0.80	0.54	0.64	0.84	0.51	0.64	0.68	0.41	0.51
Endocrine, nutritional or metabolic diseases	572	305	0.67	0.60	0.64	0.69	0.60	0.64	0.58	0.45	0.51
Mental, behavioral or neurodevelopmental disorders	1121	469	0.76	0.34	0.46	0.79	0.38	0.51	0.64	0.27	0.38
Diseases of the nervous system	233	150	0.54	0.58	0.56	0.71	0.54	0.61	0.48	0.42	0.45
Diseases of the visual system	905	634	0.85	0.81	0.83	0.90	0.80	0.85	0.73	0.68	0.70
Diseases of the ear or mastoid process	700	513	0.83	0.77	0.80	0.88	0.78	0.83	0.71	0.65	0.68
Diseases of the circulatory system	276	181	0.67	0.33	0.45	0.71	0.46	0.55	0.55	0.29	0.38
Diseases of the respiratory system	459	346	0.80	0.54	0.64	0.84	0.57	0.68	0.68	0.45	0.54
Diseases of the digestive system	671	259	0.81	0.55	0.66	0.79	0.62	0.69	0.67	0.46	0.55
Diseases of the skin	1377	1018	0.81	0.62	0.70	0.88	0.60	0.71	0.69	0.51	0.59
Diseases of the musculoskeletal system or connective tissue	1171	722	0.79	0.73	0.76	0.83	0.70	0.76	0.67	0.61	0.64
Diseases of the genitourinary system	569	334	0.76	0.59	0.66	0.79	0.67	0.73	0.65	0.49	0.56
Pregnancy, childbirth or the puerperium	65	36	0.42	0.17	0.24	0.74	0.12	0.21	0.36	0.10	0.16
Certain conditions originating in the perinatal period	39	27	0.50	0.08	0.13	0.00	0.00	0.00	0.38	0.05	0.09
Developmental anomalies	191	95	0.59	0.19	0.28	0.70	0.30	0.42	0.47	0.15	0.23
Injury, poisoning or certain other consequences of external causes	1113	636	0.67	0.67	0.67	0.73	0.70	0.71	0.58	0.55	0.56
micro average			**0.76**	**0.58**	**0.66**	**0.81**	**0.61**	**0.70**	**0.65**	**0.47**	**0.55**
macro average			**0.71**	**0.50**	**0.57**	**0.74**	**0.52**	**0.60**	**0.60**	**0.41**	**0.48**
weighted average			**0.76**	**0.58**	**0.65**	**0.81**	**0.61**	**0.69**	**0.65**	**0.47**	**0.54**

for identifying personal names (PER) such as pet, owner, and vet names. In contrast, LLaMA 3.1 achieved lower performance across all entity types, with an F1-score of 0.68 for names. For location (LOC) and organisation (ORG) entities, PetBERT outperformed BERT-base-uncased, achieving F1-scores of 0.97 and 0.97, respectively, compared to BERT-base's 0.97 and 0.98. LLaMA 3.1 showed lower performance in both entity types, with an F1 of 0.80 for LOC and 0.81 for ORG. The comparison highlights PetBERT's superior ability to process veterinary clinical text, particularly for identifying personal and organisational entities, while Llama 3.1's performance in entity recognition remained behind.

PetBERT outperformed both BERT-Base and Llama 3.1 for the disease extraction task, achieving a precision of 0.90, recall of 0.85, and F1-score of 0.87 (Table 1). BERT-Base trailed with 0.70 precision, 0.55 recall, and an F1 of 0.60, while Llama 3.1, using a few-shot prompt (Appendix), performed worst (precision: 0.60, recall: 0.35, F1: 0.40).

6 Discussion

In veterinary first-opinion clinical practice, the challenge of extracting meaningful insights from vEHRs is compounded by several notable factors. Among these is the absence of standardised data conventions within free-text inputs, and inconsistencies in spelling and abbreviations used by different clinicians (Davies et al., 2024b). This is amplified by the ambiguity surrounding the interpretation of consultation events. Specifically, the lack of diagnostic details in these narratives introduces additional layers of complexity. The moderate Cohen's kappa score of 0.7, observed between two annotators—both qualified veterinary clinicians - underscores the inherent difficulties in annotating such unstructured data. Veterinary EHRs are packed with ambiguous language, clinician-specific abbreviations, and varying documentation styles, inhibiting the ability to extract information from them effectively. Even among active clinicians, the interpretation of nuanced first-opinion notes can differ, primarily due to diagnostic uncertainties, incomplete patient histories, and the lack of standardised terminology. Despite these obstacles, the intrinsic value embedded within these clinical narratives is undeniable, with applications spanning disease outbreak detection and improving public health and animal welfare standards (Davies et al., 2024a; Farrell et al., 2023a).

Generative models, such as the LLaMA 3.1 8B applied in our baseline, exhibited relatively poor performance across tasks, particularly in NER. This highlights the ongoing challenge of designing effective prompting strategies, requiring further research. Additionally, generative models present inherent difficulties in evaluation, as their flexible outputs may not align precisely with gold-standard annotations. While our strict direct match approach may penalise performance, maintaining fidelity to the intended identifier remains a priority. Over time, we anticipate improvements in generative architectures, which may eventually surpass the limitations observed here. However, domain-adapted encoder-based models like PetBERT demonstrated superior performance across all tasks, aligning with expectations given their targeted pretraining. Beyond accuracy, their efficiency also makes them preferable for everyday deployments, especially in resource-intensive applications such as continuous disease surveillance. Given the significant environmental cost of running large LLMs (Bashir et al., 2024), there is a clear need for lightweight, domain-specific solutions that can operate effectively on consumer-level hardware, ensuring sustainability and practical usability in real-world veterinary informatics.

Strict privacy regulations in human healthcare restrict many studies to single institutions, creating discrepancies between reported performance and cross-site generalisability. PetEVAL collates from over 250 UK practices with diverse clinical approaches and provides substantial advantages for robust model evaluation. While fewer than 23% of human healthcare ML studies utilise multi-institutional data (McDermott et al., 2021), often resulting in significant biases and performance degradation when applied to external institutions (Barak-Corren et al., 2021; Burns and Kheterpal, 2020), PetEVAL's multi-institutional framework can capture practice variability and thus offers an opportunity to assess model robustness across institutions, ultimately contributing to more accurate and equitable AI-driven healthcare systems within and beyond veterinary medicine.

7 Conclusion

PetEVAL is the first benchmark dataset for veterinary EHRs, featuring expert-annotated resources across ICD-11 syndromic classifications, disease entity recognition, and anonymisation labels. Beyond addressing a critical gap in veterinary medicine, PetEVAL facilitates valuable comparative studies between animal and human health domains, promoting cross-disciplinary insights. As a foundational resource for veterinary informatics, this dataset promises to catalyse advancements in clinical decision support systems, enhance epidemiological surveillance capabilities, and strengthen WHO's One Health initiatives, ultimately advancing animal welfare and public health research outcomes.

8 Acknowledgements

We thank the data providers in veterinary practice (VetSolutions, Teleos, CVS, and other practitioners). Without their support and participation, this research would not be possible. We thank all of the annotators for their support in deriving this dataset. This work was supported by the Biotechnology and Biological Sciences Research Council [BB/T008695/1]. The carbon emissions associated with the training and evaluation of our models were entirely offset.

9 Limitations

Despite rigorous quality control, annotation errors are unavoidable due to the dataset's scale. Models trained on first-opinion vEHRs are inherently limited by the availability and accuracy of recorded information, often lacking confirmatory diagnostics due to financial constraints or resource limitations. Our evaluation method enforces strict token-level matching, penalising incomplete spans even when semantically close to the ground truth. While this is critical for anonymisation, it may be overly rigid for disease extraction. Similarly, our classification approach adheres strictly to predefined categories, which, while justified by the prompt, may overlook minor deviations. Future work could explore more flexible evaluation metrics and incorporate referral-level vEHRs to enhance diagnostic certainty.

References

John Aberdeen, Samuel Bayer, Reyyan Yeniterzi, Ben Wellner, Cheryl Clark, David Hanauer, Bradley Malin, and Lynette Hirschman. 2010. The MITRE Identification Scrubber Toolkit: Design, training, and assessment. *International Journal of Medical Informatics*, 79(12):849–859.

Karim Abouelmehdi, Abderrahim Beni-Hssane, Hayat Khaloufi, and Mostafa Saadi. 2017. Big data security and privacy in healthcare: A Review. *Procedia Computer Science*, 113:73–80.

Noor Abu-El-Rub, Jay Urbain, George Kowalski, Kristen Osinski, Robert Spaniol, Mei Liu, Bradley Taylor, and Lemuel R. Waitman. 2022.

Yuval Barak-Corren, Victor M. Castro, Solomon Javitt, Alison G. Hoffnagle, Yael Dai, Roy H. Perlis, Matthew K. Nock, Jordan W. Smoller, and Ben Y. Reis. 2017. Predicting suicidal behavior from longitudinal electronic health records. *American Journal of Psychiatry*, 174(2):154–162.

Yuval Barak-Corren, Pradip Chaudhari, Jessica Perniciaro, Mark Waltzman, Andrew M. Fine, and Ben Y. Reis. 2021. Prediction across healthcare settings: a case study in predicting emergency department disposition. *npj Digital Medicine 2021 4:1*, 4(1):1–7.

Noman Bashir, Priya Donti, James Cuff, Sydney Sroka, Marija Ilic, Vivienne Sze, Christina Delimitrou, and Elsa Olivetti. 2024. The Climate and Sustainability Implications of Generative AI. *An MIT Exploration of Generative AI*. Https://mit-genai.pubpub.org/pub/8ulgrckc.

Satesh Bidaisee and Calum N.L. Macpherson. 2014. Zoonoses and one health: A review of the literature. *Journal of Parasitology Research*, 2014.

Elena Birman-Deych, Amy D. Waterman, Yan Yan, David S. Nilasena, Martha J. Radford, and Brian F. Gage. 2005. Accuracy of ICD-9-CM codes for identifying cardiovascular and stroke risk factors. *Medical Care*, 43(5):480–485.

E. F. Bode, E. Mederska, H. Hodgkiss-Geere, A. D. Radford, and D. A. Singleton. 2022. Analysis of canine cardiovascular therapeutic agent prescriptions using electronic health records in primary care veterinary practices in the United Kingdom. *Journal of Veterinary Cardiology*, 39:35–45.

Mayla R. Boguslav, Adam Kiehl, David Kott, G. Joseph Strecker, Tracy Webb, Nadia Saklou, Terri Ward, and Michael Kirby. 2024. Fine-tuning foundational models to code diagnoses from veterinary health records.

Michael L. Burns and Sachin Kheterpal. 2020. Machine Learning Comes of AgeLocal Impact versus National Generalizability. *Anesthesiology*, 132(5):939–941.

Hui Cao, Peter Stetson, and George Hripcsak. 2003. Assessing explicit error reporting in the narrative electronic medical record using keyword searching. *Journal of Biomedical Informatics*, 36(1-2):99–105.

Richard J. Chen, Ming Y. Lu, Tiffany Y. Chen, Drew F.K. Williamson, and Faisal Mahmood. 2021. Synthetic data in machine learning for medicine and healthcare. *Nature Biomedical Engineering 2021 5:6*, 5(6):493–497.

Katharine Cheng, Timothy Baldwin, and Karin Verspoor. 2017. Automatic Negation and Speculation Detection in Veterinary Clinical Text. *In Proceedings of Australasian Language Technology Association Workshop,*, pages 70–78.

Martin R. Cowie, Juuso I. Blomster, Lesley H. Curtis, Sylvie Duclaux, Ian Ford, Fleur Fritz, Samantha Goldman, Salim Janmohamed, Jörg Kreuzer, Mark Leenay, Alexander Michel, Seleen Ong, Jill P. Pell, Mary Ross Southworth, Wendy Gattis Stough, Martin Thoenes, Faiez Zannad, and Andrew Zalewski. 2017. Electronic health records to facilitate clinical research. *Clinical Research in Cardiology*, 106(1):1–9.

Heather Davies, Goran Nenadic, Ghada Alfattni, Mercedes Arguello Casteleiro, Noura Al Moubayed, Sean Farrell, Alan D. Radford, and P.-J. M. Noble. 2024a. Text mining for disease surveillance in veterinary clinical data: part two, training computers to identify features in clinical text. *Frontiers in Veterinary Science*, 11:1352726.

Heather Davies, Goran Nenadic, Ghada Alfattni, Mercedes Arguello Casteleiro, Noura Al Moubayed, Sean O. Farrell, Alan D. Radford, and Peter John M. Noble. 2024b. Text mining for disease surveillance in veterinary clinical data: part one, the language of veterinary clinical records and searching for words. *Frontiers in Veterinary Science*, 11:1352239.

Jacob Devlin, Ming-Wei Chang, Kenton Lee, Kristina Toutanova Google, and A I Language. 2019. BERT: Pre-training of Deep Bidirectional Transformers for Language Understanding. *Proceedings of NAACL-HLT*, pages 4171–4186.

David A. Dorr, W. F. Phillips, S. Phansalkar, S. A. Sims, and J. F. Hurdle. 2006. Assessing the difficulty and time cost of de-identification in clinical narratives. *Methods of Information in Medicine*, 45(3):246–252.

Sean Farrell, Charlotte Appleton, Peter John Mäntylä Noble, and Noura Al Moubayed. 2023a. PetBERT: automated ICD-11 syndromic disease coding for outbreak detection in first opinion veterinary electronic health records. *Scientific Reports 2023 13:1*, 13(1):1–14.

Sean Farrell, John McGarry, Peter John Mäntylä Noble, Gina J. Pinchbeck, Sophie Cantwell, Alan D. Radford, and David A. Singleton. 2023b. Seasonality and other risk factors for fleas infestations in domestic dogs and cats. *Medical and veterinary entomology*, 37(2):359–370.

Ivo S. Fins, Heather Davies, Sean Farrell, Jose R. Torres, Gina Pinchbeck, Alan D. Radford, and Peter John Noble. 2024. Evaluating ChatGPT text mining of clinical records for companion animal obesity monitoring. *Veterinary Record*, 194(3):no.

Elizabeth Ford, Amanda Nicholson, Rob Koeling, A. Rosemary Tate, John Carroll, Lesley Axelrod, Helen E. Smith, Greta Rait, Kevin A. Davies, Irene Petersen, Tim Williams, and Jackie A. Cassell. 2013. Optimising the use of electronic health records to estimate the incidence of rheumatoid arthritis in primary care: what information is hidden in free text? *BMC medical research methodology*, 13(1).

Tracy D Gunter and Nicolas P Terry. 2005. The emergence of national electronic health record architectures in the United States and Australia: models, costs, and questions. *Journal of medical Internet research*, 7(1):e3.

Mika K. Hamer, Cathy J. Bradley, Richard Lindrooth, and Marcelo C. Perraillon. 2024. The Effect of Medicare Annual Wellness Visits on Breast Cancer Screening and Diagnosis. *Medical Care*, 62(8):530–537.

Mark A. Hlatky, Roberta M. Ray, Dale R. Burwen, Karen L. Margolis, Karen C. Johnson, Anna Kucharska-Newton, Joann E. Manson, Jennifer G. Robinson, Monika M. Safford, Matthew Allison, Themistocles L. Assimes, Anthony A. Bavry, Jeffrey Berger, Rhonda M. Cooper-DeHoff, Susan R. Heckbert, Wenjun Li, Simin Liu, Lisa W. Martin, Marco V. Perez, Hilary A. Tindle, Wolfgang C. Winkelmayer, and Marcia L. Stefanick. 2014. Use of Medicare Data to Identify Coronary Heart Disease Outcomes In the Women's Health Initiative (WHI). *Circulation. Cardiovascular quality and outcomes*, 7(1):157.

David C. Hsia, W. Mark Krushat, Ann B. Fagan, Jane A. Tebbutt, and Richard P. Kusserow. 2010. Accuracy of Diagnostic Coding for Medicare Patients under the Prospective-Payment System. *The New England journal of medicine*, 318(6):352–355.

Kasper Jensen, Cristina Soguero-Ruiz, Karl Oyvind Mikalsen, Rolv Ole Lindsetmo, Irene Kouskoumvekaki, Mark Girolami, Stein Olav Skrovseth, and Knut Magne Augestad. 2017. Analysis of free text in electronic health records for identification of cancer patient trajectories. *Scientific Reports 2017 7:1*, 7(1):1–12.

Zhipeng Jiang, Chao Zhao, Bin He, Yi Guan, and Jingchi Jiang. 2017. De-identification of medical records using conditional random fields and long short-term memory networks. *Journal of Biomedical Informatics*, 75:S43–S53.

Amir Kol, Boaz Arzi, Kyriacos A. Athanasiou, Diana L. Farmer, Jan A. Nolta, Robert B. Rebhun, Xinbin Chen, Leigh G. Griffiths, Frank J.M. Verstraete, Christopher J. Murphy, and Dori L. Borjesson. 2015. Companion animals: Translational scientist's new best friends. *Science translational medicine*, 7(308):308ps21.

Hyoun Joong Kong. 2019. Managing Unstructured Big Data in Healthcare System. *Healthcare Informatics Research*, 25(1):1.

Harlan M. Krumholz, Sharon Lise T. Normand, and Yun Wang. 2014. Trends in hospitalizations and outcomes for acute cardiovascular disease and stroke, 1999-2011. *Circulation*, 130(12):966–975.

Junhak Lee, Jinwoo Jeong, Sungji Jung, Jihoon Moon, and Seungmin Rho. 2022. Verification of De-Identification Techniques for Personal Information Using Tree-Based Methods with Shapley Values. *Journal of Personalized Medicine 2022, Vol. 12, Page 190*, 12(2):190.

Kahyun Lee, Nicholas J Dobbins, Bridget Mcinnes, Meliha Yetisgen, and Özlem Uzuner. 2021. Transferability of Neural Network Clinical De-identification Systems. *Journal of the American Medical Informatics Association*.

Joffrey L. Leevy, Taghi M. Khoshgoftaar, and Flavio Villanustre. 2020. Survey on RNN and CRF models for de-identification of medical free text. *Journal of Big Data*, 7(1):1–22.

Yi Liu, Jialiang Peng, James J. Q Yu, and Yi Wu. 2019. PPGAN: Privacy-preserving Generative Adversarial Network. *Proceedings of the International Conference on Parallel and Distributed Systems - ICPADS*, 2019-December:985–989.

E M Lund. 2015. Power of practice: using clinical data to advance veterinary medicine. *Veterinary Record*.

Matthew B.A. McDermott, Shirly Wang, Nikki Marinsek, Rajesh Ranganath, Luca Foschini, and Marzyeh Ghassemi. 2021. Reproducibility in machine learning for health research: Still a ways to go. *Science Translational Medicine*, 13(586).

Paul McGreevy, Peter Thomson, Navneet K. Dhand, David Raubenheimer, Sophie Masters, Caroline S. Mansfield, Timothy Baldwin, Ricardo J.Soares Magalhaes, Jacquie Rand, Peter Hill, Anne Peaston, James Gilkerson, Martin Combs, Shane Raidal, Peter Irwin, Peter Irons, Richard Squires, David Brodbelt, and Jeremy Hammond. 2017. VetCompass Australia: A National Big Data Collection System for Veterinary Science. *Animals 2017, Vol. 7, Page 74*, 7(10):74.

Mary L. McHugh. 2012. Interrater reliability: the kappa statistic. *Biochemia Medica*, 22(3):276.

Stephane M. Meystre, F. Jeffrey Friedlin, Brett R. South, Shuying Shen, and Matthew H. Samore. 2010. Automatic de-identification of textual documents in the electronic health record: A review of recent research. *BMC Medical Research Methodology*, 10(1):1–16.

Jose Antonio Miñarro-Giménez, Catalina Martínez-Costa, Daniel Karlsson, Stefan Schulz, and Kirstine Rosenbeck Gøeg. 2018. Qualitative analysis of manual annotations of clinical text with SNOMED CT. *PLOS ONE*, 13(12):e0209547.

Hiroki Nakayama. 2018. seqeval: A python framework for sequence labeling evaluation. Software available from https://github.com/chakki-works/seqeval.

Allen Nie, Ashley Zehnder, Rodney L Page, Yuhui Zhang, Arturo Lopez Pineda, Manuel A Rivas, Carlos D Bustamante, and James Zou. 2018. DeepTag: inferring diagnoses from veterinary clinical notes. *npj Digital Medicine 2018 1:1*, 1(1):1–8.

Sarah J. Price, Sal A. Stapley, Elizabeth Shephard, Kevin Barraclough, and William T. Hamilton. 2016. Is omission of free text records a possible source of data loss and bias in Clinical Practice Research Datalink studies? A case-control study. *BMJ open*, 6(5).

A. D. Radford, P. J. Noble, K. P. Coyne, R. M. Gaskell, P. H. Jones, J. G.E. Bryan, C. Setzkorn, Á Tierney, and S. Dawson. 2011. Antibacterial prescribing patterns in small animal veterinary practice identified via SAVSNET: the small animal veterinary surveillance network. *Veterinary Record*, 169(12):310–310.

I. D. Robertson, P. J. Irwin, A. J. Lymbery, and R. C.A. Thompson. 2000. The role of companion animals in the emergence of parasitic zoonoses. *International Journal for Parasitology*, 30(12-13):1369–1377.

N. J. Robinson, R. S. Dean, M. Cobb, and M. L. Brennan. 2016. Factors influencing common diagnoses made during first-opinion small-animal consultations in the United Kingdom. *Preventive Veterinary Medicine*, 131:87–94.

Royal Veterinary College (RVC). VetCompass.

Fernando Sánchez-Vizcaíno, Philip H. Jones, Tarek Menacere, Bethaney Heayns, Maya Wardeh, Jenny Newman, Alan D. Radford, Susan Dawson, Rosalind Gaskell, Peter J.M. Noble, Sally Everitt, Michael J. Day, and Katie McConnell. 2015. Small animal disease surveillance. *Veterinary Record*, 177(23):591–594.

Fernando Sánchez-Vizcaíno, Peter John M. Noble, Phil H. Jones, Tarek Menacere, Iain Buchan, Suzanna Reynolds, Susan Dawson, Rosalind M. Gaskell, Sally Everitt, and Alan D. Radford. 2017. Demographics of dogs, cats, and rabbits attending veterinary practices in Great Britain as recorded in their electronic health records. *BMC Veterinary Research*, 13(1):1–13.

Zhecheng Sheng, Emma Bollig, Jennifer Granick, Rui Zhang, and Amanda Beaudoin. 2022. Canine Parvovirus Diagnosis Classification Utilizing Veterinary Free-Text Notes. *Proceedings - 2022 IEEE 10th International Conference on Healthcare Informatics, ICHI 2022*, pages 614–615.

Gregory E. Simon, Susan M. Shortreed, R. Yates Coley, Robert B. Penfold, Rebecca C. Rossom, Beth E. Waitzfelder, Katherine Sanchez, and Frances L. Lynch. 2019. Assessing and Minimizing Re-identification Risk in Research Data Derived from Health Care Records. *eGEMs*, 7(1):6.

Jasvinder A. Singh, Aaron R. Holmgren, and Siamak Noorbaloochi. 2004. Accuracy of veterans administration databases for a diagnosis of rheumatoid arthritis. *Arthritis Care & Research*, 51(6):952–957.

David A. Singleton, Gina L. Pinchbeck, Alan D. Radford, Elena Arsevska, Susan Dawson, Philip H. Jones, Peter John M. Noble, Nicola J. Williams, and Fernando Sánchez-Vizcaíno. 2020. Factors Associated with Prescription of Antimicrobial Drugs for Dogs and Cats, United Kingdom, 2014–2016. *Emerging Infectious Diseases*, 26(8):1778.

Mike P. Starkey, Timothy J. Scase, Cathryn S. Mellersh, and Sue Murphy. 2005. Dogs really are man's best friend–canine genomics has applications in veterinary and human medicine! *Briefings in functional genomics & proteomics*, 4(2):112–128.

Amber Stubbs, Michele Filannino, and Özlem Uzuner. 2017. De-identification of psychiatric intake records: Overview of 2016 CEGS N-GRID shared tasks Track 1. *Journal of Biomedical Informatics*, 75:S4–S18.

Amber Stubbs and Özlem Uzuner. 2015. Annotating longitudinal clinical narratives for de-identification: The 2014 i2b2/UTHealth corpus. *Journal of biomedical informatics*, 58 Suppl(Suppl):S20–S29.

Zhaohao Sun, Kenneth David Strang, and Francisca Pambel. 2020. Privacy and security in the big data paradigm. *Journal of Computer Information Systems*, 60(2):146–155.

Mahanazuddin Syed, Kevin Sexton, Melody Greer, Shorabuddin Syed, Joseph VanScoy, Farhan Kawsar, Erica Olson, Karan Patel, Jake Erwin, Sudeepa Bhattacharyya, Meredith Zozus, and Fred Prior. 2022. DeIDNER Model: A Neural Network Named Entity Recognition Model for Use in the De-identification of Clinical Notes. *Biomedical engineering systems and technologies, international joint conference, BIOSTEC ... revised selected papers. BIOSTEC (Conference)*, 5:640.

Llama Team and Ai @ Meta. 2024. The Llama 3 Herd of Models.

Erik F Tjong, Kim Sang, and Fien De Meulder. 2003. Introduction to the CoNLL-2003 Shared Task: Language-Independent Named Entity Recognition. pages 142–147.

Darren J. Trott, Lucio J. Filippich, John C. Bensink, Mary T. Downs, Suzanne E. McKenzie, Kirsty M. Townsend, Susan M. Moss, and James J.C. Chin. 2004. Canine model for investigating the impact of oral enrofloxacin on commensal coliforms and colonization with multidrug-resistant Escherichia coli. *Journal of Medical Microbiology*, 53(5):439–443.

Alexander Turchin, Nikheel S. Kolatkar, Richard W. Grant, Eric C. Makhni, Merri L. Pendergrass, and Jonathan S. Einbinder. 2006. Using Regular Expressions to Abstract Blood Pressure and Treatment Intensification Information from the Text of Physician Notes. *Journal of the American Medical Informatics Association*, 13(6):691–695.

Engeline Van Duijkeren, Maurice J.H.M. Wolfhagen, Adrienne T.A. Box, Max E.O.C. Heck, Wim J.B. Wannet, and Ad C. Fluit. 2004. Human-to-Dog Transmission of Methicillin-Resistant Staphylococcus aureus. *Emerging Infectious Diseases*, 10(12):2235.

Elizabeth J. Williamson, Alex J. Walker, Krishnan Bhaskaran, Seb Bacon, Chris Bates, Caroline E. Morton, Helen J. Curtis, Amir Mehrkar, David Evans, Peter Inglesby, Jonathan Cockburn, Helen I. McDonald, Brian MacKenna, Laurie Tomlinson, Ian J. Douglas, Christopher T. Rentsch, Rohini Mathur, Angel Y.S. Wong, Richard Grieve, David Harrison, Harriet Forbes, Anna Schultze, Richard Croker, John Parry, Frank Hester, Sam Harper, Rafael Perera, Stephen J.W. Evans, Liam Smeeth, and Ben Goldacre. 2020. Factors associated with COVID-19-related death using OpenSAFELY. *Nature 2020 584:7821*, 584(7821):430–436.

World Health Organisation (WHO). 2022. International Classification of Diseases 11th Revision (ICD-11).

Judit M. Wulcan, Kevin L. Jacques, Mary Ann Lee, Samantha L. Kovacs, Nicole Dausend, Lauren E. Prince, Jonatan Wulcan, Sina Marsilio, and Stefan M. Keller. 2024. Classification performance and reproducibility of GPT-4 omni for information extraction from veterinary electronic health records. *Frontiers in Veterinary Science*, 11:1490030.

Xi Yang, Tianchen Lyu, Qian Li, Chih Yin Lee, Jiang Bian, William R. Hogan, and Yonghui Wu. 2019. A study of deep learning methods for de-identification of clinical notes in cross-institute settings. *BMC Medical Informatics and Decision Making*, 19(5):1–9.

Jinsung Yoon, Michel Mizrahi, Nahid Farhady Ghalaty, Thomas Jarvinen, Ashwin S. Ravi, Peter Brune, Fanyu Kong, Dave Anderson, George Lee, Arie Meir, Farhana Bandukwala, Elli Kanal, Sercan Arık, and Tomas Pfister. 2023. EHR-Safe: generating high-fidelity and privacy-preserving synthetic electronic health records. *npj Digital Medicine 2023 6:1*, 6(1):1–11.

Yuhui Zhang, Allen Nie, Ashley Zehnder, Rodney L Page, and James Zou. 2019. VetTag: improving automated veterinary diagnosis coding via large-scale language modeling. *npj Digital Medicine 2019 2:1*, 2(1):1–8.

10 appendices

10.1 Task 1: Anonymisation Prompt

Prompt: Extract Named Entities from Veterinary EHRs You are given short free-text veterinary electronic health records (EHRs). Your task is to extract named entities mentioned in the text. Focus on identifying Names (NAME) locations (LOC), organizations (ORG), temporal expressions (TIME), and miscellaneous named entities (MISC). Examples:

Input: "Raven GA castrate. Anaes: Premed ACP/Meth. Induced propofol maint iso/02. Good anaesthetic. Op: Routine open castrate. double ligated 2-0 polysorb. Skin closed intradermal." Output: RavenNAME = Raven LOC = ORG = TIME = MISC = Input: "Waffle/MG - back end irritation. Owner reports irritation round back end, rubbing bottom over last 2-3 weeks." Output: NAME = Waffle LOC = ORG = TIME = last 2-3 weeks MISC = Input: "Adv routine haem/biochem (est £603) owner will discuss with wife. - Prescription -. Date: Apr 3, 2002. Vet: Reese, Qualifications: MRCVS." Output: NAME = Reese LOC = ORG = TIME = Apr 3, 2002 MISC =

Guidelines: -Extract only named entities in the appropriate categories:

NAME: Pet Names, Owner Names, Cliniain names LOC: geographical locations, clinics, hospitals, animal shelters ORG: veterinary practices, laboratories, pharmaceutical companies TIME: dates, time periods, durations, temporal references MISC: animal names, medications, procedures, medical equipment, qualifications - List each entity under its proper category. - If multiple entities of the same type are mentioned, extract each one separately. - Maintain the exact form as mentioned in the text.

10.2 Task 2: Syndromic Disease Classification Prompt

You are given a free-text veterinary electronic health records (EHRs). Your task is to assign a ICD-11 chapter names based on the conditions, symptoms, and diagnoses mentioned in the text. Each assigned chapter should correspond to the primary system or disease category affected.

ICD-11 Chapters: 1. Certain infectious or parasitic diseases 2. Neoplasms 3. Diseases of the blood or blood-forming organs 4. Diseases of the immune system 5. Endocrine, nutritional, or metabolic diseases 6. Mental, behavioral, or neurodevelopmental disorders 7. Sleep-wake disorders 8. Diseases of the nervous system 9. Diseases of the eye and adnexa 10. Diseases of the ear and mastoid process 11. Diseases of the circulatory system 12. Diseases of the respiratory system 13. Diseases of the digestive system 14. Diseases of the skin 15. Diseases of the musculoskeletal system or connective tissue 16. Diseases of the genitourinary system 17. Conditions related to sexual health 18. Pregnancy, childbirth, or the puerperium 19. Certain conditions originating in the perinatal period 20. Developmental anomalies 21. Symptoms, signs, or clinical findings not elsewhere classified 22. Injury, poisoning, or certain other consequences of external causes 23. External causes of morbidity or mortality 24. Factors influencing health status or contact with health services

Examples:

1. Input: "marked signs of renal failure. not eating much. huge wt loss. not moving around much." Output: Disease of the genitourinary system

2. Input: "Bilat OE. Mild, cleaned and wax removed, no obvious sign mites. Start on ear drops, rv sooner if concerned otherwise at next vaccination on 29th." Output: Diseases of the ear and mastoid process

3. Input: "skin lesions, bloods for meds check. noticed spot like skin lesions on forehead and side of face. not rubbing/scratching. would like checked. mass on R flank, slow growing, separated masses now merged together. pulsing meloxaid for stomatogingvitis." Output: Disease of the digestive system, Disease of the skin, Neoplasms

Guidelines: - Assign at least one ICD-11 chapter name that best represents the condition(s) described. - If no condition is present then return 'None' - If multiple conditions from different systems are mentioned, include multiple ICD-11 chapter names. - Ignore non-diagnostic text (e.g., medication instructions or routine check-ups) unless relevant to a condition. - Maintain consistency in ICD-11 chapter naming as per the official classification.

10.3 Task 3: Disease Extraction Prompt

You are given a free-text veterinary electronic health records (EHRs). Your task is to **extract the disease names** mentioned in the text. Focus on identifying diseases or conditions specifically mentioned, ignoring general symptoms, treatments, or non-diagnostic text.

Examples:

1. Input: "marked signs of renal failure. not eating much. huge wt loss. not moving around much." Output: renal failure

2. Input: "Bilat OE. Mild, cleaned and wax removed, no obvious sign mites. Start on ear drops, rv sooner if concerned otherwise at next vaccination on 29th." Output: OE

3. Input: "skin lesions, bloods for meds check. noticed spot like skin lesions on forehead and side of face. not rubbing/scratching. would like checked. mass on R flank, slow growing, separated masses now merged together. pulsing meloxaid for stomatogingvitis." Output: Skinskin lesions, stomatogingvitis, mass on R flank

Guidelines: - Extract only disease names (e.g., "Renal failure", "Otitis externa", "Neoplasm"). - Do not include symptoms, treatment plans, or general findings (e.g., "not eating much", "Start on ear drops"). - If multiple diseases are mentioned, extract each disease separately. - Maintain consistency in naming diseases and conditions as per medical terminology.

Virtual CRISPR: Can LLMs Predict CRISPR Screen Results?

Steven Song[1], Abdalla Abdrabou[2], Asmita Dabholkar[3], Kastan Day[3], Pavan Dharmoju[4]
Jason Perera[2], Volodymyr Kindratenko[2,3], Aly A. Khan[1,2,†]
[1] University of Chicago, [2] Chan Zuckerberg Biohub Chicago,
[3] University of Illinois Urbana-Champaign, [4] Northwestern University
[†]**Correspondence:** aakhan@uchicago.edu

Abstract

CRISPR-Cas systems enable systematic investigation of gene function, but experimental CRISPR screens are resource-intensive. Here, we investigate the potential of Large Language Models (LLMs) to predict the outcomes of CRISPR screens *in silico*, thereby prioritizing experiments and accelerating biological discovery. We introduce a benchmark dataset derived from BioGRID-ORCS and manually curated sources, and evaluate the performance of several LLMs across various prompting strategies, including chain-of-thought and few-shot learning. Furthermore, we develop a novel, efficient prediction framework using LLM-derived embeddings, achieving significantly improved performance and scalability compared to direct prompting. Our results demonstrate the feasibility of using LLMs to guide CRISPR screen experiments.

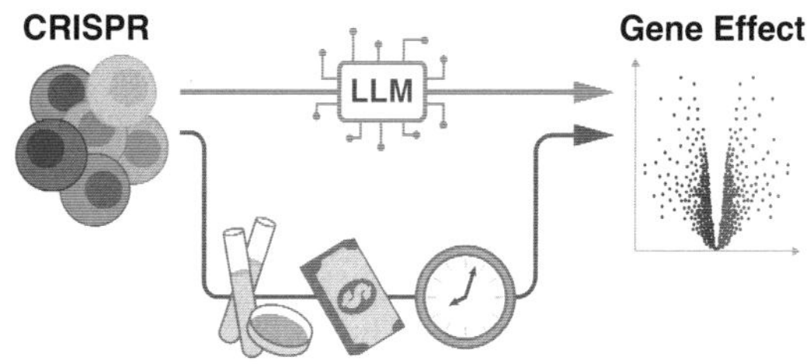

Figure 1: Conceptual motivation for LLM-driven prediction of CRISPR screen outcomes. An LLM-based approach transforms the traditionally resource-intensive experimental process of CRISPR screening into an *in silico* prediction task, where an LLM infers the phenotypic consequences of gene perturbations based on provided contextual information.

1 Introduction and Related Work

CRISPR-Cas technology has revolutionized biological research. While gene editing with CRISPR (Jinek et al., 2012) is a promising technology, genome-wide CRISPR screens have become a cornerstone of functional genomics. These screens allow researchers to systematically perturb genes and identify their causal roles in cellular processes and disease mechanisms (Shalem et al., 2014; Wang et al., 2014). However, these screens are resource-intensive, both in time, cost, and laboratory infrastructure. This can limit the scale and scope of biological investigations, hindering the discovery of novel therapeutic targets and a comprehensive understanding of complex biological systems, such as cancer progression and immune response (Doench et al., 2016; Evers et al., 2016). The ability to accurately predict the outcomes of CRISPR screens *in silico*, **before** conducting experiments, would dramatically accelerate biological discovery.

Prior work has explored computational methods for analyzing CRISPR screen data **after** experimentation. For example, MAGeCK (Li et al., 2014) and CRISPRAnalyzeR (Winter et al., 2017) provide tools for identifying essential genes and analyzing screen results. However, these methods are inherently reactive, offering insights only **after** resources have been expended on wet-lab experiments. They do not provide the capability to predict screen outcomes *a priori*. While other works have explored LLMs for CRISPR experiment *design* (e.g., guide RNA selection (Qu et al., 2024)) or discovering novel CRISPR systems (Li et al., 2024), our focus is distinctly on predicting the *phenotypic outcomes* of established screen types by leveraging an LLM's existing biological knowledge.

LLMs have demonstrated remarkable capabilities in understanding and reasoning about complex concepts across diverse domains (Brown et al., 2020; Chowdhery et al., 2023). Recent work shows promising results in applying LLMs to biological problems (Sarwal et al., 2023). For example, LLMs have been applied towards summarizing gene function (Chen and Zou, 2024), medical question answering (Singhal et al., 2023), cell-type annotation (Hou and Ji, 2024), and identifying causal genes

Proceedings of the 24th Workshop on Biomedical Language Processing (BioNLP 2025), pages 354–364
August 1, 2025 ©2025 Association for Computational Linguistics

in statistical genetics (Shringarpure et al., 2024). We hypothesize that LLMs possess the latent capacity to reason about and predict the outcomes of CRISPR screens, effectively simulating the effects of gene perturbations on cellular phenotypes (Figure 1). This would transform *in silico* biology from a primarily analytical tool to a predictive one, capable of guiding experimental design.

In this work, we investigate LLMs for CRISPR screen prediction. Our contributions are as follows:

- **Benchmark Dataset:** We introduce a new benchmark dataset for *a priori* CRISPR screen outcome prediction. It combines harmonized data from BioGRID-ORCS (Oughtred et al., 2021) with manually curated screens from recent high-impact publications, carefully selected to post-date LLM knowledge cutoffs, thus minimizing data leakage and ensuring a rigorous test of predictive capabilities.
- **Comprehensive LLM Evaluation:** We comprehensively evaluate a diverse set of LLMs (including variations of Llama-2 (Touvron et al., 2023), Llama-3.x (Dubey et al., 2024), GPT-3.5 (Brown et al., 2020), GPT-4 (Achiam et al., 2023), GPT-4o (Hurst et al., 2024), and o1 (Jaech et al., 2024)) using zero-shot, few-shot, and chain-of-thought (Wei et al., 2022) prompting strategies on our benchmark.
- **Embedding-Based Classifier:** We propose a novel, computationally efficient, and scalable CRISPR screen prediction framework. This approach leverages LLM-derived embeddings of CRISPR screen components (perturbation, gene, cell line, phenotype) as input to a multilayer perceptron (MLP) classifier, significantly outperforming direct LLM prompting, especially for complex phenotypes.

Our results show that LLMs, particularly our embedding-based model, can achieve promising performance. To our knowledge, this is the first systematic investigation of LLM potential for *a priori* CRISPR screen prediction, offering a new direction for computational biology.

2 Benchmark Dataset and Data Preparation

To systematically evaluate LLMs for *a priori* CRISPR screen prediction, we construct a new benchmark dataset. This dataset combines data

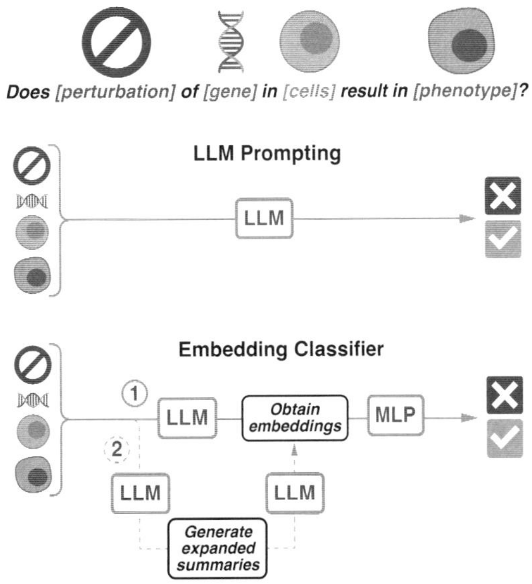

Figure 2: Two approaches for LLM-based CRISPR screen prediction: (Top) Direct prompting, where the LLM predicts a binary hit/no-hit outcome from a natural language query describing the screen. (Bottom) An embedding-based classifier, where LLM-derived embeddings of the screen components (perturbation, gene, cell line, and phenotype) are used as input to a trained MLP for outcome prediction. Embeddings are generated from either (1) raw text descriptions of each component or (2) LLM-generated summaries of these components.

from established repositories with manually curated data from recent publications. These publications were selected specifically because they were published after the training cutoffs of the LLMs we evaluate. This strategy ensures both a breadth of biological scenarios and minimizes the risk of data leakage, providing a fair assessment of the LLMs' predictive reasoning.

2.1 Data Sources

Our benchmark is built upon two primary sources. **(1) BioGRID-ORCS (v1.1.16):** The BioGRID Open Repository of CRISPR Screens (ORCS) (Oughtred et al., 2021) is the largest publicly available, harmonized database of CRISPR screens. It provides a broad foundation, encompassing a wide variety of experimental designs, cellular contexts, and observed phenotypes. **(2) Manually Curated Screens:** We complement BioGRID-ORCS with a manually curated set of screens focusing on complex phenotypes, extracted from two publications released in late 2024 (Chen et al., 2024; Skoulidis et al., 2024). This critical step ensures the dataset includes real-world screens and stringently avoids data leakage by selecting articles published after the knowledge cutoff of all evaluated LLMs.

2.2 Data Representation and High-Confidence Hit Definition

We adopt the BioGRID-ORCS harmonization approach, considering only results deemed statistically significant ("Hits") by the original study authors. This standardization minimizes inconsistencies arising from heterogeneous analysis pipelines. We focus exclusively on screens performed in human or mouse cell models, aligning genes to the GRCh38 (human) and GRCm39 (mouse) reference genomes.

Each screen result is represented as a tuple: (*perturbation method, gene, cell line, phenotype, hit/no-hit*). We refer to these tuples as gene-phenotype queries. This structure enables us to pose the prediction task as a natural language question: "Does {*perturbation method*} of {*gene*} in {*cell line*} causally result in {*phenotype*}?" The answer is binary: "Yes" (hit) or "No" (no-hit).

Crucially, we employ a high-confidence hit definition. A "Yes" (hit) indicates a statistically significant observation of the hypothesized effect. A "No" (no-hit) signifies a statistically significant effect *in the opposite direction* of the hypothesized phenotype. This is more stringent than simply the absence of the hypothesized effect. For example, if the question is "Does knockout of CD28 in T cells causally result in increased IL2 secretion?", a "No" outcome indicates that CD28 knockout *decreases* IL2 secretion significantly, not merely that it does not increase or has no effect. This strict definition is vital for evaluating LLM predictions, further detailed in Section 4.1.

2.3 Simple and Difficult Benchmarks

To evaluate LLMs across varying levels of complexity, we define two benchmarks (details in Appendix Tables 3 and 4):

Simple Benchmark: This benchmark focuses on relatively straightforward and well-understood phenotypic effects, often involving direct gene-phenotype relationships. We manually selected 4 screens from BioGRID-ORCS based on domain knowledge of the underlying biological pathways. This benchmark is comprised of 1175 gene-phenotype queries, where 41.3% of these are hits.

Model	F1 Score	FPR
Llama-2-7B	0.58 ± 0.02	0.97 ± 0.05
Llama-2-13B	0.51 ± 0.12	0.80 ± 0.32
Llama-2-70B	0.47 ± 0.20	0.71 ± 0.36
Llama-3-8B	0.48 ± 0.24	0.85 ± 0.38
Llama-3-70B	0.53 ± 0.09	0.58 ± 0.24
Llama-3.1-8B	0.39 ± 0.17	0.39 ± 0.27
Llama-3.1-70B	0.44 ± 0.14	0.38 ± 0.23
Llama-3.2-1B	0.37 ± 0.25	0.57 ± 0.43
Llama-3.2-3B	0.26 ± 0.24	0.28 ± 0.35
Llama-3.3-70B	0.40 ± 0.19	0.40 ± 0.31
o1	0.16 ± 0.02	0.03 ± 0.00
o1-mini	0.31 ± 0.04	0.10 ± 0.03
GPT-4o	**0.47 ± 0.06**	**0.22 ± 0.06**
GPT-4o-mini	0.55 ± 0.04	0.77 ± 0.16
GPT-4-turbo	0.32 ± 0.10	0.15 ± 0.08
GPT-4	0.44 ± 0.12	0.38 ± 0.21
GPT-3.5-turbo	0.42 ± 0.13	0.39 ± 0.25
Random Baseline	0.47	0.36

Table 1: Performance on the Simple CRISPR screen benchmark using LLM prompting. Results (mean and std. dev.) are aggregated across up to 10 combinations of Chain-of-Thought (CoT) and few-shot prompting strategies for each model. Highlighted model which maximizes F1 while minimizing FPR.

Difficult Benchmark: This benchmark presents more complex phenotypes requiring multi-step reasoning. For example, predicting "decreased resistance to PD1 blockade and lung carcinoma cell death" requires understanding the PD-1/PD-L1 pathway, its role in cancer cell survival, and the consequences of blocking this pathway. This benchmark is comprised of 1814 gene-phenotype queries from screens derived from the two manually curated, post-cutoff publications. Given the high-confidence hit definition and the nature of these complex screens, "hit" outcomes were rare. To address label imbalance in this dataset, we employed an inversion strategy for "hit" labels utilizing our high-confidence hit definition. For a query where the true outcome was a significant effect opposite to the hypothesized phenotype (a "no-hit" by our strict definition), we formulated an inverted query predicting this opposite phenotype and labeled this new, inverted query as a "hit". This process resulted in a balanced split of positive and negative examples for the Difficult Benchmark, totaling 907 positive and 907 negative instances. This benchmark specifically tests a model's ability to reason about more intricate biological mechanisms.

Model	AUROC	AUPRC	F1	FPR	PPV	NPV	Sens.	Spec.
Raw Emb.	**0.89**	**0.86**	**0.84**	**0.15**	**0.85**	**0.83**	**0.83**	**0.85**
Summ. Emb.	0.72	0.69	0.67	0.26	0.71	0.67	0.64	0.74
GPT-4o (prompting)	N/A	N/A	0.35 ±0.17	0.79 ±0.19	0.32 ±0.14	0.23 ±0.13	0.41 ±0.22	0.21 ±0.19

Table 2: Performance on the Difficult CRISPR screen benchmark. Compares the embedding-based classifier using embeddings of raw text (Raw Emb.) or embeddings of GPT-4o summaries (Summ. Emb.) against GPT-4o direct prompting (results aggregated across prompting strategies). Standard classification metrics reported. GPT-4o prompting gives binary outputs and thus AUROC and AUPRC are undefined. Best model highlighted.

3 Training Data for Embedding-Based Classifier

In addition to the evaluation benchmarks, we prepared a separate, larger training dataset for our embedding-based classifier. From the BioGRID-ORCS database, we selected screens performed in human or mouse cells. We excluded screens that indicated both significant positive and negative effects for the same gene-phenotype query, as this ambiguity complicates the definition of a singular phenotypic outcome from the free-text descriptions provided by BioGRID-ORCS. Instead of attempting to extract precise phenotypes from these notes, we utilize the entire phenotype note directly during the embedding process for training.

Our final training dataset consists of 1,678 screens from BioGRID-ORCS (from an initial pool of 1,924 total screens). This dataset yields approximately 22.6 million individual gene-level results across 3 perturbation methods, 40,461 unique genes (human and mouse), 133 cell lines, and 613 distinct phenotype descriptions. Within this training set, 7.74% of the gene-phenotype pairs are "hits" according to our high-confidence definition. For model development, we randomly selected 5 screens from this training dataset to serve as a validation set, used for tasks such as early stopping and hyperparameter tuning. To mitigate potential bias from dataset similarity during final evaluation, our embedding-based classifier is *exclusively* evaluated on the Difficult Benchmark (Section 2.3), which contains no data from BioGRID-ORCS.

4 CRISPR Screen Prediction Approaches

We explore two main approaches for *in silico* CRISPR screen prediction: direct LLM prompting and a scalable, embedding-based classifier (conceptualized in Figure 2).

4.1 LLM Prompting and Evaluation

We evaluated a diverse set of LLMs, including open-source Llama models and proprietary OpenAI models (see Appendix Table 7 for a complete list). We systematically tested combinations of Chain-of-Thought (CoT) (Wei et al., 2022) and few-shot prompting strategies, as well as zero-shot prompting. For CoT prompting, we instruct the model to explicitly reason through the relevant biological processes step-by-step. For few-shot prompting, we provide one or two example input-output pairs (illustrating positive and/or negative outcomes) before the target question. When combining CoT with few-shot prompting, the few-shot examples also include the CoT reasoning steps. An example prompt is provided in Appendix Table 5. Further details of our benchmarking pipeline are provided in Appendix Section A.1.

For each model and prompting strategy, we extract a binary answer ("Yes" or "No") from the generated text. Performance is primarily assessed using the F1 score and False Positive Rate (FPR). The F1 score is the harmonic mean of precision and recall, providing a balance between them. FPR (1 - Specificity) measures the proportion of actual negatives incorrectly classified as positive. We prioritize maximizing F1 and minimizing FPR due to our high-confidence hit definition, where a false positive (incorrectly predicting "Yes") means the model wrongly asserts a phenotypic effect in the opposite direction to the true significant effect.

Table 1 shows the aggregated results of LLMs on the Simple Benchmark. We report the mean and standard deviation for each model across up to 10 prompting strategy combinations. The random baseline (detailed in Appendix Section A.2) provides a performance floor. Compared to a random baseline (F1=0.47, FPR=0.36), GPT-4o (F1=0.47±0.06, FPR=0.22±0.06) is the only model which achieved a comparable F1 while attaining a notably better FPR. The generally modest performance of direct prompting, even on the Simple Benchmark, may stem from several factors. General-purpose LLMs, despite vast training data, may lack the specific, fine-grained biological nuance required even for seemingly direct gene-phenotype links. Additionally, the inherent complexity and context-dependency of biological systems mean that even "simple" effects can be modulated by cellular states or pathways not fully captured by the concise prompt. Nevertheless, GPT-4o's lower FPR suggests some capacity to avoid confident incorrect predictions of positive effects. We also observe relatively small variance in GPT-4o's performance across strategies compared to other models. The full performance metrics for GPT-4o across all prompting strategies on the Simple Benchmark are in Appendix Table 8.

4.2 Embedding-Based Classifier

Prompting LLMs for every gene in a genome-wide screen (often tens of thousands of genes) is computationally infeasible and cost-prohibitive. To address this scalability challenge, we developed a novel embedding-based classification framework. This approach leverages pre-computed embeddings of screen components, resulting in a small, efficient classifier at inference time where only cell line or phenotype descriptions may need new embeddings.

We use OpenAI's `text-embedding-3-large` model to generate embeddings for each component of the CRISPR screen tuples (perturbation method, gene symbol, cell line description, and phenotype description) from our training dataset (Section 3). We explored two strategies for generating these embeddings: (1) directly embedding the raw text terms for each component, and (2) embedding concise summaries of these terms, generated by GPT-4o, following a strategy similar to Shringarpure et al. (2024). An example prompt for summarization is in Appendix Table 6.

We then train a 5-layer MLP classifier using these concatenated LLM-derived embeddings as input. Additional training details are in Appendix Section A.3. The performance of our trained models on the Difficult Benchmark is presented in Table 2. We report Area Under the Receiver Operating Characteristic curve (AUROC), Area Under the Precision-Recall curve (AUPRC), F1 score, FPR, Positive Predictive Value (PPV, Precision), Negative Predictive Value (NPV), Sensitivity (Recall, True Positive Rate - TPR), and Specificity (True Negative Rate - TNR). Predicted probabilities are binarized using Youden's J statistic.

Our embedding-based classifiers significantly outperform direct GPT-4o prompting on the Difficult Benchmark. The model using raw text embeddings (Raw Emb.: F1=0.84, FPR=0.15) performs best, substantially exceeding GPT-4o's average prompting performance (F1=0.35±0.17, FPR=0.79±0.19). Interestingly, embeddings of raw text terms yield better results than embeddings of LLM-generated summaries (Summ. Emb.: F1=0.67, FPR=0.26). We hypothesize that the summarization process, while aiming for conciseness, may inadvertently omit subtle but critical nuances present in the original descriptions of cell lines or complex phenotypes, which are crucial for accurate prediction. This finding suggests that for tasks requiring deep, nuanced understanding, providing more complete and contextualized information to the embedding model may be beneficial.

5 Conclusion and Broader Impact

In this work, we investigated the potential of LLMs to predict CRISPR screen outcomes *a priori*. We introduced novel benchmarks designed to evaluate LLM capabilities across diverse biological contexts while mitigating data leakage. Our comprehensive evaluation of various LLMs and prompting strategies revealed inherent limitations in direct prompting for this complex task. However, our scalable and efficient embedding-based prediction framework substantially outperformed direct prompting, achieving an F1 score of 0.84 and an FPR of 0.15 on our Difficult Benchmark. This performance underscores the broad potential of LLM-driven approaches in advancing functional genomics.

Specifically, we envision LLM-guided screening as a powerful tool not only for CRISPR-based functional genomics, but also for predicting outcomes of diverse perturbation screens, thereby broadening its impact across experimental biology. Furthermore, a critical application of this framework also lies in identifying highly novel biological findings. When an LLM, drawing upon its extensive training on established knowledge, fails to predict a robust experimental hit, this discrepancy signals a result potentially unexplainable by current understanding. Such instances pinpoint exciting areas for discovering new biological mechanisms or gene functions, thereby enabling researchers to focus on novel leads and accelerating biological discovery.

Limitations and Future Work

Our study, while demonstrating promising results, has several limitations. The current prediction task is framed as a binary "hit/no-hit" classification, which simplifies the often quantitative and nuanced nature of CRISPR screen outcomes (e.g., magnitude of effect). We relied on existing general-purpose LLMs with fixed knowledge cutoffs; these models cannot dynamically incorporate the latest biological discoveries published after their training, potentially limiting predictive accuracy on cutting-edge research questions. While our Difficult Benchmark specifically used post-cutoff publications for evaluation, this is a general concern for static models. The training data for the embedding classifier, though large, may contain inherent biases (e.g., label imbalance, focus on protein-coding genes, etc). Furthermore, the size of our Difficult Benchmark test set was constrained by the availability of suitable, complex CRISPR screens published after LLM knowledge cutoffs that also lent themselves to our binary prediction framework; this reflects a necessary trade-off between test set scale and the rigor of avoiding data contamination for *a priori* evaluation.

Future work will focus on addressing these limitations and expanding the capabilities of our approach. We plan to:

- Expand our benchmark datasets to include more diverse biological contexts and screens.
- Benchmark a wider range of model families, including domain-specific LLMs pre-trained or fine-tuned on biological data, to compare against general-purpose models.
- Explore retrieval-augmented generation (RAG) techniques to enable models to incorporate the latest research findings at inference time, overcoming fixed knowledge cutoffs.
- Develop methods to predict quantitative outcomes or capture more nuanced aspects of phenotypic responses, moving beyond binary classification.
- Investigate more sophisticated prompting strategies and model architectures for improved biological reasoning.
- Conduct *in vitro* validation of the model's most confident or novel predictions to assess real-world utility.
- Develop systematic methods to analyze and prioritize discrepancies between LLM predictions and experimental outcomes to specifically flag and investigate potentially novel biological hits.
- Compare LLM-derived embeddings against other biological embedding methodologies for a broader understanding of their representational power for this task.

Ultimately, we aim to develop more robust and interpretable LLM-based tools to further accelerate biological discovery.

Ethics Statement

One potential ethical consideration is the availability of our source data. We primarily use open access data from BioGRID-ORCS. We additionally manually curated CRISPR screen data from recent publications. While one of the screens we curate for the Difficult Benchmark is derived from a publication in Nature (PMID: 39567689), which may be behind a paywall for some, the specific supplemental data containing the CRISPR screen results is publicly accessible, and the full article is available on PubMed Central. To the best of our knowledge, we have provided comprehensive descriptions, links to source code, and preprocessed data necessary to reproduce our experiments, promoting transparency and further research. Our software and benchmark data are available at: `https://github.com/czbiohub-chi/immune-llm-acl`.

Acknowledgements

The authors thank Shana Kelley, Adam Schauer, Hongyuan Mei, and Imran Razzak for their valuable feedback. S. Song is supported by NIH training grant T32GM007281. V. Kindratenko is a recipient of a Chan Zuckerberg Biohub Spoke Award. A. A. Khan is supported in part by NIH award DP2AI177884 and a Chan Zuckerberg Investigator Award.

References

Josh Achiam, Steven Adler, Sandhini Agarwal, Lama Ahmad, Ilge Akkaya, Florencia Leoni Aleman, Diogo Almeida, Janko Altenschmidt, Sam Altman, Shyamal Anadkat, et al. 2023. Gpt-4 technical report. *arXiv preprint arXiv:2303.08774*.

Tom Brown, Benjamin Mann, Nick Ryder, Melanie Subbiah, Jared D Kaplan, Prafulla Dhariwal, Arvind Neelakantan, Pranav Shyam, Girish Sastry, Amanda

Askell, et al. 2020. Language models are few-shot learners. *Advances in Neural Information Processing Systems*, 33:1877–1901.

Yiqun Chen and James Zou. 2024. Genept: a simple but effective foundation model for genes and cells built from chatgpt. *bioRxiv*, pages 2023–10.

Yu-Jung Chen, Swathi V Iyer, David Chun-Cheng Hsieh, Buren Li, Harold K Elias, Tao Wang, Jing Li, Mungunsarnai Ganbold, Michelle C Lien, Yu-Chun Peng, et al. 2024. Gliocidin is a nicotinamide-mimetic prodrug that targets glioblastoma. *Nature*, pages 1–8.

Aakanksha Chowdhery, Sharan Narang, Jacob Devlin, Maarten Bosma, Gaurav Mishra, Adam Roberts, Paul Barham, Hyung Won Chung, Charles Sutton, Sebastian Gehrmann, et al. 2023. Palm: Scaling language modeling with pathways. *Journal of Machine Learning Research*, 24(240):1–113.

John G Doench, Nicolo Fusi, Meagan Sullender, Mudra Hegde, Emma W Vaimberg, Katherine F Donovan, Ian Smith, Zuzana Tothova, Craig Wilen, Robert Orchard, et al. 2016. Optimized sgrna design to maximize activity and minimize off-target effects of crispr-cas9. *Nature Biotechnology*, 34(2):184–191.

Abhimanyu Dubey, Abhinav Jauhri, Abhinav Pandey, Abhishek Kadian, Ahmad Al-Dahle, Aiesha Letman, Akhil Mathur, Alan Schelten, Amy Yang, Angela Fan, et al. 2024. The llama 3 herd of models. *arXiv preprint arXiv:2407.21783*.

Bastiaan Evers, Katarzyna Jastrzebski, Jeroen PM Heijmans, Wipawadee Grernrum, Roderick L Beijersbergen, and Rene Bernards. 2016. Crispr knockout screening outperforms shrna and crispri in identifying essential genes. *Nature Biotechnology*, 34(6):631–633.

Wenpin Hou and Zhicheng Ji. 2024. Assessing gpt-4 for cell type annotation in single-cell rna-seq analysis. *Nature Methods*, pages 1–4.

Aaron Hurst, Adam Lerer, Adam P Goucher, Adam Perelman, Aditya Ramesh, Aidan Clark, AJ Ostrow, Akila Welihinda, Alan Hayes, Alec Radford, et al. 2024. Gpt-4o system card. *arXiv preprint arXiv:2410.21276*.

Aaron Jaech, Adam Kalai, Adam Lerer, Adam Richardson, Ahmed El-Kishky, Aiden Low, Alec Helyar, Aleksander Madry, Alex Beutel, Alex Carney, et al. 2024. Openai o1 system card. *arXiv preprint arXiv:2412.16720*.

Martin Jinek, Krzysztof Chylinski, Ines Fonfara, Michael Hauer, Jennifer A Doudna, and Emmanuelle Charpentier. 2012. A programmable dual-rna–guided dna endonuclease in adaptive bacterial immunity. *Science*, 337(6096):816–821.

Wei Li, Han Xu, Tengfei Xiao, Le Cong, Michael I Love, Feng Zhang, Rafael A Irizarry, Jun S Liu, Myles Brown, and X Shirley Liu. 2014. Mageck enables robust identification of essential genes from genome-scale crispr/cas9 knockout screens. *Genome Biology*, 15:1–12.

Wenhui Li, Xianyue Jiang, Wuke Wang, Liya Hou, Runze Cai, Yongqian Li, Qiuxi Gu, Qinchang Chen, Peixiang Ma, Jin Tang, et al. 2024. Discovering crispr-cas system with self-processing pre-crrna capability by foundation models. *Nature Communications*, 15(1):10024.

Rose Oughtred, Jennifer Rust, Christie Chang, Bobby-Joe Breitkreutz, Chris Stark, Andrew Willems, Lorrie Boucher, Genie Leung, Nadine Kolas, Frederick Zhang, et al. 2021. The biogrid database: A comprehensive biomedical resource of curated protein, genetic, and chemical interactions. *Protein Science*, 30(1):187–200.

Yuanhao Qu, Kaixuan Huang, Henry Cousins, William A. Johnson, Di Yin, Mihir Shah, Denny Zhou, Russ Altman, Mengdi Wang, and Le Cong. 2024. Crispr-gpt: An llm agent for automated design of gene-editing experiments. *bioRxiv*.

Varuni Sarwal, Viorel Munteanu, Timur Suhodolschi, Dumitru Ciorba, Eleazar Eskin, Wei Wang, and Serghei Mangul. 2023. Biollmbench: A comprehensive benchmarking of large language models in bioinformatics. *bioRxiv*, pages 2023–12.

Ophir Shalem, Neville E Sanjana, Ella Hartenian, Xi Shi, David A Scott, Tarjei S Mikkelsen, Dirk Heckl, Benjamin L Ebert, David E Root, John G Doench, et al. 2014. Genome-scale crispr-cas9 knockout screening in human cells. *Science*, 343(6166):84–87.

Suyash S Shringarpure, Wei Wang, Sotiris Karagounis, Xin Wang, Anna C Reisetter, Adam Auton, and Aly A Khan. 2024. Large language models identify causal genes in complex trait gwas. *medRxiv*, pages 2024–05.

Karan Singhal, Shekoofeh Azizi, Tao Tu, S Sara Mahdavi, Jason Wei, Hyung Won Chung, Nathan Scales, Ajay Tanwani, Heather Cole-Lewis, Stephen Pfohl, et al. 2023. Large language models encode clinical knowledge. *Nature*, 620(7972):172–180.

Ferdinandos Skoulidis, Haniel A Araujo, Minh Truong Do, Yu Qian, Xin Sun, Ana Galan Cobo, John T Le, Meagan Montesion, Rachael Palmer, Nadine Jahchan, et al. 2024. Ctla4 blockade abrogates keap1/stk11-related resistance to pd-(l) 1 inhibitors. *Nature*, 635(8038):462–471.

Hugo Touvron, Louis Martin, Kevin Stone, Peter Albert, Amjad Almahairi, Yasmine Babaei, Nikolay Bashlykov, Soumya Batra, Prajjwal Bhargava, Shruti Bhosale, et al. 2023. Llama 2: Open foundation and fine-tuned chat models. *arXiv preprint arXiv:2307.09288*.

Tim Wang, Jenny J Wei, David M Sabatini, and Eric S Lander. 2014. Genetic screens in human cells using the crispr-cas9 system. *Science*, 343(6166):80–84.

Jason Wei, Xuezhi Wang, Dale Schuurmans, Maarten Bosma, Fei Xia, Ed Chi, Quoc V Le, Denny Zhou, et al. 2022. Chain-of-thought prompting elicits reasoning in large language models. *Advances in Neural Information Processing Systems*, 35:24824–24837.

Jan Winter, Marc Schwering, Oliver Pelz, Benedikt Rauscher, Tianzuo Zhan, Florian Heigwer, and Michael Boutros. 2017. Crispranalyzer: Interactive analysis, annotation and documentation of pooled crispr screens. *BioRxiv*, page 109967.

A Appendix

A.1 Additional Benchmarking Details

All Llama models were accessed through HuggingFace and served using vLLM's OpenAI-like API server. This enabled a unified pipeline for benchmarking both Llama and OpenAI models via the OpenAI client's chat completions method. For Llama models, we used the "instruct" fine-tuned variants for Llama 3 and newer versions, and the "chat" variants for Llama 2. To ensure reproducibility, a fixed random seed was used, and the temperature parameter for LLM generation was set to 0. When benchmarking CoT prompting, additional instructions were provided to the model to explicitly use chain-of-thought reasoning to analyze the biological processes involved. For few-shot prompting, one or two manually constructed examples (positive and/or negative outcomes) were provided. These examples were interleaved between the main instructions and the final query, using transition text to clearly demarcate them as examples. For OpenAI's "o1" family of models, explicit CoT prompts were omitted as these models are designed to implicitly use CoT reasoning. All benchmarking of Llama models was conducted on a system with 2x NVIDIA A100 80GB GPUs.

A.2 Baseline Model

We compare our LLM-based approaches against a random baseline. This baseline predicts "hit" or "no-hit" outcomes randomly, with the probabilities of predicting "hit" weighted by the overall proportion of actual "hits" in the specific benchmark dataset being evaluated. This provides a simple lower-bound performance reference.

A.3 Additional Training Details for Embedding-Based Classifier

For our embedding-based classifier models, we utilized OpenAI's `text-embedding-3-large` model to compute 3072-dimensional embeddings for the raw text or summarized descriptions of CRISPR screen components (perturbation method, gene symbol, cell line, and hypothesized phenotype). For each data sample, these four embeddings were concatenated, resulting in an input vector of 12,288 dimensions (3072 * 4) for our 5-layer MLP. Each subsequent hidden layer in the MLP had half the number of neurons as the preceding layer, with a final classification layer for binary output. The hidden layer dimensions were thus [6144, 3072, 1536, 768]. The MLP was trained using a binary cross-entropy loss function. We used the AdamW optimizer with an initial learning rate of 0.001 and a weight decay of 0.01, a batch size of 8192, and a learning rate scheduler to reduce the learning rate if there was no improvement in validation loss after 5 epochs. Early stopping was triggered if validation loss did not improve for 15 consecutive epochs. The model weights corresponding to the epoch with the lowest validation loss were selected for the final model. The model trained on raw term embeddings converged after 4 epochs, while the model using summarized term embeddings trained for 3 epochs. All training was performed on a single NVIDIA A100 80GB GPU and took approximately 13 GPU hours in total for both models.

A.4 Supplemental Tables

Screen ID	Perturbation	Cell Line	Hypothesized Phenotype	Genes	Hits
1837	activation	primary CD4+ human T cells	increased TNF-alpha secretion	423	140
1885	activation	J774 macrophages	phagocytic inhibition	343	139
1835	activation	primary CD4+ human T cells	increased IL2 secretion	243	76
1733	knockout	HeLa cervical adenocarcinoma cells	increased RelA nuclear translocation	166	130

Table 3: Simple Benchmark: screens sourced from BioGRID-ORCS. "Genes" refers to the count of unique genes screened for the given phenotype in that screen after filtering for those with a significant effect, per our strict definition. "Hits" refers to the count of gene perturbations resulting in the hypothesized phenotype.

PMID	Perturbation	Cell Line	Hypothesized Phenotype	Genes	Hits
39567689	knockout	NG2-3112 mouse glioblastoma cells	increased sensitivity to gliocidin and subsequently glioblastoma cell death	881	73
39385035	knockout	3LL Lewis lung carcinoma cells	increased resistance to PD1 blockade and lung carcinoma cell survival	26	23
39567689	knockout	NG2-3112 mouse glioblastoma cells	decreased sensitivity to gliocidin and subsequently glioblastoma cell survival	881	808
39385035	knockout	3LL Lewis lung carcinoma cells	decreased resistance to PD1 blockade and lung carcinoma cell death	26	3

Table 4: Difficult Benchmark: screens derived from publications in late 2024 (Chen et al., 2024; Skoulidis et al., 2024). "Genes" refers to the count of unique genes screened for the given phenotype in that screen after filtering for those with a significant effect, per our strict definition. "Hits" refers to the count of gene perturbations resulting in the hypothesized phenotype. The two screens with high hit counts (808 and 23) correspond to the inverted phenotypes used for balancing, as described in Section 2.3.

Line	Role	Message
1	System	You are an expert in the fields of Immunology, Computational Biology, Genetics, and Molecular Biology.
2	System	Instructions: Carefully analyze the question and think through the biological process. Using the chain of thought method to clearly outline your reasoning, provide a step-by-step causal mechanistic explanation of the biological process. Give a final answer that is either "Final Answer: Yes" or "Final Answer: No". Do not give additional comments after your final answer.
3	System	Here is a positive example:
4	User	Question: Does knockout of BRCA1 in MCF-7 cells causally result in increased sensitivity to DNA-damaging agents?
5	Assistant	To address this question, let's break down the process step by step: 1. **Understanding BRCA1**: BRCA1 is a tumor suppressor gene that plays a crucial role in maintaining genome stability. It is involved in the repair of DNA double-strand breaks through homologous recombination (HR). When BRCA1 is functioning properly, it helps to repair DNA damage, thereby preventing mutations that can lead to cancer. 2. **Role of BRCA1 in DNA Repair**: In the absence of BRCA1, cells are more likely to rely on non-homologous end joining (NHEJ) for DNA repair, which is error-prone and can lead to genetic mutations. This is because BRCA1-deficient cells have impaired HR, making them more susceptible to DNA damage. 3. **Effect of BRCA1 Knockout on MCF-7 Cells**: MCF-7 cells are a human breast cancer cell line. Knocking out BRCA1 in these cells would likely impair their ability to repair DNA damage through HR, making them more reliant on NHEJ. This would increase the likelihood of genetic mutations and chromosomal instability. 4. **Sensitivity to DNA-Damaging Agents**: DNA-damaging agents, such as chemotherapy drugs, work by inducing DNA damage in cancer cells. In cells with impaired DNA repair mechanisms, such as those with a BRCA1 knockout, the DNA damage caused by these agents would be more difficult to repair. As a result, these cells would be more likely to undergo cell death, making them more sensitive to DNA-damaging agents. 5. **Causal Relationship**: The knockout of BRCA1 in MCF-7 cells would causally result in increased sensitivity to DNA-damaging agents because it would impair the cells' ability to repair DNA damage, making them more susceptible to the effects of these agents. Final Answer: Yes
6	System	Now answer the following question following the instructions and example above.
7	User	Question: Does {crispr_strategy} of {gene} in {cell_type} causally result in {phenotype}?

Table 5: Example prompt used for benchmarking LLMs, illustrating Chain-of-Thought (CoT) and 1-shot (positive example) prompting. The prompt is input to models using their respective chat templates. Placeholders {crispr_strategy}, {gene}, {cell_type}, and {phenotype} are replaced with terms specific to each screen.

Line	Role	Message
1	User	Generate a summary for the human gene {gene}. Include any relevant functions, pathways, cell types, interactions, or other important information regarding the gene.

Table 6: Example prompt used for summarizing CRISPR screen terms (specifically gene symbols in this example). Similar prompts were adapted for summarizing CRISPR perturbation methodology, cell line characteristics, and target phenotype descriptions when generating embeddings from summaries.

Model Alias	Model Version
o1	o1-2024-12-17
o1-mini	o1-mini-2024-09-12
GPT-4o	gpt-4o-2024-11-20
GPT-4o-mini	gpt-4o-mini-2024-07-18
GPT-4-turbo	gpt-4-turbo-2024-04-09
GPT-4	gpt-4-0125-preview
GPT-3.5	gpt-3.5-turbo-0125
Llama-2-7B	meta-llama/Llama-2-7b-chat-hf
Llama-2-13B	meta-llama/Llama-2-13b-chat-hf
Llama-2-70B	meta-llama/Llama-2-70b-chat-hf
Llama-3-8B	meta-llama/Llama-3-8B-Instruct
Llama-3-70B	meta-llama/Llama-3-70B-Instruct
Llama-3.1-8B	meta-llama/Llama-3.1-8B-Instruct
Llama-3.1-70B	meta-llama/Llama-3.1-70B-Instruct
Llama-3.2-1B	meta-llama/Llama-3.2-1B-Instruct
Llama-3.2-3B	meta-llama/Llama-3.2-3B-Instruct
Llama-3.3-70B	meta-llama/Llama-3.3-70B-Instruct

Table 7: List of LLMs benchmarked, with their common short name (Model Alias) and the specific version or identifier used in the experiments.

Model	Few-shot	CoT	F1	FPR	PPV	NPV	Sensitivity	Specificity
GPT-4o	0-shot	N	0.48	0.28	0.53	0.65	0.44	0.72
GPT-4o	1-shot (+)	N	0.49	0.23	0.56	0.66	0.43	0.77
GPT-4o	1-shot (-)	N	0.31	0.09	0.62	0.62	0.21	0.91
GPT-4o	2-shot (+/-)	N	0.53	0.24	0.58	0.68	0.48	0.76
GPT-4o	2-shot (-/+)	N	0.53	0.32	0.54	0.67	0.53	0.68
GPT-4o	0-shot	Y	0.53	0.25	0.58	0.68	0.49	0.75
GPT-4o	1-shot (+)	Y	0.50	0.23	0.57	0.66	0.44	0.77
GPT-4o	1-shot (-)	Y	0.46	0.20	0.57	0.65	0.39	0.80
GPT-4o	2-shot (+/-)	Y	0.44	0.18	0.58	0.64	0.36	0.82
GPT-4o	2-shot (-/+)	Y	0.45	0.21	0.56	0.65	0.38	0.79
Random	N/A	N/A	0.47	0.36	0.47	0.63	0.46	0.64

Table 8: Detailed performance of GPT-4o (model version gpt-4o-2024-11-20) on the Simple CRISPR screen benchmark across different prompting strategies. Metrics include F1 Score, False Positive Rate (FPR), Positive Predictive Value (PPV, Precision), Negative Predictive Value (NPV), Sensitivity (Recall, True Positive Rate), and Specificity (True Negative Rate). CoT indicates Chain-of-Thought prompting. (+)/(-) indicate positive/negative examples for 1-shot; (+/-) or (-/+) indicate order for 2-shot.

Overview of the BioLaySumm 2025 Shared Task on Lay Summarization of Biomedical Research Articles and Radiology Reports

Chenghao Xiao[1], **Kun Zhao**[2], **Xiao Wang**[3], **Siwei Wu**[3], **Sixing Yan**[4], **Tomas Goldsack**[5], **Sophia Ananiadou**[3], **Noura Al Moubayed**[1], **Zhan Liang**[2], **William Cheung**[4], **Chenghua Lin**[3]
[1]Durham University, [2]University of Pittsburgh, [3]The University of Manchester
[4]Hong Kong Baptist University, [5] The University of Sheffield
chenghao.xiao@durham.ac.uk chenghua.lin@manchester.ac.uk

Abstract

This paper presents the setup and results of the third edition of the BioLaySumm shared task on Lay Summarization of Biomedical Research Articles and Radiology Reports, hosted at the BioNLP Workshop at ACL 2025. In this task edition, we aim to build on the first two editions' successes by further increasing research interest in this important task and encouraging participants to explore novel approaches that will help advance the state-of-the-art. Specifically, we introduce the new task of Radiology Report Generation with Layman's terms, which is parallel to the task of lay summarization of biomedical articles in the first two editions. Overall, our results show that a broad range of innovative approaches were adopted by task participants, including inspiring explorations of latest RL techniques adopted in the training of general-domain large reasoning models.

1 Introduction

Lay Summarization describes the task of transforming a technical or specialist text that into summaries accessible to non-expert audience. By prioritizing clarity, context, and relevance over specialized terminologies, lay summaries bridge critical knowledge gaps between experts and diverse stakeholders, including practitioners, researchers in adjacent fields, patients, and the public. Despite their value in democratizing information, the creation of high-quality lay summaries remains scarce and labour-intensive, creating significant barriers to inclusive knowledge dissemination.

The need for accessible communication spans the entire biomedical ecosystem, from cutting-edge research to routine clinical care. Biomedical research publications, which contain the latest findings on prominent health-related topics, represent a key area where lay summarization is crucial. While mandatory for some journals, lay summaries are not universally adopted, leaving vital research inaccessible to non-experts. Even when required, authors who are often untrained in science communication struggle to distill their work effectively. Automatic lay summarization thus offers immense potential to scale accessibility while alleviating authorial burden, ensuring findings reach patients, policymakers, and interdisciplinary researchers

Parallel challenges exist in **clinical communication**, particularly in **radiology**. The 21st Century Cures Act (21st Century Cures Act, 2016) mandates immediate patient access to electronic health records, yet radiology reports—designed for clinicians—use highly technical language. Fewer than 4% radiology reports meet the eighth-grade reading level typical of U.S. adults (Martin-Carreras et al., 2019), causing confusion, anxiety, and poor adherence to follow-up care. Creating lay summaries of these reports is therefore not just a matter of convenience but a critical step toward a more patient-centered, transparent, and effective healthcare system.

The BioLaySumm shared task[1] is dedicated to advancing the automatic lay summarization of biomedical texts. Building on the success of the first two editions (Goldsack et al., 2023, 2024), **this year's shared task** addresses two domains: **biomedical articles** and **radiology reports**. Through this shared task, we aim to encourage the development of novel approaches and increase research interest in developing techniques for making scientific and clinical information accessible to broader audiences. In this paper, we present the results of the third edition of the BioLaySumm shared task, hosted by the BioNLP Workshop at ACL 2025. This year, we expand the scope of our challenge to include two parallel tracks: (i) the established task of Lay Summarization of Biomedical Research Articles; and (ii) a new track on the Lay Summarization of Radiology Reports.

In what remains of the paper, we address the formulation of these two tasks (§2), the datasets

[1]https://biolaysumm.org

Proceedings of the 24th Workshop on Biomedical Language Processing (BioNLP 2025), pages 365–377
August 1, 2025 ©2025 Association for Computational Linguistics

used (§3), and the evaluation procedure (§4), before providing a description of the participating systems (§5), and notable insights (§6).

2 Task Description

As part of the BioLaySumm 2025 shared task, participants developed systems capable of generating accessible summaries of biomedical content for non-expert audiences. Building upon previous editions, this year's competition introduced new challenges while maintaining core evaluation frameworks. The task was hosted using the CodaBench platform (Xu et al., 2022), with submissions automatically evaluated upon upload.

2.1 Task 1: Lay Summarization of Biomedical Articles

In Task 1, participants were required to generate plain-language summaries from technical research articles, with two distinct subtasks:

Subtask 1.1: Plain Lay Summarization required generating summaries using only the article's abstract and main text as input. As in previous editions, two separate datasets (**PLOS** and **eLife**) with notable stylistic differences were provided. Systems could employ either:

- Separate models trained independently on each dataset
- A unified model trained on both datasets

Final rankings were determined by average performance across both datasets.

Subtask 1.2: Lay Summarization with External Knowledge extended the plain summarization task by mandating incorporation of external resources to address knowledge gaps for lay audiences. Participants employed techniques such as Retrieval-Augmented Generation (RAG) or manual augmentation to integrate supplementary information (e.g., background context, terminology definitions).

2.2 Task 2: Radiology Report Generation

New in 2025, this task focused on translating medical imaging reports into patient-friendly explanations:

Subtask 2.1: Radiology Report Translation involved text-to-text simplification of professional radiology reports. Participants utilized report-layman term pairs from multiple datasets (Open-i, PadChest, BIMCV-COVID19 ± MIMIC-CXR), with separate rankings for systems using three versus four datasets.

Subtask 2.2: Multimodal Translation (optional) required generating lay summaries directly from medical images using end-to-end models (e.g., multimodal LLMs), with separate evaluation tracks based on training data scope.

Competition Framework Consistent with previous editions:

- Participants received training/validation sets with reference summaries alongside blind test sets
- For text-only tasks (Task 1 and Subtask 2.1), llama3 8B/Qwen2.5 7B will be used as the primary baseline.
- For multimodal task (Subtask 2.2), we will use finetuned Qwen-VL 7B as the finetuned baseline.

Detailed dataset characteristics appear in §3, with evaluation protocols in §4. Participants could attempt any combination of subtasks based on their research interests.

3 Datasets

The datasets used for the Task 1 are based on the previous works of Goldsack et al. (2022) and Luo et al. (2022), and are derived from two different biomedical publications: **Public Library of Science (PLOS)** and **eLife**. Each dataset consists of biomedical research articles paired with expert-written lay summaries.

As described in Goldsack et al. (2022), the lay summaries of each dataset also exhibit numerous notable differences in their characteristics, with eLife's lay summaries being longer, more abstractive, and more readable than those of PLOS.

Furthermore, for PLOS, lay summaries are author-written, and articles are derived from 5 peer-reviewed journals covering Biology, Computational Biology, Genetics, Pathogens, and Neglected Tropical Diseases. For eLife, lay summaries are written by expert editors (in correspondence with

Dataset	Task	# Train	# Val	# Test
eLife	1	4,346	241	142
PLOS	1	24,773	1,376	142*
PadChest	2	116,847	7,824	7,130
BIMCV-COVID19	2	31,364	2,042	3,221
Open-i	2	2,243	1,34	186
MIMIC-CXR	2	45,000	5,000	500

Table 1: Data split sizes for each dataset. * denotes that this split is different for each subtask.

authors), and articles are derived from the peer-reviewed eLife journal, covering all areas of the life sciences and medicine. For a more detailed analysis of dataset content, readers can refer to Goldsack et al. (2022).

For Task 2, we utilized four radiology datasets: PadChest (Bustos et al., 2020), BIMCV-COVID19+ (Vayá et al., 2020), Open-i (Demner-Fushman et al., 2012), and MIMIC-CXR (Johnson et al., 2019). The PadChest dataset comprises over 160,000 images from 67,000 patients, interpreted by radiologists at San Juan Hospital (Spain) between 2009 and 2017, and includes six positional views with supplementary acquisition and demographic metadata. The BIMCV-COVID19+ dataset contains chest X-rays (CXR/DX) and computed tomography (CT) images of COVID-19 patients, accompanied by radiographic findings, pathologies, polymerase chain reaction (PCR) tests, immunoglobulin G (IgG)/M (IgM) antibody tests, and reports from the Valencian Community Medical Image Database (BIMCV). This database includes 21,342 CR, 34,829 DX, and 7,918 CT studies. Open-i offers access to 3.7 million images from 1.2 million PubMed Central articles, including 7,470 chest X-rays with 3,955 reports. The MIMIC-CXR dataset contains 377,110 JPEG images with structured labels derived from 227,827 associated free-text reports, de-identified to comply with HIPAA Safe Harbor requirements by removing protected health information (PHI).

For the layman-style reports of Task 2, we applied the method from Zhao et al. (2025) to create the layman-style reports for all four datasets. PadChest and BIMCV-COVID19+ reports were first translated into English before transformation; Open-i and MIMIC-CXR were converted directly. A subset of MIMIC-CXR reports was selected for training and testing in this shared task.

Table 1 summarizes the data split information for all datasets of two Tasks. Note that the training and validation sets used for both datasets are identical to those published in Goldsack et al. (2022) and Zhao et al. (2025). By leveraging these datasets, we aim to develop abstractive summarization models and layman-style report generation systems capable of producing accessible summaries for unseen biomedical articles and layman-style radiology reports. This approach will facilitate effective communication of significant new publications to non-expert audiences and patients across diverse biomedical domains.

4 Evaluation

Task1: Lay Summarization For both subtasks of Task 1, we evaluate summary quality according to three criteria - *Relevance*, *Readability*, and *Factuality* - where each criterion is composed of one or more automatic metrics:

- *Relevance*: ROUGEROUGE - 1, 2, and L (Lin, 2004), *BLEU (Papineni et al., 2002), *METEOR (Banerjee and Lavie, 2005), and BERTScore (Zhang et al., 2020).
- *Readability*: Flesch-Kincaid Grade Level (FKGL), Dale-Chall Readability Score (DCRS), Coleman-Laiu Index (CLI), and LENS (Maddela et al., 2023).
- *Factuality*: AlignScore (Zha et al., 2023) and SummaC (Zha et al., 2023)

Task2: Radiology Report Generation For both subtasks of Task 2, we evaluate report quality according to three criteria - *Relevance*, *Readability*, and *Clinical* - where each criterion is composed of one or more automatic metrics:

- *Relevance*: ROUGE - 1, 2, and L (Lin, 2004), BLEU (Papineni et al., 2002), METEOR (Banerjee and Lavie, 2005), and BERTScore (Zhang et al., 2020), *Semantic Similarity scores (Pesquita et al., 2009).
- *Readability*: Flesch-Kincaid Grade Level (FKGL), Dale-Chall Readability Score (DCRS), Coleman-Laiu Index (CLI).
- *Clinical Metrics*: *CheXbert (Smit et al., 2020), and *RadGraph (Jain et al., 2021).

Here "*" indicates that the metric is newly introduced for this year's edition of the task. Specifically, the BLEU and METEOR metrics are introduced to measure how closely a system-generated summary or report matches its reference at the lexical level. To assess report quality at the semantic level, we introduce a semantic score measured based on the cosine similarity between the sentence-level embeddings of each generated report and its reference. Additionally, CheXbert and RadGraph are introduced to quantify clinical correctness, which can not be assessed by general metrics. By incorporating these two domain-aware metrics, the evaluation process could be more comprehensive.

For Task 1, The scores calculated for each metric are the average of those calculated independently for the generated lay summaries of PLOS and eLife. As for Task 2 (open track), all scores are computed on the combined public datasets — PadChest, BIMCV-COVID19, and Open-i. While for Task 2 (closed track), the reports are evaluated on open-track datasets plus MIMIC-CXR, and each metric is then averaged over the two scores.

The aim is to maximize the scores for all metrics except for FKGL, DCRS, and CLI the Readability metrics. For these metrics, the aim is to minimize scores, as lower scores are indicative of greater readability.[2]

Following the submission deadline for each subtask, an overall ranking is calculated based on the average performance of submissions across all criteria. Specifically, we first apply min-max normalization to the scores of each metric (thus establishing a common value range), before averaging across metrics within each criterion to obtain criterion-level scores. Note that, for metrics that we minimize (i.e., FKGL, DCRS, and CLI) we calculate 1 minus the mix-max normalized value. Finally, criterion-level scores are then averaged to obtain an overall score, by which submissions are then ranked.

Baselines We train Qwen2.5 (Qwen et al., 2025), LLaMA3 (Grattafiori et al., 2024), and Qwen2.5-VL (Bai et al., 2025) on the BioLaySumm 2025 training dataset as the baseline models. (1) For **Task 1 (Lay Summarization)**, we select Qwen2.5-7B-Instruct and LLaMA3-8B-Instruct as the backbone models and train them on our training data by using the whole article as input and the lay summary as output. (2) For the **Task 2 (Radiology Report Generation with Layman's Terms)**, we train Qwen2.5-7B-Instruct and LLaMA3-8B-Instruct on our training data for Subtask 2.1 (Radiology Report Translation) and train Qwen2.5-VL-7B-Instruct for the Subtask 2.2 (Multi-modal Radiology Report Translation).

[2]For these metrics, the scores are estimates of the US Grade level of education required to comprehend a given text.

5 Submissions

Out of all participating teams, 13 teams submitted system papers. Here, we provide a brief summary of the approaches taken by these teams.

AEHRC (Zhang et al., 2025) This team produced the top-ranked submission for both open-source and close-source tracks of Subtask 2.1, and provided a comparison study between encoder-decoder and decoder-only architectures. The paper presents the surprising results that a 700M T5-large-based model provides better performance than a 3B LLaMA-3.2-based model across nine out of ten metrics, including relevance, readability, and clinical accuracy, despite having significantly fewer parameters. The findings highlight the continual relevance of encoder-decoder models for lay summarization tasks in the era of LLMs.

MetninOzu (Evgin et al., 2025) This team proposes two innovative approaches, reverse data augmentation and salient sentence injection, and a detailed study of them. The authors curated a dataset of child-friendly articles with corresponding gold-standard summaries and used LLMs to rewrite them into more complex scientific variants to augment the training data beyond the shared-task training set. They also investigated whether they can insert salient sentences from the main article into the summary to enrich the input, leveraging sentence embedding models.

XSZ (Xu et al., 2025) This team investigates (i) k-shot demonstration fine-tuning with LLMs, and (ii) further employing latest reasoning-oriented RL methods to LLMs. For the first method, they use embedding models to retrieve top-K examples and fine-tune a Llama3-8B with LoRA. They then employ RL algorithms (PPO and GRPO) to further fine-tune the models. The reward function is specifically design to optimize the evaluated metrics, including factual metrics, relevant metrics and readability metrics. Although the RL results are not submitted to the competition, the paper is well-implemented and innovative, showing that RL methods are useful for lay summarization.

Rank	Team	Relevance				Readability				Factuality	
		ROUGE	BLEU	MTR	BERTS	FKGL	DCRS	CLI	LENS	AlignS	SummaC
1	**SUWMIT**	**0.370**	**10.07**	0.308	0.864	11.74	9.08	12.58	72.61	0.750	0.682
2	**Baseline-llama3-8B-sft**	0.366	9.86	**0.314**	0.863	12.20	9.25	12.98	72.86	0.722	0.644
3	**Baseline-qwen2.5-7B-sft**	0.352	8.74	0.303	0.870	12.71	9.65	13.70	60.22	0.754	0.644
4	**BDA-UCM**	0.334	8.08	0.294	**0.870**	12.32	9.26	13.20	64.07	0.691	0.590
5	**MetinOZU**	0.330	6.95	0.290	0.857	16.45	11.22	17.01	34.86	**0.881**	**0.920**
6	**MIRAGES**	0.288	4.63	0.230	0.846	11.71	**8.46**	11.99	71.27	0.681	0.605
7	**TupiQ**	0.335	7.16	0.268	0.862	13.44	10.59	13.48	43.67	0.762	0.642
8	**LaySummX**	0.321	5.44	0.253	0.855	12.33	9.51	13.38	80.46	0.675	0.521
9	**CUTN_Bio**	0.268	3.25	0.226	0.848	**10.52**	8.84	**11.43**	**84.14**	0.589	0.549
10	**Aard**	0.319	5.45	0.293	0.851	14.56	10.02	15.36	71.51	0.695	0.509
11	**LTRC**	0.288	4.27	0.222	0.850	13.36	9.30	13.29	79.34	0.601	0.476
12	**5cNLP**	0.333	6.14	0.268	0.859	16.07	10.40	15.34	76.05	0.631	0.549
13	**RainCityNLP**	0.284	4.87	0.241	0.840	16.74	11.66	16.24	9.41	0.612	0.653
14	**SXZ**	0.165	1.33	0.153	0.801	12.59	11.83	13.29	6.56	0.862	0.528
15	**demo**	0.165	1.33	0.153	0.801	12.59	11.83	13.29	6.56	0.862	0.528
16	**x2z**	0.182	1.18	0.168	0.804	12.60	8.56	12.65	63.22	0.368	0.468

(a) SubTask 1.1: Plain Lay Summarization

Rank	Team	Relevance				Readability				Factuality	
		ROUGE	BLEU	MTR	BERTS	FKGL	DCRS	CLI	LENS	AlignS	SummaC
1	**Aard**	0.292	4.32	0.262	0.848	**11.16**	**8.36**	**11.94**	**81.50**	0.614	**0.537**
2	**CUTN_Bio**	0.296	4.08	0.228	0.855	13.37	10.25	14.74	80.00	**0.689**	0.507
3	**5cNLP**	**0.335**	**5.91**	**0.275**	**0.858**	16.30	10.29	15.24	75.57	0.610	0.445
4	**LTRC**	0.215	2.01	0.169	0.818	13.71	9.66	13.60	74.48	0.378	0.429

(b) Subtask 1.2: Lay Summarization with External Knowledge

Rank	Team	Relevance					Readability			Clinical Metrics	
		ROUGE	BLEU	MTR	BERTS	SIM	FKGL	DCRS	CLI	CHEX	RG
1	**AEHRC**	**0.671**	**46.09**	**0.704**	**0.953**	**0.890**	**7.397**	9.31	**8.05**	**0.860**	**0.402**
2	**KHU_LDI**	0.529	28.66	0.577	0.935	0.843	7.528	**9.29**	8.26	0.827	0.265

(c) Subtask 2.1: Radiology Report Translation (Open Track)

Rank	Team	Relevance					Readability			Clinical Metrics	
		ROUGE	BLEU	MTR	BERTS	SIM	FKGL	DCRS	CLI	CHEX	RG
1	**AEHRC**	**0.629**	**38.99**	**0.669**	**0.948**	**0.894**	7.574	8.97	7.95	0.777	**0.377**
2	**Baseline-qwen2.5-7B-sft**	0.537	25.71	0.543	0.938	0.854	**6.440**	10.04	8.55	0.779	0.291
3	**5cNLP**	0.555	28.27	0.609	0.937	0.872	8.046	9.24	8.23	0.750	0.317
4	**Baseline-llama3-8B-sft**	0.527	25.18	0.527	0.936	0.847	6.785	**8.53**	8.67	**0.806**	0.286
5	**CUTN_Bio**	0.404	14.90	0.428	0.913	0.798	7.359	8.53	**7.36**	0.704	0.216

(d) Subtask 2.1: Radiology Report Translation (Closed Track)

Table 2: Task leaderboard - all metrics. **BertS** = BertScore, **FKGL** = Flesch-Kincaid Grade Level, **DCRS** = Dale-Chall Readability Score, **CLI** = Coleman-Liau Index, **MTR** = METOR, **SIM** = Similarity, **AlignS** = AlignScore, **CHEX** = F1 chexbert, **RG** = Radgraph.

Aard (Gupta and Krishnamurthy, 2025) This team introduced a modular and flexible system designed for generating lay summaries by leveraging large language models, a BioBERT-based named entity recognizer, and the UMLS knowledge base. For Task 1.1, they focused on summarization using only the internal content of articles, while Task 1.2 enhanced this with external biomedical knowledge like terminology definitions to improve readability and factuality. Their approach involved chunking articles, extracting key sentences, iterative rewriting, and integrating simplified definitions for complex terms. The LayForge system demonstrated strong performance, especially in readability metrics, highlighting the effectiveness of domain-specific augmentation for lay summary generation.

RainCityNLP (Wilson et al., 2025) This team utilized TF-IDF for sentence scoring and experimented with Pegasus-XSum and a Falcons.ai model pre-trained on medical data. All experiments were conducted on consumer-grade hardware, demonstrating feasibility in low-resource settings. Evaluation showed the Falcons.ai model scored highest in relevance, while Pegasus-XSum excelled in readability metrics like FKGL and LENS. The original extractive summaries outperformed others in factuality. The team also created a dictionary of medical terms translated to lay-terms for future use. Their work highlights both economic and practical accessibility in medical summarization.

TLPIQ (Bechler et al., 2025) This team focused on generating biomedical lay summaries using a fine-tuned FLAN-T5 base model, leveraging abstract and conclusion sections of articles along with expert-written lay summaries. They improved accessibility and understanding by maintaining factuality and domain relevance, despite falling short on readability compared to larger models like Llama3 and Qwen2.5. Their approach included instruction tuning with dataset tags and a specialized prompt template, achieving competitive relevance and superior factuality scores. However, the model's readability could be further enhanced through strategies such as dataset-specific training and post hoc lexical simplification.

LaySummX (Lin and Yu, 2025) This team introduced a retrieval-augmented fine-tuning framework for biomedical lay summarization, utilizing abstract-driven semantic retrieval with LoRA-tuned LLaMA3.1 models. By incorporating relevant full-text segments retrieved using the article abstracts into the fine-tuning process, they improved relevance and factuality metrics significantly compared to base models and individually tuned models, while maintaining competitive readability. Their method efficiently addresses computational constraints by segmenting articles into manageable units, demonstrating strong performance among open-source systems and closed-source models like GPT.

5cNLP (Lossio-Ventura et al., 2025) This team leveraged a combination of prompting strategies, retrieval techniques, and multimodal fusion for generating lay summaries from scientific articles and radiology reports. They utilized structured (compositional) prompting with role-based instructions to guide large language models (LLMs) like Llama-3.3-70B-Instruct and GPT-4.1 in producing summaries that are accessible to a general audience. Their method also incorporated retrieval-augmented generation (RAG) using biomedical knowledge from UMLS to enrich context understanding and employed a multimodal pipeline combining images and captions for radiology report summarization. Notably, their approach achieved second place in Subtask 2.1 close-source track and third place in Subtask 1.2, demonstrating the effectiveness of their framework in improving accessibility and understandability of complex medical information.

MIRAGES (Pong et al., 2025) The team approached the BioLaySumm 2025 task by building on an extract-then-summarize framework, emphasizing the importance of high-quality data curation for biomedical lay summarization. They experimented with various extractive summarization strategies and employed LoRA to fine-tune a Llama-3-8B to enhance readability and factual accuracy of downstream abstractive summaries. Additionally, they explored counterfactual data augmentation and post-processing definition insertion to further improve factual grounding and accessibility. Their system ranked 4th overall and achieved 2nd place in readability, demonstrating that good input design and targeted fine-tuning are critical for effective biomedical lay summarization. Their findings suggest that strategic data curation can have a more positive impact than merely increasing the volume of fine-tuning samples in domain-specific summarization tasks.

SUWMIT (Basu et al., 2025) This team developed an open-source, end-to-end pipeline for the automated generation of lay summaries from biomedical articles, achieving top scores in two out of four relevance metrics and the highest overall ranking in the plain lay summarization subtask. Their approach involved fine-tuning a Llama-3.1-8B model with LoRA, utilizing a contrastive decoding strategy known as DoLa to improve factuality and readability. They experimented with various preprocessing, extractive summarization, and abstractive summarization techniques, ultimately finding that including Flesch-Kincaid grade-level targets in system messages and applying LoRA weights during decoding were crucial for their success. Additionally, they explored different data transformation methods, including the use of BioBERT embeddings for extractive summarization, to enrich input context for improved summary quality.

KHU_LDI (Moriazi and Sung, 2025) This team explored two approaches for generating lay radiology reports: supervised fine-tuning of open-source large language models using QLoRA, and a refinement process to improve the initial generated output. They found that while the fine-tuned model outperformed the refinement approach on test data, the refinement method showed significant potential on the validation set, particularly when using GPT-4o-mini as both the feedback and refinement models. Their submission achieved second place in the open track of Subtask 2.1, highlighting the effectiveness of fine-tuning open-source models for producing patient-friendly radiology reports.

BDA-UC3M (Ramzi and Bedmar, 2025) This team focused on demonstrating that small-scale, state-of-the-art language models (4B–7B parameters) can achieve competitive performance in biomedical lay summarization. Utilizing models such as Gemma3 4B, Qwen3 4B, and GPT-4.1-mini, they employed dynamic 4-bit quantization, extractive preprocessing, prompt engineering, data augmentation, and Direct Preference Optimization to enhance efficiency and factuality. Their system ranked second in its category by generating high-quality, accurate summaries, highlighting the potential of compact models for making complex scientific content accessible to non-expert audiences without sacrificing performance.

CUTN_Bio (Sivagnanam et al., 2025) This team focused on developing a prompt-based lay summarization framework for biomedical articles and radiology reports as part of the BioLaySumm 2025 shared task. For plain lay summarization, they utilized Llama-3-8B with a Tree-of-Thought prompting strategy to generate simplified summaries. In the lay summarization with external knowledge subtask, they combined an extractive approach (LEAD-K sentence extraction) with Llama-3-8B, enriched by medical definitions from MedCAT and Wikipedia, achieving the second position in Task 2.1. For radiology report translation, they implemented a Retrieval-Augmented Generation (RAG) method using the Zephyr model, achieving third in this category. Their methodologies highlight the effectiveness of combining external knowledge, extractive summarization techniques, and instruction-tuned language models for generating accessible summaries.

6 Results Analysis

The BioLaySumm 2025 shared task revealed critical insights about biomedical lay summarization methodologies, emphasizing trade-offs, architectural innovations, and emerging trends. The analysis below synthesizes key findings from both the competition leaderboard (Table 2) and participant approaches.

Trade-offs Between Evaluation Metrics No single system dominated all evaluation dimensions (relevance, readability, factuality), revealing inherent conflicts in optimization objectives. For instance, SUWMT (1st in Subtask 1.1) excelled in relevance (ROUGE: 0.370) but produced complex text (FKGL: 11.74), while MetinOZU achieved exceptional factuality (SummaC: 0.920) at the cost of poor readability (FKGL: 16.45). Aard demonstrated balanced readability (FKGL: 11.16) and factuality (SummaC: 0.537) in Subtask 1.2 but lagged in relevance (ROUGE: 0.292). These cases illustrate how excelling in one metric often compromises others, necessitating task-specific customization.

Dominance of Retrieval-Augmented Generation Retrieval-augmented approaches emerged as a dominant trend, with 5 of 13 teams (LaySummX, BioSumEnhance, CUTN_Bio, Aard, and 5cNLP) incorporating external knowledge. This strategy proved particularly effective in Subtask 1.2 (external knowledge), where Aard and CUTN_Bio secured 1st and 2nd places with 7–9% factuality gains over non-RAG baselines. Teams leveraged UMLS, Wikipedia, and full-text segments to handle domain terminology, though sometimes at the cost of readability due to verbose outputs.

Persistence of Pipeline Approaches Pipeline frameworks remained prevalent, with 7 of 13 teams adopting multi-stage architectures rather than unified models. Examples include MIRAGES' extract-then-summarize approach using extractive summarization followed by LoRA-tuned Llama3-8B (ranking 6th with 2nd-best readability), and Aard's modular system combining BioBERT-based entity recognition with iterative rewriting. These pipelines offered interpretability advantages but introduced potential error propagation risks compared to end-to-end systems like SUWMT's top-ranked submission.

Competitiveness of Legacy Architectures Encoder-decoder models demonstrated comparable performance against larger LLMs. AEHRC's T5-large (700M parameters) outperformed 3B+ LLMs in 9 of 10 metrics for radiology report translation (Subtask 2.1), dominating both competition tracks. Similarly, TLPIQ's FLAN-T5 base model achieved competitive relevance and factuality despite its smaller size, underscoring the continued efficiency of these architectures for domain-specific generation tasks.

Emerging Methodological Innovations Several novel techniques showed promise: XSZ explored reinforcement learning (PPO/GRPO) with multi-objective rewards optimizing factuality, readability, and relevance; MetinOZU developed reverse data augmentation by generating complex scientific text from simple summaries; and BDA-UC3M implemented efficiency techniques like 4-bit quantization with Direct Preference Optimization. While not all innovations were competition submissions, they represent significant research directions.

Hardware Efficiency Demonstrations Several teams validated cost-effective approaches, most notably RainCityNLP which combined TF-IDF sentence scoring with Pegasus-XSum and medical Falcons.ai models running on consumer-grade hardware. These implementations demonstrate the feasibility of deploying lay summarization systems in resource-constrained environments while maintaining reasonable performance.

Key Gaps and Future Directions Three critical challenges emerged from the analysis: (1) The persistent conflicts between readability and factuality require new joint optimization strategies; (2) External knowledge integration through RAG sometimes disrupted narrative coherence despite improving accuracy; (3) Reinforcement learning approaches like XSZ's show untapped potential for metric-aligned reward shaping that warrants deeper exploration.

7 Conclusion

The third edition of the BioLaySumm Shared Task was hosted by the BioNLP Workshop@ACL 2025. Several changes were implemented over the previous edition, including the incorporation of the new task, lay summarization of radiology reports. The competition outcomes underscore biomedical lay summarization as a multi-faceted challenge requiring context-aware solutions. While RAG and pipeline methods dominated submissions, legacy encoder-decoder models (T5, FLAN-T5) remained surprisingly effective. Future work should prioritize hybrid approaches, particularly RAG-enhanced end-to-end models with RL fine-tuning, to better harmonize the competing demands of relevance, readability, and factuality.

References

21st Century Cures Act. 2016. 21st century cures act. An Act to accelerate the discovery, development, and delivery of 21st century cures, and for other purposes.

Shuai Bai, Keqin Chen, Xuejing Liu, Jialin Wang, Wenbin Ge, Sibo Song, Kai Dang, Peng Wang, Shijie Wang, Jun Tang, Humen Zhong, Yuanzhi Zhu, Mingkun Yang, Zhaohai Li, Jianqiang Wan, Pengfei Wang, Wei Ding, Zheren Fu, Yiheng Xu, Jiabo Ye, Xi Zhang, Tianbao Xie, Zesen Cheng, Hang Zhang, Zhibo Yang, Haiyang Xu, and Junyang Lin. 2025. Qwen2.5-vl technical report. *arXiv preprint arXiv:2502.13923*.

Satanjeev Banerjee and Alon Lavie. 2005. Meteor: An automatic metric for mt evaluation with improved correlation with human judgments. In *Proceedings of the acl workshop on intrinsic and extrinsic evaluation measures for machine translation and/or summarization*, pages 65–72.

Priyam Basu, Jose Cols, Daniel Jarvis, Yongsin Park, and Daniel Rodabaugh. 2025. Suwmit at biolaysumm2025: Instruction-based summarization with contrastive decoding. In *The 24th Workshop* on *Biomedical Natural Language Processing and BioNLP Shared Tasks*, Vienna, Austria. Association for Computational Linguistics.

Melody Bechler, Carly Crowther, Emily Luedke, Natasha Schimka, and Ibrahim Sharaf. 2025. Tlpiq at biolaysumm: Hide and seq, a flan-t5 model for biomedical summarization. In *The 24th Workshop on Biomedical Natural Language Processing and BioNLP Shared Tasks*, Vienna, Austria. Association for Computational Linguistics.

Aurelia Bustos, Antonio Pertusa, Jose-Maria Salinas, and Maria De La Iglesia-Vaya. 2020. Padchest: A large chest x-ray image dataset with multi-label annotated reports. *Medical image analysis*, 66:101797.

Dina Demner-Fushman, Sameer Antani, Matthew Simpson, and George R Thoma. 2012. Design and development of a multimodal biomedical information retrieval system. *Journal of Computing Science and Engineering*, 6(2):168–177.

Egecan Çelik Evgin, lknur Karadeniz, and Olcay Taner Yıldız. 2025. Metninozu at biolaysumm2025: Text summarization with reverse data augmentation and injecting salient sentences. In *The 24th Workshop on Biomedical Natural Language Processing and BioNLP Shared Tasks*, Vienna, Austria. Association for Computational Linguistics.

Tomas Goldsack, Zheheng Luo, Qianqian Xie, Carolina Scarton, Matthew Shardlow, Sophia Ananiadou, and Chengua Lin. 2023. Overview of the biolaysumm 2023 shared task on lay summarization of biomedical research articles. In *The 22nd Workshop on Biomedical Natural Language Processing and BioNLP Shared Tasks*, pages 468–477. Association for Computational Linguistics.

Tomas Goldsack, Carolina Scarton, Matthew Shardlow, and Chenghua Lin. 2024. Overview of the biolaysumm 2024 shared task on the lay summarization of biomedical research articles. In *Proceedings of the 23rd Workshop on Biomedical Natural Language Processing*, pages 122–131.

Tomas Goldsack, Zhihao Zhang, Chenghua Lin, and Carolina Scarton. 2022. Making science simple: Corpora for the lay summarisation of scientific literature. In *Proceedings of the 2022 Conference on Empirical Methods in Natural Language Processing*, pages 10589–10604, Abu Dhabi, United Arab Emirates. Association for Computational Linguistics.

Aaron Grattafiori, Abhimanyu Dubey, Abhinav Jauhri, Abhinav Pandey, Abhishek Kadian, Ahmad Al-Dahle, Aiesha Letman, Akhil Mathur, Alan Schelten, Alex Vaughan, Amy Yang, Angela Fan, Anirudh Goyal, Anthony Hartshorn, Aobo Yang, Archi Mitra, Archie Sravankumar, Artem Korenev, Arthur Hinsvark, Arun Rao, Aston Zhang, Aurelien Rodriguez, Austen Gregerson, Ava Spataru, Baptiste Roziere, Bethany Biron, Binh Tang, Bobbie Chern, Charlotte Caucheteux, Chaya Nayak, Chloe Bi, Chris Marra, Chris McConnell, Christian Keller, Christophe Touret, Chunyang Wu, Corinne Wong, Cristian Canton Ferrer, Cyrus Nikolaidis, Damien Allonsius, Daniel Song, Danielle Pintz, Danny Livshits, Danny Wyatt, David Esiobu, Dhruv Choudhary, Dhruv Mahajan, Diego Garcia-Olano, Diego Perino, Dieuwke Hupkes, Egor Lakomkin, Ehab AlBadawy, Elina Lobanova, Emily Dinan, Eric Michael Smith, Filip Radenovic, Francisco Guzmán, Frank Zhang, Gabriel Synnaeve, Gabrielle Lee, Georgia Lewis Anderson, Govind Thattai, Graeme Nail, Gregoire Mialon, Guan Pang, Guillem Cucurell, Hailey Nguyen, Hannah Korevaar, Hu Xu, Hugo Touvron, Iliyan Zarov, Imanol Arrieta Ibarra, Isabel Kloumann, Ishan Misra, Ivan Evtimov, Jack Zhang, Jade Copet, Jaewon Lee, Jan Geffert, Jana Vranes, Jason Park, Jay Mahadeokar, Jeet Shah, Jelmer van der Linde, Jennifer Billock, Jenny Hong, Jenya Lee, Jeremy Fu, Jianfeng Chi, Jianyu Huang, Jiawen Liu, Jie Wang, Jiecao Yu, Joanna Bitton, Joe Spisak, Jongsoo Park, Joseph Rocca, Joshua Johnstun, Joshua Saxe, Junteng Jia, Kalyan Vasuden Alwala, Karthik Prasad, Kartikeya Upasani, Kate Plawiak, Ke Li, Kenneth Heafield, Kevin Stone, Khalid El-Arini, Krithika Iyer, Kshitiz Malik, Kuenley Chiu, Kunal Bhalla, Kushal Lakhotia, Lauren Rantala-Yeary, Laurens van der Maaten, Lawrence Chen, Liang Tan, Liz Jenkins, Louis Martin, Lovish Madaan, Lubo Malo, Lukas Blecher, Lukas Landzaat, Luke de Oliveira, Madeline Muzzi, Mahesh Pasupuleti, Mannat Singh, Manohar Paluri, Marcin Kardas, Maria Tsimpoukelli, Mathew Oldham, Mathieu Rita, Maya Pavlova, Melanie Kambadur, Mike Lewis, Min Si, Mitesh Kumar Singh, Mona Hassan, Naman Goyal, Narjes Torabi, Nikolay Bashlykov, Nikolay Bogoychev, Niladri Chatterji, Ning Zhang, Olivier Duchenne, Onur Çelebi, Patrick Alrassy, Pengchuan Zhang, Pengwei Li, Petar Vasic, Peter Weng, Prajjwal Bhargava, Pratik Dubal, Praveen Krishnan, Punit Singh Koura, Puxin Xu, Qing He, Qingxiao Dong, Ragavan Srinivasan, Raj Ganapathy, Ramon Calderer, Ricardo Silveira Cabral, Robert Stojnic, Roberta Raileanu, Rohan Maheswari, Rohit Girdhar, Rohit Patel, Romain Sauvestre, Ronnie Polidoro, Roshan Sumbaly, Ross Taylor, Ruan Silva, Rui Hou, Rui Wang, Saghar Hosseini, Sahana Chennabasappa, Sanjay Singh, Sean Bell, Seohyun Sonia Kim, Sergey Edunov, Shaoliang Nie, Sharan Narang, Sharath Raparthy, Sheng Shen, Shengye Wan, Shruti Bhosale, Shun Zhang, Simon Vandenhende, Soumya Batra, Spencer Whitman, Sten Sootla, Stephane Collot, Suchin Gururangan, Sydney Borodinsky, Tamar Herman, Tara Fowler, Tarek Sheasha, Thomas Georgiou, Thomas Scialom, Tobias Speckbacher, Todor Mihaylov, Tong Xiao, Ujjwal Karn, Vedanuj Goswami, Vibhor Gupta, Vignesh Ramanathan, Viktor Kerkez, Vincent Gonguet, Virginie Do, Vish Vogeti, Vítor Albiero, Vladan Petrovic, Weiwei Chu, Wenhan Xiong, Wenyin Fu, Whitney Meers, Xavier Martinet, Xiaodong Wang, Xiaofang Wang, Xiaoqing Ellen Tan, Xide Xia, Xinfeng Xie, Xuchao Jia, Xuewei Wang, Yaelle Goldschlag, Yashesh Gaur, Yasmine Babaei, Yi Wen, Yiwen Song, Yuchen Zhang, Yue Li, Yuning Mao,

Zacharie Delpierre Coudert, Zheng Yan, Zhengxing Chen, Zoe Papakipos, Aaditya Singh, Aayushi Srivastava, Abha Jain, Adam Kelsey, Adam Shajnfeld, Adithya Gangidi, Adolfo Victoria, Ahuva Goldstand, Ajay Menon, Ajay Sharma, Alex Boesenberg, Alexei Baevski, Allie Feinstein, Amanda Kallet, Amit Sangani, Amos Teo, Anam Yunus, Andrei Lupu, Andres Alvarado, Andrew Caples, Andrew Gu, Andrew Ho, Andrew Poulton, Andrew Ryan, Ankit Ramchandani, Annie Dong, Annie Franco, Anuj Goyal, Aparajita Saraf, Arkabandhu Chowdhury, Ashley Gabriel, Ashwin Bharambe, Assaf Eisenman, Azadeh Yazdan, Beau James, Ben Maurer, Benjamin Leonhardi, Bernie Huang, Beth Loyd, Beto De Paola, Bhargavi Paranjape, Bing Liu, Bo Wu, Boyu Ni, Braden Hancock, Bram Wasti, Brandon Spence, Brani Stojkovic, Brian Gamido, Britt Montalvo, Carl Parker, Carly Burton, Catalina Mejia, Ce Liu, Changhan Wang, Changkyu Kim, Chao Zhou, Chester Hu, Ching-Hsiang Chu, Chris Cai, Chris Tindal, Christoph Feichtenhofer, Cynthia Gao, Damon Civin, Dana Beaty, Daniel Kreymer, Daniel Li, David Adkins, David Xu, Davide Testuggine, Delia David, Devi Parikh, Diana Liskovich, Didem Foss, Dingkang Wang, Duc Le, Dustin Holland, Edward Dowling, Eissa Jamil, Elaine Montgomery, Eleonora Presani, Emily Hahn, Emily Wood, Eric-Tuan Le, Erik Brinkman, Esteban Arcaute, Evan Dunbar, Evan Smothers, Fei Sun, Felix Kreuk, Feng Tian, Filippos Kokkinos, Firat Ozgenel, Francesco Caggioni, Frank Kanayet, Frank Seide, Gabriela Medina Florez, Gabriella Schwarz, Gada Badeer, Georgia Swee, Gil Halpern, Grant Herman, Grigory Sizov, Guangyi, Zhang, Guna Lakshminarayanan, Hakan Inan, Hamid Shojanazeri, Han Zou, Hannah Wang, Hanwen Zha, Haroun Habeeb, Harrison Rudolph, Helen Suk, Henry Aspegren, Hunter Goldman, Hongyuan Zhan, Ibrahim Damlaj, Igor Molybog, Igor Tufanov, Ilias Leontiadis, Irina-Elena Veliche, Itai Gat, Jake Weissman, James Geboski, James Kohli, Janice Lam, Japhet Asher, Jean-Baptiste Gaya, Jeff Marcus, Jeff Tang, Jennifer Chan, Jenny Zhen, Jeremy Reizenstein, Jeremy Teboul, Jessica Zhong, Jian Jin, Jingyi Yang, Joe Cummings, Jon Carvill, Jon Shepard, Jonathan McPhie, Jonathan Torres, Josh Ginsburg, Junjie Wang, Kai Wu, Kam Hou U, Karan Saxena, Kartikay Khandelwal, Katayoun Zand, Kathy Matosich, Kaushik Veeraraghavan, Kelly Michelena, Keqian Li, Kiran Jagadeesh, Kun Huang, Kunal Chawla, Kyle Huang, Lailin Chen, Lakshya Garg, Lavender A, Leandro Silva, Lee Bell, Lei Zhang, Liangpeng Guo, Licheng Yu, Liron Moshkovich, Luca Wehrstedt, Madian Khabsa, Manav Avalani, Manish Bhatt, Martynas Mankus, Matan Hasson, Matthew Lennie, Matthias Reso, Maxim Groshev, Maxim Naumov, Maya Lathi, Meghan Keneally, Miao Liu, Michael L. Seltzer, Michal Valko, Michelle Restrepo, Mihir Patel, Mik Vyatskov, Mikayel Samvelyan, Mike Clark, Mike Macey, Mike Wang, Miquel Jubert Hermoso, Mo Metanat, Mohammad Rastegari, Munish Bansal, Nandhini Santhanam, Natascha Parks, Natasha White, Navyata Bawa, Nayan Singhal, Nick Egebo, Nicolas Usunier, Nikhil Mehta, Nikolay Pavlovich Laptev, Ning Dong, Norman Cheng, Oleg Chernoguz, Olivia Hart, Omkar Salpekar, Ozlem Kalinli, Parkin Kent, Parth Parekh, Paul Saab, Pavan Balaji, Pedro Rittner, Philip Bontrager, Pierre Roux, Piotr Dollar, Polina Zvyagina, Prashant Ratanchandani, Pritish Yuvraj, Qian Liang, Rachad Alao, Rachel Rodriguez, Rafi Ayub, Raghotham Murthy, Raghu Nayani, Rahul Mitra, Rangaprabhu Parthasarathy, Raymond Li, Rebekkah Hogan, Robin Battey, Rocky Wang, Russ Howes, Ruty Rinott, Sachin Mehta, Sachin Siby, Sai Jayesh Bondu, Samyak Datta, Sara Chugh, Sara Hunt, Sargun Dhillon, Sasha Sidorov, Satadru Pan, Saurabh Mahajan, Saurabh Verma, Seiji Yamamoto, Sharadh Ramaswamy, Shaun Lindsay, Shaun Lindsay, Sheng Feng, Shenghao Lin, Shengxin Cindy Zha, Shishir Patil, Shiva Shankar, Shuqiang Zhang, Shuqiang Zhang, Sinong Wang, Sneha Agarwal, Soji Sajuyigbe, Soumith Chintala, Stephanie Max, Stephen Chen, Steve Kehoe, Steve Satterfield, Sudarshan Govindaprasad, Sumit Gupta, Summer Deng, Sungmin Cho, Sunny Virk, Suraj Subramanian, Sy Choudhury, Sydney Goldman, Tal Remez, Tamar Glaser, Tamara Best, Thilo Koehler, Thomas Robinson, Tianhe Li, Tianjun Zhang, Tim Matthews, Timothy Chou, Tzook Shaked, Varun Vontimitta, Victoria Ajayi, Victoria Montanez, Vijai Mohan, Vinay Satish Kumar, Vishal Mangla, Vlad Ionescu, Vlad Poenaru, Vlad Tiberiu Mihailescu, Vladimir Ivanov, Wei Li, Wenchen Wang, Wenwen Jiang, Wes Bouaziz, Will Constable, Xiaocheng Tang, Xiaojian Wu, Xiaolan Wang, Xilun Wu, Xinbo Gao, Yaniv Kleinman, Yanjun Chen, Ye Hu, Ye Jia, Ye Qi, Yenda Li, Yilin Zhang, Ying Zhang, Yossi Adi, Youngjin Nam, Yu, Wang, Yu Zhao, Yuchen Hao, Yundi Qian, Yunlu Li, Yuzi He, Zach Rait, Zachary DeVito, Zef Rosnbrick, Zhaoduo Wen, Zhenyu Yang, Zhiwei Zhao, and Zhiyu Ma. 2024. The llama 3 herd of models.

Aaradhya Gupta and Dr Parameswari Krishnamurthy. 2025. Shared task at biolaysumm2025 : Extract then summarize approach augmented with umls based definition retrieval for lay summary generation. In *The 24th Workshop on Biomedical Natural Language Processing and BioNLP Shared Tasks*, Vienna, Austria. Association for Computational Linguistics.

Saahil Jain, Ashwin Agrawal, Adriel Saporta, Steven QH Truong, Du Nguyen Duong, Tan Bui, Pierre Chambon, Yuhao Zhang, Matthew P Lungren, Andrew Y Ng, et al. 2021. Radgraph: Extracting clinical entities and relations from radiology reports. *arXiv preprint arXiv:2106.14463*.

Alistair EW Johnson, Tom J Pollard, Seth J Berkowitz, Nathaniel R Greenbaum, Matthew P Lungren, Chih-ying Deng, Roger G Mark, and Steven Horng. 2019. Mimic-cxr, a de-identified publicly available database of chest radiographs with free-text reports. *Scientific data*, 6(1):317.

Chin-Yew Lin. 2004. ROUGE: A package for automatic evaluation of summaries. In *Text Summarization Branches Out*, pages 74–81, Barcelona, Spain. Association for Computational Linguistics.

Fan Lin and Dezhi Yu. 2025. Laysummx at biolaysumm: Retrieval-augmented fine-tuning for biomedical lay summarization using abstracts and retrieved full-text context. In *The 24th Workshop on Biomedical Natural Language Processing and BioNLP Shared Tasks*, Vienna, Austria. Association for Computational Linguistics.

Juan Antonio Lossio-Ventura, Callum Chan, Arshitha Basavaraj, Hugo Alatrista-Salas, Francisco Pereira, and Diana Inkpen. 2025. 5cnlp at biolaysumm2025: Prompts, retrieval, and multimodal fusion. In *The 24th Workshop on Biomedical Natural Language Processing and BioNLP Shared Tasks*, Vienna, Austria. Association for Computational Linguistics.

Zheheng Luo, Qianqian Xie, and Sophia Ananiadou. 2022. Readability controllable biomedical document summarization. In *Findings of the Association for Computational Linguistics: EMNLP 2022*, pages 4667–4680, Abu Dhabi, United Arab Emirates. Association for Computational Linguistics.

Mounica Maddela, Yao Dou, David Heineman, and Wei Xu. 2023. LENS: A learnable evaluation metric for text simplification. In *Proceedings of the 61st Annual Meeting of the Association for Computational Linguistics (Volume 1: Long Papers)*, pages 16383–16408, Toronto, Canada. Association for Computational Linguistics.

Teresa Martin-Carreras, Tessa S Cook, and Charles E Kahn Jr. 2019. Readability of radiology reports: implications for patient-centered care. *Clinical imaging*, 54:116–120.

Nur Alya Dania binti Moriazi and Mujeen Sung. 2025. Khu_ldi at biolaysumm2025: Fine-tuning and refinement for lay radiology report generation. In *The 24th Workshop on Biomedical Natural Language Processing and BioNLP Shared Tasks*, Vienna, Austria. Association for Computational Linguistics.

Kishore Papineni, Salim Roukos, Todd Ward, and Wei-Jing Zhu. 2002. Bleu: a method for automatic evaluation of machine translation. In *Proceedings of the 40th annual meeting of the Association for Computational Linguistics*, pages 311–318.

Catia Pesquita, Daniel Faria, Andre O Falcao, Phillip Lord, and Francisco M Couto. 2009. Semantic similarity in biomedical ontologies. *PLoS computational biology*, 5(7):e1000443.

Benhamin Pong, Ju-Hui Chen, Jonathan Jiang, Abimael Hernandez Jimenez, and Melody Vahadi. 2025. Mirages at biolaysumm2025: The impact of search terms and data curation for biomedical lay summarization. In *The 24th Workshop on Biomedical Natural Language Processing and BioNLP Shared Tasks*, Vienna, Austria. Association for Computational Linguistics.

Qwen, :, An Yang, Baosong Yang, Beichen Zhang, Binyuan Hui, Bo Zheng, Bowen Yu, Chengyuan Li, Dayiheng Liu, Fei Huang, Haoran Wei, Huan Lin, Jian Yang, Jianhong Tu, Jianwei Zhang, Jianxin Yang, Jiaxi Yang, Jingren Zhou, Junyang Lin, Kai Dang, Keming Lu, Keqin Bao, Kexin Yang, Le Yu, Mei Li, Mingfeng Xue, Pei Zhang, Qin Zhu, Rui Men, Runji Lin, Tianhao Li, Tianyi Tang, Tingyu Xia, Xingzhang Ren, Xuancheng Ren, Yang Fan, Yang Su, Yichang Zhang, Yu Wan, Yuqiong Liu, Zeyu Cui, Zhenru Zhang, and Zihan Qiu. 2025. Qwen2.5 technical report.

Ilyass Ramzi and Isabel Segura Bedmar. 2025. Bdauc3m @ biolaysumm: Efficient lay summarization with small-scale sota llms. In *The 24th Workshop on Biomedical Natural Language Processing and BioNLP Shared Tasks*, Vienna, Austria. Association for Computational Linguistics.

Bhuvaneswari Sivagnanam, Rivo Krishnu C H, Princi Chauhan, and Saranya Rajiakodi. 2025. Cutn_bio at biolaysumm: Multi-task prompt tuning with external knowledge and readability adaptation for layman summarization. In *The 24th Workshop on Biomedical Natural Language Processing and BioNLP Shared Tasks*, Vienna, Austria. Association for Computational Linguistics.

Akshay Smit, Saahil Jain, Pranav Rajpurkar, Anuj Pareek, Andrew Y Ng, and Matthew P Lungren. 2020. Chexbert: combining automatic labelers and expert annotations for accurate radiology report labeling using bert. *arXiv preprint arXiv:2004.09167*.

Maria De La Iglesia Vayá, Jose Manuel Saborit, Joaquim Angel Montell, Antonio Pertusa, Aurelia Bustos, Miguel Cazorla, Joaquin Galant, Xavier Barber, Domingo Orozco-Beltrán, Francisco García-García, et al. 2020. Bimcv covid-19+: a large annotated dataset of rx and ct images from covid-19 patients. *arXiv preprint arXiv:2006.01174*.

Jen Wilson, Avery Bellamy, Rachel Edwards, Michael Pollack, and Helen Salgi. 2025. Raincitynlp at biolaysumm2025: Extract then summarize at home. In *The 24th Workshop on Biomedical Natural Language Processing and BioNLP Shared Tasks*, Vienna, Austria. Association for Computational Linguistics.

Pengcheng Xu, Sicheng Shen, Jieli Zhou, and Hongyi Xin. 2025. Team xsz at biolaysumm2025: Section-wise summarization, retrieval-augmented llm, and reinforcement learning fine-tuning for lay summaries. In *The 24th Workshop on Biomedical Natural Language Processing and BioNLP Shared Tasks*, Vienna, Austria. Association for Computational Linguistics.

Zhen Xu, Sergio Escalera, Adrien Pavão, Magali Richard, Wei-Wei Tu, Quanming Yao, Huan Zhao, and Isabelle Guyon. 2022. Codabench: Flexible, easy-to-use, and reproducible meta-benchmark platform. *Patterns*, 3(7):100543.

Yuheng Zha, Yichi Yang, Ruichen Li, and Zhiting Hu. 2023. AlignScore: Evaluating factual consistency with a unified alignment function. In *Proceedings of the 61st Annual Meeting of the Association for Computational Linguistics (Volume 1: Long Papers)*, pages 11328–11348, Toronto, Canada. Association for Computational Linguistics.

Tianyi Zhang, Varsha Kishore, Felix Wu, Kilian Q. Weinberger, and Yoav Artzi. 2020. Bertscore: Evaluating text generation with BERT. In *8th International Conference on Learning Representations, ICLR 2020, Addis Ababa, Ethiopia, April 26-30, 2020*. OpenReview.net.

Wenjun Zhang, Shekhar S. Chandra, Bevan Koopman, Jason Dowling, and Aaron Nicolson. 2025. Aehrc at biolaysumm 2025: Leveraging t5 for lay summarisation of radiology reports. In *The 24th Workshop on Biomedical Natural Language Processing and BioNLP Shared Tasks*, Vienna, Austria. Association for Computational Linguistics.

Kun Zhao, Chenghao Xiao, Sixing Yan, Haoteng Tang, William K. Cheung, Noura Al Moubayed, Liang Zhan, and Chenghua Lin. 2025. X-ray made simple: Lay radiology report generation and robust evaluation.

A Appendix

Table 3 and Table 4 present the overview and detailed metric performance after min-max normalization.

#	Team	Relevance	Readability	Factuality	Avg.
1	**SUWMIT**	**0.971**	0.816	0.616	**0.801**
2	**Baseline-llama3-8B-sft**	0.965	0.770	0.548	0.761
3	**Baseline-qwen2.5-7B-sft**	0.922	0.645	0.579	0.715
4	**BDA-UC3M**	0.892	0.726	0.406	0.675
5	**MetninOzU**	0.779	0.148	**1.000**	0.643
6	**MIRAGES**	0.531	0.886	0.466	0.628
7	**TLPIQ**	0.775	0.505	0.586	0.622
8	**LaySummX**	0.663	0.752	0.371	0.595
9	**CUTN_Bio**	0.470	**0.972**	0.316	0.586
10	**Aard**	0.708	0.505	0.376	0.530
11	**LTRC**	0.522	0.725	0.250	0.499
12	**5cNLP**	0.732	0.432	0.256	0.473
13	**RainCityNLP**	0.525	0.056	0.451	0.344
14	**sxz**	0.004	0.334	0.560	0.299
15	**demo**	0.004	0.334	0.560	0.299
16	**x2z**	0.054	0.787	0.014	0.285

(a) Subtask 1.1: Plain Lay Summarization

#	Team	Relevance	Readability	Factuality	Avg.
1	**Aard**	0.696	**1.000**	**0.879**	**0.858**
2	**CUTN_Bio**	0.667	0.382	0.861	0.637
3	**5cNLP**	**1.000**	0.039	0.447	0.495
4	**LTRC**	0.000	0.327	0.000	0.109

(b) Subtask 1.2: Lay Summarization with External Knowledge

#	Team	Relevance	Readability	Clinical	Avg.
1	**AEHRC**	**1.000**	**0.667**	**1.000**	**0.889**
2	**KHU_LDI**	0.000	0.333	0.000	0.111

(c) Subtask 2.1: Radiology Report Translation (Open Track)

#	Team	Relevance	Readability	Clinical	Avg.
1	**AEHRC**	**1.000**	0.521	**0.858**	**0.793**
2	**Baseline-qwen2.5-7B-sft**	0.567	0.384	0.601	0.517
3	**5cNLP**	0.688	0.300	0.537	0.508
4	**Baseline-llama3-8B-sft**	0.510	0.262	0.718	0.497
5	**CUTN_Bio**	0.000	**0.809**	0.000	0.270

(d) Subtask 2.1: Radiology Report Translation (Closed Track)

Table 3: Task leaderboard with min–max normalization

Rank	Team	Relevance				Readability				Factuality	
		ROUGE	BLEU	MTR	BERTS	FKGL	DCRS	CLI	LENS	AlignS	SummaC
1	**SUWMIT**	**1.000**	**1.0000**	0.964	0.921	0.804	0.815	0.793	0.851	0.745	0.487
2	**Baseline-llama3-8B-sft**	0.977	0.9770	**1.000**	0.906	0.730	0.773	0.721	0.855	0.690	0.405
3	**Baseline-qwen2.5-7B-sft**	0.909	0.8504	0.933	0.996	0.648	0.646	0.593	0.692	0.753	0.406
4	**BDA-UCM**	0.917	0.7759	0.876	**1.000**	0.710	0.763	0.688	0.742	0.631	0.181
5	**MetinOZU**	0.880	0.6490	0.854	0.812	0.046	0.183	0.000	0.365	**1.000**	**1.000**
6	**MIRAGES**	0.598	0.3886	0.482	0.656	0.809	**1.000**	0.900	0.834	0.611	0.322
7	**TupiQ**	0.829	0.6733	0.717	0.862	0.531	0.369	0.642	0.478	0.762	0.642
8	**LaySummX**	0.759	0.4798	0.623	0.793	0.718	0.690	0.650	0.953	0.600	0.142
9	**CUTN_Bio**	0.503	0.2329	0.457	0.690	**1.000**	0.888	**1.000**	**1.000**	0.431	0.202
10	**Aard**	0.749	0.4805	0.871	0.730	0.350	0.537	0.295	0.837	0.637	0.118
11	**LTRC**	0.599	0.3473	0.430	0.711	0.543	0.752	0.667	0.938	0.455	0.045
12	**5cNLP**	0.820	0.5577	0.713	0.838	0.108	0.425	0.300	0.896	0.513	0.000
13	**RainCityNLP**	0.582	0.4152	0.544	0.561	0.000	0.051	0.138	0.037	0.476	0.426
14	**sxz**	0.000	0.0169	0.000	0.000	0.668	0.000	0.669	0.000	0.964	0.157
15	**demo**	0.000	0.0169	0.000	0.000	0.668	0.000	0.669	0.000	0.964	0.157
16	**x2z**	0.085	0.0000	0.094	0.036	0.666	0.713	0.782	0.730	0.000	0.028

(a) SubTask 1.1: Plain Lay Summarization

Rank	Team	Relevance				Readability				Factuality	
		ROUGE	BLEU	MTR	BERTS	FKGL	DCRS	CLI	LENS	AlignS	SummaC
1	**Aard**	0.643	0.592	0.882	0.665	**1.000**	**1.000**	**1.000**	**1.000**	0.757	**1.000**
2	**CUTN_Bio**	0.676	0.532	0.559	0.902	0.570	0.020	0.150	0.787	**1.000**	0.722
3	**5cNLP**	**1.000**	**1.000**	**1.000**	**1.000**	0.000	0.000	0.000	0.155	0.745	0.149
4	**LTRC**	0.000	0.000	0.000	0.000	0.487	0.328	0.495	0.000	0.000	0.000

(b) Subtask 1.2: Lay Summarization with External Knowledge

Rank	Team	Relevance					Readability			Clinical Metrics	
		ROUGE	BLEU	MTR	BERTS	SIM	FKGL	DCRS	CLI	CHEX	RG
1	**AEHRC**	**1.000**	**1.000**	**1.000**	**1.000**	**1.000**	**1.000**	0.000	**1.000**	**1.000**	**1.000**
2	**KHU_LDI**	0.000	0.000	0.000	0.000	0.000	0.000	**1.000**	0.000	0.000	0.000

(c) Subtask 2.1: Radiology Report Translation (Open Track)

Rank	Team	Relevance					Readability			Clinical Metrics	
		ROUGE	BLEU	MTR	BERTS	SIM	FKGL	DCRS	CLI	CHEX	RG
1	**AEHRC**	**1.000**	**1.000**	**1.000**	**1.000**	**1.000**	0.294	0.72	0.548	0.715	**1.000**
2	**Baseline-qwen2.5-7B-sft**	0.591	0.449	0.483	0.721	0.589	**1.000**	0.06	0.094	0.733	0.468
3	**5cNLP**	0.670	0.555	0.760	0.685	0.770	0.000	0.000	0.557	0.446	0.627
4	**Baseline-llama3-8B-sft**	0.546	0.427	0.414	0.649	0.512	0.786	0.000	0.000	**1.000**	0.436
5	**CUTN_Bio**	0.404	0.427	0.000	0.000	0.000	0.428	**1.000**	**1.000**	0.000	0.000

(d) Subtask 2.1: Radiology Report Translation (Closed Track)

Table 4: Task leaderboard with min-max normalization. **BertS** = BertScore, **FKGL** = Flesch-Kincaid Grade Level, **DCRS** = Dale-Chall Readability Score, **CLI** = Coleman-Liau Index, **MTR** = METOR, **SIM** = Similarity, **AlignS** = AlignScore, **CHEX** = F1 chexbert, **RG** = Radgraph.

Overview of the ClinIQLink 2025 Shared Task on Medical Question-Answering

Brandon C. Colelough, Davis Bartels, and Dina Demner-Fushman
National Library of Medicine, National Institutes of Health
Bethesda, MD, USA
{firstname.lastname}@nih.gov

Abstract

In this paper, we present an overview of CLINIQLINK a shared task, collocated with the 24th BioNLP workshop at ACL 2025, designed to stress-test large language models (LLMs) on medically-oriented question answering aimed at the level of a General Practitioner. The challenge supplies 4 978 expert-verified, medical source-grounded question–answer pairs that cover seven formats - *true/false*, *multiple choice*, *unordered list*, *short answer*, *short-inverse*, *multi-hop*, and *multi-hop-inverse*. Participating systems, bundled in Docker or Apptainer images, are executed on the CodaBench platform or the University of Maryland's *Zaratan* cluster. An automated harness (Task 1) scores closed-ended items by exact match and open-ended items with a three-tier embedding metric. A subsequent physician panel (Task 2) audits the top model responses.

1 Introduction

LLMs have increasingly demonstrated their ability to memorize information and answer questions (Carlini et al., 2023). This has led to their increased use by consumers to ask medically relevant questions (Yun and Bickmore, 2025). However, LLMs have been shown to "hallucinate", that is, to generate factually incorrect, or even harmful answers (Singhal et al., 2023). In high-stakes domains, such as medicine, it is incredibly important to be able to evaluate the veracity of any question answering system. While there exist datasets, such as MultiMedQA (Singhal et al., 2023), designed to do just this, recent LLMs have been trained over their data. This limits their usefulness in evaluating the ability of these models to generalize to out-of-distribution data.

New datasets are necessary for the evaluation of medical question-answering systems and new systems are needed to increase accuracy and mitigate hallucinations.

To this end, we introduce the ClinIQLink shared task, inviting participants to submit question-answering systems to be evaluated on a novel dataset of medical questions. Participants are encouraged to submit systems that are capable of demonstrating medical knowledge, while mitigating hallucinations. Our dataset consists of seven question types, both closed and open ended, and a wide range of medical topics. Our task had a total of three runs from one team. Our contributions are as follows:

- A dataset of 4,978 vetted medical question-answer pairs
- Automated evaluation metrics
- A task design for participant-submitted systems
- A physician audit of system responses

2 Task Description

ClinIQLink [1] is a shared task that evaluates the ability of generative models to produce factually accurate medical information aimed at the knowledge level of a general practitioner. The submitted systems are executed in a containerized environment on CodaBench [2] or via the University of Maryland (UMD) HPC Zaratan [3] (depending on the size and model/system complexity), where the submitted systems answered a corpus of expert-curated atomic medical questions. Answers provided from the systems submitted were judged only on factual accuracy, so leaderboard ranking reflects a model's ability to retrieve correct information from its own parametric memory or any retrieval mechanism the team elected to integrate.

[1] https://cliniqlink.org/
[2] https://www.codabench.org/
[3] https://hpcc.umd.edu/hpcc/zaratan.html

Proceedings of the 24th Workshop on Biomedical Language Processing (BioNLP 2025), pages 378–387
August 1, 2025

The question sets were divided into two types (closed and open-ended QA pairs) and spanned seven modalities, including true/false, multiple choice, unordered list, short answer, short-inverse, multi-hop and multi-hop-inverse. Across all of the seven QA pair modalities, the ground truth was anchored in standard open-source medical texts, and each item targets a single, clearly defined concept such as a procedure, drug, diagnostic finding, or anatomical fact.

The challenge comprised two sequential components. Task 1 executed all baseline systems and participant submissions within our automated benchmarking harness. The script marked closed-ended items strictly for precision and evaluated open-ended answers with a semantic-similarity module that awards full or partial credit according to their closeness to the hidden ground-truth. Leaderboard rankings are derived solely from these automatic scores. Task 2 began after the leaderboard was frozen: a panel of human-expert annotators reviewed the highest-scoring outputs, ranking them from best to worst and annotating each answer on a spectrum from "good" to "bad". Participants were allowed to employ any architecture, external knowledge base, or retrieval-augmented pipeline to generate answers to questions posed, provided the final system can run end-to-end inside the supplied containerised harness. Teams were limited to three leaderboard submissions and were required to accompany their final entry with a short paper that details model design, data usage, and inference strategy for inclusion in the BioNLP 2025 proceedings. The full evaluation dataset remains private to preserve its viability for later use.

3 Dataset Description

3.1 Generation and Vetting

A neuro-symbolic pipeline was employed to produce roughly $\sim$20K atomic question–answer pairs from open-source medical texts. Each pair was linked to its supporting passage so that later reviewers could verify every biomedical fact. The QA Pairs were then ported to our online annotation portal[4], (which is now open to accredited medical schools and hospitals who wish to contribute further judgments), where human-experts (paid medical students) confirmed correctness, rated *general-practitioner (GP) relevance* on a five-point scale, and could file structured feedback or formal disputes.

3.2 Human-verification Workflow

1. **Primary review**: an expert validated factual accuracy against the source excerpt, assigned a GP-relevance score, and could flag issues or supply comments.

2. **Secondary review**: $\sim 45\%$ of items received an independent second pass; disagreements triggered adjudication. By 1 May 2025 reviewers had lodged 601 feedback notes and 461 disputes. The 1062 QA Pairs that had been flagged as feedback or disputes were not used for testing and are presently still being held for later review.

3.3 Benchmark Snapshot (1 May 2025)

At the dataset freeze the repository contained 5,118 verified QA pairs (Table 1): 5,118 had a single expert judgement and 2,505 were double-annotated. For leaderboard scoring, we retained only the 4,978 items rated maximally relevant ($score = 5$); 140 lower-relevance items were set aside for future analysis. The sample dataset plus the full evaluation architecture are available at [5].

3.4 Question Modalities

Seven formats cover both machine-gradable *closed-ended* items and semantically scored *open-ended* prompts:

- **Closed-ended**
  - True/False (TF)
  - Multiple Choice (MC) — single-best answer
  - Unordered List (LIST) — enumerate all correct elements
- **Open-ended**
  - Short Answer (SHORT) — concise factoid
  - Short-Inverse (SHORT_INV) — explain why the supplied wrong answer is incorrect
  - Multi-hop (MULTI_HOP) — required several leaps in knowledge to arrive at a final answer; models must return answer *and* knowledge leaps
  - Multi-hop Inverse (MULTI_HOP_INV) — locate the faulty step in a provided, erroneous multi-hop rationale

[4]https://bionlp.nlm.nih.gov/ClinIQLink/NIHLogin

[5]https://github.com/Brandonio-c/ClinIQLink_Sample-dataset

Table 1: ClinIQLink benchmark composition at the baseline freeze (1 May 2025). "High" denotes GP-relevance score 5 items used for leaderboard evaluation; "Low" items (< 5) were withheld.

QA Format	Counts			Subset with Two Independent Reviews			
	High	Low	Total	High	Low	Total	Percent Double
True/False (TF)	813	38	851	369	–	369	43.4%
Multiple Choice (MC)	765	29	794	346	–	346	43.6%
Unordered List (LIST)	714	28	742	341	–	341	46.0%
Short Answer (SHORT)	427	9	436	339	–	339	77.8%
Short-Inverse (SHORT_INV)	742	16	758	353	–	353	46.6%
Multi-hop (MULTI_HOP)	771	8	779	331	–	331	42.5%
Multi-hop Inverse (MULTI_HOP_INV)	746	12	758	318	–	318	42.0%
Totals	**4 978**	**140**	**5 118**	**2 497**	–	**2 505**	**48.8%**

4 Evaluation Protocol

Our assessment of CLINIQLINK was conducted in two sequential phases. First, we relied on a fully automated evaluation script (Task-1) that ingested model/participant system responses; second, we complemented the automated evaluation with an expert preference study (Task-2) in which paid medical students compared top-performing model responses.

4.1 Task-1: automatic scoring

Each submission returned answers for **seven distinct question classes**. *True/False* and single-best *multiple-choice* items were judged by straightforward **accuracy**

$$\text{Accuracy} = \frac{\#\text{correct}}{N},$$

whereas *multiple select list* questions were graded with both macro- and micro F_1 (Manning et al., 2008):

$$F_1^{\text{macro}} = \frac{1}{N}\sum_{i=1}^{N} F_1^{(i)},$$

$$F_1^{\text{micro}} = \frac{2\,\text{TP}}{2\,\text{TP} + \text{FP} + \text{FN}}.$$

All free-text tasks (`short`, `multi hop`, and their inverse variants) were assessed twice; once with the ClinIQLink **semantic-similarity** score and again with the conventional n-gram metrics BLEU (Papineni et al., 2002), ROUGE (Lin, 2004), and METEOR (Banerjee and Lavie, 2005).

ClinIQLink semantic-similarity score. The score blended *three* complementary cosine layers:

(1) **Token layer:** an IDF-weighted, greedy token-alignment F_1, rewarding exact overlap on infrequent clinical terms.

(2) **Sentence layer:** cosine similarity of SBERT–MINILM [6] CLS embeddings, capturing broader paraphrase.

(3) **Paragraph layer:** cosine similarity of the raw answer strings, offering global context.

Let $C_{\text{tok}}, C_{\text{sent}}, C_{\text{para}} \in [0, 1]$ denote these three cosines. With weights $w_{\text{tok}} = w_{\text{sent}} = 0.4$ and $w_{\text{para}} = 0.2$ the raw score is

$$S_{\text{raw}} = 0.4\,C_{\text{tok}} + 0.4\,C_{\text{sent}} + 0.2\,C_{\text{para}}.$$

Because SBERT assigns unrelated sentence pairs a baseline similarity of about $\beta = 0.25$, we subtract that offset, floor negatives, and snap near-perfect matches:

$$S = \min\Big(1,\ \max\big(0,\ S_{\text{raw}} - \beta\big)\Big),$$

$$S \geq 0.95 \implies S := 1.$$

Penalty for `multi-hop inverse`. If a model highlighted the wrong reasoning step, the semantic score was down-weighted. Let $d = |\,\text{predicted step} - \text{gold step}\,|$ be the absolute distance; then

$$\alpha(d) = \begin{cases} 1 & d = 0, \\ 0.7 & d = 1, \\ 0.3 & d = 2, \\ 0.3\,2^{-(d-2)} & d \geq 3, \end{cases} \qquad S^* = \alpha(d)\,S.$$

Hence, the final similarity S^* combined graded lexical alignment, distributional semantics, and explicit reasoning correctness, while conventional

[6]https://huggingface.co/sentence-transformers/all-MiniLM-L6-v2

BLEU/ROUGE/METEOR offered secondary diagnostics. A complete implementation of the evaluation script that implements the above can be found with the testing harness[7].

4.2 Task-2: expert preference study

We found that the automated metrics employed for analysis of the open-ended QA pairs were not effective for evaluation of model responses, nor were they effective in discriminating top-ranking model responses from mediocre model responses. Hence, to complement the automated evaluation metrics we organized a human evaluation in which we required our annotators to rank the six strongest foundation models on our public leaderboard (Falcon-10B, Llama-3.3-70B, Llama-4 Scout, Mistral-Large-2411, Microsoft Phi-4 Base, Qwen-3-32B) together with the best participant submission, *Preceptor-AI* and the ClinIQLink ground-truth answers. For every question we shuffled these seven model responses plus the ClinIQLink dataset reference solution, and asked human annotators to rank them from best to worst. Each answer also received a coarse quality tag (good/okay/bad). The annotation portal that we built for this experiment is now open to accredited medical schools and hospitals who wish to contribute further judgments [8].

5 Baseline Systems

To provide a strong reference point for future work we evaluated a broad range of publicly–available large language models on the frozen CLINIQLINK test split. For transparency, it should be noted that the LLM utilised as the "neuro" component of our neurosymbolic pipeline for data generation was **Llama 3.3-70B–Instruct**. All baseline checkpoints were used as-is and therefore reflect their pre-training and instruction-tuning quality rather than any task-specific fine-tuning.

5.1 Llama family.

The *Meta Llama 3* (Grattafiori et al., 2024) decoder-only transformer was represented by four parameter scales; 1B, 3B, 8B and 70B weights as well as an intermediate commercial variant (`llama_4-scout`, $\approx$ 45B). All are dense models built with a 32-layer architecture (70B: 80 layers) and grouped-query attention; the instruction checkpoints add a supervised fine-tuning and reinforcement learning step to the base weights.

5.2 Mistral / Mixtral family.

We included the 7-billion-parameter **Mistral-7B** (Jiang et al., 2023) dense decoder and the **Mistral-Large-Instruct-2411** release (8×22B experts, two experts routed per token, giving 47B active parameters). The **Mixtral** series consisting of Mixtral-8x7B (Jiang et al., 2024) and Mixtral-8x22B tested share the same sparse Mixture-of-Experts (MoE) scaffold, however, only two of the eight experts are selected for each input token, keeping inference costs close to their 12–13 B dense peers while exposing $>$ 140B total capacity.

5.3 Qwen3 family.

Alibaba's *Qwen3* (Yang et al., 2025) decoder stack (RoPE positional encoding, grouped-query attention) was tested at five scales: 1.7B, 3B, 4B, 8B, and 32B parameters. All checkpoints were released under an open-source licence together with alignment ("`-Instruct`") variants that follow the Supervised Fine-Tuning (SFT) + Direct Preference Optimisation recipe.

5.4 Phi family.

We evaluated Microsoft's **Phi-4** (Abdin et al., 2024) ($\sim$ 14B dense decoder) and its lightweight derivatives (`phi-4-mini-instruct` and `phi-4-mini-reasoning` (Abdin et al., 2025), $\sim$ 3.8B). This family of LLMs was designed as "small-data curriculum models" whose pre-training is dominated by synthetic textbook-style content rather than filtered web corpora.

5.5 Falcon Family

For completeness, we benchmarked **Falcon-10B-Instruct** (Almazrouei et al., 2023), an Apache–2.0 decoder model trained on the RefinedWeb dataset and alignment-tuned with RLHF.

5.6 Google Flan family.

Encoder–decoder baselines were covered by **Flan-T5-XXL** (Chung et al., 2022) (11B parameters) and **Flan-UL2** (Tay et al., 2023) (20B). Both models extend the original T5/UL2 sequence-to-sequence architecture with instruction tuning on a curated

[7] https://github.com/Brandonio-c/ClinIQLink_CodaBench_docker-setup/blob/main/submission/evaluate.py

[8] https://bionlp.nlm.nih.gov/ClinIQLink2/NIHLogin

mixture of over one thousand NLP tasks; an additional attrscore_flan_t5_xxl (Yue et al., 2023) checkpoint was tested, which augments the T5-XXL Weights with token-level attribution heads for explanation capabilities.

6 Participants and Methods

The CLINIQLINK shared task was publicly released through the Codabench evaluation platform[9], with an accompanying containerized setup for local validation and submission via Docker and Apptainer[10]. Submissions of models/systems over 10GB in size and requiring more compute than what is offered via codabench were also enabled via direct submission to organizers to be run on the University of Maryland HPC Zaratan. In total, 43 participants registered for the challenge during the initial release window. The competition remains open for new submissions on Codabench for smaller models that can run via the Codabench platform.

6.1 Preceptor AI

Although forty-three teams registered, only PRECEPTOR AI submitted runnable systems. They provided three containerised runs, v001, v002, and v003, but discuss only v001 in their participant paper.

v001 – VeReaFine (Verifier-augmented RAG). v001 is an iterative, evidence-seeking pipeline that couples a Qwen-7B-Instruct generator with a separately fine-tuned Qwen-8B medical-reasoning verifier. For each question the system:

1. retrieves up to 20 passages from a ColBERT (Khattab and Zaharia, 2020) + BM25 (Robertson et al., 2009) hybrid index built over *PubMed* abstracts and *StatPearls*;
2. drafts an answer *with inline citations*;
3. scores every generated claim with the verifier's token-level entailment head;
4. if any claim falls below a 0.8 confidence threshold, expands the evidence pool and repeats steps (1)–(3) (max. four rounds).

The loop stops when all claims are verified or the round limit is reached, after which the final answer and citation list are emitted. This design yields strong gains on all four open-ended modalities (top-10 P75 recall) but was *not* tuned for the closed-ended formats, explaining its low rank on multiple-choice and true/false items (see Table 2).

v002 and v003. The team also submitted v002 (a retrieval-free Qwen-32B classifier optimised for closed-ended questions) and an ablation run v003. Because their accompanying paper focuses on the verifier-augmented strategy, only v001 is analysed in detail there; we include the headline numbers for all three runs in the leaderboard for completeness.

7 Results

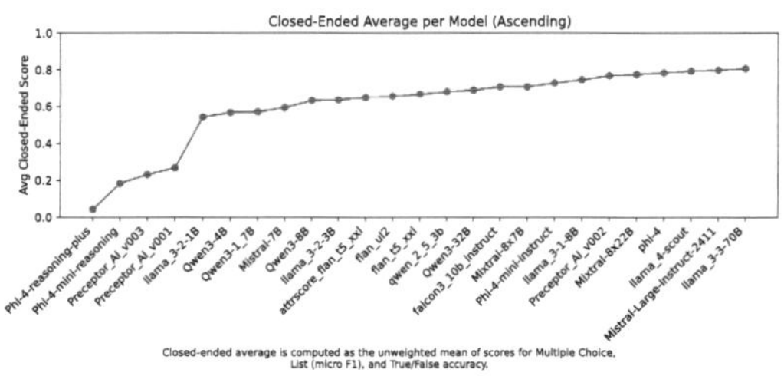

Figure 1: Average performance on closed-ended tasks (True/False accuracy, multiple-choice accuracy and list F_1).

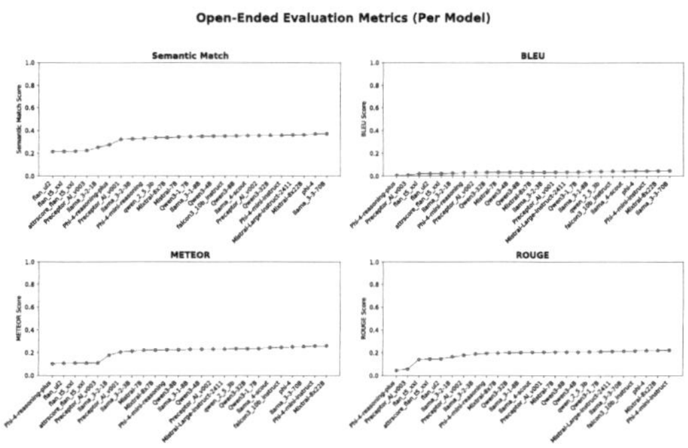

Figure 2: Distributions of individual n-gram scores (BLEU, ROUGE, METEOR) and semantic similarity for each open-ended question type.

Figure 1 summarised mean performance on the three closed-ended tasks. The spread between True/False, multiple-choice and list accuracy was modest, indicating that the leading models handled discrete answer formats with broadly comparable competence.

Open-ended behaviour was more nuanced. The per-task distributions in Figure 2 showed markedly heavier tails for semantic-similarity than for surface n-gram metrics, confirming that several systems produced answers that were lexically novel yet semantically similar. This pattern was especially

[9]https://www.codabench.org/competitions/5117/
[10]https://github.com/Brandonio-c/ClinIQLink_CodaBench_docker-setup

Table 2: ClinIQLink leaderboard snapshot (higher is better). Models were evaluated across all seven modalities of the ClinIQLink challenge. All models were retrieved from their public Hugging Face repositories, except for `Preceptor_AI`, which is private.

Rank	Model	Overall	MC Acc	TF Acc	List F1	Short	S-Inv	MHop	MH-Inv
1	`llama_3-3-70B`	0.541	0.796	0.822	0.682	0.235	0.488	0.313	0.450
2	`Mistral-Large-Instruct-2411`	0.530	0.797	0.822	0.645	0.260	0.472	0.313	0.398
3	`phi-4`	0.528	0.775	0.790	0.658	0.229	0.493	0.311	0.440
4	`llama_4-scout`	0.524	0.776	0.822	0.652	0.238	0.492	0.302	0.388
5	`Mixtral-8x22B`	0.521	0.752	0.800	0.643	0.227	0.491	0.306	0.428
6	`Preceptor_AI_v002`	0.512	0.762	0.817	0.583	0.213	0.479	0.298	0.430
7	`llama_3-1-8B`	0.499	0.720	0.765	0.613	0.223	0.479	0.293	0.396
8	`Phi-4-mini-instruct`	0.498	0.672	0.745	0.636	0.222	0.485	0.299	0.424
9	`falcon3_10b_instruct`	0.482	0.673	0.760	0.538	0.219	0.487	0.302	0.396
10	`Qwen3-32B`	0.477	0.737	0.803	0.373	0.233	0.474	0.307	0.415
11	`Mixtral-8x7B`	0.474	0.656	0.750	0.570	0.213	0.472	0.304	0.353
12	`qwen_2_5_3b`	0.461	0.629	0.726	0.535	0.216	0.484	0.292	0.347
13	`Qwen3-8B`	0.454	0.722	0.748	0.293	0.223	0.477	0.316	0.397
14	`llama_3-2-3B`	0.436	0.502	0.733	0.517	0.200	0.463	0.271	0.369
15	`Mistral-7B`	0.427	0.425	0.701	0.491	0.216	0.483	0.295	0.378
16	`Qwen3-4B`	0.423	0.515	0.752	0.294	0.212	0.470	0.310	0.408
17	`Qwen3-1_7B`	0.419	0.393	0.681	0.484	0.206	0.483	0.299	0.390
18	`flan_t5_xxl`	0.390	0.599	0.705	0.558	0.220	0.420	0.220	0.005
19	`flan_ul2`	0.383	0.567	0.695	0.556	0.205	0.430	0.223	0.003
20	`attrscore_flan_t5_xxl`	0.383	0.571	0.680	0.552	0.214	0.428	0.227	0.005
21	`llama_3-2-1B`	0.354	0.379	0.610	0.477	0.181	0.450	0.269	0.111
22	`Preceptor_AI_v001`	0.295	0.047	0.713	0.021	0.163	0.482	0.277	0.363
23	`Phi-4-mini-reasoning`	0.249	0.095	0.068	0.256	0.196	0.456	0.281	0.389
24	`Preceptor_AI_v003`	0.221	0.000	0.581	0.074	0.111	0.286	0.233	0.263
25	`Phi-4-reasoning-plus`	0.167	0.000	0.000	0.070	0.206	0.470	0.290	0.135

pronounced for the `multi-hop` and `multi-hop inverse` questions, where BLEU occasionally under-estimated quality relative to the embedding-based score.

To illustrate model-specific traits, Figures 3–5 present the full metric dashboards for three representative baselines. The FLAN-UL2 run exhibited tight clustering around mid-range similarity values and an extreme outlier for the multi-hop inverse modality.

LLAMA-3 70B displayed a broader inter-quartile range on semantic scores but maintained competitive n-gram fidelity, suggesting flexible paraphrasing capabilities.

Similarly, the PHI-4-REASONING-PLUS submission produced a long tail of semantically similar scores when evaluated with the CLinIQLInk semantic similarity metric, but low scoring across all the n-gram scoring metrics utilised; further inspection of the model responses revealed that, despite using the prescribed stop tokens and output template, the model frequently emitted extensive chain-of-thought traces capped by an ambiguous or missing "final answer" cue. Our automated evaluation script extracted only the required answers utilising pre-determined queues (i.e. the prompt templates used explicitly constrained models to provide list-type responses as comma-separated lists, etc.) and as such, the digressions observed from the Phi-4-Reasoning-Plus (amongst others) translated into poor task compliance rather than genuine comprehension deficits.

A consolidated leaderboard is provided in Table 2. The ranking served solely as an empirical reference from the evaluation metrics gathered from task 1 automated evaluation script.

8 Discussion

8.1 Closed-ended tasks (Figure 1).

Table 2 confirms what is visually apparent in the right-hand side of Figure 1, which is that single-labelled questions (e.g., t/f, MC, etc.) are close to saturation for modern LLMs. The top five systems tested (`llama_3-3-70B`, `Mistral-Large`, `phi-4`, `llama_4-scout`, and `Mixtral-8x22B`) all scored between 0.75 – 0.80 on **multiple-choice** and 0.79 – 0.82 on **True/False**. By contrast, **list** questions remained challenging with macro–micro F_1 not exceeding 0.68. List answers required both recognition of all correct options *and* rejection of distractors, and as such, the metric penalised even minor hallucinations; consequently, models whose generation style tended to "hedge" with ex-

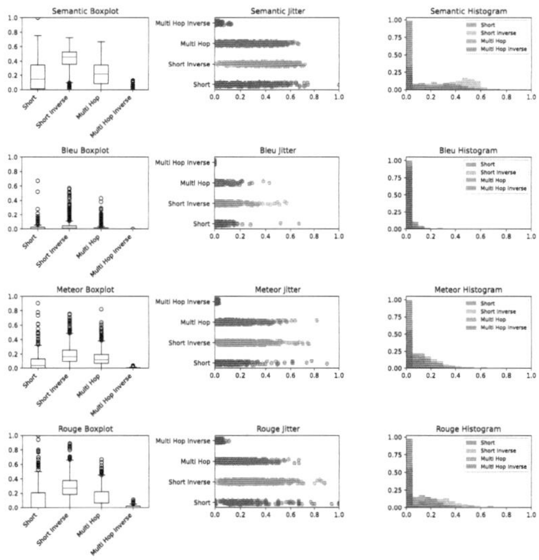

Figure 3: Comprehensive dashboard for FLAN-UL2 showing boxplots, jitter plots and histograms across semantic and n-gram metrics.

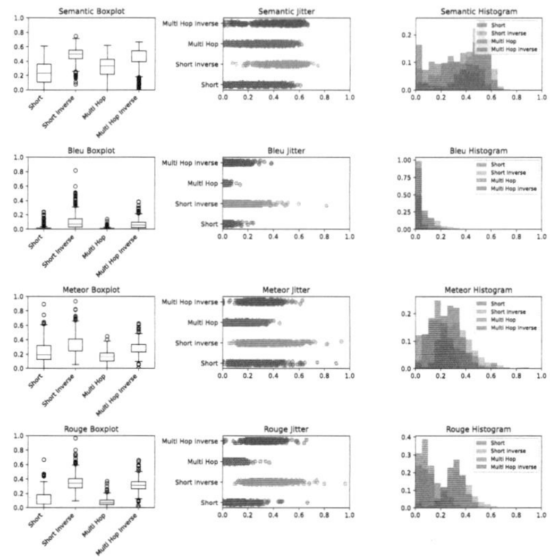

Figure 4: Comprehensive dashboard for LLaMA-70B showing boxplots, jitter plots and histograms across semantic and n-gram metrics.

tra choices (e.g. `falcon3_10b_instruct`) underperformed relative to their multiple-choice score. The tight inter-quartile ranges on True/False and multiple-choice further suggest that most contemporary LLMs share a common ceiling on purely factual one-shot classification, leaving little room for architectural distinctions to distinguish in these settings.

8.2 Aggregate open-ended behavior

Figure 2 shows that, across all models evaluated, the `short inverse` distributions peaked around 0.50 semantic similarity, while the forward `short` items clustered near 0.25, indicating that simply *critiquing* providing an answer was easier than generating an answer from scratch. The gap between semantic and n-gram scores widens for larger checkpoints. Mixtral-8×22B and LLaMA 3.3 70B frequently achieved high semantic similarity scores (above 0.60) despite very low BLEU scores (below 0.1), indicating that their correct answers were often paraphrased rather than copied verbatim, supporting the long-tailed distribution of paraphrastic responses seen in Figure 2. Inspection of the model responses for multi-hop inverse QA types also revealed answers that often diagnosed the wrong knowledge hop step, which in turn attracted the multiplicative penalties. Traditional n-gram metrics failed to flag these omissions, underscoring the necessity of the custom semantic evaluation platform.

8.3 Model-specific open-ended evaluations

Figures 3–5 illustrate how aggregate patterns materialised at the system level. The model-specific open-ended evaluations are shown for only the highest performing model across the board (llama-3.3 70B and the lowest performing model for closed and open-ended metrics (Phi-4-reasoning and FLAN-UL2, respectively).

- **FLAN-UL2** Figure 3 reveals that FLAN-UL2's outputs cluster tightly between 0.20 and 0.60 for the three forward-facing open-ended tasks, yet its `multi-hop inverse` scores collapse toward the origin on *all four* axes—semantic similarity, BLEU, ROUGE, and METEOR rarely rise above 0.05. The dashboard traces that floor effect to the model's habit of supplying only a step label (e.g., "Step 5") with no explanatory text, which earns minimal credit under the step-penalised rubric. Elsewhere, list questions are answered with bare option letters (e.g. "B, C, D"), boosting recall but cutting precision to roughly 0.33–0.50, while short prompts receive one or two-word noun phrases, driving n-gram metrics to zero even when the semantics are acceptable. These abrupt, template-bound behaviours keep variance low and prevent catastrophic errors, but they also cap the weighted open-ended average at 0.14 and hold FLAN-UL2 in 18th place despite competent

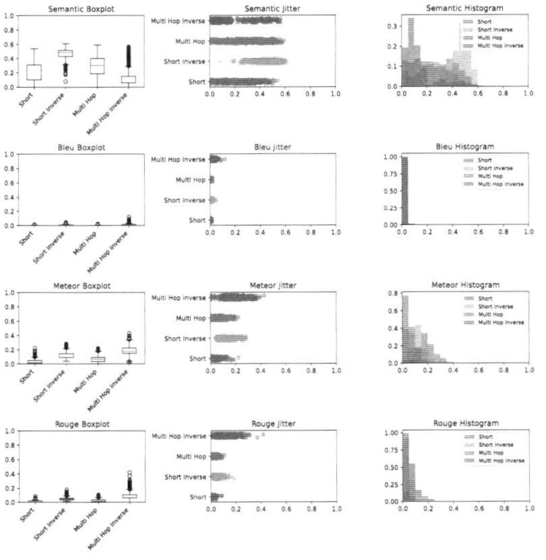

Figure 5: Comprehensive dashboard for Phi-4-reasoning-plus showing boxplots, jitter plots and histograms across semantic and n-gram metrics.

closed-ended performance.

- **LLaMA-3-3-70B** Figure 4 shows that LLaMA-3 70 B's open-ended answers cluster in the mid-range for every metric, not at the extremes. Its *semantic-similarity* box-plot sits roughly between 0.35 and 0.55, with whiskers reaching only the mid-0.70s; BLEU, ROUGE, and METEOR centre much lower (BLEU's median is barely above 0.04, ROUGE around 0.20, METEOR around 0.25). The small band of higher-value semantic outliers (around 0.65–0.75) is confined to `short inverse` and `multi-hop inverse` items in which the model repeated key medical terms but re-ordered the surrounding sequence of words, so n-gram overlap stayed muted. Conversely, many `short` replies are abrupt noun-phrases, depressing all four metrics and keeping the inter-quartile ranges tight.

- **Phi-4-Reasoning-Plus** (Figure 5). The cloud at the extreme lower-left of the dashboard mirrors the 624 malformed list entries and 813 invalid True/False lines produced by this model. Extensive "chain-of-thought" preambles obscured the required delimiters, so the automated evaluation script extracted empty or partial lines. BLEU/ROUGE medians (around 0.04) remained higher than the semantic median (around 0.02) because the responses still shared surface n-grams with the references.

8.4 Cross-metric contrasts.

1. The ClinIQLink Semantic similarity metric displayed higher variance than any n-gram metric across every model dashboard, reflecting sensitivity to both omissions *and* verbose digressions.

2. The gap between ClinIQLink Semantic similarity metric and BLEU was inversely correlated with parameter count; smaller Qwen checkpoints recycled reference wording, whereas 70-B LLaMAs paraphrased aggressively.

3. `Multi hop inverse` was the most discriminative sub-task; its step-penalty compressed medians for every system (lowest boxes in Figure 2), frequently reshuffling neighbouring ranks in Table 2.

8.5 Findings

- High closed-ended scores hide residual hallucinations. Even with a vocabulary capped at just true/false or four choice-letters, every model occasionally invented an out-of-range option, proving that 0.75–0.82 headline accuracies do not equal flawless control.

- List questions are the singular closed-ended format that is still able to effectively discriminate model effectiveness because they demand selecting all true items while rejecting distractors, and as such, macro–micro F_1 was found to be spread from 0.30 to 0.68. Those wider answer sets surface the hallucinated extras that multiple-choice and true/false conceal.

- "Critique" is easier than "generate". Across the board, `short inverse` prompts (spot the error) cluster around 0.50 semantic similarity which is roughly double the median for forward `short` prompts that require composing a fresh answer.

- Multi-hop-inverse is the most discriminative open-ended task. Its step-distance penalty drags every model's median to the bottom of Figure 2, reshuffling several adjacent leaderboard positions and exposing brittle reasoning chains.

- Embedding-level similarity scores for LLM evaluation tasks are now required as the minimum standard. High-ranked systems such as Mixtral 22B and LLaMA-3.3 70B often score > 0.60 on the semantic metric while BLEU, ROUGE and METEOR sit < 0.05, confirming that lexically novel yet faithful paraphrases fool token-overlap measures. Conversely, runs that recycle reference text earn decent n-gram scores but remain low on the embedding metric, demonstrating that overlap alone no longer tracks answer fidelity.

- Note on open-ended evaluation challenges. Despite impressive progress in embedding-based metrics (e.g., BERTScore, Sentence-Mover, BLEURT, COMET etc.) and NLI-based metrics (e.g., MENLI, UniEval), no single method can yet (a) decide with high confidence that two free-form LLM responses convey the same meaning, while also (b) grounding that decision in consistent entity and relation alignment across passages. Embedding similarity captures distributional closeness but is blind to logical entailment; NLI classifiers reason over sentence-level entailment yet lack explicit entity grounding and scale poorly beyond short contexts, and recent surveys and benchmark studies conclude that integrating these complementary views into a robust, scalable metric remains an unsolved problem and a key direction for future work (Ito et al., 2025; Croxford et al., 2025).

9 Conclusion

The CLINIQLINK evaluation shows that modern LLMs reach impressive headline scores on tightly constrained *True/False* and single-letter *multiple-choice* items, yet every model evaluated still sporadically produces out-of-vocabulary or otherwise invalid answers; unordered *list* questions, with their wider response space, remain the only closed-ended format able to expose this fragility. On open-ended tasks, embedding-based semantic similarity distinguishes genuinely informative paraphrases from superficial n-gram overlap. Conventional n-gram indices systematically mis-score open responses, rewarding superficial token overlap while penalising lexically novel yet factually correct paraphrases; embedding-based similarity aligns far more closely with clinical accuracy and, through the step-penalised *multi-hop-inverse* task, reveals brittle reasoning chains. More work is required to produce an effective semantic similarity scoring metric with explicit reasoning validation into a composite metric that more rigorously captures factuality, logical coherence, entity relationship framing, and schema compliance. To support this goal, future iterations of CLINIQLINK will link each question–answer pair to a machine-readable knowledge graph for graph-based verification of multi-step rationales and will introduce multimodal variants that couple text queries with images, thereby challenging models to ground their answers in heterogeneous clinical evidence.

Acknowledgments

This research was supported in part by the Division of Intramural Research (DIR) of the National Library of Medicine (NLM), National Institutes of Health. This work utilized the computational resources of the NIH HPC Biowulf cluster (`https://hpc.nih.gov`) and the University of Maryland's Zaratan cluster (`https://hpcc.umd.edu/hpcc/zaratan.html`). This research was also partly supported by the U.S. Fulbright program, enabling international collaborative efforts.

References

Marah Abdin, Sahaj Agarwal, Ahmed Awadallah, Vidhisha Balachandran, Harkirat Behl, Lingjiao Chen, Gustavo de Rosa, Suriya Gunasekar, Mojan Javaheripi, Neel Joshi, Piero Kauffmann, Yash Lara, Caio César Teodoro Mendes, Arindam Mitra, Besmira Nushi, Dimitris Papailiopoulos, Olli Saarikivi, Shital Shah, Vaishnavi Shrivastava, and 4 others. 2025. Phi-4-reasoning technical report. *Preprint*, arXiv:2504.21318.

Marah Abdin, Jyoti Aneja, Harkirat Behl, Sébastien Bubeck, Ronen Eldan, Suriya Gunasekar, Michael Harrison, Russell J. Hewett, Mojan Javaheripi, Piero Kauffmann, James R. Lee, Yin Tat Lee, Yuanzhi Li, Weishung Liu, Caio C. T. Mendes, Anh Nguyen, Eric Price, Gustavo de Rosa, Olli Saarikivi, and 8 others. 2024. Phi-4 technical report. *Preprint*, arXiv:2412.08905.

Ebtesam Almazrouei, Hamza Alobeidli, Abdulaziz Alshamsi, Alessandro Cappelli, Ruxandra Cojocaru, Mérouane Debbah, Étienne Goffinet, Daniel Hesslow, Julien Launay, Quentin Malartic, Daniele Mazzotta, Badreddine Noune, Baptiste Pannier, and Guilherme Penedo. 2023. The falcon series of open language models. *Preprint*, arXiv:2311.16867.

Satanjeev Banerjee and Alon Lavie. 2005. METEOR: An automatic metric for MT evaluation with improved correlation with human judgments. In *Proceedings of the ACL Workshop on Intrinsic and Extrinsic Evaluation Measures for Machine Translation and/or Summarization*, pages 65–72, Ann Arbor,

Michigan. Association for Computational Linguistics.

Nicholas Carlini, Daphne Ippolito, Matthew Jagielski, Katherine Lee, Florian Tramer, and Chiyuan Zhang. 2023. Quantifying memorization across neural language models. *Preprint*, arXiv:2202.07646.

Hyung Won Chung, Le Hou, Shayne Longpre, Barret Zoph, Yi Tay, William Fedus, Yunxuan Li, Xuezhi Wang, Mostafa Dehghani, Siddhartha Brahma, Albert Webson, Shixiang Shane Gu, Zhuyun Dai, Mirac Suzgun, Xinyun Chen, Aakanksha Chowdhery, Alex Castro-Ros, Marie Pellat, Kevin Robinson, and 16 others. 2022. Scaling instruction-finetuned language models. *Preprint*, arXiv:2210.11416.

Emma Croxford, Yanjun Gao, Nicholas Pellegrino, Karen Wong, Graham Wills, Elliot First, Frank Liao, Cherodeep Goswami, Brian Patterson, and Majid Afshar. 2025. Current and future state of evaluation of large language models for medical summarization tasks. *npj Health Systems*, 2(1).

Aaron Grattafiori, Abhimanyu Dubey, Abhinav Jauhri, Abhinav Pandey, Abhishek Kadian, Ahmad Al-Dahle, Aiesha Letman, Akhil Mathur, Alan Schelten, Alex Vaughan, Amy Yang, Angela Fan, Anirudh Goyal, Anthony Hartshorn, Aobo Yang, Archi Mitra, Archie Sravankumar, Artem Korenev, Arthur Hinsvark, and 542 others. 2024. The llama 3 herd of models. *Preprint*, arXiv:2407.21783.

Takumi Ito, Kees van Deemter, and Jun Suzuki. 2025. Reference-free evaluation metrics for text generation: A survey. *arXiv preprint*.

Albert Q. Jiang, Alexandre Sablayrolles, Arthur Mensch, Chris Bamford, Devendra Singh Chaplot, Diego de las Casas, Florian Bressand, Gianna Lengyel, Guillaume Lample, Lucile Saulnier, Lélio Renard Lavaud, Marie-Anne Lachaux, Pierre Stock, Teven Le Scao, Thibaut Lavril, Thomas Wang, Timothée Lacroix, and William El Sayed. 2023. Mistral 7b. *Preprint*, arXiv:2310.06825.

Albert Q. Jiang, Alexandre Sablayrolles, Antoine Roux, Arthur Mensch, Blanche Savary, Chris Bamford, Devendra Singh Chaplot, Diego de las Casas, Emma Bou Hanna, Florian Bressand, Gianna Lengyel, Guillaume Bour, Guillaume Lample, Lélio Renard Lavaud, Lucile Saulnier, Marie-Anne Lachaux, Pierre Stock, Sandeep Subramanian, Sophia Yang, and 7 others. 2024. Mixtral of experts. *Preprint*, arXiv:2401.04088.

Omar Khattab and Matei Zaharia. 2020. Colbert: Efficient and effective passage search via contextualized late interaction over bert. *Preprint*, arXiv:2004.12832.

Chin-Yew Lin. 2004. ROUGE: A package for automatic evaluation of summaries. In *Text Summarization Branches Out*, pages 74–81, Barcelona, Spain. Association for Computational Linguistics.

Christopher D. Manning, Prabhakar Raghavan, and Hinrich Schütze. 2008. *Introduction to Information Retrieval*. Cambridge University Press.

Kishore Papineni, Salim Roukos, Todd Ward, and Wei-Jing Zhu. 2002. Bleu: a method for automatic evaluation of machine translation. In *Proceedings of the 40th Annual Meeting of the Association for Computational Linguistics*, pages 311–318, Philadelphia, Pennsylvania, USA. Association for Computational Linguistics.

Stephen Robertson, Hugo Zaragoza, and 1 others. 2009. The probabilistic relevance framework: Bm25 and beyond. *Foundations and Trends® in Information Retrieval*, 3(4):333–389.

Karan Singhal, Shekoofeh Azizi, Tao Tu, S. Sara Mahdavi, Jason Wei, Hyung Won Chung, Nathan Scales, Ajay Tanwani, Heather Cole-Lewis, Stephen Pfohl, Perry Payne, Martin Seneviratne, Paul Gamble, Chris Kelly, Abubakr Babiker, Nathanael Schärli, Aakanksha Chowdhery, Philip Mansfield, Dina Demner-Fushman, and 13 others. 2023. Large language models encode clinical knowledge. *Nature*, 620(7972):172–180.

Yi Tay, Mostafa Dehghani, Vinh Q. Tran, Xavier Garcia, Jason Wei, Xuezhi Wang, Hyung Won Chung, Siamak Shakeri, Dara Bahri, Tal Schuster, Huaixiu Steven Zheng, Denny Zhou, Neil Houlsby, and Donald Metzler. 2023. Ul2: Unifying language learning paradigms. *Preprint*, arXiv:2205.05131.

An Yang, Anfeng Li, Baosong Yang, Beichen Zhang, Binyuan Hui, Bo Zheng, Bowen Yu, Chang Gao, Chengen Huang, Chenxu Lv, Chujie Zheng, Dayiheng Liu, Fan Zhou, Fei Huang, Feng Hu, Hao Ge, Haoran Wei, Huan Lin, Jialong Tang, and 41 others. 2025. Qwen3 technical report. *Preprint*, arXiv:2505.09388.

Xiang Yue, Boshi Wang, Ziru Chen, Kai Zhang, Yu Su, and Huan Sun. 2023. Automatic evaluation of attribution by large language models. *Preprint*, arXiv:2305.06311.

Hye Sun Yun and Timothy Bickmore. 2025. Online health information–seeking in the era of large language models: Cross-sectional web-based survey study. *J Med Internet Res*, 27:e68560.

SMAFIRA Shared Task at the BioNLP'2025 Workshop: Assessing the Similarity of the Research Goal

Mariana Neves[1] **Iva Sovadinova**[2] **Susanne Fieberg**[1] **Céline Heinl**[1]
diana Rubel[1] **Gilbert Schönfelder**[1,3] **Bettina Bert**[1]
[1]German Centre for the Protection of Laboratory Animals (Bf3R),
German Federal Institute for Risk Assessment (BfR), Berlin, Germany
[2]RECETOX, Masaryk University, Faculty of Science, Brno, Czech Republic
[3]Institute of Clinical Pharmacology and Toxicology,
Charité - Universitätsmedizin Berlin, Berlin, Germany

Abstract

We organized the SMAFIRA Shared in the scope of the BioNLP'2025 Workshop. Given two articles, our goal was to collect annotations about the similarity of their research goal. The test sets consisted of a list of reference articles and their corresponding top 20 similar articles from PubMed. The task consisted in annotating the similar articles regarding the similarity of their research goal with respect to the one from the corresponding reference article. The assessment of the similarity was based on three labels: "similar", "uncertain", or "not similar". We released two batches of test sets: (a) a first batch of 25 reference articles for five diseases; and (b) a second batch of 80 reference articles for 16 diseases. We collected manual annotations from two teams (RCX and Bf3R) and automatic predictions from two large language models (GPT-4omini and Llama3.3). The preliminary evaluation showed a rather low agreement between the annotators, however, some pairs could potentially be part of a future dataset.

1 Introduction

Many countries require the researchers to ask for a permission before they carry out an animal experiment (Vasbinder and Locke, 2017). Some countries, e.g., Germany, require a through search of the scientific literature in order to certify that no alternative methods are already available.

We recently developed the SMAFIRA tool[1] (Butzke et al., 2024) to support the above task. The input to the tool is a PubMed identifier (PMID) of an animal experiment, hereafter called "reference article". From PubMed, the tool retrieves the similar articles to the reference article, for which it performs two automatic tasks: (a) classification of the methods (Neves et al., 2023a), and (b) re-ranking of the retrieved similar articles.

For the latter, the goal is to rank the similar articles according the similarity of their research goal, i.e., with respect to the research goal of the corresponding reference article. Previously, we created the SMAFIRA-c dataset (Butzke et al., 2020), for which we annotated the top 100 (approximately) for four reference articles (cf. Section 2). Based on this dataset, we recently performed an evaluation of various similarity methods (Neves et al., 2023b). However, the dataset is rather small for training or even for a comprehensive evaluation of various methods.

The SMAFIRA Shared Task[2] is a collaborative effort that aimed to collect additional data for this task. We released a list of various reference articles, grouped according to some pre-selected diseases (MeSH terms). Participants were asked validate the top 20 similar articles for any number of reference articles. The similarity was assessed in terms of three labels, namely, "similar", "uncertain", and "not similar". The annotations could be performed either automatically, with any system of their choice, or manually using the SMAFIRA tool.

We describe the shared task in the next section of this publication, including the test sets, annotation tasks, guidelines, and the available dataset. In Section 3 we list the various teams (manual and automatic annotations), including how we retrieved automatic annotations from two large language models (LLMs). We give and overview of the annotations that we obtained in Section 4, as well as the computation of the agreement. Finally, we present an analysis of the annotations in Section 5.

2 SMAFIRA Shared Task

2.1 Test Sets

We compiled a list of reference articles for various disease categories. We started with a list of 23 diseases from the "Diseases (C)" category in the MeSH terms[3]. For each sub-category (MeSH

[1]https://smafira.bf3r.de/

[2]https://smafira-bf3r.github.io/smafira-st/

[3]https://meshb.nlm.nih.gov/treeView

Proceedings of the 24th Workshop on Biomedical Language Processing (BioNLP 2025), pages 388–395
August 1, 2025 ©2025 Association for Computational Linguistics

batch1					
Infections [C01]	36159784	36577999	32485164	37071015	31689515
Neoplasms [C04]	34233949	33320838	36311701	37429473	35623658
Nervous System Diseases [C10]	35709748	37084732	37339207	37749256	37126714
Cardiovascular Diseases [C14]	33635944	37010266	37380648	37268711	35917178
Immune System Diseases [C20]	34503569	36179018	37079985	37256935	37168850
batch2					
Musculoskeletal Diseases [C05]	37775153	36328744	36209953	36661300	36302840
Digestive System Diseases [C06]	26313006	34089528	36717026	30974318	34774008
Stomatognathic Diseases [C07]	32541832	34190354	33673616	35082168	37143319
Respiratory Tract Diseases [C08]	31694835	33524990	33166988	32707078	37730992
Otorhinolaryngologic Diseases [C09]	38531465	35331657	38608332	31570054	30970038
Eye Diseases [C11]	37345657	32721019	32341164	37429715	37757825
Urogenital Diseases [C12]	36581059	37324943	35264456	38688639	34270549
Hemic and Lymphatic Diseases [C15]	32001657	32494068	33639162	31797883	38713510
Congen., Heredit., and Neonatal Dis. and Abnorm. [C16]	33922602	31476705	34533563	38891999	33729473
Skin and Connective Tissue Diseases [C17]	32440554	33391503	34078596	38361478	31481954
Nutritional and Metabolic Diseases [C18]	33762572	38263084	36463128	37245586	36854163
Endocrine System Diseases [C19]	21211517	1617104	23777580	26517045	37480416
Pathological Conditions, Signs and Symptoms [C23]	33744277	32544087	26667043	38690023	24286894
Occupational Diseases [C24]	34139709	27775689	38669965	33705732	28762870
Chemically-Induced Disorders [C25]	23449255	7236062	28263289	31641018	36162952
Wounds and Injuries [C26]	26123115	31111883	29603350	19841895	16929202

Table 1: List of reference articles (test sets) for batch1 and batch2.

term) from the list, we queried PubMed with the corresponding term and for animal models[4]. Subsequently, we filtered for articles with available abstract and that were published in the last five years.

For each disease, we screened the list of results and selected five reference articles that described an animal experiment. We skipped surveys and review articles and checked that the reference article contained a pre-compiled list of similar articles. We aimed at selecting reference articles that referred to distinct diseases, e.g., distinct cancer types for the category "Neoplasms". From the original list of 23 categories, we ended up with 21 categories. We could not find five interesting animal experiments for two categories, namely, "Disorders of Environmental Origin [C21]" and "Animal Diseases [C22]".

We split the above reference articles into two groups: "batch1" and "batch2". Batch1 was released in February/2025 and contains five preselected disease categories, namely, "Infections [C01]", "Neoplasms [C04]", "Nervous System Diseases [C10]", "Cardiovascular Diseases [C14]", and "Immune System Diseases [C20]". Batch2 contains the remaining 16 disease categories and was released in the end of April/2025. Table 1 shows all reference articles for both batches.

[4]e.g., "(Infections[MeSH Major Topic]) AND (Models, Animal[MeSH Major Topic])"

2.2 Annotation Tasks

We proposed two annotation tasks: manual and automatic annotation. For both tasks, for any reference article, the top 20 similar articles should be annotated. The annotation should be based on the similarity of the research goal (cf. Section 2.3), and over three possible values for the similarity: "similar", "uncertain", and "not similar".

For the manual annotation, the task should be carried out in the SMAFIRA tool. Participants should enter one of the reference articles (cf. Table 1) in the input field and the tool retrieves the list of similar articles as available in PubMed. The top 20 similar articles should be annotated based on the SMAFIRA-Rank option (the default option). After the annotation, participants have two possibilities to submit their annotations to us per e-mail: (a) share their session URL, or (b) export the annotations into a file. More details about the annotation with the SMAFIRA tool is available on the web site of the shared task.

For the automatic annotation, we provide the reference articles, their corresponding top 20 similar articles, and all titles and abstracts, which were retrieved using the TeamTat tool (Islamaj et al., 2020). This data is available for download in the JSON format in the GitHub repository[5]. There is one folder for each of the batches, in which we released the following files:

[5]https://github.com/smafira-bf3r/smafira-st

- (a) "batch1.json" or "batch2.json": complete JSON file with all reference articles, their respective top 20 similar articles, as well as title and abstracts for all PMIDs;
- (b) (optional) "batch1_teamtat.zip" or "batch2_teamtat.zip": zip file with all articles as exported by TeamTat;
- "sample_submission.json": sample submission file that include all reference articles and their similar articles, but not the labels.

2.3 Guidelines

For each pair, i.e., a reference article and one of the similar articles, our goal is to assess their similarity based on three labels: "similar", "uncertain", or "not similar". We decided some simple aspects that should be taken into account during the annotation:

- The assessment should only be based on the title and the abstract, thus, the annotator should not consider the full text of the article.
- The methods should not be considered, since two research goals can be similar even if, for instance, one article describes an *in vivo* experiment and the other an *in vitro* experiment.

The actual decision of the label for a particular pair is very subjective and dependent on the opinion of the annotator. The SMAFIRA-c dataset (cf. below) has some examples that can be used for better understanding the various similarity situations. Further, we give some examples on the web site based on three aspects that were curated in the JRC's reports (e.g., (Commission et al., 2020)), namely, application, disease, and disease feature.

The application refers to the the main scientific aim of the article or the application of the described model or method, e.g., whether the article describes the mechanism of the disease or the development of a new treatment. When addressing a certain disease, an article usually describe which specific aspects of the latter are under study, e.g., the progression of the tumor into an invasive form.

The assessment of the similarity could be based on these three aspects, though this is not mandatory. Pair of articles in which all these aspects are equal (or very similar) could certainly be tagged as "similar". Since our annotation is based on the list of similar articles, all articles are somehow similar to the reference article. For instance, the disease is usually the same, and exceptions to this usually constitute a good reason for tagging an article as "not similar". However, the disease feature is often not the same, or more than one are described, and their similarity (or lack of similarity) is usually the main aspect to be observed when deciding about the label. Finally, the application is also usually the same across the articles and exceptions could also be tagged as "not similar".

2.4 Available data

Previously, we have annotated (approximately) the top 100 similar articles for four reference articles, namely, the SMAFIRA-c dataset[6] (Butzke et al., 2020). This data could be used for manually checking some annotated examples, e.g., for training purposes. Further, for automatic methods, it could be used for few-shot strategies or for the evaluation. However, given its small size, it might not be appropriate for supervised learning purposes. The mapping between the annotations in SMAFIRA-c ("Equivalence" column) and the three labels used in the shared task is shown below:

- "similar": equivalent "++", partially equivalent "+(+)" or "+", noteworthy "n"
- "uncertain": limbo "L"
- "not similar": not equivalent "-"

3 Teams and Systems

In this section we give details of the participants of the shared task. For the sake of simplicity, we will sometimes refer to all of participants, whether manual annotators or automatic systems, as "teams" throughout this publication.

For the manual annotation, we had the participation of two teams:

- "RCX" (RECETOX, Faculty of Science, Masaryk University, Czechia);
- "Bf3R" (German Centre for the Protection of Laboratory Animals, Germany).

The annotations from "Bf3R" were carried out by five experts. Some of them annotated the same reference article in order to compare their results, but they did not try to reach a consensus. In these cases, we selected one of them as the official submission of the team.

[6]https://github.com/SMAFIRA/c_corpus

For the automatic annotations, we relied on a zero-shot approach with two LLMs: (a) the GPT-4o-mini model using the OpenAI API[7]; and (b) Llama3.3 (llama-3.3-70b-versatile) using the Groq API[8]. We provided the two texts (title and abstract) in the prompt, i.e., first the one for the reference article and then the one for one of the similar article, followed by the questions with detailed instruction on how to assess the similarity. We used the following user message:

You are a helpful assistant designed to evaluate the similarity between two texts.

and the following user content:

Text 1: REF_ARTICLE_TEXT
Text 2: SIMILAR_ARTICLE_TEXT
Are the the research goals of the two texts above similar? You should compare the research goal based on four aspects: (1) Are the disease(s) addressed in the texts the same? (2) Do they address the same characteristic symptom/feature of the disease? (3) Do they refer to the same biological endpoints, e.g., the same disease mechanism, gene/protein or chemical coumpounds? (4) Is the scientific aim or the future application of the results the same, e.g., for drug development, model development, disease treatment or diagnosis? Answer with either 'similar', 'uncertain', or 'not similar'. The answer is:

We evaluated our prompts with the cases studies of the SMAFIRA-c dataset (cf. Section 2.4) and show the statistics of the corpus (cf. Table 3) and the results (cf. Table 4) in the Appendix A. The same prompt was used for both LLMs when obtaining annotations for the shared task, as well as for the evaluation of the SMAFIRA-c dataset. We retrieved annotations from the two LLMs for all reference articles in batch1.

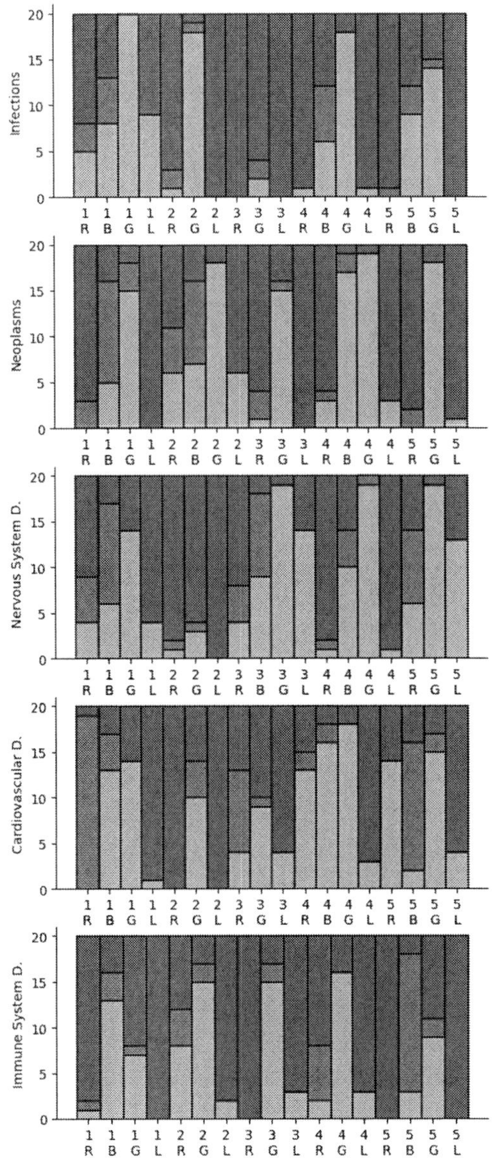

Figure 1: Overview of the annotations per disease in terms of number of annotations (y-axis). The x-axis shows the five reference article (in the order shown in Table 1) and the teams: (R)CX, (B)f3R, (G)PT-4o-mini, and (L)lama3.3. The three-value similarity is the following (from darker to lighter color, from top to bottom in each graph): "similar": dark blue (top color), "uncertain": dark blue/green (middle color), "not similar": light green (bottom color).

4 Results

4.1 Overview of the annotations

We describe the annotations that we obtained from two participants and from two LLMs. In this publication, we present results only for batch1.

We obtained manual annotations for all 25 reference articles from RCX and for 14 reference articles from Bf3R. Further, we collected annotations for all reference articles from the two LLMs. Regarding the three similarity labels, we obtained the following number of annotations (from a total of 1,780):

[7]https://openai.com/

[8]https://console.groq.com/docs/model/llama-3.3-70b-versatile

948 (53%) for "similar", 202 (11%) for "uncertain", and 630 (35%) for "not similar". All annotations are available in our GitHub repository.

We depict the number of annotations for each label in Figure 1, from which we can observe some differences across the teams. On the one hand, RCX and Llama3.3 frequently assigned the "similar" label to all (or most) of the similar articles of some reference articles. On the other hand, GPT-4o-mini frequently assigned the "not similar" label for all (or most) of the similar articles of some reference articles. Further, the "uncertain" label was more frequently assigned by human annotators, but rarely returned by GPT-4o-mini, and never by Llama3.3.

We did not observe considerable differences across the diseases. For all five diseases, the "similar" label was the most frequent one (45% to 62%), followed by "not similar" (28% to 41%) and "uncertain" (7% to 16%).

4.2 Agreement between teams

We analyzed the agreement of the annotations in various ways, e.g., pairwise comparison between two teams, or multiple comparison across all teams. We present the results below.

Agreement for manual annotation. For the RCX and Bf3R teams and for the 14 reference articles annotated by both teams, we observed the following (cf. Figure 1):

- Three cases with good agreement: "36159784" (no. 1) of "Infections", "33320838" (no. 2) of "Neoplasms" and "37268711" (no. 4) of "Cardiovascular Diseases".
- Two cases had some agreement: "35709748" (no. 1) and "37339207" (no. 3) of "Nervous System Diseases".
- Four cases in which one assigned mostly the "uncertain" label, which might overlap with the "similar" or "not similar" labels from the other: "34233949" (no. 1) of "Neoplasms", "33635944" (no. 1) and "35917178" (no. 5) of "Cardiovascular Diseases", and "37168850" (no. 5) of "Immune System Diseases".
- Five cases with very bad agreement: "37071015" (no. 4) and "31689515" (no. 5) of "Infections", "37429473" (no. 4) of "Neoplasms", "37749256" (no. 4) of "Nervous System Diseases", and "34503569" (no. 1) of "Immune System Diseases".

In general, the agreement for the manual annotation was rather good for the reference articles in the "Cardiovascular Diseases". However, this comparison did not consider the labels for each particular article, nor agreements that might have occurred by chance.

Pairwise agreement. We computed the kappa score[9] (McHugh, 2012) for all pairwise comparison between the teams and plotted a heatmap in Figure 2. From a total of 114 pairs, 26 of them were negative, (no agreement), 42 between zero and 0.2 (slight agreement), 12 between 0.2 and 0.4 (fair agreement), four between 0.4 and 0.6 (moderate agreement), and none above these values (substantial or perfect agreement). As already observed above, there are less negative scores for the "Cardiovascular Diseases". From the 14 reference articles annotated by human annotators, seven of them had a negative agreement. The three highest scores, namely, 0.52, 0.51, and 0.49 were obtained between RCX and Llama3.3, followed by a good agreement (0.44) by Bf3R and GPT-4o-mini.

Multiple agreement. For each reference article, we also computed the krippendorff's alpha score[10] across annotations from all teams (whenever available). We plot the scores on Figure 3. From the 25 reference article, 22 of them were negative, which mean a systematic disagreement. The highest (and positive) scores, i.e., 0.056, 0.051, and 0.035, were obtained by the following reference articles, respectively: "37380648" and "35917178" of "Cardiovascular Diseases", and "37126714" from "Nervous System Diseases".

5 Discussion

5.1 Analysis of the annotations

We analyzed the articles that could already be part a future dataset, as well as potential articles that could be included after an additional round of consensus. Further, we also analyzed whether articles tagged as "similar" were usually ranked higher in the top 20 list.

Pairs of articles with high agreement. The aim of this shared task was to build a dataset for pairs of articles with respect to the similarity of their research goal. Our previous effort, i.e., SMAFIRA-c,

[9] https://scikit-learn.org/stable/modules/generated/sklearn.metrics.cohen_kappa_score.html

[10] https://pypi.org/project/krippendorff/

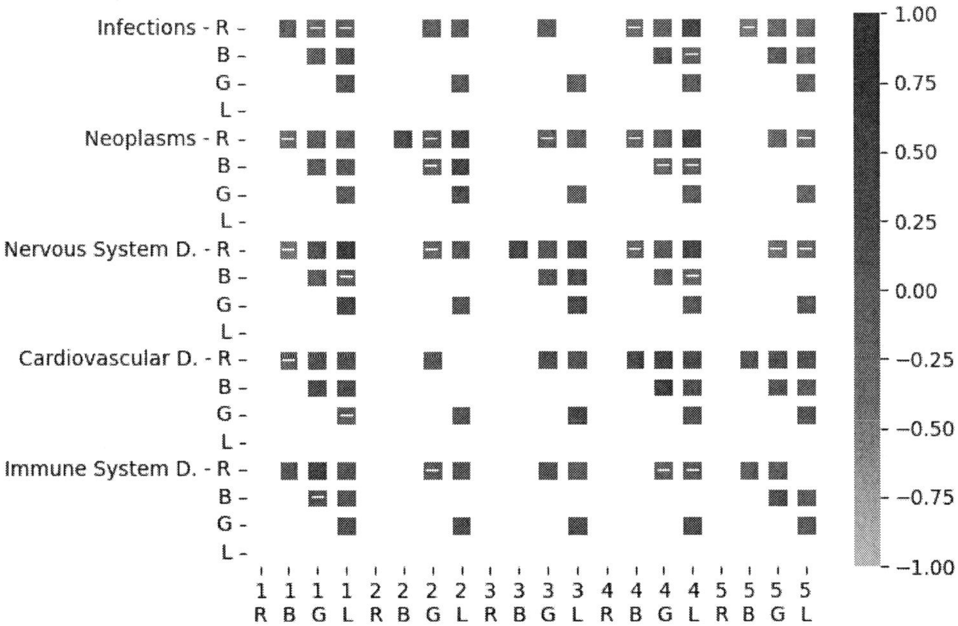

Figure 2: Cohen's kappa scores for each pair of teams. The x-axis shows the five reference article (in the order shown in Table 1) and the teams: (R)CX, (B)f3R, (G)PT-4o-mini, and (L)lama3.3. The y-axis depicts the teams along with the five diseases. Cells with negative scores are depicted with a minus ("-").

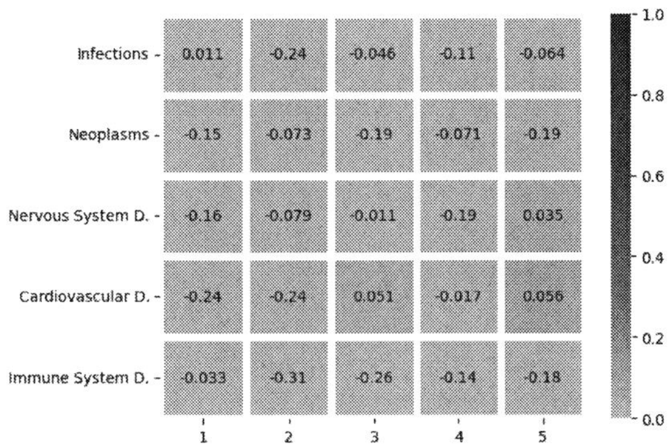

Figure 3: Krippendorff' alpha scores for each reference article across all teams. The x-axis shows the five reference articles (in the order shown in Table 1). The y-axis depicts the five diseases.

is a rather large dataset (around 400 articles), but includes only four reference articles. For all reference articles in batch1, the low kappa and krippendorff's alpha scores showed above indicate that such as dataset should not include all articles from top 20 list. Therefore, we identified the articles (PMIDs) that have high agreement across the teams and that could potentially be included in a dataset. We only considered PMIDs with four equal votes of the same label, i.e., agreement across all teams. We obtained 28 PMIDs from all five diseases with unanimous agreement, 14 "similar" pairs and 14 "not similar" (cf. Table 2).

Pairs of articles with good agreement. Many pairs have a good agreement, even though they have no agreement across the four teams. We identified 55 articles with three unanimous labels, i.e., from the RCX team and the two LLMs. From these, 50 of them were tagged as "similar" and 5 of them as "not similar". These articles come from 11 reference articles, and these are the ones whose annotation from team Bf3R should be prioritized, especially those with already many unanimous labels from the three teams, namely, reference articles "37084732" from "Nervous System Disease" and "32485164" from "Infections". Further, from the 14 reference articles with annotations from the four teams, we identified 77 articles with just one different annotation, e.g., three "similar" annotations and one "not similar". These constitute potential additional 42 "similar" articles and 35 "not similar" articles. A consensus round of annotation could potentially solve these disagreements.

Cases with very low agreement. From the reference articles annotated by all four teams, two of them had no article with an unanimous label,

Diseases	Ref. articles	Articles	Label
Infections	37071015	35605915 36441775	similar
	31689515	28456941 26920550 35798933 26189763	similar
	36159784	34228857	not similar
Neoplasms	34233949	35027827	similar
	33320838	36339405 37376562 35995402 35507699	not similar
	37429473	36740846	not similar
Nervous S.	35709748	25362208	not similar
	37339207	31010153 27174093 27045344 34788059	not similar
Cardiov. S.	37268711	31140393	similar
		36674651 31780864 36990303	not similar
	35917178	23563994	similar
Immune. S.	34503569	35325396 23335001 32693359	similar
	37168850	22673798 25778936	similar

Table 2: Selected unanimous pairs for each disease.

i.e., namely "37749256" of "Nervous System Disease" and "33635944" of "Cardiovascular Diseases". However, some articles in these reference articles had three votes of the same label (cf. above). Further, the reference article "36179018" from "Immune System Disease" was annotated by three teams and did not obtain any article with unanimous label. Finally, in general, the "uncertain" label had a very low agreement, and no article obtained an unanimous label of this type, not even three unanimous labels (cf. above).

Ranks of the articles. For the articles with full agreement across the four teams (cf. above), we checked whether articles tagged as "similar" were usually on the top of the list, and those tagged as "not similar" were rather at the bottom of the list. For the 14 articles tagged as "similar", their positions in the list varied from 1 to 14 (average of 5.5). For the 14 articles tagged as "not similar", their positions in the list varied from 6 to 19 (average of 11.6). On the one hand, and even if the sample is rather small for significant insights, it seems that "similar" articles were actually found in rather higher ranks and "not similar" ones in rather lower ranks. On the other hand, there are some cases of "not similar" ones in the top 10, namely, positions 6, 7, and 9, and "similar" ones below the top 10, namely positions 14 and 15.

6 Conclusion

We proposed the SMAFIRA Shared Task with the aim to collect data for a dataset about the similarity of the research goal between two articles, namely, a reference article and one candidate article from the list of similar articles. We released two batches of references articles: (i) a first one related to five diseases, five reference articles each; (ii) a second one with 16 diseases, also five reference articles each. For any reference article in these batches, we asked the participants to annotate the top 20 similar articles, as available in PubMed. The annotation consisted on assessing the similarity in terms of three labels, namely, "similar", "uncertain", and "not similar".

For the first batch, we collected annotations from two teams that performed manual annotation and two LLMs for automatic annotations. For each reference article, we presented a detailed analysis based on the number of the labels, as well as agreement based on the kappa and the kippendorff's alpha scores. These scores were very low (often negative) for most reference articles, which means that there is a systematic disagreement across most articles on the top 20.

In spite of the above, there are some articles with high agreement and which could already be part of a dataset. Additionally, some more articles received three equal labels (out of four teams) and could also possibly be included after a consensus round. Finally, some more articles have three unanimous labels and could also potentially be selected after an annotation round from the Bf3R teams. Finally, the RCX team has already agreed to further annotate the second batch, and we could also obtain annotations from the two LLMs, as well as some additional ones from the Bf3R team.

Finally, a preliminary analysis of the ranks of the articles tagged as "similar" or "not similar" confirmed that some articles could have been pushed higher in the top 20 list. Therefore, we need better methods for assessing the similarity of the articles' research goals. However, a comprehensive dataset is essential for a reliable evaluation of these methods, and for training few-shot approaches.

Acknowledgments

We would like to thank Sebastian Dunst, Daniel Butzke, and Ines Schadock for their feedback on the shared task.

References

Daniel Butzke, Bettina Bert, Konrad Gulich, Gilbert Schönfelder, and Mariana Neves. 2024. SMAFIRA: a literature-based web tool to assist researchers with retrieval of 3R-relevant information. *Laboratory Animals*, 58(4):369–373. PMID: 38872231.

Daniel Butzke, Nadine Dulisch, Sebastian Dunst, Matthias Steinfath, Mariana Neves, Brigitte Mathiak, and Barbara Grune. 2020. Smafira-c: A benchmark text corpus for evaluation of approaches to relevance ranking and knowledge discovery in the biomedical domain.

European Commission, Joint Research Centre, I Adcock, T Novotny, M Nic, K Dibusz, J Hynes, L Marshall, and L Gribaldo. 2020. *Advanced non-animal models in biomedical research : respiratory tract diseases*. Publications Office of the European Union.

Rezarta Islamaj, Dongseop Kwon, Sun Kim, and Zhiyong Lu. 2020. Teamtat: a collaborative text annotation tool. *Nucleic Acids Research*, 48(W1):W5–W11.

M. L. McHugh. 2012. Interrater reliability: the kappa statistic. *Biochemia Medica*, 22:276 – 282.

Mariana Neves, Antonina Klippert, Fanny Knöspel, Juliane Rudeck, Ailine Stolz, Zofia Ban, Markus Becker, Kai Diederich, Barbara Grune, Pia Kahnau, Nils Ohnesorge, Johannes Pucher, Gilbert Schönfelder, Bettina Bert, and Daniel Butzke. 2023a. Automatic classification of experimental models in biomedical literature to support searching for alternative methods to animal experiments. *Journal of Biomedical Semantics*, under review.

Mariana Neves, Ines Schadock, Beryl Eusemann, Gilbert Schönfelder, Bettina Bert, and Daniel Butzke. 2023b. Is the ranking of PubMed similar articles good enough? an evaluation of text similarity methods for three datasets. In *The 22nd Workshop on Biomedical Natural Language Processing and BioNLP Shared Tasks*, pages 133–144, Toronto, Canada. Association for Computational Linguistics.

Mary Ann Vasbinder and Paul Locke. 2017. Introduction: Global Laws, Regulations, and Standards for Animals in Research. *ILAR Journal*, 57(3):261–265.

A Evaluation of the LLMs with SMAFIRA-c dataset

We show the statistics of the annotation of the corpus are in Table 3. Further, we evaluated both

ref. PMIDs	similar	uncertain	not similar
16850029	11	14	71
19735549	12	5	81
21494637	5	42	56
24204323	26	0	76

Table 3: Number of annotations in the SMAFIRA-c dataset.

	GPT-4omini			Llama3.3		
16850029	P	R	F1	P	R	F1
similar	0.31	0.90	0.46	0.15	1.00	0.26
not similar	0.84	0.75	0.79	0.89	0.35	0.51
uncertain	0.33	0.07	0.12	0	0	0
overall	0.66	0.66	0.66	0.37	0.37	0.37
19735549	P	R	F1	P	R	F1
similar	0.46	0.55	0.50	0.13	1.00	0.23
not similar	0.87	0.86	0.87	1.00	0.17	0.29
uncertain	0	0	0	0	0	0
overall	0.78	0.78	0.78	0.26	0.26	0.28
21494637	P	R	F1	P	R	F1
similar	0.42	0.56	0.48	0.42	1.00	0.59
not similar	0.58	0.43	0.49	1.00	0.07	0.13
uncertain	0	0	0	0	0	0
overall	0.46	0.46	0.46	0.44	0.44	0.44
24204323	P	R	F1	P	R	F1
similar	0.52	0.48	0.50	0.39	0.92	0.55
not similar	0.84	0.82	0.83	0.95	0.53	0.68
uncertain	0	0	0	0	0	0
overall	0.73	0.73	0.73	0.62	0.62	0.62

Table 4: Evaluation of the LLMs on the SMAFIRA-c dataset.

LLMs in the SMAFIRA-c dataset with the corresponding mapping for the labels (cf. Section 2.4). We show results in Table 4.

Overview of the ArchEHR-QA 2025 Shared Task on Grounded Question Answering from Electronic Health Records

Sarvesh Soni, Soumya Gayen, Dina Demner-Fushman
Division of Intramural Research
National Library of Medicine, National Institutes of Health, Bethesda, MD, USA
sarvesh.soni@nih.gov, ddemner@mail.nih.gov

Abstract

This paper presents an overview of the ArchEHR-QA 2025 shared task, which was organized with the 24th BioNLP Workshop at ACL 2025. The goal of this shared task is to develop automated responses to patients' questions by generating answers that are grounded in key clinical evidence from patients' electronic health records (EHRs). A total of 29 teams participated in the task, collectively submitting 75 systems, with 24 teams providing their system descriptions. The submitted systems encompassed diverse architectures (including approaches that select the most relevant evidence prior to answer generation), leveraging both proprietary and open-weight large language models, as well as employing various tuning strategies such as fine-tuning and few-shot learning. In this paper, we describe the task setup, the dataset used, the evaluation criteria, and the baseline systems. Furthermore, we summarize the methodologies adopted by participating teams and present a comprehensive evaluation and analysis of the submitted systems.

1 Introduction

The volume of messages received through patient portals is on the rise, which includes requests from patients for medical information (Holmgren et al., 2023; Martinez et al., 2024). This is one of the main contributors to increasing clinician burden. One promising strategy to address this challenge is to assist clinicians in formulating responses to patient inquiries. To this end, automatically generating answers to questions from patients considering their medical records is important.

While there is extensive work on answering general health-related queries from patients (Welivita and Pu, 2023), relatively little focuses on addressing patient questions specifically about their own medical records. Within the work on patient portal messages, most research has focused on message triage (Ren et al., 2023; Liu et al., 2024c) or on helping patients formulate their questions (Liu et al., 2024b). Efforts to automatically generate answers to patient questions rarely incorporate relevant information from the patient's medical record (Liu et al., 2024a; Chen et al., 2024). Among the few that do, none evaluate how effectively the generated responses leverage that evidence (Small et al., 2024; Garcia et al., 2024).

Grounding an answer in evidence is the process of citing or referencing specific segments of the input evidence to support the generated response (Chandu et al., 2021). This practice is especially critical in medicine, where accuracy and traceability are paramount–particularly when the target users are not proficient in medical knowledge (Haug and Drazen, 2023). Although grounding has been extensively studied in open-domain (Wang et al., 2025; Sung et al., 2025), its application in the clinical domain remains relatively underexplored.

To foster research in these underexplored areas of clinical natural language processing (NLP), we introduced the ArchEHR-QA (pronounced *"Archer"*) shared task[1]. The goal of the task is to develop automated systems that generate answers to patients' questions, grounded in key clinical evidence from their electronic health records (EHRs). Participants were provided with patient-posed questions, their clinician-interpreted versions, and corresponding clinical notes. Systems were expected to produce answers accompanied by sentence-level citations to the relevant sentences of the clinical note.

2 ArchEHR-QA 2025 Task Description

Given a patient-posed natural language question, the corresponding clinician-interpreted question, and the patient's clinical note excerpt, the task is to generate a natural language answer with sentence-

[1] archehr-qa.github.io

Proceedings of the 24th Workshop on Biomedical Language Processing (BioNLP 2025), pages 396–405
August 1, 2025 ©2025 Association for Computational Linguistics

Patient Question	I had severe abdomen pain and was hospitalised for 15 days in ICU, diagnosed with CBD sludge thereafter on udiliv. Doctor advised for ERCP. My question is if the sludge was there does not the medication help in flushing it out? Whether ERCP was the only cure?
Clinician Question	Why was ERCP recommended to him over continuing a medication-based treatment?
Clinical Note Excerpt	1: During the ERCP a pancreatic stent was required to facilitate access to the biliary system (removed at the end of the procedure), and a common bile duct stent was placed to allow drainage of the biliary obstruction caused by stones and sludge. 2: However, due to the patient's elevated INR, no sphincterotomy or stone removal was performed. 3: Frank pus was noted to be draining from the common bile duct, and post-ERCP it was recommended that the patient remain on IV Zosyn for at least a week. 4: The Vancomycin was discontinued. 5: On hospital day 4 (post-procedure day 3) the patient returned to ERCP for re-evaluation of her biliary stent as her LFTs and bilirubin continued an upward trend. 6: On ERCP the previous biliary stent was noted to be acutely obstructed by biliary sludge and stones. 7: As the patient's INR was normalized to 1.2, a sphincterotomy was safely performed, with removal of several biliary stones in addition to the common bile duct stent. 8: At the conclusion of the procedure, retrograde cholangiogram was negative for filling defects.
Clinician Answer	An endoscopic retrograde cholangiopancreatography, ERCP, was recommended to place a common bile duct stent [1]. This stent was placed to allow drainage of the biliary obstruction which was caused by stones and sludge [1]. Due to no improvement in liver function, the patient needed a repeat ERCP [5]. The repeat ERCP showed that the biliary stent placed in the first ERCP was obstructed by stones and sludge [6]. The stones and stent were successfully removed during this procedure by performing a sphincterotomy [7].
Sample System Answer	ERCP was recommended as medication alone may not effectively clear the biliary obstruction caused by stones and sludge [1]. The procedure allowed for direct visualization, stenting, and subsequent removal of biliary stones and sludge [1,7,8]. Medication, such as IV Zosyn, was used adjunctively to manage infection but was not sufficient to flush out the sludge and stones on its own [3].

Table 1: Example annotated patient case from the ArchEHR-QA dataset. Focus areas in the patient question are underlined, which are taken into account when creating the clinician question. In the clinical note excerpt, the sentences with identifiers [1,5,6,7] are essential, [2,8] are supplementary, and [3,4] are not-relevant.

level citations to the specific clinical note sentences.

3 Data Description

We used a subset of 120 patient cases from the ArchEHR-QA dataset (Soni and Demner-Fushman, 2025). Each patient case in the dataset comprises a hand-curated, realistic patient question (reflective of patient portal messages), relevant focus areas identified within the question (as determined by a clinician), corresponding clinician-rewritten version (crafted to aid in formulating responses), and note excerpt providing essential clinical context (Table 1). The dataset was curated by aligning real patient questions posted to public health forums with clinical notes from publicly accessible EHR databases, namely, MIMIC-III and MIMIC-IV (Johnson et al., 2016, 2023). Each sentence in the note excerpt is manually annotated to mark its importance in answering the question as "essential" (must be cited in the answer), "supplementary" (may be cited to provide support), or "not-relevant" (should not be cited). For more details about the dataset curation process, please refer to the dataset paper (Soni and Demner-Fushman, 2025).

A total of 20 patient cases were provided to the participants with sentence relevance keys for the development and validation of systems. The remaining 100 patient cases were used for testing the participant systems and released to the participants closer to the final submission date without the sentence relevance labels. Tables 2 and 3 provide the dataset statistics.

4 Evaluation

4.1 Metrics

Submissions were evaluated based on their use of clinical evidence for grounding (*"Factuality"*) and the relevance of the generated answers (*"Relevance"*). The scoring script is available on GitHub[2].

Factuality is assessed by calculating Precision, Recall, and F1 Scores between the cited evidence sentences in the generated answers (i.e., predicted as "essential") and the manually anno-

[2]github.com/soni-sarvesh/archehr-qa

	Patient Question		Clinician Question		Note Excerpt		Clinician Answer	
	Dev	**Test**	**Dev**	**Test**	**Dev**	**Test**	**Dev**	**Test**
Mean	85.2	92.3	10.8	10.6	320.8	380.4	73.6	72.4
Median	81.0	74.5	10.0	10.0	320.5	345.0	74.0	73.0
S.D.	35.1	62.4	2.8	3.9	174.6	213.3	2.3	3.6
Min	40.0	33.0	7.0	3.0	109.0	76.0	66.0	55.0
Max	170.0	440.0	17.0	21.0	678.0	1028.0	78.0	76.0

Table 2: Word count statistics by dataset split. *Dev*: development; *S.D.*: standard deviation.

	Sentences	
Relevance	**Dev**	**Test**
`all`	21.4 (100%)	26.0 (100%)
`essential`	6.0 (28.3%)	6.6 (25.3%)
`supplementary`	1.3 (6.1%)	5.5 (21.3%)
`not-relevant`	14.1 (65.7%)	13.9 (53.4%)

Table 3: Average sentence counts by relevance and dataset split. *Dev*: development.

tated ground truth sentence relevance labels. Two variations of Citation F1 Scores are calculated. In the *"strict"* variation, only `essential` sentences are considered as answers. In the *"lenient"* variation, both `essential` and `supplementary` sentences are considered as answers.

Relevance is evaluated by comparing the generated answer text with the ground truth answer. Two variations of ground truth answers were used for relevance computations: clinician-authored answer and a concatenation of `essential` note sentences with patient and clinician questions. A suite of text and semantics based relevance metrics are used to compare the predicted and ground truth text: BLEU (Papineni et al., 2002), ROUGE-LSum (Lin, 2004), SARI (Xu et al., 2016), BERTScore (Zhang et al., 2019), AlignScore (Zha et al., 2023), and MEDCON (Yim et al., 2023).

4.2 Baseline

As a simple yet strong baseline, we prompted the LLaMa 3.3 70B model (Grattafiori et al., 2024) in a zero-shot setting to generate an answer using both the patient and clinician questions, along with the note excerpt as input. We provided the note sentence identifiers to the model and instructed it to cite the specific sentence IDs within its generated answer. In instances where the model failed to follow the required formatting or citation guidelines, we iteratively prompted the model with specific feedback from the previous attempt (e.g., an *invalid citation*) up to five times.

5 Participation

5.1 Participating Teams

We used the Codabench platform[3] to facilitate shared task submission process (Xu et al., 2022). In total, 29 teams participated in the task and submitted a total of 75 systems. Of these, 24 teams provided a description of their submitted system. We report the evaluation scores exclusively for those submissions accompanied by a system description.

5.2 Results

Participants were provided with a preliminary version of sentence relevance keys during the development phase, where note excerpts and questions were used to compute the relevance scores. Table 4 presents the submission results on the test set (with hidden keys) using this setup. In this setting, DMIS Lab (Hwang et al., 2025) achieved the highest overall score of 53.7% with a strict micro F1 score of 58.6% and an average relevance score of 48.8%. This was followed by Neural (Bogireddy et al., 2025) and LAILab (Le et al., 2025), which attained overall scores of 51.5% and 51.0%, respectively. Notably, ArgHiTZ (Cortes et al., 2025) obtained the highest strict micro F1 score of 60.5%.

Upon completion of the annotation reconciliation process, we recalculated the evaluation metrics using revised sentence relevance keys and clinician-authored reference answers. These results are reported in Table 5. While the overall score range remained relatively stable, there were substantial changes in the ranking of individual systems, and we observed a general drop in the overall scores for

[3]codabench.org/competitions/5302

Rank			Team	Factuality						Relevance							
				Lenient			Strict			Text			Semantics				
O	F	R	ID	P	R	F1	P	R	F1	BL	RG	SA	BS	AS	MD	Avg	OS
1	4	1	DMIS Lab	61.2	59.2	60.2	57.9	59.3	58.6	14.3	46.5	36.7	53.9	92.4	49.3	48.8	53.7
2	3	2	Neural	58.4	63.7	60.9	55.4	63.8	59.3	8.5	34.1	73.1	39.1	67.3	40.0	43.7	51.5
3	2	4	LAILab	59.7	66.0	62.7	56.0	65.5	60.4	6.5	32.7	69.2	37.4	65.3	38.4	41.6	51.0
4	6	5	LAMAR	64.0	53.5	58.3	60.6	53.6	56.9	6.0	32.1	65.8	36.4	64.3	43.6	41.4	49.1
5	14	3	ssagarwal	71.7	35.6	47.6	68.8	36.2	47.5	4.7	31.1	70.0	36.9	74.9	38.0	42.6	45.0
6	7	8	LIMICS	63.6	49.6	55.8	59.9	49.4	54.2	3.0	26.2	61.2	31.2	52.3	39.4	35.5	44.9
7	10	7	cuni-a	60.2	48.1	53.5	56.9	48.1	52.1	5.1	26.5	63.2	32.0	58.2	37.7	37.1	44.6
8	1	22	ArgHiTZ	58.9	65.8	62.1	55.8	65.9	60.5	0.9	21.1	48.1	22.1	42.3	30.9	27.6	44.0
9	5	16	Loyola	51.1	70.5	59.3	48.3	70.5	57.3	2.9	25.5	54.4	26.1	42.4	30.8	30.4	43.9
10	8	11	unibuc-sd	66.5	47.2	55.2	62.7	47.0	53.8	1.4	22.2	53.3	27.5	53.4	38.2	32.7	43.2
11	15	6	SzegedAI	69.7	37.0	48.4	65.6	36.9	47.2	3.2	27.8	63.6	32.9	64.2	37.8	38.2	42.7
12	11	12	KRLabs	50.7	56.6	53.5	48.1	56.8	52.1	2.0	21.4	57.9	26.3	49.0	35.2	31.9	42.0
13	12	10	FK	70.0	37.9	49.2	66.7	38.2	48.6	2.0	25.4	54.4	28.2	55.8	36.8	33.8	41.2
14	9	20	UTSA-NLP	47.0	68.4	55.7	43.7	67.2	53.0	0.7	17.8	56.6	22.7	40.4	29.4	27.9	40.4
15	17	13	UIC	70.4	35.2	46.9	67.3	35.6	46.5	0.7	19.4	55.6	24.6	57.7	31.4	31.6	39.0
16	13	21	utsamuel	55.1	45.3	49.7	51.4	44.7	47.8	0.6	20.0	56.7	24.2	35.4	29.6	27.8	37.8
17	16	23	aehrc	55.5	42.0	47.8	52.9	42.4	47.1	0.6	19.0	48.4	22.5	41.9	30.3	27.1	37.1
18	18	19	unibuc-sb	61.7	35.9	45.4	58.7	36.1	44.7	0.6	19.9	49.0	23.9	43.0	32.4	28.1	36.4
19	20	17	HurLab	52.9	34.8	41.9	49.3	34.3	40.4	1.8	24.0	47.3	24.9	48.1	34.5	30.1	35.2
20	19	18	JUNLP	57.5	32.4	41.4	54.2	32.3	40.5	1.5	22.8	49.3	24.4	49.1	30.9	29.6	35.1
21	21	9	WisPerMed	59.1	27.1	37.1	55.4	26.9	36.2	2.0	22.6	61.0	29.5	62.3	25.9	33.9	35.0
22	22	15	DKIT	59.9	23.1	33.4	56.5	23.1	32.7	1.7	23.6	49.8	26.2	47.8	33.9	30.5	31.6
23	23	14	heiDS	71.2	16.0	26.2	67.7	16.1	26.0	0.7	18.1	53.6	22.2	61.0	29.9	30.9	28.5
24	24	24	razreshili	39.7	8.4	13.9	36.8	8.2	13.5	0.4	16.8	45.8	19.9	43.9	24.5	25.2	19.3
-	-	-	baseline	77.0	22.3	34.6	71.6	21.9	33.6	0.1	15.2	47.8	20.5	57.7	25.6	27.8	30.7

Table 4: Submission scores using the preliminary version of answer keys, with note excerpts and questions used for evaluating relevance. Factuality scores are reported at the micro level. O, F, R: Rank using Overall, Factuality (Strict F1), and the average Relevance score. ID: Team identifier; P: Precision; R: Recall; F1: F1 Score; BL: BLEU; RG: ROUGE; SA: SARI; BS: BERTScore; AS: AlignScore; MD: MEDCON; Avg: Overall Relevance Score; OS: Overall Score. All scores are percentages.

most submissions. Under this revised evaluation, LAMAR (Yoadsanit et al., 2025) achieved the highest overall score and strict micro F1 score of 46.9% and 58.8%, respectively. FK and unibuc-sd (Ghinea and Rîncu, 2025) followed closely, securing the second and third positions with overall scores of 46.6% and 45.6%, respectively.

5.3 Approaches

Table 6 summarizes the key characteristics of the systems submitted to the shared task. The majority of teams (20 [83.3%]) adopted a two-stage pipeline in which relevant evidence was first identified from the note excerpts, followed by answer generation in a subsequent stage. Several teams also incorporated additional post-generation steps, such as citation assignment (5 [20.8%]) or answer reformulation (8 [33.3%]) with an aim to further enhance the quality of responses.

All participating teams utilized language models as part of their systems. Over half of the teams (14 [58.3%]) employed proprietary models, such as OpenAI's GPT, while 11 teams (45.8%) used open-weight large language models (LLMs), e.g., Meta's LLaMA. Additionally, 9 teams (37.5%) integrated small language models (SLMs), such as BERT, into their systems. Model tuning strategies varied, with fine-tuning being the most common (6 [25.0%]), followed by few-shot learning (5 [20.8%]), the use of synthetic data (3 [12.5%]), and hyperparameter tuning (2 [8.3%]). Postprocessing steps to refine the generated answers were also reported, with some teams leveraging the language model itself for editing (5 [20.8%]) and some applying heuristic-based approaches (4 [16.7%]).

Among the top-scoring systems, answer reformulation emerged as a common component, so did the use of proprietary LLMs. Notably, the leading

Rank			Team	Factuality						Relevance							
				Lenient			Strict			Text			Semantics				
O	*F*	*R*	ID	P	R	F1	P	R	F1	BL	RG	SA	BS	AS	MD	Avg	OS
1	1	2	LAMAR	72.9	57.4	64.2	49.6	72.0	58.8	8.4	24.8	55.9	40.8	36.9	43.2	35.0	46.9
2	3	1	FK	78.8	40.3	53.3	59.1	55.6	57.3	7.4	24.6	53.9	40.2	47.9	41.6	35.9	46.6
3	2	6	unibuc-sd	75.4	50.4	60.4	53.0	65.2	58.4	4.2	21.3	53.0	41.0	34.6	41.9	32.7	45.6
4	5	5	ssagarwal	79.0	37.0	50.4	58.3	50.3	54.0	8.1	24.1	54.3	36.8	33.7	39.3	32.7	43.4
5	10	3	UIC	77.6	36.6	49.7	55.5	48.2	51.6	4.5	22.8	54.5	41.7	40.5	38.9	33.8	42.7
6	9	4	SzegedAI	78.4	39.3	52.3	54.4	50.2	52.2	7.1	23.4	54.1	39.3	35.0	39.1	33.0	42.6
7	6	10	LIMICS	71.2	52.3	60.3	46.9	63.5	54.0	5.5	22.2	54.3	38.6	25.7	39.3	30.9	42.5
8	4	18	Neural	67.5	69.3	68.4	42.8	81.0	56.0	6.3	20.7	53.1	30.6	25.9	33.3	28.3	42.2
9	7	13	LAILab	65.7	68.4	67.0	40.6	77.8	53.3	7.4	22.0	53.3	33.9	26.8	34.4	29.6	41.5
10	13	7	JUNLP	64.9	34.5	45.0	49.5	48.5	49.0	5.2	21.7	51.6	38.7	39.9	37.8	32.5	40.8
11	11	16	cuni-a	65.7	49.4	56.4	43.3	60.0	50.3	4.4	19.1	52.3	31.6	30.7	34.8	28.8	39.6
12	14	12	utsamuel	62.8	48.7	54.9	41.6	59.4	49.0	4.3	23.2	53.4	39.5	23.3	34.5	29.7	39.3
13	16	8	unibuc-sb	66.9	36.6	47.3	45.6	46.0	45.8	5.1	22.9	53.5	40.8	31.2	39.8	32.2	39.0
14	12	21	ArgHiTZ	64.7	68.1	66.3	38.0	73.6	50.1	2.9	18.4	48.5	34.9	25.8	32.8	27.2	38.6
15	17	9	KRLabs	57.6	60.6	59.1	34.3	66.4	45.2	5.5	23.4	53.8	38.2	27.8	42.8	31.9	38.6
16	15	19	Loyola	54.6	70.9	61.7	32.8	78.4	46.2	6.2	21.8	50.6	31.5	24.3	34.3	28.1	37.2
17	19	15	aehrc	57.7	41.2	48.1	37.5	49.2	42.5	2.8	20.6	51.3	38.5	28.5	33.4	29.2	35.9
18	18	22	UTSA-NLP	53.2	72.9	61.5	30.4	76.7	43.6	2.5	17.5	51.5	33.9	22.7	30.9	26.5	35.0
19	8	24	DMIS Lab	68.3	62.3	65.2	42.1	70.7	52.8	0.6	12.7	34.8	19.1	9.2	16.2	15.4	34.1
20	20	17	HurLab	56.2	34.8	43.0	36.5	41.6	38.9	4.6	21.0	48.6	37.3	26.5	33.2	28.5	33.7
21	22	11	heiDS	79.0	16.7	27.6	63.0	24.6	35.4	4.4	18.7	51.5	36.1	37.4	33.6	30.3	32.8
22	21	14	DKIT	64.6	23.5	34.5	44.2	29.6	35.5	5.1	21.3	49.8	37.2	27.9	35.1	29.4	32.4
23	23	23	WisPerMed	63.7	27.5	38.4	40.0	31.8	35.4	4.2	18.8	51.7	29.3	24.7	26.0	25.8	30.6
24	24	20	razreshili	40.5	8.1	13.5	30.2	11.1	16.2	2.9	19.4	48.7	32.7	29.4	31.9	27.5	21.9
-	-	-	baseline	83.7	22.9	35.9	65.3	32.8	43.7	2.4	21.0	49.2	39.3	47.0	36.7	32.6	38.1

Table 5: Submission scores using the reconciled answer keys, with clinician-authored answers used for evaluating relevance. Factuality scores are reported at the micro level. *O*, *F*, *R*: Rank using Overall, Factuality (Strict F1), and the average Relevance score. ID: Team identifier; P: Precision; R: Recall; F1: F1 Score; BL: BLEU; RG: ROUGE; SA: SARI; BS: BERTScore; AS: AlignScore; MD: MEDCON; Avg: Overall Relevance Score; OS: Overall Score. All scores are percentages.

systems favored few-shot learning paradigms or the incorporation of synthetic data generated by LLMs over traditional fine-tuning setup. For example, LAMAR (Yoadsanit et al., 2025) created synthetic examples using an LLM to facilitate few-shot prompting with a separate LLM, which was employed to identify relevant note sentences. These sentences were subsequently leveraged to generate the final answer text. In contrast, some systems opted to utilize pre-trained models directly without substantial modifications (e.g., FK).

6 Conclusion

We presented an overview of the ArchEHR-QA Shared Task organized at the BioNLP Workshop in ACL 2025. We discussed the proposed task, the dataset used, the evaluation metrics, and a summary of the baseline and participants' systems. The shared task attracted significant interest, with 29 teams submitting a total of 75 systems and 24 teams providing their system descriptions. Our analysis indicated that systems leveraging proprietary language models achieved higher overall performance, and that top-performing approaches favored few-shot learning strategies over traditional fine-tuning. Additionally, system architectures incorporating an answer reformulation step demonstrated notable improvements in answer quality. The strong interest and competitive submissions underscore the growing momentum in this field. We believe that the insights and resources provided by the ArchEHR-QA Shared Task will promote further advancements in the development and evaluation of EHR-based question answering systems for patient-centered applications.

Rank			Team	Components				Model			Adaptation				Postprocess	
O	F	R	ID	ES →	AG →	CA →	AR	Pty	O-LLM	O-SLM	FT	FS	SD	HT	Mod	Heur
1	1	2	LAMAR	✓	✓			✓				✓	✓			
2	3	1	FK		✓			✓								
3	2	6	unibuc-sd	✓	✓		✓		✓							✓
4	5	5	ssagarwal	✓	✓		✓		✓			✓			✓	
5	10	3	UIC	✓	✓		✓	✓								✓
6	9	4	SzegedAI	✓	✓		✓	✓		✓					✓	
7	6	10	LIMICS		✓	✓	✓	✓							✓	
8	4	18	Neural	✓	✓			✓								
9	7	13	LAILab	✓	✓			✓	✓					✓		
10	13	7	JUNLP		✓			✓				✓				
11	11	16	cuni-a	✓	✓		✓	✓		✓		✓	✓		✓	
12	14	12	utsamuel	✓	✓			✓								
13	16	8	unibuc-sb	✓	✓	✓	✓		✓	✓	✓					✓
14	12	21	ArgHiTZ	✓	✓	✓		✓		✓						✓
15	17	9	KRLabs	✓	✓		✓		✓						✓	
16	15	19	Loyola	✓	✓	✓		✓		✓				✓		
17	19	15	aehrc		✓				✓		✓					
18	18	22	UTSA-NLP	✓	✓				✓			✓				
19	8	24	DMIS Lab	✓	✓			✓								
20	20	17	HurLab	✓	✓				✓	✓	✓					
21	22	11	heiDS	✓	✓				✓							
22	21	14	DKIT	✓	✓				✓	✓	✓					
23	23	23	WisPerMed	✓	✓	✓		✓		✓	✓		✓			
24	24	20	razreshili	✓	✓				✓	✓	✓					

Table 6: Characteristics of the submitted systems with their rankings based on reconciled keys and human answers. O, F, R: Rank using Overall, Factuality (Strict F1), and Relevance score. ID: Team identifier. Broad categories of system components comprise Evidence Selection (ES), Answer Generation (AG), Citation Assignment (CA), and Answer Reformulation (AR). Different types of models employed can be categorized into Proprietary model (Pty), Open-weight large language model (O-LLM), and Open-weight small language model (O-SML). Adaptation or learning methods employed were: Fine Tuning (FT), Few-shot Learning (FS), Use of Synthetic Data (SD), and Hyperparameter Tuning (HT). Postprocessing was performed using the Model itself (Mod) or using Heuristics (Heur).

Limitations

The primary evaluation of system submissions in this shared task relied on automated metrics, which serve as practical proxies for system performance. While such metrics offer scalability and efficiency, they may not fully capture the nuances of answer quality, especially in the clinical domain. Human evaluation that assesses system-generated answers considering the input question and the corresponding clinical note remains the gold standard for determining answer relevance, accuracy, and evidence grounding. However, due to the intensive time and resource requirements, as well as the limited window between the submission deadline and the proceedings release, comprehensive manual evaluation was not feasible within the scope of the shared task. To address this limitation, we plan to conduct a thorough manual assessment of the top submissions from each participating team, focusing on three key criteria: (i) whether the system response adequately answers the question, (ii) whether it leverages relevant clinical evidence, and (iii) whether it uses general knowledge. We anticipate that this forthcoming analysis will provide deeper insights into system performance and help inform and accelerate future development of patient-centered EHR question answering systems.

Acknowledgments

This research was supported by the Division of Intramural Research (DIR) of the National Library of Medicine (NLM), National Institutes of Health, and utilized the computational resources of the NIH HPC Biowulf cluster (`http://hpc.nih.gov`).

References

Sai Prasanna Teja Reddy Bogireddy, Abrar Majeedi, Viswanath Reddy Gajjala, Zhuoyan Xu, Siddhant Rai, and Vaishnav Potlapalli. 2025. Neural at archehr-qa 2025: Agentic prompt optimization for evidence-grounded clinical question answering. In *The 24th Workshop on Biomedical Natural Language Processing and BioNLP Shared Tasks*, Vienna, Austria. Association for Computational Linguistics.

Khyathi Raghavi Chandu, Yonatan Bisk, and Alan W Black. 2021. Grounding 'Grounding' in NLP. In *Findings of the Association for Computational Linguistics: ACL-IJCNLP 2021*, pages 4283–4305, Online. Association for Computational Linguistics.

Shan Chen, Marco Guevara, Shalini Moningi, Frank Hoebers, Hesham Elhalawani, Benjamin H. Kann, Fallon E. Chipidza, Jonathan Leeman, Hugo J. W. L. Aerts, Timothy Miller, Guergana K. Savova, Jack Gallifant, Leo A. Celi, Raymond H. Mak, Maryam Lustberg, Majid Afshar, and Danielle S. Bitterman. 2024. The effect of using a large language model to respond to patient messages. *The Lancet Digital Health*, 0(0).

Adrian Cuadron Cortes, Aimar Sagasti, Maitane Urruela, Iker De la Iglesia, Ane García Domingo-Aldama, Aitziber Atutxa Salazar, Josu Goikoetxea, and Ander Barrena. 2025. Arghitz at archehr-qa 2025: A two-step divide and conquer approach to patient question answering for top factuality. In *The 24th Workshop on Biomedical Natural Language Processing and BioNLP Shared Tasks*, Vienna, Austria. Association for Computational Linguistics.

Patricia Garcia, Stephen P. Ma, Shreya Shah, Margaret Smith, Yejin Jeong, Anna Devon-Sand, Ming Tai-Seale, Kevin Takazawa, Danyelle Clutter, Kyle Vogt, Carlene Lugtu, Matthew Rojo, Steven Lin, Tait Shanafelt, Michael A. Pfeffer, and Christopher Sharp. 2024. Artificial Intelligence–Generated Draft Replies to Patient Inbox Messages. *JAMA Network Open*, 7(3):e243201.

Dragos Dumitru Ghinea and Ștefania Rîncu. 2025. Unibuc-sd at archehr-qa 2025: Prompting our way to clinical qa with multi-model ensembling. In *The 24th Workshop on Biomedical Natural Language Processing and BioNLP Shared Tasks*, Vienna, Austria. Association for Computational Linguistics.

Aaron Grattafiori, Abhimanyu Dubey, Abhinav Jauhri, Abhinav Pandey, Abhishek Kadian, Ahmad Al-Dahle, Aiesha Letman, Akhil Mathur, Alan Schelten, Alex Vaughan, Amy Yang, Angela Fan, Anirudh Goyal, Anthony Hartshorn, Aobo Yang, Archi Mitra, Archie Sravankumar, Artem Korenev, Arthur Hinsvark, and 542 others. 2024. The Llama 3 Herd of Models. *Preprint*, arXiv:2407.21783.

Charlotte J. Haug and Jeffrey M. Drazen. 2023. Artificial Intelligence and Machine Learning in Clinical Medicine, 2023. *New England Journal of Medicine*, 388(13):1201–1208.

A. Jay Holmgren, Maria E. Byron, Carrie K. Grouse, and Julia Adler-Milstein. 2023. Association Between Billing Patient Portal Messages as e-Visits and Patient Messaging Volume. *JAMA*, 329(4):339–342.

Hyeon Hwang, Hyeongsoon Hwang, JongMyung Jung, Jaehoon Yun, Minju Song, Yein Park, Dain Kim, Taewhoo Lee, Jiwoong Sohn, Chanwoong Yoon, Sihyeon Park, Jiwoo Lee, Heechul Yang, and Jaewoo Kang. 2025. Dmis lab at archehr-qa 2025: Evidence-grounded answer generation for ehr-based qa via a multi-agent framework. In *The 24th Workshop on Biomedical Natural Language Processing and BioNLP Shared Tasks*, Vienna, Austria. Association for Computational Linguistics.

Alistair E. W. Johnson, Lucas Bulgarelli, Lu Shen, Alvin Gayles, Ayad Shammout, Steven Horng, Tom J. Pollard, Sicheng Hao, Benjamin Moody, Brian Gow, Li-wei H. Lehman, Leo A. Celi, and Roger G. Mark. 2023. MIMIC-IV, a freely accessible electronic health record dataset. *Scientific Data*, 10(1):1.

Alistair E. W. Johnson, Tom J. Pollard, Lu Shen, Li-wei H. Lehman, Mengling Feng, Mohammad Ghassemi, Benjamin Moody, Peter Szolovits, Leo Anthony Celi, and Roger G. Mark. 2016. MIMIC-III, a freely accessible critical care database. *Scientific Data*, 3(1):160035.

Tuan Dung Le, Thanh Duong, Shohreh Haddadan, Behzad Jazayeri, Brandon Manley, and Thanh Thieu. 2025. Lailab at archehr-qa 2025: Test-time scaling for evidence selection in grounded question answering from electronic health records. In *The 24th Workshop on Biomedical Natural Language Processing and BioNLP Shared Tasks*, Vienna, Austria. Association for Computational Linguistics.

Chin-Yew Lin. 2004. ROUGE: A Package for Automatic Evaluation of Summaries. In *Text Summarization Branches Out*, pages 74–81, Barcelona, Spain. Association for Computational Linguistics.

Siru Liu, Allison B McCoy, Aileen P Wright, Babatunde Carew, Julian Z Genkins, Sean S Huang, Josh F Peterson, Bryan Steitz, and Adam Wright. 2024a. Leveraging large language models for generating responses to patient messages—a subjective analysis. *Journal of the American Medical Informatics Association*, page ocae052.

Siru Liu, Aileen P Wright, Allison B Mccoy, Sean S Huang, Julian Z Genkins, Josh F Peterson, Yaa A Kumah-Crystal, William Martinez, Babatunde Carew, Dara Mize, Bryan Steitz, and Adam Wright. 2024b. Using large language model to guide patients to create efficient and comprehensive clinical care message. *Journal of the American Medical Informatics Association*, 31(8):1665–1670.

Vincent X Liu, Pamela Kaercher, Jennifer Manickam, Eric Smallberg, Kanishka Bhutani, Michelle Mancha, and Kristine Lee. 2024c. Content of Patient Electronic Messages to Physicians in a Large Integrated System. *JAMA Network Open*, 7(4):e244867.

Kathryn A. Martinez, Rebecca Schulte, Michael B. Rothberg, Maria Charmaine Tang, and Elizabeth R. Pfoh. 2024. Patient Portal Message Volume and Time Spent on the EHR: An Observational Study of Primary Care Clinicians. *Journal of General Internal Medicine*, 39(4):566–572.

Kishore Papineni, Salim Roukos, Todd Ward, and Wei-Jing Zhu. 2002. BLEU: A method for automatic evaluation of machine translation. In *Proceedings of the 40th Annual Meeting on Association for Computational Linguistics*, pages 311–318. Association for Computational Linguistics.

Yang Ren, Dezhi Wu, Aditya Khurana, George Mastorakos, Sunyang Fu, Nansu Zong, Jungwei Fan, Hongfang Liu, and Ming Huang. 2023. Classification of Patient Portal Messages with BERT-based Language Models. In *2023 IEEE 11th International Conference on Healthcare Informatics (ICHI)*, pages 176–182.

William R. Small, Batia Wiesenfeld, Beatrix Brandfield-Harvey, Zoe Jonassen, Soumik Mandal, Elizabeth R. Stevens, Vincent J. Major, Erin Lostraglio, Adam Szerencsy, Simon Jones, Yindalon Aphinyanaphongs, Stephen B. Johnson, Oded Nov, and Devin Mann. 2024. Large Language Model–Based Responses to Patients' In-Basket Messages. *JAMA Network Open*, 7(7):e2422399.

Sarvesh Soni and Dina Demner-Fushman. 2025. A Dataset for Addressing Patient's Information Needs related to Clinical Course of Hospitalization. *Preprint*, arXiv:2506.04156.

Mujeen Sung, Song Feng, James Gung, Raphael Shu, Yi Zhang, and Saab Mansour. 2025. Structured List-Grounded Question Answering. In *Proceedings of the 31st International Conference on Computational Linguistics*, pages 8347–8359, Abu Dhabi, UAE. Association for Computational Linguistics.

Jui-I Wang, Hen-Hsen Huang, and Hsin-Hsi Chen. 2025. MESAQA: A Dataset for Multi-Span Contextual and Evidence-Grounded Question Answering. In *Proceedings of the 31st International Conference on Computational Linguistics*, pages 10891–10901, Abu Dhabi, UAE. Association for Computational Linguistics.

Anuradha Welivita and Pearl Pu. 2023. A survey of consumer health question answering systems. *AI Magazine*, 44(4):482–507.

Wei Xu, Courtney Napoles, Ellie Pavlick, Quanze Chen, and Chris Callison-Burch. 2016. Optimizing Statistical Machine Translation for Text Simplification. *Transactions of the Association for Computational Linguistics*, 4:401–415.

Zhen Xu, Sergio Escalera, Adrien Pavão, Magali Richard, Wei-Wei Tu, Quanming Yao, Huan Zhao, and Isabelle Guyon. 2022. Codabench: Flexible, easy-to-use, and reproducible meta-benchmark platform. *Patterns*, 3(7):100543.

Wen-wai Yim, Yujuan Fu, Asma Ben Abacha, Neal Snider, Thomas Lin, and Meliha Yetisgen. 2023. Aci-bench: A Novel Ambient Clinical Intelligence Dataset for Benchmarking Automatic Visit Note Generation. *Scientific Data*, 10(1):586.

Seksan Yoadsanit, Nopporn Lekuthai, Watcharitpol Sermsrisuwan, and Titipat Achakulvisut. 2025. Lamar at archehr-qa 2025: Clinically aligned llm-generated few-shot learning for ehr-grounded patient question answering. In *The 24th Workshop on Biomedical Natural Language Processing and BioNLP Shared Tasks*, Vienna, Austria. Association for Computational Linguistics.

Yuheng Zha, Yichi Yang, Ruichen Li, and Zhiting Hu. 2023. AlignScore: Evaluating Factual Consistency with A Unified Alignment Function. In *Proceedings of the 61st Annual Meeting of the Association for Computational Linguistics (Volume 1: Long Papers)*, pages 11328–11348, Toronto, Canada. Association for Computational Linguistics.

Tianyi Zhang, Varsha Kishore, Felix Wu, Kilian Q. Weinberger, and Yoav Artzi. 2019. BERTScore: Evaluating Text Generation with BERT. In *International Conference on Learning Representations*.

A Appendix

Tables 7 and 8 provide the factuality scores both at the macro level (averaging per-case F1 scores) and the micro level (aggregating true positives, false positives, and false negatives across all cases).

Rank	Team	Micro						Macro					
		Lenient			Strict			Lenient			Strict		
O	ID	P	R	F1	P	R	F1	P	R	F1	P	R	F1
1	DMIS Lab	61.2	59.2	60.2	57.9	59.3	58.6	66.6	67.1	63.2	62.1	69.0	61.2
2	Neural	58.4	63.7	60.9	55.4	63.8	59.3	68.1	69.8	64.8	62.7	71.3	62.6
3	LAILab	59.7	66.0	62.7	56.0	65.5	60.4	67.2	72.1	64.6	62.1	72.8	61.5
4	LAMAR	64.0	53.5	58.3	60.6	53.6	56.9	70.0	62.2	61.8	65.4	64.0	60.2
5	ssagarwal	71.7	35.6	47.6	68.8	36.2	47.5	77.8	44.9	52.1	72.9	46.5	51.4
6	LIMICS	63.6	49.6	55.8	59.9	49.4	54.2	71.0	58.2	59.4	66.6	59.8	57.4
7	cuni-a	60.2	48.1	53.5	56.9	48.1	52.1	66.0	54.2	55.6	61.1	56.0	53.6
8	ArgHiTZ	58.9	65.8	62.1	55.8	65.9	60.5	62.4	69.1	61.9	57.0	69.5	58.5
9	Loyola	51.1	70.5	59.3	48.3	70.5	57.3	56.2	72.9	60.4	52.1	74.0	57.6
10	unibuc-sd	66.5	47.2	55.2	62.7	47.0	53.8	70.8	55.7	58.4	65.7	56.5	56.2
11	SzegedAI	69.7	37.0	48.4	65.6	36.9	47.2	73.6	46.1	53.1	68.3	47.1	51.4
12	KRLabs	50.7	56.6	53.5	48.1	56.8	52.1	60.4	60.6	56.2	55.8	62.3	54.3
13	FK	70.0	37.9	49.2	66.7	38.2	48.6	74.9	49.8	54.5	70.8	51.3	53.4
14	UTSA-NLP	47.0	68.4	55.7	43.7	67.2	53.0	49.6	77.4	56.7	45.1	77.3	52.6
15	UIC	70.4	35.2	46.9	67.3	35.6	46.5	79.1	42.1	51.2	74.7	44.1	51.4
16	utsamuel	55.1	45.3	49.7	51.4	44.7	47.8	57.0	55.4	51.8	52.2	56.0	49.0
17	aehrc	55.5	42.0	47.8	52.9	42.4	47.1	65.4	48.0	50.4	61.4	49.5	49.1
18	unibuc-sb	61.7	35.9	45.4	58.7	36.1	44.7	68.5	41.4	47.8	63.6	42.7	46.4
19	HurLab	52.9	34.8	41.9	49.3	34.3	40.4	61.2	42.0	44.8	56.7	43.0	42.7
20	JUNLP	57.5	32.4	41.4	54.2	32.3	40.5	62.4	43.6	46.9	58.4	45.0	45.8
21	WisPerMed	59.1	27.1	37.1	55.4	26.9	36.2	59.5	33.9	39.9	54.0	34.0	37.7
22	DKIT	59.9	23.1	33.4	56.5	23.1	32.7	63.4	31.1	36.5	60.0	32.4	35.9
23	heiDS	71.2	16.0	26.2	67.7	16.1	26.0	73.9	22.5	30.7	69.7	24.0	30.7
24	razreshili	39.7	8.4	13.9	36.8	8.2	13.5	53.8	13.6	19.1	49.6	14.5	19.0
-	baseline	77.0	22.3	34.6	71.6	21.9	33.6	83.0	30.8	39.9	77.4	31.5	39.0

Table 7: Factuality scores using the preliminary version of answer keys, with both micro and macro level calculations. *O*: Rank using Overall score. ID: Team identifier; P: Precision; R: Recall; F1: F1 Score. All scores are percentages.

Rank	Team	Micro						Macro					
		Lenient			Strict			Lenient			Strict		
O	ID	P	R	F1	P	R	F1	P	R	F1	P	R	F1
1	LAMAR	72.9	57.4	64.2	49.6	72.0	58.8	78.3	67.2	68.4	56.7	75.5	61.9
2	FK	78.8	40.3	53.3	59.1	55.6	57.3	81.8	52.5	58.7	64.3	60.8	59.8
3	unibuc-sd	75.4	50.4	60.4	53.0	65.2	58.4	80.2	59.8	65.2	60.7	69.1	62.3
4	ssagarwal	79.0	37.0	50.4	58.3	50.3	54.0	84.6	47.5	55.9	67.0	55.2	56.8
5	UIC	77.6	36.6	49.7	55.5	48.2	51.6	86.8	43.7	54.1	68.7	51.1	54.5
6	SzegedAI	78.4	39.3	52.3	54.4	50.2	52.2	80.7	47.9	56.2	58.5	53.8	53.4
7	LIMICS	71.2	52.3	60.3	46.9	63.5	54.0	78.6	61.5	64.4	55.5	67.2	57.5
8	Neural	67.5	69.3	68.4	42.8	81.0	56.0	76.5	75.6	72.3	54.6	82.8	62.1
9	LAILab	65.7	68.4	67.0	40.6	77.8	53.3	73.3	74.3	68.9	51.9	79.9	57.8
10	JUNLP	64.9	34.5	45.0	49.5	48.5	49.0	68.9	46.5	51.0	53.6	54.4	52.0
11	cuni-a	65.7	49.4	56.4	43.3	60.0	50.3	72.5	56.6	59.3	54.5	64.3	54.8
12	utsamuel	62.8	48.7	54.9	41.6	59.4	49.0	63.8	58.0	56.6	45.1	63.8	50.6
13	unibuc-sb	66.9	36.6	47.3	45.6	46.0	45.8	72.5	43.2	49.5	54.1	48.9	48.4
14	ArgHiTZ	64.7	68.1	66.3	38.0	73.6	50.1	68.4	71.1	65.6	45.8	74.4	52.8
15	KRLabs	57.6	60.6	59.1	34.3	66.4	45.2	67.4	64.9	62.0	48.0	69.0	51.9
16	Loyola	54.6	70.9	61.7	32.8	78.4	46.2	59.0	73.9	62.4	40.1	78.4	49.9
17	aehrc	57.7	41.2	48.1	37.5	49.2	42.5	68.7	47.6	51.4	51.5	52.7	47.8
18	UTSA-NLP	53.2	72.9	61.5	30.4	76.7	43.6	56.4	80.3	61.7	35.5	79.8	46.2
19	DMIS Lab	68.3	62.3	65.2	42.1	70.7	52.8	73.9	70.3	68.6	50.5	74.7	57.2
20	HurLab	56.2	34.8	43.0	36.5	41.6	38.9	63.9	41.3	45.4	45.6	44.5	41.1
21	heiDS	79.0	16.7	27.6	63.0	24.6	35.4	82.4	24.5	32.9	66.3	28.5	36.7
22	DKIT	64.6	23.5	34.5	44.2	29.6	35.5	68.1	32.3	38.3	51.1	34.6	37.6
23	WisPerMed	63.7	27.5	38.4	40.0	31.8	35.4	62.9	34.8	40.8	39.7	35.1	35.3
24	razreshili	40.5	8.1	13.5	30.2	11.1	16.2	57.8	13.4	18.3	47.4	14.6	20.2
-	baseline	83.7	22.9	35.9	65.3	32.8	43.7	89.2	32.4	42.2	74.2	38.6	47.3

Table 8: Factuality scores using the reconciled answer keys, with both micro and macro level calculations. *O*: Rank using Overall score. ID: Team identifier; P: Precision; R: Recall; F1: F1 Score. All scores are percentages.

Author Index

Association for Computational Linguistics
209 N. Eighth Street
Stroudsburg, Pennsylvania 18360

ISBN 979-8-3313-2402-5

9 798331 324025

CURRAN ASSOCIATES INC.
proceedings.com

First Workshop on Natural Language Processing for Political Sciences (PoliticalNLP)

Held at LREC 2022

Marseille, France
20 – 25 June 2022